Aquatic Biology

Aquatic Biology

Editor: Luke Holt

Callisto Reference

www.callistoreference.com

Callisto Reference,
118-35 Queens Blvd., Suite 400,
Forest Hills, NY 11375, USA

Visit us on the World Wide Web at:
www.callistoreference.com

ISBN: 978-1-63239-942-7 (Hardback)

The publisher's policy is to use permanent paper from mills that operate a sustainable forestry policy. Furthermore, the publisher ensures that the text paper and cover boards used have met acceptable environmental accreditation standards.

Printed in China.

Cataloging-in-Publication Data

Aquatic biology / edited by Luke Holt.
p. cm.
Includes bibliographical references and index.
ISBN 978-1-63239-942-7
1. Aquatic biology. 2. Aquatic ecology. 3. Aquatic sciences. I. Holt, Luke.
QH90 .A68 2018
578.76--dc23

Table of Contents

Preface

Freshwater biology studies freshwater ecosystems which includes lakes, streams, springs and wetlands. As a subject, it studies the relationship between organisms found in the freshwater and their environment. It is mostly used in industrial processes such as sewage treatment, etc. This book is a valuable compilation of topics, ranging from the basic to the most complex advancements in the field of freshwater biology. It attempts to understand the multiple branches that full under this discipline and how such concepts have practical application. For someone with an interest and eye for detail, this book covers the most significant topics in the field of freshwater biology.

The information contained in this book is the result of intensive hard work done by researchers in this field. All due efforts have been made to make this book serve as a complete guiding source for students and researchers. The topics in this book have been comprehensively explained to help readers understand the growing trends in the field.

I would like to thank the entire group of writers who made sincere efforts in this book and my family who supported me in my efforts of working on this book. I take this opportunity to thank all those who have been a guiding force throughout my life.

Editor

1

Methane-Derived Carbon in the Benthic Food Web in Stream Impoundments

John Gichimu Mbaka*, Celia Somlai, Denis Köpfer, Andreas Maeck, Andreas Lorke, Ralf B. Schäfer

Institute for Environmental Sciences, University of Koblenz-Landau, Landau, Rhineland-Palatinate State, Germany

Abstract

Methane gas (CH_4) has been identified as an important alternative source of carbon and energy in some freshwater food webs. CH_4 is oxidized by methane oxidizing bacteria (MOB), and subsequently utilized by chironomid larvae, which may exhibit low $\delta^{13}C$ values. This has been shown for chironomid larvae collected from lakes, streams and backwater pools. However, the relationship between CH_4 concentrations and $\delta^{13}C$ values of chironomid larvae for in-stream impoundments is unknown. CH_4 concentrations were measured in eleven in-stream impoundments located in the Queich River catchment area, South-western Germany. Furthermore, the $\delta^{13}C$ values of two subfamilies of chironomid larvae (i.e. Chironomini and Tanypodinae) were determined and correlated with CH_4 concentrations. Chironomini larvae had lower mean $\delta^{13}C$ values (−29.2 to −25.5 ‰), than Tanypodinae larvae (−26.9 to −25.3 ‰). No significant relationships were established between CH_4 concentrations and $\delta^{13}C$ values of chironomids ($p>0.05$). Mean $\delta^{13}C$ values of chironomid larvae (mean: −26.8‰, range: −29.2‰ to −25.3‰) were similar to those of sedimentary organic matter (SOM) (mean: −28.4‰, range: −29.3‰ to −27.1‰) and tree leaf litter (mean: −29.8 ‰, range: −30.5‰ to −29.1‰). We suggest that CH_4 concentration has limited influence on the benthic food web in stream impoundments.

Editor: David William Pond, Scottish Association for Marine Science, United Kingdom

Funding: JGM was funded by the German Academic Exchange service (DAAD) (Grant number: A/12/91652). The funder had no role in study design, data collection and analysis, decision to publish, or preparation of the manuscript.

Competing Interests: The authors have declared that no competing interests exist.

* Email: mbakagichimu@gmail.com

Introduction

Allochthonous and autochthonous plant organic matter are major sources of carbon and energy for freshwater ecosystems [1]. Recent studies have revealed that also methane, which can be produced by microbial degradation of organic matter under anoxic conditions, can significantly contribute to the carbon budget of freshwater ecosystems [2]. Part of this gas is released to the atmosphere, where it contributes to the pool of green house gases [3], or is oxidized by methane oxidizing bacteria (MOB) [4]. The biogenic methane in MOB can contribute to the biomass of chironomid larvae [5]. Chironomids are one of the most dominant invertebrate groups in the soft sediments in freshwater ecosystems and their larvae feed mainly on algae or allochthonous organic material and associated microroganisms [6]. CH_4 derived organic carbon may constitute a crucial source of carbon and energy for chironomids, in comparison to other aquatic invertebrates, because their burrowing habit creates and exposes them to oxyclines at the sediment-water interface, where MOB density and CH_4 oxidation rates are usually high [7–10]. For example, chironomid larvae collected from some lakes were sustained (up to 70%) by CH_4 derived carbon [11].

These quantitative estimates are based on the stable carbon isotope signature ($\delta^{13}C$) of CH_4 which is highly depleted due to carbon isotopic fractionation related to methanogenesis [12]. Additionally, MOB that oxidize CH_4 are usually characterized by further depletion in $\delta^{13}C$ [13]. Therefore, organisms that consume MOB have lower $\delta^{13}C$ values (typically <−40‰; [14]), in comparison to organisms that feed on plant organic matter (−32 to −21 ‰; [15]). Bunn & Boon [16] determined the $\delta^{13}C$ values of invertebrates in backwater pools and found Chironominae larvae to have $\delta^{13}C$ values (<−35‰) that were lower than for particulate organic matter (−29 to −25 ‰). Kiyashko et al. [18] and Jones & Grey [19] also reported lower (−64 to −55 ‰) $\delta^{13}C$ values for some chironomid larvae than for particulate organic matter in lakes. The observed differences between $\delta^{13}C$ values of chironomid larvae and potential food resources led to the conclusion that the chironomid larvae might have fed on MOB, which has very low $\delta^{13}C$ values. The carbon isotope composition of consumers (e.g. insects) is determined by their diet and usually potrays an enrichment by about 1 ‰, even though the $\delta^{13}C$ can deviate from −3 ‰ to +3 ‰ [20]. Given that methane is isotopically very distinct, stable carbon isotopes are particularly useful for tracing methane derived carbon [17].

Although most existing studies on the importance of CH_4 derived carbon in freshwater food webs mainly focused on lakes [2], a wide array of anoxic habitats with high potential for CH_4 production also exist in rivers and streams [21–23]. Particularly, impoundments increases the residence time of water, promotes accumulation of organic matter and sediment, and have been identified as hot spots of CH_4 emissions [24]. Maeck et al. [24] measured CH_4 concentrations in riverine and impoundment reaches and found sediment accumulation in dams to be the main source of CH_4 emissions. Guérin et al. [25] reported an increase in CH_4 emissions at the downstream sides of impoundments as a result of release of water enriched with CH_4.

In shallow aquatic systems such as rice paddies and small lakes, CH_4 has been shown to be an important source of energy in the benthic food webs [26–28]. In spite of the high abundance of small in-stream ponds in smaller streams [29], the relationship between CH_4 concentrations and stable carbon isotope ratios ($\delta^{13}C$) of chironomid larvae in such systems has not been examined. Globally, there exist millions of small impoundments (height < 15 m; [30]). Within this study, we assessed (i) CH_4 concentrations in stream and pore-water and (ii) the relationship between CH_4 concentrations and $\delta^{13}C$ values of chironomid larvae in impoundments located in the Queich River catchment area, South-western Germany. We hypothesized that CH_4 would have a significant influence on the $\delta^{13}C$ of chironomid larvae.

Methods

Ethics statement

This study was conducted in the Queich River catchment area (see coordinates below) and was not conducted in an area requiring research permit (e.g. national park) or private land. This study did not involve endangered or protected species.

Study area and sites

The study was conducted in the Queich River catchment area, Rhineland-Palatinate State, South-western Germany. The Queich River (length: 52 km) originates from the Palatinate forest (49°10′6°N 7°50′48°E) and flows (mean discharge: 1.31 $m^3\ s^{-1}$; www.geoportal-wasser.rlp.de) through the upper Rhine Valley to its confluence with the Rhine River in Germersheim (49°13′39°N 8°23′4°E). The catchment (area: 271 km^2) is primarily covered by sandstone and is between 100 m and 673 m above sea level. The Rhineland-Palatinate region has dry climate conditions in summer.

Typical for most stream networks in central Europe, 67 small in-stream impoundments (www.geoportal-wasser.rlp.de) have been constructed on the main stem and tributaries of the Queich River, South-western Germany, for various purposes such as hydropower generation and flood control. Here, we selected eleven study sites (e.g. Figure S1 in File S1) located from the downstream to the upstream reaches of the Queich River catchment area (Figure 1, Table S1 in File S1). The study sites were located at five different streams. The sampled impoundments were approximately 0.5–2.0 metres deep and had water bypasses that transported water to the downstream reaches.

Water chemistry and physical characteristics

Field measurements and sampling were conducted between 9th and 24th June, 2013. Data collection was done between 9 a.m. and 4 p.m. Electrical conductivity, temperature, dissolved oxygen concentration and pH of stream water were measured *in situ* with a WTW Multi 340i/SET (Wissenschaftlich Werkstätten GmbH, Weilheim, Germany). Average water depth was computed from three measurements taken on a transect across the river channel and current speed was estimated by timing a float over a distance of 5 metres [31]. Water discharge was calculated from velocity, width and depth [31]. Water residence time was calculated as follows:

$$T = \frac{V}{Q} \qquad \text{(i)}$$

where: T is the water residence time, V is the volume of water stored in the impoundment, and Q is the water discharge [32]. Nitrate and phosphate concentrations in stream water were determined in the laboratory using Macherey-Nagel viscolor kits (Macherey-Nagel, Düren, Germany).

CH_4 concentrations

Concentrations of dissolved CH_4 in stream and pore-water were measured at the impoundments. Water samples were collected from each study impoundment using 20 mL serum bottles. The samples for stream water CH_4 analysis were collected by filling water to the sample bottles from the bottom to top, and overflowing the sample bottles several times over. Three bottles were completely filled with water at each sampling site and several drops (250 µL) of mercuric chloride were added to each bottle as preservative [33]. The bottles were capped and sealed and transported to the laboratory. A headspace was prepared by replacing 10% of the bottle (i.e. 2 mL) with nitrogen gas. To generate the headspace, each sample bottle was held upside down and a 20 gauge needle was inserted through the septum. Then 2 mL nitrogen gas was added to each bottle using a syringe, while the replaced water sample escaped through the needle. The samples were manually shaken, for 1 minute, to equilibrate the gas between the headspace and the water [34]. The samples were analyzed using a CH_4 analyser (Los Gatos Research Inc., Mountain view, CA, U.S.A.). A closed loop was created between the gas inlet and outlet of the analyser. A gas tight syringe was then used to inject 0.5 mL gas sample into the closed loop. The CH_4 concentrations were averaged over 30 seconds before and after gas injection. Concentration of the injected gas was computed as:

$$c_{sample} = \frac{\Delta c_{LosGatos}(V_{LosGatos} + V_{Injection})}{V_{Injection}} \qquad \text{(ii)}$$

where: c_{sample} is the mol fraction of the sampled gas in parts per million, $\Delta c_{LosGatos}$ is the change in mole fraction before and after gas sample injection, $V_{LosGatos} = 92.5$ mL and $V_{Injection} = 0.5$ mL.

Sediment samples for pore-water CH_4 analysis were obtained at the impoundments, where fine sediment accumulated. Pore-water CH_4 concentrations were assessed from sediment cores (1 core per site). Cores were taken at location of soft sediment using a piston corer and analyzed for porosity, carbon:nitrogen (C:N) ratio, total organic carbon (TOC) content and pore-water CH_4 concentration. Cut-off syringes (3 mL) were used to extract sediment subsamples which were immediately placed into crimp capped 20 mL vials containing 3 mL of 2.5% NaCl solution for conservation of the CH_4. Pore-water was sampled in the cores from the homogenized upper (0–10 cm) sediment layer, where chironomid larvae are found [35]. The pore-water CH_4 samples were measured as described for water samples. For C:N ratio, TOC and porosity, three sediment sub-samples were extracted from the cores (0–10 cm) and placed into glass tubes before analysis in the laboratory.

Chironomid sampling and processing

Chironomid samples were collected from the deepest point in each impoundment using an Ekman grab sampler (Hydro-Bios, Kiel, Germany). Sediments were sieved by passing them through

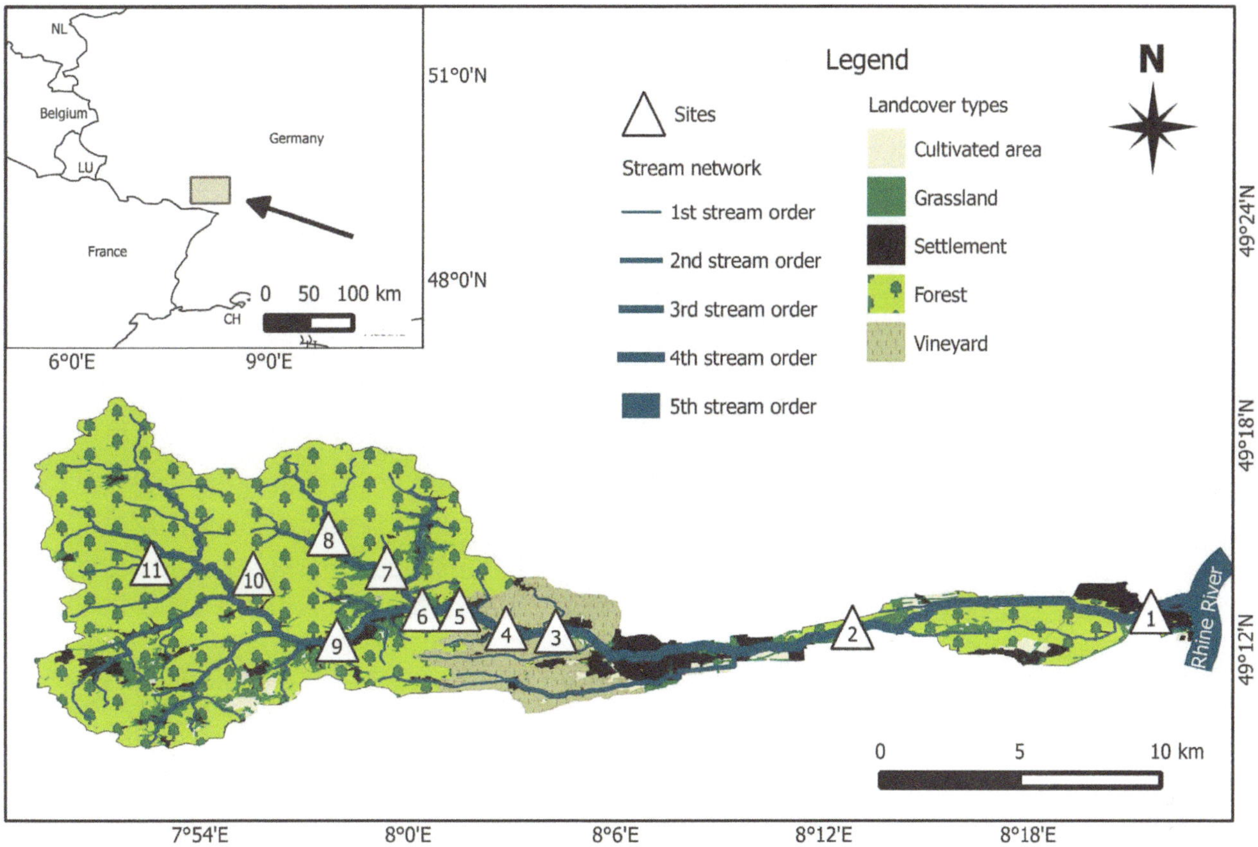

Figure 1. Locations of the study impoundments in the Queich River catchment area.

two metal sieves (mesh size: 500 µm and 2 mm). Materials such as stones and large pieces of organic matter (>5 cm) were removed and chironomid larvae were picked from a sorting tray using forceps and placed into 500 mL sample bottles containing river water. In some sites only few chironomids (<5 chironomids) were found. The chironomid samples were transported to the laboratory and transferred to sample bottles containing clean tap water for 24 hours to allow gut clearance. Faecal materials were periodically removed to prevent ingestion by chironomids [36]. Chironomids were sorted by tribe and subfamily [37–40] and size (instar) [41,42] under a dissecting microscope (magnification: x 40-100). Chironomids of the same tribe, subfamily and size were pooled to obtain sufficient mass (0.5–1 mg) for isotope analysis [43]. Sorting of chironomids by size was done to detect the potential effect of body size on the isotopic signal as demonstrated by Grey et al. [44]. In most cases the limited number of specimens excluded replicate analyses. Therefore, we collected individuals from a site with a high abundance of chironomids to exemplarily determine the $\delta^{13}C$ variability from 9 replicates. Second instar larvae were discarded as they were too small for identification and their mass was insufficient for isotope analysis. Before isotope analysis, chironomids were placed into glass tubes, oven dried at 60°C for 24 hours and subsequently stored in a desiccator.

Stable isotope, TOC and C:N analyses

Three replicate sediment samples, from each site were analysed, for $\delta^{13}C$ of sedimentary organic matter (SOM), TOC and C:N ratios. They were rinsed with a 2.5% HCL solution for four hours to remove carbonates [45], rinsed three times with demineralised water, oven dried at 60°C for 24 hours and ground using a mortar and pestle. Leaves, for analysis of $\delta^{13}C$ of potential allochthonous food resources, were collected from trees near the impoundments, washed with demineralised water, rinsed, oven dried and ground before analysis.

Sediment samples for C:N ratios and TOC were weighed into tin cups (15–20 g) and analysed using a Vario Microcube elemental analyser (Elementar Analysensysteme, Hanau, Germany). The chironomid, SOM and leaf litter samples for stable isotope analysis were also weighed (approximately 0.5–1.0 mg for chironomids and 5–20 mg for SOM and leaf litter) into tin cups before their combustion in an isotope ratio mass spectrometer (ThermoScientific, Bremen, Germany). Stable isotope ratios were expressed in per mille (‰).

Statistical analysis

Relations between variables were tested using Spearman's rank correlation test [46]. Comparisons of CH_4 concentrations in pore and stream water, and between $\delta^{13}C$ values of chironomids, and leaf litter and SOM were done using paired t-test. A value of $p < 0.05$ was considered as statistically significant. Homogeneity of variances was examined with Bartlett's test and data were square root transformed to improve normality. Statistical analyses were done using the R statistical package [47] and all data are provided in File S1.

Results

Water chemistry and physical characteristics

Water residence time varied from 0.5 to 6.0 minutes. Nutrient concentrations in stream water ranged from 3.5 to 5.0 mg NO_3 L^{-1}, and from 0.1 to 0.3 mg PO_4 L^{-1}. Water temperature ranged from 10.2 to 18.6°C, whereas electrical conductivity, dissolved oxygen concentrations and water discharge ranged from 61 to 380 μS cm^{-1}, 8.5 to 11.9 mg L^{-1} and 0.2 to 7.8 m^3 s^{-1} (Table S1 in File S1).

Methane gas concentrations

Average values of dissolved CH_4 concentrations in stream water ranged from 0.07 μmol L^{-1} at Site 6 to 0.7 μmol L^{-1} at Site 10 (Table 1). Pore-water CH_4 concentrations ranged from 0.3 μmol L^{-1} at Site 7 to 1657.5 μmol L^{-1} at Site 9, and were statistically significantly (t-value $=3.1$, $p=0.005$) higher than the stream water CH_4 concentrations. Some pore-water CH_4 measurements at sites 3 and 6 differed greatly from the other measurements, either due to disturbance during sampling or gas leakage during analysis, and were therefore taken to be unreliable and excluded from further analysis.

$\delta^{13}C$ of SOM and leaf litter, C:N ratios and TOC

$\delta^{13}C$ values of SOM ranged from −29.3‰ at site 7 to −27.1‰ at site 3 (Table 1), whereas $\delta^{13}C$ values of leaf litter ranged from −30.5 ‰ to −29.1‰. C:N ratios of sediment and the TOC ranged from 12.0 at site 3 to 23.2 at site 2, and from 0.2% at site 3 to 3.1% at site 2, respectively (Table 1).

$\delta^{13}C$ of chironomids

The chironomids were identified as Chironomini, Tanypodinae and *Chironomus* sp. *Chironomus* sp. were only found at site 10 and were not analyzed because they did not provide an adequate pooled mass for isotope analysis. The lowest $\delta^{13}C$ value, −29.2 ‰, was measured in a third instar Chironomini larvae collected from site 8, whereas the highest $\delta^{13}C$ values, −25.3 ‰, were measured in third instar Tanypodinae larvae collected from Sites 9 and 11 (Table S2 in File S1). Generally, the highest mean $\delta^{13}C$ values were measured in third (−25.3±0.01‰) and fourth (−26.3±0.14 ‰) instar Tanypodinae, whereas slightly lower mean $\delta^{13}C$ values were measured in fourth (−27.2±0.16 ‰) and third (−26.9±0.25 ‰) instar Chironomini. $\delta^{13}C$ values did not differ significantly between third and fourth instar Chironomini larvae (t-value $=1.4$, $p=0.17$). The analysis of 9 replicates of fourth instar Chironomini larvae from site 4 had a standard deviation as low as 0.19 ‰ for $\delta^{13}C$ (Table S2 in File S1). No significant correlations were observed between CH_4 concentrations and $\delta^{13}C$ values of chironomids (Figure 2, Table 2). However, $\delta^{13}C$ values of chironomid larvae were significantly correlated to those of the SOM and to each other within site ($p<0.05$) (Table 2). Mean $\delta^{13}C$ of chironomid larvae (−26.8 ‰) was more similar to that of the SOM (−28.4 ‰) than leaf litter (−29.8 ‰) and there were significant differences ($p<0.05$) between $\delta^{13}C$ values of chironomids, and SOM and leaf litter.

Discussion

Methane gas concentrations

CH_4 concentrations measured in this study are comparable to those measured in other aquatic ecosystems [48–50]. Pore-water CH_4 concentrations showed a highly variable pattern among the impoundments (Table 1). These differences can be attributed to heterogeneity in the distribution of sedimentary organic materials within the impoundments. Sanders et al. [51] found pore-water CH_4 concentrations to be influenced by sediment heterogeneity.

Table 1. Average CH_4 concentrations (μmol L^{-1}) and bulk sediment characteristics (TOC, C:N ratio, porosity and $\delta^{13}C$) of samples from the studied impoundments (in parentheses are standard errors, ±SE (when applicable).

Name	Code	Pore-water CH_4	Dissolved CH_4	TOC (%)	C:N ratio	Porosity	$\delta^{13}C$ (‰)
Germersheim	1	27.8 (4.6)	0.2 (0.006)	2.8 (0.1)	17.6 (0.8)	0.6 (0.02)	−28.7 (0.01)
Fuchsbach	2	658.1 (244.4)	0.4 (0.004)	3.1 (0.1)	23.2 (0.9)	0.7 (0.02)	−28.2 (0.3)
Godramstein	3	142.5	0.3 (0.04)	0.2 (0.02)	12.0 (1.0)	0.4 (0.02)	−27.1 (0.02)
Siebeldingen	4	1499.6 (784.6)	0.1 (0.06)	2.8 (0.03)	17.1 (0.6)	0.7 (0.003)	−28.8 (0.04)
Albersweiler Pfalz	5	199.6 (8.2)	0.1 (0.004)	0.6 (0.04)	16.1 (1.2)	0.5 (0.02)	−28.9 (0.2)
Rosenfeldt Mill	6	263.0	0.07 (0.001)	1.8 (0.04)	16.4 (0.6)	0.5 (0.03)	−28.5 (0.1)
Eußerbach	7	0.3 (0.01)	0.3 (0.01)	0.3 (0.03)	15.2 (1.4)	0.4 (0.03)	−29.3 (0.2)
Eisbach	8	45.2 (8.7)	0.2 (0.03)	1.6 (0.2)	14.7 (0.6)	0.6 (0.02)	−28.6 (0.1)
Annweiler Am Trifels	9	1657.5 (128.9)	0.1 (0.002)	2.4 (0.2)	19.5 (1.2)	0.6 (0.01)	−28.7 (0.1)
Langenbächel	10	4.6 (2.2)	0.7 (0.07)	0.9 (0.1)	17.9 (1.5)	0.6 (0.02)	−28.9 (0.04)
Modenbach	11	105.9 (18.3)	0.3 (0.01)	0.5 (0.02)	16.1 (0.6)	0.3 (0.01)	−27.3 (0.1)

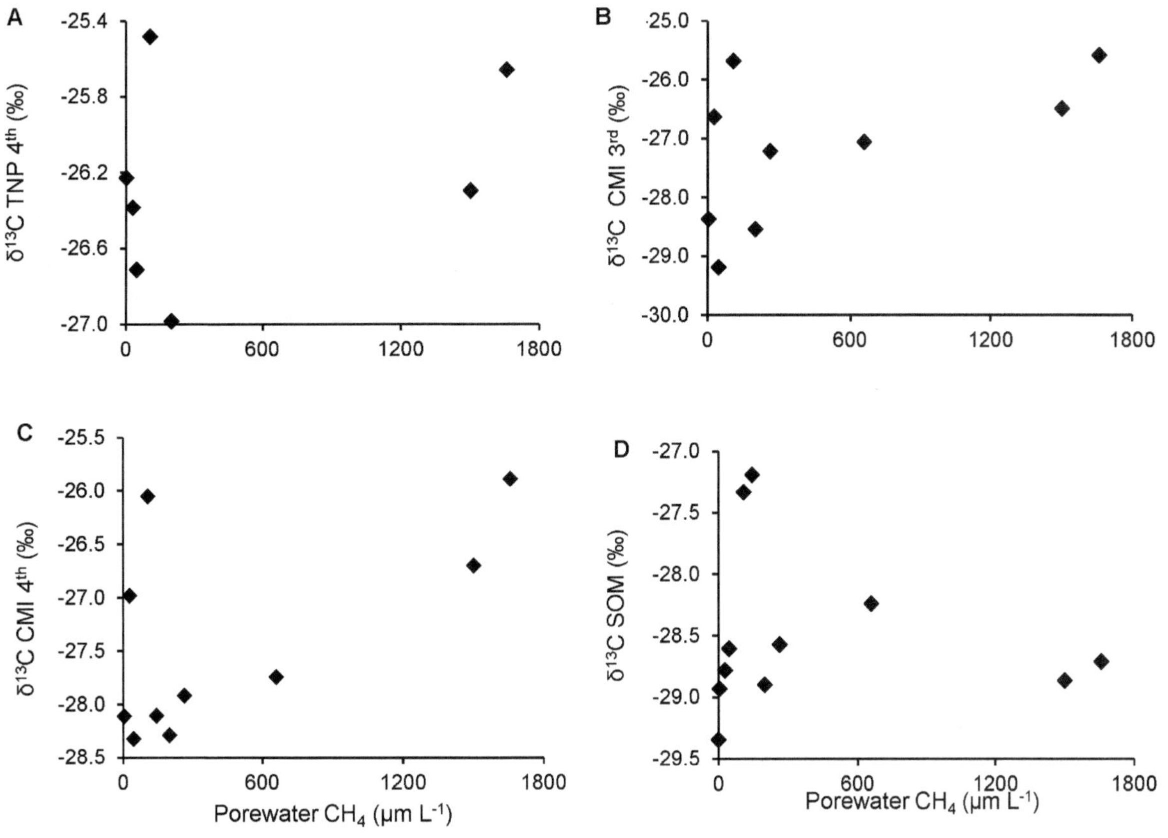

Figure 2. Relationships between pore-water CH_4 concentrations and $\delta^{13}C$ values. TNP 4th, CMI 3rd, CMI 4th and SOM stands for Tanypodinae fourth generation, Chironomini third and fourth generation, and sedimentary organic matter, respectively.

The authors reported that enhanced retention of sediment by macrophytes (*Ranunculus penicillatus*) increased pore-water CH_4 production in streams. Significantly higher mean CH_4 concentrations were recorded in pore than stream water. This can be explained by the fact that CH_4 is usually produced in sediments, where anoxic conditions are likely to develop, whereas the stream water is rather well oxygenated (Table S1 in File S1) [21].

Relationship between $\delta^{13}C$ of chironomids and methane gas

Few studies examined CH_4 as a source of carbon and energy for stream invertebrates (e.g. [52]). In these studies, mean $\delta^{13}C$ values of invertebrates supported by CH_4 derived carbon were lower (<−40 ‰) than those of potential photoautotrophic food resources, indicating ingestion and assimilation of MOB, which had oxidized isotopically light CH_4. This was also demonstrated for lake invertebrates [7,43,53], where significant negative relationships between CH_4 concentrations and $\delta^{13}C$ values of some invertebates indicated ingestion and assimilation of MOB.

Mean $\delta^{13}C$ of chironomid larvae ranged from −27.2 (fourth instar Chironomini) to −25.3 ‰ (fourth instar Tanypodinae). The mean $\delta^{13}C$ values of potential food resources ranged from −28.4 (SOM) to −29.8 ‰ (leaf litter) and were significantly correlated to those of the chironomids. The similarity of $\delta^{13}C$ of Chironomini larvae and SOM can be attributed to the fact that Chironomini larvae are either filterers or gathering-collectors, feeding on fine particulate organic matter in aquatic systems [54]. In comparison to Chironomini, Tanypodinae utilize different types of food (e.g. detritus, oligochaetes, diatoms; [55]) and their $\delta^{13}C$ values may be difficult to interpret when compared with the other chironomids. The differences between the $\delta^{13}C$ values of chironomids, SOM and leaf litter were within the reported range (±3 ‰) of $\delta^{13}C$ values for consumers and their food resources [20].

In the current study, no significant relationships were established between CH_4 concentrations in stream and pore-water and $\delta^{13}C$ values of chironomid larvae. In the UK, Trimmer et al. [52] found river water to have higher mean (0.16 µmol L^{-1}) CH_4 concentrations than pore-water (0.07 µmol L^{-1}). Highly depleted mean $\delta^{13}C$ values (<−42 ‰) of Trichoptera larvae (*Agapetus fuscipes*), relative to those of potential food sources (−38 ‰), were only measured in areas with low pore-water CH_4 concentrations. In comparison to Trimmer et al., we recorded higher mean CH_4 concentrations in stream and pore-water and the mean $\delta^{13}C$ values of chironomid larvae were not low. The type of food consumed by chironomids and presence of MOB influence the $\delta^{13}C$ values of chironomid larvae. The short water residence times and the shallow nature of the studied impoundments could have enhanced water turn over rates, mixing of the entire water column and supply of organic matter or periphyton into the benthic zone. Thus, in the case of chironomid larvae feeding on sedimenting organic matter, we would anticipate their $\delta^{13}C$ to be similar to that of their food resources. Mixing of the water column may also increase oxygen concentration and CH_4 oxidation rates, and reduce CH_4 production and the biomass of MOB available to chironomids [9]. For example, Eller et al. [9] recorded two fold higher MOB density in the anoxic waters (0.1 mg L^{-1} O_2) of a stratified lake than in the well mixed and oxygenated waters (9 mg L^{-1} O_2) of a polymictic lake. Additionally, the contribution of

Table 2. Correlations (*rho*-values) between $\delta^{13}C$ values, C:N ratio of bulk sediment, water residence time (min) and CH_4 concentrations ($^*p<0.05$).

	$\delta^{13}C_{SOM}$	Dissolved methane	Pore-water methane	CMI 4th	TNP 4th	Water residence time	CMI 3rd
$C:N_{SOM}$	−0.16	−0.28	0.35	−0.46	−0.15	−0.55	−0.29
$\delta^{13}C_{SOM}$		−0.80*	0.30	0.66*	0.92*	−0.28	0.76*
Dissolved CH_4			−0.78*	−0.11	−0.57	0.22	−0.28
Pore-water CH_4				−0.50	−0.05	0.21	−0.35
CMI 4th					0.87*	−0.41	0.97*
TNP 4th						−0.53	0.95*
Water residence time							−0.55

$\delta^{13}C_{SOM}$, $\delta^{13}C$ of bulk sediment; CMI 4th, fourth instar Chironomini; CMI 3rd, third instar Chironomini; TNP 4th, fourth instar Tanypodinae.

MOB to chironomid larvae biomass was higher in the anoxic than oxygenated waters. Jones et al. [11] found that the contribution of MOB to the chironomid larvae biomass was highest at sites with low dissolved oxygen content (2–4 mg L^{-1}). Within the above mentioned studies, the isotopic values of chironomids were more similar to those of the SOM, when they were collected from sediments overlaid by well oxygenated waters. Use of MOB was particularly pronounced under hypoxia or post mixing following on from stratification.

$\delta^{13}C$ values of chironomid larvae (mean = ~−27 ‰; −29 to −25 ‰) were similar to those of SOM (mean = ~−28 ‰; −29 to −27 ‰) and leaf litter (mean = ~−29 ‰; −30 to −29 ‰). Additionally, $\delta^{13}C$ values of chironomid larvae were significantly correlated to each other. Given that the $\delta^{13}C$ values varied little between consumers and their food sources [56], the chironomid larvae collected from the sampled impoundments most likely obtained their carbon through ingestion and assimilation of SOM or allochthonous leaf litter. Other studies using stable carbon isotope analysis also demonstrated SOM and allochthonous plant organic matter as significant sources of carbon and energy for freshwater invertebrates [57]. The C:N ratio can be used to determine the source of organic matter in aquatic ecosystems because autochthonous organic matter generally has lower C:N ratios (e.g. algae: 4–10; [58]) than allochthonous organic matter. The measured C:N ratios (12.0–23.2) of sediments indicated elevated proportions of allochthonous organic matter.

Although the $\delta^{13}C$ values of chironomid larvae did not indicate utilization of methane derived carbon, other invertebrates could have used it as a source of energy. For example, Kohzu et al. [59] found coleopterans collected from backwater pools to have lower mean (−40 to −67 ‰) $\delta^{13}C$ values than the other invertebrates (e.g. chironomids; −36 ‰), suggesting increased utilization of methane derived carbon. In summary, this study reveals that methane derived carbon did not contribute substantially to chironomid larval biomass in small impoundments, rather that allochthonous organic matter was the main source of energy. Future studies assessing the role of methane derived carbon in stream impoundments should include MOB community characterization, CH_4 oxidation rates and fluxes, and $\delta^{13}C$ values of other invertebrates.

Supporting Information

File S1 Supporting figure and tables. Figure S1, Example of a weir that impounded the studied rivers. The white arrow shows the direction of water movement. Table S1, Location and environmental characteristics recorded from the studied impoundments. Table S2, $\delta^{13}C$ values (‰) of chironomid larvae from the studied impoundments. For each of the impoundments, between 1 and 9 replicates (*n*) were made from pooled samples.

Acknowledgments

We thank Vivian Rhee, Sebastian Geissler, Zeyad Alshboul and Anne König for invaluable practical assistance in the field. We also thank Markus Kurtz for C:N and TOC analyses and Andreas Hirsch for stable isotope analysis.

Author Contributions

Conceived and designed the experiments: JGM RS AL. Performed the experiments: JGM CS AM DK. Analyzed the data: JGM RS. Contributed reagents/materials/analysis tools: RBS AL. Wrote the paper: JGM RBS AL AM CS.

References

1. Meili M (1992) Sources, concentrations and characteristics of organic matter in softwater lakes and streams of the Swedish forest region. Hydrobiologia 229: 23–41.
2. Jones RI, Grey J (2011) Biogenic methane in freshwater food webs. Freshwater Biol 56: 213–229.
3. Bastviken D, Tranvik LJ, Downing JA, Crill PM, Enrich-Prast A (2011) Freshwater methane emissions offset the continental carbon sink. Science 331: 50.
4. Kajan R, Frenzel P (1999) The effect of chironomid larvae on production, oxidation and fluxes of methane in a flooded rice soil. FEMS Microbiol Ecol 28: 121–129.
5. Kiyashko SI (2001) Contribution of methanotrophs to freshwater macroinvertebrates: evidence from stable isotope ratios. Aquat Microb Ecol 24: 203–207.
6. Berg H (1995) Larval food and feeding behaviour. In: Armitage, P.D., Cranston, P.S. & Pinder, L.C.V. (eds), *The Chironomidae: biology and ecology of non biting midges*. Chapman& Hall, London. pp 136–168.
7. Deines P, Grey J (2006) Site-specific methane production and subsequent midge mediation within Esthwaite Water, UK. Arch Hydrobiol 167: 317–334.
8. Sanseverino AM, Bastviken D, Sundh I, Pickova J, Enrich-Prast A (2012) Methane carbon supports aquatic food webs to the fish level. PLoS ONE 7: e42723–e42723.
9. Eller G, Deines P, Grey J, Richnow H, Krager M (2005) Methane cycling in lake sediments and its influence on chironomid larval $\delta^{13}C$. FEMS Microbiol Ecol 54: 339–350.
10. Yasuno N, Shikano S, Shimada T, Shindo K, Kikuchi E (2013) Comparison of the exploitation of methane-derived carbon by tubicolous and non-tubicolous chironomid larvae in a temperate eutrophic lake. Limnology 14: 239–246.
11. Jones RI, Carter CE, Kelly A, Ward S, Kelly DJ, et al. (2008) Widespread contribution of methane-cycle bacteria to the diets of lake profundal chironomid larvae. Ecology 89: 857–864.
12. Teh YA, Silver WL, Conrad ME, Borglin SE, Carlson CM (2006) Carbon isotope fractionation by methane-oxidizing bacteria in tropical rain forest soils. J Geophys Res-Biogeo 111: G2.
13. Summons RE, Jahnke LL, Roksandic Z (1994) Carbon isotopic fractionation in lipids from methanotrophic bacteria: relevance for interpretation of the geochemical record of biomarkers. Geochim Cosmochim Acta 58: 2853–2863.
14. Zemskaya TI, Sitnikova TY, Kiyashko SI, Kalmychkov GV, Pogodaeva TV, et al. (2012) Faunal communities at sites of gas-and oil-bearing fluids in Lake Baikal. Geo-Marine Lett 32: 437–451.
15. Hershey A, Binkley E, Fortino K, Keyse M, Medvedeff C, et al. (2010) Use of allochthonous and autochthonous carbon sources by Chironomus in arctic lakes. Verh Internat Verein Limnol 30: 1321–1325.
16. Bunn SE, Boon PI (1993) What sources of organic carbon drive food webs in billabongs? A study based on stable isotope analysis. Oecologia 96: 85–94.
17. Zanden MJV, Rasmussen JB (1999) Primary consumer $\delta^{13}C$ and $\delta^{15}N$ and the trophic position of aquatic consumers. Ecology 80: 1395–1404.
18. Kiyashko S, Narita T, Wada E (2001) Contribution of methanotrophs to freshwater macroinvertebrates: evidence from stable isotope ratios. Aquat Microb Ecol 24: 203–207.
19. Jones RI, Grey J (2004) Stable isotope analysis of chironomid larvae from some Finnish forest lakes indicates dietary contribution from biogenic methane. Boreal Environ Res 9: 17–24.
20. DeNiro MJ, Epstein S (1978) Influence of diet on the distribution of carbon isotopes in animals. Geochim Cosmochim Acta 42: 495–506.
21. Trimmer M, Maanoja S, Hildrew AG, Pretty JL, Grey J (2010) Potential carbon fixation via methane oxidation in well-oxygenated riverbed gravels. Limnol Oceanogr 55: 560–568.
22. Shelley F, Grey J, Trimmer M (2014) Widespread methanotrophic primary production in lowland chalk rivers. Proc R Soc Lond B: Biol Sci 281: 20132854.
23. Crawford JT, Stanley EH, Spawn SA, Finlay JC, Loken LC, et al. (2014) Ebullitive methane emissions from oxygenated wetland streams. Glob Change Biol Doi:10.1111/gcb.12614.
24. Maeck A, DelSontro T, McGinnis DF, Fischer H, Flury S, et al. (2013) Sediment trapping by dams creates methane emission hot spots. Environ Sci Technol 47: 8130–8137.
25. Guérin F, Abril G, Richard S, Burban B, Reynouard C, et al. (2006) Methane and carbon dioxide emissions from tropical reservoirs: significance of downstream rivers. Geophys Res Lett 33.
26. Jones RI, Grey J (2004) Stable isotope analysis of chironomid larvae from some Finnish forest lakes indicates dietary contribution from biogenic methane. Boreal Environmen Res 9: 17–24.
27. Khalil M, Shearer MJ, Rasmussen R, Duan C, Ren L (2008) Production, oxidation, and emissions of methane from rice fields in China. J Geophys Res-Biogeo 113: G00A04.
28. Minami K, Neue HU (1994) Rice paddies as a methane source. Climate Change 27: 13–26.
29. Van Looy K, Tormos T, Souchon Y (2014) Disentangling dam impacts in river networks. Ecol Ind 37: 10–20.
30. WCD (2000) Dams and development. A new framework for decision making, World commision on Dams. Earthscan publications ltd. London Sterling, VA. 399 pages.
31. Gordon ND, McMahon TA, Finlayson BL (1994) Stream hydrology, an introduction for ecologists. Wiley, New York. 526 pages.
32. Rueda F, Moreno-Ostos E, Armengol J (2006) The residence time of river water in reservoirs. Ecol model 191: 260–274.
33. Christian JR, Karl DM (1995) Measuring bacterial ectoenzyme activities in marine waters using mercuric chloride as a preservative and a control. Mar Ecol Prog Ser 123: 217–224.
34. Baird AJ, Stamp I, Heppell CM, Green SM (2010) CH_4 flux from peatlands: a new measurement method. Ecohydrol 3: 360–367.
35. Deines P, Grey J, Richnow H-H, Eller G (2007) Linking larval chironomids to methane: seasonal variation of the microbial methane cycle and chironomid $\delta^{13}C$. Aquat Microb Ecol 46: 273–282.
36. Feuchtmayr H, Grey J (2003) Effect of preparation and preservation procedures on carbon and nitrogen stable isotope determinations from zooplankton. Rapid Commun Mass Spectrom 17: 2605–2610.
37. Ashe P, O'Connor JP (2009) A world catalogue of chironomidae (Diptera). Part 1. Buchonomyiinae, chilenomyiinae, Podominae, Aphoratemiinae, Tanypodinae, Usambaromyiinae, Dimesinae, Prodiamesinae and Telmatogetominae. Irish Biogeographical Society & National Museum of Ireland, Dublin, 445 pages.
38. Orendt C, Dettinger-klemm A, Spies M (2012) Chironomidae larvae in brakish waters of Germany and adjacent areas. Federal Environment Agency, Berlin, 214 pages.
39. Orendt C, Spies M (2012) Chironomini (Diptera: Chironomidae). Keys to Central European larvae using mainly macroscopic characters. 2nd revised edition, 64 pages.
40. Orendt C, Spies M (2012) *Chironomus* (Meigen). Key to the larvae of importance to biological water analysis in Germany and adjacent areas. 2nd revised edition, 24 pages.
41. Dillon PM (1985) Chironomid larval size and case presence influence capture success achieved by dragonfly larvae. Freshwat Biol 4: 22–29.
42. Frouz J, Ali A, Lobinske RJ (2002) Suitability of morphological parameters for instar determination of pestiferous midges *Chironomus crassicaudatus* and *Glyptotendipes paripes* (Diptera: Chironomidae) under laboratory conditions. J Am Mosq Control 18: 222–227.
43. van Hardenbroek M, Lotter AF, Bastviken D, Duc NT, Heiri O (2012) Relationship between $\delta^{13}C$ of chironomid remains and methane flux in Swedish lakes. Freshw Biol 57: 166–177.
44. Grey J, Kelly A, Jones RI (2004) High intraspecific variability in carbon and nitrogen stable isotope ratios of lake chironomid larvae. Limnol Oceanogr 49: 239–244.
45. Kelly A, Jones RI, Grey J (2004) Stable isotope analysis provides fresh insights into dietary separation between *Chironomus anthracinus* and *C. plumosus*. JNABS 23: 287–296.
46. Hauke J, Kossowski T (2011) Comparison of Values of Pearson's and Spearman's Correlation Coefficients on the Same Sets of Data. Quaestiones Geographicae 30: 87–93.
47. R Core Team (2013) R: A language and environment for statistical computing. R Foundation for Statistical Computing, Vienna, Austria. ISBN 3-900051-07-0, Available: http://www.R-project.org/. Accessed 2014 Jan 6.
48. Bastviken D, Santoro AL, Marotta H, Pinho LQ, Calheiros DF, et al. (2010) Methane emissions from Pantanal, South America, during the low water season: toward more comprehensive sampling. Environ Sci Technol 44: 5450–5455.
49. Deborde J, Anschutz P, Guérin F, Poirier D, Marty D, et al. (2010) Methane sources, sinks and fluxes in a temperate tidal Lagoon: The Arcachon lagoon (SW France). Estuar Coast Shelf S 89: 256–266.
50. Hofmann H, Federwisch L, Peeters F (2010) Wave-induced release of methane: Littoral zones as a source of methane in lakes. Limnol Oceanogr 55: 1990–2000.
51. Sanders I, Heppell C, Cotton J, Wharton G, Hildrew A, et al. (2007) Emission of methane from chalk streams has potential implications for agricultural practices. Freshw Biol 52: 1176–1186.
52. Trimmer M, Hildrew AG, Jackson MC, Pretty JL, Grey J (2009) Evidence for the role of methane-derived carbon in a free-flowing, lowland river food web. Limnol Oceanogr 54: 1541.
53. van Hardenbroek M, Heiri O, Parmentier FJW, Bastviken D, Ilyashuk BP, et al. (2013) Evidence for past variations in methane availability in a Siberian thermokarst lake based onδ13C of chitinous invertebrate remains. Quat Sci Rev 66: 74–84.
54. Armitage PD, Cranston P, Pinder L (1995) The Chironomidae: biology and ecology of non-biting midges. Chapman and Hall, London. 572 pages.
55. Baker A, McLachlan A (1979) Food preferences of Tanypodinae larvae (Diptera: Chironomidae). Hydrobiol 62: 283–288.
56. Peterson BJ, Fry B (1987) Stable Isotopes in Ecosystem Studies. Annu Rev Ecol Evol Syst 18: 293–320.
57. Fuentes N, Güde H, Straile D (2013) Importance of allochthonous matter for profundal macrozoobenthic communities in a deep oligotrophic lake. Int Rev Hydrobiol 98: 1–13.
58. Meyers PA (1994) Preservation of elemental and isotopic source identification of sedimentary organic matter. Chem Geol 114: 289–302.
59. Kohzu A, Kato C, Iwata T, Kishi D, Murakami M, et al. (2004) Stream food web fueled by methane-derived carbon. Aquat Microb Ecol 36: 189–194.

2

High Genetic Diversity and Novelty in Eukaryotic Plankton Assemblages Inhabiting Saline Lakes in the Qaidam Basin

Jiali Wang[1,2], Fang Wang[1*¤], Limin Chu[1,3], Hao Wang[1], Zhiping Zhong[4], Zhipei Liu[4], Jianyong Gao[5], Hairong Duan[5]

1 State Key Laboratory of Simulation and Regulation of Water Cycle in River Basin, China Institute of Water Resources and Hydropower Research, Beijing, China, **2** Institute of Shandong River Wetlands, Laiwu, China, **3** College of Hydrology and Water Resource, Hohai University, Nanjing, China, **4** Institute of Microbiology, Chinese Academy of Sciences, Beijing, China, **5** Genewiz, Inc, Beijing, China

Abstract

Saline lakes are intriguing ecosystems harboring extremely productive microbial communities in spite of their extreme environmental conditions. We performed a comprehensive analysis of the genetic diversity (18S rRNA gene) of the planktonic microbial eukaryotes (nano- and picoeukaryotes) in six different inland saline lakes located in the Qaidam Basin. The novelty level are high, with about 11.23% of the whole dataset showing <90% identity to any previously reported sequence in GenBank. At least 4 operational taxonomic units (OTUs) in mesosaline lakes, while up to eighteen OTUs in hypersaline lakes show very low CCM and CEM scores, indicating that these sequences are highly distantly related to any existing sequence. Most of the 18S rRNA gene sequence reads obtained in investigated mesosaline lakes is closely related to Holozoa group (48.13%), whereas Stramenopiles (26.65%) and Alveolates (10.84%) are the next most common groups. Hypersaline lakes in the Qaidam Basin are also dominated by Holozoa group, accounting for 26.65% of the total number of sequence reads. Notably, Chlorophyta group are only found in high abundance in Lake Gasikule (28.00%), whereas less represented in other hypersaline lakes such as Gahai (0.50%) and Xiaochaidan (1.15%). Further analysis show that the compositions of planktonic eukaryotic assemblages are also most variable between different sampling sites in the same lake. Out of the parameters, four show significant correlation to this CCA: altitude, calcium, sodium and potassium concentrations. Overall, this study shows important gaps in the current knowledge about planktonic microbial eukaryotes inhabiting Qaidam Basin (hyper) saline water bodies. The identified diversity and novelty patterns among eukaryotic plankton assemblages in saline lake are of great importance for understanding and interpreting their ecology and evolution.

Editor: Melanie R. Mormile, Missouri University of Science and Technology, United States of America

Funding: Independent Study Project of State Key Laboratory of Simulation and Regulation of Water Cycle in River Basin (NO. 2013ZY06). The funders had no role in study design, data collection and analysis, decision to publish, or preparation of the manuscript.

Competing Interests: The authors declare that JG and HD are employed by commercial companies: Genewiz, Inc, Beijing, 102206, China. They do not have any other relevant declarations relating to employment, consultancy, patents, products in development or marketed products, etc.

* Email: wangf@iwhr.com

¤ Current address: A-1 Fuxing Road, Haidian District, Beijing, China

Introduction

Saline lakes usually occur in endorheic drainage basins, which span approximately 1/10 of the Earth's surface area [1]. Inland saline lakes represent approximately 5% of modern drylands [2]; these lakes are numerous and are distributed worldwide in semi-arid or arid areas [3]. Inland saline lakes and freshwater lakes from humid areas account for similar proportions of global water, approximately 0.008% and 0.009%, respectively [4–5]. Saline lakes are important reservoirs of largely unseen microbial biodiversity with high phylogenetic richness and novelty [5]. Saline lakes at high altitudes are also productive and represent an important and extreme ecosystem harboring many novel prokaryotic microorganisms [6–7]. Small-sized planktonic microorganisms are critical for aquatic systems, mostly as major contributors to production and biomass and as key players driving carbon and nutrient cycles [8–9]. The genetic diversity of microbial communities in saline lakes has been studied in different areas of the world, including the USA [10–11], Mongolia [12], China [7], Iran [13], Australia [14], Spain [5,15], and the Andean Altiplano [16]. However, our current knowledge on microorganisms isolated in culture does not completely represent the microbial diversity in saline systems [5,7,15,17].

Salinity is an important factor that selects and structures microbial assemblages globally [18–20], and microorganisms inhabiting high salinity environments, mostly prokaryotes, have developed several salinity-stress adaptation strategies [21]. Eukaryotes might have greater difficulty in coping with the selective effect of high salinity [21–22], resulting in large decreases in the number of species as salinity increases [23]. This hypothesis might explain why eukaryotes are poorly represented in high-salinity environments compared to prokaryotes. Description of the molecular diversity of small marine eukaryotes through rRNA gene cloning and sequencing has revealed a large diversity of ribosomal types and identified novel lineages within microbial eukaryotes [24–25]. However, there are few studies analyzing the genetic diversity of eukaryotic assemblages in high-salt environments at high altitudes, although consistent changes in eukaryotic community composition and richness have been observed along salinity gradients [26]. Sequence analysis of selected major denaturing gradient gel electrophoresis (DGGE) bands revealed many sequences (largely protist) that are not related to any known cultures but that are related to uncultured eukaryotic picoplankton and unidentified eukaryotes in Eastern Tibetan Lakes [7]. High-salinity water bodies in inland saline ponds contain an unexpected large genetic diversity of novel protists [15], but the number of such eukaryotic microbial species in these environments remains to be elucidated [5,27].

Traditionally, studies on the diversity of eukaryotic assemblages (protist) have largely relied on morphological surveys using different microscopic techniques [28–30], and some important components of the microbial diversity in environmental samples have remained undetected using traditional methods [5,15]. Microscopy approaches have difficulties in identifying small cells (<10 μm), and thus, this fraction is understudied [25]. Recently, the development of high-throughput next-generation sequencing (NGS) technology for DNA sequencing [5,27,30–33] has facilitated extensive sequence-based characterization of diverse natural microbial communities and has allowed an assessment of microbial communities at high resolution based on deep taxon sampling [34]. Because millions of sequence reads are generated in a single experiment, NGS has revolutionized surveys of microbial diversity. Compared to microscopy, NGS-based amplicon sequencing is superior in detecting rare species [35], and it is now possible to recognize and identify nano- and picophytoplankton such as unicellular cyanobacteria and small flagellates, which cannot be discriminated based on morphological features [25,27,30,33].

The 18S rRNA gene is a widely used and valuable 'bar-code' to analyze eukaryotic diversity, because it is universally present in living organisms, and there are significant sequence data for comparison in public databases such as GenBank [30,33,36]. Recently, this gene marker is commonly used for the next-generation sequencing [25,41–42]. The repetitive arrangement of rRNA genes within the genome provides large amounts of template DNA for PCR, even in the smallest organisms. Comparative studies using 18S rRNAs have observed that the major length-variable regions are distributed on the surface of the molecules [43–45], whereas the intronic splicing sites are clustered in the inner region [45]. However, some sources of bias have been identified, including pyrosequencing errors [46–47], inappropriate clustering approaches [48] or inconsistent results from different targeted 18S rDNA regions [42]. More recently, Hadziavdic [33] et al. (2014) completely characterized the variable and conserved regions in the 18S rRNA gene, and their results suggested that the V2, V4, and V9 regions are best suited for biodiversity analysis. Furthermore, algorithms to remove pyrosequencing errors have been developed to reduce these overestimations, such as single-linkage preclustering (SLP) approach [48]. The 18S rDNA amplicon pyrosequencing has become a widespread approach for microbial community diversity studies. However, to the best of our knowledge, no similar studies have been performed on eukaryotic plankton in high-salt environments at high altitudes.

The Qaidam Basin is located in the northeastern section of the Plateau of Tibet, occupying the northwestern part of Qinghai province, western China. The basin is almost entirely an area of interior drainage, with rivers discharging either into Koko Nor or into one of the numerous salt lakes and saline swamps in the basin's central area, such as Tuosu, Dasugan, Gahai, Gasikule and Xiaochaidan. In the present work, we have analyzed the genetic diversity of planktonic microbial eukaryotes (size range 0.2–20 μm) along a salinity gradient in six water bodies in the Qaidam Basin using next generation sequencing (NGS). Samples were obtained from different geographic regions and covered a wide range of environmental conditions such as salinity (concentration and composition), in situ temperatures, trophic status, water and connectivity regimes, and altitude, which captured part of the high variety of saline habitats present in continental areas.

Materials and Methods

Sample collection and DNA extraction

No specific permissions are required for six lakes, located in publicly accessible areas. The field studies did not involve endangered or protected species. Overall, 13 samples (Lake Keluke 1, Lake Keluke 2, Lake Dasugan 10, Lake Dasugan 11, Lake Tuosu 3, Lake Tuosu 11, Lake Tuosu 12, Xiaochaidan 12, Xiaochaidan 13, Lake Gahai 1, Lake Gahai 7, Lake Gsikule 1 and Lake Gsikule 3, hereafter referred to as K11, K13, D10, D11, T3, T11, T12, M1, M7, X12, X13, G1, G3, respectively) were analyzed (Table 1). Six lakes located on the Qaidam Basin at altitudes ranging from 2,792 to 3,170 m above sea level were investigated (Figure 1). The lakes were chosen in order to cover a salinity gradient from 0 to 466 g/L. Water samples were collected from surface waters (top 30 cm) with a 5 L schindler sampler in August 2013 and were first condensed by 0.2-μm pore-size filters. Plankton samples (2000–2500 ml water) for NGS analyses were further filtered through a 20 μm mesh sieve immediately to remove most of the large particles. Water temperatures, pH and dissolved oxygen levels were measured on board using a Hydrolab sensor (Austin, TX, USA). Concentrations of the six major ions potassium (K^+), sodium (Na^+), calcium (Ca^{2+}), magnesium (Mg^{2+}), chloride (Cl^-), and sulfate (SO_4^{2-}), as well as the concentration of total nitrogen (TN) and total phosphorus (TP) were measured according to the standard methods [5,7,49] after transportation of samples to the laboratory. The total dissolved solid (TDS) of the investigated habitats was determined by a conventional conductivity meter (Table 1). Filters for extraction of DNA were centrifuged to the bottom of the eppendorf tube (3200 g 5 min, Eppendorf 5810R) and stored at −80°C in an ultra low-temperature freezer for DNA extraction. DNA was extracted after the cetyltrimethylammonium bromide extraction procedure [7]. The final ethanol-rinsed DNA pellets were dried and resuspended in 100 μL of 1× TE buffer (10 mM Tris–HCl, 1 mM EDTA) and stored at −80°C. DNA concentration and quality were determined with a NanoDrop 1000 spectrophotometer (Wilmington, DE, USA).

Next-generation sequencing

PCR was performed using 454 sequencing adaptor-linked primers flanking V4 region of the 18S rRNA gene: A-528F (5′-gcctccctcgcgccatcag-GCGGTAATTCCAGCTCCAA-3′) and B-1055R (5′-gccttgccagcccgctcag-ACGGCCATGCACCACCACC-CAT-3′) (adaptor sequences shown in lowercase) [34] by GeneWiz, Inc. (Beijing, China). PCR mixtures (50 μl) were prepared in duplicate and each contained 2 μl of DNA template, 5 μl of 10× PCR buffer (50 mM KCl, 10 mM Tris-HCl and

Table 1. The geographical, physical, and chemical characteristics of the investigated lakes. *ND* not determined, *n.d.* not detectable.

	Sample ID	Longitude	Latitude	TN mg/L	NH_3-N mg/L	PH	D0 mg/L	Sampling depth	Tempe-rature°C	Altitude m	TDS g/L	Na^+ g/L	K^+ g/L	Mg^{2+} g/L	Ca^{2+} g/L	Cl^- g/L	SO_4^{2-} g/L
Fresh water lake	Keluke11	96°51′39.05″	37°17′01.98″	2.21	0.40	8.83	8.7	0.3	15.1	2807	*ND*	*n.d*	*n.d*	*n.d*	*n.d*	*n.d*	*n.d*
	Keluke13	96°53′57.04″	37°17′12.23″	1.57	0.33	8.89	9.7	0.3	15.6	2816	*ND*	*n.d*	*n.d*	*n.d*	*n.d*	*n.d*	*n.d*
Mesosaline lake	Tuosu3	96°57′26.00″	37°08′59.00″	2.45	1.26	8.63	5.7	0.3	17.9	2803	30.9	7.47	0.22	1.5	0.04	12.36	6.52
	Tuosu11	96°54′54.27″	37°08′58.91″	2.74	0.36	8.82	6.5	0.3	17.1	2802	30.3	7.37	0.22	1.31	0.03	12.09	6.39
	Tuosu12	96°53′34.00″	37°19′58.00″	3.60	0.23	8.9	6.6	0.3	15.3	2801	30.5	*ND*	*ND*	1.25	0.04	10.42	6.37
	Dasugan10	93°56′02.39″	38°52′48.02″	0.55	0.40	8.95	6.4	0.3	13.2	2796	31.2	6.94	0.39	1.16	0.09	9.29	9.41
	Dasugan11	93°57′21.25″	38°52′48.02″	4.19	0.94	8.98	7	0.3	12.8	2792	30.7	6.57	0.37	1.08	0.11	9.16	9.28
Hypersaline lake	Gahai1	97°31′32.45″	37°08′41.78″	2.05	0.46	8.19	2.8	0.3	19.3	2853	96.9	24.5	0.39	1.06	0.27	45.65	11.63
	Gahai7	97°31′52.61″	37°08′36.19″	3.11	0.4	8.19	2.2	0.3	19.3	2851	93.6	25.02	0.41	1.23	0.27	45.02	11.52
	Xiaochadan12	95°30′19.03″	37°28′31.77″	3.27	1.08	8.31	4.9	0.3	10.8	3170	96.7	31.43	0.56	1.3	0.39	44.68	18.19
	Xiaochadan15	95°33′9.10″	37°28′52.57″	3.29	1.07	8.32	4.9	0.3	10.5	3168	96.2	*ND*	*ND*	1.71	0.41	42.76	4.08
	Gasikule1	90°49′01.89″	38°09′43.03″	8.9	1.18	6.73	1.4	0.3	16	2858	348.3	74.65	2.46	2.96	0.16	134.28	10.75
	Gasikule3	90°46′59.72″	38°07′58.62″	6.6	1.36	7.29	1.5	0.3	15.5	2858	466	75.5	2.32	1.97	0.18	156.1	14.48

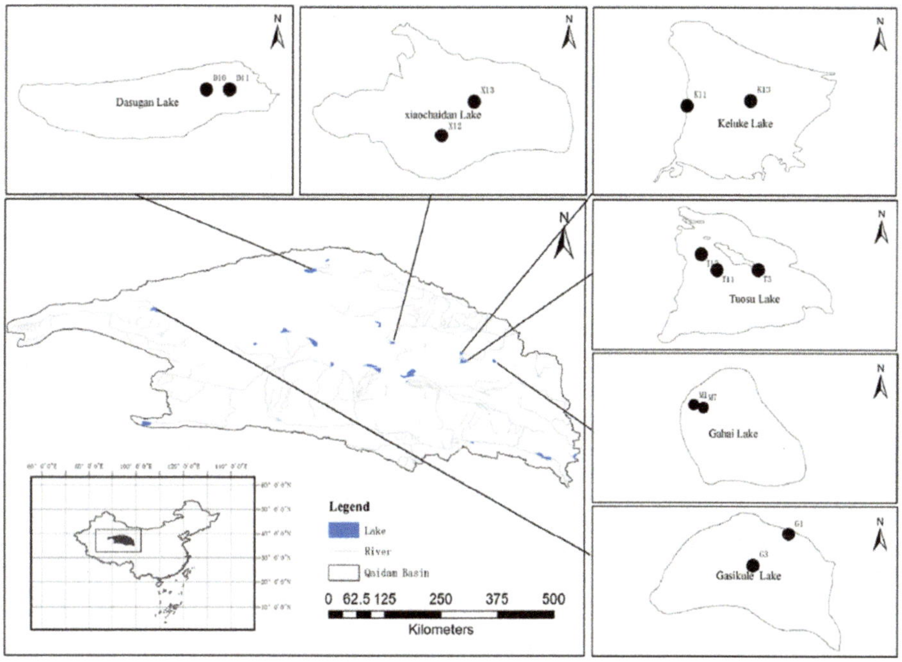

Figure 1. Locations of investigated lakes in Qaidam Basin. In Figure, K, T, D, M, X and G referred to Lake Keluke, Lake Tuosu, Lake Dasugan, Lake Gahai, Lake Xiaochaidan and Lake Gsikule, respectively.

1.5 mM $MgCl_2$), 200 μM of dNTP, 0.2 μM of each primer and 2.5 U Taq polymerase (Promega, Madison, WI, USA). The PCR thermal regime consisted of an initial denaturation of 3 min at 94°C, followed by 30 cycles of 30 s at 94°C, 30 s at 60°C, 1 min at 72°C and a final cycle of 5 min at 72°C. PCR products were pooled and purified with the Qiaquick gel purification kit according to the manufacturer's instructions (Gene Company Limited, Beijing, China).

Analysis of 454 pyrosequencing data and clustering of sequence reads into OTUs

The 454 sequencing reads were processed using the Qiime (Quantitative Insights into Microbial Ecology, v1.3.0) pipeline as flowgrams (.sff files). The data was processed following the steps recommended in the Qiime processing 18S data tutorial. In order to increase the quality of the raw sequences, reads shorter than 300 bp were excluded from the analysis to guarantee the analysis of the whole V4 region. Sequences longer than 670 bp (expected amplicon size) and those with more than one uncertain base (N) were further removed. The following parameters settings were used, namely an operational taxonomic unit (OTU) threshold of 0.97, 2 as maximum number of primer mismatches, 0 ambiguous bases, a maximum length of homopolymer run of 10, 300 nucleotides as a minimum sequence length and 670 as a maximum sequence length. The multiplexed reads were assigned to samples based on their nucleotide barcode. The data were then de-noised using the de-noiser wrapper within Qiime to remove the sequence errors characteristic of 454 sequencing machines. Chimeras were identified using ChimeraSlayer [50] and rejected from the dataset before construction of the OTU table. The OTUs were assigned using UCLUST [51], an open reference OTU picker within Qiime, a representative set of sequences was then generated and these sequences were assigned taxonomy using the SILVA release 115 database.

Another clustering process was to apply the NDIST module of AmpliconNoise [52] with default parameters to generate the distance matrix of all Needleman–Wunsch pairwise sequence alignments. Then, the average linkage option in the AmpliconNoise FCluster module was used to cluster reads into OTUs. In addition, we ran CD-HIT-OTU, a recent clustering algorithm for 18S rDNA pyrosequence data, with default parameters (including –e 0.0025) on the OTU finder web server (http://weizhong-lab.ucsd.edu/metagenomic-analysis/server/cd-hit-otu/).

The Closest environmental match (CEM) and the Closest cultured match (CCM) available

We explored the 18S rRNA gene novelty of the dataset by BLAST identity search against GenBank sequences (search May 2014). The identity of each single sequence was related to both the closest environmental match (CEM), and the closest cultured match (CCM) available in GenBank. Histograms [15,24,53] were used to assess the degree of novelty comparing salinity gradient (Hypersaline lake vs. Mesosaline lake).

Phylogenetic analysis

Sequence alignment of amplicon reads was performed with MegAlign implemented in DNAStar 6.0 software package (DNASTAR, Madison, USA) and then was confirmed visually by BioEdit 7.0.9 [54]. The ambiguous regions of alignment were discarded and eventually 733 nucleotide bases were obtained. Phylogenetic tree was reconstructed based on the full alignment of 733 sequences by using approximately Maximum Likelihood (ML) analysis in FastTree 2.1.3 [55]. For FastTree 2 analysis, a heuristics search strategy was employed with an estimated rate of evolution for each site (the "CAT" approximation), minimum-evolution subtree-pruning regrafting (SPRs), and maximum-likelihood nearest-neighbor interchanges (NNIs). Bootstrap ML analysis was carried out using 1000 pseudo-replicates. Trees were edited with the online tool iTOL [56].

Other Statistics

For statistical analysis, the environmental parameters were transformed to avoid skewed data distributions: ion concentrations were arcsine transformed, other chemical parameters were log10 transformed; pH, altitude, and latitude were not transformed. We used principal component analysis (PCA) on chemical parameters (ion concentrations, pH, and conductivity) to display the main gradients in chemical parameters by CANOCO program (57). Significant marginal effects were analyzed by running separately a canonical correspondence analysis (CCA) on the OTU using square root transformation for each of the environmental factors separately (i.e. ion percentages, pH, altitude, TP, TN, TDS) by CANOCO program [57]. The data set generated in this study has been deposited at GenBank's Short Read Archive (SRA) under Accession No. SRA178606.

Results

Characteristics of the studied lakes

The surveyed systems covered a wide range of variability in environmental conditions, including up to 40-fold differences in the salinity gradient (from 0 to approximately 46% salinity), altitude (from 2792 m to 3170 m), and in situ temperature (from 10.8 to 19.3°C) (Figure 1, Table 1). The Gahai, Gasikule, and Xiaochaidan lakes are more saline (hypersaline lakes >50 g/L) compared to the lakes Tuosu and Dasugan (mesosaline lakes, 20–50 g/L). The Lake Keluke is the only freshwater lake in this study. These six lakes are located in four different regions, with a minimum distance of 5 kilometers between them (Figure 1). Lake Gasikule is located in the northwestern part of the Qaidam Basin, where less precipitation and high evaporation have resulted in the highest salinity. Lake Dasugan, in which the ground water is considered the primary source of runoff, is located in the northern part of the Qaidam Basin. Lake Xiaochaidan is located in the northeastern part of the Qaidam Basin, whereas the other three lakes are located in close proximity to each other in the eastern part of the Qaidam Basin and have significantly different physical and chemical conditions (Table 1). Over the last century, Lake Xiaochaidan has been completely drained to exploit mineral resources and has been recovered by the reinjection of water. Lake Tuosu is connected to the freshwater lake Keluke by a short river; Lake Tuosu is expanding, resulting in altered salinities and phytoplankton communities. Table 1 lists the sampling sites, collected samples and physicochemical parameters that were measured. PCA of the lake environmental factors revealed that the first two components accounted for 79.44% of the total variation (PC1, 63.48%; PC2, 15.96%). There was a significant salinity gradient among the lakes, as observed by the high correlation between PCA axis 1 (Figure 2) and the chemical parameters (Table S1).

Composition of the eukaryotic (nano- and picoeukaryotes) plankton community in saline lake

After quality filtering and preprocessing, 350 640 reads were obtained from the 13 sequenced samples in the study; approximately eighty-one percent, or a total of 286 360 reads, could be assigned to eukaryotic assemblages. Sequencing yielded highly variable results among the samples, ranging from 15 482 to 30 625 total reads per sample. A ML phylogenetic tree with all sequence reads provided a detailed picture of the diversity of Qaidam Basin lake eukaryotes (Figure 3A). To our surprise, at least 11.23% in total number of sequence reads could not be precisely assigned to any known eukaryotic taxonomic group. Further analysis showed that the 18S rRNA gene amplicon sequence were distributed among thirteen high-rank taxonomic groups and matched 70 eukaryal classes in freshwater lake (indigo in the outer ring) (Figure 3A, Table S2). Notably, at least 30 classes might be phylogenetically novel at least at the class level in freshwater lake. Most sequence reads from freshwater lake were affiliated with class Intramacronucleata (Alveolata), accounting for 29.50% of the total number of sequence reads. Eukaryotes from mesosaline lakes (lime in the outer ring) were distributed among eleven high-rank taxonomic groups and 38 eukaryal classes, whereas at least 16 classes could not be clearly attributed to known eukaryotic taxonomic groups (Figure 3A, Table S2). Craspedida (Holozoa) were the most prevalent class in mesosaline lakes, accounting for 26.52% of all sequence reads. The 18S rRNA gene sequences obtained from hypersaline lakes (seagreen in the outer ring) were distributed among eleven high-ranking taxonomic groups and matched 53 classes. The most prevalent class in hypersaline lakes was affiliated with Animalia (Holozoa), accounting for 21.98% of the total number of sequence reads. Notably, as was shown in Table S2, at least 35.85% of the total numbers of classes likely represented novel phylogenetic lineages at the class level.

Holozoa were the most common kingdom in mesosaline Lake Tuosu, accounting for 44.50% of the total number of sequence reads (Figure 4). Stramenopiles were the second most prevalent kingdom, accounting for 24.2% of the total number of sequence reads. Approximately 10.97% of the sequence reads were new sequence types, defined as unclassified eukaryal clusters. The most prevalent kingdom in mesosaline Lake Dasugan was affiliated with Holozoa, accounting for 51.75% of the total number of sequence reads. Figure 4 showed that approximately 7.1% of the sequence reads were not classified into any of the known eukaryal clusters in Lake Dasugan.

There were approximately 14.85%, 2.85%, and 29.85% sequence reads defined as unclassified eukaryal clusters in the hypersaline lakes Xiaochaidan, Gasikule, and Gahai, respectively. Most sequences from Lake Xiaochaidan were affiliated with Stramenopiles (34.10%) (Figure 4). Haptophyta are the next most prevalent kingdom in Lake Xiaochaidan, accounting for 31.60% of all sequence reads. However, the most abundant sequence reads were mainly represented by Holozoa (36.35%), Chloroplastida (28.00%), Alveolata (17.25%), and Stramenopiles (10.20%) in Lake Gasikule. Holozoa is also the most common kingdom in Lake Gahai, accounting for 30.15% of the total number of sequence reads. The qualitative species composition of the abundant biosphere of Lake Gahai is shown in Figure 4. Alveolata were the third most prevalent kingdom, accounting for 19.00% of the total number of sequence reads.

Taxonomic Composition of investigated lakes was shown in Table 2. Notably, the composition of the eukaryotic plankton community in different sampling sites of the same lake was significantly different (Figure 5, Table 2). Intramacronucleata (Alveolata) were the dominant class in sampling sites T3 (37.10%), whereas the most prevalent class in sampling sites T11 was Craspedida (Holozoa) (20.80%) (Table S2).

Most abundant taxa in different sampling sites

As shown in Figure 5, an *uncultured_freshwater cercozoan* exhibited the highest proportion in K13, accounting for 10.45% of the total number of amplicon reads. However, the most common taxa in the K11 sampling point were *an uncultured_freshwater Eukaryote*, accounting for 6.89% of the total number of amplicon reads. The most common taxa in the D10 sampling point belong to the class Craspedida, accounting for 50.89% of the total number of amplicon reads (Figure 5). In contrasts, another species that affiliated with the class Craspedida was prevalent at sample

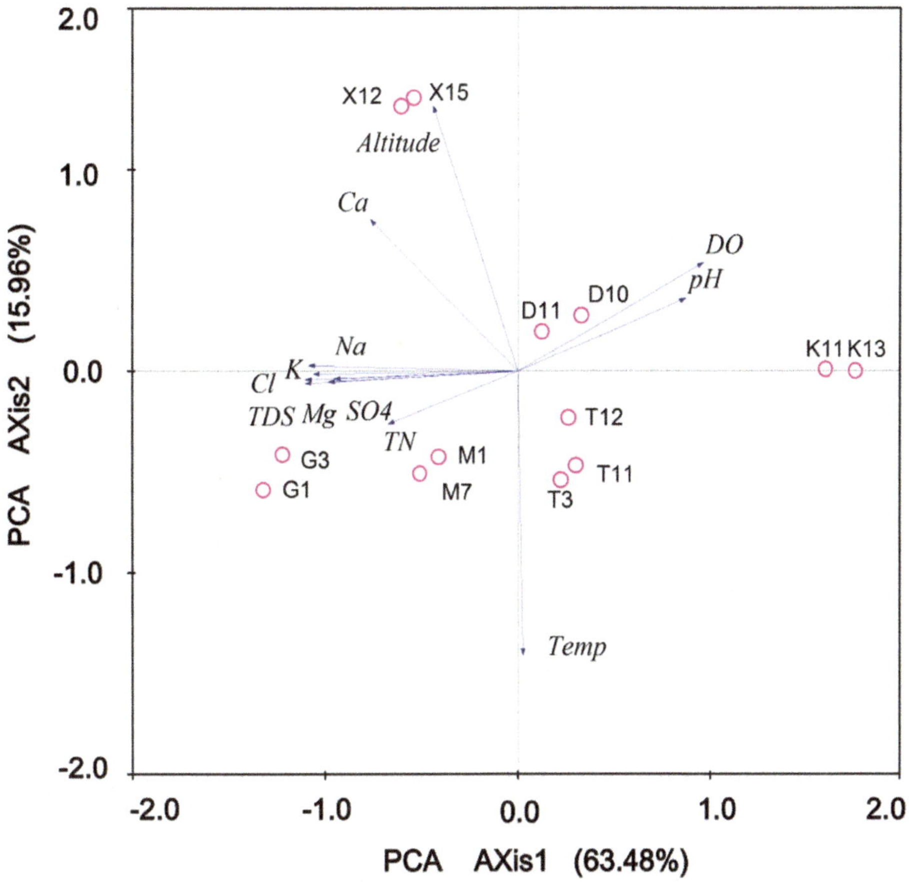

Figure 2. PCA analyses on DO, pH, TN and ion percentages of investigated Tibetan lakes. DO dissolved oxygen, TN total nitrogen, TDS total dissolved solid. K, Na, Ca, Mg, Cl, SO4 was represent for ions potassium (K^+), sodium (Na^+), calcium (Ca^{2+}), magnesium(Mg^{2+}), chloride (Cl^-), and sulfate (SO_4^{2-}), respectively. The numbers 1 to 13 refer to investigated lakes, Lake Keluke (G1, G3), Lake Gahai (M1, M7), Lake Dasugan (D10, D11), Lake Tuosu (T3, T11, T12), and Lake Keluke (K11, K13), respectively.

site D11, accounting for 30.19% of the total number of amplicon reads. At stations T3 and T11, the eukaryal communities consisted mainly of uncultured_Scuticociliatia, and Monosiga brevicollis, respectively (Figure 5).

Lake Gahai is located quite far from Lake Gasikule. However, the compositions of the most common taxa in these lakes were quite similar. As shown in Figure 5, one species that affiliated with the Mammalia family was prevalent at sample site G1 and M1, accounting for 28.81% and 23.35% of the total number of amplicon reads, respectively. However, in another hypersaline lake (X15), the most abundant sequences were affiliated with a species affiliated with phylum Bicosoecida (31.20%), which was also the next most common taxa in X12 sample site (Figure 5). As was shown in Figure 5, the eukaryotic plankton assemblage was dominated by a species from the phylum Pavlovophyceae at the X12 sample site, accounting for 33.43% of the total number of amplicon reads.

Abundance tables of OTU were used to calculate Shannon's diversity index and Simpson's evenness. As indicated by the Observed species rarefaction curves, (Figure S1), the shape of the curves indicates a trend of diminishing chance of finding new phylotypes as sampling continues. The estimated number of observed species varied between 86.20 and 250.80 (Figure S1). The highest diversity index was observed in Lake Keluke (Observed Species = 250.80), and the lowest diversity index was in Lake Dasugan (Observed Species = 86.20). The Chao1 estimator was calculated to predict the total number of OTUs (richness) in the water samples. The taxonomic richness estimated for the six lakes was 267.50, 208.14, 199.21, 101.75, 117.80, and 119.35 for Lakes Keluke, Gasikule, Gahai, Dasugan, Tuosu, and Xiaochaidan, respectively (Figure S1). Considering the mean estimated richness in DNA datasets, Lake Keluke is the most diverse lake, closely followed by Lake Gsikule. Notably, eukaryotic plankton communities were more diverse in the hypersaline lakes compared to the mesosaline lakes.

Cluster Analysis

Phylogenetic tree reconstruction using the maximum likelihood (ML) with all sequence reads elucidated the diversity of eukaryotic plankton assemblages in lakes located in the Qaidam Basin (Figure 3A). There were approximately 1.90%, 9.04% and 15.85% OTUs defined as unclassified eukaryal clusters in fresh water lake, mesosaline lakes and hypersaline lakes, respectively. This highly supported tree was pivotal to place very divergent sequences that could not be classified into any known eukaryal class that could not be identified by BLAST. Interestingly, some novel OTUs affiliated within a given taxonomic group. As shown in Figure 3A, about twelve novel OTUs were related to Holozoa groups. Nevertheless, at least four OTUs in mesosaline lake and eighteen OTUs in hypersaline lake could still not be related to any eukaryal taxonomic group, not even to a supergroup, and occupy

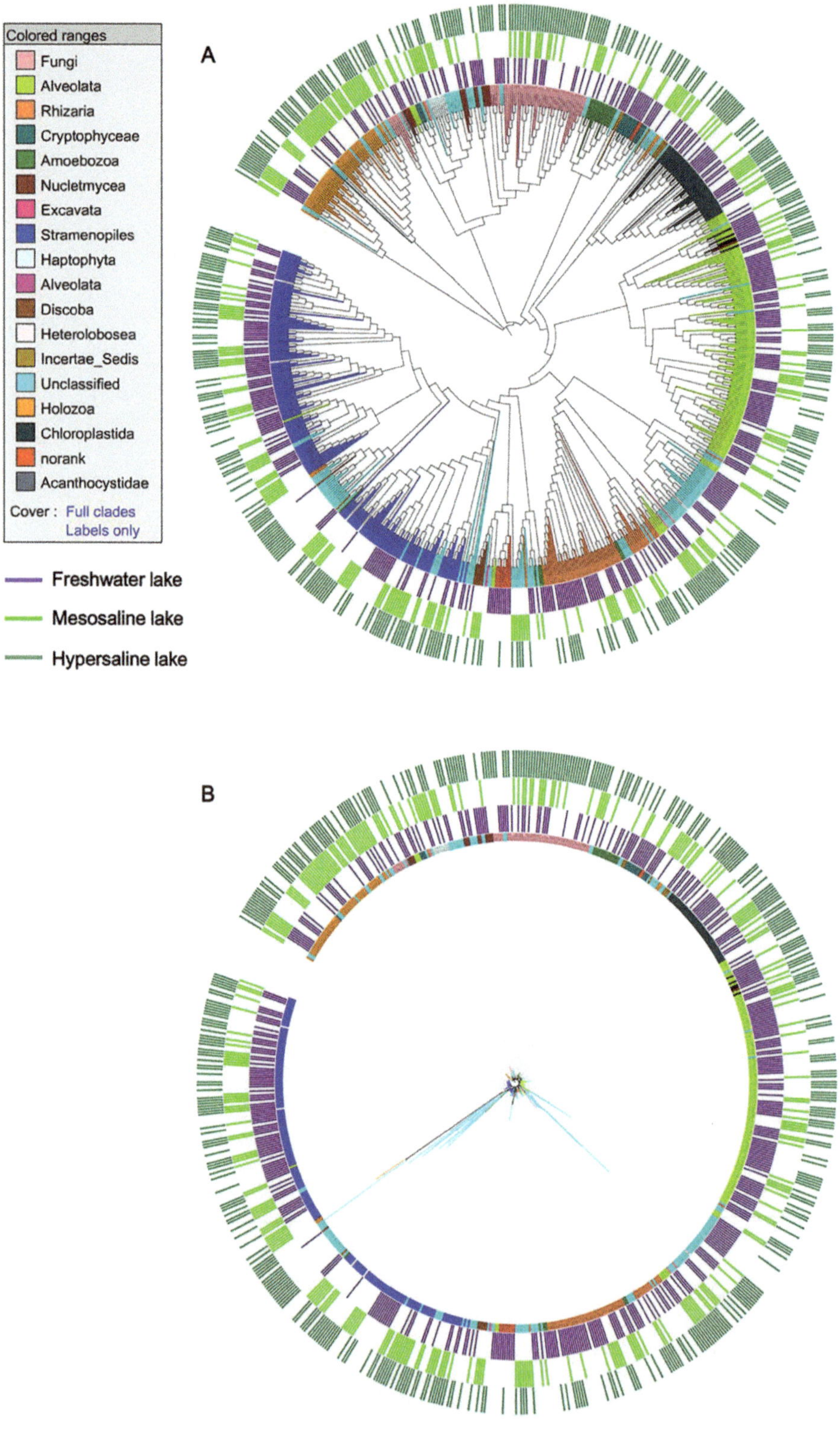

Figure 3. ML phylogenetic tree with 18S rDNA sequences of eukaryotes retrieved from the investigated lakes in the Qaidam Basin. The tree was constructed using 733 sequences. (A) Tree ignoring branch lengths and overlaid with colors with independent taxonomic assignations of sequences to an eukaryotic supergroup. The branches with cyan colors correspond to unclassified eukaryotes. (B) Same tree showing branch lengths, colored as before. The scale bar indicates 0.1 substitutions per position.

highly unique branches in this phylogenetic analysis (Figure 3B). Phylogenetic analysis indicated that most OTUs from hypersaline, mesosaline, and freshwater lakes tended to form separate groups and shared rather low similarities (Figure 3).

Taxonomy	Fresh water lake	Mesosaline lake		Hypersaline lake		
	Keluke	Dasugan	Tuosu	Gahai	Xiaochaidan	Gasikule
Other	1.90%	7.10%	10.97%	29.85%	14.85%	2.85%
Alveolata	40.95%	7.50%	14.17%	19.00%	3.20%	17.25%
Amoebozoa	1.20%	0.00%	0.03%	0.05%	0.00%	0.25%
Centrohelida	0.05%	0.00%	0.00%	0.00%	0.00%	0.00%
Chloroplastida	0.55%	0.10%	0.37%	0.50%	1.15%	28.00%
Cryptophyceae	6.00%	0.00%	0.03%	0.35%	0.00%	0.00%
Discicristoidea	0.00%	0.00%	0.00%	0.00%	0.00%	0.00%
Discoba	0.05%	0.00%	0.03%	0.15%	0.00%	0.00%
Excavata	0.00%	0.00%	0.00%	0.00%	0.00%	0.00%
Fungi	1.20%	0.20%	0.77%	1.85%	1.25%	4.40%
Haptophyta	0.15%	0.05%	0.03%	1.00%	31.60%	0.00%
Holozoa	4.95%	51.75%	44.50%	30.15%	11.70%	36.35%
Incertae_Sedis	0.75%	0.45%	0.50%	0.00%	0.00%	0.00%
Nucletmycea	2.40%	0.00%	0.00%	0.00%	0.00%	0.25%
Rhizaria	19.80%	2.35%	4.87%	9.05%	0.25%	0.35%
Stramenopiles	19.50%	30.60%	22.70%	6.60%	34.10%	10.20%
norank	0.55%	0.00%	1.03%	1.30%	1.90%	0.00%

Figure 4. Distribution matrix with the proportion of sequence reads in investigated lakes with different physical and chemical conditions. Taxa were defined at Kingdom level. "Incertae sedis" is a term used for a taxonomic group where its broader relationships are unknown or undefined. "other" means group that could not be precisely assigned to any known eukaryotic taxonomic group. "norank" means group that could not be precisely assigned to eukaryotic taxonomic group at least at the kingdom level.

Novelty in each eukaryal class

After water samples were obtained and sequenced, the identity of the different microorganisms was analyzed by BLAST search. For each OTUs, the similarity to the CEM and the CCM was recorded (Figure 6). The results were further classified according to salinity concentration (i.e., mesosaline vs. hypersaline, Figure 6A, B). The average CEM similarity (83%) was slightly lower than the average CCM similarity (83.87%) in hypersaline lakes. However, the average CEM similarity (89.96%) was much higher than the average CCM similarity (86.40%) in mesosaline lakes. We then analyzed the novelty of the taxonomic composition of the different assemblages. Each phylogenetic group exhibits a different novelty pattern, as observed with the supergroups Alveolates and Stramenopiles (Figure 6). In the kingdom Stramenopiles, the median genetic diversity values were ≥75% for both cultured and environmental matches, with similar sequence identity distributions in hypersaline lakes. Conversely, only approximately 35% of the Stramenopiles group in mesosaline lake matched both cultured and environmental sequences. As shown in Figure 6, dots below 80% similarity in both axes indicated highly divergent novel sequences. Notably, we observed that in mesosaline lakes (Figure 6A), at least four sequences had no similarity to CCM or CEM data, whereas in hypersaline lakes, up to eighteen OTUs yield generally low similarity values (Figure 6B). These OTUs were either not successfully cultured in the laboratory or have not been observed in nature.

Distribution of OTUs and Influence of Physicochemical Parameters

Lake Keluke contained the highest number of OTUs unique to one lake, which account for approximately 59.92% of all OTUs (Figure S2). Venn diagrams comparing the number of common OTUs among samples showed that the hypersaline lake Gasikule and the freshwater Lake Keluke had a relatively large overlap (approximately 137 OTUs) compared to other lakes. The smallest proportions of OTUs were shared between the freshwater Lake Keluke and the mesosaline Lake Dasugan (approximately 47 OTUs). OTUs that were present in all conditions were rare: only six such OTUs were observed in our data set.

To analyze the influence of eukaryotic plankton community structure and other measured physicochemical parameters, a distinct CCA was generated from pooled habitat datasets. CCA analysis of chemical variables yielded four clusters, separated mainly by altitude and the percentage of calcium ions. The first cluster contained hypersaline lake sample points (X12 and X15); the second contained large hypersaline lake sample points (M1, M7, G1 and G3); the third cluster contained freshwater lake sample points (K11 and K13), and the fourth cluster contained mesosaline lake sample points (D10, D11, T3, T11 and T12). Altitude and the calcium-ion percentage affected the eukaryotic community structure in Lake Xiaochaidan, whereas the calcium ion percentage and TN were major factors in Lakes Gahai and Gasikule. pH and dissolved oxygen (DO) were the most important factors influencing the distribution of eukaryotic plankton assemblages in Lakes Dasugan and Tuosu (Figure 7).

Discussion

The high-rank diversity observed in this study, both in terms of the eukaryotic supergroups detected and the presence and relative abundance of specific lineages, was typical of molecular surveys of eukaryotes [5,7,15,25]. Shannon rarefaction curves reached plateau in the present study, indicates that more intensive sampling is likely to yield only few additional species. As not expected, the lake with the most extreme conditions (hypersaline

Table 2. Statistics of Taxonomic Composition of investigated lakes.

	Samples	Kingdom	Phylum	Class	Order	Family	Genus	OTU
Fresh water lake	K11	12	15	50	86	98	112	343
	K13	11	13	33	53	59	71	165
Mesosaline lake	T3	8	12	30	43	50	59	128
	T11	10	12	28	40	46	49	111
	T12	7	9	22	32	35	40	95
	D10	5	9	22	27	30	30	75
	D11	7	11	24	35	40	42	105
Hypersaline lake	M1	9	12	36	62	73	86	238
	M7	8	12	26	38	40	42	101
	X12	7	10	24	34	40	43	103
	X15	7	10	24	31	35	38	95
	G1	7	13	31	56	69	78	144
	G3	8	12	32	59	74	85	174

lake Gasikule) had the highest OTU diversity, followed by the lake with the second highest salinity (Gahai). Although the actual numbers have to be regarded with caution, it seems clear that this study (and the many more to come with improved technologies) will significantly raise the higher limit of eukaryotes diversity.

Taxonomic groups detected

Picoeukaryotes are probably the most abundant eukaryotes on earth. They are found in all lakes and oceans at densities from 10^2 to 10^4 cells/mL [58–61]. Alveolates, Fungi and Stramenopiles represent 65% of the total diversity and differ from the dominant groups known from microscopic studies in freshwaters Lake Pavin [62]. Recent research showed that most of the 18S rRNA gene sequences affiliate with Stramenopiles, Cryptophyta and Alveolata in several alpine oligotrophic lakes of the Central Pyrenees (Spain) [15,63]. The taxonomic composition of the eukaryotic plankton assemblages, collected from freshwater Lake Keluke, differed from that reported in June 2012 from several alpine oligotrophic lakes of the Central Pyrenees (Spain) [15], where Stramenopiles, Cryptophyta and Alveolata dominated the 18S rRNA gene sequences. In the present study, Alveolata group represented a greater proportion of the community. Recent experiments by Mangot et al. (2013) showed that Almost 27% of the OTUs are affiliated with the Alveolates and more precisely the *Ciliophora* and *Perkinsozoa* taxa in freshwater Lake Geneva [64]. The dominance of Intramacronucleata within kingdom Alveolata in Lake Keluke highlights a difference in the composition of microbial eukaryotic communities between this lake and other lake.

From the general ecological principles it can be established that a more extreme environment is expected to be less species rich [65]. To our surprise, all the investigated saline lakes harbored remarkably diverse eukaryotic microbial communities, considering their higher salt concentrations. Mesosaline lake presented a different picture of the eukaryotic microorganisms community with freshwater lake. At least 7.10% of all sequences were related to unidentified eukaryal lineages in mesosaline lakes. Most 18S rRNA gene sequences obtained in mesosaline lakes were closely related to Holozoa (mainly Craspedida), differed from that reported in freshwater lake [58–61]. Genetic diversity of eukaryotic plankton assemblages has been well investigated in marine ecosystems. However, there is very few published studies utilised cloning-free high throughput sequencing to estimate the eukaryotic diversity of mesosaline lake. 454 pyrosequencing analysis reveal that Stramenopiles, Alveolata, Ciliates, and Prasinophytes were the dominant picoeukaryote groups in subtropical coastal waters [41]. The Stramenopiles constitute one of the major eukaryotic branches and include a vast number of heterotrophic and autotrophic groups with large ecological importance in the oceans [9]. Unlike in mesosaline lakes located in Qaidam Basin, in the Indian Ocean world, Alveolata and Stramenopiles are the most common groups, whereas Rhizaria are the next most abundant [24]. Recent experiments by Wolf et al. (2014) showed Alveolata (mainly Dinoflagellates) group are dominant in the sequence assemblage in the sub-Antarctic Zone [25]. These were not in agreement with the present result that Holozoa (mainly Craspedida) group was most prevalent in mesosaline lakes located in Qaidam Basin.

Hypersaline systems are overwhelming dominated by prokaryotes, with communities dominated by the haloarchaea [66]. However, eukaryotic organisms are also commonly present and might be play important ecological roles. In a study of 34 sites in different inland and coastal environments in Spain, most sequences are affiliated with Chloroplastida (mainly Chlorophyta), Alveolata, Stramenopiles, Opisthokonta, and Rhizaria, whereas

A

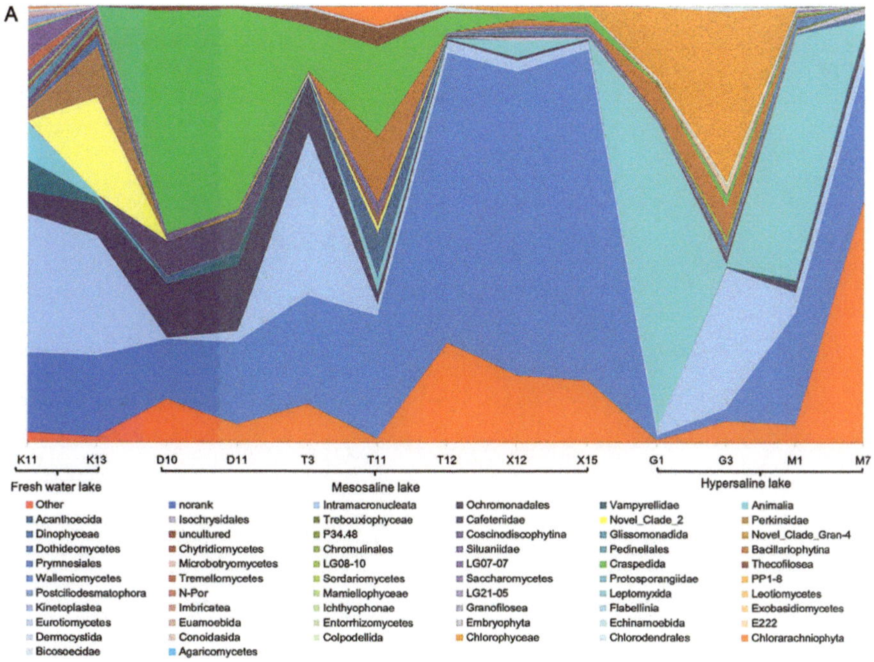

B

Lake type	Sample ID	Species	Number	Ratio in total reads(%)	OTU
Fresh water lake	K13	Rhizaria; s__uncultured_freshwater_cercozoan	Species	10.45	OTU116
		Rhizaria; s__ uncultured_freshwater_cercozoan	Species	10.23	OTU317
	K11	Rhizaria; s__uncultured_freshwater_eukaryote	Species	6.89	OTU677
		Alveolata; g__Halteria	Genus	6.52	OTU353
Mesosaline lake	T3	Alveolata; s__uncultured_Scuticociliatia	Species	30.91	OTU729
		Holozoa; s__uncultured_eukaryote	Species	8.56	OTU688
	T11	Holozoa; s__Monosiga_brevicollis	Species	19.03	OTU557
		Stramenopiles; s__Stramenopile_sp._ME8	Species	11.35	OTU305
	T12	Holozoa; c__norank	Class	26.68	OTU667
		Holozoa; c__norank	Class	25.80	OTU104
	D10	Holozoa; c__Craspedida	Class	50.89	OTU117
		Stramenopiles; s__Paraphysomonas_foraminifera	Species	6.91	OTU648
	D11	Holozoa; c__Craspedida	Class	30.19	OTU304
		Holozoa; c__Craspedida	Class	15.40	OTU117
Hypersaline lake	M1	Holozoa; f__Mammalia	Famliy	23.35	OTU620
		Holozoa; f__Mammalia	Famliy	13.55	OTU428
	M7	Eukaryota	Not classified	46.38	OTU535
		Alveolata; s__Oxyrrhis_marina	Species	7.13	OTU327
	X15	Stramenopiles; c__norank	Class	31.20	OTU670
		Haptophyta; c__norank	Class	20.68	OTU192
	X12	Haptophyta; c__norank	Class	33.43	OTU192
		Stramenopiles; c__norank	Class	12.60	OTU670
	G1	Holozoa; f__Mammalia	Famliy	28.81	OTU620
		Holozoa; f__Mammalia	Famliy	16.82	OTU428
	G3	Alveolata; f__Hypotrichia	Famliy	29.76	OTU576
		Chloroplastida; g__Dunaliella	Genus	29.27	OTU33

Figure 5. Taxa assignments at Class rank or below of investigated sample sits and the most abundant taxa in each habitat. (A) Taxa assignments at Class rank or below of investigated sample sits. (B) Taxa assignments at Class rank or below in each habitat. The class "other" means group that could not be precisely assigned to any known eukaryotic taxonomic group. The class "norank" means group that could not be precisely assigned to eukaryotic taxonomic group at least at the class level within kingdom. "Incertae sedis" is a term used for a taxonomic group where its broader relationships are unknown or undefined.

Centroheliozoa, Haptophyceae, and Telonemida are less well represented [15]. Chlorophyta have adapted to specialised and extreme environments, such as deserts, arctic environments, hypersaline habitats [15,67]. A study of high mountain saline lakes of the Eastern Tibet Plateau showed that most sequences are affiliated with Chloroplastida (mainly Chlorophyta) and Alveolata (mainly Dinophyceae) [7]. As we not expect, Chloroplastida group was only observed in high abundance in Lake Gasikule, whereas was less represented in other hypersaline lakes such as Gahai and Xiaochaidan. Our finding that Holozoa (mainly Animalia) were the most common kingdom in hypersaline lakes in the Qaidam Basin was contrast with a recent research that water samples were dominated (91%) by a novel cluster of the Alveolate in hypersaline Lake Tyrrell (Australia) [66]. These results were in agreement with previous reports that distinct biogeographical patterns are defined by the environmental conditions in particular regions [25]. The

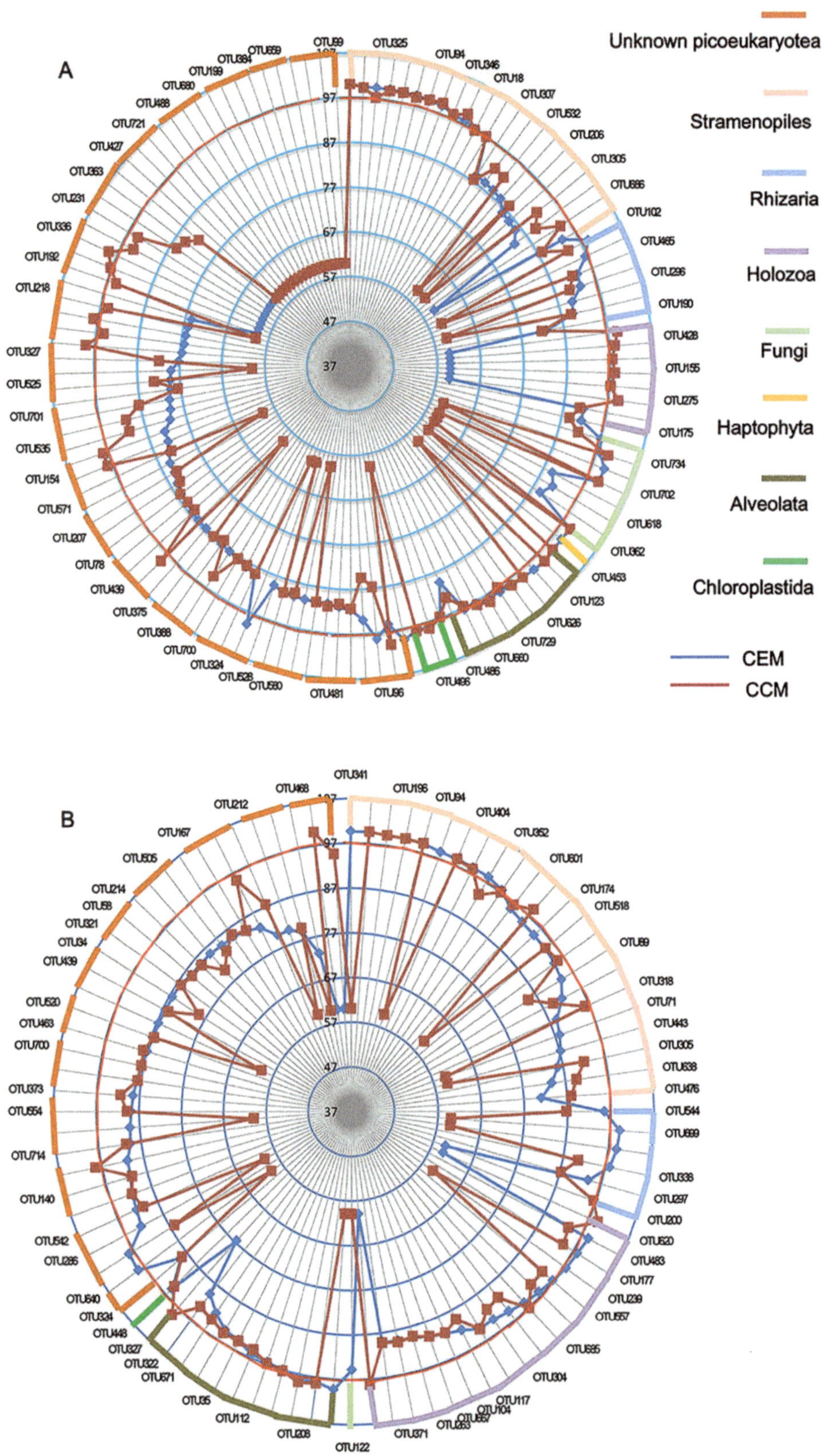

Figure 6. Novelty pattern plot for the different eukaryal classes found in mesosaline lake and hypersaline lake. (A) mesosaline saline, (B) hypersaline. Closest environmental match (CEM) and the closest cultured match (CCM) available in GenBank (BLAST search, May 2014). Dots below 61% similarity in both axes indicated no similarity sequences.

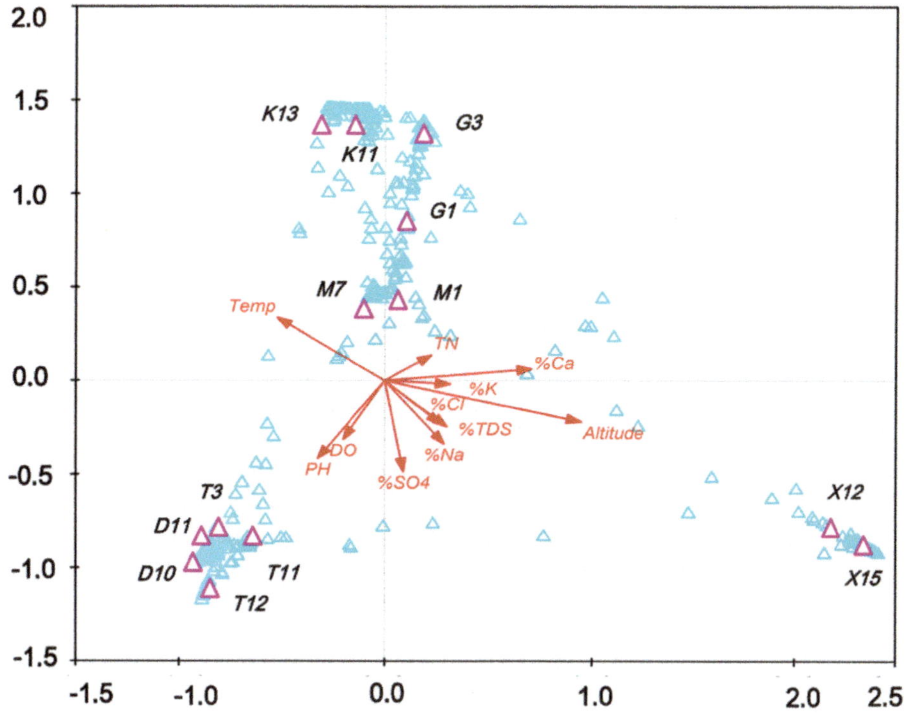

Figure 7. CCA biplots based on OTU and geographical or selected chemical parameters. Chemical parameters yield in a very similar spreading of sampling sites; DO dissolved oxygen, TN total nitrogen, TDS total dissolved solid. K, Na, Ca, Cl, SO4 was represent for ions potassium (K^{+}), sodium (Na^{+}), calcium (Ca^{2+}), chloride (Cl^{-}), and sulfate (SO_4^{2-}), respectively. In Figure, K, T, D, M, X and G referred to Lake Keluke, Lake Tuosu, Lake Dasugan, Lake Gahai, Lake Xiaochaidan and Lake Gsikule, respectively.

dominance of Holozoa (mainly Animalia) in water samples highlights a difference in the composition of eukaryotes between hypersaline lakes in the Qaidam Basin and other hypersaline environments.

It is of interest that Lake Xiaochaidan has the same salinity as Lake Gahai, but these lakes differed in the quantitative composition of their eukaryotic communities. Haptophyta were in high abundance in Lake Xiaochaidan, whereas present in lower numbers throughout coastal environments in Spain [15], Lake Gahai and Lake Gasikule. Several reasons might explain the different distributions of taxonomic groups in Lake Xiaochaidan compared to hypersaline lakes. First, Lake Xiaochaidan has been completely drained to exploit mineral resources in the last century, which has eliminated certain species, decreasing the abundance of the Holozoa group. Second, some species such as Alveolata, which are commonly observed in the freshwater Lake Keluke, might thrive because of freshwater injection. More interestingly, recent experiments by Charvet et al. (2014) showed that eukaryotic plankton assemblages in fresh waters were related to the Alveolata and in the lower saline waters were related to Stramenopiles [61]. However, further studies are required to confirm this hypothesis. Nevertheless, these data are consistent with previous conclusions that human activity is one of the greatest threats to these ecologically valuable habitats [3].

Novelty analysis of environmental sequences

The novelty analysis revealed highly divergent sequences that appeared in the area of the dispersion plot with very low CCM and/or CEM values (Figure 6). A recent reports showed that 10% of the entire dataset showing <90% identity to any previously reported sequence in GenBank in 34 different coastal and inland saline ponds [15]. Notably, approximately 32% and 37.98% of the entire dataset in mesosaline lakes and hypersaline lakes, respectively, had <90% identity to previously reported sequences in GenBank (Figure 6). Most of these sequences formed long branches in the ML phylogenetic tree, although some could be confidently placed in a taxonomic group based on the dendrogram (Figure 3). The novelty level was also not equally distributed among the different taxa. In this study, Alveolata and Rhizaria contained the highest and lowest novelty, respectively, in hypersaline lakes. These results are not consistent with previous observations that Opisthokonta and Rhizaria contained the highest novelty and Chlorophyta and Alveolata contained the lowest novelty in 34 different coastal and inland saline ponds [15].

Furthermore, at least four OTUs in mesosaline lakes and eighteen OTUs in hypersaline lakes showed very low CCM and CEM scores, indicating that these OTUs are highly distantly related to any existing OTUs. As expected, phylogenetic analysis further showed that these OTUs occupy highly unique branches in phylogenetic tree (Figure 3B). Molecular approaches have been successfully used to estimate the diversity of eukaryotic plankton assemblages (largely, the protistan assemblages and picoeukaryotic communities) in marine ecosystems and have revealed a large and novel diversity of protists and other eukaryotic microorganisms [68–69]. Over a 4-year period, eukaryotic molecular surveys were performed and thousands of sequences were deposited [5,7,15,25]. However, in this study, some OTUs still had no similarity to CCM or CEM data, yielding generally low similarity values. These OTUs could represent high-rank novel phylogenetic lineages and are obvious candidates for further research. Our data highlight incomplete culturing of the dominant eukaryotes in hypersaline environments in Qaidam Basin. From an evolutionary perspective, we are faced with very divergent sequences that could account for new, unexpected and fascinating evolutionary lineages [24].

Influence of Salinity on Eukaryotic Plankton Assemblages

Understanding the spatial distribution of aquatic microbial diversity and the underlying mechanisms causing differences in community composition is a challenging task. Recent research suggested that distinct protist community assemblages are present in different large-scale water masses in the Southern Ocean [25]. Our data are consistent with this observation and have revealed the distinct distribution of eukaryotic plankton assemblages along a salinity gradient in the Qaidam Basin lakes. Notably, the eukaryotic plankton assemblage compositions were most variable between different sampling sites in the same lake (Figure 5). CCA analysis indicated that pH and dissolved oxygen (DO) were significantly correlated with the distribution of the eukaryotic plankton assemblages in Lakes Dasugan and Tuosu, whereas TN was a major factor influencing the taxonomic composition of eukaryotic plankton assemblages in Lakes Gahai and Gasikule (Figure S2). The present result confirmed that environmental factor strongly influences the taxonomic composition of eukaryotic plankton assemblages in inland waters [7,60,64].

The most important environmental gradients among the studied lakes were TDS in the present study. Previous reports also suggest that salinity and oxygen are important factors shaping the microbial composition in aquatic habitats [7,56,70]. Our results suggest that TDS defined distinct eukaryotic plankton assemblages among lakes, whereas other factors affected the distribution of eukaryotic plankton assemblages within lakes. This hypothesis consist with previous reports that chlorine and carbonate ion percentages were the most important chemical variables potentially structuring the diversity of eukaryotic plankton assemblages in Eastern Tibetan lakes [7]. Oxygen (presence or absence), pH, Na^+, and K^+ concentrations were significantly correlated with the OTU composition in Ethiopian soda lakes [27].

Inland saline lakes are extremely responsive to changes in climatic conditions and have significant ecological, economic, and cultural value; however, these lakes are threatened worldwide by diversion and pollution of their inflows, the introduction of exotic species, and economic development with changes in land use. This study shows that saline lakes in the Qaidam Basin, which are considered to be less species-rich most likely, harbored remarkably diverse eukaryotic microbial communities. Therefore, it is important to perform detailed studies and to develop active conservation strategies to preserve the microbial biodiversity in these areas. This result further highlights the major gap existing between the well-defined diversity and classification inferred from cultivated microorganisms, and their significantly larger natural diversity that is not yet well understood. These data provide the baseline information needed to further study the ecological context of the roles of microbial eukaryotes in the system.

Supporting Information

Figure S1 Estimating species richness of the investigated lakes.

Figure S2 Venn diagram showing the distribution of shared OTUs across lakes in the Qaidam Basin.

Table S1 shows Cumulative fit per species as fraction of variance of species.

Table S2 shows Taxa assignments at class level of investigated lakes.

Acknowledgments

We would like to thank Lanping Zhou (Haixi Water Conservancy Bureau, Delingha, China), Qiang Zhang and Peng Zhang for assistance with the collection of sample in water of saline lake.

Author Contributions

Conceived and designed the experiments: JLW LMC FW HW ZPL. Performed the experiments: JLW LMC ZPZ. Analyzed the data: JLW JYG HRD. Contributed reagents/materials/analysis tools: JLW FW. Contributed to the writing of the manuscript: JLW HW FW.

References

1. Waiser MJ, Robarts RD (2009) Saline Inland Waters, In G.E. Likens (Ed.) Encyclopedia of Inland Waters. Elsevier, Amsterdam.
2. Bryant RG (1996) Validated linear mixture modeling of Landsat TM data for mapping evaporite minerals on a playa surface: methods and applications. International Journal of Remote Sensing 17: 315–330.
3. Williams WD (2002) Environmental threats to salt lakes and the likely status of inland saline ecosystems in 2025. Environmental Conservation 29: 154–167.
4. Ramsar Convention Secretariat (2010) Designating Ramsar Sites: Strategic Framework and guidelines for the future development of the List of Wetlands of International Importance, Ramsar handbooks for the wise use of wetlands, 4th edition, vol. 17. Ramsar Convention Secretariat, Gland, Switzerland.
5. Casamayor, Emilio O, Xavier Triadó-Margarit, Carmen Castañeda (2013) Microbial biodiversity in saline shallow lakes of the Monegros Desert, Spain. FEMS microbiology ecology 85: 503–518.
6. Massana R, Balague V, Guillou L, Pedros-Alio C (2004) Picoeukaryotic diversity in an oligotrophic coastal site studied by molecular and culturing approaches, FEMS microbiology ecology 50: 231–243.
7. Wu QL, Chatzinotas A, Wang J, Boenigk J (2009) Genetic diversity of eukaryotic plankton assemblages in Eastern Tibetan Lakes differing by their salinity and altitude. Microbial Ecology 58: 569–581.
8. Caron D, Countway P, Jones A, Kim D, Schnetzer A (2012) Marine protistan diversity. Annual Review of Marine Science 4: 6.1–6.27.
9. Logares R, Lindström ES, Langenheder S, Logue JB, Paterson H, et al. (2013) Biogeography of bacterial communities exposed to long-term environmental change. The ISME Journal 5: 937–948.
10. Dillon JG, McMath LM, Trout AL (2009) Seasonal changes in bacterial diversity in the Salton Sea. Hydrobiol 632: 49–64.
11. Parnell JJ, Rompato G, Latta LC IV, Pfrender ME, Van Nostrand JD, et al. (2010) Functional Biogeography as Evidence of Gene Transfer in Hypersaline Microbial Communities. PloS One 5: e12919. doi:10.1371/journal.pone.0012919.
12. Pagaling E, Wang H, Venables M, Wallace A, Grant WD, et al. (2009) Microbial biogeography of six salt lakes in inner Mongolia, China, and a salt lake in Argentina. Applied and Environmental Microbiology 75: 5750–5760.
13. Makhdoumi-Kakhki A, Amoozegar MA, Kazemi B, Pasic L, Ventosa A (2012) Prokaryotic Diversity in Aran-Bidgol Salt Lake, the Largest Hypersaline Playa in Iran. Microbes and Environments 27: 87–93.
14. Narasingarao P, Podell S, Ugalde JA, Brochier-Armanet C, Emerson JB, et al. (2012) De novo metagenomic assembly reveals abundant novel major lineage of Archaea in hypersaline microbial communities. The ISME Journal 6: 81–93.
15. Triadó-Margarit X, Casamayor EO (2013) High genetic diversity and novelty in planktonic protists inhabiting inland and coastal high salinity water bodies. FEMS Microbiology Ecology 85: Early View. doi:10.1111/1574-6941.12095.
16. Demergasso C, Escudero L, Casamayor EO, Chong G, Balagué V, et al. (2008) Novelty and spatio-temporal heterogeneity in the bacterial diversity of hypersaline Lake Tebenquiche (Salar de Atacama). Extremophiles 12: 491–504.
17. Oren A, Baxter BK, Weimer BC (2009). Microbial communities in salt lakes: Phylogenetic diversity, metabolic diversity, and in situ activities. Natural Res & Environ Iss 15: article 51.
18. Lozupone CA, Knight R (2007). Global patterns in bacterial diversity. Proceedings of the National Academy of Sciences of the United States of America 104: 11436–11440.
19. Auguet JC, Barber_an A, Casamayor EO (2010) Global ecological patterns in uncultured Archaea. The ISME Journal 4: 182–190.
20. Barberán A, Casamayor EO (2010) Global phylogenetic community structure and beta-diversity patterns of surface bacterioplankton metacommunities. Aquatic Microbial Ecology 59: 1–10.
21. Oren A (2002) Diversity of halophilic microorganisms: environments, phylogeny, physiology, and applications. Journal of Industrial Microbiology and Biotechnology 28: 56–63.

22. Pedrós-Alió C, Calderón-Paz JI, MacLean MH, Medina G, Marrasé C, et al. (2000) The microbial food web along salinity gradients. FEMS Microbiology Ecology 32: 143–155.
23. Hauer G, Rogerson A (2005) Heterotrophic protozoa from hypersaline environments. Adaptation to Life at High Salt Concentrations in Archaea, Bacteria and Eukarya (Gunde- Cimerman N, Oren A & Plemenitas A, eds), 519–539. Springer, Dordrecht.
24. Massana R, Pernice M, Bunge JA, del Campo J (2011) Sequence diversity and novelty of natural assemblages of picoeukaryotes from the Indian Ocean. The ISME Journal 5: 184–195.
25. Wolf C, Frickenhaus S, Kilias ES, PeekenI, Metfies K (2014) Protist community composition in the Pacific sector of the Southern Ocean during austral summer 2010. Polar Biology 1–15 37: 375–389.
26. Casamayor EO, Massana R, Benlloch S, Ovreas L, Diez B, et al. (2002) Changes in archaeal, bacterial and eukaryal assemblages along a salinity gradient by comparison of genetic fingerprinting methods in a multipond solar saltern. Environmental Microbiology 4: 338–348.
27. Lanzen A, Simachew A, Gessesse A, Chmolowska D, Jonassen I, et al. (2013) Surprising Prokaryotic and Eukaryotic Diversity, Community Structure and Biogeography of Ethiopian Soda Lakes. PloS One 8: e72577.
28. Beaver JR, Crisman TL (1989) The role of ciliated protozoa in pelagic freshwater ecosystems. Microbial Ecology 17: 111–136.
29. Gaedke U, Wickham S (2004) Ciliate dynamics in response to changing biotic and abiotic conditions in a large, deep lake (L Constance). Aquatic Microbial Ecology 34: 247–261.
30. Eiler A, Drakare S, Bertilsson S, Pernthaler J, Peura S, et al. (2013) Unveiling distribution patterns of freshwater phytoplankton by a next generation sequencing based approach. PloS One 8: e53516.
31. Sogin ML, Morrison HG, Huber JA, Mark Welch D, Huse SM, et al. (2006) Microbial diversity in the deep sea and the underexplored "rare biosphere". Proceedings of the National Academy of Sciences of the United States of America 32: 12115–12120.
32. Andersson AF, Riemann L, Bertilsson S (2010) Pyrosequencing reveals contrasting seasonal dynamics of taxa within Baltic Sea bacterioplankton communities. The ISME Journal 4: 171–181.
33. Hadziavdic K, Lekang K, Lanzen A, Jonassen I, Thompson EM, et al. (2014) Characterization of the 18S rRNA Gene for Designing Universal Eukaryote Specific Primers. PloS One 9: e87624.
34. Kilias E, Wolf C, Nöthig EM, Peeken I, Metfies K (2013) Protist distribution in the Western Fram Strait in summer 2010 based on 454-pyrosequencing of 18S rDNA. Journal of Phycology 49: 996–1010.
35. Medinger R, Nolte V, Pandey RV, Jost S, Ottenwaelder B, et al. (2010) Diversity in a hidden world: potential and limitation of next-generation sequencing for surveys of molecular diversity of eukaryotic microorganisms. Molecular Ecology 19: 32–40.
36. Ebenezer V, Medlin LK, Ki JS (2012) Molecular Detection, Quantification, and Diversity Evaluation of Microalgae. Marine Biotechnology 14: 129–142.
37. Ki JS (2012) Hypervariable regions (V1–V9) of the dinoflagellate 18S rRNA using a large dataset for marker considerations. Journal of Applied Phycology 24: 1035–1043.
38. Hosoi-Tanabe S, Sako Y (2005) Rapid detection of natural cells of Alexandrium tamarense and A-catenella (Dinophyceae) by fluorescence in situ hybridization. Harmful Algae 4: 319–328.
39. Metfies K, Medlin LK (2008) Feasibility of transferring fluorescent in situ hybridization probes to an 18S rRNA gene phylochip and mapping of signal intensities. Applied and Environmental Microbiology 74: 2814–2821.
40. Diez B, Pedros-Alio C, Massana R (2001) Study of genetic diversity of eukaryotic picoplankton in different oceanic regions by small-subunit rRNA gene cloning and sequencing. Applied and Environmental Microbiology 67: 2932–2941.
41. Cheung MK, Au CH, Chu KH, Kwan HS, Wong CK (2010) Composition and genetic diversity of picoeukaryotes in subtropical coastal waters as revealed by 454 pyrosequencing. The ISME Journal 4: 1053–1059.
42. Stoeck T, Bass D, Nebel M, Christen R, Jones MDM, et al. (2010) Multiple marker parallel tag environmental DNA sequencing reveals a highly complex eukaryotic community in marine anoxic water. Molecular Ecology 19: 21–31.
43. Spahn CMT, Beckmann R, Eswar N, Penczek PA, Sali A, et al. (2001) Structure of the 80S ribosome from Saccharomyces cerevisiae–tRNA-ribosome and subunitsubunit interactions. Cell 107: 373–386.
44. Wuyts J, Van de Peer Y, De Wachter R (2001) Distribution of substitution rates and location of insertion sites in the tertiary structure of ribosomal RNA. Nucleic Acids Research 29: 5017–5028.
45. Chandramouli P, Topf M, Ménétret JF, Eswar N, Cannone JJ, et al. (2008) Structure of the mammalian 80S ribosome at 8.7 A resolution. Structure 16: 535–548.
46. Huse SM, Huber JA, Morrison HG, Sogin ML, Welch DM(2007) Accuracy and quality of massively parallel DNA pyrosequencing. Genome Biology 8: R143.
47. Kunin V, Engelbrektson A, Ochman H, Hugenholtz P (2010) Wrinkles in the rare biosphere: pyrosequencing errors lead to artificial inflation of diversity estimates. Environmental Microbiology 12: 118–123.
48. Huse SM, Welch DM, Morrison HG, Sogin ML (2010) Ironing out the wrinkles in the rare biosphere through improved OTU clustering. Environmental Microbiology 12: 1889–1898.
49. Greenberg AE, Clesceri LS, Eaton AD (1992) Standard methods for the examination of water and wastewater. American Public Health Association, Washington.
50. Haas BJ, Gevers D, Earl AM, Feldgarden M, Ward DV, et al. (2011) Chimeric 16S rRNA sequence formation and detection in Sanger and 454-pyrosequenced PCR amplicons. Genome Research 21: 494–504.
51. Edgar RC, Haas BJ, Clemente JC, Quince C, Knight R (2011) UCHIME improves sensitivity and speed of chimera detection. Bioinformatics 27: 2194–2200.
52. Quince C, Lanzen A, Curtis TP, Davenport RJ, Hall N, et al. (2009) Accurate determination of microbial diversity from 454 pyrosequencing data. Nature Methods 6: 639–641.
53. Del Campo J, Massana R (2011) Emerging diversity within chrysophytes, choanoflagellates and bicosoecids based on molecular surveys. Protist 162: 435–448.
54. Hall TA (1999) Bioedit: a user-friendly biological sequence alignment editor and analysis program for windows 95/98/Nt. Nucleic Acids Symposium Series 41: 95–98.
55. Price MN, Dehal PS, Arkin AP (2010) FastTree 2–Approximately Maximum-Likelihood Trees for Large Alignments. Plos One 5: e9490.
56. Letunic I, Bork P (2007) Interactive Tree Of Life (iTOL): an online tool for phylogenetic tree display and annotation. Bioinformatics 23: 127–128.
57. Ter Braak CJF, Verdonschot PFM (1995) Canonical correspondence analysis and related multivariate methods in aquatic ecology. Aquatic Science 57: 255–289.
58. Lefranc M, Thénot A, Lepère C, Debroas D (2005) Genetic diversity of small eukaryotes in lakes differing by their trophic status. Applied and Environmental Microbiology 71: 5935–5942.
59. Lepere C, Domaizon I, Debroas D (2008) Unexpected Importance of Potential Parasites in the Composition of the Freshwater Small-Eukaryote Community. Applied and Environmental Microbiology 74: 2940–2949.
60. Lepère C, Domaizon I, Taïb N, Mangot JF, Bronner G, et al. (2013) Geographic distance and ecosystem size determine the distribution of smallest protists in lacustrine ecosystems. FEMS Microbiology Ecology 85: 85–94.
61. Charvet S, Vincent WF, Lovejoy C (2014) Effects of light and prey availability on Arctic freshwater protist communities examined by high-throughput DNA and RNA sequencing. FEMS Microbiology Ecology 88: 550–564.
62. Lefèvre E, Roussel B, Amblard C, Sime-Ngando T (2008) The molecular diversity of freshwater picoeukaryotes reveals high occurrence of putative parasitoids in the plankton. PloS One 3: e2324.
63. Triadó-Margarit X, Casamayor EO (2012) Genetic diversity of planktonic eukaryotes in high mountain lakes (Central Pyrenees, Spain). Environmental Microbiology 14: 2445–2456.
64. Mangot JF, Domaizon I, Taib N, Marouni N, Duffaud E, et al. (2013) Short-term dynamics of diversity patterns: evidence of continual reassembly within lacustrine small eukaryotes. Environmental Microbiology 15: 1745–1758.
65. Frontier S (1985) Diversity and structure in aquatic ecosystems. Oceanography and Marine Biology-An Annual Review 23: 253–312.
66. Heidelberg KB, Nelson WC, Holm JB, Eisenkolb N, Andrade K, et al. (2013) Characterization of eukaryotic microbial diversity in hypersaline Lake Tyrrell, Australia. Frontiers in Microbiology 4: 1–14.
67. Buchheim MA, Sutherland DM, Buchheim JA, Wolf M (2013) The blood alga: phylogeny of Haemotococcus (Chlorophyceae) inferred from ribosomal RNA gene sequence data. European Journal of Phycology 48: 318–329.
68. Countway PD, Gast RJ, Savai P, Caron DA (2005) Protistan diversity estimates based on 18S rDNA from seawater incubations in the Western North Atlantic. Journal of Eukaryotic Microbiology 52: 95–106.
69. Countway PD, Gast RJ, Dennett MR, Savai P, Rose JM, et al. (2007) Distinct protistan assemblages characterize the euphotic zone and deep sea (2500 m) of the western North Atlantic (Sargasso Sea and Gulf Stream). Environmental Microbiology 9: 1219–32.
70. Herlemann DP, Labrenz M, Jürgens K, Bertilsson S, Waniek JJ, et al. (2011) Transitions in bacterial communities along the 2000 km salinity gradient of the Baltic Sea. The ISME Journal 5: 1571–1579.

3

Integrating Bacterial and Viral Water Quality Assessment to Predict Swimming-Associated Illness at a Freshwater Beach

Jason W. Marion[1¤a], Cheonghoon Lee[1¤b], Chang Soo Lee[1¤c], Qiuhong Wang[2], Stanley Lemeshow[3], Timothy J. Buckley[1¤d], Linda J. Saif[2], Jiyoung Lee[1,4*]

1 Division of Environmental Health Sciences, College of Public Health, The Ohio State University, Columbus, Ohio, United States of America, **2** Food Animal Health Research Program, Ohio Agricultural Research and Development Center, Department of Veterinary Preventive Medicine, The Ohio State University, Wooster, Ohio, United States of America, **3** Division of Biostatistics, College of Public Health, The Ohio State University, Columbus, Ohio, United States of America, **4** Department of Food Science and Technology, The Ohio State University, Columbus, Ohio, United States of America

Abstract

Background & Objective: Recreational waters impacted by fecal contamination have been linked to gastrointestinal illness in swimmer populations. To date, few epidemiologic studies examine the risk for swimming-related illnesses based upon simultaneous exposure to more than one microbial surrogate (e.g. culturable *E. coli* densities, genetic markers). We addressed this research gap by investigating the association between swimming-related illness frequency and water quality determined from multiple bacterial and viral genetic markers.

Methods: Viral and bacterial genetic marker densities were determined from beach water samples collected over 23 weekend days and were quantified using quantitative polymerase chain reaction (qPCR). These genetic marker data were paired with previously determined human exposure data gathered as part of a cohort study carried out among beach users at East Fork Lake in Ohio, USA in 2009. Using previously unavailable genetic marker data in logistic regression models, single- and multi-marker/multi-water quality indicator approaches for predicting swimming-related illness were evaluated for associations with swimming-associated gastrointestinal illness.

Results: Data pertaining to genetic marker exposure and 8- or 9-day health outcomes were available for a total of 600 healthy susceptible swimmers, and with this population we observed a significant positive association between human adenovirus (HAdV) exposure and diarrhea (odds ratio = 1.6; 95% confidence interval: 1.1–2.3) as well as gastrointestinal illness (OR = 1.5; 95% CI: 1.0–2.2) upon adjusting for culturable *E. coli* densities in multivariable models. No significant associations between bacterial genetic markers and swimming-associated illness were observed.

Conclusions: This study provides evidence that a combined measure of recreational water quality that simultaneously considers both bacterial and viral densities, particularly HAdV, may improve prediction of disease risk than a measure of a single agent in a beach environment likely influenced by nonpoint source human fecal contamination.

Editor: Andrew C. Singer, NERC Centre for Ecology & Hydrology, United Kingdom

Funding: This study was funded by the Public Health Preparedness for Infectious Diseases program (http://phpid.osu.edu) and The Ohio State University Targeted Investment in Excellence program sponsored by the Offices of Academic Affairs, the President, and Research. The human health and exposure data collection effort was funded by the Ohio Water Development Authority. The funders had no role in study design, data collection and analysis, decision to publish, or preparation of the manuscript.

Competing Interests: The authors have declared that no competing interests exist.

* Email: lee.3598@osu.edu

¤a Current address: Department of Environmental Health Science, Eastern Kentucky University, Richmond, Kentucky, United States of America
¤b Current address: Department of Environmental Health and Institute of Health and Environment, Graduate School of Public Health, Seoul National University, Seoul, Korea
¤c Current address: Environmental Biotechnology Research Center, Korea Research Institute of Bioscience and Biotechnology, Yuseong-gu, Daejeon, Korea,
¤d Current address: U.S. Environmental Protection Agency, National Exposure Research Laboratory, Research Triangle Park, North Carolina, United States of America

Introduction

Beach water advisories are issued to discourage human contact (e.g., swimming, wading, etc.) when water is potentially harmful to human health. In the United States, many of these advisories are issued at freshwater beaches when densities of fecal indicator bacteria (*E. coli* or enterococci) are observed or predicted to be in excess of single-day maximum criteria. Prior to November 26,

2012, the U.S. Environmental Protection Agency (USEPA) had established single-day maximum criteria, which were derived from several large epidemiological studies performed in the early 1980s [1]. In 2003, the 1986 *E. coli* criteria were supported by a comprehensive meta-analysis demonstrating *E. coli* was the most consistent indicator for predicting gastrointestinal illness in freshwater [2]. This same study indicated that methods for quantifying *E. coli* were problematic for same-day water quality advisories. The method required 18–24 h incubation and therefore was untimely for communication of risk to susceptible beachgoers [2].

Since the 2003 meta-analysis, several epidemiological studies evaluating associations between illness and more rapid measures of fecal indicators have been performed with an emphasis on genetic markers of bacterial or viral contamination, which have demonstrated the effectiveness of rapidly measured *Enterococcus* via quantitative polymerase chain reaction (qPCR) for predicting human illness at sewage-impacted Great Lakes and marine beaches [3,4]. However, in epidemiology studies on swimmers at non-point source-impacted beaches, associations between qPCR-based bacterial markers and gastrointestinal (GI) illness were not observed [5,6]. A potential reason for the lack of association in both studies could have related to the source of fecal contamination, which was likely dominated by avian sources [7], potentially presenting less risk for swimming-associated gastrointestinal (GI) illness than equivalent amounts of human-associated fecal contamination [8]. The diffuse nature of the source(s) of fecal contamination has been proposed as a possible explanation for the lack of association between bacterial genetic markers and human illness [9–11].

To date, few epidemiological studies on beachgoers have been performed at areas dominated by human-associated non-point source contamination, particularly in freshwater environments. Several peer-reviewed epidemiological studies have evaluated illness associations with human enteric viruses in European recreational waters [12–14] and U.S. waters [6,15,16]. Epidemiological studies pertaining to viruses are highly relevant as viruses are known to have a broad distribution in the aquatic environment [17], and are the most observed etiological agents in disease outbreaks associated with untreated U.S. recreational waters [18]. Recent epidemiological studies on swimmers [9–11] have suggested that viruses were the primary etiologic agents associated with recreational water illnesses based upon relatively short incubation periods following swimming exposure [10,19]. Similarly with regards to incubation periods, among limited-contact recreational users of freshwater in Chicago, Illinois (U.S.A.), most gastrointestinal illnesses generally developed within three days of water exposure [19].

To date, the majority of recreational water-related epidemiological studies have been performed to identify effective single indicators, often bacterial, for practical application among those making beach management decisions. Multiple fecal indicators and multiple genetic markers have been evaluated in recreational waters with respect to their association with each other as well as GI illness frequency among swimmers. There is some inconsistency in the findings of studies that have studied associations (or the lack thereof) between fecal indicator densities and measured pathogen and/or viral levels in freshwater [20] and coastal nvironments [21,22], and legitimate concerns remain regarding the use of single bacterial indicators for predicting and communicating swimming-associated illness risks to the general public [23–25].

To address these knowledge gaps, this study examined previously undescribed samples which were archived during our prospective cohort study [26] that gathered culture-based *E. coli* and human exposure data. In our 2010 study [26], new GI illnesses among swimmers were associated with increasing culture-based *E. coli* densities at East Fork Lake (Ohio, USA); however, no assessment of associations between genetic marker densities and human health occurred. In 2012, the U.S. Environmental Protection Agency (EPA) emphasized a need for additional evaluation of rapid molecular methods for the timely determination of water quality to protect recreational water users [27]. Building upon this need, we examined associations between GI illness incidence among swimmers exposed to multiple genetic markers targeting both viruses and bacteria in this non-point source human-impacted beach environment. This study presented a unique opportunity to successfully use our historical exposure and health data coupled with our previously unexplored genetic material to do a timely assessment of the effectiveness of viral and bacterial molecular markers for predicting GI illness risk among swimmers who used the East Fork beach in 2009.

Materials and Methods

Overall Approach

The exposure and health data were collected using the prospective cohort study approach adopted by Wade et al. in their epidemiological studies related to recreational water [3,4]. This study builds upon our epidemiological investigations of recreational water-related illnesses by evaluating human illness associations with our recently qPCR-determined densities of bacteria and viruses. In brief, beach water samples, beach water quality data, and beach user exposure/behavior data were collected at East Fork State Park (39.0198° N; 84.1432° W) near Cincinnati, Ohio, U.S.A. Approvals to use this location for water sample collection and beachgoer interviews were kindly provided orally and in writing by Chris Dauner (Park Manager, East Fork State Park), and Dan West (Chief, Ohio Department of Natural Resources, Division of Parks and Recreation). Complete health and exposure data were collected from 891 beachgoers from 278 households. As described in our 2010 study [26], participants reported their health status via a telephone questionnaire after an eight- to nine-day follow-up period. More recently we quantified densities of four viral (human adenovirus (HAdV), human enterovirus (HEntV), and human norovirus genogroup I (HNoV GI) and genogroup II (HNoV GII)) and four bacterial (*E. coli* by *uidA* and 23S, Enterococci by 23S, *Bacteroides-Prevotella* by HuBac) markers from 23 samples collected over 23 weekend days from 2009 using qPCR and paired them with human exposure data.

Ethics Statement

Approvals for the study design, questionnaires, verbal consent procedures, and related materials were obtained from the Institutional Review Board (IRB) at The Ohio State University (IRB Protocol #2009H0107). For this study, adult participants (18 years of age or older) capable of providing verbal consent were welcomed into the study by investigators or key personnel trained and authorized by the university to obtain consent. The written consent waiver permitting verbal consent was granted given the low study risk and because written consent would have required surname information and the gathering of consent documentation in a periodically windy and wet beach environment. Thus, written consent procedures would have increased the likelihood of losing or damaging documentation which would have then included participant surnames that could potentially be matched with telephone numbers and health data. Verbal consent was recorded on the datasheet for each household interviewed. In addition, participants were provided with a card sharing contact information for the principal investigator and the ethics coordinator in the Office of Responsible Research Practices at the university.

Enrollment and Exposure Survey

The East Fork Beach Study [26] recruited beach user households into the study at the beach and interviewed them upon their departure to ascertain information regarding their activities at the beach on that same day. For each household in the study, an adult spokesperson provided investigators with information requested from a standard script. The obtained information was used for classifying exposure status and obtaining information related to potential confounding variables (age, gender, ethnicity, food consumption, source of food/drink, number of beach visits per year, duration of time at the beach, hand hygiene, extent of sand exposure, distance traveled, and number of other beach users). More specifically, individuals were asked whether any household members had: (1) no water contact; (2) waded, played or swam in the water; and/or (3) immersed their head in the water. Questions pertaining to any recent or on-going illnesses were also asked. Individual beach users were dichotomously classified as having or not having each of the above-mentioned exposures as well as any prevalent symptoms of gastrointestinal illness. These exposure data were recently paired with date-specific *E. coli* genetic marker density data to further establish our exposure classification of each beach user.

Health Outcomes Survey

Health outcome data were obtained by Marion et al. [26] in which enrolled households were contacted eight to nine days after their beach visit via telephone. Only adult household members who were at the beach at the time of enrollment were eligible for interview participation. Interviews were conducted using a standard questionnaire inquiring about illness among household members at the beach at the time of enrollment. Reported illnesses were recorded as a yes/no response for a variety of symptoms (e.g., stomach cramps, nausea, diarrhea, vomiting, headache, fever, etc.) for each person who visited the beach. For comparing with other studies, persons were further classified by gastrointestinal illness. Similar to the Wade et al. studies [3,4], persons were defined to have experienced "GI illness" if they were reported to have had any one of the following: nausea, stomach ache or stomach cramps, diarrhea (three or more loose or watery stools in a 24-hour period), or vomiting. With respect to nausea, all persons in this study who reported nausea were considered positive for GI illness. This GI illness definition is slightly different than the Wade et al. studies [3,4], which only included nausea cases into their GI illness definition when the condition interfered with daily activities. In our study, we were unable to make this distinction as we did not ascertain if the nausea interfered with daily activities. Beyond the "GI Illness" definition, using the highly credible gastrointestinal illness (HCGI) definition employed by Colford [5] as "HCGI-1", we likewise coded individuals as positive for HCGI if they were reported to have experienced any of the following: (1) vomiting; (2) diarrhea and fever, (3) stomach-ache and fever, or (4) nausea and fever.

Sample Collection and Water Analysis

Beach water samples used for obtaining genetic marker data were collected on the same day as enrollment and administration of the exposure questionnaire on summer weekends during the time of day generally used for swimming. In total, 91% of samples were collected at the median time of 13:45±3 h. Two samples (9%) were collected after 16:00 on days when daily swimming attendance was low (n = 4, n = 28) in anticipation of more swimmers late in the day. For each day, 3 L was collected into multiple autoclaved 500 mL bottles (Nalgene, Rochester, NY, USA) and sterile Whirl-Pak bags (Nasco, Fort Atkinson, WI, USA) over 23 weekend days. Samples were collected near the beach center in water with an approximate depth of one meter by sweeping containers 30 cm below the water surface. Within 30 minutes after collection, a 1 L sample aliquot was used for membrane filtration for culturable *E. coli*, a 1 L sample aliquot was temporarily stored (24–48 h) at 4°C and then immediately filtered for bacterial marker analysis, and a 1 L aliquot was stored at −20°C for virus analysis.

E. coli densities were quantified for three separate filtration volumes (20, 50, and 100 mL) for each sample using the culture-based method described in EPA Method 1603 [27] and Marion et al. [26]. Other water quality parameters including temperature, pH, dissolved oxygen, and turbidity were also measured using a YSI 600XL data sonde (Yellow Springs Instruments, Yellow Springs, Ohio, U.S.A.) and a Hach 2100P Turbidimeter (Loveland, Colorado, U.S.A.) as described in Marion et al. [26]. UV index data were obtained from the National Oceanic and Atmospheric Administration [28]. Rainfall, lake stage (surface elevation), and lake inflow data were obtained from the U.S. Army Corps of Engineers [29].

A detailed description of sample preparation, water concentration, and quantification of the bacterial and viral marker densities by qPCR is provided as (**Text S1**), including a description of the primers and probes used in this study (**Table S1**). In brief, samples were filtered to capture bacteria for quantifying the following bacterial markers: *uid*A genes (uidA) and 23S rRNA genes (23S *E. coli*) of *E. coli*, 16S rRNA genes of *Bacteroides-Prevotella* (HuBac), and 23S rRNA genes of *Enterococcus* spp. (23S *Enterococcus*). For concentrating viruses (HEntV, HAdV, HNoV GI, and HNoV GII), membrane filtration was performed using cation (Al^{3+})-coated membranes. DNA or RNA was then extracted from the microbes captured on the membranes using a QIAmp DNA Stool Kit or the RNeasy Mini Kit (Qiagen, Valencia, CA, USA), respectively. RNA extracts for HEntV were amplified by reverse transcription (RT)-PCR and quantified using TaqMan real-time RT-PCR (RT-qPCR) with the QIAGEN OneStep RT-PCR Kit (Qiagen, Valencia, CA, USA). DNA extracts for HAdV and all bacterial markers were amplified by PCR and quantified using a TaqMan-based real-time qPCR system. All RT-qPCR and qPCR reactions except HNoV GI and HNoV GII were carried out in an ABI 48-well StepOne Real Time System (Applied Biosystems, Foster City, CA). RT-qPCR assays for HNoV GI and HNoV GII were carried out with an Eppendorf Mastercycler RealPlex^2 (Eppendorf, Germany). Quantification of the markers was then determined using standard curves generated by plotting Ct values (y) versus the log gene copy numbers of the target microbes (x) and each limit of detection (LOD) was determined based on the lowest gene copy number from which above 90% of the replicates could be amplified in each qPCR assay. Additionally, with regards to these samples, the presence/absence of qPCR inhibition was previously evaluated for East Fork samples [20] using the Sketa22 qPCR assay as described in Haugland et al. [30].

Data Analysis

The analysis focused on healthy susceptible individuals, and persons who were classified as prevalent or existing cases based upon responses to the enrollment questionnaire were excluded in analyses using incident cases. Logistic regression was employed to evaluate crude associations between genetic marker exposure and new illness among swimmers reporting head immersion. The data for the microbiological terms, including *E. coli* density data, were skewed, accordingly they were log (base 10)-transformed. Prior to log-transforming *E. coli* results, samples with 0 CFU/100 mL were assigned a value of 1 CFU/100 mL. In scatterplots and models containing genetic marker densities, for markers not detected after 45 cycles, densities were assigned a value of half the detection limit in log CFUs per 100 mL or log gene equivalents (GE) per 100 mL for bacterial and viral markers, respectively.

Multivariable logistic regression was used for estimating odds ratios for any GI illness, diarrhea, and HCGI among swimmers and non-swimmers (negative controls) as recommended in observational studies [31] and done by Colford et al. [9] in their beach user health study. The swimmer population was defined as beach users who reported immersing their head in the beach water. Non-swimmers were defined as beach users reporting no water contact or limited water contact that did not include head immersion in the beach water. All swimmers were assumed to have been exposed to the same density of *E. coli* and the various genetic markers observed from the single sample collection for that day. Densities of *E. coli* and the various genetic markers were determined from the single mid-day water sample collection.

The multivariable logistic regression models considered potential confounders and/or modifying influences, including covariates related to demographic, exposure, meteorological, and water quality factors. Since data were obtained from households, for the purpose of statistical analysis, the data were treated as clustered by household [32]. For constructing multivariable models, a backward selection approach was employed whereby covariates were dropped stepwise based on the highest p-value until arriving at the most parsimonious model [32]. The rule of ten events (cases) per covariate was used in logistic regression modeling [33]. Due to the low frequency of HCGI cases, HCGI model construction was limited to only two microbiological terms by relaxing the rule of ten events per covariate [34].

Multivariable logistic regression models were evaluated for fit via the Hosmer-Lemeshow Goodness-of-Fit Test [32] and for model discrimination via the area under the receiver-operating-characteristic curve (AUC) [32]. The modeling efforts and assessments of logistic model fit and discrimination were performed using Stata 11 (Stata Corporation, College Station, TX, USA). A time-series plot was generated with Minitab 16 (Minitab Inc., State College, PA, USA).

Results

Water Quality

Among the 23 weekend water samples collected for qPCR analysis, bacterial markers were more readily detected than viruses (Table 1). HEntV and HAdV were detected in 22% (5/23) and 35% (8/23) of all samples, respectively, whereas, HNoV GI and GII were not detected (0/23). As previously reported [20], qPCR inhibition was not observed in any of the 23 samples. Culturable *E. coli* densities greater than 0 CFU/100 mL were observed in 91% (21/23) of all samples. The time-series plot of culturable *E. coli* and HAdV densities by day (Figure 1) demonstrate the lack of association between the two water quality parameters, which was confirmed by correlation analysis (Spearman's $\rho = 0.147$; $p = 0.503$). Overall, culturable *E. coli* densities exceeded Ohio single-day maximum criteria (>235 CFU/100 mL) in 8.6% (2/23) of our samples; however, no beach advisories were issued by local or state authorities due to their adherence to the one sample per week guideline used by the Ohio Department of Health for inland waters. Other water quality parameters are summarized in Marion et al. [26].

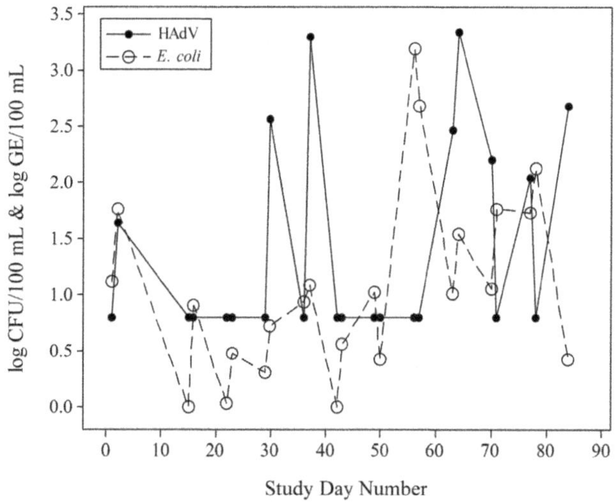

Figure 1. Time series plot for HAdV and *E. coli* densities measured during the 2009 swimming season at the study beach (East Fork Lake, Ohio).

Study Population

A total of 891 individuals were included in this study (from 278 households). The study population was 93% white, 2.5% black, 1% Hispanic and 3.5% other. The average study participant age was 24 years (median =21 years). Young children (≤5 years) represented 13% of the participants on whom data were available. Females were more represented (56%) than males (44%). The population described in this report is a subset of the 965 individuals from 300 households in our previous study [26], which was conducted over 26 weekend days. Genetic marker data were only obtained for 23 of 26 days of the 2009 East Fork Beach Study [26], and accordingly, this study reports the results of the investigation using genetic marker data. The collection of sufficient water for obtaining qPCR data was an additional task added to the previous study on the third sample day.

Among the participants in this study, 618 (69%) individuals reported head immersion in East Fork beach waters. Prevalent GI illness was reported for 18 (2.9%) of these swimmers at enrollment (Table 2). Chronic GI problems were self-reported at enrollment in 9 (50%) of these 18 individuals. All 18 individuals with prevalent GI illness (including chronic GI problems) were excluded from the exposure-related illness models. The prevalence of fever, nausea, and stomach cramping among participants was not ascertained at enrollment.

Water Quality and Human Illness

The qPCR-based densities for bacteria and viruses do not show any significant associations ($p<0.05$) with HCGI, GI, and diarrhea in single marker models using univariable and multivariable logistic regression when treating marker densities as continuous terms or binary terms in models (Table 3). Among all the continuous terms, HAdV had the strongest association with HCGI (crude OR (cOR) =1.6; 95% CI 0.90–2.9) and diarrhea (adjusted OR (aOR) =1.5; 95% CI 0.98–2.4); however, these associations were not significant but were noteworthy ($p = 0.105$; $p = 0.066$, respectively). When evaluating the association of either HEntV or HAdV detection with HCGI, GI, and diarrhea

Table 1. Microbial water quality measured by various genetic markers and *E. coli* at East Fork Lake, Ohio (N = 23).

Genetic Marker or Fecal Indicator	No. Samples Above LOD[a]	Detection Limit	Median	Range
HAdV (log gene equivalents/100 mL)	8	1.6	BD[b]	BD-3.34
HEntV (log gene equivalents/100 mL)	5	1.9	BD	BD-2.13
uidA (log copies/100 mL)	19	1.5	2.08	1.50–2.49
23S *E. coli* (log copies/100 mL)	19	2.46	2.94	2.46–3.19
HuBac (log copies/100 mL)	18	2.05	3.04	2.05–3.39
23S *Enterococcus* (log copies/100 mL)	17	1.93	2.19	1.93–2.88
E. coli (log CFU/100 mL)	21	ND[c]	1.01	LOD-3.19

[a]Limit of Detection or 0 CFU/100 mL for *E. coli*.
[b]Below Detection.
[c]Not Determined.

incidence, no association was observed ($p>0.05$). The covariates used in the illness models are summarized in **Table S2** and **Table S3**, respectively. The most significant covariate, food consumption at the beach (see **Table S2** and **Table S3**), was used in all genetic marker and *E. coli* models except the HCGI model, as food consumption at the beach was common to all HCGI cases. Beyond food consumption at the beach, specific conductivity and previous 72-hour UV results were the most significant covariates in all of the GI illness models. With respect to the diarrhea illness model, food consumption and previous 72-hour UV data were used as covariates, as they were identified as most significant through the backward elimination approach. In evaluating HCGI models relying on two microbiological terms, the only two-term model achieving significance was a combined model using HAdV and culturable *E. coli* ($p = 0.003$; likelihood ratio test). None of the bacterial markers were significant predictors of HCGI in models using an additional marker.

Using this same bacteria and virus approach for GI illness, we observed a similar association. Since more GI cases were detected than HCGI, the model was able to contain an additional two covariates. In this model, significant ($p<0.05$) or near significant odds ratios ($p = 0.056$) were observed for all covariates in this model (Table 4), including food consumption, specific conductivity, *E. coli*, and HAdV. Increases in water conductivity were associated with protective effects (aOR<1); whereas, food consumption, increasing HAdV, and increasing *E. coli* were identified as risk factors for GI (aOR>1). The use of two microbial terms was supported in model building as the removal of the HAdV term resulted in significantly lower log-likelihood values for HCGI, GI, and diarrheal models ($p<0.05$), and the two microbial terms had the low p-values throughout the entire backward selection procedure allowing their retention in the model. The final four-term model was evaluated using a negative control group (the non-swimmer group), and no significant associations were observed for any model terms, suggesting no day-effects, and implying a water quality-related association in our exposed group. Lastly, all possible models for predicting GI illness using a single bacterial marker in tandem with either culturable *E. coli* or viral marker levels were developed, and no bacterial markers were associated with illness.

Models constructed for predicting diarrhea using *E. coli* and HAdV preliminarily indicate that HAdV is associated with diarrheal disease incidence as the aOR for HAdV was significant ($p = 0.023$) when adjusting *E. coli* density by including *E. coli* density in the model. Compared to the HCGI and GI models, the diarrheal model provided poor discrimination (AUC <0.70) with

Table 2. Health status among study participants at enrollment and follow-up.

Reported Health Outcomes	Swimmers (n = 618)		Non-Swimmers (n = 274)	
	Health Status at Beach Enrollment (No. (%))	Health Status at Follow-up[a] (No. (%))	Health Status at Beach Enrollment (No. (%)	Health Status at Follow-up[a] (No. (%))
No Reported GI Problems	600 (97)	562 (94)	259 (95)	251 (97)
Any GI Illness	18 (2.9)	38 (6.3)	15 (5.5)	8 (3.1)
Chronic GI Problems	9 (1.5)	ND[b]	8 (2.9)	ND[b]
Diarrhea	6 (0.97)	28 (4.6)	5 (1.8)	8 (3.1)
Fever	ND	10 (1.6)	ND	2 (0.73)
Stomach Cramps	ND	8 (1.2)	ND	8 (1.2)
Nausea + Other GI Illness	ND	7 (1.2)	ND	1 (0.36)
Nausea Only	ND	3 (0.50)	ND	0 (0.00)
Vomiting	6 (0.97)	10 (1.6)	3 (1.1)	1 (0.36)
HCGI	ND	13 (2.2)	267 (97)	1 (0.37)

[a]Health status at telephone follow-up excluding positive cases at enrollment.
Not determined as information was not collected by the questionnaire.

Table 3. Odds ratios for HCGI, GI, and diarrhea associated with exposure to varying levels of various molecular genetic markers and/or fecal indicators at East Fork Lake, Ohio.

Molecular Marker	HCGI		Any GI Illness		Diarrhea	
	cOR[a] (95% CI)	Wald (*p*)	aOR[b] (95% CI)	Wald (*p*)	aOR (95% CI)	Wald (*p*)
HEntV (+)[c]	1.6 (0.34–7.6)	0.552	0.76 (0.21–2.7)	0.674	0.17 (0.02–1.3)	0.081
HAdV (+)	2.1 (0.65–7.1)	0.212	1.3 (0.45–3.5)	0.654	2.2 (0.65–7.6)	0.202
uidA *E. coli* (+)	1.1 (0.22–5.4)	0.892	0.78 (0.27–2.2)	0.637	0.88 (0.24–3.2)	0.843
23S *E. coli* (+)	0.69 (0.18–2.7)	0.595	0.78 (0.30–2.0)	0.599	1.1 (0.31–3.8)	0.905
HuBac (+)	1.5 (0.32–7.1)	0.604	1.1 (0.38–3.2)	0.846	0.86 (0.25–2.9)	0.815
23S *Enterococcus* (+)	0.70 (0.21–2.4)	0.586	0.77 (0.27–2.2)	0.613	0.58 (0.18–1.9)	0.363
Log HAdV	1.6 (0.90–2.9)	0.105	1.2 (0.77–1.9)	0.399	1.5 (0.97–2.4)	0.066
Log uidA *E. coli*	1.0 (0.38–2.9)	0.927	0.76 (0.38–1.5)	0.442	0.98 (0.34–2.9)	0.974
Log 23S *E. coli*	0.78 (0.38–1.6)	0.489	0.85 (0.50–1.4)	0.526	1.2 (0.48–3.2)	0.648
Log HuBac	1.5 (0.63–3.7)	0.355	1.1 (0.67–1.8)	0.7	0.78 (0.49–1.2)	0.289
Log 23S *Enterococcus*	0.96 (0.38–2.4)	0.921	0.89 (0.49–1.6)	0.714	0.67 (0.37–1.2)	0.207

[a]Crude odds ratio.
[b]Adjusted odds ratios, see Table S2 and S3, to see the covariates used for adjustment.
[c](+), Positive detection by qPCR, binary term.

respect to the ability of the model to properly classify diarrhea cases (Table 5). The Goodness-of-Fit test results (Table 5) also demonstrate the GI and HCGI models have better fits, as the p-values for those models are higher, indicating that the modeled results are not significantly different than the observed GI and HCGI results from the health survey.

For considering possible interaction effects, a water quality index was constructed based upon the detection limit of HAdV and the median *E. coli* density. Table 6 presents the four water quality index values, and the respective *E. coli* and HAdV densities represented by each water quality index level. When evaluating associations for this model, the reference was set as the

Table 4. Multivariable logistic regression models for predicting new HCGI[a], GI, and diarrhea among swimmers immersing their head and beachgoers not immersing their head in beach water (East Fork Lake, Ohio).

Model & Exposure	Covariate	β	SE_β	Wald (*p*)	AOR (95% CI)
HCGI Swimmers	Log HAdV (gene equivalents/100 mL)	0.8191	0.2864	0.005	2.3 (1.3–3.9)
	Log *E. coli* (CFU/100 mL)	1.101	0.3571	0.002	3.0 (1.5–6.1)
	Constant Term	−7.003			
GI Swimmers	Log HAdV (gene equivalents/100 mL)	0.421	0.1978	0.034	1.5 (1.0–2.2)
	Log *E. coli* (CFU/100 mL)	0.6515	0.2473	0.009	1.9 (1.2–3.1)
	Consumed Food at the Beach	1.543	0.7019	0.029	4.7 (1.2–19)
	Specific Conductivity (μS)	−0.092	0.0477	0.056	0.91 (0.83–1.0)
	Constant Term	19.37			
GI Non-Swimmers	Log HAdV (gene equivalents/100 mL)	0.0788	0.5481	0.886	1.1 (0.37–3.2)
	Log *E. coli* (CFU/100 mL)	0.2675	0.3346	0.425	1.3 (0.67–2.5)
	Consumed Food at the Beach	0.4254	0.8889	0.633	1.5 (0.26–8.9)
	Specific Conductivity (μS)	0.04	0.0279	0.154	1.0 (0.98–1.1)
	Constant Term	−15.33			
Diarrhea Swimmers	Log HAdV (gene equivalents/100 mL)	0.453	0.1977	0.023	1.6 (1.1–2.3)
	Log *E. coli* (CFU/100 mL)	0.3275	0.2466	0.186	1.4 (0.85–2.3)
	Constant Term	−4.226			
Diarrhea Non-Swimmers	Log HAdV (gene equivalents/100 mL)	0.0034	0.4905	0.994	1.0 (0.38–2.6)
	Log *E. coli* (CFU/100 mL)	0.2309	0.336	0.493	1.3 (0.65–2.4)
	Constant Term	−3.735			

[a]A model could not be constructed for HCGI among non-swimmers due to the low sample size resulting in an insufficient number of HCGI cases for model development.

Table 5. Model diagnostics for models from Table 4, pertaining to discrimination (AUC) and model calibration (goodness-of-fit).

Model Name for Swimmers	AUC[a]	Goodness-of-Fit[b] (*p*)
Multivariable HCGI Model	0.75	0.895
Multivariable GI Model	0.753	0.958
Multivariable Diarrhea Model	0.64	0.206

[a]Area under the receiver-operator-characteristic curve (AUC).
[b]Hosmer-Lemeshow Goodness-of-Fit Test [32].

group of days in which *E. coli* densities were lowest and HAdV was not detected. Here we observe multiple associations with increasing viral and/or bacteria densities and both GI and diarrheal illness incidence. Swimmers with the presumed greatest exposure to *E. coli* and HAdV had the greatest odds of reporting new GI illness (OR = 6.1; 95% CI 1.5–25) compared to swimmers with the least exposure to *E. coli* and HAdV. Exposure to the highest levels of *E. coli* when HAdV levels were low also presented increased GI illness risk (OR = 5.4; 95% CI 1.5–19) compared to those swimmers with low *E. coli* and low HAdV exposure. In the diarrheal illness model, significantly increased odds of diarrhea were only observed in the group who swam on days with elevated *E. coli* and HAdV compared to the reference group (OR = 5.2; 1.3–22). Model discrimination for each model (GI and diarrhea) was evaluated and the AUC values were 0.63 and 0.65, respectively.

Discussion

This prospective cohort study from an inland U.S. beach demonstrates the predictive potential of an integrative, multi-microbial approach for estimating recreational waterborne disease risk from viral and bacterial indicators. The term 'indicator' used here does not imply fecal indicator, but instead refers to 'health-relevant' indicators. Our results showed a positive association between increasing densities of individual health-relevant indicators (HAdV and *E. coli*) and increased odds of GI and HCGI among swimmers at the studied beach. More importantly, the two-indicator model (HAdV and *E. coli*) represented a significant improvement over single health indicator approaches with this data. Although speculative, particularly due to the small sample size of the study, the findings of this two-indicator approach are etiologically plausible as a combined viral + bacterial (ViBac) approach may account for GI and HCGI illness associations with viral and bacterial pathogens. Based upon our cohort study results, we have some reason to speculate that a multiple health-indicator approach may be beneficial for natural recreational waters such as the East Fork beach, which is part of a watershed comprised of mixed land uses and a high density of septic systems, all of which may promote a more diffuse type of fecal contamination than observed in point-source impacted waters. Within the beach watershed of interest, two municipal wastewater treatment plants (Williamsburg and Bethel WWTPs) and several smaller WWTPs operate, while malfunctioning septic systems are presumed to be a very significant source of contamination [35]. In the East Fork of the Little Miami River watershed, which includes the study beach, the septic system density in sub-catchments has been directly linked to human-associated fecal contamination through the use of human-associated genetic markers [35]. The only Harsha Lake (East Fork Lake) tributary investigated by Peed et al. [35] had the highest mean human-specific marker (HF183) density and the second highest density of septic systems ($41/km^2$) among tributaries investigated suggesting diffuse septic system-associated fecal contamination impacts this beach.

In health studies of swimmers in waters impacted by nonpoint source and/or diffuse fecal contamination, the lack of association between bacteria-based genetic markers and human illness has been observed [5,6,10,16]. Similarly, we did not observe associations between qPCR-based bacterial densities and human illness at our inland freshwater beach. The association of culturable *E. coli* with HCGI and GI was already observed [26], but the lack of association of GI and HCGI with human-specific genetic markers (e.g., HuBac) was unknown in this study population. Two possible reasons for our findings are as follows: (1) transport and fate varies considerably between bacterial genetic markers and viable waterborne pathogens, including viruses; and/or (2) our sample size was not big enough to detect significant associations.

With respect to viruses at the beaches with diffuse contamination, male-specific coliphage density has been associated with HCGI and GI illness in marine studies [5,16]. Similarly, our study observed an association between a virus (HAdV) and GI illness; however, unlike the marine studies, our study did not evaluate

Table 6. Odds ratios for GI and diarrhea associated with exposure to varying levels of HAdV and *E. coli* at East Fork Lake, Ohio.

Water Quality Index	Any GI Illness			Diarrhea		
	Cases/*n* (%)	OR (95% CI)	Wald (*p*)	Cases/*n* (%)	OR (95% CI)	Wald (*p*)
Group 1: Low HAdV & Low *E. coli* (HAdV <DL[a], log *E. coli* <1.05)	3/187 (1.6)	Referent		3/189 (1.6)	Referent	
Group 2: Low HAdV & High *E. coli* (HAdV <DL, log *E. coli*>1.05)	16/199 (8.0)	**5.4 (1.5–19)**	**0.011**	9/199 (4.5)	2.9 (0.74–12)	0.126
Group 3: High HAdV & Low *E. coli* (HAdV> DL, log *E. coli* <1.05)	11/126 (8.7)	5.9 (0.88–39)	0.067	9/127 (7.1)	4.7 (0.56–40)	0.153
Group 4: High HAdV & High *E. coli* (HAdV> DL, log *E. coli*>1.05)	8/88 (9.0)	**6.1 (1.5–25)**	**0.011**	7/90 (7.8)	**5.2 (1.3–22)**	**0.023**

[a]Detection Limit.

coliphages, focusing solely on viral pathogens. Like the associations between coliphages and illness [5,16], our finding of HAdV associations with HCGI and GI incidence in the adjusted models [Table 4] preliminarily suggests a possible benefit of using a viral indicator for evaluating recreational water illness risk in certain environments.

Similar to the Colford et al. study [5], human noroviruses (HNoV GI and HNoV GII) were not detected; however, human adenoviruses (HAdV) were detected from eight (35%) samples at East Fork versus one detection in the Colford et al. study [5]. The infrequent detection of HEntV (22%) in our study is not entirely unexpected as enteroviruses have been observed at densities below 1 PFU/100 mL in European waters [36] and have been infrequently detected (9% of samples) elsewhere [37]. These findings are similar to earlier findings from Fleisher et al. [14] that determined positive enterovirus detection (by cell culture) was less likely than negative detection in United Kingdom waters, which was also described as a limitation of the utility of the enterovirus assay for swimming advisory determination. The low detection frequency of viruses coupled with practicality of measurement presents challenges for establishing their use unless monitored in tandem with other indicators of water quality. Their use in a monitoring scheme may complement fecal indicator bacteria which have not effectively predicted viral densities in freshwater and marine environments [17,21,38,39]. Among viruses, adenovirus has shown promise as a potential human-associated marker of fecal contamination, whereby adenovirus presence has been linked to human-associated fecal problems as reflected by human-associated bacterial markers [40,41].

Despite a lack of association between viruses and bacteria, swimming advisories are primarily bacteria-based for practical reasons. Furthermore, there appears to be growing consensus among experts suggesting viruses are responsible for many water-related illnesses [18,19,42]. It is becoming more widely understood that enteric viruses (e.g., noroviruses, rotaviruses) are responsible for a substantial number of GI illnesses among U.S. beach users [18]. With viruses gaining attention and being more frequently measured, approaches for monitoring viruses to support integrative health risk models for water-associated illnesses are meaningful. Viral indicators of GI illness, particularly enteroviruses and bacteriophages, were described as promising predictors in 2003 by Wade et al. [2], but were also described as being limited by the difficulty and time for cultivation and enumeration. Now, with increasing use of molecular approaches such as qPCR, viral detection and enumeration is becoming less difficult and more rapid. In less contaminated or nonpoint source impacted waters, the use of a more commonly detected virus, like HAdV, may serve as an effective viral marker over noroviruses or enteroviruses for beach monitoring, since the detection frequency of adenovirus (36%) has been observed to be much greater than norovirus detection (9%) in European water samples [41,43].

The low detection frequency of norovirus in our samples and elsewhere may involve RNA virus instability, unstable DNA amplification potential from single-stranded RNA, and virus seasonality since the most prevalent human norovirus (GII.4) has clear winter peak seasonality [44] and therefore is less likely to be observed in summer surface waters. Unlike HNoV and HEntV, HAdV is a double-stranded DNA virus presumed to have better stability in environments and greater opportunities for successful detects. Furthermore, adenoviruses are believed to be up to 60 times more resistant to damage from ultraviolet irradiation than RNA viruses [45] likely affording HAdV particles greater integrity and viability leading to more undamaged infectious particles [46]. Given detection frequencies, Wyn-Jones et al. [41] encourages the consideration of adenovirus over norovirus as a recreational water quality indicator

Accordingly, water quality advisories should be protective for preventing human illness from all potential infectious agents as deemed practical with the available methods. The need for bacteria-based indicators of health risk are warranted, as the historical paradigm supports the notion that individual fecal indicators like *E. coli* and enterococci as well as human-specific markers explain the potential pathogenic bacterial genera (*Shigella, Camplyobacter, Salmonella*, etc.) as well as gastrointestinal illness among swimmers. However, the need for data attempting to address and/or quantify viral-associated human health risks is significant, and some attempt has recently been made using a site-specific quantitative microbial risk assessment (QMRA) for adenovirus illness [47]. Although speculative, virus monitoring may be beneficial in recreational waters where the contamination sources of the aquatic system are complex and impacted by a combination of non-point source human-associated contamination as well as agricultural runoff, as in Kundu et al. [47]. The concept of 'know your beach' appears to be particularly important as it relates to water quality when fecal indicator bacteria are low, and some consideration to broadening the paradigm to include multiple microbiological terms may be warranted with further research, with a particular emphasis on viruses capable of predicting recreational water illnesses more efficiently. Similar approaches demonstrating the value of a holistic design, like our combined virus-bacteria model, for enhancing risk assessments pertaining to water are already gaining attention in microbial source tracking studies using toolbox approaches [22]. Beyond water quality analysis, additional assessments of other known exposures presenting health risks, such as sand quality [48] and hydrology [49] are also recommended to be considered for more holistic beach condition modeling [25]. In this study, our integrative risk model using two microbiological terms, a fecal indicator bacteria and a viral genetic marker, provided a better assessment of human health risk than any single indicator/marker approach with limited data. Additional epidemiological studies using HAdV or other indicators of viral contamination that control for illnesses explained by bacteria are warranted for improving our predictive capability of swimming-related illness as well as our characterization of water quality at inland beaches.

Study Limitations

Future studies with larger populations are needed for drawing less speculative conclusions. The total swimmer population after exclusion is relatively small (n = 600) for rigorous analysis and may be leading us to spurious conclusions. Given the low number of cases presented, model construction was limited to two to four covariates, which therefore limited our ability to adequately assess multiple confounders, interaction effects, and linearity. The cohort study approach used here and in other U.S. studies [1,3,4] also presents several methodological limitations [50] related to potential misclassification of exposure and disease among study participants. We acknowledge a variety of confounders (assessed and not assessed) may be responsible for GI illness cases beyond adverse water quality conditions, such as food consumption at the beach. For example, personal risk perceptions were not measured in this study, which have been linked to self-reported illness among swimmers in other studies [15]. To reduce self-selection bias among swimmers, this study performed modeling comparing swimmers in various exposure groups to a low exposure swimmer group. Exposure misclassification is also relevant, particularly if beach water quality varies significantly throughout each study day which would limit reliable exposure estimates from a single daily sampling effort. Lastly, illness misclassification relying on self-reported information was more likely in this study than in studies using direct follow-up examinations.

Despite this sample size limitation, several constructed models did provide acceptable discrimination (AUC>0.70; Table 5) and acceptable model fit, particularly the HCGI and GI models from Table 4. Although the study presents several limitations, the study likely represents the only prospective epidemiological investigation exploring HAdV associations with recreational water-associated illness at this important beach type (inland lake recreational water). Future studies with larger samples sizes are warranted, particularly in beach environments where fecal contamination is human-associated and from dispersed sources or a mixture of point and dispersed sources.

Conclusions

- No association was observed between exposure to various qPCR-measured bacterial markers and recreational water-associated illness among study participants.
- Increasing levels of qPCR-measured HAdV were associated with increased odds of recreational water-associated gastrointestinal illness among the swimmers.
- Combined measurement of HAdV and culturable *E. coli* densities potentially enhances recreational water illness prediction among swimmers.
- Future studies with larger sample sizes enabling adjustment for additional confounding variables are needed to permit more robust investigations of human illness associations with exposures to HAdV and other viral markers in recreational water environments.

Supporting Information

Table S1 Primers and probes used in this study.

Table S2 Adjusted odds ratios for covariates used in GI illness models from Table 3.

Table S3 Adjusted odds ratios for covariates used in diarrheal illness models from Table 3.

Text S1 Detailed methods for sample processing and qPCR procedures.

Acknowledgments

The authors are grateful to the East Fork beach users, as well as Scott Fletcher, Chris Dauner, Bob Bowman, and Dan West from the Ohio Division of Parks and Recreation. The authors thank James Rosenblum, Michael Lofton, Matthew Farley, Sara Zolinski, and Kaedra Wetzel for assisting in the administration of the beach user questionnaire. The authors are thankful to Mary K. Estes at the Baylor College of Medicine for kindly providing the RNA transcript of Hu/NoV/GI/Norwalk.

Author Contributions

Conceived and designed the experiments: JL JWM. Performed the experiments: JWM CSL CL. Analyzed the data: JWM JL SL CL CSL QW. Contributed reagents/materials/analysis tools: JL LJS QW. Wrote the paper: JWM JL TJB SL CL CSL QW LJS.

References

1. Dufour A (1984) Health Effects Criteria for Fresh Recreational Waters. U.S. Environmental Protection Agency Report EPA-600-1-84-2004. Cincinnati, OH: U.S. Environmental Protection Agency, Office of Research and Development Available: http://www.epa.gov/nerlcwww/documents/frc.pdf. Accessed 2014 June 19.
2. Wade TJ, Pai N, Eisenberg JN, Colford JMJ (2003) Do U.S. Environmental Protection Agency water quality guidelines for recreational waters prevent gastrointestinal illness? A systematic review and meta-analysis. Environ Health Perspect 111: 1102–1109.
3. Wade TJ, Calderon RL, Sams E, Beach M, Brenner KP, et al. (2006) Rapidly measured indicators of recreational water quality are predictive of swimming-associated gastrointestinal illness. Environ Health Perspect 114 (1): 24–28. doi: 10.1289%2Fehp.8273.
4. Wade TJ, Sams E, Brenner KP, Haugland R, Chem E, et al. (2010) Rapidly measured indicators of recreational water quality and swimming-associated illness at marine beaches: a prospective cohort study. Environmental Health 9:66. doi: 10.1186/1476-069X-9-66.
5. Colford JM, Wade TJ, Schiff KC, Wright CC, Griffith JF, et al. (2007) Water quality indicators and the risk of illness at beaches with nonpoint sources of fecal contamination. Epidemiology . 18 (1), 27–35. doi: 10.1097/01.ede.0000249425.32990.b9.
6. Sinigalliano CD, Fleisher JM, Gidley ML, Solo-Gabriele HM, Shibata T, et al. (2010) Traditional and molecular analyses for fecal indicator bacteria in non-point source recreational marine waters. Water Res 44 (13): 3763–3772. doi: dx.doi.org/10.1016/j.watres.2010.04.026.
7. Gruber SJ, Kay LM, Kolb R, Henry K (2005) Mission Bay bacterial source identification study. Stormwater 6: 40–51.
8. Soller JA, Schoen ME, Bartrand T, Ravenscroft J, Ashbolt NJ (2010) Estimated human health risks from exposure to recreational waters impacted by human and non-human sources of faecal contamination. Water Res 44 (16): 4674–4691. doi: 10.1016/j.watres.2010.06.049.
9. Colford JM, Schiff KC, Griffith JF, Yau V, Arnold BF, et al. (2012) Using rapid indicators for Enterococcus to assess the risk of illness after exposure to urban runoff contaminated marine water. Water Res 46 (7): 2176–2186. doi: 10.1016/j.watres.2012.01.033.
10. Arnold BF, Schiff KC, Griffith JF, Gruber JS, Yau V, et al. (2013) Swimmer illness associated with marine water exposure and water quality indicators: impact of widely used assumptions. Epidemiology 24 (6): 845–853. doi: 10.1097/01.ede.0000434431.06765.4a.
11. Yau VM, Schiff KC, Arnold BF, Griffith JF, Gruber JS, et al. (2014) Effect of submarine discharge on bacterial indicators and swimmer health at Avalon Beach, CA, USA. Water Res 59: 23–36. doi: 10.1016/j.watres.2014.03.050.
12. Fewtrell L, Jones F, Tech B, Kay D, Wyer MD, et al. (1992) Health effects of white-water canoeing. Lancet 339 (8809): 1587–1589. doi:10.1016/0140-6736(92)91843-W.
13. Pike E (1994) Health Effects of Sea Swimming, Phase III – Final Report to the Department of the Environment. WRC Report: DoE 3412/2. Medmenham, U.K. UK: Water Research Center. pp 9–24.
14. Fleisher JM, Kay D, Salmon RL, Jones F, Wyer MD, et al. (1996) Marine waters contaminated with domestic sewage: nonenteric illnesses associated with swimmer exposure in the United Kingdom. Am J Public Health 86 (9): 1228–1234.
15. Fleisher JM, Fleming LE, Solo-Gabriele HM, Kish JK, Sinigalliano D, et al. (2010) The BEACHES Study: health effects and exposures from non-point source microbial contaminants in subtropical recreational marine waters. Int J Epidemiol 39 (5): 1291–1298. doi: 10.1093/ije/dyq084.
16. Abdelzaher AM, Wright ME, Ortega C, Hasan AR, Shibata T, et al. (2011) Daily measures of microbes and human health at a non-point source marine beach. J Water Health 9 (3): 443–457. DOI:10.2166/wh.2011.146Griffen DW, Donaldson KA, Paul JH, Rose JB (2003) Pathogenic human viruses in coastal waters. Clin Microbiol Rev 16 (1): 129–143. doi: 10.1128/CMR.16.1.129-143.2003.
17. Hlavsa MC, Roberts VA, Anderson AR, Hill VR, Kahler AM, et al. (2011) Surveillance for Waterborne Disease Outbreaks and Other Health Events Associated with Recreational Water -United States, 2007-2008. MMWR 60 (ss12): 1–32.
18. Soller JA, Bartrand T, Ashbolt NJ, Ravenscroft J, Wade TJ (2010). Estimating the primary etiologic agents in recreational freshwaters impacted by human sources of fecal contamination. Water Res 44 (16): 4736–4747. doi: 10.1016/j.watres.2010.07.064.
19. Dorevitch S, Pratap P, Wroblewski M, Hryhorczuk DO, Li H, et al. (2012) Health risks of limited-contact water recreation. Environ Health Perspect 120 (2): 192–197. doi: 10.1289/ehp.1103934.
20. Lee CS, Lee C, Marion JW, Wang Q, Saif L, et al. (2014) Occurrence of human enteric viruses at freshwater beaches during swimming season and its link to water inflow. Sci Total Environ 472: 757–766. doi: 10.1016/j.scitotenv.2013.11.088.
21. Noble RT, Fuhrman JA (2001) Enteroviruses detected by reverse transcriptase polymerase chain reaction from the coastal waters of Santa Monica Bay,

California: low correlation to bacterial indicator levels. Hydrobiologia 460: 175–184. doi: 10.1023/A:1013121416891.

22. McQuaig S, Griffith J, Harwood VJ (2012) Association of fecal indicator bacteria with human viruses and microbial source tracking markers at coastal beaches impacted by nonpoint source pollution. Appl Environ Microbiol 78 (18): 6423–6432. doi: 10.1128/AEM.00024-12.
23. Ashbolt NJ, Grabow WOK, Snozzi M (2001) Indicators of microbial water quality. In: Fewtrell L, Bartram J, editors. Water Quality: Guidelines, Standards and Health. London: World Health Organization: pp.289–315.
24. Boehm AB, Ashbolt NJ, Colford JM, Dunbar LE, et al. (2009) A sea of change ahead for recreational water quality criteria. J Water Health 7 (1): 9–20. doi: 10.2166/wh.2009.122.
25. Abdelzaher AM, Solo-Gabriele HM, Phillips MC, Elmir SM, Fleming LE (2013) An alternative approach to water regulations for public health protection at swimming beaches. J. Environ Public Health 2013: 138521. DOI: 10.1155/2013/138521.
26. Marion JW, Lee J, Lemeshow S, Buckley TJ (2010) Association of gastrointestinal illness and recreational water exposure at an inland U.S. beach. Water Res 44 (16): 4796–4804. doi: 10.1016/j.watres.2010.07.065.
27. U.S. Environmental Protection Agency (2002) Method 1603: *Escherichia coli* (*E. coli*) in Water by Membrane Filtration Using Modified membrane-Thermotolerant *Escherichia coli* Agar (Modified mTEC). Office of Water Report EPA-821-R-02-023. Washington D.C. Available: http://www.epa.gov/nerlcwww/documents/1603sp02.pdf. Accessed 2014 June 19.
28. National Oceanic and Atmospheric Administration (2013) Database: UV Index Bulletins Archives. http://www.cpc.ncep.noaa.gov/products/stratosphere/uv_index/uv_archive.shtml. Accessed 2014 June 19.
29. U.S. Army Corps of Engineers, Louisville District (2013) Database: Yearly Lake Reports. http://www.lrl.usace.army.mil/wcds/dlbrpt.htm. Accessed 2014 June 19.
30. Haugland RA, Siefring SC, Wymer LJ, Brenner KP, Dufour AP (2005) Comparison of *Enterococcus* measurements in freshwater at two recreational beaches by quantitative polymerase chain reaction and membrane filter culture analysis. Water Res 39 (4): 559–568. doi: 10.1016/j.watres.2004.11.011.
31. Lipsitch M, Tchetgen ET, Cohen T (2010). Negative controls: a tool for detecting confounding and bias in observational studies. Epidemiology 21 (3): 383–388. doi: 10.1097/EDE.0b013e3181d61eeb.
32. Hosmer DW, Lemeshow S (2000) Applied logistic regression, 2nd ed. New York: Wiley and Sons. 375 p.
33. Peduzzi P, Concato J, Kemper E, Holford TR, Feinstein AR (1996). A simulation of the number of events per variable in logistic regression analysis. J Clin Epidemiol 49 (12): 1373–1379.
34. Vittinghoff E, McCulloch CE (2007). Relaxing the rule of ten events per variable in logistic and cox regression. Am J Epidemiol 165 (6): 710–718. doi: 10.1093/aje/kwk052.
35. Peed LA, Nietch CT, Kelty CA, Meckes M, Mooney T, et al. (2011) Combining land use information and small stream sampling with PCR-based methods for better characterization of diffuse sources of human fecal pollution. Environ Sci Technol 45 (13): 5652–5659. doi: 10.1021/es2003167.
36. Lodder WJ, de Roda Husman AM (2005) Presence of noroviruses and other enteric viruses in sewage and surface waters in the Netherlands. Appl Environ Microbiol 71 (3): 1453–1461. doi: 10.1128/AEM.71.3.1453-1461.2005.
37. Haramoto E, Katayama H, Oguma K, Ohgaki S (2005) Application of cation-coated filter method to detection of noroviruses, enteroviruses, adenoviruses, and torque teno viruses in the Tamagawa River in Japan. Appl Environ Microbiology 71 (5): 2403–2411. doi: 10.1128/AEM.71.5.2403-2411.2005.
38. Jiang S, Noble R, Chu W (2001) Human adenoviruses and coliphages in urban-runoff impacted coastal waters of Southern California. Appl Environ Microbiol 67 (1): 179–184. doi: 10.1128/AEM.67.1.179-184.2001.
39. Payment P, Lemieux M, Trudel M (1982) Bacteriological and virological analysis of water from four fresh water beaches. Water Res 16 (6): 939–943. doi: 10.1016/0043-1354(82)90026-4.
40. Newton RJ, VandeWalle JL, Borchardt MA, Gorelick MH, McLellan SL (2011) Lachnospiraceae and Bacteroidales alternative fecal indicators reveal chronic human sewage contamination in an urban harbor. Appl Environ Microbiol 77 (19): 6972–6981. doi: 10.1128/AEM.05480-11.
41. Wyn-Jones AP, Carducci A, Cook N, D'Agostino M, Divizia M, et al. (2011) Surveillance of adenoviruses and noroviruses in European recreational waters. Water Res 45 (3): 1025–1038. doi: 10.1016/j.watres.2010.10.015.
42. Sinclair RG, Jones EL, Gerba CP (2009) Viruses in recreational water-borne disease outbreaks: a review. J Appl Microbiol 107 (6): 1769–1780. doi: 10.1111/j.1365-2672.2009.04367.x.
43. Albinana-Gimenez N, Miagostovich MP, Calgua B, Huguet JM, Matia L, et al. 2009. Analysis of adenoviruses and polyomaviruses quantified by qPCR as indicators of water quality in source and drinking-water treatment plants. Water Res 43 (7): 2011–2019. DOI: 10.1016/j.watres.2009.01.025.
44. Leshem E, Wikswo M, Barclay L, Brandt E, Storm W, et al. (2013) Effects and clinical significance of GII.4 Sydney Norovirus, United States, 2012–2013. Emerg Infect Dis 19 (8): 1231–1238. doi: 10.3201/eid1908.130458.
45. Gerba CP, Gramos DM, Nwachuku N (2002) Comparative inactivation of enteroviruses and adenovirus 2 by UV light. Appl Environ Microbiol 68: 5167–5169. doi: 10.1128/AEM.68.10.5167-5169.2002.
46. Fongaro G, Nascimento MA, Rigotto C, Ritterbusch G, da Silva AD, et al. (2013) Evaluation and molecular characterization of human adenovirus in drinking water supplies: viral integrity and viability assays. J Virol 10: 166. doi: 10.1186/1743-422X-10-166.
47. Kundu A, McBride G, Wuertz S (2013) Adenovirus-associated health risks for recreational activities in a multi-use coastal watershed based on site-specific quantitative microbial risk assessment. Water Res 47 (16): 6309–6325. doi: 10.1016/j.watres.2013.08.002.
48. Heaney CD, Sams E, Dufour AP, Brenner KP, Haugland RA, et al. (2012) Fecal indicators in sand, sand contact, and risk of enteric illness among beachgoers. Epidemiol 23 (1): 95–106.
49. Ge Z, Whitman RL, Nevers MB, Phanikumar MS (2012) Wave-induced mass transport affects daily *Escherichia coli* fluctuations in nearshore water. Environ Sci Technol 46 (4): 2204–2211. doi: 10.1021/es203847n.
50. Fleisher JM, Jones J, Kay D, Stanwell-Smith R, Wyer M, et al. (1993). Water and non-water related risk factors for gastroenteritis among swimmers exposed to sewage-contaminated marine waters. Int J Epidemiol 22 (4): 698–708. doi: 10.1093/ije/22.4.698.
51. Lee C, Agidi S, Marion JW, Lee J (2012) *Arcobacter* in Lake Erie beach waters: an emerging gastrointestinal pathogen linked with human-associated fecal contamination. Appl and Environ Microbiol 78 (16): 5511–5519. doi: 10.1128/AEM.08009-11.
52. Lee J, Deininger RA (2004) Detection of *E. coli* in beach water within 1 hour using immunomagnetic separation and ATP bioluminescence. Luminescence 19 (1): 31–36. doi: 10.1002/bio.753.
53. Haramoto E, Katayama H, Oguma K, Ohgaki S (2005) Application of cation-coated filter method to detection of noroviruses, enteroviruses, adenoviruses, and torque teno viruses in the Tamagawa River in Japan. Appl Environ Microbiol 71 (5): 2403–2411. doi: 10.1128%2FAEM.71.5.2403-2411.2005.
54. Monpoeho S, Dehee A, Mignotte B, Schwartzbrod L, Marechal V, et al. (2000) Quantification of enterovirus RNA in sludge samples using single tube real-time RT-PCR. Biotechniques 29 (1): 88–93.
55. Kageyama T, Kojima S, Shinohara M, Uchida K, Fukushi S, et al. (2003) Broadly reactive and highly sensitive assay for Norwalk-like viruses based on real-time quantitative reverse transcription-PCR. J Clin Microbiol 41 (4): 1548–1557. doi: 10.1128/JCM.41.4.1548-1557.2003.
56. Heim A, Ebnet C, Harste G, Pring-Akerblom P (2003) Rapid and quantitative detection of human adenovirus DNA by real-time PCR. J Med Virol 70 (2): 228–239. doi: 10.1002/jmv.10382.
57. Chern EC, Brenner K, Wymer L, Haugland RA (2009) Comparison of fecal indicator bacteria densities in marine recreational waters by QPCR. Water Qual Expo Healt 1 (3–4): 203–214.
58. Chern EC, Siefring S, Paar J, Doolittle M, Haugland RA (2011) Comparison of quantitative PCR assays for *Escherichia coli* targeting ribosomal RNA and single copy genes. Lett Appl Microbiol 52 (3): 298–306. doi: 10.1111/j.1472-765X.2010.03001.x.
59. Bernhard AE, Field KG (2000) Identification of nonpoint sources of fecal pollution in coastal waters by using host-specific 16S ribosomal DNA genetic markers from fecal anaerobes. Appl Environ Microbiol 66 (4): 1587–1594. doi: 10.1128/AEM.66.4.1587-1594.2000.
60. Okabe S, Okayama N, Savichtcheva O, Ito T (2007) Quantification of host-specific *Bacteroides-Prevotella* 16S rRNA genetic markers for assessment of fecal pollution in freshwater. Appl Microbiol Biotechnol 74 (4): 890–901. doi: 10.1007/s00253-006-0714-x.
61. Ludwig W, Schleifer KH (2000) How quantitative is quantitative PCR with respect to cell counts? Syst Appl Microbiol 23 (4): 556–562. doi: 10.1016/S0723-2020(00)80030-2.
62. U.S. Environmental Protection Agency (2010) Method A: Enterococci in Water by TaqMan Quantitative Polymerase Chain Reaction (qPCR) Assay. EPA-821-R-10-004. Washington, D.C. Available: http://water.epa.gov/scitech/methods/cwa/bioindicators/upload/rapid1.pdf. Accessed 2014 June 20.
63. Lee CS, Lee J, (2010) Evaluation of new gyrB-based real-time PCR system for the detection of *B. fragilis* as an indicator of human-specific fecal contamination. J Microbiol Methods 82 (3): 311–318. doi: 10.1016/j.mimet.2010.07.012.

4

Evaluating Primers for Profiling Anaerobic Ammonia Oxidizing Bacteria within Freshwater Environments

Puntipar Sonthiphand, Josh D. Neufeld*

Department of Biology, University of Waterloo, Waterloo, Ontario, Canada

Abstract

Anaerobic ammonia oxidizing (anammox) bacteria play an important role in transforming ammonium to nitrogen gas and contribute to fixed nitrogen losses in freshwater environments. Understanding the diversity and abundance of anammox bacteria requires reliable molecular tools, and these are not yet well established for these important Planctomycetes. To help validate PCR primers for the detection of anammox bacteria within freshwater ecosystems, we analyzed representative positive controls and selected samples from Grand River and groundwater sites, both from Ontario, Canada. The objectives of this study were to identify a suitable anammox denaturing gradient gel electrophoresis (DGGE) fingerprint method by using GC-clamp modifications to existing primers, and to verify the specificity of anammox-specific primers used for DGGE, cloning and qPCR methods. Six primer combinations were tested from four published primer sets (i.e. A438f/A684r, Amx368f/Amx820r, An7f/An1388r, and Pla46/1392r) for both direct and nested PCR amplifications. All PCR products were run subsequently on DGGE gels to compare the resulting patterns. Two anammox-specific primer combinations were also used to generate clone libraries and quantify anammox bacterial 16S rRNA genes with qPCR. The primer set A438f/A684r was highly specific to anammox bacteria, provided reliable DGGE fingerprints and generated a high proportion of anammox-related clones. A second primer set (Amx368f/Amx820r) was anammox specific, based on clone library analysis, but PCR products from different candidate species of anammox bacteria resolved poorly using DGGE analysis. Both DGGE and cloning results revealed that *Ca.* Brocadia and an uncharacterized anammox bacterial cluster represented the majority of anammox bacteria found in Grand River sediment and groundwater samples, respectively. Together, our results demonstrate that although Amx368f/Amx820r was useful for anammox-specific qPCR and clone library analysis, A438f/A684r was the most suitable primer set for multiple molecular assessments of anammox bacteria in freshwater environments.

Editor: Mark R. Liles, Auburn University, United States of America

Funding: This research was supported by a Strategic Projects Grant from the National Science and Engineering Research Council of Canada (NSERC). The funders had no role in study design, data collection and analysis, decision to publish, or preparation of the manuscript.

Competing Interests: The authors have declared that no competing interests exist.

* E-mail: jneufeld@uwaterloo.ca

Introduction

Even before the discovery of anaerobic ammonia oxidizing (anammox) bacteria as Planctomycetes [1] and elucidation of their metabolic pathways, physiology and morphology [2–3], the anammox process was recognized by nutrient profiles and thermodynamic calculations [4–6]. Anammox bacteria have the ability to transform ammonium into nitrogen gas by using nitrite as an electron acceptor under anoxic conditions. Anammox bacteria were first discovered in a laboratory-scale reactor in 1995 [7]. Since then, many reports have demonstrated the widespread occurrence and contribution of anammox bacteria in a variety of natural ecosystems [8–12]. Anammox bacteria are important in the global nitrogen cycle, and responsible for high nitrogen losses: ~50% in marine sediments [13–16], ~40% in contaminated groundwater [17] and 4–37% in terrestrial habitats [18]. These findings demonstrate the important role of anammox bacteria in natural environments.

There are still no pure culture isolates of anammox bacteria due to their extremely slow growth rates, relatively low biomass yields and inactivation by low concentrations of oxygen and nitrite [19–21]. Culture-independent methods such as 16S rRNA gene-based analysis [22–24] and fluorescence *in situ* hybridization (FISH) [2,25–26] are the common methods used for anammox bacterial community analysis. The 16S rRNA genes of known anammox bacteria show several phylogenetically distinct *Candidatus* genera including *Ca.* Brocadia, *Ca.* Scalindua, *Ca.* Kuenenia, *Ca.* Anammoxyglobus and *Ca.* Jettenia. The average sequence similarity between *Ca.* Scalindua and *Ca.* Brocadia or *Ca.* Kuenenia clusters is only 85% [3,26]. Thus, it is very challenging to design primer sets that target all known anammox genera. Although several anammox-specific primers have been used for 16S rRNA gene amplification, reported problems include low recovery efficiencies of anammox-related clones, non-specific amplification and an inability to target all anammox bacterial clusters [8,22,24,27–32]. Anammox bacteria from groundwater have recently been profiled by comparing bacteria-specific denaturing gradient gel electrophoresis (DGGE) patterns (341f-GC/518r) with "anammox specific" patterns from a nested PCR protocol (An7f/An1388r followed by 341f-GC/518r; [17]). The intense bands that appeared in DGGE fingerprints were confirmed to be related to anammox bacteria.

Since identifying the contribution of anammox bacteria to fixed nitrogen losses in natural ecosystems [33], much research has focused on studying anammox bacteria in marine environments [34]. Since anammox bacteria were first reported in freshwater environments [35], there have been only five known studies characterizing anammox bacterial communities in freshwater habitats [17,22,29,35–36]. Thus, information on the diversity, abundance and activity of anammox bacteria in freshwater ecosystems is still scarce. In this study, we focused on PCR primer-based detection methods for anammox bacteria within freshwater environments, including samples taken from the Grand River and from a previously studied groundwater site.

The primer set, A438f and A684r, successfully quantified anammox bacteria 16S rRNA gene copies in wetland soils [37] but was not tested for DGGE prior to this study. Another primer set, Amx368f and Amx820r, was commonly used to detect anammox bacteria in various environments. Both primers were originally designed for FISH probes and were then applied as forward and reverse PCR primers for detecting anammox bacteria by clone library analysis in freshwater, terrestrial and marine ecosystems, such as Lake Kitaura [36], groundwater [17], fertilized paddy soil [18], constructed wetland [38], peat soil [39], Cape Fear River estuary [23], coastal marine sediment [27], the Jiaojiang estuary [40], the South China Sea [41], and a high-temperature petroleum reservoir [42]. In this study, these existing PCR primers were modified with GC clamps and tested for the ability to generate anammox bacterial 16S rRNA gene DGGE fingerprints. The other two primer sets investigated here (An7f/An1388r and Pla46/1392r) amplified a large amplicon (~1,400 bp). The primers An7f and An1388r were designed originally to target anammox bacteria in freshwater and marine sediments [22]. The Planctomycetes-specific FISH probe, Pla46, has been used as a forward primer with the reverse universal primer 1392R to obtain PCR products that were subsequently used for nested PCR templates in both natural [12,28,30,43] and artificial [44–45] environments.

The main objectives of this study were (i) to identify suitable PCR primer combinations for DGGE assessment of anammox bacterial communities and (ii) to compare the efficiency and specificity of the existing primer sets for DGGE, clone library and qPCR assays. The results provide important experimental validation for using specific primer sets for investigating the diversity and abundance of anammox bacteria within freshwater environments.

Methods

Sampling site description and sample collection

The Grand River is located in southwestern Ontario, Canada. This large river and its tributaries receive high nitrogen inputs, mainly from agricultural runoff and wastewater discharge. Two sampling sites along the Grand River (Bridgeport and Blair), located in the city of Waterloo, were chosen as representative freshwater environmental sites. Sediment (SedBr and SedBl), epilithic biofilm (EpBr and EpBl) and water (WaBr and WaBl) samples were collected from each sampling site in June 2010. In addition, groundwater samples were collected from the Zorra township, Ontario (site details and sampling information were previously described in [17]). Both groundwater (GW) and groundwater sediment core (GS) samples were collected at a 7.5-m depth in July 2009. All water samples of approximately 300 ml were filtered onsite onto 0.22-μm Sterivex filters (Millipore, USA). All samples were stored on dry ice during transportation and kept at −80°C until DNA extraction. The environmental chemistry analyses from each site were shown in Table 1.

Table 1. Water chemistry for each sample site.

Sampling site	NH_4^+N (mg NH_4^+ L^{-1})	NO_2^-N (mg NO_2^- L^{-1})	NO_3^-N (mg NO_3^- L^{-1})	DO (mg L^{-1})	pH
Bridgeport	0.05	0.04	2.06	7.9	8.14
Blair	0.47	0.46	1.80	7.3	7.91
Zorra*	ND	NA	10	1.97	7.04

NA = Not available; ND = Not detected;
*Samples from Zorra site were collected in July, 2010 but all parameters reported were measured in August 2010.

DNA extraction

Genomic DNA was extracted from Grand River sediment and epilithic biofilm samples using the MoBio PowerSoil DNA kit (MoBio Laboratories, USA), following the manufacturer's protocol. Nucleic acids from all Sterivex filters were extracted according to a previously published protocol [46]. The quality and quantity of extracted DNA were measured by agarose gel electrophoresis and spectrophotometry (NanoDrop Spectrophotometer ND-100; Thermo Fisher Scientific, USA), respectively. Extracted DNA was then diluted to 5–10 ng μl^{-1} for use as PCR template.

Denaturing gradient gel electrophoresis (DGGE)

All samples were PCR amplified with bacteria-specific primers (341f-GC/518r), targeting the bacterial 16S rRNA gene. This PCR, in addition to a nested PCR approach for detecting anammox bacteria, followed previously published protocols [17]. Briefly, for the nested PCR, template was amplified with anammox bacteria-specific primers An7f/An1388r, followed by amplification by the bacteria-specific DGGE primers 341f-GC/518r. Additional published primers were also modified with GC-clamps for DGGE assessment [47]. These anammox-specific nested PCR amplifications involved either An7f/An1388r or Pla46/1392r for the first round of PCR, followed by anammox-specific A438f-GC/A684r or Amx368f-GC/Amx820r for a second reaction. The PCR components contained 2.5 μl of 10× ThermoPol Reaction Buffer, 0.05 μl of forward and reverse primer (100 μM stocks), 0.05 μl of dNTPs (100 mM stock), 0.1 μl of *Taq* DNA polymerase (5 U μl^{-1} stock), 1.5 μl of bovine serum albumin (10 mg ml^{-1} stock) and 1 μl of DNA template (representing 5–10 ng of genomic DNA) in a total reaction volume of 25 μl. All PCR amplifications were carried out with an initial denaturation at 95°C for 5 min, followed by primer-set-specific thermal cycling conditions (Table 2) with a total of 30–35 cycles for the first round PCR and a final extension of 72°C for 10 min to complete the reaction. The first round PCR products were diluted 100-fold to serve as template for the nested PCR. Nested PCR conditions and thermal cycle profiles of each primer set were the same as previously described, except for the number of PCR cycles. All nested PCR were run for a total of 20–25 cycles. After each amplification, PCR products were verified by agarose gel electrophoresis to confirm amplicon size. The 341f-GC/518r and A438f-GC/A684r PCR products were run on 10% acrylamide gels, with 30%-70% denaturing gradients. The Amx368f-GC/Amx820r PCR products were profiled on 8% acrylamide gels, with 30%-70% denaturing gradients. All DGGE gels were run for 15 h at 85 V and at 60°C, using a DGGEK-2401 (CBS Scientific Company, USA). The DGGE gels were stained with SYBR green (Invitrogen, USA) and scanned with a Pharos FXTM Plus

Table 2. Summary of the PCR primers and conditions used in this study.

Primer[1]	Specificity	*E. coli* position	PCR conditions			Reference
			Denaturation	Annealing	Extension	
A438f[2]	All anammox bacteria	438–455	95°C, 30 sec	55°C, 30 sec	72°C, 30 sec	[37]
A684r[2]		667–684				
Amx368f	All anammox bacteria	368–385	95°C, 45 sec	59°C, 45 sec	72°C, 45 sec	[26]
Amx820r	*Ca.* Kuenenia, *Ca.* Brocadia	820–841				[25]
An7f An1388r	*Ca.* Kuenenia, *Ca.* Brocadia, *Ca.* Scalindua	7–26 1372–1388	95°C, 45 sec	63°C, 1 min	72°C, 1 min	[22]
Pla46	Planctomycetes	46–63	95°C, 45 sec	59°C, 1 min	72°C, 1 min	[62]
1392r	Universal bacteria	1392–1406				[63]
341f	Universal bacteria	341–357	95°C, 30 sec	55°C, 30 sec	72°C, 30 sec	[64]
518r		518–534				

[1]For DGGE, a GC-clamp was attached to the forward primers for PCR (A348f-GC, Amx368f-GC and 341f-GC).
[2]The qPCR conditions were exactly the same (primer concentrations, annealing temperature and without additional BSA) as the original condition [37].

Molecular Imager (Bio-Rad, USA). Representative bands were excised and sequenced by the corresponding anammox-specific primers at Beckman Coulter Genomics using an ABI 3730XL sequencer.

Clone library analysis

Three representative samples (SedBr, SedBl and GW) were selected to generate clone libraries with three anammox primer sets (An7f/An1388r, A438f/A684r and Amx368f/Amx820r) to compare the efficiency and specificity of each primer set. The PCR conditions for cloning were the same as those described for DGGE (Table 2). The PCR products were ligated and transformed using a TOPO TA Cloning kit and One Shot TOP10 Chemically Competent cells (Invitrogen, USA), respectively, according to manufacturer's protocols. Between 30–70 white colonies were selected from each library and screened for the presence of inserts by each anammox-specific PCR primer set prior to being sequenced at Beckman Coulter Genomics, USA as previously mentioned in the DGGE section.

Quantitative real-time PCR (qPCR)

Anammox bacterial 16S rRNA genes were quantified by two specific primer sets (A438f/A684r and Amx368f/Amx820r) for comparison. Total bacterial abundance was also investigated by primers 341f/518r as a reference. The qPCR master mix contained 5 µl of SsoAdvanced SYBR Green Supermix (Bio-Rad, USA), 0.03 µl of each primer (100 µM stocks), 0.02 µl of bovine serum albumin (10 mg ml^{-1} stock) and 1 µl of genomic DNA template (5–10 ng stock) in a total volume of 10 µl. All qPCR amplifications were performed in duplicate on a CFX96 real-time system (Bio-Rad, USA). Although specific annealing temperatures were used (Table 2), qPCR thermal programs were common for all three primer sets. An initial denaturation at 98°C for 2 min was followed by 35 cycles of 98°C for 5 sec, annealing at the primer-specific annealing temperature for 30 sec and 72°C for 30 sec, with a plate read after each cycle. Following PCR, melt curves were generated between 65°C–95°C in 0.5°C increments to ensure PCR specificity. Reference freshwater samples with high anammox abundance were amplified by each specific primer set, pooled by primer set, then purified to serve as anammox bacterial standard templates for qPCR. For general bacterial qPCR, the standard curves were constructed from *Escherichia coli* genomic DNA. Each PCR product was purified using a MinElute kit (Qiagen, USA) and quantified by the NanoDrop Spectrophotometer ND-100. Ten-fold serial dilutions were applied to the standard DNA PCR product template to create the qPCR standard curves, which was linear between 10^1–10^7 copies, with efficiencies of 84–93% and coefficients of determination (R^2)≥0.996 for all standard curves. The specificity of qPCR amplification was confirmed by melting curve analysis and agarose gel electrophoresis of all products after each run.

Statistical and phylogenetic analysis

All analyzed sequences from clone libraries were clustered into operational taxonomic units (OTUs), based on 1% and 3% dissimilarity cut-off settings for 16S rRNA gene nucleotide similarity by AXIOME version 1.5.0 [48]. All clone sequences showing 97% and 99% identical sequences were grouped together before constructing two phylogenetic trees, based on differing distance levels in nucleotide sequences, to compare anammox bacterial clusters and tree topologies. All sequences from DGGE and clone library analysis were compared to the Genbank non-redundant database with the Basic Local Alignment Search Tool (BLAST) to identify related sequences. All DGGE bands and representative clones from each library were aligned with selected uncultured anammox bacteria sequences and reference *Ca.* anammox bacteria species using MUSCLE [49]. A phylogenetic tree was constructed in PhyML v.3.0.1 [50–51], with the GTR model. The approximate likelihood ratio test (aLRT) statistic was used to calculate branch support values. Phylogenetic trees were run with five random starts to optimise the tree topology.

Nucleotide accession number

All anammox bacterial 16S rRNA gene sequences were deposited in GenBank with accession numbers JX392915-JX392948.

Results

Anammox bacterial primers for DGGE

In a previous study, DGGE, qPCR and Illumina 16S rRNA gene data indicated that anammox bacteria were numerically important community members of an ammonium-contaminated groundwater site (Zorra, Ontario). Our initial benchmarking experiment for this study was to repeat the DGGE protocol with DNA extracts from this groundwater at a 7.5-m depth, and include additional sediment, epilithic biofilm and water samples from two representative sites within the Grand River, which are not as strongly dominated by anammox bacteria [52]. As positive controls, we used plasmids carrying anammox 16S rRNA genes associated with *Ca.* Jettenia, *Ca.* Brocadia and *Ca.* Scalindua; genomic DNA from *E. coli* was used as a negative control. The results demonstrated that in all cases, patterns generated by bacteria-specific primers (341f-GC/518r) were distinct from those generated by the nested anammox PCR protocol for DGGE (Fig. 1). Of the 20 bands selected for sequencing, all three analyzed bands from SedBr, one band from SedBl and three bands from GW were affiliated with anammox bacteria (Fig. 1), demonstrating that this nested PCR design resulted in the enrichment of anammox bacterial amplicons. Genbank BLAST analysis results showed that the sequences were 97%–100% identical to previously reported sequences recovered from sediment in wetlands (JQ762203) and ammonium-contaminated groundwater from the same site analyzed in this study (HQ595700 and HQ595667). Phylogenetic analysis revealed that all six bands indicated by a yellow triangle were associated with Unknown 1 cluster, while another band indicated by a purple triangle fell into Unknown 2 cluster (Fig. 1 and Fig. 2). Importantly, non-specific amplification of the *E. coli* 16S rRNA gene was observed (Fig. 1), which was a non-specific amplification problem also seen in most of the freshwater samples included in this study. Bands indicated by black triangles were not related to anammox bacteria. These bands were 97%–100% identical to Actinobacteria, Chloroflexi and Cyanobacteria. These results demonstrate that a coupling of an anammox-specific amplification using An7f/An1388r with bacteria-specific PCR for DGGE (341f-GC/518r) was not suitable for targeting anammox bacteria in the environmental samples at the expense of all other bacteria, especially if anammox bacterial abundances were relatively low, such as in the Grand River epilithic biofilm and water column samples. *In silico* analysis (Table S1) revealed that An7f showed two mismatches at the 5′end and one, none and three mismatches at 3′end against Actinobacteria, Chloroflexi, and *E. coli* 16S rRNA gene sequences, respectively. Although, An1388r showed many mismatches for Actinobacteria, Chloroflexi, and *E. coli*, it was highly specific for both anammox and non-anammox of Planctomycetales 16 rRNA gene sequences. This analysis supported that An7f/An1388r could amplify non-anammox bacterial sequences. Note that band positions on a DGGE gel were different for the three anammox *Ca.* genera (Fig. 1), which is useful for distinguishing different anammox populations if non-specific amplification is not a concern.

Because many bands from the previously published nested DGGE approach were not affiliated with anammox bacteria in the Grand River epilithic biofilm and water samples (Fig. 1), we combined alternative published primer sets to generate DGGE profiles from the selected freshwater environments. The two tested primer sets were A438f-GC/A684r and Amx368f-GC/Amx820r; neither primer set had been tested with GC clamps for DGGE prior to this study. A nested PCR technique was also included in this comparison to identify whether this approach, useful for samples with low target abundance, could increase the sensitivity of anammox bacterial 16S rRNA gene signals without altering the profiles generated. We used primers An7f/An1388r, which targets a near full-length (~1,400 bp) region of the anammox bacterial 16S rRNA gene associated with *Ca.* Scalindua, *Ca.* Brocadia and *Ca.* Kuenenia genera [22]. Primers Pla46/1392r, targeting all bacterial 16S rRNA genes within the Planctomycetes phylum (~1,400 bp), were also used for the first PCR amplification for comparison. In all nested PCR assays, after generating a larger amplicon from the initial PCR, a shorter fragment was amplified by more specific anammox bacterial primers with the GC-clamp. All freshwater samples and both positive and negative controls were amplified by direct PCR using the two main primer sets (Fig. 3A and 3D) and the four additional combinations for nested PCR approaches (Fig. 3B, 3C, 3E and 3F) to compare the DGGE patterns, anammox-specific bands and diversity of anammox bacteria detected by each set.

The results demonstrated that the primer set A438f-GC/A684r generated reproducible patterns that were unique for the different samples included in this study. DGGE fingerprints generated directly by primers A438f-GC/A684r (Fig. 3A) were similar to those from the nested PCR techniques (Fig. 3B and 3C). The DGGE fingerprints produced by both nested PCR amplifications were highly similar to each other. Seven of eight samples showed positive anammox signals from direct PCR amplification with A438f-GC/A684r (Fig. 3A). Only one sample, GS, was not amplified by direct PCR, but a nested PCR generated an anammox band for this sample. Intense DGGE bands from each sample were sequenced and included in a phylogenetic analysis (Fig. 2). All analyzed bands from primer set A438f-GC/A684r were related to anammox sequences. The BLAST results indicated them to be 97%–100% identical to uncultured anammox bacteria recovered from range of freshwater habitats. Phylogenetic analysis demonstrated that bands identified as 2-1 were associated with *Ca.* Scalindua-like sequences (Fig. 3). They were 96% and 94% identical to *Ca.* Scalindua brodae and *Ca.* Scalindua wagneri, respectively. These two identical bands, found in samples EpBr and WaBr, were at the same position as *Ca.* Scalindua control plasmid (Fig. 3A). Bands labeled as 2-2 fell into the Unknown 1 cluster, which was an anammox cluster previously reported [17]. All of the anammox bands from GW and GS were related to this Unknown 1 cluster. The majority of anammox bands found across our samples, corresponding to bands labeled 2-3, were closely related to *Ca.* Brocadia-like sequences (Fig. 2). They were 98% identical to *Ca.* Brocadia caroliniensis and *Ca.* Brocadia fulgida.

All retrieved anammox bands from the Grand River samples from nested PCR DGGE fingerprints fell into the *Ca.* Brocadia cluster. This demonstrates the potential for the direct PCR method to generate higher anammox diversity representation than that detected by either nested PCR approach for the Grand River samples (Fig. 3A, 3B and 3C). All three bands detected in WaBr (bands 2-1, 2-2, and 2-3) fell into *Ca.* Scalindua, Unknown 1 and *Ca.* Brocadia clusters, respectively, whereas all four bands from the same samples, generated by both nested PCR amplifications, were all *Ca.* Brocadia-like phylotypes. The nested PCR approach, due to probable PCR bias, underrepresented anammox bacterial diversity in our freshwater samples but can nonetheless increase amplification of anammox template from samples with low anammox bacterial abundance.

Apart from the *Ca.* Scalindua plasmid template, all PCR amplicons, with or without a nested PCR design, migrated similarly on the DGGE gel for primer set Amx368f-GC/Amx820r (Fig. 3D, 3E, 3F). In addition, only three samples, SedBr, SedBl and GW, could amplify product by direct PCR with primers Amx368f-GC/Amx820 (Fig. 3D); anammox bands from all samples can be captured by both nested PCR conditions (Fig. 3E and 3F). The DGGE patterns from this primer set were clearly different from those from the previous set, A438f-GC/A684r. All sequenced bands were indicated by the triangles. They were 99%–100% identical to the previously reported anammox-like sequences retrieved from various freshwater environments. Phylogenetic analysis revealed that anammox sequences from GW and GS still

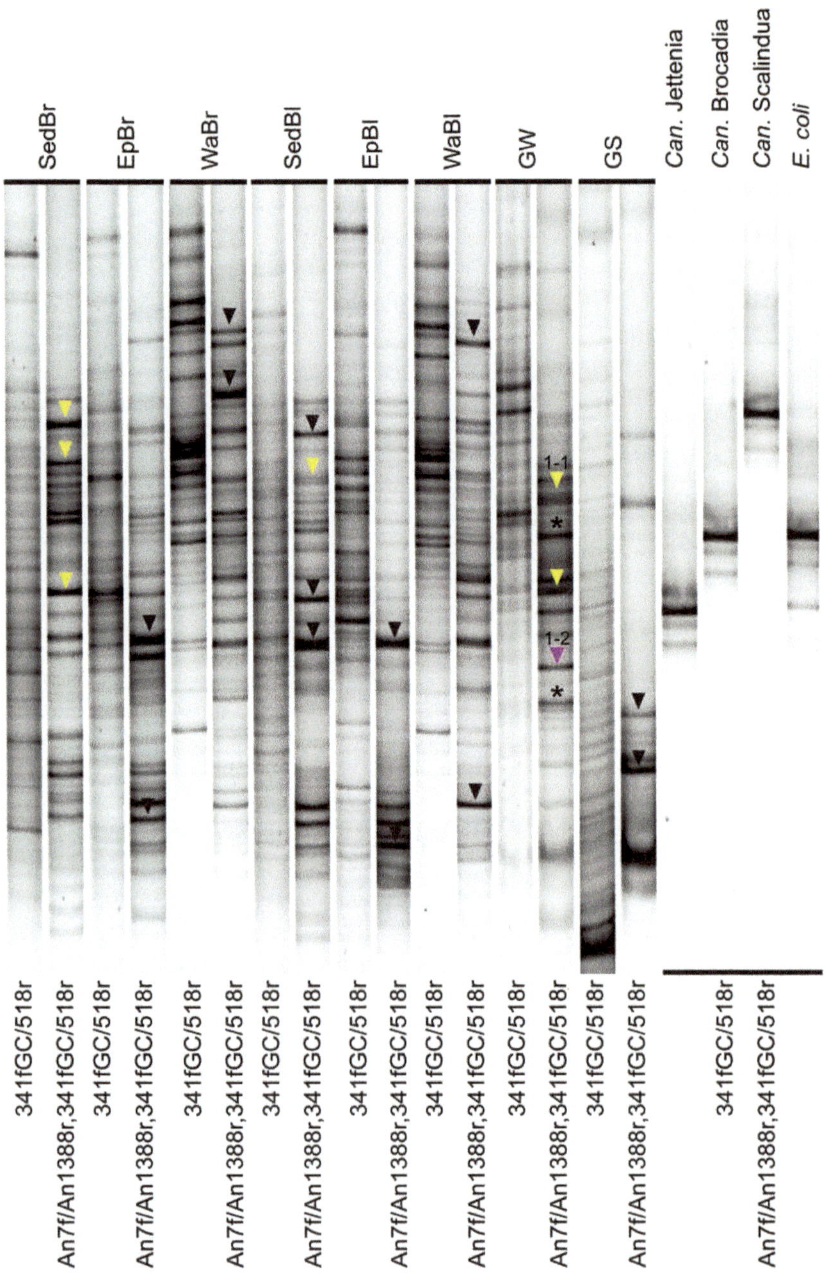

Figure 1. Denaturing gradient gel electrophoresis (DGGE) profiles of bacterial and anammox bacterial 16S rRNA genes in comparison. Together with six environmental samples, positive control template (*Ca.* Jettenia, *Ca.* Brocadia and *Ca.* Scalindua) and a negative control (*E. coli*) were included. Triangles indicate sequenced bands. Band 1-1, indicated by a yellow triangle, was associated with Unknown anammox cluster 1 [17] and band 1-2, indicated by a purple triangle, was affiliated with Unknown anammox cluster 2 [17]. Bands indicated by black triangles were not affiliated with anammox bacteria. A star indicates a band with low quality sequence, which was excluded from subsequent phylogenetic analysis.

grouped together and fell into Unknown 1 cluster; all sequenced bands from the Grand River samples were affiliated with a *Ca.* Brocadia-like phylotype (Fig. 2).

All positive controls were amplified by the two primer sets A438f-GC/A684r and Amx368f-GC/Amx820, with or without nested PCR. The results were quite similar for both sets; *Ca.* Brocadia and *Ca.* Jettenia controls were close to each other on DGGE gels, but further from the *Ca.* Scalindua control. The negative control, *E. coli*, showed no signal for all primer combinations, except sets An7f/An1388r nested by both A438f-GC/A684r and Amx368f-GC/Amx820 (Fig. 3B and 3E).

In addition to the two main primer sets, A438f-GC/A684r and Amx368f-GC/Amx820r, two additional primer combinations, A438f-GC/Amx820r and Amx368f-GC/A684r were tested with and without nested PCR amplifications, following the same pattern as previously described (Fig. S1). The gradient PCR program was run at annealing temperatures between 51–60°C; the optimum temperature (sharp and bright band on agarose gel) was 55°C for both primer sets (data not shown). The primer set A438f-

Westlake freshwater sediment (HQ540388)
Qiantang River freshwater sediment (HM537196)
An7f_SedBr_OTU1 (1)
Fuhe River soil (HM565015)
Candidatus Brocadia sp. 40 (AM285341)
Upper Cape Fear River sediment (HM851588)
Lake Ormajarvi (HE583334)
A438f_SedBl_OTU1 (50)
Lake Rassnitzer, 29 m depth (FJ830386)
North canal freshwater wetland (GU083981)
Amx368f_GW_OTU4 (1)
anammox reactor (JN043329)
A438f_SedBr_OTU5 (1)
Amx368f_SedBr_OTU2 (31)
Lake Rassnitzer, 29 m depth (FJ830384)
Amx368f_SedBl_OTU2 (41)
Yangcheng Lake sediment (JF261087)
An7f_SedBl_OTU1 (1)
sediment of eutrophic freshwater lake (HQ847792)
An7f_SedBr_OTU2 (2)
Yangtze Delta sediment (JQ688400)
Songhua Delta freshwater wetland (GU084045)
A438f_SedBr_OTU2 (27)
wetland sediment (JQ762202)
A438f_SedBr_OTU3 (15)
DGGE_band_2-3
Jiaojiang estuarine sediment (JN051511)
wetland sediment (JQ762038)
Amx368f_SedBl_OTU1 (5)
Amx368f_SedBr_OTU1 (14)
A438f_SedBr_OTU1 (1)
lake sediment (AB509315)
Qiantang River freshwater sediment (HM537207)
A438f_SedBl_OTU4 (6)
Candidatus Brocadia fulgida (DQ459989)
Candidatus Brocadia caroliniensis (JF487828)
Candidatus Brocadia sinica (AB56547)
Candidatus Brocadia anammoxidans (AF375994)
Amx368f_GW_OTU5 (1)
A438f_GW_OTU4 (1)
Amx368f_GW_OTU1 (34)
A438f_GW_OTU3 (14)
ammonium-contaminated groundwater (HQ595700)
An7f_GW_OTU3 (5)
lake sediment (AB509308)
Amx368f_SedBr_OTU3 (2)
ammonium-contaminated groundwater (HQ595493)
Kunming Lake freshwater wetland (GU083955)
An7f_GW_OTU2 (1)
groundwater (AJ871740)
groundwater (AJ871736)
paddy field soil (AB602675)
reductisol (FN691943)
A438f_SedBl_OTU3 (2)
A438f_SedBr_OTU4 (3)
wetland sediment (JQ762116)
Lake Paajarvi (HE583366)
DGGE_band_2-2
DGGE_band_1-1
North Carolina groundwater (HM851669)
A438f_GW_OTU2 (35)
Candidatus Anammoxoglobus propionicus (DQ317601)
SBR for anammox enrichment (JN006728)
A438f_SedBl_OTU2 (3)
A438f_GW_OTU1 (1)
Candidatus Jettenia asiatica (DQ301513)
Amx368f_GW_OTU2 (2)
Songhua Delta freshwater wetland (GU084034)
ammonium-contaminated groundwater (HQ595667)
An7f_GW_OTU1 (9)
DGGE_band_1-2
Amx368f_GW_OTU3 (4)
Candidatus Kuenenia stuttgartiensis (AF375995)
Candidatus Scalindua sinooilfield (HM208769)
Candidatus Scalindua wagneri (AY254882)
DGGE_band_2-1
channel sediment (AB518012)
Candidatus Scalindua arabica (EU478624)
Candidatus Scalindua zhenghei (GQ331167)
Candidatus Scalindua brodae (AY254883)
Candidatus Scalindua sorokinii (AY257181)
Pirellula sp. (EU703486)
0.1
Candidatus Brocadia
Unknown 1
Candidatus Anammoxoglobus
Candidatus Jettenia
Unknown 2
Candidatus Kuenenia
Candidatus Scalindua

Figure 2. Phylogenetic tree of anammox bacterial 16S rRNA genes retrieved from both DGGE and clone library methods (shown in bold). All DGGE band sequences are indicated by "DGGE" with representative band ID, related to Fig. 1 and 3. The number of individual DGGE bands is indicated by the triangles with corresponding colour. Clone sequences with 97% identity from each library were grouped together; the representative clones from each operational taxonomic unit (OTU) were included in the analysis. The number of sequences belonging to each OTU is indicated in parentheses. The maximum likelihood tree was constructed with the GTR model. Branch support values (aLRT) greater than 50% are indicated at the nodes.

GC/Amx820r generated a dominant band across all samples, although Amx368f-GC/A684r produced additional DGGE bands (Fig. S1). Several representative bands from each position were sequenced and included in phylogenetic analysis (data not shown). All analyzed bands were anammox-related sequences and fell into *Ca.* Brocadia, Unknown 1 and Unknown 2 groups.

Diversity of anammox bacteria within freshwater environments

To confirm the specificity of An7f/An1388r, A438f/A684r and Amx368f/Amx820, three representative samples were selected to construct clone libraries from each pair. Based on DGGE profiles, two sediments from the Grand River (SedBr and SedBl) and Zorra groundwater (GW) samples were included in the clone library analysis. The cloning results showed that the ratio between the numbers of colonies containing inserts and the total number of selected colonies was lower for primers An7f/An1388r (Table 3). Using this primer set, the ratio was ~0.6 for the SedBr and SedBl libraries, but it increased to 0.9 for the GW library. The other two primer sets produced higher insert ratios, with >90% of screened colonies containing inserts (Table 3). The sequencing results revealed that only 3 out of 37 and 1 out of 36 analyzed sequences from SedBr and SedBl, constructed by primers An7f/An1388r, were anammox-related sequences, respectively. The GW library, constructed by the same primers, yielded a better result because 15 out of 25 were closely related to anammox bacteria (Table 3). The other primer sets (A438f/A684r and Amx368f/Amx820) revealed 100% specificity; all analyzed clones from all libraries were associated with anammox-related sequences. The BLAST results indicated that all clone sequences were 92%–100% identical to previously reported anammox bacteria found in various ecosystems. The number of OTUs and anammox bacterial sequences were similar for A438f/A684r and Amx368f/Amx820 libraries (Table 3). The clone sequences showing as 99% and 97% identical

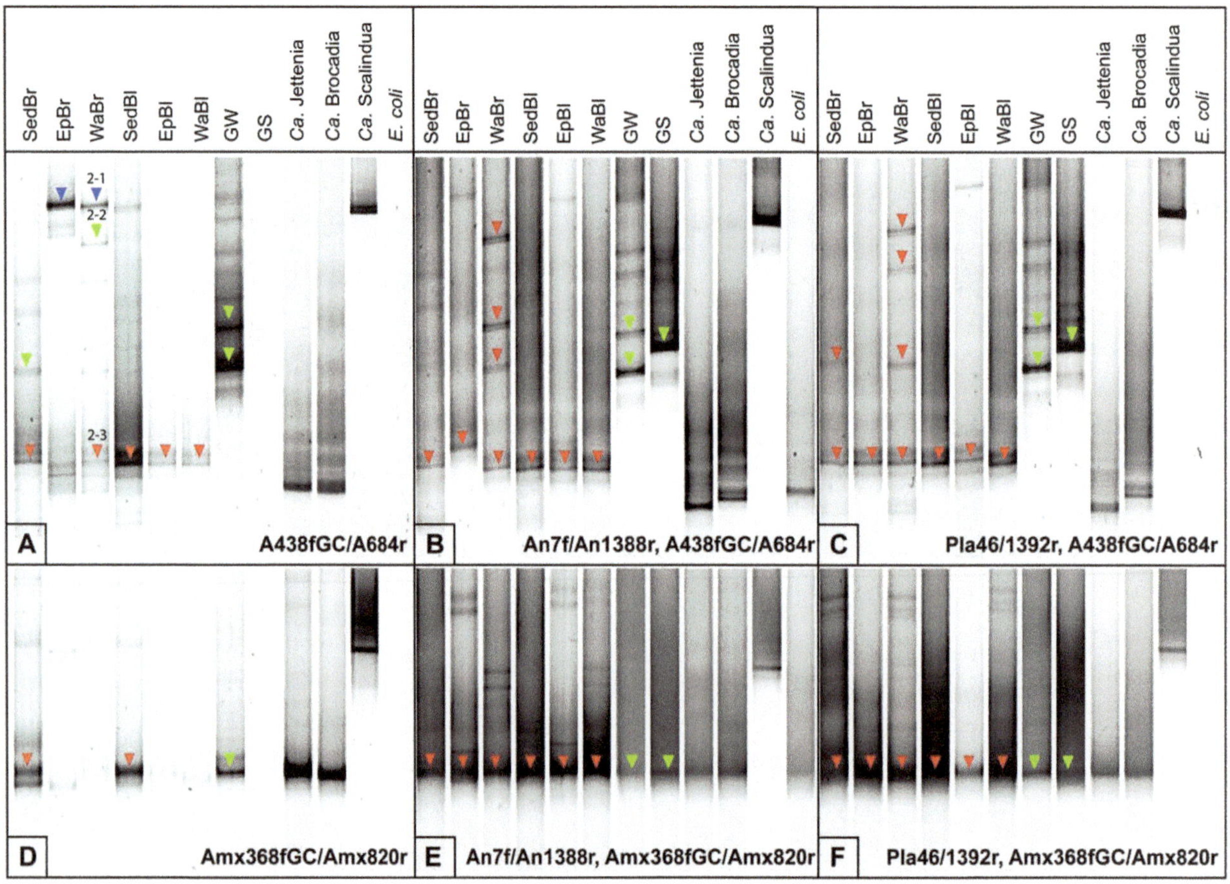

Figure 3. Denaturing gradient gel electrophoresis (DGGE) profiles of anammox bacterial 16S rRNA genes. The two main primer sets were used to generate the anammox-fingerprints (A and D) by direct PCR amplification. The additional four patterns (B, C, E and F) were produced by the nested PCR assay, using other two primer sets and followed by the two main sets. Triangles indicate a total of 55 bands, associated with anammox bacteria. Only three representative bands were shown in the phylogenetic tree. Bands 2-1, 2-2 and 2-3 were indicated by the blue, green and red triangles, respectively. Each coloured triangle indicates exactly the same phylogeny shown in Fig. 2.

from each library grouped together. Representative clones from each OTU were included in phylogenetic analysis (Fig. 2). Note that phylogenetic trees based on both 1% and 3% cut-off nucleotide sequences exhibited the same anammox grouping and tree topology (data not shown). The resulting phylogeny of the clone libraries constructed by primers An7f/An1388r revealed that *Ca.* Brocadia-like sequences made up the majority of anammox bacteria found in SedBr and SedBl; GW contained anammox bacterial sequences associated with Unknown 1 and 2 clusters (Fig. 2). For primers A438f/A684r, both Grand River sediment and GW were dominated by *Ca.* Brocadia and Unknown 1 clusters, respectively (Fig. 2). Although *Ca.* Jettenia-like sequences are targeted by this primer set (Table S1), only a few clones from SedBl and GW samples fell into this cluster. Fewer clones recovered from SedBr and SedBl were closely related to Unknown 1 cluster, compared to the GW library. The expected diversity of anammox bacteria was supported by Amx368f/Amx820r libraries because the majority of anammox bacteria found in SedBr and SedBl were related to *Ca.* Brocadia-like sequences, whereas Unknown 1 cluster dominated the GW library (Fig. 2). Other minor clone library OTUs detected from GW were related to Unknown 2 and *Ca.* Brocadia clusters.

To test spatial and temporal changes of anammox bacterial diversity, representative sediment samples from Bridgeport (SedBr) were collected at three time points, including Summer 2010, Fall 2010 and Summer 2012. All samples were amplified by primer set A438f-GC/A684r and profiled by DGGE (Fig. S2). Overall, anammox patterns were consistent across the three time points. One additional band was apparent in the pattern from Summer 2012 and its sequence clustered with *Ca.* Brocadia phylotypes.

Abundance of anammox bacteria in freshwater environments

DGGE and cloning results demonstrated that the primer sets A438f/A684r and Amx368f/Amx820 were specific for detecting anammox bacteria within freshwater environments. We used these existing two primer sets to assess the qPCR method. Total bacterial 16S rRNA gene copies were also quantified for comparison. The results showed that the measured bacterial 16S rRNA gene copies were consistent for all analyzed samples (Fig. 4). The abundance of anammox bacterial 16S rRNA genes in GW was the highest ($\sim 10^3$–10^4 copies per ng of genomic DNA), whereas those in SedBr and SedBl were lower and similar to each other ($\sim 10^2$–10^3 copies per ng of genomic DNA). The qPCR with Amx368f/Amx820r generated bacterial 16S rRNA gene abundance estimates that were approximately four times higher than for A438f/A684r in all analyzed samples (Fig. 4). Consequently, caution must be taken in using Amx368f/Amx820r to quantify anammox abundance due to possible overestimation.

Discussion

The anammox-DGGE method developed by Moore and colleagues [17], using a nested amplification beginning with An7f/An1388r, was able to generate DGGE patterns with confirmed anammox bands for only three samples in this study (SedBr, SedBl and GW). Of 20 bands sequenced, 7 sequences were associated with anammox bacteria (Fig. 1). The remaining bands showed no relationship with any reported anammox bacterial sequences in the Genbank database. However, this method provided a reliable result for all previously analyzed groundwater samples [17], obtained from the same site as GW. Improved specificity for anammox bacterial template may be explained by a high sample-specific relative abundance of anammox bacteria,

Table 3. Summary of the cloning results using three primer sets and three sample sites.

Primer set	Site	Total screened clone (A)	Total inserted clone (B)	B/A ratio	Anammox sequence (% recovery)	Number of OTUs		Cluster[1]
						1%	3%	
An7f/An1388r	SedBr	62	38	0.61	3/37 (8)	3	2	*Ca.* Brocadia (3/3)
	SedBl	62	39	0.63	1/36 (3)	1	1	*Ca.* Brocadia (1/1)
	GW	30	27	0.90	15/25 (60)	6	3	Unknown 1 (6/15), Unknown 2 (9/15)
A438f/A684r	SedBr	55	53	0.96	47/47 (100)	13	5	*Ca.* Brocadia (44/47), Unknown 1 (3/47)
	SedBl	70	66	0.94	61/61 (100)	11	4	*Ca.* Brocadia (56/61), *Ca.* Jettenia (3/61), Unknown 1 (2/61)
	GW	58	54	0.93	51/51 (100)	10	4	*Ca.* Jettenia (1/51), Unknown 1 (50/51)
Amx368f/Amx820r	SedBr	52	48	0.92	47/47 (100)	10	3	*Ca.* Brocadia (45/47), Unknown 1 (2/47)
	SedBl	51	48	0.94	46/46 (100)	4	2	*Ca.* Brocadia (46/46)
	GW	52	48	0.92	42/42 (100)	12	5	*Ca.* Brocadia (1/42), Unknown 1 (35/42), Unknown 2 (6/42)

[1]The number of clones out of the total number of sequences affiliated with each cluster is listed in parentheses.

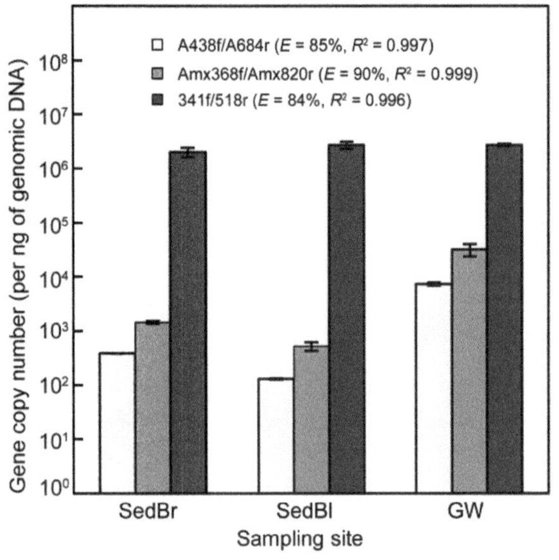

Figure 4. Abundance of anammox bacterial 16S rRNA genes quantified by two anammox specific primer sets, and general bacterial qPCR data for comparison. The qPCR efficiency (E) and coefficient of determination (R^2) of each primer set are shown in parentheses.

which is supported by anammox bacterial bands appearing in both bacterial and anammox-specific DGGE profiles from the GW site. Our qPCR results demonstrated that most Grand River samples contained anammox bacterial 16S rRNA genes at $\leq 10^2$ copies per ng of genomic DNA (data not shown), except for GW ($\sim 10^4$ gene copies; Fig. 4). Clone library analysis was consistent with this finding because lower recovery frequencies of anammox-related clones were obtained from the SedBr (8%) and SedBl (3%) libraries, yet higher for GW (60%; Table 3). Consistent with this observation, a previous marine sediment clone library constructed by primers An7f/An1388r showed 12% anammox-related sequences [8]. Our previous groundwater libraries showed a high proportion of anammox clones with this primer set, in the range of 86%–100% [17]. Overall results suggest that primers An7f/An1388r may be specific only for anammox samples with high proportions of anammox bacteria.

To improve the detection efficiency of anammox bacteria present in low abundance, different primer combinations, with or without a nested PCR step, were tested to enhance the specificity of DGGE (Fig. 3). In this study, primer sets A438f-GC/A684r and Amx368f-GC/Amx820 were confirmed as specific for detecting anammox bacteria within freshwater environments. The A438f-GC/A684r primer pair was superior for DGGE based on well-resolved bands for both samples and controls. The resulting phylogeny revealed that direct PCR amplification by A438f-GC/A684r primers captured the most diverse anammox bacterial groups. *Ca.* Scalindua-like sequences were retrieved from picked bands only by these primers in this study. Although this cluster was normally found in marine and estuary environments and proposed to be a marine anammox-specific cluster [9,10,41,53–56], the *Ca.* Scalindua genus has been associated with freshwater habitats such as Lake Tanganyika [35], Wintergreen Lake [22] and Lake Rassnitzer [57]. This anammox cluster was also retrieved previously from ammonium-contaminated groundwater samples [17].

Both the A438f/A684r and Amx368f/Amx820r primer sets were highly specific based on clone library analysis; all clones were affiliated with anammox bacterial sequences (Table 3). The Amx368f/Amx820r primer pair has been used for cloning in many previous studies. Reported recoveries of anammox-related clones with this primer set vary depending on the sampling sites. Clone libraries constructed from coastal marine sediment and eutrophic freshwater lake yielded 98% and 90%–100% anammox bacterial sequences, respectively [27,36]. However, low recovery frequencies of 12%–59% were previously reported from deep-sea subsurface sediment libraries [41]. There is no prior information on the specificity of anammox bacterial community libraries generated by the primer set A438f/A684r. *In silico* analysis revealed that these four existing primers (Amx368f, Amx820r, A438f and A684r) showed high specificity and were selective for capturing known anammox *Ca.* species (Table S1). In this study, the anammox bacterial diversity recovered from cloning was in general agreement with those of the DGGE method.

The majority of anammox bacterial 16S rRNA genes found in the Grand River samples were similar to those of previously studied environments in identifying that *Ca.* Brocadia-like phylotypes were dominant in freshwater ecosystems [17,29,36]. The dominant anammox bacteria found in groundwater-related samples were Unknown 1 and 2 clusters. This result agreed with previous findings, showing that these unknown and uncultured groups have a potential to be a specific anammox cluster present in groundwater sites [17]. Sequences from a reductisol and ammonium contaminated aquifer also fell into a distinct group without any affiliation to known anammox clusters, being named "cluster II" in the study of Humbert and colleague (2010). These sequences were related to the Unknown 1 cluster in this study. *Ca.* Jettenia cluster was a minority population found in both the Grand River sediment and groundwater samples. The *Ca.* Jettenia lineage does not normally dominate in any specific habitat previously reported but was detectable in various terrestrial habitats [24], peat soil [39], paddy soil [18], estuarine sediments [40] and groundwater [11,17]. Due to limited studies on the distribution of anammox bacteria in freshwater environments, future research will include more freshwater aquifers to explore their anammox bacterial communities.

The abundance of anammox bacterial 16S rRNA genes in GW was in the same range as in groundwater samples from the Zorra site previously reported [17]. In the case of SedBr and SedBl, anammox bacterial 16S rRNA genes were present in the range of 10^2–10^3 copies per ng of genomic DNA. River estuary sediments also contained anammox bacterial 16S rRNA in this range [23]. The qPCR results revealed that Amx368f/Amx684r captured more anammox bacterial 16S rRNA genes in all analyzed samples. Primer pair A438f/A684r could provide more accurate results than Amx368f/Amx820 because of potential false positive amplification by Amx368f/Amx820 in samples with low anammox bacterial abundance [37]. These findings are obtained mainly from the anammox bacterial 16S rRNA genes within freshwater environments; future research will include a broad range of environmental samples such as marine, terrestrial and engineered systems to evaluate the efficiency and specificity of primers for targeting anammox bacteria. Other than anammox 16S rRNA genes, functional genes such as the hydrazine oxidoreductase (*hzo*) gene have been used for anammox bacterial detection in marine sediment [8–10], aquatic ecosystems [11] and mangrove sediment [58]. Another functional gene marker is the nitrite reductase (*nirS*) gene, which has been used to detect anammox bacteria in the ocean [59–60]. The hydrazine synthase (*hzsA*) gene has also been tested as a unique biomarker for detecting anammox bacteria in both natural and built environments [61]. However, primers targeting these functional genes are still limited. Our results demonstrate that primer sets should be evaluated in a range of environments and with a careful selection of positive and negative controls to avoid false positive amplification, as seen here with An7f/An1388r. We recommend primers A438f/A684r or Amx368f/Amx820 for clone library or qPCR analyses, but only A438f-GC/A684r for DGGE-based analyses of freshwater anammox communities.

Supporting Information

Table S1 Alignment of anammox primer sequences against known anammox Ca. species, non-anammox species of Planctomycetales and non-Planctomycetales1. The direction of all sequences is 5′-3′.

Acknowledgments

We thank Eduardo Cejudo and Tara Moore for their assistance with the sampling at the Grand River and Zorra sites, together with Richard Elgood for providing all chemistry measurements at both sites. We also would like to thank Laura A. Sauder for her constructive comments on this study, Michael D. J. Lynch and Mike W. Hall for their assistance with phylogenetic and AXIOME analyses, respectively.

Author Contributions

Conceived and designed the experiments: PS JDN. Performed the experiments: PS. Analyzed the data: PS. Wrote the paper: PS JDN.

References

1. Strous M, Fuerst JA, Kramer EHM, Logemann S, Muyzer G, et al. (1999) Missing lithotroph identified as new planctomycete. Nature 400: 446–449.
2. Schmid MC, Maas B, Dapena A, van de Pas-Schoonen K, van de Vossenberg J, et al. (2005) Biomarkers for in situ detection of anaerobic ammonium-oxidizing (anammox) bacteria. Appl Environ Microbiol 71: 1677–1684.
3. Jetten MSM, Sliekers O, Kuypers MMM, Dalsgaard T, van Niftrik L, et al. (2003) Anaerobic ammonium oxidation by marine and freshwater planctomycete-like bacteria. Appl Microbiol Biotechnol 63: 107–114.
4. Broda E (1977) Two kinds of lithotrophs missing in nature. Z Allg Mikrobiol 17: 491–493.
5. Richards FA (1965) Anoxic basins and fjords. In: Ripley JP, Skirrow G, editors. Chemical Oceanography. pp. 611–645.
6. Helmer C, Kunst S, Juretschko S, Schmid MC, Schleifer KH, et al. (1999) Nitrogen loss in a nitrifying biofilm system. Water Sci Technol 39: 13–21.
7. Mulder A, van de Graaf AA, Robertson LA, Kuenen JG (1995) Anaerobic ammonium oxidation discovered in a denitrifying fluidized bed reactor. FEMS Microbiol Ecol 16: 177–183.
8. Li M, Hong Y, Klotz MG, Gu JD (2010) A comparison of primer sets for detecting 16S rRNA and hydrazine oxidoreductase genes of anaerobic ammonium-oxidizing bacteria in marine sediments. Appl Microbiol Biotechnol 86: 781–790.
9. Dang H, Chen R, Wang L, Guo L, Chen P, et al. (2010) Environmental factors shape sediment anammox bacterial communities in hypernutrified Jiaozhou Bay, China. Appl Environ Microbiol 76: 7036–7047.
10. Hong YG, Yin B, Zheng TL (2011) Diversity and abundance of anammox bacterial community in the deep-ocean surface sediment from equatorial Pacific. Appl Microbiol Biotechnol 89: 1233–1241.
11. Hirsch MD, Long ZT, Song B (2011) Anammox bacterial diversity in various aquatic ecosystems based on the detection of hydrazine oxidase genes (*hzoA/hzoB*). Microb Ecol 61: 264–276.
12. Mohamed NM, Saito K, Tal Y, Hill RT (2010) Diversity of aerobic and anaerobic ammonia-oxidizing bacteria in marine sponges. ISME J 4: 38–48.
13. Thamdrup B, Dalsgaard T (2002) Production of N_2 through anaerobic ammonium oxidation coupled to nitrate reduction in marine sediments. Appl Environ Microbiol 68: 1312–1318.
14. Engstrom P, Penton CR, Devol AH (2009) Anaerobic ammonium oxidation in deep-sea sediments off the Washington margin. Limnol Oceanogr 54: 1643–1652.
15. Brandsma J, van de Vossenberg J, Risgaard-Petersen N, Schmid MC, Engström P, et al. (2011) A multi-proxy study of anaerobic ammonium oxidation in marine sediments of the Gullmar Fjord, Sweden. Environ Microbiol Rep 3: 360–366.
16. Trimmer M, Nicholls JC (2009) Production of nitrogen gas via anammox and denitrification in intact sediment cores along a continental shelf to slope transect in the North Atlantic. Limnol Oceanogr 54: 577–589.
17. Moore TA, Xing Y, Lazenby B, Lynch MD, Schiff S, et al. (2011) Prevalence of anaerobic ammonium-oxidizing bacteria in contaminated groundwater. Environ Sci Technol 45: 7217–7225.
18. Zhu G, Wang S, Wang Y, Wang C, Risgaard-Petersen N, et al. (2011) Anaerobic ammonia oxidation in a fertilized paddy soil. ISME J 5: 1905–1912.
19. Strous M, Heijnen JJ, Kuenen JG, Jetten MSM (1998) The sequencing batch reactor as a powerful tool for the study of slowly growing anaerobic ammonium-oxidizing microorganisms. Appl Microbiol Biotechnol 50: 589–596.
20. Manz W, Amann RI, Ludwig W, Wagner M, Schleifer KH (1992) Phylogenetic oligodeoxynucleotide probes for the major subclasses of proteobacteria: problems and solutions. Syst Appl Microbiol 15: 593–600.
21. Schmid MC, Schmitz-Esser S, Jetten MSM, Wagner M (2001) 16S-23S rDNA intergenic spacer and 23S rDNA of anaerobic ammonium-oxidizing bacteria: implications for phylogeny and in situ detection. Environ Microbiol 3: 450–459.
22. Penton CR, Devol AH, Tiedje JM (2006) Molecular evidence for the broad distribution of anaerobic ammonium-oxidizing bacteria in freshwater and marine sediments. Appl Environ Microbiol 72: 6829–6832.
23. Dale OR, Tobias CR, Song B (2009) Biogeographical distribution of diverse anaerobic ammonium oxidizing (anammox) bacteria in Cape Fear River Estuary. Environ Microbiol 11: 1194–1207.
24. Humbert S, Tarnawski S, Fromin N, Mallet MP, Aragno M, et al. (2010) Molecular detection of anammox bacteria in terrestrial ecosystems: distribution and diversity. ISME J 4: 450–454.
25. Schmid MC, Twachtmann U, Klein M, Strous M, Juretschko S, et al. (2000) Molecular evidence for genus level diversity of bacteria capable of catalyzing anaerobic ammonium oxidation. Syst Appl Microbiol 23: 93–106.
26. Schmid MC, Walsh K, Webb R, Rijpstra WI, van de Pas-Schoonen K, et al. (2003) *Candidatus* "Scalindua brodae", sp. nov., *Candidatus* "Scalindua wagneri", sp. nov., two new species of anaerobic ammonium oxidizing bacteria. Syst Appl Microbiol 26: 529–538.
27. Amano T, Yoshinaga I, Okada K, Yamagishi T, Ueda S, et al. (2007) Detection of anammox activity and diversity of anammox bacteria-related 16S rRNA genes in coastal marine sediment in Japan. Microbes Environ 22: 232–242.
28. Rich JJ, Dale OR, Song B, Ward BB (2008) Anaerobic ammonium oxidation (anammox) in Chesapeake Bay sediments. Microb Ecol 55: 311–320.
29. Zhang Y, Ruan XH, Op den Camp HJM, Smits TJM, Jetten MSM, et al. (2007) Diversity and abundance of aerobic and anaerobic ammonium-oxidizing bacteria in freshwater sediments of the Xinyi River (China). Environ Microbiol 9: 2375–2382.
30. Tal Y, Watts JEM, Schreier HJ (2005) Anaerobic ammonia-oxidizing bacteria and related activity in Baltimore inner harbor sediment. Appl Environ Microbiol 71: 1816–1821.
31. Tal Y, Watts JEM, Schreier HJ (2006) Anaerobic ammonium-oxidizing (anammox) bacteria and associated activity in fixed-film biofilters of a marine recirculating aquaculture system. Appl Environ Microbiol 72: 2896–2904.
32. Kirkpatrick J, Oakley B, Fuchsman C, Srinivasan S, Staley JT, et al. (2006) Diversity and distribution of Planctomycetes and related bacteria in the suboxic zone of the Black Sea. Appl Environ Microbiol 72: 3079–3083.
33. Kuypers MMM, Sliekers AO, Lavik G, Schmid MC, Jørgensen BB, et al. (2003) Anaerobic ammonium oxidation by anammox bacteria in the Black Sea. Nature 422: 608–611.
34. Dalsgaard T, Thamdrup B, Canfield DE (2005) Anaerobic ammonium oxidation (anammox) in the marine environment. Res Microbiol 156: 457–464.
35. Schubert CJ, Durisch-Kaiser E, Wehrli B, Lam P, Kuypers MMM (2006) Anaerobic ammonium oxidation in a tropical freshwater system (Lake Tanganyika). Environ Microbiol 8: 1857–1863.
36. Yoshinaga I, Amano T, Yamagishi T, Okada K, Ueda S, et al. (2011) Distribution and diversity of anaerobic ammonium oxidation (anammox) bacteria in the sediment of a eutrophic freshwater lake, Lake Kitaura, Japan. Microbes Environ 26: 189–197.
37. Humbert S, Zopfi J, Tarnawski SE (2012) Abundance of anammox bacteria in different wetland soils. Environ Microbiol Rep 4: 484–490.
38. Zhu G, Wang S, Feng X, Fan G, Jetten MSM, et al. (2011) Anammox bacterial abundance, biodiversity and activity in a constructed wetland. Environ Sci Technol 45: 9951–9958.
39. Hu BL, Rush D, van der Biezen E, Zheng P, van Mullekom M, et al. (2011) New anaerobic, ammonium-oxidizing community enriched from peat soil. Appl Environ Microbiol 77: 966–971.
40. Hu B, Shen L, Du P, Zheng P, Xu X, et al. (2012) The influence of intense chemical pollution on the community composition, diversity and abundance of anammox bacteria in the Jiaojiang Estuary (China). PloS ONE 7: e33826.
41. Hong YG, Li M, Cao H, Gu JD (2011) Residence of habitat-specific anammox bacteria in the deep-sea subsurface sediments of the South China Sea: analyses of marker gene abundance with physical chemical parameters. Microb Ecol 62: 36–47.
42. Li H, Chen S, Mu BZ, Gu JD (2010) Molecular detection of anaerobic ammonium-oxidizing (anammox) bacteria in high-temperature petroleum reservoirs. Microb Ecol 60: 771–783.

43. Woebken D, Fuchs BM, Kuypers MMM, Amann R (2007) Potential interactions of particle-associated anammox bacteria with bacterial and archaeal partners in the Namibian upwelling system. Appl Environ Microbiol 73: 4648–4657.
44. van der Star WRL, Miclea AI, van Dongen UGJM, Muyzer G, Picioreanu C, et al. (2008) The membrane bioreactor: A novel tool to grow anammox bacteria as free cells. Biotechnol Bioeng 101: 286–294.
45. Bae H, Park KS, Chung YC, Jung JY (2010) Distribution of anammox bacteria in domestic WWTPs and their enrichments evaluated by real-time quantitative PCR. Process Biochem 45: 323–334.
46. Neufeld JD, Schafer H, Cox MJ, Boden R, McDonald IR, et al. (2007) Stable-isotope probing implicates *Methylophaga* spp. and novel *Gammaproteobacteria* in marine methanol and methylamine metabolism. ISME J 1: 480–491.
47. Green SJ, Leigh MB, Neufeld JD (2010) Denaturing gradient gel electrophoresis (DGGE) for microbial community analysis. In: Timmis K, editor. Hydrocarbon Microbiology. Berlin: Springer. pp. 4137–4158.
48. Lynch MDJ, Masella AP, Hall MW, Bartram AK, Neufeld JD (2013) AXIOME: automated exploration of microbial diversity. Gigascience. In press.
49. Edgar RC (2004) MUSCLE: multiple sequence alignment with high accuracy and high throughput. Nucl Acids Res 32: 1792–1797.
50. Guindon S, Gascuel O (2003) A simple, fast, and accurate algorithm to estimate large phylogenies by maximum composite likelihood. Syst Biol 52: 696–704.
51. Galtier N, Gouy M, Gautier C (1996) SEAVIEW and PHYLO_WIN: two graphic tools for sequence alignment and molecular phylogenies. Bioinformatics 12: 543–548.
52. Sonthiphand P, Cejudo E, Schiff S, Neufeld JD (2012) Microbiology of ammonia oxidation in the Grand River: disentangling molecular signatures of wastewater effluent from autochthonous prokaryotic communities. Environ Microbiol (submitted).
53. Lam P, Jensen MM, Lavik G, McGinnis DF, Muller B, et al. (2007) Linking crenarchaeal and bacterial nitrification to anammox in the Black Sea. Proc Natl Acad Sci USA 104: 7104–7109.
54. Schmid MC, Risgaard-Petersen N, van de Vossenberg J, Kuypers MMM, Lavik G, et al. (2007) Anaerobic ammonium-oxidizing bacteria in marine environments: widespread occurrence but low diversity. Environ Microbiol 9: 1476–1484.
55. Ward BB, Devol AH, Rich JJ, Chang BX, Bulow SE, et al. (2009) Denitrification as the dominant nitrogen loss process in the Arabian Sea. Nature 461: 78–81.
56. Woebken D, Lam P, Kuypers MMM, Naqvi SW, Kartal B, et al. (2008) A microdiversity study of anammox bacteria reveals a novel *Candidatus* Scalindua phylotype in marine oxygen minimum zones. Environ Microbiol 10: 3106–3119.
57. Hamersley MR, Woebken D, Boehrer B, Schultze M, Lavik G, et al. (2009) Water column anammox and denitrification in a temperate permanently stratified lake (Lake Rassnitzer, Germany). Syst Appl Microbiol 32: 571–582.
58. Li M, Hong YG, Cao HL, Gu JD (2011) Mangrove trees affect the community structure and distribution of anammox bacteria at an anthropogenic-polluted mangrove in the Pearl River Delta reflected by 16S rRNA and hydrazine oxidoreductase (HZO) encoding gene analyses. Ecotoxicology 20: 1780–1790.
59. Li M, Ford T, Li XY, Gu JD (2011) Cytochrome *cd1*-containing nitrite encoding gene *nirS* as a new functional biomarker for detection of anaerobic ammonium oxidizing (Anammox) bacteria. Envion Sci Technol 45: 3547–3553.
60. Lam P, Lavik G, Jensen MM, van de Vossenberg J, Schmid MC, et al. (2009) Revising the nitrogen cycle in the Peruvian oxygen minimum zone. Proc Natl Acad Sci USA 106: 4752–4757.
61. Harhangi HR, Roy ML, van Alen T, Hu BL, Groen J, et al. (2012) Hydrazine synthase, a unique phylomarker with which to study the presence and biodiversity of anammox bacteria. Appl Environ Microbiol 78: 752–758.
62. Neef A, Arnann R, Schlesner H, Schleiferl KH (1998) Monitoring a widespread bacterial group: in situ detection of planctomycetes with 16S rRNA-targeted probes. Microbiology 144: 3257–3266.
63. Ferris M, Muyzer G, Ward D (1996) Denaturing gradient gel electrophoresis profiles of 16S rRNA-defined populations inhabiting a hot spring microbial mat community. Appl Environ Microbiol 62: 340–346.
64. Muyzer G, de Waal EC, Uitterlinden AG (1993) Profiling of complex microbial populations by denaturing gradient gel electrophoresis analysis of polymerase chain reaction-amplified genes coding for 16S ribosomal RNA. Appl Environ Microbiol 59: 695–700.

5

The Application of Predictive Modelling for Determining Bio-Environmental Factors Affecting the Distribution of Blackflies (Diptera: Simuliidae) in the Gilgel Gibe Watershed in Southwest Ethiopia

Argaw Ambelu[1]*, Seblework Mekonen[1], Magaly Koch[2], Taffere Addis[3], Pieter Boets[4], Gert Everaert[4], Peter Goethals[4]

1 Department of Environmental Health Sciences and Technology, Jimma University, P.O. Box 378, Jimma, Ethiopia, **2** Center for Remote Sensing, Boston University, 725 Commonwealth Avenue, Boston, Massachusetts, United States of America, **3** Ethiopian Institute of Water Resources, Addis Ababa University, Akaki Kifleketema, Addis Ababa, Ethiopia, **4** Laboratory of Environmental Toxicology and Aquatic Ecology, Ghent University, J. Plateaustraat 22, 9000 Ghent, Belgium

Abstract

Blackflies are important macroinvertebrate groups from a public health as well as ecological point of view. Determining the biological and environmental factors favouring or inhibiting the existence of blackflies could facilitate biomonitoring of rivers as well as control of disease vectors. The combined use of different predictive modelling techniques is known to improve identification of presence/absence and abundance of taxa in a given habitat. This approach enables better identification of the suitable habitat conditions or environmental constraints of a given taxon. Simuliidae larvae are important biological indicators as they are abundant in tropical aquatic ecosystems. Some of the blackfly groups are also important disease vectors in poor tropical countries. Our investigations aim to establish a combination of models able to identify the environmental factors and macroinvertebrate organisms that are favourable or inhibiting blackfly larvae existence in aquatic ecosystems. The models developed using macroinvertebrate predictors showed better performance than those based on environmental predictors. The identified environmental and macroinvertebrate parameters can be used to determine the distribution of blackflies, which in turn can help control river blindness in endemic tropical places. Through a combination of modelling techniques, a reliable method has been developed that explains environmental and biological relationships with the target organism, and, thus, can serve as a decision support tool for ecological management strategies.

Editor: Arash Rashed, University of Idaho, United States of America

Funding: This research was supported by Jimma University, Ethiopia and VLIR-UOS, Belgium. The funders had no role in study design, data collection and analysis, decision to publish, or preparation of the manuscript.

Competing Interests: The authors have declared that no competing interests exist.

* Email: aambelu@yahoo.com

Introduction

It is important to investigate the ecological factors affecting the distribution of blackflies in order to understand blackfly ecology and their environmental dynamcis [1,2]. Blackflies are one of the most frequently occurring aquatic taxa in tropical countries such as Ethiopia [3,4]. These organisms are important pollution indicators of running water habitats [5,6]. Because of their sensitivity to different environmental changes, they have been used to assess the impact of climate change and other anthropogenic activities [7]. Some species of blackflies (e.g. *Simulium damnosum*) are also known vectors of river blindness (onchocerciasis) in sub-Saharan Africa [8].

Predictive models are often applied to assess, monitor and control environmental factors of a given taxon [9,10]. Predictive modelling is one of the most essential steps in the development of a standard habitat assessment protocol that links organisms and habitat information to environmental data [11–13]. Effective habitat models need to be simple, robust and at the same time biologically meaningful [14]. The goal of applying different predictive models is to simplify complex systems and to enable reliable predictions [15].

Generalized additive models (GAMs) [16] and classification trees (CTs) [17] are widely used predictive models because they are fairly simple and transparent to understand, which allow easy application into an environmental decision support system [10,18,19]. Such models can be useful for policy and decision-makers to improve the effectiveness of monitoring and assessment activities in different ecosystems [20].

Although linear models are attractive because of their simplicity, they often fail in addressing natural relationships between a species and biotic and abiotic variables because of their nonlinear nature [21]. Non-linear and non-monotonic relationships between the outcome and the set of explanatory variables can be meaningfully modelled using GAMs. The model accommodates non-normal data by clearly constructing the distribution as a member of the exponential family and map the relationship between the predictor and the mean of the data [22]. The main advantage of GAMs is their ability to deal with non-linear and non-monotonic relationships between the predictor and response variables because of the capability to model non-linear data using non-parametric smoothers [23–25].

CTs are used as an effective habitat suitability modelling technique to determine the presence/absence and abundance of species [9,10,18]. Genetic algorithms (GA) are one of known techniques to boost model performance and improve the accuracy and predictive power by minimizing number of irrelevant attributes [10,26]. GA is widely used optimisation method for predictive models in the field of aquatic ecology [9,10,27]. Reliable CT models having best performance can be constructed when it is combined with GA [10].

The use of CT combined with GA and the application of GAMs can help to identify the major variables predicting the occurrence of Simuliidae larvae by minimizing model uncertainty [28]. In addition to the model combination, the use of environmental as well as biological predictors in the model construction is known to minimize prediction errors and ensure reliable model output [29]. Our main aim is to identify biological determinants in terms of other macroinvertebrate groups and environmental parameters, which are crucial for the presence/absence and abundance of blackflies, using GAMs and CTs combined with GAs in order to fill current knowledge gaps on the blackfly ecology, thus, leading to a better understanding of the underlying environmental factors.

Methods

Study area

The study was performed in the Gilgel Gibe watershed, which is part of the Omo-Gibe River basin situated in southwest Ethiopia. Simuliidae larvae are found in most of the study sites where their abundance is indicated as a bar graph in Fig. 1. The area is bounded by latitudes 7°25′ and 7° 55′ North and longitudes 36°30′ and 37° 22′ East. The watershed is mainly located in the Jimma administrative zone, which has an estimated population of over 2.5 million people (CSA, 2007). The study area receives annual rainfall in the range of 1200–2800 mm, while the altitude ranges from 1096 to 3259 m above sea level. The Gilgel Gibe watershed is located in the tropical afro-alpine ecological region.

The river basin has a catchment area of about 5371 km^2 [10] and the sampling points are distributed along a total length of 186 km from the source to an area further downstream of the Gilgel Gibe hydropower reservoir. During the last 20 years, the Gilgel Gibe river basin has received increased attention from the Ethiopian government for implementing development projects, specifically for hydropower generation [10]. The Gilgel Gibe watershed has many rivers and streams from fast flowing forest streams to stagnant waters and even marshlands. Jimma region is known to have a high forest cover compared to other parts of the country though this is currently dramatically changing due to resettlement and agricultural expansion [30]. The sampling sites and the distribution of Simuliidae larvae are shown in Fig. 1.

Data collection

Data was collected from different rivers in the Gilgel Gibe river basin. About 180 samples were collected from 34 study sites in five sampling campaigns. The governing authority for rivers in Ethiopia is the Ministry of Water, Irrigation and Energy. However, to undertake this study, permission from the Ministry was not required because none of the sampling sites were protected or needed special permission. Therefore, obtaining the permission from Jimma University was sufficient to collect samples from each of the sites as they are authorized to undertake such activities.

Coordinate points of each of the sampling sites are 36°39′ 7.853"E & 7°33′46.697"N, 36°40′12.455"E & 7°34′51.858"N, 36°40′12.455"E & 7°34′51.858"N, 36°40′52.675"E & 7°36′ 2.288"N, 36°43′59.173"E & 7°36′51.886"N, 36°44′42.679"E & 7°36′14.303"N, 36°43′43.812"E & 7°34′4.363"N, 36°46′ 28.268"E & 7°36′9.447"N, 36°45′53.931"E & 7°42′5.768"N, 36°49′19.443"E & 7°42′55.442"N, 36°49′9.523"E & 7°39′ 57.136"N, 36°50′3.572"E & 7°40′16.315"N, 36°50′33.787"E & 7°38′50.039"N, 36°50′44.889"E & 7°38′50.549"N, 36°49′ 51.072"E & 7°36′50.987"N, 36°51′20.99"E & 7°34′52.595"N, 36°54′31.28"E & 7°41′22.464"N, 36°53′40.939"E & 7°37′ 38.038"N, 37°0′30.67"E & 7°43′58.139"N, 36°59′12.895"E & 7°31′37.667"N, 36°59′16.273"E & 7°29′11.822"N, 37°3′ 18.938"E & 7°32′38.206"N, 37°4′16.022"E & 7°41′49.515"N, 37°4′42.857"E & 7°42′43.66"N, 37°6′23.256"E & 7°41′ 58.186"N, 37°8′16.153"E & 7°34′46.464"N, 37°9′50.927"E & 7°45′13.589"N, 37°12′8.41"E & 7°47′45.456"N, 37°11′ 36.415"E & 7°45′46.457"N, 37°14′17.279"E & 7°41′30.491"N, 37°14′28.6"E & 7°39′31.672"N, 37°17′24.483"E & 7°51′ 14.785"N, 37°19′24.012"E & 7°49′54.218"N, and 37°20′ 26.46"E & 7°49′19.344"N 36°50′44.889"E 7°38′50.549"N.

Each campaign was carried out at a six-month interval and samples were taken during dry and wet seasons. The study sites were selected *a priori* based on the criteria of accessibility, geographical distribution, and existing variations of natural and anthropogenic activities. The collected data are categorized into three parts: a) physical-chemical data, b) macroinvertebrate data, and c) physical habitat (physiographic) data (e.g. water depth, water width, river bed, vegetation cover, etc).

Physical-chemical parameters. Temperature (°C), conductivity (µS.cm^{-1}), pH (-), oxygen saturation (%) and turbidity (NTU) were measured onsite at each sampling location using hand electrodes. Five day biochemical oxygen demand (BOD_5) (mg.L^{-1}), nitrate-nitrogen (described as nitrate) (mg.L^{-1}), ammonium-nitrogen (described as ammonium) (mg.L^{-1}) and orthophosphate-phosphorus (described as phosphate) (mg.L^{-1}) were analysed in the laboratory according to standard methods [31].

Physiographic and habitat data. The water body width, water depth and flow velocity were assessed according to Ambelu (2009). The riparian vegetation, river sinuosity, river bank status and embeddedness were estimated using US-EPA habitat assessment protocol [32].

Biological data. Larvae of Simuliidae and other macroinvertebrates were collected using the kick-sampling technique which consists of a D-frame net having a mesh size of 300 µm diameter (Ambelu et al., 2010). Kick sampling was performed along a 10 meter stretch of the river for five minutes including all the microhabitats within the sampling reach [33]. During sampling, the river bed was thoroughly disturbed by kicking with the feet in order to dislodge the macroinvertebrates from the substrate. All substrates in the sampling reach were thoroughly checked to capture organisms attached to it. Within the five minutes of kick sampling, all the possible areas of pool, riffle, edge and center were sampled. After sampling, macroinvertebrates were sorted alive onsite and preserved in 70% ethanol. In the laboratory, the sorted macroinvertebrates were identified to family level using a stereo-microscope and the identification keys [34,35].

Modelling procedures

The modelling was performed using two groups of predictors, namely environmental and macroinvertebrate data. The summary statistics of the response variables in relation to Simuliidae larvae

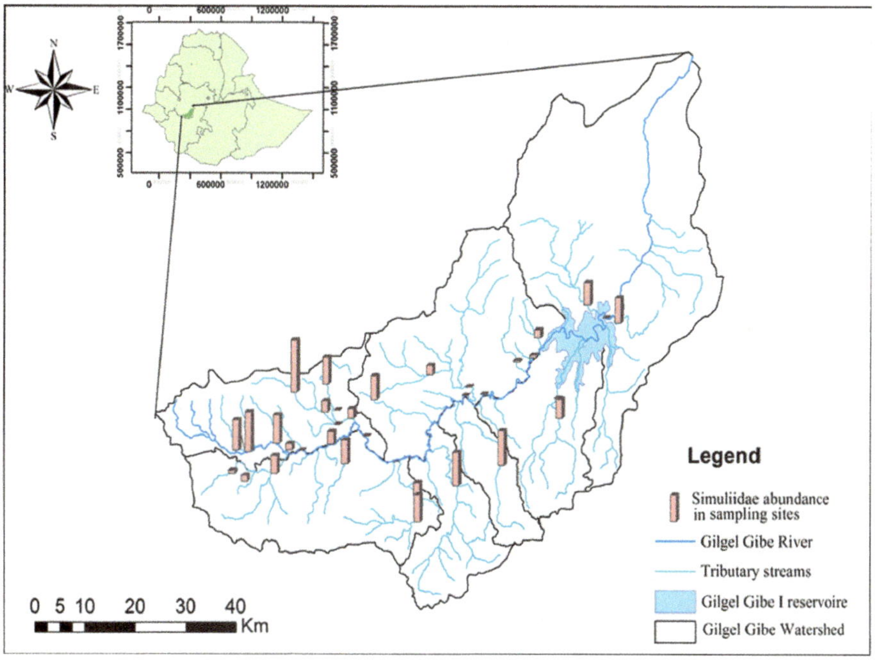

Figure 1. Location of the study area with bar graphs showing the abundance of Simuliidae larvae at each sampling site. The longest bar represents 33 Simuliidae individuals and the shortest one represents zero individuals.

are presented in Table 1 and Table 2. All the environmental variables used were log transformed (except pH) and a square root transformation was done for all macroinvertebrate data. For the application of GAM, a transformation was necessary in order to achieve a uniform distribution [24].

Generalized additive models. GAMs were applied in order to define the set of the environmental parameters and macroinvertebrate taxa that best described the habitat condition of Simuliidae and presence-absence. Additive models are a nonparametric alternative for the more conventionally used generalized linear models (GLMs). GLMs have been frequently used in ecology (Guisan et al., 2006) and are defined by

$$Y_i = \alpha + \beta_1 X_{1i} + \beta_2 X_{2i} + \ldots + \beta_j X_{ji} + \varepsilon_i \quad \text{Where } \varepsilon_i \sim N(0, \sigma^2) \quad (1)$$

The Y_i is the response variable, and X_i represent the explanatory variable(s). The residuals (ε_i) capture the unexplained variation in the data, which is assumed to be normally distributed with a mean value of 0 and variance σ^2. The parameters α and β represent the intercept and slope of the regression respectively. If multiple explanatory variables are used, the number of products between β and X_i is equal to the number of explanatory variables. (1) can be further conceptualized as

$$g^{-1}[E(Y_i|X_{ij})] = \alpha + \Sigma(\beta_j X_{ij}) \quad (2)$$

Where $g^{-1}(\cdot)$ is a local scoring algorithm that specifies the link function between the expected value of Y_i and the explanatory variables. A GAM is defined by

$$g^{-1}[E(Y_i|X_{ij})] = \alpha + \Sigma f_j(x_{ji}) \quad (3)$$

The Y_i is the response variable, $X_i\beta$ represents the intercept of the regression equation, $f_j(x_{ji})$ is a smooth function of the jth explanatory variable, $i = 1, \ldots, n$, is the number of observations.

The number of knots affects the amount of smoothing applied to the data [36]. A smoother with two knots is linear, has little variability and may be biased since there is only one piecewise function [37]. Increasing the number of knots allows more flexibility, but may result in over-fitting. For smaller data sets, below 30, three knots is a good starting point. [37] report that a number of four to five knots is appropriate for most applications. In our analysis, the number of knots for the smoothing curves was fixed to five for macroinvertebrate analysis and 10 for environmental variables as the number of records per substance in our training dataset varied from below 30 to more than 100.

The 'mgcv' library in the R statistical software [38] was used to select the GAMs smoothing predictors following the method proposed by Wood and Augustin [36]. The individual models cannot be tested for significance using the P-values provided by 'mgcv' library since the true number of degrees of freedom is unknown (Giannoulaki et al., 2008; Wood, 2012). Each model fit was analyzed by the level of deviance explained (0–100%; the higher the better), and the unbiased risk estimator (UBRE) in which the lowest value is considered as the best model performance indicator. The degree of smoothing was also chosen based on the observed data and the generalized cross validation method suggested by [25] and incorporated in the 'mgcv' library. To avoid the over-fitting problem, the effective degree of freedom of each model count in the GCV score was increased by a factor of $\gamma = 1.4$ [39].

To increase the model performance and decrease the collinearity problem, independent variables were eliminated [22,23,25]

Table 1. The minimum (Min), 1st quartile (1st Qu), median, mean, 3rd quartile (3rd Qu), maximum (Max) and standard deviation (StDv) of environmental predictors used to analyse Simuliidae occurrence.

Environmental variables	Min	1st Qu	Median	Mean	3rd Qu	Max	StDv
Altitude (m)	1625	1698	1742	1772	1788	2488	121.43
Vegetation (score out of 20)	2	6	10	10.3	13	20	5.3
Water temperature (°C)	13.9	18	19.68	19.8	21.1	27.5	2.42
Width (m)	0.6	3	6	8.7	10	43	9.24
Depth (m)	0.01	0.25	0.43	0.6	0.7	2.5	0.43
Velocity (m/s)	0.005	0.22	0.44	0.5	0.7	1.8	0.32
Flow rate (m3/s)	0.001	0.26	1.05	2.8	3.15	27.36	3.84
Sinuosity (score out of 20)	6	10	14	14	18	20	4.13
Distance from source (m)	2	12.5	19	29.8	29	154	34.9
Embeddedness (score out of 20)	3	10	16	14.4	18	20	4.91
River bank status (score out of 20)	4	12	15	14.2	18	20	4.4
pH	5.3	7.02	7.4	7.4	7.7	8.5	0.47
Conductivity (µS/cm)	27.1	80	100	114	130	455	57.9
DO (mg/L)	0.34	5.83	6.7	6.4	7.31	9.3	1.54
BOD (mg/L)	0.21	1.6	2.5	4.1	3.6	80	6.34
Phosphate (mg/L)	0	0.03	0.16	0.4	0.5	4.47	0.57
Nitrate (mg/L)	0.01	0.402	1.2	1.4	1.9	6.156	1.13
Ammonium (mg/L)	0.002	0.05	0.22	0.5	0.8	3.13	0.62
Simuliidae (count)	0	0	0	5.297	4	150	16.08

Table 2. The median, mean, 3rd quartile (3rd Qu), maximum (Max) and standard deviation (StDv) of macroinvertebrate (MI) variables used to predict Simuliidae abundance and presence-absence.

MI variables	Median	Mean	3rd Qu	Max	StDv
Aeshnidae	0	1	0	10	1
Anthomyidae	0	9	0	74	19
Baetidae	5	15	19	150	25
Belostomatidae	0	1	0	27	3
Caenidae	4	11	14	155	21
Chironomidae	6	11	12	125	17
Coenagrionidae	4	11	13	88	17
Corduliidae	0	1	0	20	2
Corixidae	0	2	2	50	6
Dytiscidae	0	4	2	150	17
Elmidae	0	1	1	43	3
Ephemerellidae	0	1	0	53	3
Glossiphonidae	0	1	0	47	3
Glossosomatidae	0	1	0	62	5
Gomphidae	0	1	2	22	3
Gyrinidae	0	1	0	23	2
Heptagenidae	0	3	2	110	10
Hydrophilidae	0	1	1	26	2
Hydropsychidae	3	15	19	150	26
Libellulidae	1	5	4	100	11
Naucoridae	0	1	1	31	3
Nepidae	0	0	0	4	1
Notonectidae	0	1	0	82	5
Protoneuridae	0	3	3	37	6
Sphaeriidae	0	1	1	41	4
Tipulidae	0	0	0	8	1
Unionidae	0	1	0	21	3

The minimum and the 1st quartile values are not presented in the table because all were zero.

and the best model was chosen based on a stepwise backward selection method. Specifically, models were compared using the estimated UBRE and percent deviance explained, the environmental variables were ranked and selection of the final model was based on the minimization of the above criteria. Following the recommendation forwarded by Wood (2001), during model fitting manual elimination of attributes was done when all of the following three criteria are met: the estimated degree of freedom of the model term is closer to 1; the plotted confidence band from the model term include zero everywhere; and URBE score is dropped when the model term (attribute) is eliminated.

The relationship between Simuliidae larvae and the predicting variables (e.g. pH) with the ith observation in the data, smooth function $s()$, constant a, and residual error i is represented by:

$$[\text{Simuliidae}]\text{i} = \text{a} + \text{s}(\text{pH}) + \varepsilon\text{i} \qquad (4)$$

Therefore a model with n smooth functions (predictor variables) in this relationship can be generalized to:

$$Ai = \Sigma_{\text{j}=1}^{\text{n}} sj(xi) + a + \varepsilon i \qquad (5)$$

The ith Simuliidae abundance in the data set is Ai. $sj(xi)$ is the smooth for the jth variable and gives the value of this smooth for the ith observation. i is the residual error for this observation and a is a constant.

Classification tree combined with genetic algorithms. First, the model was developed based on CT using all input predictors, while in a next steps the CT was combined with a genetic algorithm, which was used to select the most relevant input variables. CTs [17,40] predict the value of a discrete dependent variable with a finite set of values (called classes) from the values of a set of independent variables (called attributes), which may be either continuous or discrete. The J48 algorithm with binary splits was applied to induce CT. There are a variety of algorithms to build classification trees that share the desirable quality of interpretability. A well-known and frequently used algorithm is the C4.5 which is a java reimplementation of the J48 algorithm in the WEKA machine learning package [41]. The dependent variable (output value) consisted of the presence-absence of

Simuliidae larvae whereas the independent parameters were the physical-chemical and MI larvae predictors (Table 1 and 2).

Different folds of cross-validation were tested for the training and validation of CTs. The maximum stability and model performance of CT was maximized using a 10-fold cross-validation in terms of percentage of correctly classified instances (%CCI) and Kappa statistics (*k*). In the 10-fold cross-validation, the original data were randomly partitioned into 10 subsamples of approximately equal size using WEKA default settings. In addition, the default values of the J48 algorithm with binary split were used to find the most important explanatory variables for the prediction of Simuliidae.

The next step was the application of the GA search method on the CT to select the best explanatory variables for the Simuliidae larvae. GAs follow the principle of "*survival of the fittest*" which begin with a population of randomly generated chromosomes that advances towards the selection of better chromosomes [42–44]. Following the principle of natural selection, the population undergoes evolution with successive generations. During this process, chromosomes in the population are rated for their fitness and consequently a new population of chromosomes are formed depending on the applied selection method.

During CT model development, wrapper subset evaluator was used on J48 learning algorithm in which the attributes (variables) are evaluated by using accuracy estimations [45]. During GA application, we used 20 chromosomes as initial population that evolved through a maximum of 20 generations [26]. Default settings of Weka machine learning algorithm was used for crossover and mutation probability which is 60% and 3.3%, respectively. Before the GA application, the dataset was randomized and then attributes were selected. After the selection of the successful chromosomes, CTs were run seven to ten times to each subset (chromosome) after randomization to check the stability of the model. The subsets of selected attributes by GAs (chromosomes) that showed the lowest standard deviation, based on %CCI and *K*, were retained. In addition, attributes that appeared most frequently in the subsequent GA application were finally used for CT-GA model development.

Results

GAM output

Using the abundance of the response variable (Simuliidae), 11 environmental predictor variables were obtained from the model after a backward stepwise elimination of the terms. The selected variables significantly contributed to the prediction of the Simuliidae larvae (Table 2). All eliminated variables had a very low value of estimated degrees of freedom and had non-significant p-values. The GAM has an adjusted R^2 of 0.62 and the total deviance explained was 62% and the un-biased risk estimator (UBRE) score was 0.345. The relationship between environmental attributes selected by GAMs and the Simuliidae larvae is shown in Fig. 2.

However, when GAM prediction of the Simuliidae larvea with its presence-absence data is made, only three environmental predictors (distance, flow velocity and water depth) were selected with significant prediciton (p-value<0.01). The estimated degrees of freedom for the three environmental predictors were 2.43, 2.06 and 1.51, respectively. The EBRE score, adjusted R^2, and percent deviance explained were respectively −0.462, 0.323 and 33.9.

Among the 27 macroinvertebrate predictors, eight were selected by the GAMs. After backward stepwise selection of the predicting variables of macroinvertebrate families, those which showed significant predicting power were fitted as shown in Fig. 2. The presence-absence of the Simuliidae larvae was also predicted with GAMs and only four macroinvertebrate predictors (Beatidae, Dytiscidae, Hydropsychidae and Libeluliidae) were selected as important variables. All four variables showed a significant (p-value<0.05) contribution to the model and have an R^2 adjusted = 0.58, percent deviance explained = 63, and UBRE score = 0.243.

CT-GA output

Classification tree models were built using a genetic search algorithm. Prior to the selection of the environmental attributes, the classification tree was built. The tree size was 67 with 34 leaves whereas the %CCI and *k* were 69.4±1.3 and 0.38±0.03, respectively. During the application of the genetic search algorithm, the distance of the sampling site from the source of the river appears in all the successful chromosomes. Whereas the flow velocity and embeddedness appears nine times, river bank status and DO appear seven times, BOD and ammonium appear four times, electrical conductivity (EC), flow rate and water depth appear three times, pH and nitrate appear only one time from the ten independently identified subset of attributes (chromosomes). Finally, using the most frequently selected attributes (four to ten times), a classification tree model was built. The model indicated that the presence or absence of Simuliidae is primarily determined by the distance of the site from the stream source. According to the model, the Simuliidae community are often absent for sites which are 32 km far from the source. In addition, Simuliidae is absent for sites whose flow velocity is 0.125 m/s (Fig. 3).

Before the application of GA on the CT, all 28 macroinvertebrate variables were used and the average performance in terms of %CCI and *K* was 78.26±0.02 and 0.53±0.02, respectively. After the application of GA, each chromosome or group of successful macroinvertebrate variables picked by the GA showed an average %CCI and *K* of 80.2–82.46 and 0.60–0.65, respectively (Fig. 4).

In each chromosome five to nine macroinvertebrates were chosen by the GA to predict the presence of Simuliidae. Corixidae (9 times), Hydropsychidae (9 times), Protoneuridae (8 times), Chironomidae (8 times), and Elmidae (6 times) were the most frequently selected macroinvertebrate variables. Glossosomatidae, Aeshnidae, Gyrinidae, Libellulidae, Nepidae, Belostomatidae, Caenidae, Dytiscidae, Hydrophilidae, Sphaeriidae, Tipulidae, Ephemerellidae, and Anthomyidae appeared rarely (one to two times) among the ten selected chromosomes. The other macroinvertebrates were not selected by GA. The CT model, constructed with the most frequently selected macroinvertebrate predictors by GA, is shown in Fig. 3. The model indicated that among the macroinvertebrate communities, Hydropsychidae, Corixidae, Protoneuridae, Chironomidae and Elmidae were the major determinants of the presence and absence of Simuliidae larvae.

Discussion

Bio-environmental factors that are influencing blackflies distribution in the Gilgel Gibe watershed has been identified using combined modelling techniques. This approach enabled us a better identification of the suitable habitat conditions or environmental constraints for Simuliidae larvae. Characterizing and modelling the distribution and abundance of taxa is one of the major tasks of ecologists [46]. The availability of reliable environmental dataset obtained from wider area of sampling sites for an extended period of time often encourages prediction of taxa to identify the environmental requirements so that their distribu-

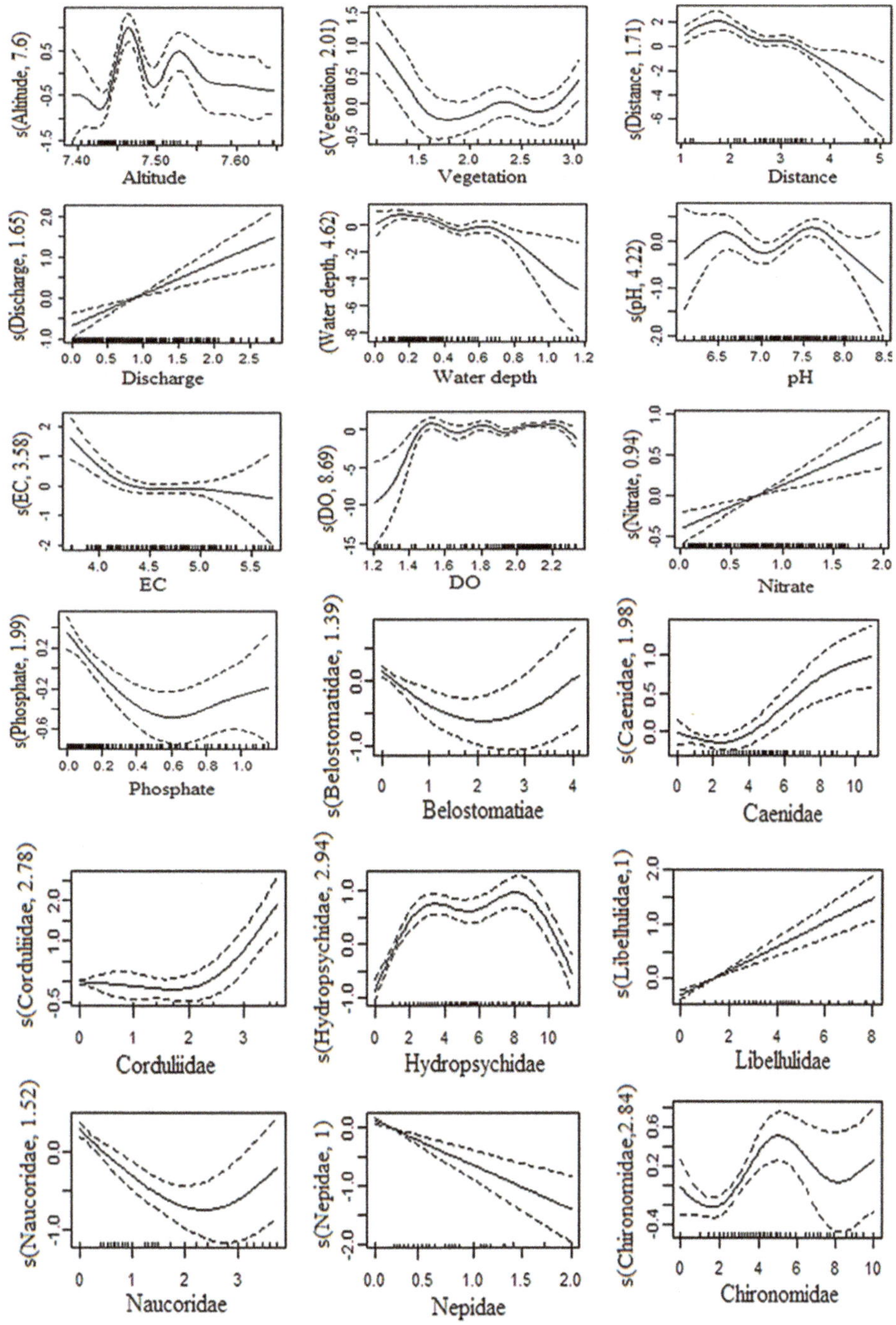

Figure 2. Smooth plot of the GAM output of the selected environmental and macroinvertebrate predictors showing their relationship with Simuliidae larvae and the fitted nonparametric terms with 95% confidence interval (dashed lines). The y-axis is scaled to zero and the rug plot on the x-axis indicates number of observations.

tion can be inferred. This is especially helpful for the prediction of species distribution over large unsampled areas and for reducing sampling costs. In addition, the model output could provide important information for decision support of environmental management systems. Here, we have used two well-established habitat suitability modelling techniques in order to identify important predictors that can explain the abundance and occurrence of Simuliidae larvae.

Simultaneous modelling of Simuliidae using GAMs and CTs has enabled the identification of the most important environmental and macroinvertebrate predictors. Among the environmental predictors, distance from the source, river discharge, water depth, river bank status, electrical conductivity and nitrate concentration were selected by both modelling techniques as important variables determining the occurrence and abundance of black flies in the region.

The GAM outputs indicate that the model performance indicators between the presence-absence of Simuliidae larvae significantly differ from the abundance prediction. The number of selected predicting variables (both environmental and macroin-

```
distance <= 1.518514
| Velocity <= 0.10721
| | distance <= 1.30103
| | | BOD <= 0.579784: a (24.0/2.0)
| | | BOD > 0.579784
| | | | Velocity <= 0.071882
| | | | | Embededness <= 1.146128: a (10.0/1.0)
| | | | | Embededness > 1.146128
| | | | | | Velocity <= 0.051153: a (5.0/1.0)
| | | | | | Velocity > 0.051153: p (3.0)
| | | | Velocity > 0.071882: p (7.0)
| | distance > 1.30103: a (33.0/1.0)
| Velocity > 0.10721
| | Embededness <= 0.845098: a (10.0/1.0)
| | Embededness > 0.845098: p (179.0/57.0)
distance > 1.518514: a (54.0/3.0)
```

[A]

```
Hydropsychidae <= 4.0
| Hydropsychidae <= 0.0: a (120.0/12.0)
| Hydropsychidae > 0.0
| | Protoneuridae <= 19.0
| | | Chironomidae <= 14.0: a (27.0/4.0)
| | | Chironomidae > 14.0: p (5.0/1.0)
| | Protoneuridae > 19.0: p (3.0)
Hydropsychidae > 4.0
| Corixidae <= 0.0
| | Protoneuridae <= 0.0
| | | Elmidae <= 2.0: p (79.0/17.0)
| | | Elmidae > 2.0
| | | | Chironomidae <= 7.0: a (7.0/1.0)
| | | | Chironomidae > 7.0: p (2.0)
| | Protoneuridae > 0.0
| | | Elmidae <= 4.0
| | | | Chironomidae <= 10.0: a (14.0/1.0)
| | | | Chironomidae > 10.0
| | | | | Chironomidae <= 17.0: p (3.0)
| | | | | Chironomidae > 17.0: a (3.0)
| | | Elmidae > 4.0: p (3.0)
| Corixidae > 0.0
| | Protoneuridae <= 18.0: p (34.0/2.0)
| | Protoneuridae > 18.0: a (3.0)
```

[B]

Figure 3. Classification tree constructed by the most frequently selected environmental [A] and macroinvertebrate [B] predictors using genetic algorithm predicting the presence (p) and absence (a) of Simuliidae larvae.

vertebrate predictors) was fewer for presence-absence compared with the Simuliidae abundance. Except for flow velocity, the other environmental presence-absence predictors were also identified by GAM during abundance prediction. A previous study done by Barry and Welsh (2002) also has indicated that the model pattern of presence-absence of a species, conditional on the covariates, is markedly different from the pattern of abundance.

We therefore determined that the abundance of Simuliidae increases with increasing river flow rate, nitrate concentration and flow velocity. Nevertheless, Simuliidae abundance regularly decreases with increasing distance of the sampling site from the source, electrical conductivity of the water, water depth and phosphate concentration. The other environmental predictors like altitude, vegetation cover, river bank status and DO concentrations show irregular patterns with regard to the abundance of

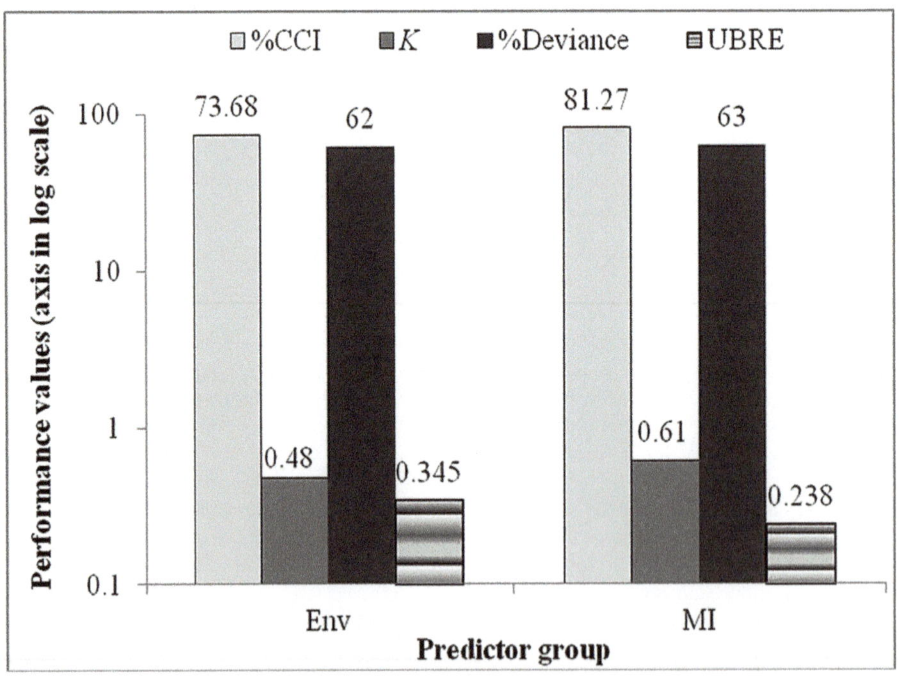

Figure 4. Model performances of GAMs and classification trees based on environmental (Env) and macroinvertebrate (MI) predictors. %CCI = percent correctly classified instances, K = kappa statistics, UBRE = unbiased risk estimator.

Simuliidae. The optimum pH condition for Simuliidae larvae abundance was found to be approximately between 6.5 and 7.7. Regarding the selected macroinvertebrate predictors, the occurrence of Libellulidae, Baetidae, and Caenidae promotes the availability of Simuliidae larvae in the river system. However, higher abundances of Hydropsychidae, Belostomatidae, Naucoridae and Nepidae could reduce the availability of the dependent variable, i.e. Simuliidae larvae. It has been found that the GAMs prediction using macroinvertebrate communities showed better performance (in terms of UBRE, adjusted R^2 and percent deviance explained) than the environmental predictors.

Clear model results were obtained when classification tree models were supported by a genetic search algorithm to select environmental and macroinvertebrate predictors of Simullidae larvae. The application of GA to CT significantly improved the model performance as well as the clarity of the decision tree. The decision tree model without the application of GA was complicated to understand and describe due to its large tree size. Recently [10,19,26] have also improved clarity of their classification tree models by applying GAs. However, those authors and many others [4,26,47–50] are often using model boosting mechanisms such as bagging and boosting, together with the use of attribute selection tools (GA, greedy stepwise algorism) rather than combining the model with robust statistical techniques like that of GAMs. Based on the given data set, the CT-GA has given clear environmental predictor values for which the Simuliidae larvae could be present or absent. The majority of environmental and macroinvertebrate predictors selected by GAM were also identified by GA as important predictors of the presence-absence of Simuliidae larvae. The two modelling techniques (GAMs and CT-GAs) showed reliable predictors which can be very useful for understanding the distribution of Simuliidae larvae and, thus, controlling the vector of onchocerchiasis. On the other hand, both the GAMs and CT-GAs models have indicated that Simuliidae larvae may be an important water quality indicator in head waters (with shorter distances from the source), shallow and fast flowing rivers.

Vector control and patient treatment is a major component of the Onchocerciasis control program and is based on routine aerial application of larvicides. This is found to be very expensive to implement in many developing countries like Ethiopia and Ghana where the disease is endemic [51]. Therefore, our model outputs could indicate an alternative means to control the disease vector larvae based on environmental management and biological control mechanisms. Environmental management and biological control of the disease vector may be a much more effective strategy than the use of pesticides to overcome the residual effects of chemical applications to the different environmental compartments. The GAMs and CT-GA have been successfully applied to identify the environmental variables and macroinvertebrates that can play a detrimental role in the elimination of Simuliidae larvae from the river system. GAMs and classification trees can even indicated which areas should be focused on for insecticide application if it becomes a choice of vector control. Based on the selected variables it should be possible to map the sites where Simuliidae is present. Such mapping has been proposed by [51] to help control the occurrence of onchocerchiasis.

According to GAMs, one of the major environmental management strategies that could be applied is minimizing the flow velocity and increasing the water depth so that the abundance of Simuliidae larvae would be minimized or eliminated. This could be achieved by slowing down the flow in the highlands which would reduce the flow velocity and increase the water depth. This procedure could benefit communities affected by Onchocerchiasis because they could utilize the additional water for irrigation to ensure food security. This is a very relevant issue in arid tropical countries where farmers cannot dependant on rain water only but need river water for irrigation. The model outputs based on macroinvertebrate variables could be an important indication for when biological control methods need to be applied to Simuliidae. However, it is recommended to further study the biological relationship of the identified macroinvertebrates and Simuliidae to effectively apply such biological control of Simuliidae.

Conclusion

In conclusion, the combination of GAMs and CT-GA techniques has led to the identification of suitable habitat conditions of Simuliidae larvae and the macroinvertebrate families, which are crucial for their existence or disappearance. Such models are important for conservation purposes as well as for disease vector control in the tropics because they can be used to eliminate the suitable environmental conditions of the target organism [52]. Accurate representation of species distribution models derived from sampled data is essential for tropical ecosystem management purposes. Effective prediction of the habitat suitability of Simuliidae larvae has been obtained by the combined application of GAMs and CT-GAs. Through this modelling approach, a more reliable ecological assessment and Onchocerchiasis disease vector control could be achieved based on environmental management and biological control techniques. The results may lead to improved vector control methods using habitat modification techniques and site specific application of pesticides.

Acknowledgments

We would like to thank all people who have contributed in the sample collection and laboratory analyses.

Author Contributions

Conceived and designed the experiments: AA. Analyzed the data: AA GE. Contributed reagents/materials/analysis tools: SM TA. Wrote the paper: AA MK PB GE PG. Edited the English language: MK.

References

1. Figueiró R, Gil-Azevedo LH (2010) The role of Neotropical blackflies (Diptera: Simuliidae) as vectors of the Onchocerciasis: a short overview of the Ecology behind the disease. Oecologia Australis 14: 745–755. doi:10.4257/oeco.2010.1403.10.
2. Lock K, Adriaens T, Goethals P (2014) Effect of water quality on blackflies (Diptera: Simuliidae) in Flanders (Belgium). Limnologica - Ecology and Management of Inland Waters 44: 58–65. doi:10.1016/j.limno.2013.08.001.
3. Figueiró R, Gil-Azevedo LH, Maia-Herzog M, Monteiro RF (2012) Diversity and microdistribution of black fly (Diptera: Simuliidae) assemblages in the tropical savanna streams of the Brazilian cerrado. Mem Inst Oswaldo Cruz 107: 362–369.
4. Mereta ST, Boets P, Ambelu Bayih A, Malu A, Ephrem Z, et al. (2012) Analysis of environmental factors determining the abundance and diversity of macroinvertebrate taxa in natural wetlands of Southwest Ethiopia. Ecological Informatics 7: 52–61. doi:10.1016/j.ecoinf.2011.11.005.
5. Kazanci N, Ertunc Ö (2010) Use of Simuliidae (Insecta, Diptera) species as indicators of aquatic habitat quality of Yeşilırmak River Basin (Turkey). Review of Hydrobiology 3: 27–36.
6. Lechthaler W, Car M (2004) Digital keys for biological indicators, part I, Simuliidae: presentation of a key to larvae and pupae from central and western Europe. Acta Zoologica Universitatis Comenianae 46: 39–44.
7. Kazanci N (2006) Ordination of Simuliidae and climate change impact. Acta entomologica serbica Suplement: 69–76.

8. Gibbins EG (2009) further studies on Ethiopian Simuliidae*. Transactions of the Royal Entomological Society of London 82: 51–97. doi:10.1111/j.1365-2311.1934.tb00028.x.
9. D'heygere T, Goethals PLM, De Pauw N (2006) Genetic algorithms for optimisation of predictive ecosystems models based on decision trees and neural networks. Ecological Modelling 195: 20–29. doi:10.1016/j.ecolmodel.2005.11.005.
10. Ambelu A, Lock K, Goethals P (2010) Comparison of modelling techniques to predict macroinvertebrate community composition in rivers of Ethiopia. Ecological Informatics 5: 147–152. doi:10.1016/j.ecoinf.2009.12.004.
11. Reid JL, Harris JBC, Zahawi RA (2012) Avian Habitat Preference in Tropical Forest Restoration in Southern Costa Rica. Biotropica 44: 350–359. doi:10.1111/j.1744-7429.2011.00814.x.
12. Planque B, Bellier E, Lazure P (2007) Modelling potential spawning habitat of sardine (Sardina pilchardus) and anchovy (Engraulis encrasicolus) in the Bay of Biscay. Fisheries Oceanography 16: 16–30. doi:10.1111/j.1365-2419.2006.00411.x.
13. Austin MP, Smith TM, Van Niel KP, Wellington AB (2009) Physiological responses and statistical models of the environmental niche: a comparative study of two co-occurring *Eucalyptus* species. Journal of Ecology 97: 496–507. doi:10.1111/j.1365-2745.2009.01494.x.
14. Hilborn R, Mangel M (1997) The ecological detective: confronting models with data. Princeton, NJ: Princeton University Press.
15. Cheaib A, Badeau V, Boe J, Chuine I, Delire C, et al. (2012) Climate change impacts on tree ranges: model intercomparison facilitates understanding and quantification of uncertainty. Ecology Letters 15: 533–544. doi:10.1111/j.1461-0248.2012.01764.x.
16. Giannoulaki M, Pyrounaki MM, Liorzou B, Leonori I, Valavanis VD, et al. (2011) Habitat suitability modelling for sardine juveniles (Sardina pilchardus) in the Mediterranean Sea. Fisheries Oceanography 20: 367–382. doi:10.1111/j.1365-2419.2011.00590.x.
17. Quinlan JR (1986) Induction of decision trees. Machine Learning 1: 81–106. doi:10.1007/BF00116251.
18. Zarkami R, Sadeghi R, Goethals P (2012) Use of fish distribution modelling for river management. Ecological Modelling 230: 44–49. doi:10.1016/j.ecolmodel.2012.01.011.
19. Zarkami R, Goethals Peter, De Pauw, Neils (2010) Use of classification tree methods to study the habitat requirements of tench (Tinca tinca) (L., 1758). CJES 8: 55–63.
20. Guisan A, Zimmermann NE (2000) Predictive habitat distribution models in ecology. Ecological Modelling 135: 147–186. doi:10.1016/S0304-3800(00)00354-9.
21. Hastie T, Tibshirani R (1986) Generalized Additive Models. 1: 297–310.
22. Guisan A, Edwards TC, Hastie T (2002) Generalized linear and generalized additive models in studies of species distributions: setting the scene. Ecological Modelling 157: 89–100. doi:10.1016/S0304-3800(02)00204-1.
23. Giannoulaki M, Valavanis VD, Palialexis A, Tsagarakis K, Machias A, et al. (2008) Modelling the presence of anchovy Engraulis encrasicolus in the Aegean Sea during early summer, based on satellite environmental data. Hydrobiologia 612: 225–240. doi:10.1007/s10750-008-9498-6.
24. Hastie T, Tibshirani R (1999) Generalized additive models. Boca Raton, Fla: Chapman & Hall/CRC. 335 p.
25. Wood SN (2006) Generalized additive models: an introduction with R. Boca Raton, FL: Chapman & Hall/CRC. 391 p.
26. Sadeghi R, Zarkami R, Sabetraftar K, Van Damme P (2013) Application of genetic algorithm and greedy stepwise to select input variables in classification tree models for the prediction of habitat requirements of Azolla filiculoides (Lam.) in Anzali wetland, Iran. Ecological Modelling 251: 44–53. doi:10.1016/j.ecolmodel.2012.12.010.
27. Hoang TH, Lock K, Mouton A, Goethals PLM (2010) Application of classification trees and support vector machines to model the presence of macroinvertebrates in rivers in Vietnam. Ecological Informatics 5: 140–146. doi:10.1016/j.ecoinf.2009.12.001.
28. Barry S, Elith J (2006) Error and uncertainty in habitat models. Journal of Applied Ecology 43: 413–423. doi:10.1111/j.1365-2664.2006.01136.x.
29. Wooley JC, Lin HS (2005) Catalyzing inquiry at the biocomp interface. Washington, DC: National Academies of Sciences.
30. Ambelu A (2009) Biological monitoring based on macroinvertebrates for decision support of water management in Ethiopia Ghent: Ghent University.
31. American Public Health Association, American Water Works Association, Water Environment Federation (2005) Standard methods for the examination of water & wastewater. Washington, D.C.: American Public Health Association.
32. Barbour MT, Gerritsen J, Snyder BD, Stribling JB (1999) Rapid Bioassessment Protocols for Use in Streams and Wadeable Rivers: Periphyton, Benthic Macroinvertebrates and Fish. Second Edition. Washington, D.C.: United States Environmental Protection Agency.
33. IBN (1984) Biological water quality: determination of the biotic index based on aquatic macroinvertebrates, NBN T92–402.
34. Bouchard JRW (2004) Guide to Aquatic Invertebrates of the Upper Midwest. University Of Minnesota. 207 p.
35. Voshell JR (2002) A guide to common freshwater invertebrates of North America. Blacksburg, Va: McDonald & Woodward Pub. 442 p.
36. Wood SN, Augustin NH (2002) GAMs with integrated model selection using penalized regression splines and applications to environmental modelling. Ecological Modelling 157: 157–177. doi:10.1016/S0304-3800(02)00193-X.
37. Keele L (2008) Semiparametric regression for the social sciences. Chichester, England; Hoboken, NJ: Wiley. 213 p.
38. R Core Team (2012) R: A language and environment for statistical computing. Vienna, Austria: R Foundation for Statistical Computing. Available: http://www.R-project.org/.
39. Katsanevakis S, Maravelias CD, Damalas D, Karageorgis AP, Tsitsika EV, et al. (2009) Spatiotemporal distribution and habitat use of commercial demersal species in the eastern Mediterranean Sea. Fisheries Oceanography 18: 439–457. doi:10.1111/j.1365-2419.2009.00523.x.
40. Breiman L (1993) Classification and regression trees. New York, N.Y.: Chapman & Hall.
41. Witten IH, Frank E, Hall MA (2011) Data mining: practical machine learning tools and techniques. 3rd ed. Burlington, MA: Morgan Kaufmann. 629 p.
42. Goldberg DE (1989) Genetic algorithms in search, optimization, and machine learning. Reading, Mass: Addison-Wesley Pub. Co. 412 p.
43. Holland JH (1992) Adaptation in natural and artificial systems: an introductory analysis with applications to biology, control, and artificial intelligence. 1st MIT Press ed. Cambridge, Mass: MIT Press. 211 p.
44. Vose MD (1999) The simple genetic algorithm: foundations and theory. Cambridge, Mass: MIT Press. 251 p.
45. Hall MA, Holmes G (2003) Benchmarking attribute selection techniques for discrete class data mining. IEEE Transactions on Knowledge and Data Engineering 15: 1437–1447. doi:10.1109/TKDE.2003.1245283.
46. Andrewartha HG, Birch C (1986) The ecological web: more on the distribution and abundance of animals. Chicago: University of Chicago Press.
47. Dakou E, D'heygere T, Dedecker AP, Goethals PLM, Lazaridou-Dimitriadou M, et al. (2006) Decision Tree Models for Prediction of Macroinvertebrate Taxa in the River Axios (Northern Greece). Aquatic Ecology 41: 399–411. doi:10.1007/s10452-006-9058-y.
48. Jowett IG, Davey AJH (2007) A Comparison of Composite Habitat Suitability Indices and Generalized Additive Models of Invertebrate Abundance and Fish Presence–Habitat Availability. Transactions of the American Fisheries Society 136: 428–444. doi:10.1577/T06-104.1.
49. Goethals PLM, Dedecker AP, Gabriels W, Lek S, Pauw N (2007) Applications of artificial neural networks predicting macroinvertebrates in freshwaters. Aquatic Ecology 41: 491–508. doi:10.1007/s10452-007-9093-3.
50. Sadeghi R, Zarkami R, Sabetraftar K, Van Damme P (2012) Application of classification trees to model the distribution pattern of a new exotic species Azolla filiculoides (Lam.) at Selkeh Wildlife Refuge, Anzali wetland, Iran. Ecological Modelling 243: 8–17. doi:10.1016/j.ecolmodel.2012.06.011.
51. Taylor MJ, Awadzi K, Basáñez M-G, Biritwum N, Boakye D, et al. (2009) Onchocerciasis Control: Vision for the Future from a Ghanian perspective. Parasites & Vectors 2: 7. doi:10.1186/1756-3305-2-7.
52. Palialexis A, Georgakarakos S, Karakassis I, Lika K, Valavanis VD (2011) Prediction of marine species distribution from presence–absence acoustic data: comparing the fitting efficiency and the predictive capacity of conventional and novel distribution models. Hydrobiologia 670: 241–266. doi:10.1007/s10750-011-0673-9.

6

Taxonomic and Functional Diversity Provides Insight into Microbial Pathways and Stress Responses in the Saline Qinghai Lake, China

Qiuyuan Huang[1,4], Brandon R. Briggs[1], Hailiang Dong[1,2,3*], Hongchen Jiang[3], Geng Wu[3], Christian Edwardson[4], Iwijn De Vlaminck[5], Stephen Quake[5]

1 Department of Geology and Environmental Earth Science, Miami University, Oxford, Ohio, United States of America, **2** State Key Laboratory of Biogeology and Environmental Geology, China University of Geosciences, Beijing, China, **3** State Key Laboratory of Biogeology and Environmental Geology, China University of Geosciences, Wuhan, China, **4** Department of Microbiology, University of Georgia, Athens, Georgia, United States of America, **5** Departments of Bioengineering and Applied Physics, Stanford University and the Howard Hughes Medical Institute, Stanford, California, United States of America

Abstract

Microbe-mediated biogeochemical cycles contribute to the global climate system and have sensitive responses and feedbacks to environmental stress caused by climate change. Yet, little is known about the effects of microbial biodiversity (i.e., taxonmic and functional diversity) on biogeochemical cycles in ecosytems that are highly sensitive to climate change. One such sensitive ecosystem is Qinghai Lake, a high-elevation (3196 m) saline (1.4%) lake located on the Tibetan Plateau, China. This study provides baseline information on the microbial taxonomic and functional diversity as well as the associated stress response genes. Illumina metagenomic and metatranscriptomic datasets were generated from lake water samples collected at two sites (B and E). Autotrophic *Cyanobacteria* dominated the DNA samples, while heterotrophic *Proteobacteria* dominated the RNA samples at both sites. Photoheterotrophic *Loktanella* was also present at both sites. Photosystem II was the most active pathway at site B; while, oxidative phosphorylation was most active at site E. Organisms that expressed photosystem II or oxidative phosphorylation also expressed genes involved in photoprotection and oxidative stress, respectively. Assimilatory pathways associated with the nitrogen cycle were dominant at both sites. Results also indicate a positive relationship between functional diversity and the number of stress response genes. This study provides insight into the stress resilience of microbial metabolic pathways supported by greater taxonomic diversity, which may affect the microbial community response to climate change.

Editor: Yiguo Hong, CAS, China

Funding: This research was supported by the National Natural Science Foundation of China (Grant Nos. 41030211 and 41002123), the Scientific Research Funds for the 1000 "Talents" Program Plan from China University of Geosciences (Beijing), State Key Laboratory of Biogeology and Environmental Geology, China University of Geosciences (No. GBL11201), and the Fundamental Research Funds for National University, China University of Geosciences (Wuhan). This work was fostered by NSF IOS grant 1238801. The funders had no role in study design, data collection and analysis, decision to publish, or preparation of the manuscript.

Competing Interests: The authors have declared that no competing interests exist.

* Email: dongh@miamioh.edu

These authors contributed equally to this work.

Introduction

Microorganisms have been key respondents to and drivers of global climate change by affecting the atmospheric concentrations of greenhouse gases [1]. The microbial response to future global climate change is likely controlled by the biodiversity (i.e. taxonomic and functional diversity) of the ecosystem [2,3,4]. For example, ecosystems with higher biodiversity are more likely to be stable against environmental change because of a greater likelihood of having key functioning species [5]. However, ecosystems with low diversity can be stable if the organisms have mechanisms to respond to stress [6]. For example, *Synechococcus* has multiple protective mechanisms to cope with UV stress and can maintain photosynthesis, while *Procholorcoccus* lacks these protective mechanisms and shuts down several key metabolic processes under similar UV stress [7]. Thus, studying the biodiversity and potential stress response mechanisms can aid in understanding microbial community response to stress.

Baseline information on the microbial biodiversity and stress response mechanisms is needed in ecosystems that are sensitive to climate change. For example, the Tibetan Plateau has experienced significant warming in recent decades and is considered to be a sensitive indicator of regional and global climate change [8]. Temperature has increased 0.28°C per decade since the early 1960 s [9] causing 82% of the 46,000 glaciers to retreat [10]. The melting glaciers have caused numerous floods and altered salinity and water levels in most of the Tibetan lakes [11]. The fragility and sensitivity of the Tibetan Plateau's ecosystem to these environmental changes have resulted in loss of habitats and extinctions of endemic macrobiota [12].

Qinghai Lake, located on the Tibetan Plateau, is characterzied by oligotrophy, low temperature, moderate salinity, and high UV radiation, making it a unique ecosystem for studying microbial response to global climate change [13–15]. Previous studies on Qinghai Lake have detected novel archaea commonly found in marine environments [16], and the microbial diversity, composition, and lipid profiles all showed a response to salinity change [14,17,18]. However, the following key questions remain unanswered: (1) what is the microbial taxonomic and functional diversity for the Qinghai Lake water column, (2) what is the metabolic potential and active metabolisms related to the carbon and nitrogen cycles, and (3) what stress response genes are present in organisms involved in the carbon and nitrogen cycles? Answering these questions can provide baseline knowledge that can be used to understand the effect of biodiversity on the microbial community response to environmental stress.

An integrated approach including geochemical, metagenomic and metatranscriptomic analyses were used to answer these questions. The metagenomic and metatranscriptomic reads were annotated with both taxonomic and functional information. The synthesized cDNA was compared to the DNA retrieved from the same sample to assess the relative activity of different populations and functional gene transcription in the microbial community. In addition, the relationship between species richness and the number of stress response genes was assessed. Microbial processes in Qinghai Lake are involved in both the carbon and nitrogen biogeochemical cycles. However, certain processes (e.g., photosynthesis, denitrification) had lower diversity and fewer stress response genes, which may make them more susceptible to environmental stresses.

Materials and Methods

Site description

Qinghai Lake is a perennial lake located on the Tibetan Plateau at an elevation of 3196 m above sea level. The lake is located in a structural intermontane depression at the north-eastern corner of the Qinghai-Tibetan Plateau (Figure 1) [19]. The lake has a surface area of 4300 km^2 and lies within a catchment of limestone, sandstone, and shale. The average water depth is 19.2 m with the maximum of 28.7 m. The evaporation of the lake (~1400 mm/year) is in excess of mean annual precipitation (~400 mm/year), resulting in a mesohaline lake. Qinghai Lake is separated into two subbasins by a normal faulting horst in the middle of the lake. The northern subbasin is more dynamic than the southern subbasin because of riverine input. The southern subbasin water is stratified in winter due to ice cover, but in summer, the water column geochemistry is fairly uniform [14,16]. No specific permits were required for the described field studies because no animal or human subjects were involved in this research. The sampling locations are not privately owned or protected in any way. The field studies did not involve endangered or protected species.

Sampling and geochemical analysis

Water samples were collected from the southern subbasin midday in August 2011 at sites B and E (Figure 1). The water samples were pumped from water depths of 12.5 m and 13.6 m below the lake surface from sites B and E, respectively. These two water depths were selected as representatives of the microbial community in the water column. Ten to 12 L of lake water was filtered through a polyethersulfone filter with a pore size of 0.2 μm (Supor; Pall Life Sciences, Ann Arbor, MI, USA). A Horiba multiparameter meter (W-20XD Series, HORIBA, Kyoto, Japan) was used to measure *in-situ* environmental parameters of the water including; temperature, pH, conductivity, dissolved oxygen (DO), depth, chloride (Cl^-), and salinity at both sites. Concentrations of major ions (i.e., sulfide, sulfate, silica, nitrite, nitrate, ferrous iron, and ammonia) in the filtered water of both sites were measured using a HACH colorimeter (model CEL 850, HACH Chemical Co., Iowa, USA) as previously described [20]. Filters collected for microbial analysis were placed on dry ice immediately after filtering of lake water from each site, and stored at −80°C in the laboratory until DNA and RNA extractions were conducted. A portable global positioning system (GPS) unit (eTrex H, Garmin) was used to indicate the location of each sampling site (Figure 1).

RNA and DNA extractions

Total RNA was extracted from one half of the filter from each site following a modified version of the RNeasy kit (Qiagen, Valencia, CA, USA) as previously described [21]. Briefly, half of each frozen filter was thawed and vortexed for 10 min with 8 ml of RLT lysis buffer and 3 g of low protein binding zirconium beads (200 μm, OPS Diagnostics, Lebanon, NJ, USA). RNA was then extracted using the RNeasy kit according to manufacturer's instructions. A Turbo DNA-free kit (Ambion, Austin, TX, USA) was used to remove any residual DNA. The resulting RNA was purified and concentrated using the RNeasyMinElute Cleanup kit (Qiagen, Valencia, CA, USA) [22]. Genomic DNA was extracted from the other half of the filter using the FastDNA Spin Kit (MP Biomedical, OH, USA) as previously described [20]. The amount of DNA that was extracted from sites B and E was 14.6 and 13.8 ng μl^{-1}, indicating that DNA was extracted from roughly the same amount of biomass.

Metatranscriptomic sample preparation

Ribosomal RNA (rRNA) was removed from total RNA via a subtractive hybridization process using sample-specific biotinylated rRNA probes [23]. rRNA-subtracted RNA was amplified linearly using the MessageAmp II-Bacteria kit (Ambion, Austin, TX, USA). Amplified RNA was converted to double-stranded cDNA using the Universal RiboClone cDNA synthesis system (Promega, Madison, WI, USA) using random hexamer primers [23]. The synthesized cDNA was purified with the QIAquick PCR purification kit (Qiagen, Valencia, CA, USA).

Library preparation and sequencing

The extracted DNA and synthesized cDNA was sheared to 500 base pairs (bp) using a Covaris ultrasonicator (Covaris Inc., Woburn, MA, USA) according to manufacturer's recommendations. The sheared DNA was end-repaired, adaptor-ligated with multiplexing, and purified using the Ovation SP ultralow DR multiplex system (NuGEN Technologies Inc., CA, USA). The prepared library was then sequenced using an Illumina MiSeq (250 bp paired-end reads).

Data analysis

Sequences were paired, trimmed, and filtered using the CLC Genomics Workbench version 6 (CLC Bio, Aarhus, Denmark). Paired reads were assembled together if 25 bp overlapped. Reads were trimmed based on the length (minimum length 50 bp) and quality (quality score ≥20) [24]. Sequences were uploaded to the metagenomics RAST (MG-RAST) server [25] for annotation and are available under MG-RAST ID 4532866.3, 4532865.3, 4522126.3, and 4522125.3 for the samples B_DNA, B_RNA, E_DNA, and E_RNA, respectively.

Sequences for DNA and RNA libraries were assigned a taxonomy and function using BLASTX [26] on MG-RAST

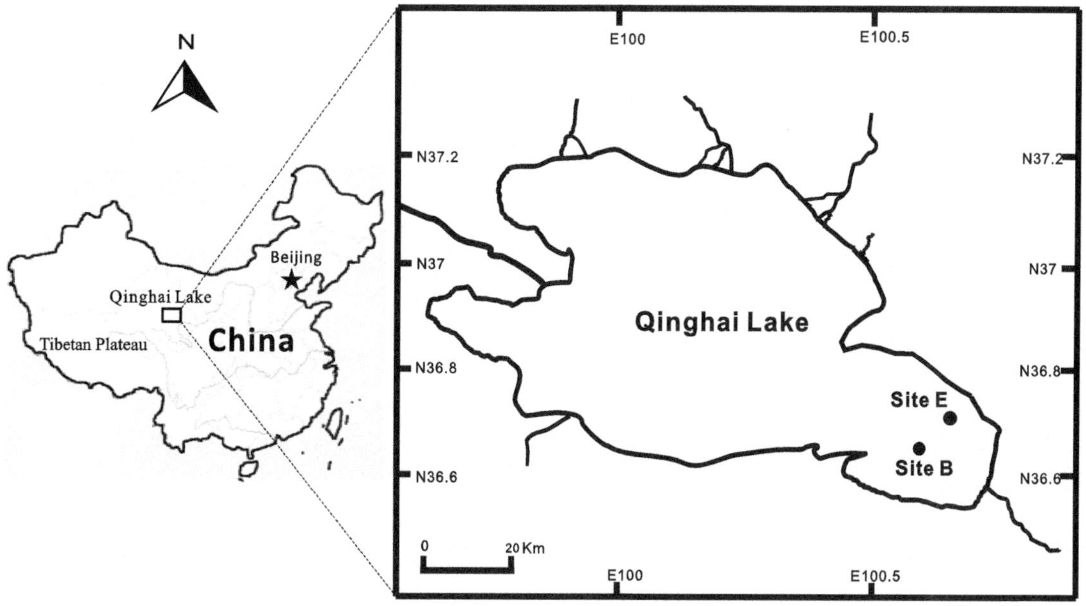

Figure 1. A geographical map showing the sampling sites in Qinghai Lake, China.

v3.3.6 against M5NR database, which integrates multiple databases (e.g., NCBI-nr, KEGG, SEED, and *etc.*). Sequences were given an annotation if it had at least a bit score cut-off of 50, E-values of 1×10^{-5}, and a minimum alignment of 15 amino acids [24]. DNA and RNA reads that were annotated with taxonomy and function were downloaded from MG-RAST and a custom R script was used to search for functional annotations involved in the carbon and nitrogen cycles, and stress responses [27]. KEGG categories were used to determine annotations that were involved in stress response. The richness of each functional gene was determined by counting the number of species or phyla that contained that particular functional gene.

Rarefaction curves were created for each of the samples to determine whether the sequencing depth was sufficient to detect the majority of species containing a functional gene. This was performed using the "rarecurve" function in the R package "Vegan" [28]. A network of correlated genes (i.e. functional genes found [DNA] or expressed [RNA] in the same organisms as stress response genes) was created using the WGCNA package [29] and the igraph package [30] in R [27], and viewed in Cytoscape [31]. Species with only one functional gene and functional genes in only a single organism were removed to reduce the complexity of the network [32]. A Spearman correlation between each functional gene and a stress response gene was calculated with a p-value adjusted for multiple comparisons. Correlations with a p-value greater than 0.05 were removed from the network.

amoA amplification

Jiang et al. [19] detected abundant ammonia oxidizing bacteria (AOB) and ammonia oxidizing archaea (AOA) in Qinghai Lake; however, the ammonia monooxygenase (*amoA)* gene was not detected in our metagenomic or metatranscriptomic datasets (see results below). Therefore, the *amoA* gene was amplified from the same DNA pool used for metagenomics. PCR amplifications were performed using FailSafe PCR System (Epicentre Biotechnologies, Madison, WI), AOB specific primer set: amoA-1F (5′-GGGGTTTCTACTGGTGGT-3′) and amoA-2R (5′-CCCCTCKGSAAAGCCTTCTTC −3′), and with the same conditions as described previously [33]. The amplicons were stained with EtBr and visualized on a 1% agarose gel.

Results and Discussion

Water geochemistry

Sites B and E shared similar geochemical profiles for most parameters (Table 1). For example, both sites had 1.4% salinity and a pH of ~9.1. The DO content was 8.9 and 8.7 ppm for sites B and E, respectively. Site E was about 1 m deeper and 2°C warmer and had a higher ammonia concentration (1 mg/L) than site B (below detection).

Descriptive statistics of DNA and RNA

After quality control of sequence reads, a total of 10,514,407 and 4,035,731 reads were retrieved in the DNA libraries for sites B and E, respectively (Table 2). The RNA libraries had a total of 1,728,111 and 2,893,577 reads for sites B and E, respectively (Table 2). Twenty-eight percent to 81% of the DNA and RNA reads that were identified to have an open reading frame were annotated with both function and organism (Table 2). Protein and rRNA features were predicted and identified for RNA and DNA libraries from both sample sites (Table 2).

Taxonomic diversity

Protein-coding genes, while not as phylogenetically robust as 16S rRNA genes, can be used to identify approximate taxonomic affiliations [34,35]. A total of 832 genera within 31 phyla were detected. The DNA samples from both sites contained more genera than were detected in the RNA samples, indicating some genera had no or low activities (Table 2). Bacterial sequences dominated both locations, with each sample containing fewer than 1% archaeal sequences. Eukaryotic sequences were removed for further analysis because of the unreliability of FragGeneScan (prokaryotic gene calling algorithm used by MG-RAST) to identify eukaryotic open reading frames [36] and our hypotheses were specific to prokaryotes.

Table 1. Sample locations and geochemistry.

Parameters	B	E
GPS location (N, E)	36.66, 100.60	36.74, 100.69
Temperature (°C)	13.7	15.6
pH	9.1	9.2
Depth (m)	12.5	13.6
Conductivity (s/m)	2.28	2.26
DO (ppm)	8.9	8.7
Salinity (%)	1.4	1.4
Cl^- (mg/L)	2930	3420
NH_4^+ (mg/L)	ND	1
NO_3^- (mg/L)	0.39	0.4
NO_2^- (mg/L)	0.13	NA
Si (mg/L)	1.43	NA
S^{2-} (mg/L)	0.05	ND
SO_4^- (mg/L)	>80	>80
Fe^{2+} (mg/L)	0.065	NA

ND: not detected. NA: not available.

At the phylum level, *Cyanobacteria* dominated the B_DNA and E_DNA samples with 52% and 63% of the total abundance, respectively (Figure 2). *Synechococcus* was the dominant genus, which has been previously detected in clone libraries from Qinghai Lake [14,15,37] and is known to be important contributors to carbon fixation in freshwater and marine ecosystems [38,39]. The second most abundant phylum in the DNA samples was *Proteobacteria,* with 28 and 24% of total abundance in B_DNA and E_DNA, respectively (Figure 2). Within the *Proteobacteria*, *Loktanella* was the dominant genus. It is an aerobic heterotroph that can supplement its energy requirements through aerobic anoxygenic photosynthesis [40,41]. Other major phyla that were detected in the DNA samples included *Actinobacteria, Bacteroidetes, Planctomycetes*, and *Verrucomicrobia* (Figure 2).

In the RNA samples, *Proteobacteria* were the most abundant (B_RNA: 47%, E_RNA: 70%) and *Cyanobacteria* were the second most abundant (B_RNA: 35%, E_RNA: 8%) (Figure 2). Other major phyla detected in the RNA samples were *Firmicutes*, *Actinobacteria*, *Bacteroidetes*, and *Verrucomicrobia* (Figure 2). Differences in RNA degradation rates and possible mis-annotations make it difficult to access true activity levels, but the relative activity levels can be defined as the ratio of RNA to DNA [42]. With this assumption, *Proteobacteria* were 1.6 and 2.8 fold more active than *Cyanobacteria* in samples B_RNA and E_RNA, respectively.

Metabolic Pathways in the Carbon Cycle

Functional identification of protein-coding genes identified autotrophic and heterotrophic carbon metabolisms (Figure 3). At both sites carbon fixation was performed by ribulose bis-phosphate carboxylase (RuBisCo), which is indicative of the Calvin-Benson-Bassham (CBB) cycle [43]. However, activity levels of RuBisCo (RNA:DNA 0.46 [B] and 0.2 [E]) were low compared to heterotrophic processes (see below).

Table 2. Statistic results of metagenome and metatranscriptome sequences from sites B and E in Qinghai Lake.

	B		E	
	RNA	DNA	RNA	DNA
Number of reads	1,728,111	10,514,407	2,893,577	4,035,731
Mean sequence length (bp)	211±60	194±55	200±59	185±59
Total Mbp	365	2,049	581	749
Mean GC content (%)	49±8	60±8	44±9	63±5
Reads with ORF[a]	1,463,714	10,161,795	2,098,251	3,901,755
Identified protein features	288,106	3,233,826	304,159	945,021
Identified rRNA features	413,195	10,573	113,464	2,307
% annotated reads[b]	65.0	81.2	28.3	77.3

[a]Open reading frame.
[b]% of reads with an open reading frame that were annotated by function and taxonomy.
base pairs (bp).

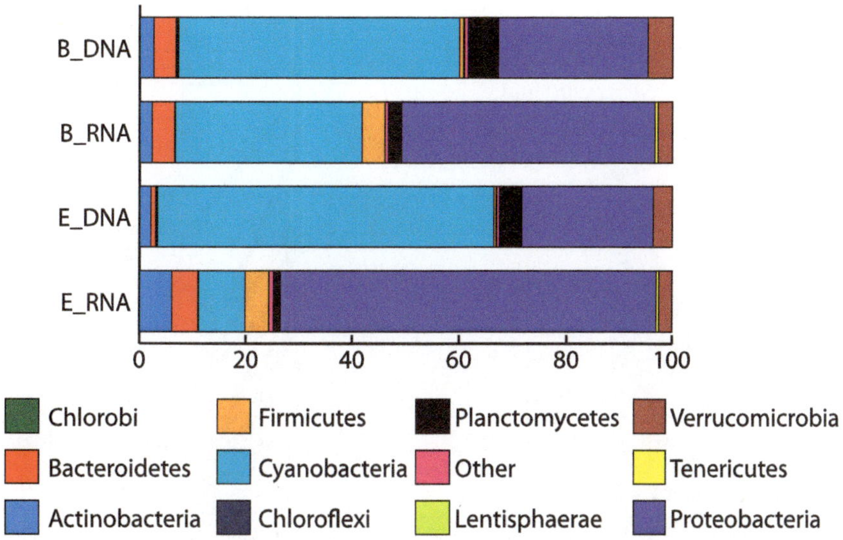

Figure 2. Distribution of phyla detected in the DNA or RNA samples from sites B and E determined by taxonomic assignment of metagenomic or metatranscriptomic reads. Phyla with <1% abundance were grouped into "other".

Photosystems I and II were detected, indicating the reliance of carbon fixation on light energy (Figure 3). Photosystem II (RNA:DNA 1.86 [B] and 0.65 [E]) was more highly expressed than photosystem I (RNA:DNA 0.52 [B] and 0.07 [E]) (Figure 3). The higher expression of photosystem II than photosystem I indicate that the *Cyanobacteria* are under UV stress [44], which is expected in Qinghai Lake because of its high UV irradiance due to the high elevation. In addition, network analysis revealed that organisms expressing photosystems I and II were also expressing orange carotenoid protein (Figure 4), which plays an important role in protecting photosynthetic organisms (such as *Cyanobacteria*) from solar radiation [45,46]. These data suggested that the photoautotrophs in Qinghai Lake had low activities and were under UV stress.

Heterotrophic remineralization of organic carbon proceeded through the glycolysis and tricarboxylic acid cycle (TCA). The key enzymes for glycolysis (phosphofructokinase- RNA:DNA 0.67 [B] and 0.76 [E]) and the TCA cycle (pyruvate dehydrogenase- RNA:DNA 0.62 [B] and 0.66 [E]) had slightly higher RNA:DNA ratios than RuBisCo. Oxidative phosphorylation, identified by cytochrome C oxidase, was the most abundant and active (RNA:DNA 1.53 [B] and 1.99 [E]) energetic pathway. Aerobic anoxygenic phototrophic *puf* operon was also detected indicating that some energy generation came from photoheterotrophy

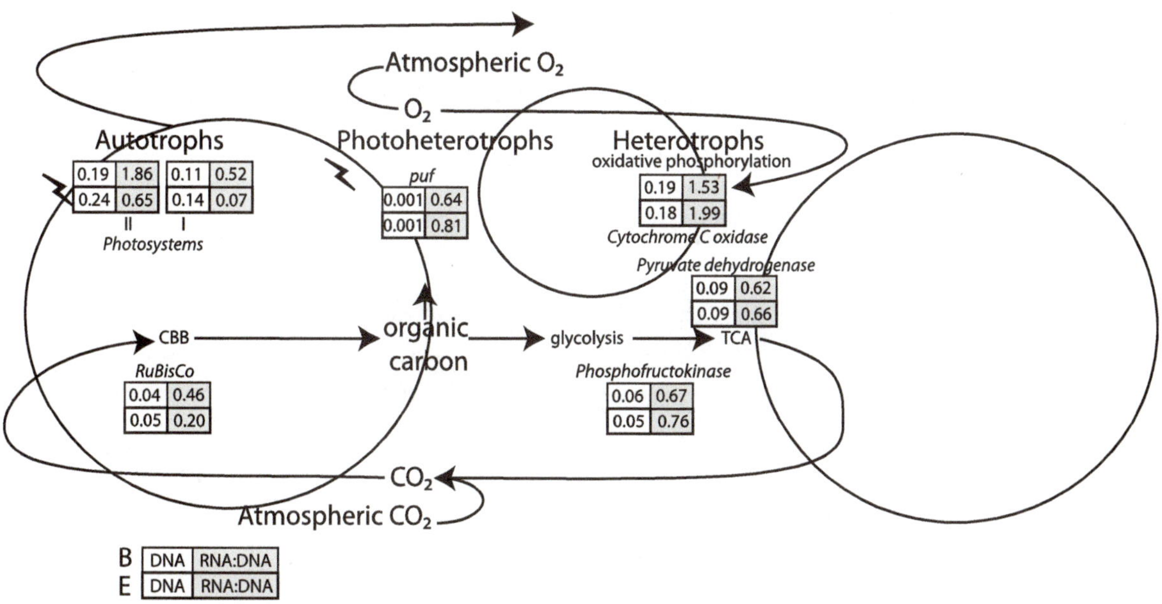

Figure 3. The carbon cycle depicted by a generalized autotroph, photohetertroph, and heterotroph in Qinghai Lake. The numbers in boxes represent the percentage and the RNA:DNA ratio of reads that were annotated within each metabolic pathway for sites B and E. The key genes used to identify a pathway was Ribulose-bisphosphate carboxylase (RuBisCo): Calvin-Benson-Bassham cycle (CBB), D-glucose 6-phosphotransferase: glycolysis, pyruvate dehydrogenase: tricarboxylic acid cycle (TCA), and cytochrome C oxidase: oxidative phosphorylation.

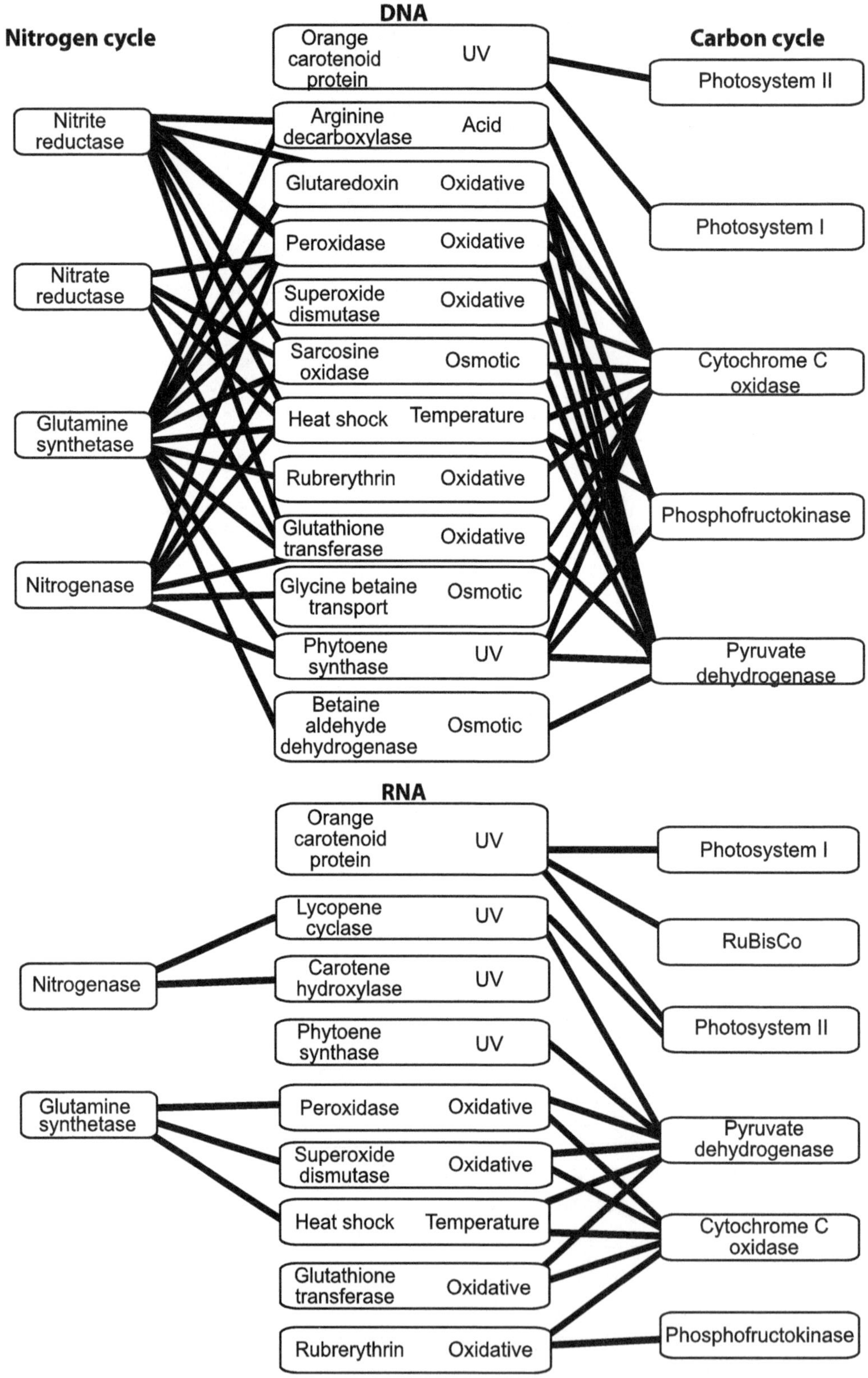

Figure 4. Network showing correlations of carbon and nitrogen cycle-related genes with stress response genes for DNA and RNA samples. Lines represent genes that are directly correlated (i.e. detected (DNA) or expressed (RNA) in the same species). Only significant correlations ($p<0.05$) are shown. The type of stress that each gene is involved in is also listed.

(RNA:DNA 0.64 [B] and 0.81 [E]). The high concentration of DO (8.7–8.9 ppm) suggests aerobic respiration was the most prominent. Network analysis showed that these heterotrophs were also expressing genes involved in osmotic and oxidative stress

(Figure 4), again consistent with the saline and oxic nature of the water in the lake.

Metabolic Pathways in the Nitrogen Cycle

Nitrogen is biologically available in the form of ammonia. Ammonia assimilation into organic molecules occurred primarily via the glutamine synthetase pathway (Figure 5). The high functional activity of the ammonia assimilatory pathway observed probably led to the low concentration of ammonia measured at both sites (≤1 mg/L). Active nitrogen fixation genes were also detected, which is possibly driven by low levels of ammonia in the lake (Figure 5). Denitrification pathways (i.e., nitric oxide reductase and nitrous oxide reductase) had very low abundance but were highly transcriptionally active (Figure 5). Denitrification requires anoxic conditions, so it was unexpected to detect denitrification transcripts in the oxic water column. It is possible that localized anoxic conditions can exist because of oxygen microgradients, shown in marine snow [48] and microbial mats [49]. These small patches of anoxic water could harbor a relatively low abundance of active anaerobic denitrifiers.

Biological ammonia oxidation is the first step in the nitrification process ($NH_3 \rightarrow NO_2^-$) and is mainly carried out by AOB and AOA that contain the *amoA* gene [50]. Putative AOB and AOA were detected at both sites. AOB genera that were detected included *Nitrosospira*, *Nitrosomonas* and *Nitrosococcus* and AOA genera that were detected included *Nitrosocaldus* and *Nitrosopumilus*. The AOB were more abundant than AOA in all samples; however, AOB and AOA comprised less than 0.1% and 0.01% of total reads, respectively. Although putative AOB- and AOA-like sequences were identified, the *amoA* gene was not detected in our metagenomic or metatranscriptomic samples. However, the *amoA* gene was detected using primers specific for *amoA*. These results indicate that, while present, the abundance of the *amoA* gene is low at the time of sampling.

In contrast, Jiang et al. [19] found that the microbial community in Qinghai lake was dominated by AOA and AOB. This discrepancy is likely because the Jiang et al., samples were collected in early summer (July 2005 and 2007). The lake is frozen during the winter and the ice cover causes stratification of the water column, resulting in gradients of salinity, light availability, and oxygen. In spring, inflow glacial meltwater and thawing permafrost reduces the lake salinity and increases the flow of organic matter and nutrients into the lake [10,51,52]. The early summer samples had a DO concentration of 5.4 ppm at the surface and 7 ppm at the bottom, salinity was 1.25%, and the ammonia concentration was 5.6 mM. High ammonia concentrations in early spring water are conducive for AOA and AOB. Late summer samples described here (late August 2011) had higher DO (8.9 ppm), higher salinity (1.4%), and little measureable ammonia. This suggests that, similar to many inland lakes and oceans [53–56], seasonal changes in the physicochemical conditions and microbiology occurs in Qinghai Lake.

Comparison of sites B and E

An additional 100 genera were detected in the B_DNA sample compared to the E_DNA sample. Both sites had similar rarefaction curve asymptotes (Figure S1), suggesting that the difference in genera is likely because more reads were obtained for the B_DNA sample (~6 million more reads) (Table 2). This also influenced the number of ORFs and protein features detected between the two sites (Table 2). Sites B and E were very similar in the taxonomy and functional genes that were detected; however, slight differences were observed in some RNA:DNA ratios. For example, differences were observed in RNA:DNA ratios of nitric oxide reductase and nitrous oxide reductase. It is difficult to conclude if this is biological or a sampling artifact. Small changes in the RNA reads annotated to these genes could translate to large RNA:DNA ratio changes. For example, nitrous oxide reductase at site E had nine DNA reads annotated and would only need four

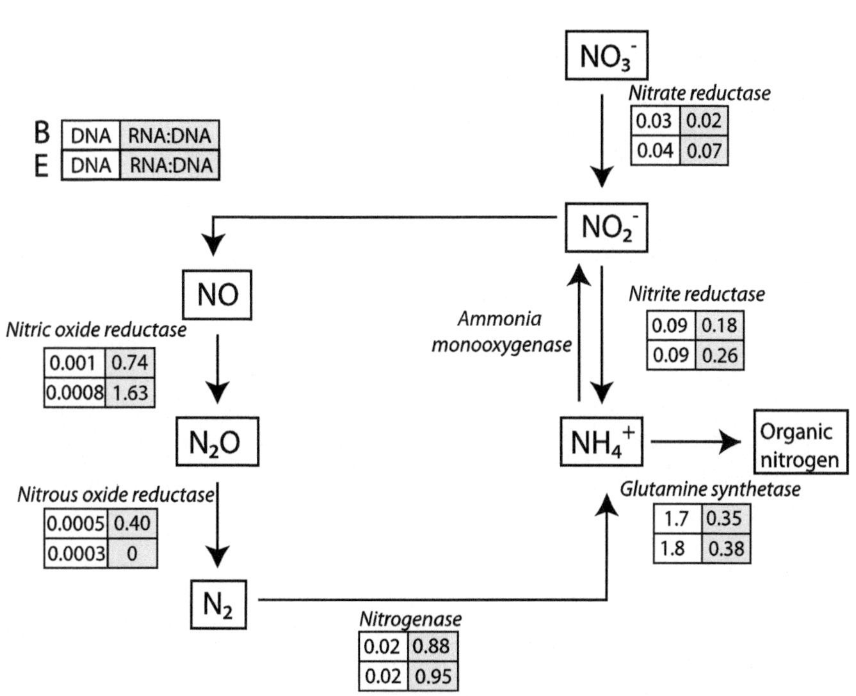

Figure 5. DNA and RNA reads detected and annotated within the nitrogen cycle. The numbers in boxes represent either the percent or the RNA:DNA ratio of reads that were annotated as genes within each pathway for sites B and E.

RNA reads to obtain a RNA:DNA ratio similar to site B. Furthermore, site E had less photosynthetic activity than site B. This is more likely a biological phenomenon because of the high abundance of reads annotated to photosystem II and I. While both sites had similar physicochemical conditions (Table 1), a possible reason for this difference is that site E was 1 m deeper than site B and as a consequence receives less photosynthetically active radiation (PAR) [7] because of the attenuation of PAR in water [47].

Biodiversity

The ability to assign both taxonomy and function to a sequence allowed for identification of the number of organisms that contained a particular functional gene. The richness of each functional gene was calculated as the number of taxa in which the gene was identified. The diversity was sufficiently sampled to detect the majority of organisms that contained a functional gene involved in the carbon and nitrogen cycle (Figure 6). The most diverse functional genes were genes that were part of the heterotrophic pathway: cytochrome C oxidase, glutamine synthetase, phosphofructokinase, and pyruvate dehydrogenase; while photosystem I, photosystem II, and genes involved in the denitrification process had low diversity (Figure 6).

Generally, there is a positive relationship between ecosystem function and taxonomic diversity, because a more diverse community increases the likelihood of finding high-performance key species or the probability of functional redundancy [57,58]. For example, in soils with a high taxonomic diversity, denitrification rates were higher and decreased more slowly with increased salinity [6]. Similar results were found for respiration [59–61], photosynthesis [62–64], and nitrogen fixation [65]. This suggests that, as a consequence of the greater diversity, the heterotrophic pathway in Qinghai Lake would be more resilient to stress.

While the greater diversity increases the probability of stress resilience, a mechanism to respond to stress is needed to continue functioning. Stress response genes are one mechanism that can be detected with metagenomics and metatranscriptomics. Genes were found that respond to UV, osmotic, and oxidative stress (Figure 4). The detection of these stress genes in Qinghai Lake was expected because of the intense UV radiation and oxidative stress (related to the high elevation of the lake) and its salinity. In addition, samples were retrieved mid-day when UV radiation was at its highest. Furthermore, the number of stress gene types increased logarithmically with the species richness for each functional gene involved in the carbon and nitrogen cycles (Figure 7). For example, in the 365 species that had a cytochrome c oxidase, there were a total of 20 different stress response genes. This observation provides a possible mechanism as to why higher biodiversity is more resilient to stress because one of the species will be able to respond to a stress.

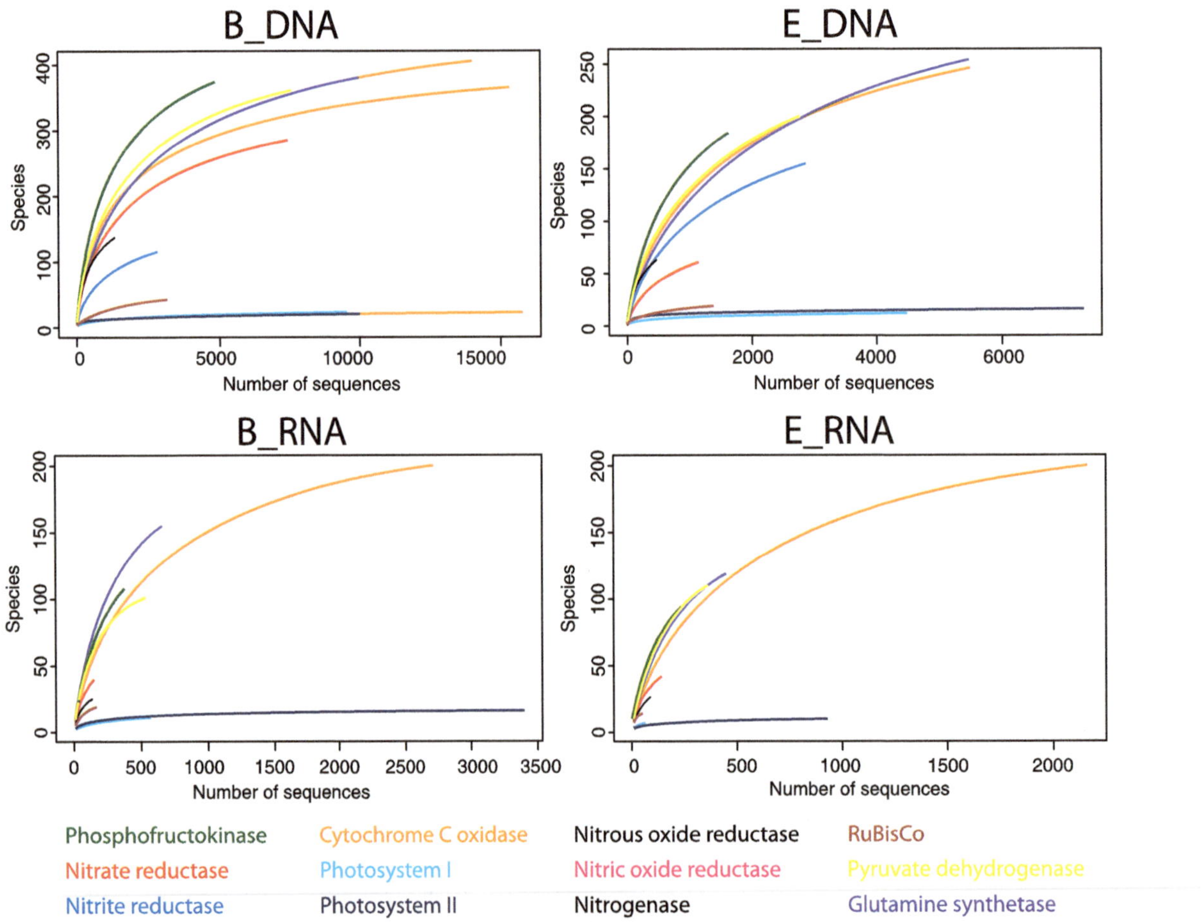

Figure 6. Rarefaction curves for the detected genes related to carbon and nitrogen cycle. Highly saturated curves (e.g. photosystem I and II) indicate that sequencing depth was sufficient to capture most species containing the corresponding gene.

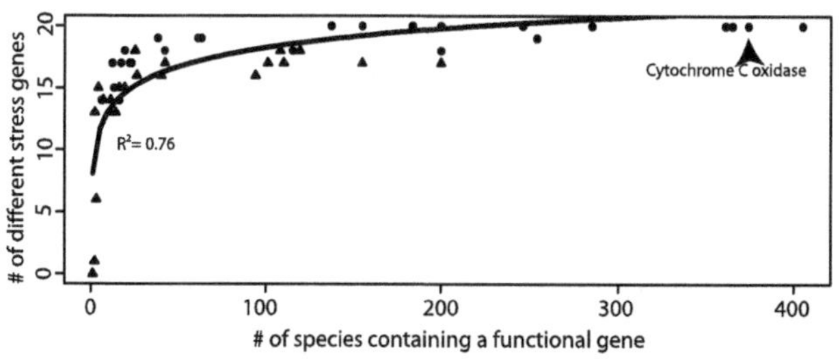

Figure 7. Plot showing the number of species that contained a particular functional gene in the carbon and nitrogen cycle and the number of stress genes that were found in all species with that functional gene (Cytochrome C oxidase is labeled as an example). Triangle symbols represent RNA samples and circle symbols represent DNA samples. Best-fit curve followed a logarithmic function.

Conclusions

A metagenomic and metatranscriptomic survey of two locations in Qinghai Lake identified microbial processes involved in the carbon and nitrogen cycle. While photoautotrophic organisms were the most abundant in the DNA samples, heterotrophic organisms were the most active at both sites. Energy generation was also supplemented by photoheterotrophy. Coupled with previous reports from early summer sampling, our data suggest that successional changes occur in the microbiota throughout the summer. Our data also shows a positive relationship between the number of stress gene types and the species richness for each functional gene in the carbon and nitrogen cycles, suggesting that microbial processes with higher taxonomic diversity would have an increased ability to respond to a variety of stresses. Heterotrophic respiration and ammonia assimilation had the highest richness and would be expected to be more resistant to environmental change.

Acknowledgments

We thank Dr. James Hollibaugh for providing technical support and reviewing of the manuscript. We also thank Dr. Tammie Gerke and Dr. Jennifer Biddle for reviewing the manuscript. This work was fostered by NSF IOS grant 1238801. We are grateful to two anonymous reviewers whose comments improved the quality of the manuscript.

Author Contributions

Conceived and designed the experiments: BRB HD HJ. Performed the experiments: QH BRB CE IV GW. Analyzed the data: QH BRB. Contributed reagents/materials/analysis tools: CE IV SQ. Contributed to the writing of the manuscript: QH BRB HD HJ GW CE IV SQ.

References

1. Canfield DE, Glazer AN, Falkowski PG (2010) The evolution and future of Earth's nitrogen cycle. Science 330: 192–196.
2. Bardgett RD, Freeman C, Ostle NJ (2008) Microbial contributions to climate change through carbon cycle feedbacks. ISME J 2: 805–814.
3. Singh BK, Bardgett RD, Smith P, Reay DS (2010) Microorganisms and climate change: terrestrial feedbacks and mitigation options. Nat Rev Microbiol 8: 779–790.
4. Heimann M, Reichstein M (2008) Terrestrial ecosystem carbon dynamics and climate feedbacks. Nature 451: 289–292.
5. Griffiths BS, Philippot L (2013) Insights into the resistance and resilience of the soil microbial community. FEMS Microbiol Rev 37: 112–129.
6. Hallin S, Welsh A, Stenstrom J, Hallet S, Enwall K, et al. (2012) Soil functional operating range linked to microbial biodiversity and community composition using denitrifiers as model guild. PLoS One 7: e51962.
7. Mella-Flores D, Six C, Ratin M, Partensky F, Boutte C, et al. (2012) *Prochlorococcus* and *Synechococcus* have evolved different adaptive mechanisms to cope with light and UV stress. Front Microbiol 3: 285.
8. Liu X, Chen B (2000) Climatic warming in the Tibetan Plateau during recent decades. Int J Climatol 20: 1729–1742.
9. Guo D, Wang H, Li D (2012) A projection of permafrost degradation on the Tibetan Plateau during the 21st century. J Geophys Res Atmos 117: D05106.
10. Qiu J (2008) China: the third pole. Nature 468: 141–142.
11. Lei Y, Yao T, Yi C, Wang W, Sheng Y, et al. (2012) Glacier mass loss induced the rapid growth of Linggo Co on the central Tibetan Plateau. J Glaciology 58: 177–184.
12. Yang SJ, Dong H, Lei FM (2009) Phylogeography of regional fauna on the Tibetan Plateau: A review. Progr Natural Sci 19: 789–799.
13. Dong H, Jiang H, Yu B, Liu X (2010) Impacts of environmental change and human activity on microbial ecosystems on the Tibetan Plateau, NW China. GSA Today 20: 4–10.
14. Dong H, Zhang G, Jiang H, Yu B, Chapman LR, et al. (2006) Microbial diversity in sediments of saline Qinghai Lake, China: linking geochemical controls to microbial ecology. Microb Ecol 51: 65–82.
15. Xing P, Hahn MW, Wu QL (2009) Low taxon richness of bacterioplankton in high-altitude lakes of the eastern tibetan plateau, with a predominance of *Bacteroidetes* and *Synechococcus spp*. Appl Environ Microbiol 75: 7017–7025.
16. Jiang H, Dong H, Yu B, Ye Q, Shen J, et al. (2008) Dominance of putative marine benthic Archaea in Qinghai Lake, north-western China. Environ Microbiol 10: 2355–2367.
17. Jiang H, Dong H, Deng S, Yu B, Huang Q, et al. (2009) Response of archaeal community structure to environmental changes in lakes on the Tibetan Plateau, Northwestern China. Geomicrobiol J 26: 289–297.
18. Wang HY, Liu WG, Zhang CLL, Jiang HC, Dong HL, et al. (2013) Assessing the ratio of archaeol to caldarchaeol as a salinity proxy in highland lakes on the northeastern Qinghai-Tibetan Plateau. Org Geochem 54: 69–77.
19. Jiang H, Dong H, Yu B, Lv G, Deng S, et al. (2009) Diversity and abundance of ammonia-oxidizing archaea and bacteria in Qinghai Lake, Northwestern China. Geomicrobiol J 26: 199–211.
20. Huang Q, Jiang H, Briggs BR, Wang S, Hou W, et al. (2013) Archaeal and bacterial diversity in acidic to circumneutral hot springs in the Philippines. FEMS Microbiol Ecol 85: 452–464.
21. Poretsky RS, Hewson I, Sun S, Allen AE, Zehr JP, et al. (2009) Comparative day/night metatranscriptomic analysis of microbial communities in the North Pacific subtropical gyre. Environ Microbiol 11: 1358–1375.
22. Hollibaugh JT, Gifford S, Sharma S, Bano N, Moran MA (2011) Metatranscriptomic analysis of ammonia-oxidizing organisms in an estuarine bacterioplankton assemblage. ISME J 5: 866–878.
23. Stewart FJ, Ottesen EA, DeLong EF (2010) Development and quantitative analyses of a universal rRNA-subtraction protocol for microbial metatranscriptomics. ISME J 4: 896–907.
24. Jiménez DJ, Andreote FD, Chaves D, Montaña JS, Osorio-Forero C, et al. (2012) Structural and functional insights from the metagenome of an acidic hot spring microbial planktonic community in the Colombian Andes. PLoS One 7: e52069.
25. Meyer F, Paarmann D, D'Souza M, Olson R, Glass E, et al. (2008) The metagenomics RAST server - a public resource for the automatic phylogenetic and functional analysis of metagenomes. BMC Bioinformatics 9: 386.
26. Altschul SF, Madden TL, Schäffer AA, Zhang J, Zhang Z, et al. (1997) Gapped BLAST and PSI-BLAST: a new generation of protein database search programs. Nucleic Acids Res 25: 3389–3402.
27. R Development Core Team (2012) R: A language and environment for statistical computing. Vienna: R Found. for Stat. Comput.
28. Oksanen J, Blanchet FG, Kindt R, Legendre P, Minchin PR, et al. (2011) vegan: Community ecology package. http://cran.r-project.org/, http://vegan.r-forge.r-project.org/.
29. Langfelder P, Horvath S (2008) WGCNA: an R package for weighted correlation network analysis. BMC Bioinformatics 9: 559.
30. Csardi G, Nepusz T (2006) The igraph software package for complex network research. InterJournal, Complex Systems 1695.
31. Shannon P, Markiel A, Ozier O, Baliga NS, Wang JT, et al. (2003) Cytoscape: a software environment for integrated models of biomolecular interaction networks. Genome Res 13: 2498–2504.
32. Kara EL, Hanson PC, Hu YH, Winslow L, McMahon KD (2013) A decade of seasonal dynamics and co-occurrences within freshwater bacterioplankton communities from eutrophic Lake Mendota, WI, USA. ISME J 7: 680–684.

33. Rotthauwe J-H, Witzel K-P, Liesack W (1997) The ammonia monooxygenase structural gene *amoA* as a functional marker: molecular fine-scale analysis of natural ammonia-oxidizing populations. Appl Environ Microbiol 63: 4704–4712.
34. Stewart FJ, Ulloa O, DeLong EF (2012) Microbial metatranscriptomics in a permanent marine oxygen minimum zone. Environ Microbiol 14: 23–40.
35. Urich T, Lanzen A, Qi J, Huson DH, Schleper C, et al. (2008) Simultaneous assessment of soil microbial community structure and function through analysis of the meta-transcriptome. PLoS One 3: e2527.
36. Rho M, Tang H, Ye Y (2010) FragGeneScan: predicting genes in short and error-prone reads. Nucleic Acids Res 38: e191.
37. Jiang H, Dong H, Zhang G, Yu B, Chapman LR, et al. (2006) Microbial diversity in water and sediment of Lake Chaka, an athalassohaline lake in northwestern China. Appl Environ Microbiol 72: 3832–3845.
38. Joint IR (1986) Physiological ecology of picoplankton in various oceanographic provinces. Photosynthetic picoplankton, Can Bull Fish Aquat Sci 214: 287–309.
39. Fahnenstiel GL, Carrick HJ, Iturriaga R (1991) Physiological characteristics and food-web dynamics of *Synechococcus* in Lakes Huron and Michigan. Limnol 36: 219–234.
40. Newton RJ, Griffin LE, Bowles KM, Meile C, Gifford S, et al. (2010) Genome characteristics of a generalist marine bacterial lineage. ISME J 4: 784–798.
41. Jiang H, Dong H, Yu B, Lv G, Deng S, et al. (2009) Abundance and diversity of aerobic anoxygenic phototrophic bacteria in saline lakes on the Tibetan plateau. FEMS Microbiol Ecol 67: 268–278.
42. Yu K, Zhang T (2012) Metagenomic and metatranscriptomic analysis of microbial community structure and gene expression of activated sludge. PLoS One 7: e38183.
43. Berg IA (2011) Ecological aspects of the distribution of different autotrophic CO_2 fixation pathways. Appl Environ Microbiol 77: 1925–1936.
44. Campbell D, Eriksson M-J, Öquist G, Gustafsson P, Clarke AK (1998) The cyanobacterium *Synechococcus* resists UV-B by exchanging photosystem II reaction-center D1 proteins. Proc Natl Acad Sci USA 95: 364–369.
45. Armstrong GA, Alberti M, Leach F, Hearst JE (1989) Nucleotide sequence, organization, and nature of the protein products of the carotenoid biosynthesis gene cluster of *Rhodobacter capsulatus*. Mol Gen Genet 216: 254–268.
46. Moeller RE, Gilroy S, Williamson CE, Grad G, Sommaruga R (2005) Dietary acquisition of photoprotective compounds (mycosporine-like amino acids, carotenoids) and acclimation to ultraviolet radiation in a freshwater copepod. Limnol Oceanogr 50: 427–439.
47. Miller CB (2004) Biological oceanography. Malden, MA.: Blackwell Publishing. 416 p.
48. Ploug H (2001) Small-scale oxygen fluxes and remineralization in sinking aggregates. Limnol Ocean 46: 1624–1631.
49. Cravo-Laureau C, Duran R (2014) Marine coastal sediments microbial hydrocarbon degradation processes: contribution of experimental ecology in the omics'era. Front Microbiol 5: 39.
50. Thamdrup B (2012) New pathways and processes in the global nitrogen cycle. Annu Rev Ecol, Evol, and Syst 43: 407–428.
51. Li X, Cheng GD, Jin HJ, Kang E, Che T, et al. (2008) Cryospheric change in China. Global Planet Change 62: 210–218.
52. Liu J, Wang S, Yu S, Yang D, Zhang L (2009) Climate warming and growth of high-elevation inland lakes on the Tibetan Plateau. Global Planet Change 67: 209–217.
53. Pernthaler J, Glöckner F-O, Unterholzner S, Alfreider A, Psenner R, et al. (1998) Seasonal community and population dynamics of pelagic bacteria and archaea in a high mountain lake. Appl Environ Microbiol 64: 4299–4306.
54. Yannarell AC, Kent AD, Lauster GH, Kratz TK, Triplett EW (2003) Temporal patterns in bacterial communities in three temperate lakes of different trophic status. Microb Ecol 46: 391–405.
55. Gilbert JA, Steele JA, Caporaso JG, Steinbrück L, Reeder J, et al. (2012) Defining seasonal marine microbial community dynamics. ISME J 6: 298–308.
56. Treusch AH, Vergin KL, Finlay LA, Donatz MG, Burton RM, et al. (2009) Seasonality and vertical structure of microbial communities in an ocean gyre. ISME J 3: 1148–1163.
57. Yachi S, Loreau M (1999) Biodiversity and ecosystem productivity in a fluctuating environment: the insurance hypothesis. Proc Natl Acad Sci U S A 96: 1463–1468.
58. Philippot L, Spor A, Henault C, Bru D, Bizouard F, et al. (2013) Loss in microbial diversity affects nitrogen cycling in soil. ISME J 7: 1609–1619.
59. Chowdhury N, Marschner P, Burns R (2011) Response of microbial activity and community structure to decreasing soil osmotic and matric potential. Plant Soil 344: 241–254.
60. Setia R, Marschner P, Baldock J, Chittleborough D, Smith P, et al. (2011) Salinity effects on carbon mineralization in soils of varying texture. Soil Biol Biochem 43: 1908–1916.
61. Langenheder S, Lindstrom ES, Tranvik LJ (2005) Weak coupling between community composition and functioning of aquatic bacteria. Limnol Ocean 50: 957–967.
62. Allakhverdiev SI, Murata N (2008) Salt stress inhibits photosystems II and I in cyanobacteria. Photosynth Res 98: 529–539.
63. Murata N, Takahashi S, Nishiyama Y, Allakhverdiev SI (2007) Photoinhibition of photosystem II under environmental stress. Biochim Biophys Acta 1767: 414–421.
64. Takahashi S, Murata N (2008) How do environmental stresses accelerate photoinhibition? Trends Plant Sci 13: 178–182.
65. Severin I, Confurius-Guns V, Stal LJ (2012) Effect of salinity on nitrogenase activity and composition of the active diazotrophic community in intertidal microbial mats. Arch Microbiol 194: 483–491.

7

Vegetation Controls on Weathering Intensity during the Last Deglacial Transition in Southeast Africa

Sarah J. Ivory[1]*, Michael M. McGlue[2], Geoffrey S. Ellis[3], Anne-Marie Lézine[4], Andrew S. Cohen[5], Annie Vincens[6]

1 Brown University, Providence, Rhode Island, United States of America, **2** University of Kentucky, Lexington, Kentucky, United States of America, **3** U.S. Geological Survey, Denver, Colorado, United States of America, **4** LOCEAN, CNRS, Paris, France, **5** University of Arizona, Tucson, Arizona, United States of America, **6** CEREGE, CNRS, Aix-en-Provence, France

Abstract

Tropical climate is rapidly changing, but the effects of these changes on the geosphere are unknown, despite a likelihood of climatically-induced changes on weathering and erosion. The lack of long, continuous paleo-records prevents an examination of terrestrial responses to climate change with sufficient detail to answer questions about how systems behaved in the past and may alter in the future. We use high-resolution records of pollen, clay mineralogy, and particle size from a drill core from Lake Malawi, southeast Africa, to examine atmosphere-biosphere-geosphere interactions during the last deglaciation (~18–9 ka), a period of dramatic temperature and hydrologic changes. The results demonstrate that climatic controls on Lake Malawi vegetation are critically important to weathering processes and erosion patterns during the deglaciation. At 18 ka, afromontane forests dominated but were progressively replaced by tropical seasonal forest, as summer rainfall increased. Despite indication of decreased rainfall, drought-intolerant forest persisted through the Younger Dryas (YD) resulting from a shorter dry season. Following the YD, an intensified summer monsoon and increased rainfall seasonality were coeval with forest decline and expansion of drought-tolerant miombo woodland. Clay minerals closely track the vegetation record, with high ratios of kaolinite to smectite (K/S) indicating heavy leaching when forest predominates, despite variable rainfall. In the early Holocene, when rainfall and temperature increased (effective moisture remained low), open woodlands expansion resulted in decreased K/S, suggesting a reduction in chemical weathering intensity. Terrigenous sediment mass accumulation rates also increased, suggesting critical linkages among open vegetation and erosion during intervals of enhanced summer rainfall. This study shows a strong, direct influence of vegetation composition on weathering intensity in the tropics. As climate change will likely impact this interplay between the biosphere and geosphere, tropical landscape change could lead to deleterious effects on soil and water quality in regions with little infrastructure for mitigation.

Editor: Olivier Boucher, Université Pierre et Marie Curie, France

Funding: Lake Malawi Drilling Project-Earth System History Program (NSF-EAR-0602404) funded field operations, logistics, and some laboratory analysis. NSF Graduate Research Fellowship (2009078688) provided student salary and tuition and some travel support for laboratory analysis. The funders had no role in study design, data collection and analysis, decision to publish, or preparation of the manuscript.

Competing Interests: The authors have declared that no competing interests exist.

* Email: sarah_ivory@brown.edu

Introduction

Tropical climate is rapidly changing resulting in altered atmospheric and oceanic circulation, as well as changing variability of important climatic modes like the El Niño–South Oscillation [1]. Much work is being done within the framework of the IPCC to better understand climate sensitivity and climate-induced changes to the biosphere; however, integrating the effects on and feedbacks from the geosphere has been largely unexplored [2]. Alterations to the geosphere, an important component of the Earth's critical zone, could have costly, hazardous implications for water and soil quality, as well as a host of ecosystems services provided by inland waters (e.g., serving as fisheries, rookeries, and zones of stormwater retention).

Weathering and climate have hypothetically been linked via feedback loops that regulate the Earth system [3]. However, the low resolution of most marine sedimentary records and the strong influence of long time-scale (10^6–10^7 yrs) processes such as orogeny do not provide a scalable framework for evaluating future changes in the critical zone. Additionally, although climate has long been thought to play a strong, direct role, a number of datasets indicate that the role of climate for weathering and erosion may in fact be indirect, mediated by vegetation and soil storage of organic acids [4–7].

Some studies have generated geochemical indicator records of weathering or produced denudation rates for tropical watersheds based on mass balance models; however, such studies are unable to determine how these rates vary over long time-scales [8–11]. Due to the strong interrelation of vegetation and climate, it has been particularly difficult to separate the individual influences of each mechanism. However, it is expected that vegetation plays an important role. Other studies that have examined the relationship between land use and sediment yield in tropical catchments have demonstrated that heavily forested areas generate less sediment than grasslands or areas modified for agriculture or ranching (e.g., [12]). Furthermore, human land-use change could have strong but unclear effects on weathering and the carbon cycle via alteration of vegetation with no climatic change [13]. To answer the question of how climate change influences weathering and erosion in the tropics, well-dated, high-resolution paleo-records from the critical zone are needed. However, records of tropical lowland vegetation and climate are exceedingly rare, despite the great need to better understand these regions and their growing populations, as articulated by the IPCC [1]. Existing Quaternary paleoenvironmental records from tropical Africa attest to the fact that both vegetation and climate have varied greatly over the last 20 ka [14]. However, it is as yet uncertain what the relative importance of their effects on the geosphere may have been.

Our study seeks to address this important knowledge gap using a scientific drill core from Lake Malawi, southeast Africa (Figure 1). The lake is situated at the current southern extent of the Intertropical Convergence Zone (ITCZ) and has been found to be climatically and ecologically sensitive, making it a remarkable natural laboratory to investigate the relationships and feedbacks among vegetation, climate, weathering, and erosion during the last deglaciation (18–9 ka). This period is of interest because of high-amplitude changes in both climate and vegetation that are both progressive and abrupt. This record thus has the potential to assess the influence of climate change on weathering in the tropics.

Modern Setting

Lake Malawi is the southernmost rift lake in the western branch of the East African Rift System (Figure 1). The physical geography of Lake Malawi and its watershed is controlled by Cenozoic extensional tectonics and volcanism [15]. The lake occupies a series of half-graben basins that are linked *en echelon*, such that steeply dipping border faults with opposing polarity are connected by accommodation zones [16]. Subsidence and deformation patterns within individual half-graben basins control lake bathymetry, with maximum depths (~700 m) achieved adjacent to border faults [17].

The drill core used in this study was collected from the northern basin of the lake (Figure 1). Onshore, distinct topography and drainage patterns follow major tectonic features. A border fault margin to the east of the lake is characterized by high-elevation escarpments which form the Livingstone Mountains. The topography of this faulted margin precludes all but short, steep river systems from forming [18]. In contrast, flexural margins and accommodation zones exhibit comparatively lower altitudes and relief, and large river systems are common in these settings [19]. North of the lake, a well-developed axial delta system building along low-gradient plains is sourced by the Songwe-Kiwira River, which drains the Rungwe highlands. Because of the sloping topography and presence of the delta, Johnson and McCave [20] considered the Songwe-Kiriwa River to be the only significant sediment source to the northern basin. The Songwe-Kiriwa delta has a relatively gentle subaqueous slope (1:70) with a broad shelf, reaching 2–3 km offshore [19]. The drill core site is located ~40 km from the mouth of the river. Although the nearest shoreline is 20 km from the drill site, a sub-lacustrine trough to the east probably prevents significant clastic input from the eastern part of the watershed [20]. Offshore, the lake deepens adjacent to the border fault to ~500 m, and sedimentary processes are dominated by debris and turbidity flows [21].

The smaller watersheds that drain into the lake consist primarily of Neogene alkaline volcanic bedrock, with abundant olivine and alkali basalts, phonolites, trachytes, and nephelinites [22]. Permo-Triassic and Cretaceous sedimentary rocks also crop out north and west of the lake (Figure 1). Soils in the woodlands and forests of the lowlands dominantly consist of pellic vertisols and mollic andosols [23]. The higher-gradient watersheds along the northeastern margin of the basin draining the Livingstone Mountains consist primarily of the Neoproterozoic Mozambique belt, with abundant biotite-horneblende-pyroxene gneiss, charnockites, and minor schists and quartzites [22]. Soils in this area are typically thin and weakly developed, consisting of lithosols, chromic cambisols and dystric regosols [23].

Climate within the watershed is primarily controlled by the yearly passage of the ITCZ, imparting a highly seasonal rainfall regime in which rainfall occurs from November to March. A single, long dry season lasts from April to October when little to no rainfall occurs and prevailing winds are strong southeasterlies [24]. Mean annual precipitation varies from ~800 mm/yr in the lowlands to over 2400 mm/yr in the Rungwe highlands [25]. In addition to a marked N-S gradient of precipitation, the rift escarpments bordering the lake create a pronounced local orographic effect. This results in substantial variability in local rainfall over relatively short spatial scales throughout the watershed.

Vegetation within the watershed is controlled by rainfall and dry season length, although temperature plays a role in the subalpine and alpine zones of the highlands [26–28]. Four principal biome types are observed within the Malawi watershed. In the lowlands (<1500 masl), Zambezian miombo woodlands, a low-diversity deciduous tropical woodland, dominate the landscape. These woodlands primarily comprise *Uapaca*, *Brachystegia*, *Isoberlinia*, *Julbernardia*, and species of Combretaceae and are relatively open, with 30–70% canopy cover. Tropical seasonal forests, closed-canopy semi-deciduous forests— with trees such as *Myrica*, *Macaranga*, Ulmaceae, and Moraceae—are not common in the lowlands; however, they are typically found in areas with locally moist conditions and as riparian corridors along streams and rivers. In the highlands (>1500 masl), afromontane vegetation is found. Today, this region consists of discontinuous patches of afromontane forest separated by high-elevation grasslands. Within the forests, composition is controlled by elevation and rainfall with lower montane forests having moister forest taxa such as *Olea capensis* (1500–2000 mm/yr; 0–3 dry months) from 1500 to 2500 m, whereas *Podocarpus*, *Juniperus*, Ericaceae are common above 2000 m in drier sites (800–1700 mm/yr; ~4 dry months) [27]. The most open biome occurs in the southernmost part of the watershed, where rainfall is the lowest (<800 mm/yr) and contains wooded grasslands with Zambezian affinities.

Methods

Core MAL05-2A was collected from the northern basin of Lake Malawi (10°1.1′S, 34°11.2′E; 359 m water depth) during the Lake Malawi Drilling Project (LMDP) in 2005 (permits issued by Malawi Geological Survey) [29]. Coring site, stratigraphy, and age model details can be found in Figures 1 and 2 as well as in Brown et al. [11] and Scholz et al. [29]. We studied a total of 40 samples, which were taken every ~5 cm from a 3-meter core section (~6–9 meters below lake floor [mblf]). The only gap occurred at ~8.1 mblf, where not enough sediment was available in the core archive to maintain the routine sample interval. The age model of the upper 22 m of the core is based on 24 calibrated accelerator mass spectrometry ^{14}C dates fit with a second-order polynomial

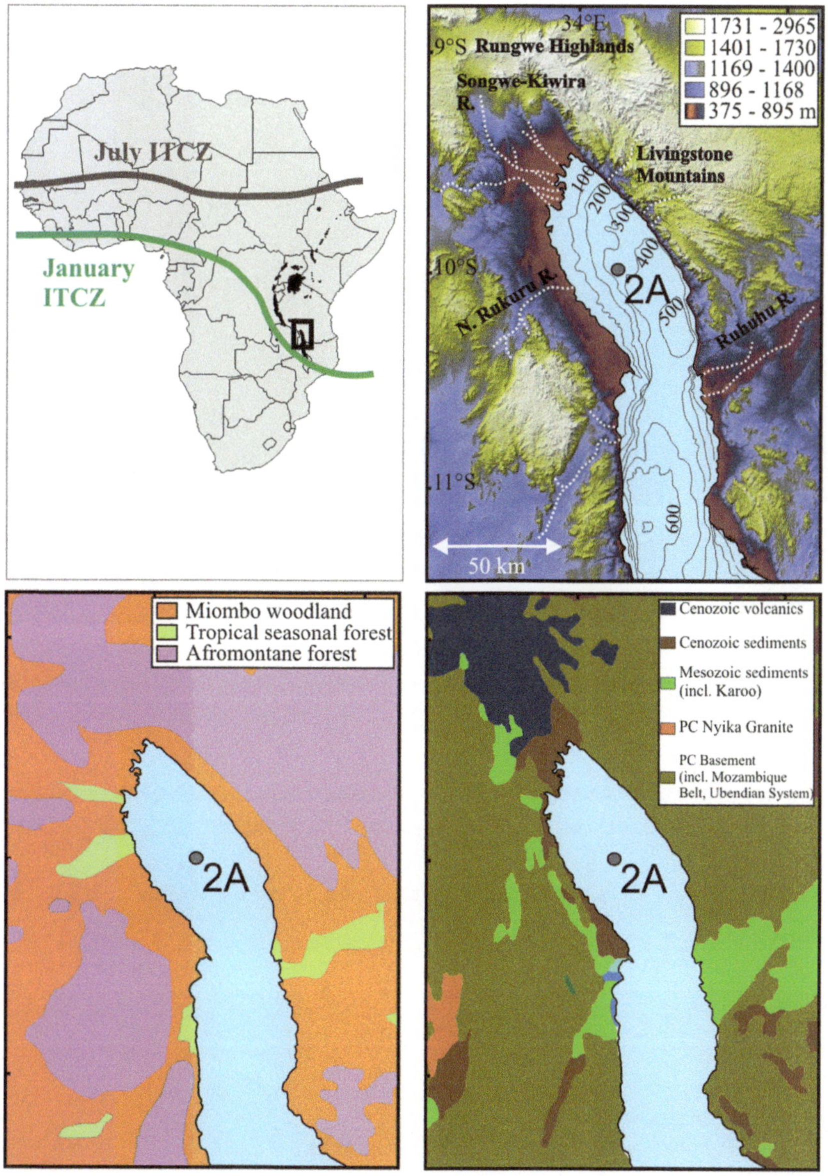

Figure 1. Geography, vegetation, and geology of the study site. (A) showing the July and January positions of the Intertropical Convergence Zone (ITCZ); rectangle represents inset area for watershed maps (modified from Nicholson, 1996). (B) map of the northern basin of Lake Malawi showing topography and bathymetry (isobaths [in meters] modified from Scholz et al., 1989), (C) modern potential vegetation distribution (modified from White, 1983), and (D) bedrock geology (modified from Schlüter, 2006). Scale for panels C and D is the same as for panel B. 2A identifies the location of drill core MAL05-2A.

(calibrated with the "Fairbanks 0107" calibration curve [30] so as to make ages comparable to those of previous studies). The average temporal resolution within the studied section is 208 years, but is ~100 years from 14–11 ka.

Terrigenous mass accumulation rates (TMAR) are a metric of the amount of sediment entering the lake from the watershed and provide insights on the erosion of Lake Malawi's northern watershed. Calculation of TMAR relied on several datasets. Linear sedimentation rates (cm/yr) were calculated using the radiocarbon age model presented in earlier LMDP publications [11]. Dry bulk density was calculated for MAL05-2A sediment horizons using the formulas presented in Dadey et al. [31]. We calculated water content for this purpose and used the gamma-ray density curves from GEOTEK multi-sensor logging of the core [32]. Sediment mass accumulation rates ($g/cm^2/yr$) were calculated by multiplying the linear sedimentation rate and dry bulk density. Weight percent terrigenous sediment was determined from X-ray patterns analyzed using the RockJock quantitative mineralogy computer program [33].

Terrigenous particle size data (sand: >62.5 μm; silt: 3.9–62.5 μm; clay: <3.9 μm), interpreted in conjunction with core lithostratigraphy, provide an indicator of landscape sediment flushing, sub-lacustrine hydrodynamic energy, and depositional processes [e.g., [34]]. Particle size analysis was conducted on the terrigenous fraction of the core MAL05-2A samples using a Malvern laser-diffraction particle size analyzer coupled to a Hydro 2000S dispersion bench (see Methods S1).

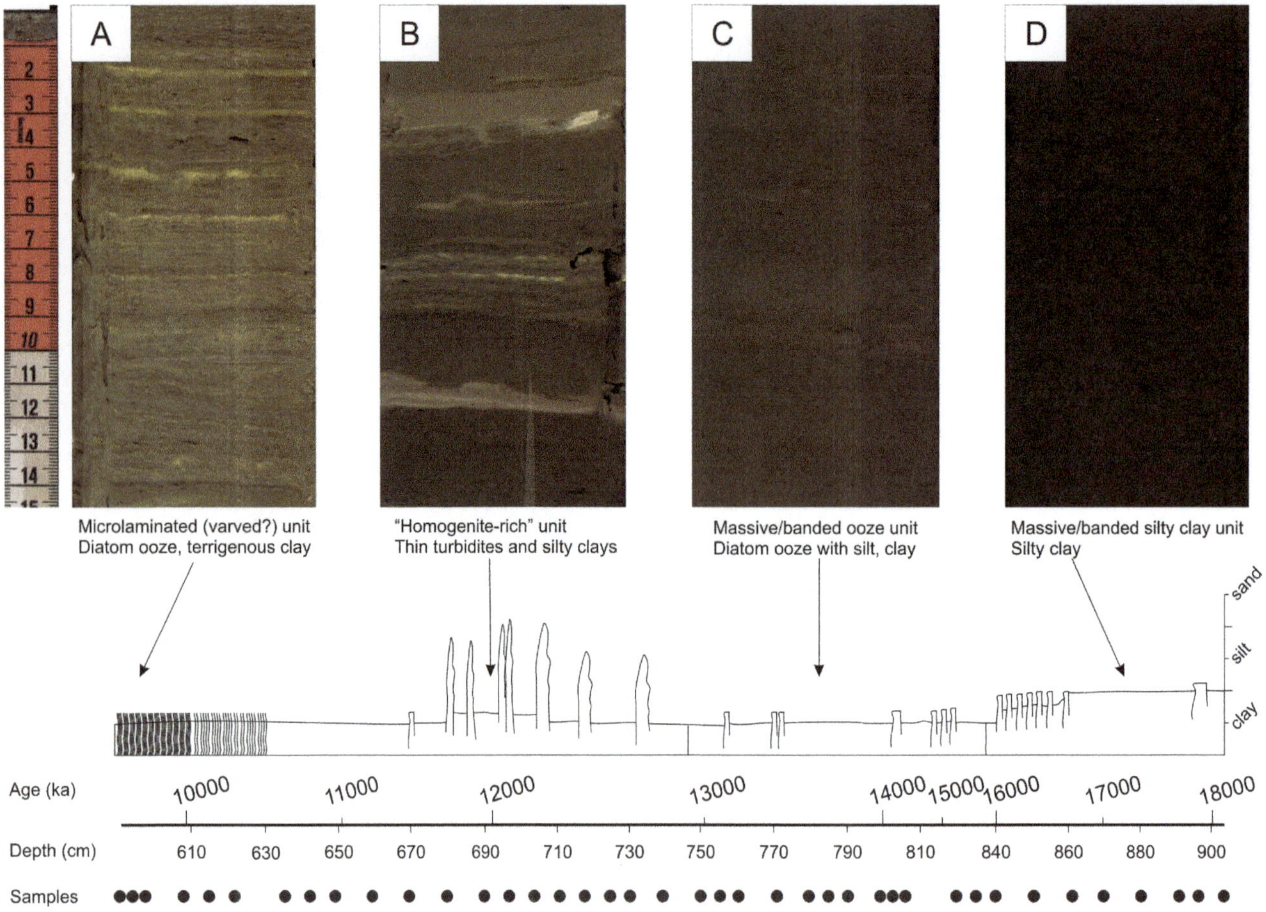

Figure 2. Core MAL05-2A lithostratigraphy within the study interval. Sediment composition and structures vary significantly over the deglacial interval.

Detrital clay minerals in lake sediments provide evidence for the alteration of watershed parent lithologies by physical (disintegration) and chemical (compositional alteration) weathering processes. Quantitative mineralogy was determined using powder X-ray diffraction (XRD) and the computer program RockJock v.11, which has been successfully used for the analysis of fine-grained Cenozoic sediments at a number of locales globally (e.g., [35–37]). For more information on the preparation and processing of these samples, see Methods S1.

In our interpretation of clay mineralogy, we focus on the ratio of kaolinite to smectite (K/S) as a proxy for chemical weathering intensity, which has been used in other similar studies in Africa [38–39]. Kaolinite is produced by the process of leaching in tropical regions marked by high rainfall [40]. In contrast, smectite is typical of tropical semiarid regions and points to increased rainfall seasonality, as it frequently forms during the dry season from the concentration of chemical elements transported to downstream areas by runoff [41–43]. We interpret elevated percentages of micas (illite plus chlorite) to reflect soil formation influenced by physical weathering processes [44].

Pollen samples were taken at the same depths as those for particle size and clay mineralogy and were processed following the standard methods of Faegri and Iverson [45]. More details about sample preparation and pollen identification can be found in Ivory et al. [46]. Vegetation groupings presented in this study are based on biomes of White [27], as well as prior pollen studies within the watershed by Debusk [25] and Vincens et al. [47].

Results

Vegetation

Detailed descriptions of the ecological dynamics surrounding vegetation change in the Malawi watershed during the last deglaciation appear in Ivory et al. [46]. In this study, we present pollen taxa grouped into biomes (afromontane forest, tropical seasonal forest, Zambezian miombo woodland) in order to present the composition and physiognomy of vegetation on the landscape over time. Following the Last Glacial Maximum (LGM) at 18.1–16.4 ka, afromontane forest taxa percentages were relatively stable at around 20% (Figure 3). Their subsequent decline began in a stepwise manner until 14 ka, at which time, the high-elevation arboreal taxa stabilized at percentages of 7 to 10%. Coeval with the decline of montane taxa, percentages of both lowland arboreal vegetation types increased. Beginning at 14 ka, both tropical seasonal forest and miombo woodland increased progressively in abundance until ~11.8 ka from percentages of <5% up to ~13%. Similarly, a progressive increase in grasses is observed in the record from around 30% to over 45% following the decline of afromontane forest; however, grasses once again returned to lower values around 30% relatively abruptly at 13 ka before the increase of the lowland arboreal vegetation types. The lowland arboreal vegetation continued to increase in a similar manner to a maximum value of 13.5% until 11.8 ka when trends in the tropical seasonal forest and miombo woodland diverged. Forest taxa began a slow decline until a minimum of 3.5 to 7% at 10 ka. At this same time, grasses began to increase at the expense of lowland trees with abundances of nearly 60% by 11 ka. Miombo woodland continued to increase until the end of the record at a maximum value of around 15%.

Mineralogy and Particle Size

Qualitative analysis of X-ray diffraction patterns generated from oriented clay mounts indicated the presence of smectite, kaolinite, illite, and minor chlorite in MAL05-2A (for more information, see the Methods S1). These results are in accord with the findings of Kalindekafe et al. [48] and Branchu et al. [49] for the modern clay mineralogy of northern Lake Malawi. Using RockJock, we determined that smectite was the most abundant clay in the deglacial sequence at all times, ranging from 9 to 29%. Kaolinite ranges from 6 to 13%, and illite ranges from 4 to15%. At the end of the LGM, kaolinite:smectite (K/S) within the lake sediments is relatively high, with values varying from 0.55 to 0.70 from 18.1 to 15 ka. The maximum value (0.79) of K/S within the lake is reached at 14.5 ka. After this time, sediment K/S begins a two-step decline with values of 0.55 to 0.65 from 14.5 to 13 ka, followed by highly variable values from 0.45 to 0.70 from 13 to 11.8 ka. It is only at the onset of the early Holocene at 11.8 ka that stable minimum values are reached (0.40 to 0.50). In addition, illite plus chlorite (maximum = 18%) increased sharply for the first time in the record at 11.8 ka and remained at stable values for the rest of the record.

Over the entire record (~18.1 to 9.5 ka), there is a decreasing trend in the ratio of silt to clay (silt/clay) which is marked by a two-step transition towards finer material. In the interval from ~18.1 to 14.6 ka, we observe high relative silt/clay values, with a mean of 1.69. Here, particles in the silt-size class (~4.0 to 62.5 μm) dominated the terrigenous fraction. The form of the silt/clay curve is blocky during this early interval, with prolonged periods of invariant conditions on the order of 1000–2000 years in duration. Values of silt/clay from ~18.1 to 16.9 ka range from ~1.84 to 1.95, whereas values from ~16.9 to 14.6 ka range from 1.48 to 1.59. A transition in the record after ~14.6 ka is marked by the saw-tooth form of the silt/clay curve until ~11.9 ka. During this more variable interval, values of silt/clay range from 0.87 to 1.72 with a mean of 1.21. The highest silt/clay values in this interval correlate with high relative values of sand (up to ~3.0%); this was the only instance very fine sand was detected by our analysis. The second important transition in the record took place after ~11.9 ka, where silt/clay ranges from 0.68 to 1.0, with a mean of ~0.82. Over this interval, particles in the clay-size fraction (<4.0 μm) generally exceeded 50% of the terrigenous fraction.

Overall, the record of TMAR is marked by a major transition from low to relatively high rates after 14.3 ka. From ~18.1 to 14.3 ka, the average was ~0.004 gm/cm^2/yr, and TMAR never exceeded ~0.006 gm/cm^2/yr. The transition after ~14.3 ka was initially marked by an abrupt two-fold increase in TMAR. At a finer scale, the TMAR record shows a stepwise change from ~14.3 to 9.5 ka. The initial step, from ~14.3 to 12.8 ka, was characterized by TMAR values ranging from ~0.007 to 0.012 gm/cm^2/yr. This period is followed by brief decline to lower values from ~12.8 to 11.9 ka, when the average TMAR was ~0.006 gm/cm^2/yr. The second TMAR increase was after ~11.9 ka, with a mean value of 0.012 gm/cm^2/yr until ~10.1 ka.

Interpretation

Dense forest and chemical weathering

At the end of the LGM, paleoclimate records from throughout Africa and at Lake Malawi suggest cooler temperatures and aridity [14,50–51]. Although it might be expected that these environmental conditions would slow reaction rates and mineral transformation during the early deglacial period, this does not seem to be the case in northern Malawi. Throughout this phase, K/S is relatively high in our core record (Figure 3). Higher proportions of kaolinite, a clay mineral leached of mobile cations typically associated with soils of the humid tropics, suggests that chemical weathering was relatively intense. The increased delivery of heavily weathered clays may appear counterintuitive given the comparatively cool temperatures and aridity during this time as recorded in the same core based on TEX-86 and leaf wax $\delta^{13}C$, respectively [50–51]; however, the clay mineralogy record is similar to changes in vegetation during the early deglacial period.

At the end of the LGM, many studies have suggested that East African afromontane forest communities expanded to lower altitudes than at present because of cooler temperatures and low atmospheric CO_2 [52–54]. This was also the case within the Lake Malawi watershed, where afromontane taxa, dominated by *Juniperus*, Ericaceae, and *Podocarpus*, were found in high percentages at the earliest part of our record beginning around 18.1 ka and lasting until ~15.5 ka (Figure 3) [46]. Although it is unlikely that the afromontane forest reached the lakeshore, Ivory et al. [46] suggest that these normally high-altitude communities expanded down to ~900 masl within several kilometers of the lakeshore. Dense forest communities must have been prevalent within the watershed, given the presence of taxa typical of the lower montane moist forest (*Olea* spp., *Myrica*). Furthermore, grass percentages during this interval are very low, the lowest of the record (Figure 3). In modern lake-floor sediment samples, grass percentages within Lake Malawi contribute ~55% of the total pollen [25], supporting the idea of a much denser canopy cover and larger contribution of the arboreal taxa throughout the watershed and into the lowlands. This may suggest that the presence of dense vegetation on the landscape as a result of lowering of the afromontane belt resulted in higher concentrations of organic acids and increased soil moisture, counteracting any slowdown in weathering due to climate.

The first marked change in both the vegetation and sedimentary records began at 15.5 ka, a period which is coeval with the end of Heinrich Stadial 1 (H1) [55]. Within Africa and at Lake Malawi, this period was characterized by progressive warming [50–51]. During this phase from 15.5 to 14.3 ka, K/S reached the highest values of the record (Figure 3). This suggests that, as in the previous phases of afromontane forest dominance, chemical weathering within the northern Malawi watershed remained relatively intense. Although some wetting occurs in East Africa during this post-H1 period, organic geochemical analysis from leaf wax $\delta^{13}C$ implies that Malawi remained relatively arid with respect to modern at this time, despite indications of warming [50–51]. Once again, although chemical weathering reactions would be expected to slow during an arid period, intense weathering continued in the northern Lake Malawi watershed.

Following H1, all of the principal afromontane arboreal taxa began to decline [46]. This decline has been attributed to increasing temperatures which forced afromontane forest to retreat to higher elevations [25]. However, the transition was stepwise. Although the principal afromontane forest taxa that were dominant just following the LGM (Ericaceae, *Juniperus*, and *Podocarpus*) began to decline, from 15.5 to 14.3 ka, the dense

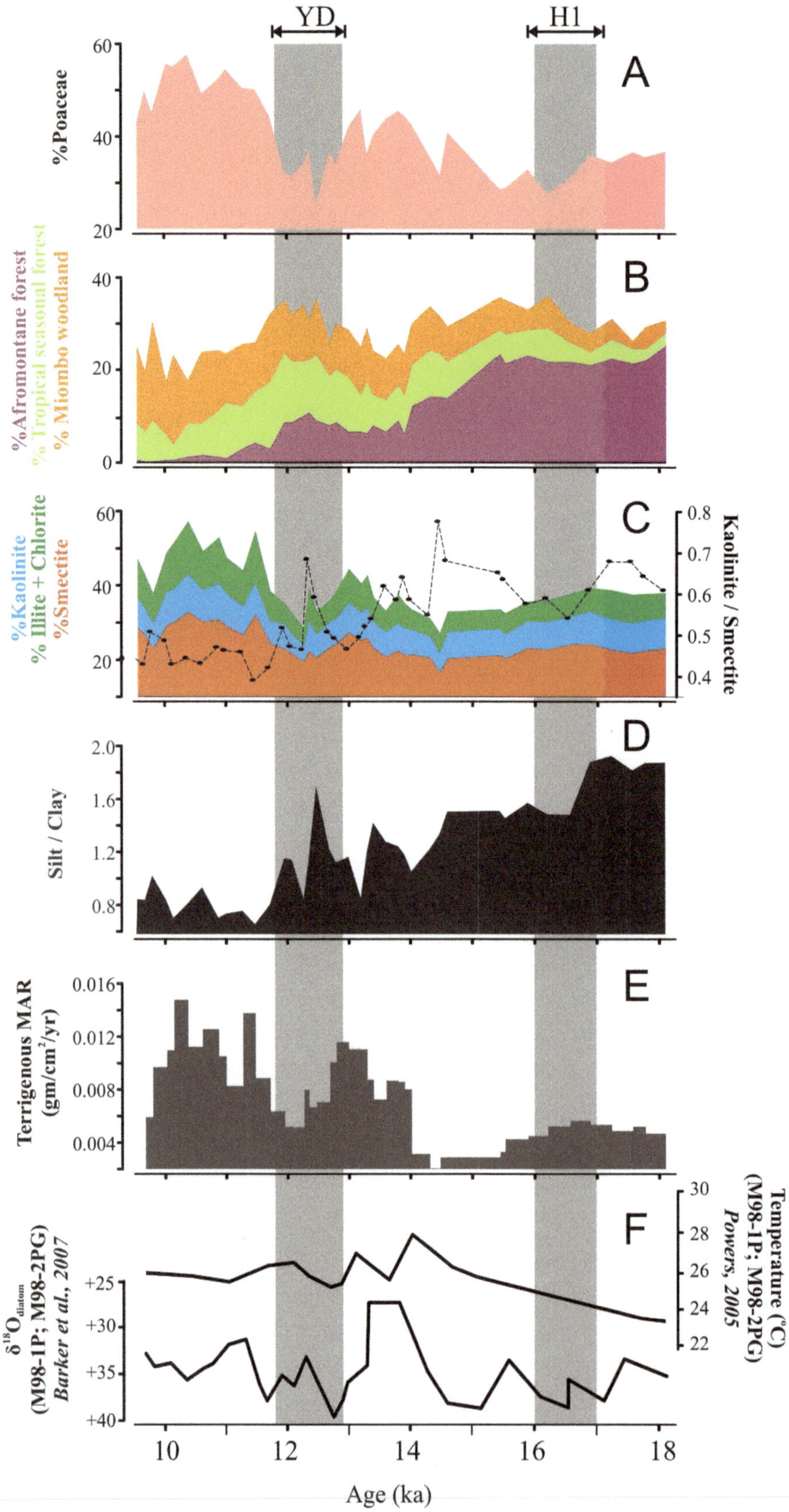

Figure 3. Vegetation and weathering indicators from drill core MAL05-2A.YD, Younger Dryas. H1, Heinrich event 1. (A) pollen percentages of Poaceae and (B) arboreal vegetation types (afromontane forest, tropical seasonal forest, miombo woodland), (C) XRD clay mineral percentages of smectite, kaolinite, and illite+chlorite, ratio of kaolinite to smectite (K/S), (D) ratio of silt to clay (silt/clay), and (E) terrigenous mass

accumulation rate, and (F) $\delta^{18}O$ from diatoms and TEX-86 temperature from piston cores M98-1P and M98-2PG from the northern basin of Lake Malawi from Barker et al. (2007) and Powers (2005), respectively.

lower montane moist forest taxa, such as *Myrica* and *Olea*, known pioneer taxa, continued to expand [46]. The presence of dense forest in the lowlands is the best explanation for the continued high values of K/S at this time.

Throughout the early deglacial period, minimum values of TMAR suggest that erosion was low (Figure 3). Johnson and McCave [20] suggest that TMAR in the northern basin of Lake Malawi is indicative of moisture, and therefore low values would suggest a reduced riverine transport. However, our dataset makes clear that a more likely mechanism for abating the flushing of weathered parent material from the landscape is the presence of dense lowland forests. During the LGM, a slight lake-level regression at Malawi of ~40 to 100 m would have decreased the distance from the mouth of the Songwe-Kiriwa River and other important drainages to the core site [56]. A basinward progradation of the delta during a lake-level lowstand might be expected to show an increase in the amount of terrigenous material deposited; however, we observe the opposite. Thus, despite noted cooler temperature and aridity relative to modern, intense weathering and low erosion are observed over this period by K/S and TMAR. This regime strongly suggests that dense moist forest in the watershed had a strong control on clay transformation and delivery of terrigenous siliciclastic material to the basin.

Transition to the Holocene (14.3 to 11.8 ka)

At 14.3 ka, a transitional period began that was coeval with the onset of the Bolling-Allerod (BA) [57]. In contrast with the predominantly silty detritus in our record lower in the core, a broad range of grain sizes produces highly variable values of silt/clay (1.8 to 0.6) during this time (Figure 3). Over this interval, lake levels were highly variable. Evidence from diatom assemblages implies high-frequency changes on the order of 50 to 100 m [56]. Sand is present in the samples from this interval, but only in significant percentages in one sample. However, there is clear evidence of turbidity flows here as early as ~13.2 ka (Figure 2). The presence of time-equivalent turbidites has also been described in other cores in northern Lake Malawi ("homogenites" of Barry et al. [58]). Thus, the variable silt content is expected to be caused by a change in sub-lacustrine depositional processes and energy. Thin, dominantly fine-grained turbidites may have been produced by wave remobilization of distal deltaic sediments, which move downslope at low velocity under the force of gravity as benthic nepheloid plumes [20].

At this time, the moist montane forest taxa became less abundant in the lowlands, and the afromontane forest as a biome reached very low percentages values, suggesting trees may have retreated to higher elevations by this time (Figure 3). Within the lowlands, however, arboreal taxa of the tropical seasonal forest began to increase in abundance. Miombo woodland abundances began increasing as well; however, the response time is much slower than the other arboreal taxa and values remained relatively low. Although the lowland closed-canopy tropical seasonal forest began expanding at this time, the percentages remained at <10%, values which are not significantly larger than those observed today, suggesting that dense forests were restricted largely to riparian corridors [25]. Ivory et al. [46] attribute this rise of lowland dense forest to the higher temperatures and reinforced monsoon recorded at the time in the basin. The lack of other arboreal vegetation on the landscape away from river banks in the lowlands indicates a marked change of physiognomy from the previous phase following the LGM to more open vegetation. This openness in the lowlands is reflected in the high percentages of grass pollen during this phase (Figure 3).

Within the basin, the increase in grass is tracked by rising TMAR values in core MAL05-2A (Figure 3). The increase in terrigenous material as the landscape opens supports the argument that TMAR represents erosion from the hinterland into the basin when vegetation is open and forests are scarce. This suggests that the relatively open character of vegetation caused by the retreat of montane forests prevented water storage on the landscape and promoted flushing of clastic material into the basin.

The presence of turbidites makes the clay record in this interval more complex. From 14.3 to 12.7 ka, an overall declining trend in K/S is observed, followed by an abrupt increase at 12.7 to 12.3 ka. Dense tropical seasonal forest and miombo woodland continued to expand in the lowlands at Lake Malawi between 13 and 11.7 ka, and tropical seasonal forest reached its maximum extent at the end of this interval (Figure 3). The beginning of this zone is coeval with the inception of the Younger Dryas (YD) in the Northern Hemisphere [59]. Although many other sites in East Africa record a retreat or slowdown of deglacial forest expansion during this time due to increased aridity, the dense lowland forest became more prevalent at Lake Malawi [60-64].

This indication of dense forest in conjunction with higher K/S values could possibly suggest a return to more intense chemical weathering beginning around 12.7 ka. However, independent particle size evidence, which shows elevated silt/clay and minor sand in the core during the period from 12.7 to 12.3 ka, suggests instead that the elevated K/S may be related to minor reworking by turbidites. Therefore, we interpret the multiple sedimentary indicators from this interval to be influenced by mass wasting. This may be the result of flushing from enhanced precipitation promoting sediment transport (and hyperpycnal sub-lacustrine flows) rather than chemical weathering.

Early Holocene open woodlands

Although records from farther north in tropical and subtropical Africa show an abrupt resumption of the African monsoon and significant wetting during the early Holocene, including at Lake Rukwa (~200 km north of Lake Malawi), the situation at Lake Malawi is slightly different [14,65]. Organic geochemical studies of $\delta^{13}C$ agree that the watershed was wetter than during the late Pleistocene; however, a ~100 m lake-level regression occurred at this time, suggesting a change in hydrology [66-67]. Pollen analysis on a core from Lake Masoko, a small maar lake within the Malawi watershed, suggests that the expansion of open miombo woodlands indicates high summer rainfall, but heightened rainfall seasonality [47]. This conclusion is in agreement with Zr:Ti and biomarker studies from Lake Malawi, which demonstrate wind direction and air mass changes at this time consistent with major reorganization of tropical circulation and a more northerly ITCZ than during the late Pleistocene [11,68].

At this time, total abundance of clays was very high, resulting in a net increase in all dominant clay minerals; however, K/S reached a minimum (Figure 3). Additionally, for the first time in this record, illite plus chlorite reached higher abundances than kaolinite (Figure 3). The dominance of detrital clays retaining mobile cations suggests a very strong decrease in the intensity of chemical weathering beginning at 11.8 ka. Furthermore, the importance of illite, a clay mineral that is only present in low abundances in the modern lake today, is suggestive of increasing physical weathering and reworking during the early Holocene despite increased wetness [48]. In addition, TMAR during this interval reaches maximum values. The agreement of these two independent datasets seems to indicate peak erosion from 11.8 to 10 ka.

This transition in clay mineralogy at 11.8 ka is coeval with the opening of the lowland vegetation (Figure 3). At 11.8 ka, near the onset of the Holocene, an abrupt transition occurred from denser lowland forests to more open miombo woodland in the watershed (Figure 3). Additionally, for the first time in the record, afromontane forest tree pollen is nearly absent. Given the high abundance of woodland and grasses, it is likely that this early Holocene transition represented that largest change in vegetation physiognomy in Malawi during the studied interval.

This increase in erosion was likely caused by the more intense summer rains of the African monsoon and the open character vegetation, conditioning the landscape for the flushing of physically weathered material into the basin. Cecil and Edgar [69] noted a similar relationship between greater coarse siliciclastic sediment transport and strongly seasonal rainfall.

Discussion

Traditional models for the tropics suggest strong climatic controls such as precipitation and temperature on chemical and physical weathering and erosion [44,70]. However, our record from Lake Malawi suggests a more indirect role of climate. During the early part of the last deglaciation, climate within the watershed varied dramatically in mean state and variability, including cooler temperatures and relative aridity from 18.1 to 14.5 ka. This was followed by successive wetting and warming from 14.5 to 12.5 ka, a return to slightly drier conditions with much reduced rainfall seasonality during the YD, and finally an increase in rainfall seasonality at the time of the resumption of the African monsoon at the early Holocene. However, despite the changes in climate, both highland and lowland forests were common in the watershed throughout this time until 11.8 ka. Although changes in rainfall and temperature lead to compositional vegetation changes during this period as high-altitude forest was replaced successively by lowland forest, the physiognomy of the landscape, with dominance of arboreal taxa, remained until the early Holocene. Similarly, although chemical weathering should be more intense during the moister periods, instead we see relative stability in the clay minerals with higher values of kaolinite throughout this period regardless of climatic regime. During the early Holocene, the physiognomic change to more open woodland is coeval with higher mean annual precipitation because of heightened seasonality following the resumption of the African monsoon. It might be expected that this high temperature and high rainfall should have resulted in intense chemical weathering, but the available data do not support this idea. At this time, kaolinite decreases, and high abundance of illite suggests not only less-intense chemical weathering, but higher physical weathering and transport of less-weathered micas into the basin.

The strong, direct control of vegetation on mineral transformation and erosion is not surprising for several reasons as many processes mediated by plants influence chemical and physical breakdown of bedrock. First, although heat and moisture are necessary for driving chemical breakdown of aluminosilicate rocks, organic acids, particularly low-weight organic acids such as those produced by plants and mycorrhizae associated with plant roots, are essential for biological mediation of weathering in soils [71–72]. These acids not only provide H+ for the transformation of primary materials, but also act as ligands that form strong complexes with trivalent cations such as Al^{3+} and Fe^{3+} [6]. This function of organic acids has the potential to dramatically intensify weathering rates by increasing the solubility of minerals containing these elements. Second, root networks, particularly in dense, old-growth forests, are critical for the mechanical breakdown of bedrock [73]. Furthermore, the removal of trees results in an increased response of erosion rates and closer control of abiotic erosion mechanisms as seen in our record.

Perhaps the most important biotic control on weathering is the role of forests and large forest trees in regulating the hydrologic cycle locally within a watershed and at the scale of a soil [74]. In dense forests, canopies intercept rainfall. During intense monsoon rains, this function acts to reduce runoff and increase storage on the landscape, increasing the residence time of moisture available for chemical reactions in soils. Additionally, evapotranspiration within a forest often creates a microclimate that results in higher local precipitation over dark, dense trees as well as more stored water on the landscape and cooler temperatures from evaporation [75]. More importantly however, evapotranspiration controlled by forest trees limits the drainage depth within the weathering profile of a soil, reducing physical weathering processes and allowing time for chemical alteration [76]. Vegetation transitions, as we see in the Malawi record, however, are key as long-lived dense forests, such as those found in the Congo Basin or in the Amazon exhibit very little chemical weathering [77–78]. Early stages of succession, such as the period of intense chemical weathering observed during the deglaciation following the retreat of the afromontane forest as lowland forest colonized the lake shore, are in fact typically the period of most intense chemical weathering as large trees colonize previously unweathered substrate [79–80].

TMAR values show a positive correlation with grass abundances throughout our record. Although TMAR has been interpreted as an indication of moisture availability, we instead suggest that delivery of terrigenous material to the basin is strongly regulated by vegetation on the landscape. This means that more open, grassy woodlands are less effective at trapping material on the landscape leading to greater erosion and deposition of terrigenous material than when forests are dominant in the lowlands [12,81]. This first order control of erosion on the landscape is in agreement with cosmogenic nuclide studies in the Alps by Vanacker et al. [82] who show exponential increases in denudation rates due to modern land use change. Although this suggests rapid landscape conversion once forest is gone, this study, as well as that of Vanacker et al. [82], suggests that once forest returns, stabilization is possible as denudation rates return to background levels. Reducing the residence time of water on the landscape likewise influences chemical weathering intensity, and we note that the inverse correlation of K/S with TMAR and grass percentages, which appears to support this hypothesis.

In our analysis of core records from Lake Malawi, we have observed coarser sediments (high silt/clay or % sand) around the inception of lake-level highstands, which also show high ratios of trees to herb pollen. Several factors may influence this particle size trend. For example, storage release (i.e., flushing) of coarser sediment from low-to-moderate gradient flexural and axial margins concomitant with transgression likely helps to explain some of the variability we have observed. Additionally, increased fluvial discharge implied by the presence of afromontane forest is consistent with what is known about the efficacy of coarse siliciclastic transport [34]. Another potential influence relates to the shortening of fluvial transport networks associated with the higher base level (i.e., less reduction of siliciclastic particle size by transport abrasion and downstream fining), but because fining upward patterns are characteristic of the highstand strata we have examined, the absolute influence of this mechanism is not well known. By contrast, during lowstand intervals such as the early Holocene, when dry season length was long, fluvial networks lengthen and appear to generate finer siliciclastic detritus. Physical weathering of parent rocks and soils, captured in the clay mineral record, may serve as a feedback on reducing particle size during these intervals. Indeed, the lack of forest cover to intercept rainfall during these intervals may also serve to help flush fine sediment from hillslopes and floodplains into river channels. Variability in depositional processes can complicate the interpretation of silt/clay, which is why evidence of mass wasting is critical to identify in the lithostratigraphy. For example, the transition from lowstand to highstand in rift lakes occurs rapidly (10^2 to 10^5 yrs) [83], and as such flexural margins can become prone to gravity flows as pore pressures change and slopes destabilize. Therefore, our study employed particle size and stratigraphic data in support of pollen, clay minerals, and TMAR to provide the most rigorous assessment of weathering and erosion patterns for the deglacial period yet developed for this region of East Africa.

Conclusions

Vegetation composition and structure at Lake Malawi and elsewhere does not unequivocally track simple precipitation amount [28,46–47]. Our record attests to a strong negative relationship between increased rainfall seasonality and vegetation density that results in a specific depositional signature within the lake. When dense forests occupy the watershed, chemical weathering is intense and erosion is low. Leaching within soils leads to generation of highly altered clay minerals like kaolinite. However, during times of open vegetation, chemical weathering is less intense, but erosion is increased. Smectite dominates during these periods, but flashy precipitation on the open landscape leads to flushing of siliciclastics into the lake and thus high relative TMAR.

Finally, these results have important implications for better understanding how weathering and erosion may change in the future. Although little work has been done to quantify potential future alterations in weathering, most models do not take ecosystem change related to climatic change into account, leading to significant potential biases. Only in the last few years has vegetation been added to projections of soil weathering, thus leading to the suggestion of high sensitivity of weathering to future climate states [84–86]. Although paleostudies hint at the role of biological mediation in weathering over long time scales, no other studies have used paleo-records to quantify the effects of vegetation on weathering processes in southeast Africa [7]. This study points to the importance of vegetation for mediating weathering in the past; however, more and longer records are needed in order to better quantify this effect over a larger range of climatic and vegetation variability in watersheds throughout the tropics.

Acknowledgments

Disclaimer: Any use of trade, product, or firm names is for descriptive purposes only and does not imply endorsement by the U.S. Government.

Thanks to Jean-Pierre Cazet, Guillaume Buchet, Owen Davis, and John Logan for help with sample processing and pollen identification, William Benzel and Adam Boehlke for assistance with x-ray diffraction, and the National Lacustrine Core Repository (LacCore) at the University of Minnesota for sub-sampling and core curation. We would also like to thank three reviewers for helpful comments that improved the manuscript.

Author Contributions

Conceived and designed the experiments: SJI MMM. Performed the experiments: SJI MMM. Analyzed the data: SJI MMM. Contributed reagents/materials/analysis tools: SJI MMM GE AC AML AV. Wrote the paper: SJI MMM.

References

1. IPCC (2013) Climate Change 2013: The Physical Science Basis. Contribution of Working Group I to the Fifth Assessment Report of the Intergovernmental Panel on Climate Change [Stocker TF, Qin D, Plattner GK, Tignor M, Allen SK, et al. (eds.)]. Cambridge University Press, Cambridge, United Kingdom and New York, NY, USA, 1535 pp.
2. Brantley SL, Megonigal JP, Scatena FN, Balogh-Brunstad Z, Barnes RT, et al. (2011) Twelve testable hypotheses on the geobiology of weathering. Geobiology 9: 140–165.
3. Walker JC, Hays PB, Kasting JF (1981) A negative feedback mechanism for the long-term stabilization of Earth's surface temperature. J Geophys Res: Oceans 86: 9776–9782.
4. Langbein WB, Schumm SA (1958) Yield of sediment in relation to mean annual precipitation. Eos 39: 1076–1084.
5. Egli M, Mirabella A, Sartori G (2008) The role of climate and vegetation in weathering and clay mineral formation in late Quaternary soils of the Swiss and Italian Alps. Geomorphology 102: 307–324.
6. Goddéris Y, Roelandt C, Schott J, Pierret M.C, François LM (2009) Towards an integrated model of weathering, climate, and biospheric processes. Rev Mineral Geochem 70: 411–434.
7. Dosseto A, Hesse PP, Maher K, Fryirs K, Turner S (2010) Climatic and vegetation control on sediment dynamics during the last glacial cycle. Geology 38: 395–398.
8. Einsele G, Hinderer M (1998) Quantifying denudation and sediment–accumulation systems (open and closed lakes): basic concepts and first results. Palaeogeogr Palaeoclimatol Palaeoecol 140: 7–21.
9. Sémah AM, Sémah F, Moudrikah R, Fröhlich F, Djubiantono T (2004) A late Pleistocene and Holocene sedimentary record in Central Java and its palaeoclimatic significance. Modern Quaternary Research in SE Asia, Balkema 19: 63–88.
10. Felton AA, Russell JM, Cohen AS, Baker ME, Chesley JT, et al. (2007) Paleolimnological evidence for the onset and termination of glacial aridity from Lake Tanganyika, Tropical East Africa. Palaeogeogr Palaeoclimatol Palaeoecol 252: 405–423.
11. Brown ET, Johnson TC, Scholz CA, Cohen AS, King JW (2007) Abrupt change in tropical African climate linked to the bipolar seesaw over the past 55,000 years. Geophys Res Lett 34.
12. Dunne T (1979) Sediment yield and land use in tropical catchments. J Hydrol 42: 281–300.
13. Dixon RK, Solomon AM, Brown S, Houghton RA, Trexier MC, et al. (1994) Carbon Pools and Flux of Global Forest Ecosystems. Science 263: 185–190.
14. Gasse F (2000) Hydrological changes in the African tropics since the Last Glacial Maximum. Quat Sci Rev 19: 189–211.
15. Ebinger CJ (1989) Tectonic development of the western branch of the East African rift system. Geol Soc Am Bull 101: 885–903.
16. Rosendahl BR (1987) Architecture of continental rifts with special reference to East Africa. Annu Rev Earth Planet Sci 15: 445–503.
17. Scholz CA, Rosendahl BR, Versfelt JW, Kaczmarick J, Woods LD (1989) Seismic Atlas of Lake Malawi (Nyasa), East Africa. Project PROBE Geophysical Atlas Series, Vol. 2. Duke University, Durham, NC.
18. Soreghan MJ, Scholz CA, Wells JT (1999) Coarse-grained, deep-water sedimentation along a border fault margin of Lake Malawi, Africa: seismic stratigraphic analyses. J Sediment Res 69: 832–846.
19. Scholz CA (1995) Deltas of the Lake Malawi rift, East Africa: seismic expression and exploration implications. AAPG Bull 79: 1679–1697.

20. Johnson T, McCave I (2008) Transport mechanism and paleoclimatic significance of terrigenous silt deposited in varved sediments of an African rift lake, Limnol Oceanogr 5: 1622–1632.
21. Scott DL, Ng'ang'a P, Johnson TC, Rosendahl BR (1991) High-resolution acoustic character of Lake Malawi (Nyasa), East Africa, and its relationship to sedimentary processes. Special Publication of the International Association of Sedimentologists 13: 129–145.
22. Schlüter T (2006) Geological Atlas of Africa. Springer, Berlin. 272 pp.
23. FAO (1988) Soil Map of the World, Revised legend, World Resources Report, 60, FAO, Rome, Italy.
24. Malawi Government (1983) The National Atlas of Malawi. National Atlas Committee and Department of Surveys of Malawi. Natl. Atlas Comm. Dep. Surv. Malawi.
25. DeBusk GH (1994) Transport and stratigraphy of pollen in Lake Malawi, Africa. PhD Thesis, Duke University.
26. Polhill RM (1966) Flora of Tropical East Africa. In: Hubbard, C.E., Milne-Redhead, E. (Eds.), Crown Agents for Oversea Governments and Administrations, London.
27. White F (1983) Vegetation of Africa—a descriptive memoir to accompany the Unesco/ AETFAT/UNSO vegetation map of Africa. Natural Resources Research Report XX. UNESCO, Paris, France.
28. Hély C, Bremond L, Alleaume S, Smith B, Sykes M, Guiot J (2006) Sensitivity of African biomes to changes in the precipitation regime. Global Ecol Biogeogr 15: 258–270.
29. Scholz CA, Johnson TC, Cohen AS, King J, Peck J, et al. (2007) East African megadroughts between 135 and 75 thousand years ago and bearing on early-modern human origins. PNAS 104: 16416–16421.
30. Fairbanks R, Mortlock R, Chiu T, Cao L, Kaplan A, et al. (2005) Radiocarbon calibration curve spanning 0 to 50,000 years BP based on paired 230Th/234U/238U and 14C dates on pristine corals. Quat Sci Rev 25: 1781–1796.
31. Dadey KA, Janecek T, Klaus A (1992) Dry-bulk density: its use and determination. In Proceedings of the Ocean Drilling Program, Scientific Results (Vol. 126, pp. 551–554) College Station, TX: Ocean Drilling Program.
32. Lyons WB, Carey AE, Hicks DM, Nezat CA (2005) Chemical weathering in high-sediment yielding watersheds, New Zealand. J Geophys Res: Earth Surface (2003–2012) 110.
33. Eberl DD (2003) User's guide to RockJock – A program for determining quantitative mineralogy from powder X-ray diffraction data: U.S. Geological Survey Open-File Report 2003–78, 47 p.
34. Heins Keiro (2007), in Sedimentary Provenance and Petrogenesis: Perspectives from petrography and geochemistry, in Arribas, J, Critelli, S. and Johnson, M., eds., GSA Special paper 420, p. 345–379
35. Andrews JT (2008) The role of the Iceland Ice Sheet in the North Atlantic during the late Quaternary: a review and evidence from Denmark Strait. J Quat Sci 23: 3–20.
36. Refsnider KA (2010) Dramatic increase in late Cenozoic alpine erosion rates recorded by cave sediment in the southern Rocky Mountains. Earth Planet Sci Lett 297: 505–511.
37. Dühnforth M, Anderson RS, Ward D J, Blum A (2012) Unsteady late Pleistocene incision of streams bounding the Colorado Front Range from measurements of meteoric and in situ 10Be. J Geophys Res: Earth Surface (2003–2012) 117.
38. Pastouret L, Chamley H, Delibrias G, Duplessy JC, Thiede J (1978) Late Quaternary climatic changes in western tropical Africa deduced from deep-sea sedimentation off Niger Delta. Oceanol Acta 1: 217–232.
39. Lézine AM, Duplessy JC, Cazet JP (2005) West African monsoon variability during the last deglaciation and the Holocene: Evidence from fresh water algae, pollen and isotope data from core KW31, Gulf of Guinea. Palaeogeogr Palaeoclimatol Palaeoecol 219: 225–237.
40. Birkeland PW, (1984) Soils and Geomorphology. Oxford University Press, Oxford. 310 pp.
41. Chamley H (1989) Clay sedimentology (Vol. 623) New York: Springer-Verlag. 623 pp.
42. Weaver CE (1989) Clays, muds, and shales. Elsevier. 818 pp.
43. Alizai A, Hillier S, Clift PD, Giosan L, Hurst A, et al. (2012) Clay mineral variations in Holocene terrestrial sediments from the Indus Basin. Quat Res 77: 368–381.
44. Thiry M (2000) Palaeoclimatic interpretation of clay minerals in marine deposits; an outlook from the continental origin. Earth Sci Rev 49: 201–221
45. Faegri K, Iversen J (1989) Textbook of Pollen Analysis, Wiley, Chichester, UK. 328 pp.
46. Ivory S, Lézine AM, Vincens A, Cohen A (2012) Effect of aridity and rainfall seasonality on vegetation in the southern tropics of East Africa during the Pleistocene/Holocene transition, Quat Res 77: 77–76.
47. Vincens A, Garcin Y, Buchet G (2007) Influence of rainfall seasonality on African lowland vegetation during the late Quaternary: pollen evidence from Lake Masoko, Tanzania. J Biogeog 34: 1274–1288.
48. Kalindekafe LN, Dolozi MB, Yuretich R (1996) Distribution and origin of clay minerals in the sediments of Lake Malawi, in Johnson, T.C., and Odada, E., eds., The Limnology, Climatology and Paleoclimatology of the East African Lakes: Amsterdam, Gordon and Breach, p. 443–460.
49. Branchu P, Bergonzini L, Delvaux D, De Batist M, Golubev V, et al. (2005) Tectonic, climatic and hydrothermal control on sedimentation and water chemistry of northern Lake Malawi (Nyasa), Tanzania. J African Earth Sci 43: 433–446.
50. Powers LA, Johnson TC, Werne JP, Castañeda I, Hopmans E, et al. (2005) Large temperature variability in the southern African tropics since the Last Glacial Maximum. Geophys Res Lett 32: L08706.
51. Castañeda IS, Werne JP, Johnson TC (2007) Wet and arid phases in the southeast African tropics since the Last Glacial Maximum. Geology 35: 823–826.
52. Street-Perrott FA, Huang Y, Perrott RA, Eglinton G, Barker P, et al. (1997) Impact of lower atmospheric carbon dioxide on tropical mountain ecosystems. Science 278: 1422–1426.
53. Wu H, Guiot J, Brewer S, Guo Z (2007) Climatic changes in Eurasia and Africa at the last glacial maximum and mid-Holocene: reconstruction from pollen data using inverse vegetation modelling. Clim Dyn 29: 211–229.
54. Woltering M, Johnson TC, Werne JP, Schouten S, Sinninghe Damsté JS (2011) Late Pleistocene temperature history of southeast Africa: A TEX86 temperature record from Lake Malawi. Palaeogeogr Palaeoclimatol Palaeoecol 303: 93–102.
55. Hemming SR (2004) Heinrich events: Massive late Pleistocene detritus layers of the North Atlantic and their global climate imprint. Rev Geophys 42.
56. Stone JR, Westover KS, Cohen AS (2011) Late Pleistocene paleohydrography and diatom paleoecology of the central basin of Lake Malawi, Africa. Palaeogeogr Palaeoclimatol Palaeoecol 303: 51–70.
57. Wohlfarth B (1996) The chronology of the last termination: A review of radiocarbon date, high resolution terrestrial stratographies. Quat Sci Rev 15: 267–284.
58. Barry SL, Filippi ML, Talbot MR, Johnson TC (2002) Sedimentology and geochronology of late Pleistocene and Holocene sediments from northern Lake Malawi. In The East African Great Lakes: Limnology, Palaeolimnology and Biodiversity, Springer Netherlands, 369–391.
59. Alley RB, Meese DA, Shuman CA, Gow AJ, Taylor KC, et al. (1993) Abrupt increase in Greenland snow accumulation at the end of the Younger Dryas event. Nature 362: 527–527.
60. Vincens A (1991) Late quaternary vegetation history of the South-Tanganyika basin. Climatic implications in south central Africa. Palaeogeogr Palaeoclimatol Palaeoecol 86: 207–226.
61. Vincens A (1993) Nouvelle sequence pollinique du Lac Tanganyika: 30,000 ans d'histoire botanique et climatique du Bassin Nord. Rev Palaeobot Palyno 78: 381–394.
62. Bonnefille R, Riollet G, Buchet G, Icole M, Lafont R, et al. (1995) Glacial-interglacial record from intertropical Africa, high resolution pollen and carbon data at Rusaka, Burundi. Quat Sci Rev 14: 917–936.
63. Beuning KR, Talbot MR, Kelts K (1997) A revised 30,000-year paleoclimatic and paleohydrologic history of Lake Albert, East Africa. Palaeogeography, Palaeoclimatology, Palaeoecology 136: 259–279.
64. Ryner M, Gasse F, Rumes B, Verschuren D (2007) Climatic and hydrological instability in semi-arid equatorial East Africa during the late glacial to Holocene transition: a multi-proxy reconstruction of aquatic ecosystem response in northern Tanzania. Palaeogeogr Palaeoclimatol Palaeoecol 248: 440–458.
65. Cohen AS, Van Bocxlaer B, Todd JA, McGlue M, Michel E, et al. (2013) Quaternary ostracodes and molluscs from the Rukwa Basin (Tanzania) and their evolutionary and paleobiogeographic implications. Palaeogeogr Palaeoclimatol Palaeoecol 392: 79–97.
66. Johnson TC, Brown ET, McManus J, Barry S, Barker P, et al. (2002) A high-resolution paleoclimate record spanning the past 25,000 years in southern East Africa. Science 296: 113–132.
67. Barker PA, Leng MJ, Gasse F, Huang Y (2007) Century-to-millennial scale climatic variability in Lake Malawi revealed by isotope records. Earth Planet Sci Lett 261: 93–103.
68. Konecky BL, Russell JM, Johnson TC, Brown ET, Berke MA, et al. (2011) Atmospheric circulation patterns during late Pleistocene climate changes at Lake Malawi, Africa. Earth Planet Sci Lett 312: 318–326.
69. Cecil CB, Edgar NT (2003) Climate Controls on Stratigraphy: Society for Sedimentary Geology Special Publication 77, 275 p.
70. Burnett AP, Soreghan MJ, Scholz CA, Brown ET (2011) Tropical East African climate change and its relation to global climate: a record from Lake Tanganyika, Tropical East Africa, over the past 90+ kyr. Palaeogeogr Palaeoclimatol Palaeoecol 303: 155–167.
71. Landeweert R, Hoffland E, Finlay RD, Kuyper TW, van Breemen N (2001) Linking plants to rocks: ectomycorrhizal fungi mobilize nutrients from minerals. Trends Ecol Evol 16: 248–254.
72. Bonneville S, Smits MM, Brown A, Harrington J, Leake JR, et al. (2009) Plant-driven fungal weathering: early stages of mineral alteration at the nanometer scale. Geology 37: 615–618.
73. Roering JJ, Marshall J, Booth AM, Mort M, Jin Q (2010) Evidence for biotic controls on topography and soil production. Earth Planet Sci Lett 298: 183–190.
74. Rodríguez-Iturbe I, Porporato A (2004) Ecohydrology of water-controlled ecosystems. In Soil Moisture and Plant Dynamics.
75. Spracklen DV, Arnold SR, Taylor CM (2012) Observations of increased tropical rainfall preceded by air passage over forests. Nature 489: 282–285.
76. Roelandt C, Goddéris Y, Bonnet MP, Sondag F (2010) Coupled modeling of biospheric and chemical weathering processes at the continental scale. Global Biogeochem Cy 24: GB2004.

77. Gaillardet J, Dupré B, Louvat P, Allegre CJ (1999) Global silicate weathering and CO2 consumption rates deduced from the chemistry of large rivers. Chem Geo159: 3–30.
78. Moquet JS, Crave A, Viers J, Seyler P (2011) Chemical weathering and atmospheric/soil CO2 uptake in the Andean and foreland Amazon basins. Chem Geo 287: 1–26.
79. Moulton KL, West J, Berner RA (2000) Solute flux and mineral mass balance approaches to the quantification of plant effects on silicate weathering. Am J Sci 300: 539–570.
80. Goddéris Y, Donnadieu Y, Tombozafy M, Dessert C (2008) Shield effect on continental weathering: implication for climatic evolution of the Earth at the geological timescale. Geoderma 145: 439–448.
81. Hay WW (1998) Detrital sediment fluxes from continents to oceans. Chem Geo 145: 287–323.
82. Vanacker V, von Blanckenburg F, Govers G, Molina A, Poesen J, et al. (2007) Restoring dense vegetation can slow mountain erosion to near natural benchmark levels. Geology 35: 303–306.
83. McGlue MM, Lezzar KL, Cohen AS, Russell JM, Tiercelin JJ, et al. (2008) Seismic records of late Pleistocene aridity in Lake Tanganyika, tropical East Africa. J Paleolimn 40: 635–653.
84. Beaulieu E, Goddéris Y, Labat D, Roelandt C, Oliva P, et al. (2010) Impact of atmospheric CO2 levels on continental silicate weathering. Geochem Geophys Geosyst 11.
85. Beaulieu E, Goddéris Y, Donnadieu Y, Labat D, Roelandt C (2012) High sensitivity of the continental-weathering carbon dioxide sink to future climate change. Nature Climate Change 2: 346–349.
86. Goddéris Y, Brantley S, François LM, Schott J, Pollard D, et al. (2013) Rates of consumption of atmospheric CO 2 through the weathering of loess during the next 100 yr of climate change. Biogeosciences 10: 135–148.

8

Monitoring Urban Greenness Dynamics Using Multiple Endmember Spectral Mixture Analysis

Muye Gan[1], Jinsong Deng[1,2]*, Xinyu Zheng[1], Yang Hong[2,3], Ke Wang[1]*

1 College of Environmental and Resource Sciences, Zhejiang University, Hangzhou, Zhejiang Province, China, **2** School of Civil Engineering and Environmental Sciences and School of Meteorology, University of Oklahoma, Norman, Oklahoma, United States of America, **3** State Key Laboratory of Hydroscience and Engineering, Department of Hydraulic Engineering, Tsinghua University, Beijing, China

Abstract

Urban greenness is increasingly recognized as an essential constituent of the urban environment and can provide a range of services and enhance residents' quality of life. Understanding the pattern of urban greenness and exploring its spatiotemporal dynamics would contribute valuable information for urban planning. In this paper, we investigated the pattern of urban greenness in Hangzhou, China, over the past two decades using time series Landsat-5 TM data obtained in 1990, 2002, and 2010. Multiple endmember spectral mixture analysis was used to derive vegetation cover fractions at the subpixel level. An RGB-vegetation fraction model, change intensity analysis and the concentric technique were integrated to reveal the detailed, spatial characteristics and the overall pattern of change in the vegetation cover fraction. Our results demonstrated the ability of multiple endmember spectral mixture analysis to accurately model the vegetation cover fraction in pixels despite the complex spectral confusion of different land cover types. The integration of multiple techniques revealed various changing patterns in urban greenness in this region. The overall vegetation cover has exhibited a drastic decrease over the past two decades, while no significant change occurred in the scenic spots that were studied. Meanwhile, a remarkable recovery of greenness was observed in the existing urban area. The increasing coverage of small green patches has played a vital role in the recovery of urban greenness. These changing patterns were more obvious during the period from 2002 to 2010 than from 1990 to 2002, and they revealed the combined effects of rapid urbanization and greening policies. This work demonstrates the usefulness of time series of vegetation cover fractions for conducting accurate and in-depth studies of the long-term trajectories of urban greenness to obtain meaningful information for sustainable urban development.

Editor: Clinton N. Jenkins, Instituto de Pesquisas Ecológicas, Brazil

Funding: This work was supported by the Survey and Evaluation of National Ecological Environment (2000-2010) using Remote Sensing Program of Chinese Ministry of Environmental Protection (No. STSN-05-11). The funders had no role in study design, data collection and analysis, decision to publish, or preparation of the manuscript.

Competing Interests: The authors have declared that no competing interests exist.

* Email: kwang@zju.edu.cn (KW); jsong_deng@zju.edu.cn (JSD)

Introduction

The rapid urbanization in China in recent decades, accompanied by the continuous increase in urban population and the unprecedented growth of cities, has put significant pressure on urban environments. This pressure has led to severe environmental issues, such as urban heat islands, air pollution, and urban flooding [1]. Moreover, improved economic conditions have lead residents to demand better living environments and a better quality of life.

Urban greenness, represented in this study by all vegetation cover in and around cities, e.g., street plantation, lawns, parks, gardens, crops, wetlands, and forests, contributes valuable ecosystem services and plays an irreplaceable role in the improvement of the urban environment [2]. Urban vegetation improves air quality [3], intercepts storm water runoff [4], reduces energy demands [5], and provides numerous psychological benefits, aesthetic views and restorative opportunities for city dwellers who may otherwise have limited exposure to natural environments [6,7].

Rapid urbanization has resulted in significant alterations in the quality and quantity of urban greenness [8]. In a large-sample survey involving 386 European cities, green space was found to increase with the city area and slightly decline with the population density [9]. Dallimer et al. indicated that nine out of thirteen cities in England exhibited declining urban green spaces between 2000 and 2008 [10]. Infill development during urbanization is considered to be a major cause of green space loss. Similarly, a statistically significant decline in tree cover was found in many major U.S. cities, along with increased population densities and likely increased development pressure between 2003 and 2009 [11]. Unlike the declining trend reported by the above studies, green space coverage may increase with urbanization intensity. Green space coverage in the built-up areas of Chinese cities, for example, has increased steadily during the last two decades [12]. Yang et al. confirmed the greening trend in larger cities, and they note that rapid urbanization caused a dramatic turnover in vegetation cover [13]. Accurate and timely monitoring of the status of urban greenness is essential for the protection and management of urban environments. Furthermore, characterizing and understanding the trends in urban vegetation cover change provide insights into the relationship between urban greenness and urbanization and can help guide sustainable urban development [14].

As an important source of urban information, remote sensing data provide a spatially consistent coverage of large areas with both high spatial detail and high temporal frequency [15]. With the increased availability of current and historical remotely sensed data and new analytical techniques, it is now possible to quantify urban greenness in a timely and cost-effective manner [16,17]. Despite the increasing application of high-resolution images, medium-resolution imagery (e.g., Landsat and SPOT) is still the preferred data source. Specifically, medium-resolution imagery is globally available, and it uniquely composes the only long-term, consistent digital dataset. However, commonly used detection techniques, such as the calculation of classification and vegetation indices from medium spatial resolution imagery, may be ineffective in quantifying physically fine-resolution information related to urban greenness. The urban environment is highly heterogeneous and complex, resulting in the mixed pixel problem [18,19]. Because the status and variation of urban greenness typically occur at finer spatial scales than most moderate resolution imagery, the vegetation in a pixel can only be recorded as either present or absent with traditional hard classifiers, resulting in a loss of information on isolated vegetation and within-class variation [20,21]. However, the status of and subtle change in urban greenness are highly related to the urban environment and significantly affect local residents' quality of life. The specific characteristics of the urban landscape require further consideration regarding the capabilities and limitations of remote sensing data and the use of appropriate analysis techniques [22,23].

A variety of approaches has been developed to overcome the mixed pixel problem in urban landscapes. Spectral mixture analysis techniques (SMA and MESMA) [24] have been widely utilized to estimate the proportion of representative urban land cover within each pixel because of their ability to support the repeatable and accurate extraction of quantitative subpixel information with physical meaning [25,26]. To date, a considerable number of studies have utilized this method to characterize urban compositions [18,27,28], map urban vegetation [16,19,29], and monitor urban changes [30,31,32,33]. However, only a few multi-temporal studies have employed these techniques to monitor long-term changes in urban greenness [34]. Given the sharp growth in urban areas in China in terms of area and intensity, it is imperative to analyze the pattern of vegetation cover change within Chinese cities.

Taking Hangzhou, one of the most rapidly urbanizing cities in eastern China, this paper proposes a feasible and cost-effective greenness change detection method in highly-fragmented urban environment by integrating multi-date remote sensing, MESMA and GIS spatial analysis. Specifically, this paper explores the spatiotemporal dynamics and evolution of greenness in response to the rapid urbanization process in the past two decades (1990–2010) aiming to improve the understanding of the effects of urbanization on greenness patterns and provide basic information for appropriate decision-making towards urban sustainable development.

Methods

Study site

Hangzhou, the capital of Zhejiang Province, is located near the eastern edge of the Qiantang River in the southern Yangtze River Delta (Figure 1). This study was conducted in the area within the administrative boundaries of Hangzhou, the city proper, which covers an area of 728 km^2 and has approximately 3.56 million registered permanent residents. At the end of 2010, the local GDP per capita was approximately 109,708 RMB Yuan (equivalent to 17,981 US dollars) [35]. As a notable international garden city, as well as a vigorous and economically competitive key city, Hangzhou has experienced significant economic development and tremendous population growth over the past decades [15]. The urbanization process has significantly modified the vegetation cover in the wetland, forest, and agriculture ecosystems in and around the city.

Data and preprocessing

The Landsat TM images (Path: 119, Row: 39) used to conduct the study were obtained on October 8, 1990, September 23, 2002, and May 24, 2010 (i.e., a total of 20 years). The level 1T data were carefully selected from the USGS GLOVIS portal [36], with special consideration of phenology and cloudlessness. Thermal bands were excluded from the analysis. FLAASH (Fast Line-of-sight Atmospheric Analysis of Spectral Hypercubes) was used for atmospheric correction of these TM data. The study area was then extracted using the boundaries of Hangzhou.

In addition to Landsat imagery, the following ancillary data were collected for analytical purposes: (1) two historical true color aerial photographs with a 0.5-m resolution, acquired on September 26, 2002 and May 20, 2011; and (2) land-use maps derived from the National Detailed Land-use Inventory for 2010. All the auxiliary data were registered to the same projection as the TM images: WGS 84, UTM map projection Zone 50, with a root mean squared error (RMSE) of less than 15 m.

Greenness mapping requires special consideration of vegetation phenology. Generally, images from the same season with full vegetation growth are the most appropriate for detecting changes. However, because a high-quality image was not available for the late summer of 2010, a TM image acquired on May 24 was used. No significant bias in the vegetation cover was observed in late May or late September (except for cropland), according to our pre-experiment using a MODIS NDVI time series and a comparison between the fraction results from a Landsat 5 TM image acquired on May 24, 2010 and a Landsat 7 ETM+ SLC-off image acquired on September 21, 2010. To minimize the information error caused by the crop phenology, a common cropland mask was developed based on the 2010 land-use map and visual interpretation of three TM images. The common cropland pixels for the three dates were extracted, and the vegetation fraction values were set to a consistent value of 1 (Figure 2c, yellow circles). Thus, we assumed that the phenological effect of the vegetation in our study had no effect on our results during processing.

Multiple endmember spectral mixture analysis

The SMA approach assumes that a landscape is formed from continuously varying proportions of idealized land cover types or spectrally 'pure' materials called endmembers. In linear SMA, the reflectance P' measured at pixel i represents a linear combination of N endmembers, weighted by their areal fractions f_{ki}, within the pixel [24,25,33] as follows:

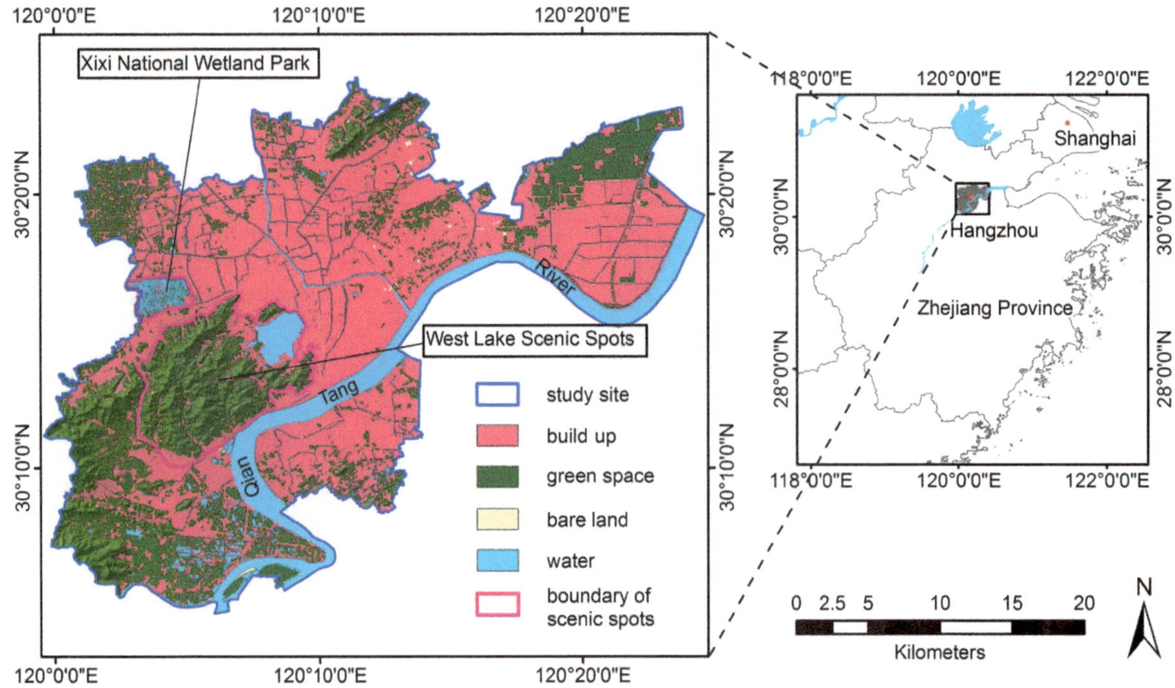

Figure 1. A map of the study area, i.e., Hangzhou city proper.

$$P'_{i\lambda} = \sum_{k=1}^{N} f_{ki} \times P_{k\lambda} + e_{i\lambda} \text{ and } \sum_{k=1}^{N} f_{ki} = 1 \quad (1)$$

where $P_{k\lambda}$ is the reflectance of endmember k for a specific band (λ) and $e_{i\lambda}$ is a residual term indicating the unmodeled portions of the spectrum. The model fit is assessed via an RMSE metric [24,37], calculated as follows:

$$\text{RMSE}_i = \sqrt{\sum_{\lambda=1}^{M} (e_{i\lambda})^2 / M} \quad (2)$$

where M is the number of bands [38]. The standard SMA has two major limitations. First, only one spectrum is allowed for each endmember, i.e., every image pixel is interpreted using the same endmember spectra. Therefore, the method is unable to account for within-class spectral variability. This is particularly problematic in Chinese cities where a large variation of material spectra exists due to rapid development and ubiquitous urban reconstruction. Second, the standard SMA is most often implemented with a fixed number of endmembers for the entire image, regardless of whether the ground components represented by the endmembers are present in the pixel. This may result in the decreased accuracy of the estimated fractions [18,25,39]. MESMA, which is an extension of the standard SMA approach, allows the number and type of endmembers to vary on a per pixel basis to better represent the spectral variability of land cover; thus, it is more suitable for urban landscapes [25,27].

In this study, MESMA was conducted according to the following steps: (1) endmember selection, (2) spectral unmixing modeling, and (3) accuracy assessment. These steps are each detailed in turn. MESMA was implemented using VIPER Tools Version 1.5, which is a plug-in to the ENVI software package [40].

Endmember selection. The careful selection of representative endmembers is essential to all applications of SMA [20,41,42]. In this study, the endmembers were extracted from the 1990, 2002, and 2010 Landsat TM images. When building a spectral library for MESMA, the number of spectra needed for adequately representing the spectral variation in the materials and for attaining computational efficiency should be considered [25,43,44]. Therefore, the endmembers were organized into five groups: vegetation, impervious surfaces, soil, water, and shade. Each group included several subsets, e.g., vegetation (urban forest, primary forest, grass, and crops), impervious surfaces (urban center and suburbs), soil (mine soil, farmland soil, and construction sites), and water (lakes and rivers). An endmember for shade (zero value in all bands) was included in all of the models [41,45].

First, the spectrally "pure" pixels were collected from the image based on the pixel purity index (PPI) [46] and high-resolution aerial photographs to build a candidate spectral library for each endmember. Pixels with high PPI scores, representing pure land cover in the high-resolution aerial photograph and having reasonable spectral signatures, were chosen as candidate spectra and grouped into the corresponding subclasses. The representative spectra for each subclass were then selected from the candidate spectra using the endmember average root mean squared error (EAR) [41], count-based endmember selection (CoB) [47], and minimum average spectral angle (MASA) [38]. The optimal set of spectra for each endmember was iteratively selected by adding spectra with a high CoB index, low EAR, or low MASA to the library and assessing the model performance using RMSE images and a visual comparison with the high-resolution aerial photographs [16]. Ultimately, libraries with 47, 53, and 69 spectra were used for the 1990, 2002, and 2010 images, respectively, to run the unmixing models.

Spectral unmixing modeling. Previous research has shown that natural systems are usually best modeled by two endmembers, while disturbed regions require three and urban landscapes

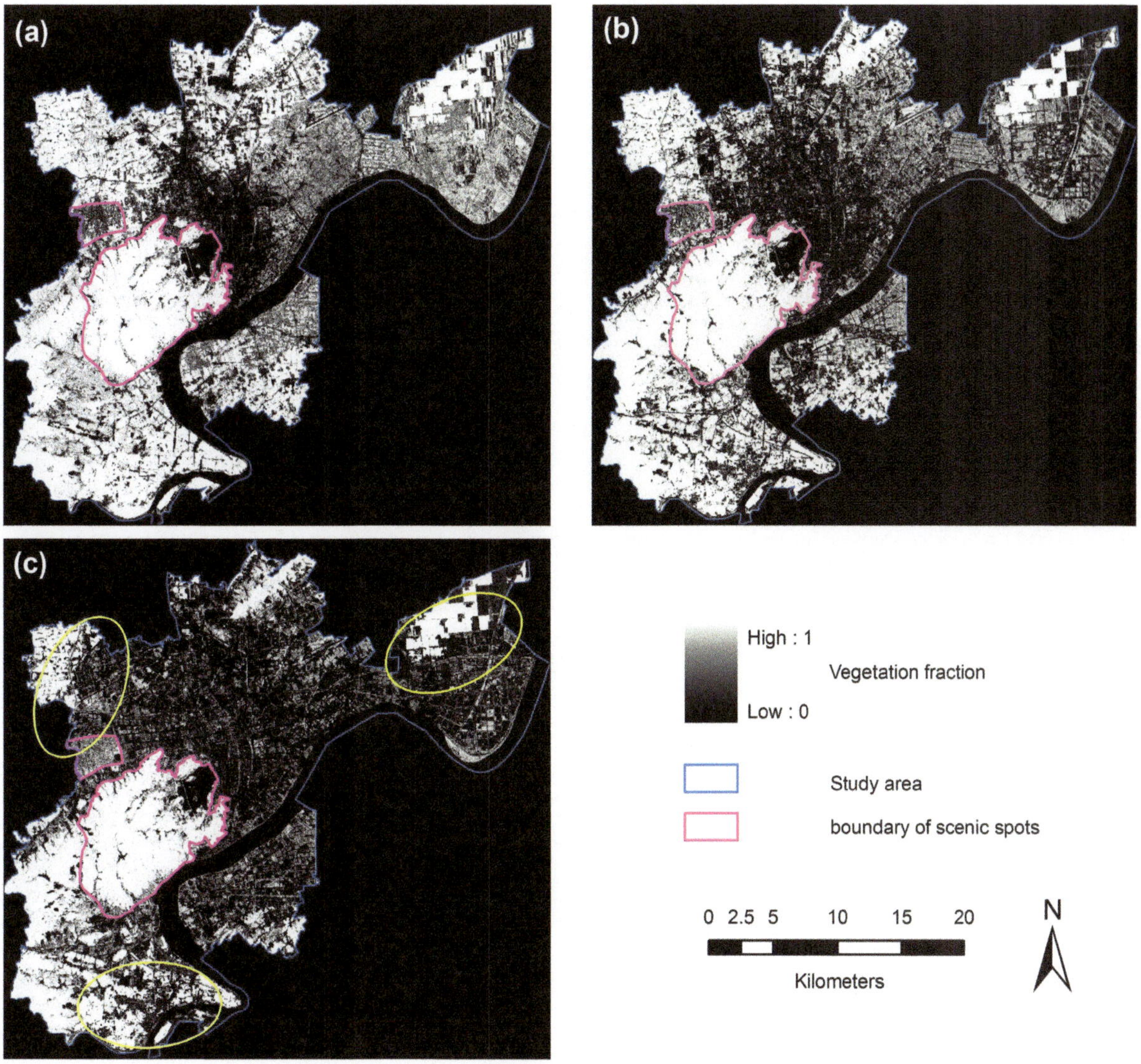

Figure 2. Vegetation fraction maps of Hangzhou for (a) 1990, (b) 2002, and (c) 2010. The vegetation fractions are indicated by the grayscale, where 1 represents pure vegetation (light gray) and 0 represents no vegetation (dark gray) in each pixel.

require four [48]. Therefore, two-endmember models with all the spectra were first applied to map each pixel in the image (Table 1). Then, only spectra that effectively represented the pure land cover within the urban area in the two-endmember models were retained to run three- and four-endmember models. Intra-class spectral mixtures were only considered for impervious surfaces [48]. Every model was evaluated for every pixel in the image. Constraints were determined based on iterative experimentation and the findings of previous studies [37,41,45]. The non-shade fractions were confined to the range of −0.05 to 1.05, the maximum allowable shade fraction was set to 0.5, and a maximum RMSE of 2.5% in reflectance was applied.

After the models with the different levels of spectral complexity (i.e., two-, three-, and four-endmember models) were applied, the model with the lowest RMSE at each level was compared to determine the best model for each pixel. Generally, a lower RMSE in a more complex model does not necessarily indicate better modeling of the true land cover component [28,49]. In addition, we noticed that the introduction of a water endmember significantly improved the model performance in the water-containing and water-edge pixels (e.g., wetlands), but it also caused a mis-unmixing between water and shade classes in pixels affected by topographic shadowing within mountainous regions [50]. Based on previous findings, simpler endmember models are preferred over more complex endmember models, except in cases in which adding an endmember significantly improves the RMSE beyond a specified threshold [30,47,50]. The RMSE threshold value between the two- and three-endmember models was 0.2% for the reflectance, the RMSE threshold between the three- and four-endmember models was 0.4%, and the RMSE threshold between the two- and four-endmember models was 0.6%. When a water endmember was used, the corresponding threshold values were set to 0.7%, 0.4% and 1.1%, respectively, for the reflectance.

The fractions produced by the optimal models were then shade-normalized [24], rescaled [26], and combined to generate fractional maps of each component. Vegetation cover fraction (VCF) maps were retained to investigate the vegetation cover change.

Accuracy assessment. Evaluating vegetation fraction estimates can be challenging due to the difficulty of obtaining

Table 1. Combinations of endmember models by land cover class.

2-endmember	3-endmember	4-endmember
Imp+SHD	Imp_a+Imp_b+SHD	Imp+V+S+SHD
V+SHD	Imp+V+SHD	Imp+V+W+SHD
S+SHD	Imp+S+SHD	S+V+W+SHD
W+SHD	Imp+W+SHD	Imp_a+Imp_b+V+SHD
	V+S+SHD	Imp_a+Imp_b+S+SHD
	V+W+SHD	Imp_a+Imp_b+W+SHD
	S+W+SHD	

V refers to green vegetation, Imp to impervious surface, S to soil, W to water, and SHD to shade. Endmembers with subscripts (a, b) were used multiple times. The same endmember spectra were not used in one combination.

reference data, especially for historical datasets. Because of the lack of high-quality reference images from the 1990s, accuracy assessments for only two dates were performed using the aerial photographs acquired in 2002 and 2011 for the relevant TM images. We assumed that the result of the 1990 fraction image would have acceptable accuracy if the vegetation fraction images in 2002 and 2010 were reasonably accurate. The lack of an accuracy assessment for the 1990 fraction image is a limitation of this study.

Two hundred samples were used to assess the modeled VCF after removing the samples with apparent land cover changes. Each sample had a window size of 3×3 pixels to reduce the impact of geometric errors between the aerial photographs and the TM images [27,51]. The 'true' proportion of vegetation was then calculated by dividing the area of vegetation cover measured through visually interpreting the aerial photograph by the total sampling area (i.e., 8,100 m^2). The accuracy of the vegetation cover fraction was assessed using three metrics: the coefficient of determination (R^2), the root mean squared error (RMSE), and the systematic error (SE) [52] as follows:

$$\mathrm{RMSE} = \sqrt{\sum_{i=1}^{S} (\hat{V}_i - V_i)^2 / S} \quad (3)$$

$$\mathrm{SE} = \sum_{i=1}^{S} (\hat{V}_i - V_i) / S \quad (4)$$

where $\hat{V}_i$ is the modeled fractional value of vegetation measured at sample i, V_i is the reference fractional value, and S is the number of samples. The RMSE is a measure of the overall estimation accuracy for all samples, and the SE is a measure of the effects of systematic errors (e.g., overestimation or underestimation) [53].

RGB-vegetation fraction model

An RGB-vegetation fraction model based on additive color theory was applied to the multi-temporal vegetation fraction images to interpret change events in our study area. Using this technique, multiple dates of vegetation fraction images could be viewed at one time, and the changes were highlighted. Vegetation fraction images in 1990, 2002, and 2010 were placed chronologically by red, green, and blue colors, respectively. Any combination of primary colors of similar brightness produces a complementary color that can identify the direction and magnitude of the vegetation cover change.

Change intensity analysis

To study the spatial distribution of the vegetation change intensity, fraction change images from 1990 to 2002 (T1) and from 2002 to 2010 (T2) were produced by subtracting two sequential fraction images. The intensity of the fraction change was divided into 4 intensity categories: extreme decrease (−1 to −0.5), moderate decrease (−0.5 to −0.2), moderate increase (0.2 to 0.5) and extreme increase (0.5 to 1). The relationship between the change intensity and the distance to the urban center was analyzed by counting the percentage of pixels of a certain intensity category at a certain distance.

Concentric analysis

A concentric analysis technique was used to obtain insight into the relationship between urban vegetation status and urban expansion. Twenty concentric belts were identified, radiating from the urban center to the fringe (Figure 3). Each belt spanned 0.5 km, approximately 16 pixels in the TM image, which is approximately two times the minimum resolution of assessment suggested by Powell et al. [18]. To account for the multi-nuclei expansion pattern in Hangzhou [54], we only applied the concentric analysis to the main urban area rather than to the entire study area. Furthermore, two famous tourist attractions, the West Lake Scenic Spots (WLSS) and the Xixi National Wetland Park (XNWP), were excluded from the concentric analysis because they were subject to a different land use policy than the rest of the study area.

MESMA provided subpixel information with physical meaning, where a fraction value of 1 represented full coverage of vegetation within a pixel. The vegetation fraction images were reclassified into 4 density categories: low coverage (0 to 0.2), middle coverage (0.2 to 0.5), middle-high coverage (0.5 to 0.8), and high coverage (0.8 to 1). The overall average vegetation fraction and the area percentages of the pixels within the 4 density categories were calculated for each belt to estimate the general pattern of urban greenness.

Results

Accuracy assessment

The R^2, RMSE, and SE were calculated for the 2002 and 2010 vegetation fraction images to evaluate the modeled results quantitatively. Scatter plots of the accuracy assessment results are shown in Figure 4. The RMSE was 7.78% for the 2010 image

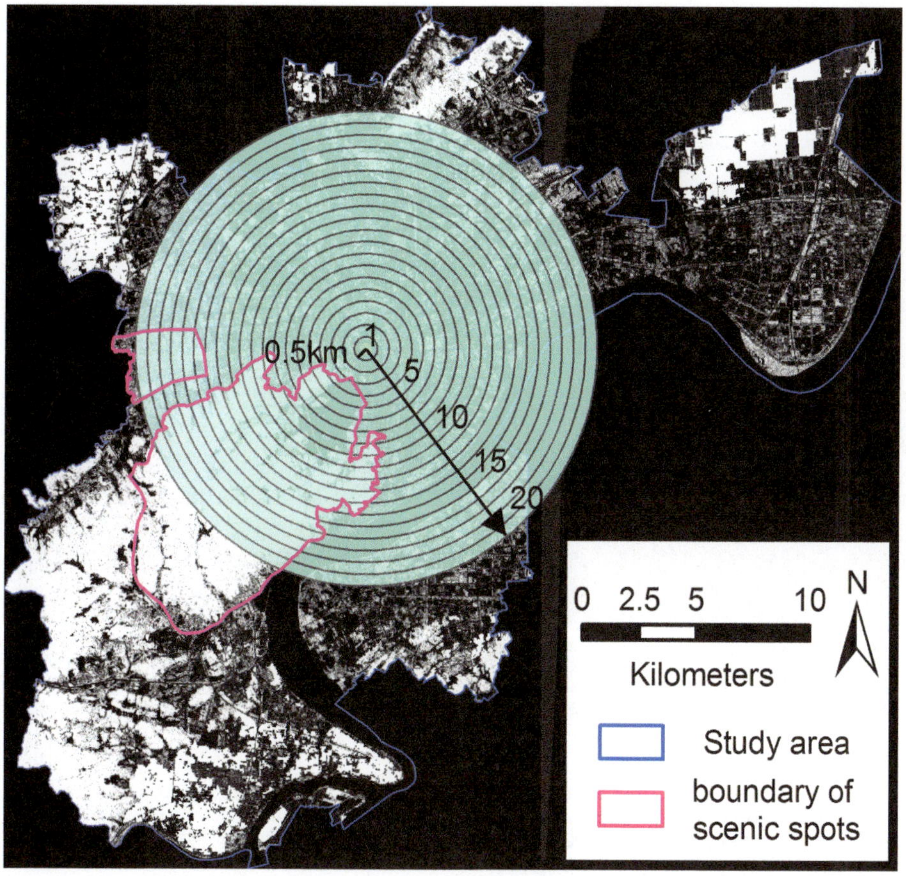

Figure 3. The layout of the concentric belts. Twenty concentric belts radiating from the city center, where each belt spans 0.5 km.

and 10.38% for the 2002 image. The R^2 values were relatively high, ranging from 0.929 to 0.881, indicating a good fit between the 'true' fraction and the fraction derived from MESMA.

The analysis of the residuals, in terms of the SE (−0.800% for 2010 and 0.320% for 2002), indicated that there was no significant underestimation or overestimation relative to the reference fraction.

Spatiotemporal analysis of urban greenness change

To construct vegetation cover images for corresponding years, the amount of vegetation in each pixel was represented by the assigned gray value. Brighter areas signify higher proportions of vegetation, and darker areas signify lower proportions of vegetation.

Overall, the urban greenness of Hangzhou has decreased steadily over the past two decades (Figures 2a, b, and c). In 1990, most of the high-density green space was distributed in natural land-use areas (i.e., agricultural areas, primary forests, and wetlands) around the built-up area, and the vegetation cover within the city was relatively sparse. Along with the urban expansion, an obvious loss of greenness occurred in the urban fringe and the suburban area. The urban greenness became more fragmented and isolated, except in mountainous areas and in the remaining cropland. The total percentage of the vegetated area of Hangzhou decreased greatly from 64.61% to 42.76% (Table 2), especially in T2, nearly twice the rate observed in T1.

Using the RGB-vegetation fraction model, distinctly different patterns were demonstrated among the various segments of the urban area (Figure 5). The core built-up area exhibited an increasing percentage of vegetation cover in T1 and T2, dominated by cyan and blue colors (Figure 5c). The other colors, such as red, yellow and green, revealed a complex reconfiguration of urban greenness in the core built-up area over the past two decades due to urban reconstruction and the urban greening project. In contrast, the urban peripheral areas showed dramatic declines in vegetation cover in the two periods. Vast fields of red pixels were observed near the core built-up area, indicating vegetation loss occurred during the urban expansion in T1 (Figure 5b). Two large blocks of vegetation reduction occurred in the eastern and southern parts of the city due to the construction of two new subcities (Xiasa and Jiangnan). Furthermore, a more severe decrease in the greenness in T2 could be confirmed by the widespread yellow color around the existing built-up areas in the main city and in the two subcities. In Xiasa, a suburban college town, road networks and communities constructed in T1 were depicted in red (Figure 5e). The remaining vegetation along the road networks was replaced by new construction in T2, indicated by yellow. Meanwhile, some of the roads and communities represented by magenta suggest vegetation recovery in the form of road and community greening after 2002. In most WLSS regions, the vegetation coverage remained consistently high, as indicated in white. However, a much more complex color composition of blue, green, purple and magenta was observed in XNWP and reflects a complex but overall subtle increase in wetland vegetation coverage over the past two decades.

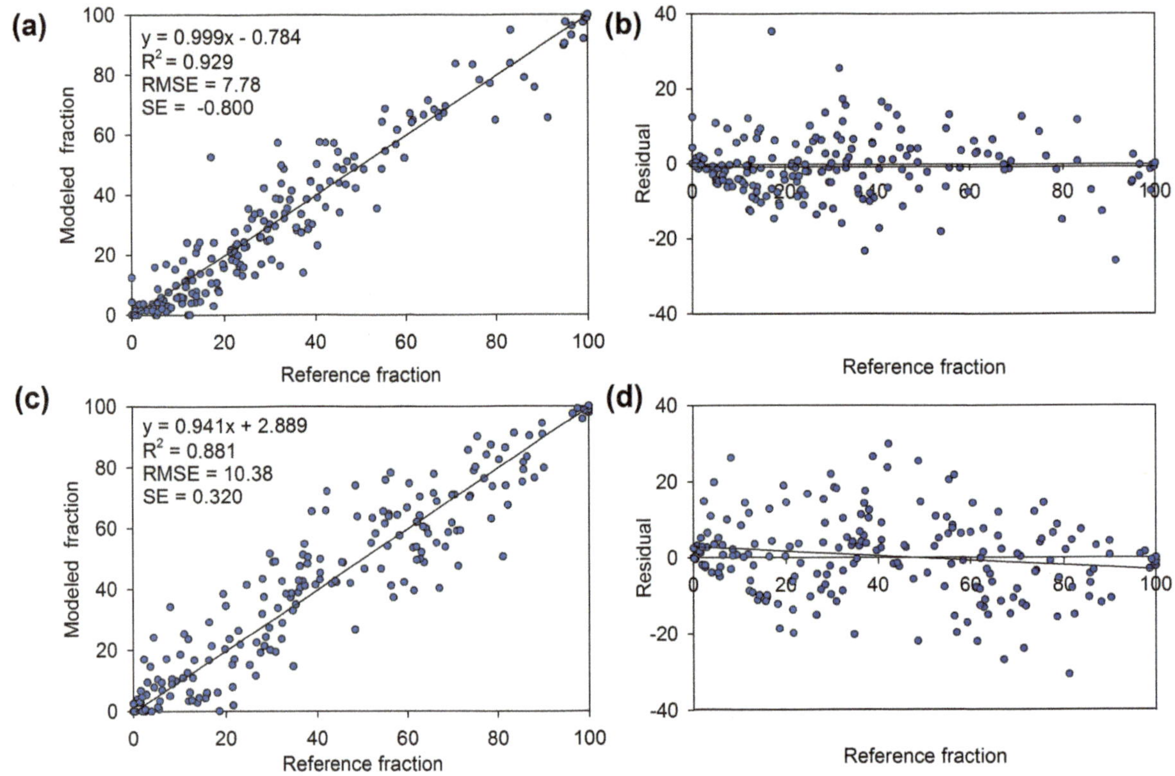

Figure 4. Comparisons between the reference and modeled fractions. (**a**) Vegetation fraction scatter plot for 2010, (**b**) vegetation fraction residuals for 2010, (**c**) vegetation fraction scatter plot for 2002, and (**d**) vegetation fraction residuals for 2002. Perfect agreement, represented by the 1:1 line, is displayed in the scatter plots. The best-fit line is displayed in the residual plots to indicate the general trends of overestimation and underestimation.

A clear spatial disparity in vegetation coverage change intensity was found (Figure 6) by analyzing the relationships between the change intensity and the distance to the urban center. In T1, the vegetation change pattern was dominated by a moderate increase within 2 km. At 2 km, the extreme decrease soared to over 20% and became the major change process. Two peaks appeared at approximately 4 km and 19 km, reflecting the vegetation loss due to the expansion of the core built-up area and construction of the new subcities. In T2, a greater intensity and extent of moderate increases was observed; this suggests that vegetation coverage more vigorously recovered in the core built-up area. However, a more severe extreme decrease occurred in the outer region of the core built-up area and subcities and peaked at approximately 11 km and 15 km. Therefore, the hotspot of the vegetation decline moved to the suburban area between the core built-up area and the subcities. The conversion, both negative and positive, was much more significant in the later decade. A pattern similar to the class of the extreme decreases could be found in the class of moderate decreases, while the class of extreme increases remained consistently low and showed no distinct pattern related to distance in either period.

General pattern of urban greenness accompanied by urban expansion

Twenty concentric belts radiating from the city center to the fringe were analyzed to provide insights into the spatial patterns of urban greenness change within and around the city during the two periods. In 1990, the average vegetation fraction remained low in belts 1–5 but increased significantly in belt 6; this highlights the difference in vegetation cover between the core built-up area and

Table 2. Total percent vegetated area and changes from 1990 to 2010 for the study area and the two scenic spots.

	1990	2002	2010	1990–2002 (per year)	1990–2002 (per year)
TPVA	64.61%	54.51%	42.76%	−0.84%	−1.53%
TPVA in XNWP	52.39%	54.44%	58.92%	0.17%	0.58%
TPVA in WLSS	85.51%	84.70%	83.13%	−0.07%	−0.20%

TPVA is the total percent of vegetated area, XNWP is the Xixi National Wetland Park, and WLSS is the West Lake Scenic Spots.

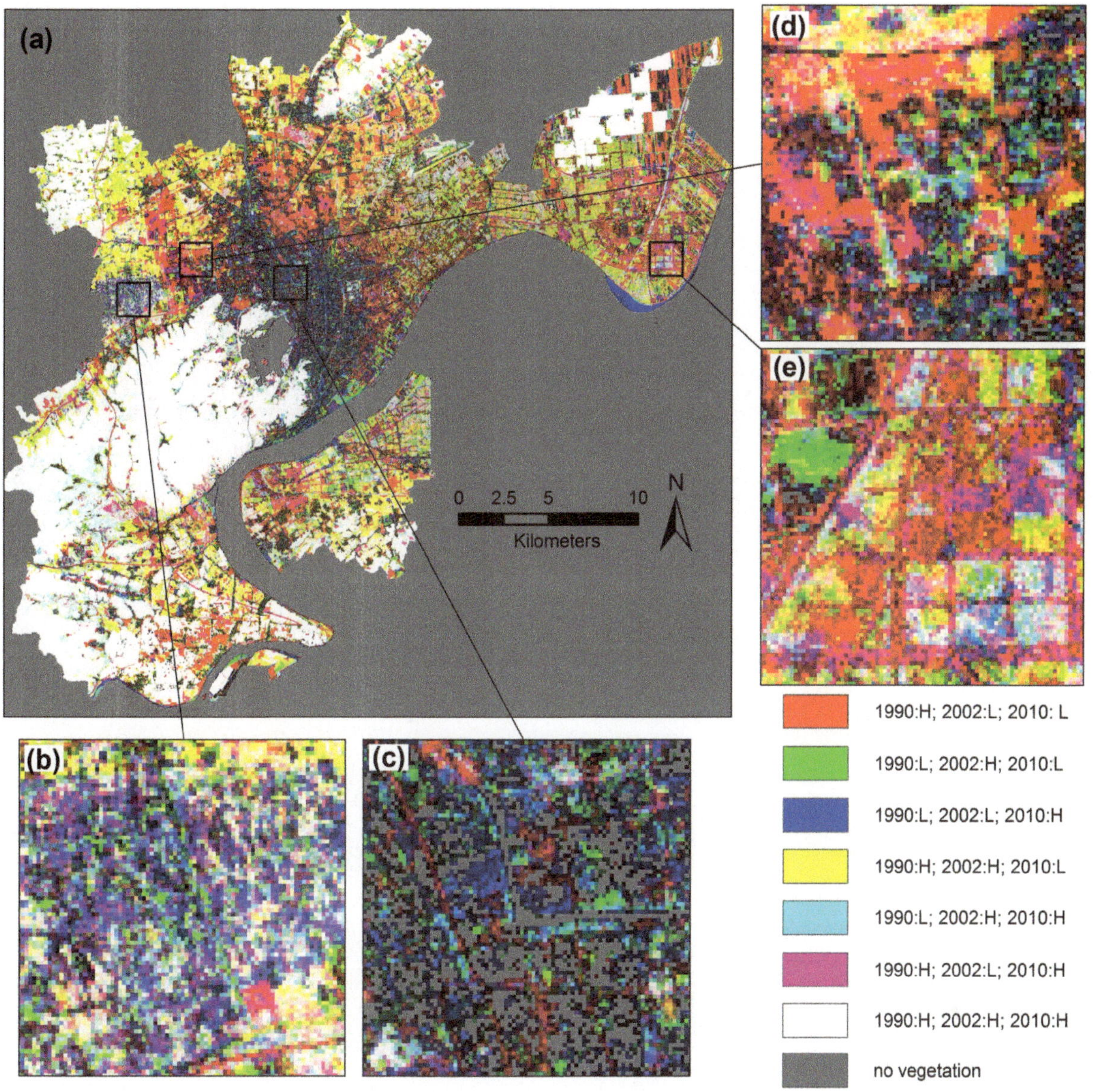

Figure 5. RGB color composite using vegetation fraction maps of 1990 (R), 2002 (G) and 2010 (B). The typical examples demonstrated are (a) the study area, (b) Xixi National Wetland Park, (c) the urban center, (d) residential communities and (e) the Xiasha suburban college town. Colors for the typical compositions of the vegetation fractions on the three dates are illustrated. H represents high vegetation fraction and L represents low vegetation fraction.

the suburbs. Because of urban expansion and sprawl, the boundary between the core built-up area and the suburbs was extrapolated to location near belt 9 in 2002. Meanwhile, a slight increase was detected for belts 1–5, indicating an improvement in the vegetation cover in the existing urban area. By 2010, a more obvious increase occurred in belts 1–9, along with a more severe decrease in belts 10–20. Additionally, the correlation between the vegetation abundance and the distance from the city center became weaker. These changes occurred because nearly all the regions within the twenty belts were highly urbanized, and only sparse native vegetation remained in belts 16–20. With the expansion and sprawl of built-up areas, urban greenness declined sharply in the urban fringe and the suburbs. A subsequent recovery of the vegetation cover occurred in the existing urban area, along with a decrease in the outer belts. This reflects the main process of conversion from natural vegetation to artificial urban greening because of urban expansion.

Urban greenness in the existing urban area contributes valuable ecosystem services which are related closely with the daily life of the local residents [55]. The lowest vegetation fraction was found in 1990 (Figure 7a: belts 1–5, 1990). Since then, the vegetation cover continued to increase in the older neighborhoods over the past two decades (belts 1–5 in T1 and belts 1–9 in T2), indicating sustained improvement in the urban environment can be a component of the urbanization. In 2010, the average vegetation fraction of the urban area was nearly double that in 1990. In addition, an obvious increase in the middle cover pixels occurred in belts 1–10, especially in belts 1–5 (Figure 7d), along with a decrease in the low cover pixels (Figure 7e). The result of change intensity analysis also showed that the changing pattern in urban center was dominated by moderate

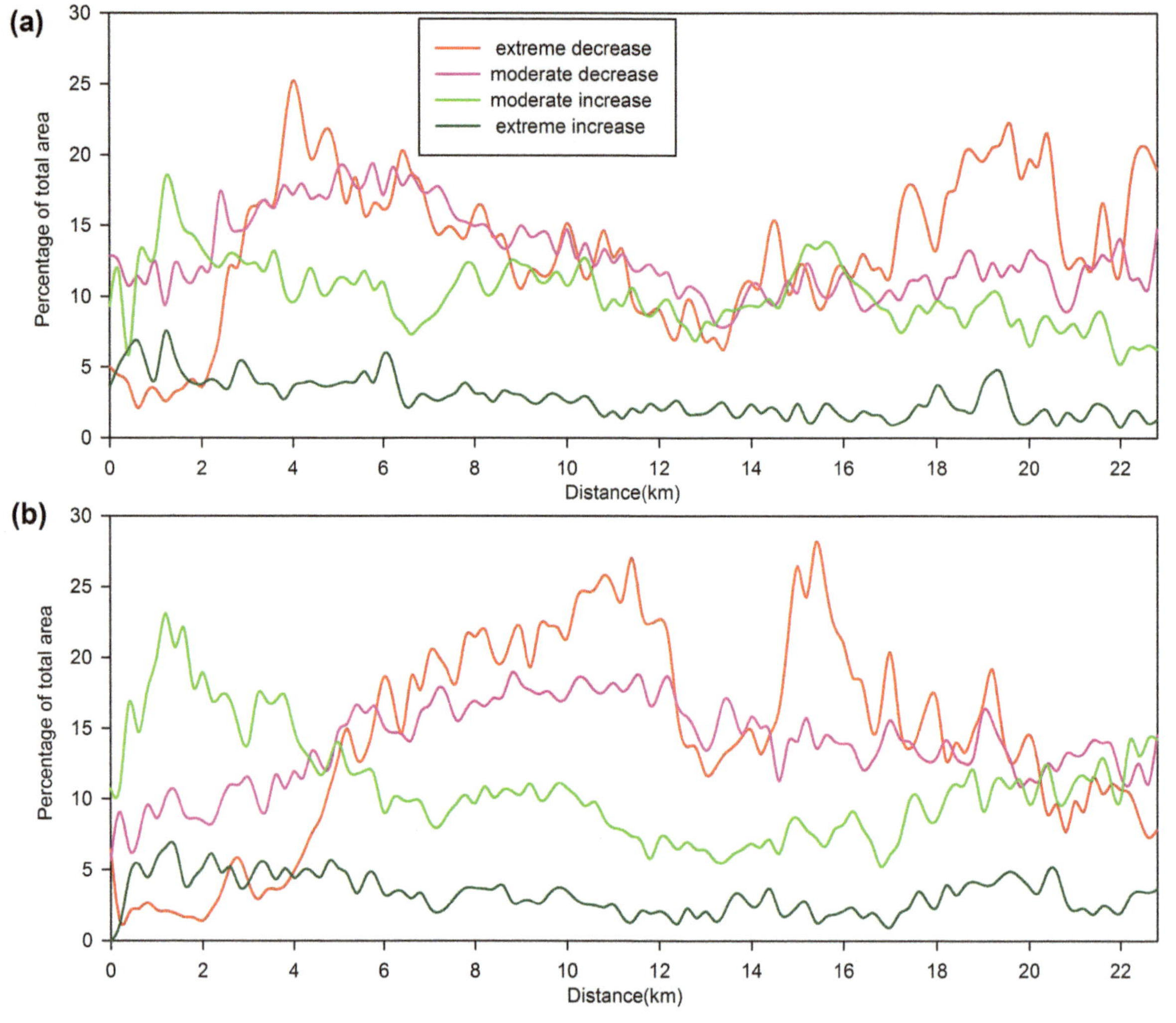

Figure 6. Spatial characteristics of the vegetation fraction change from the urban center over the two periods. (a) 1990–2002 and (b) 2002–2010. The vegetation change categories are shown by the different colors in the legend.

increase. These characteristics of vegetation fraction within a pixel further indicated those small green patches with size smaller than the pixel size of the Landsat data has mainly contributed to the increase of green space in existing urban area.

Discussion

MESMA and integration of the multiple analysis technique

The accuracy of the information acquired on vegetation cover has a direct influence on how well the trends and processes of greenness change are understood, especially for a diverse urban area. Urban greenness is generally highly fragmented in Chinese cities, resulting in a large number of small patches. Li et al. indicated that 86% of the patches of urban greenness were smaller than 900 m^2, and 37% were smaller than 100 m^2 in their study area [56]. Likewise, our study highlights the importance of the increasing area of small green patches, which may be difficult to detect by traditional classification techniques. An analysis based on directly classifying medium-resolution data would lead to an underestimation of the amount of urban greenness, which is crucial to urban areas with generally limited amounts of vegetation cover. This study demonstrated the advantage of MESMA in providing accurate and reliable subpixel information with physical meaning. However, this method still failed to acquire accurate information on the species and structure of urban greenness in multispectral images. Moreover, the method was unable to resolve nonlinear mixing effects, and it generally needed more supervision in the unmixing process. By integrating RGB-vegetation fraction model, the change intensity analysis and concentric analysis, detailed information on the direction, magnitude, location and pattern of the variation of urban greenness were able to be visually and quantitatively analyzed. These information have improved our understanding of how urban greenness evolved under rapid urbanization. The strength of MESMA is its ability to represent a wide variety of surface reflectance types as simple combinations of endmember abundances. Furthermore, because of the global availability and rich historical data archive of the Landsat data, analysis based on these techniques can be easily expanded to other cities around the world.

Greenness change in response to urbanization and greening policies

Despite the proven benefits of urban green spaces [3,4,5,6,7], widespread losses of these spaces have been reported in American

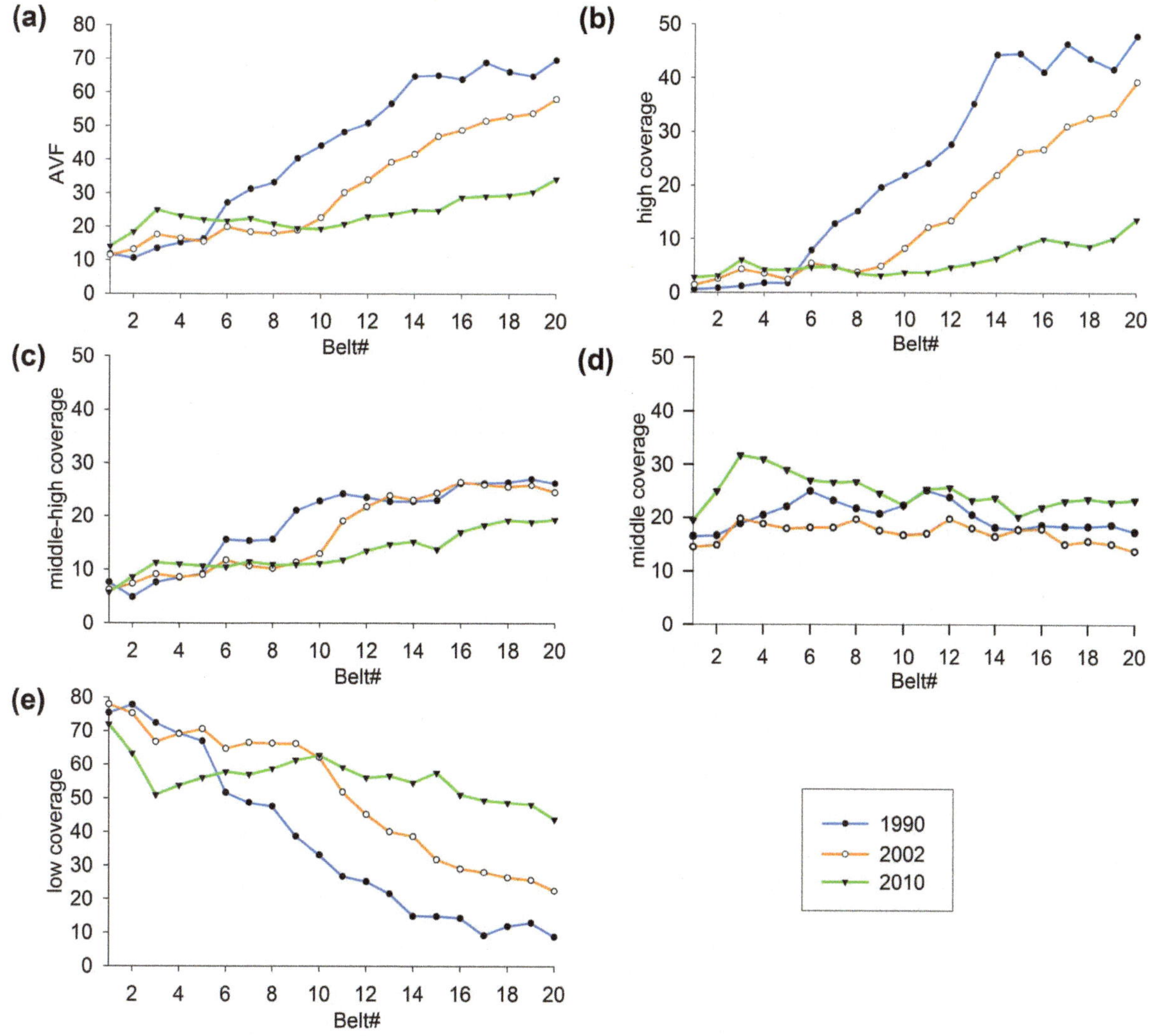

Figure 7. Concentric vegetation coverage analysis in each belt for each year. (a) Average vegetation fraction (AVF), (b) percentage of area of high coverage pixels, (c) percentage of area of middle-high coverage pixels, (d) percentage of area of middle coverage pixels, and (e) percentage of area of low coverage pixels.

and European cities [9,11]. In our study area, one-third of the vegetation cover that was present in 1990 was gradually lost over these two decades. The RGB-vegetation fraction model showed that most of the vegetation loss occurred in the urban fringe or in the new subcities. This trend agrees with the regional decline in vegetation cover caused by urbanization, as observed by Sun et al. [57]. In contrast to the overall decline throughout the study area, both the RGB-vegetation fraction model and concentric analysis demonstrated a consistent increase in urban greenness in the existing urban area. This confirmed the finding that Chinese cities have undergone a steady increase in urban greenness in the built-up areas [12]. We further demonstrated the contribution to the recovery of urban greenness by the increasing area of small green patches. The vegetation cover of the scenic areas remained relatively stable, even with a slight increase in XNWP, which reflects the effect of a scenic protection policy and an ecological renewal project. The changing pattern of greenness in the study area from 1990 to 2010 was a response to the combined effects of rapid urbanization and greening policies. Because of China's unique institutional character, cities have developed into modern metropolitan areas in a much different way than other regions. The government plays a key role in guiding urban planning. Greenness was very limited in the urban area in 1990, and most of the greenness was located in the West Lake Scenic Spots. This uneven distribution limited the access of urban residents to the ecological benefits of urban green space. The high degree of urbanization in the city center and the rapid growth of urban population exacerbated the demand for urban greenness but also placed limits on its development. The implementation of new urbanization strategies since 2000, such as "leapfrog development along the Qiantang River" and "crossing the Qiantang River and developing southward", triggered eastward sprawl and the construction of two subcities [54], resulting in the use of more space to reduce the population and building density and the availability of space for the creation and development of urban green space system. A green network system called "Two rings, Two axles, Six ecological zones" and a series of greening projects, e.g., the West Lake comprehensive protection project, the Grand Canal comprehensive protection project, and the Xixi wetland comprehensive protection project, have been conducted. These projects contributed to the accelerated recovery of urban greenness from 2002 to 2010 within the city. Meanwhile, innovative efforts such as retrofitting green space alongside formerly dilapidated canals, underneath and alongside main roads and railway lines were done to activate neglected spaces to restore lost green space [58]. In addition, the urban greening management regulation of Hangzhou set the minimum greening cover for new residential area to 30%. These factors explain the remarkable increase in the number of middle cover pixels.

Conclusions

Acquiring accurate and area-wide information remains a challenge in urban environments due to the extremely heterogeneous spatial patterns within cities. Subpixel methods can overcome this problem and provide valuable quantitative information. In this paper, we investigated the patterns of change in urban greenness in Hangzhou, China, over the past two decades using Landsat imagery from 1990, 2002, and 2010. Multiple endmember spectral mixture analysis (MESMA) was used to derive vegetation cover fractions (VCFs) for various segments of the urban area. Despite the complex spectral confusion of different land cover types in the urban area, MESMA could provide accurate and reliable subpixel information of the vegetation cover fraction. An RGB-vegetation fraction model, change intensity analysis and the concentric technique were integrated to characterize the detailed, spatial characteristics and overall pattern of the urban greenness change over time.

The study area has experienced a drastic loss of urban greenness. Nearly one-third of the vegetation coverage that was present in 1990 was gradually lost over the past two decades. Most of this loss is located in the urban fringe and suburban areas due to urban expansion and the construction of new subcities. No significant change occurred in scenic spots, except for a recovery of vegetation coverage in the Xixi National Wetland Park from 2002 to 2010. Meanwhile, a remarkable recovery in the vegetation coverage occurred in the existing urban area that was mainly attributed to the continuous increase in small green patches. These changing patterns were more obvious from 2002 to 2010 than from 1990 to 2002. The characteristics of the changing pattern of urban greenness over time and in different parts of the urban environment reveal the combined effects of rapid urbanization and greening policies.

This study demonstrated the unique role of medium-resolution satellite images in the longitudinal analysis of the evolution of urban greenness. Subpixel methods provided abundant meaningful information, and more analysis techniques could be developed based on such continuous information. Furthermore, regional comparisons of cities with different economic and natural conditions should be conducted using these techniques to investigate the relationships between urban greenness and urbanization.

Acknowledgments

We thank Shahtahmassebi Amirreza for his help during the study and the anonymous reviewers and the editor for their valuable comments, which helped improve our manuscript.

Author Contributions

Conceived and designed the experiments: MYG JSD KW. Performed the experiments: MYG XYZ. Analyzed the data: MYG XYZ. Contributed reagents/materials/analysis tools: MYG XYZ. Contributed to the writing of the manuscript: MYG XYZ. Designed/drew figures: MYG XYZ. Revised the manuscript: MYG JSD YH KW.

References

1. Grimm NB, Faeth SH, Golubiewski NE, Redman CL, Wu JG, et al. (2008) Global change and the ecology of cities. Science 319: 756–760.
2. Robinson SL, Lundholm JT (2012) Ecosystem services provided by urban spontaneous vegetation. Urban Ecosystems 15: 545–557.
3. Davies RG, Barbosa O, Fuller RA, Tratalos J, Burke N, et al. (2008) City-wide relationships between green spaces, urban land use and topography. Urban Ecosystems 11: 269–287.
4. Walsh CJ, Fletcher TD, Burns MJ (2012) Urban Stormwater Runoff: A New Class of Environmental Flow Problem. PLoS ONE 7: e45814.
5. Akbari H, Pomerantz M, Taha H (2001) Cool surfaces and shade trees to reduce energy use and improve air quality in urban areas. Solar Energy 70: 295–310.
6. Donovan GH, Butry DT, Michael YL, Prestemon JP, Liebhold AM, et al. (2013) The Relationship Between Trees and Human Health: Evidence from the Spread of the Emerald Ash Borer. American Journal of Preventive Medicine 44: 139–145.
7. Everson-Rose SA, Lewis TT (2004) Psychosocial factors and cardiovascular diseases. Annual Review of Public Health 26: 469–500.
8. Tan PY, Wang J, Sia A (2013) Perspectives on five decades of the urban greening of Singapore. Cities 32: 24–32.
9. Fuller RA, Gaston KJ (2009) The scaling of green space coverage in European cities. Biology Letters 5: 352–355.
10. Dallimer M, Tang ZY, Bibby PR, Brindley P, Gaston KJ, et al. (2011) Temporal changes in greenspace in a highly urbanized region. Biology Letters 7: 763–766.
11. Nowak DJ, Greenfield EJ (2012) Tree and impervious cover change in US cities. Urban Forestry & Urban Greening 11: 21–30.
12. Zhao JJ, Chen SB, Jiang B, Ren Y, Wang H, et al. (2013) Temporal trend of green space coverage in China and its relationship with urbanization over the last two decades. Science of the Total Environment 442: 455–465.
13. Yang J, Huang C, Zhang Z, Wang L (2013) The temporal trend of urban green coverage in major Chinese cities between 1990 and 2010. Urban Forestry & Urban Greening.
14. Zhou XL, Wang YC (2011) Spatial-temporal dynamics of urban green space in response to rapid urbanization and greening policies. Landscape and Urban Planning 100: 268–277.
15. Deng JS, Wang K, Hong Y, Qi JG (2009) Spatio-temporal dynamics and evolution of land use change and landscape pattern in response to rapid urbanization. Landscape and Urban Planning 92: 187–198.
16. Liu T, Yang XJ (2013) Mapping vegetation in an urban area with stratified classification and multiple endmember spectral mixture analysis. Remote Sensing of Environment 133: 251–264.
17. He C, Convertino M, Feng Z, Zhang S (2013) Using LiDAR Data to Measure the 3D Green Biomass of Beijing Urban Forest in China. PLoS ONE 8: e75920.
18. Powell RL, Roberts DA, Dennison PE, Hess LL (2007) Sub-pixel mapping of urban land cover using multiple endmember spectral mixture analysis: Manaus, Brazil. Remote Sensing of Environment 106: 253–267.
19. Small C (2001) Estimation of urban vegetation abundance by spectral mixture analysis. International Journal of Remote Sensing 22: 1305–1334.
20. Myint SW (2006) Urban vegetation mapping using sub-pixel analysis and expert system rules: A critical approach. International Journal of Remote Sensing 27: 2645–2665.
21. Foody GM, Boyd DS (1999) Detection of partial land cover change associated with the migration of inter-class transitional zones. International Journal of Remote Sensing 20: 2723–2740.
22. Herold M, Couclelis H, Clarke KC (2005) The role of spatial metrics in the analysis and modeling of urban land use change. Computers, Environment and Urban Systems 29: 369–399.
23. Patino JE, Duque JC (2013) A review of regional science applications of satellite remote sensing in urban settings. Computers, Environment and Urban Systems 37: 1–17.
24. Adams JB, Smith MO, Gillespie AR (1993) Imaging spectroscopy: Interpretation based on spectral mixture analysis. In: CM P, P E, ^editors. Remote geochemical analysis: elemental and mineralogical composition: New York: Cambridge Univ. pp. 145–166.
25. Roberts DA, Gardner M, Church R, Ustin S, Scheer G, et al. (1998) Mapping chaparral in the Santa Monica Mountains using multiple endmember spectral mixture models. Remote Sensing of Environment 65: 267–279.
26. Smith MO, Ustin SL, Adams JB, Gillespie AR (1990) Vegetation in deserts: I. A regional measure of abundance from multispectral images. Remote Sensing of Environment 31: 1–26.
27. Lu DS, Weng QH (2004) Spectral mixture analysis of the urban landscape in Indianapolis with landsat ETM plus imagery. Photogrammetric Engineering & Remote Sensing 70: 1053–1062.
28. Rashed T, Weeks JR, Roberts D, Rogan J, Powell R (2003) Measuring the physical composition of urban morphology using multiple endmember spectral

mixture models. Photogrammetric Engineering & Remote Sensing 69: 1011–1020.
29. Small C, Lu JWT (2006) Estimation and vicarious validation of urban vegetation abundance by spectral mixture analysis. Remote Sensing of Environment 100: 441–456.
30. Michishita R, Jiang Z, Xu B (2012) Monitoring two decades of urbanization in the Poyang Lake area, China through spectral unmixing. Remote Sensing of Environment 117: 3–18.
31. Powell RL, Roberts DA (2010) Characterizing urban land-cover change in Rondônia, Brazil: 1985 to 2000. Journal of Latin American Geography 9: 183–211.
32. Weng QH, Lu DS (2009) Landscape as a continuum: an examination of the urban landscape structures and dynamics of Indianapolis City, 1991–2000, by using satellite images. International Journal of Remote Sensing 30: 2547–2577.
33. Rashed T, Weeks JR, Stow D, Fugate D (2005) Measuring temporal compositions of urban morphology through spectral mixture analysis: toward a soft approach to change analysis in crowded cities. International Journal of Remote Sensing 26: 699–718.
34. Tang JM, Chen F, Schwartz SS (2012) Assessing spatiotemporal variations of greenness in the Baltimore-Washington corridor area. Landscape and Urban Planning 105: 296–306.
35. Hangzhou almanac 2011 (2012) Beijing: Fangzhi Press. 546 p.
36. USGS Global Visualization Viewer. Available: http://glovis.usgs.gov/. Accessed 2013 Oct 17.
37. Roberts DA, Batista G, Pereira J, Waller E, Nelson B (1999) Change Identification Using Multitemporal Spectral Mixture Analysis: Applications in Eastern Amazonia. In: Lunetta R, Elvidge C, ^editors. London: Taylor & Francis. pp. 318, 16.
38. Dennison PE, Halligan KQ, Roberts DA (2004) A comparison of error metrics and constraints for multiple endmember spectral mixture analysis and spectral angle mapper. Remote Sensing of Environment 93: 359–367.
39. Song CH (2005) Spectral mixture analysis for subpixel vegetation fractions in the urban environment: How to incorporate endmember variability? Remote Sensing of Environment 95: 248–263.
40. Roberts DA, Halligan K, Dennison PE (2007) VIPER Tools user manual (Version1.5) University of California at Santa Barbar a, 91P. Available: http://www.vipertools.org/. Accessed 2013 Aug 28.
41. Dennison PE, Roberts DA (2003) Endmember selection for multiple endmember spectral mixture analysis using endmember average RMSE. Remote Sensing of Environment 87: 123–135.
42. Tompkins S, Mustard JF, Pieters CM, Forsyth DW (1997) Optimization of endmembers for spectral mixture analysis. Remote Sensing of Environment 59: 472–489.
43. Okin GS, Roberts DA, Murray B, Okin WJ (2001) Practical limits on hyperspectral vegetation discrimination in arid and semiarid environments. Remote Sensing of Environment 77: 212–225.
44. Somers B, Asner GP, Tits L, Coppin P (2011) Endmember variability in Spectral Mixture Analysis: A review. Remote Sensing of Environment 115: 1603–1616.
45. Powell R (2011) Characterizing Urban Subpixel Composition Using Spectral Mixture Analysis. In: Yang X, edito. Urban Remote Sensing: Monitoring, Synthesis and Modeling in the Urban Environment. chichester,UK: Wiley-blackwell. pp. 111–128.
46. Boardman JW, Kruse FA, Green RO (1995) Mapping target signatures via partial unmixing of AVIRIS data. In AVIRIS Airborne Geoscience Workshop Proceedings. Pasadena.
47. Roberts DA, Dennison PE, Gardner ME, Hetzel Y, Ustin SL, et al. (2003) Evaluation of the potential of Hyperion for fire danger assessment by comparison to the Airborne Visible/Infrared Imaging Spectrometer. IEEE Transactions on Geoscience & Remote Sensing 41: 1297–1310.
48. Powell RL, Roberts DA (2008) Characterizing Variability of the Urban Physical Environment for a Suite of Cities in Rondonia, Brazil. Earth Interactions 12.
49. Dennison PE, Roberts DA (2003) The effects of vegetation phenology on endmember selection and species mapping in southern California chaparral. Remote Sensing of Environment 87: 295–309.
50. Michishita R, Gong P, Xu B (2012) Spectral mixture analysis for bi-sensor wetland mapping using Landsat TM and Terra MODIS data. International Journal of Remote Sensing 33: 3373–3401.
51. Lu DS, Moran E, Hetrick S (2011) Detection of impervious surface change with multitemporal Landsat images in an urban-rural frontier. ISPRS Journal of Photogrammetry and Remote Sensing 66: 298–306.
52. Wu CS, Yuan F (2007) Seasonal sensitivity analysis of impervious surface estimation with satellite imagery. Photogrammetric Engineering & Remote Sensing 73: 1393–1401.
53. Wu CS (2004) Normalized spectral mixture analysis for monitoring urban composition using ETM plus imagery. Remote Sensing of Environment 93: 480–492.
54. Wu K, Zhang H (2012) Land use dynamics, built-up land expansion patterns, and driving forces analysis of the fast-growing Hangzhou metropolitan area, eastern China (1978–2008). Applied Geography 34: 137–145.
55. Jim CY, Chen WY (2009) Ecosystem services and valuation of urban forests in China. Cities 26: 187–194.
56. Li XM, Zhou WQ, Ouyang ZY (2013) Relationship between land surface temperature and spatial pattern of greenspace: What are the effects of spatial resolution? Landscape and Urban Planning 114: 1–8.
57. Sun J, Wang X, Chen A, Ma Y, Cui M, et al. (2011) NDVI indicated characteristics of vegetation cover change in China's metropolises over the last three decades. Environmental monitoring and assessment 179: 1–14.
58. Wolch JR, Byrne J, Newell JP (2014) Urban green space, public health, and environmental justice: The challenge of making cities 'just green enough'. Landscape and Urban Planning.

9

Hydrologic Landscape Regionalisation Using Deductive Classification and Random Forests

Stuart C. Brown[1]*, Rebecca E. Lester[1], Vincent L. Versace[2,3], Jonathon Fawcett[4], Laurie Laurenson[1]

1 School of Life and Environmental Sciences, Deakin University, Warrnambool, Victoria, Australia, **2** Greater Green Triangle University, Department of Rural Health, Flinders University, Warrnambool, Victoria, Australia, **3** Centre for Environmental Management, Faculty of Science and Technology, Federation University Australia, Ballarat, Victoria, Australia, **4** Sinclair-Knight-Merz, Bendigo, Victoria, Australia

Abstract

Landscape classification and hydrological regionalisation studies are being increasingly used in ecohydrology to aid in the management and research of aquatic resources. We present a methodology for classifying hydrologic landscapes based on spatial environmental variables by employing non-parametric statistics and hybrid image classification. Our approach differed from previous classifications which have required the use of an *a priori* spatial unit (e.g. a catchment) which necessarily results in the loss of variability that is known to exist within those units. The use of a simple statistical approach to identify an appropriate number of classes eliminated the need for large amounts of *post-hoc* testing with different number of groups, or the selection and justification of an arbitrary number. Using statistical clustering, we identified 23 distinct groups within our training dataset. The use of a hybrid classification employing random forests extended this statistical clustering to an area of approximately 228,000 km^2 of south-eastern Australia without the need to rely on catchments, landscape units or stream sections. This extension resulted in a highly accurate regionalisation at both 30-m and 2.5-km resolution, and a less-accurate 10-km classification that would be more appropriate for use at a continental scale. A smaller case study, of an area covering 27,000 km^2, demonstrated that the method preserved the intra- and inter-catchment variability that is known to exist in local hydrology, based on previous research. Preliminary analysis linking the regionalisation to streamflow indices is promising suggesting that the method could be used to predict streamflow behaviour in ungauged catchments. Our work therefore simplifies current classification frameworks that are becoming more popular in ecohydrology, while better retaining small-scale variability in hydrology, thus enabling future attempts to explain and visualise broad-scale hydrologic trends at the scale of catchments and continents.

Editor: Rachata Muneepeerakul, Arizona State University, United States of America

Funding: The authors received no specific funding for this work. Jonathon Fawcett is employed by Sinclair-Knight-Merz. Sinclair-Knight-Merz provided support in the form of salary for author JF, but did not have any additional role in the study design, data collection and analysis, decision to publish, or preparation of the manuscript. The specific role of this author is articulated in the 'author contributions' section.

Competing Interests: Jonathon Fawcett is employed by Sinclair-Knight-Merz. There are no patents, products in development or marketed products to declare.

* Email: stuart.brown@research.deakin.edu.au

Introduction

Flow variability and ecological controls

Long-term trends in flow variability in streams have the ability to create and maintain ecosystem dynamics for a range of ecologically-important conditions [1,2] and can therefore influence biotic communities and abiotic conditions [1] at local to regional scales, both temporally and spatially [1,3]. These long-term trends are controlled by the same factors influencing the hydrologic cycle in a landscape and ultimately influence physical habitat and refuge availability, food distribution and abundance, and opportunities for migration, reproduction and recruitment [4]. Given this ability for hydrologic variability to control the ecological and biophysical attributes of in-stream and riparian systems, landscapes that have similar hydrologic properties should have similar biological and ecological assemblages [5]. Furthermore, if the same or similar hydrologic landscapes can exist in multiple spatial locations within bioregions, it stands to reason that the ecology of these systems should also be similar, regardless of spatial location. The ability to identify, classify, and validate spatial patterns in hydrologic landscapes is an important step in creating a solid foundation to assess the impact of natural flow variability, associated ecological conditions and management of water resources across a range of spatial scales. As such, hydrologic classification has been identified as a critical step in providing a spatially-explicit understanding of the magnitude and timing of flow regime variation within and between rivers and regions [2,6].

Landscape and hydrologic units

Landscape characteristics affecting the quality, quantity, and movement of water are extremely complex [7]. The earth is made up of a number of different landforms, geological settings and climatic conditions, and the idea of a simple, unifying conceptual hydrologic framework may seem impossible to achieve [7]. However, landscapes that appear unique and diverse often actually have a common set of attributes (e.g. governing the movement of water). Winter [7] introduced the concept of hydrologic landscape units, which suggests that the complete hydrologic system (i.e. incorporating surface runoff, groundwater flow and atmospheric water) interacts with simple physiographic features, and that these features then become the building blocks of all hydrologic landscapes. Therefore, by this rationale, the movement, storage and release of surface and subsurface water are controlled by a common set of physical principles regardless of the geographic location of the landscape [8]. Winter [7] termed these 'fundamental hydrologic landscape units' (FHLU), and defined the conceptual unit as a land surface form which includes an upland, an adjacent lowland and the valley side that separates them. The hydrologic system of an FHLU consists of: 1) the movement of surface water (controlled by the slopes and permeability of the landscape); 2) the movement, storage and release of groundwater (a function of the geologic setting); and 3) atmospheric water exchange (controlled by climate) [7]. Much peer-reviewed research supports the idea that all hydrologic landscapes can be considered to be variations and multiples of FHLUs, and that these can then be used to describe major, spatially-contiguous and discrete landscape types that should have similar hydrologic conditions (e.g. [6,8,9]). Since the concept was first introduced, further research has been conducted to delineate hydrologic landscape regions based on a number of different approaches and across a variety of scales (see Olden et al. [9] and Kennard et al. [6] for an extensive list of examples).

Deductive and inductive landscape classification

Classification is the process of systematically placing objects into classes that are similar with respect to a set of variables or characteristics. Hydrologic classification is therefore the process of systematically arranging streams, rivers or catchments into classes that are similar with respect to their flow regime [6,9]. While hydrologic classification can refer to a broad assortment of methods, a review by Olden et al. [9] recognises two broad approaches to hydrologic classification; deductive and inductive approaches (not to be confused with top-down and bottom-up logic; see below). The inductive approach uses the emergent properties of discharge time series data to generate classes. In contrast, the deductive approach to classification is used when attempting to describe broad spatial patterns in flow regime variability where there is a lack of gauged or modelled streamflow data available. Deductive methods of environmental classification are commonly used when the objective is to quantify and describe spatial variation in flow regime attributes. This approach to classification identifies groups on the basis of physical and climatic attributes that, over broad scales, produce similar hydrologic responses in stream systems [9]. The increased availability of high-quality, hydrologically-relevant spatial datasets (e.g. climate, topography, land use) makes deductive reasoning an appealing method when attempting to define spatial similarities or dissimilarities in hydrological characteristics [9]. It has been demonstrated that the deductive approach to hydrological classification can help in the prediction of streamflow metrics [10,11], and that it improves predictive streamflow models when those models are stratified by hydrologic regions [12]. However, some facets of flow regimes (e.g. low flow magnitude and duration) are difficult to accurately characterise and quantify with this approach due to limitations in data quality and conceptual knowledge of the systems, and spatial variability of hydrological processes in many regions [6,9,13].

Wolock et al. [8] used the concept of hydrologic landscapes introduced by Winter [7] to classify nearly 44,000 catchments (~200 km^2 in area each) using a combination of multivariate ordination and cluster analyses. Kennard et al. [6] presented a method combining non-hierarchical clustering of climate, topography, soils and geology, vegetation and flow data to group Australian streams at a continental scale with mixed success. Sawicz et al. [14] employed the use of precipitation-temperature-streamflow signatures and Bayesian clustering to characterise 280 non-contiguous catchments located in eastern USA so as to understand similarities in climatic and landscape attributes across the region. Their work found that signatures which vary along climatic gradients exerted a stronger influence on cluster separation than those signatures which may vary as a result of geology or land cover. It has also been shown by McManamay et al. [13] that hydrological regionalisations [15] can be severely lacking in their ability to explain variation in a number of streamflow metrics.

The approaches by Wolock et al. [8], Kennard et al. [6], Sawicz et al. [14] and others all require the use of catchments or some choice of arbitrary spatial unit (e.g. eco/bio-region) to delineate and display the results of the clustering. However, there is evidence of significant flow variability within river catchments [6,16] and significant spatial variability in climate and land use within sub-catchments that affect wetland extent [17]. The approach of delineating spatial units *a priori* leads to a loss of spatial variability, particularly as the catchment or spatial units become larger. Olden et al. [9] state that while deductive classification is common in the literature, hydrologic landscape regions and other similar concepts that are founded on physical principles have rarely been tested with this approach. The *a priori* (or 'top-down') specification of boundaries between classes has been criticised, while alternative 'bottom-up' approaches, where groups are developed as an emergent property of the data [18] (not to be confused with inductive reasoning which relies on time series hydrologic data) have been considered to be in keeping with physical ecohydrological principles [9]. Using a bottom-up approach, spatial and group clustering patterns are generated based on the analysis of a large number of units, such as pixels or micro-catchments. These units are then allocated into clusters based on their multivariate similarity [18]. However, a number of subjective choices as to which datasets to include, classification strategies and the number of groups in the classification process still need to be made. Such decisions could affect the quality and repeatability of the classification process when applied to different regions and datasets [9,14,18,19]. Despite the potential limitations, the routine availability of these datasets and the application of statistical clustering and analyses have allowed scientists to begin to link spatial patterns to ecohydrological processes.

Statistical clustering and multivariate analyses

Statistical clustering and multivariate analyses are important and powerful tools in the identification of spatial and temporal gradients. There is a multitude of variations on the theme of statistical clustering [20–22], but the most commonly used are hierarchical agglomerative methods [23] which fuse individual samples into like groups, gradually increasing the similarity within groups while lowering the similarity level between groups; i.e. each sample starts as its own group and pairs of groups are merged moving up a hierarchy. The process is considered complete when all samples are contained within a single group or cluster. Unlike hierarchical clustering, non-hierarchical clustering places samples into groups that are not related hierarchically, but differ from each other significantly in multivariate space. Described simply, non-hierarchical clustering tends to work by assigning each sample (n) into a pre-defined number of clusters (k) and then cluster membership of the samples is iteratively reassessed, usually with the criterion of maximising between-cluster variance while simultaneously minimising within-cluster variance. The most common example of non-hierarchical clustering is the *k*-means algorithm [24]. In some instances, the groups extracted by hierarchical and non-hierarchical algorithms do not differ significantly [25], but non-hierarchical methods can be much more efficient at extracting groups from large datasets [26].

Once statistical clustering has occurred, an analysis of the performance of the clustering can be conducted through the use of multivariate statistics and, in particular, ordination plots. There are a number of ordination techniques (e.g. principal components analysis (PCA), and multidimensional scaling (MDS)) that, broadly speaking, reduce multidimensional space so that objects can be compared graphically in two or three dimensions, without a significant loss of explanatory information. It is also possible to use the clustering information to train predictive models to classify samples not included in the original classification. This is where supervised classification algorithms, such as random forests (RF) [27,28], coupled with geographic information systems (GIS) and image processing software can be applied to extend the applicability of deductive landscape classification approaches to regionalisation studies.

Supervised classification of landscapes

One of the most common applications of remotely-sensed images and data is the creation of maps of vegetation type, soil properties or other discrete classes. In supervised classification, the location of known classes on those maps (i.e. training sites) is used by the software to determine the spectral signature of the pixels belonging to each of those classes. Each pixel in the image (i.e. outside the training sites) is then assigned, based on its spectral signature, to the class it most closely matches. Supervised classification can be applied at the individual pixel level or to groups of adjacent, similar pixels for the creation of contiguous regions. However, for the classification to work effectively, *a priori* knowledge of where the classes of interest (e.g. land cover types) are located is required. When supervised classification is combined with, for example, an unsupervised statistical classification, the process is referred to as hybrid (or semi-supervised) classification [29]. A major benefit of hybrid classifications for landscape regionalisation is that they permit the bottom-up approach to deductive classification as recommended by Mackey et al. [18] and Olden et al. [9]. This eliminates the need for survey approaches to develop *a priori* knowledge of the location of classes of interest which require expert opinion and substantial amounts of qualitative evidence which is not always available or suitable. The hybrid approach also eliminates the need to define a spatial unit a priori (e.g. a catchment) and allows small-scale (e.g. intra-catchment) variability to be identified and preserved where it may otherwise be lost.

Aim of study

The aim of this study was to create a hydrologic landscape regionalisation using deductive reasoning and a bottom-up approach to statistical clustering combined with a hybrid classification. The regionalisation was then assessed based on its ability to discriminate between groups (regions) based on a number of streamflow indices. In this research, we used unsupervised classification (i.e. the statistical clustering) to first determine class membership based on multivariate space and then used supervised image classification to classify the remaining pixels from a number of ecohydrologically-important layers into the classes of interest as defined by the statistical clustering. This approach will permit the regionalisation of spatially non-contiguous regions, while maintaining small-scale intra-catchment variability that would be lost using catchments as the unit of classification as has often been done in the past. The assessment of the ability of the regions to differentiate among streams based on a number of flow indices provides insight into the utility of the method in predicting streamflow characteristics in ungauged catchments.

Materials and Methods

To clarify the process used in the creation of the hydrological regionalisation and the validation and training methods for the RF models, a graphical overview of the methods is presented in Figure 1.

Site description

Victoria is the southernmost state of mainland Australia, comprising an area of 227,594 km^2, and bordered by the southern bank of the Murray River to the north, South Australia to the west and separated from Tasmania by Bass Strait to the south. Topographically, geologically, and climatically, Victoria is diverse, varying from wet temperate climates in the southeast to alpine areas rising to ~2000 m altitude in the northeast (Figure 2). To the west and northwest are extensive, flat areas of semi-arid plains, while most of the rest of the state experiences a Mediterranean climate consisting of hot, dry summer and cool, wet winters [30]. Median annual rainfall in Victoria exceeds 2,500 mm in some parts of the mountainous northeast but is less than 300 mm in a large part of the west and northwest [31]. Generally, snowfall is only observed in the mountains and hills to the east and centre of the state. Victoria has an extensive wetland system, with nearly 17,000 wetlands larger than 0.01 km^2 in surface area [32], and a large river network, with the largest being the Murray River system.

Site description of case-study area in western Victoria

The Glenelg-Hopkins region of western Victoria covers approximately 27,000 km^2 and the regional cities of Warrnambool, Ararat, Hamilton, Portland and the western fringes of Ballarat are within its boundary. The region contains the Grampians Ranges in the north but is generally a low-lying series of catchments across three major catchments – Glenelg, Hopkins and Portland. The region has been previously studied with respect to land-use and land-cover changes [33] and the associated impacts on nutrient exports, in-stream salinity and dryland salinity [34,35], while recent work has examined the spatio-temporal variability between land cover, climate and wetland extent [17], the impact of land-cover changes on groundwater levels [36] and empirically modelled streamflow response to land-use change [37].

Variable selection and Processing

The first phase of the classification involved selecting suitable variables upon which to base our classification. Steps associated with variable selection and processing, described in this section, are outlined in green in Figure 1. Based on the concept of FHLUs, 25 variables were chosen that could explain the storage, movement, and quality of surface water, groundwater, and atmospheric water. A full list and brief description of each of the variables are presented in Table S1. All raster calculations and raster analysis for the processing of variables was conducted in ArcGIS 10.1 [38].

The raster datasets employed in the study covered a wide range of resolutions (30 m–10 km). Typically, with GIS, analyses are only considered to be suitable if all rasters are resampled to the coarsest resolution. However, this can result in the loss of a

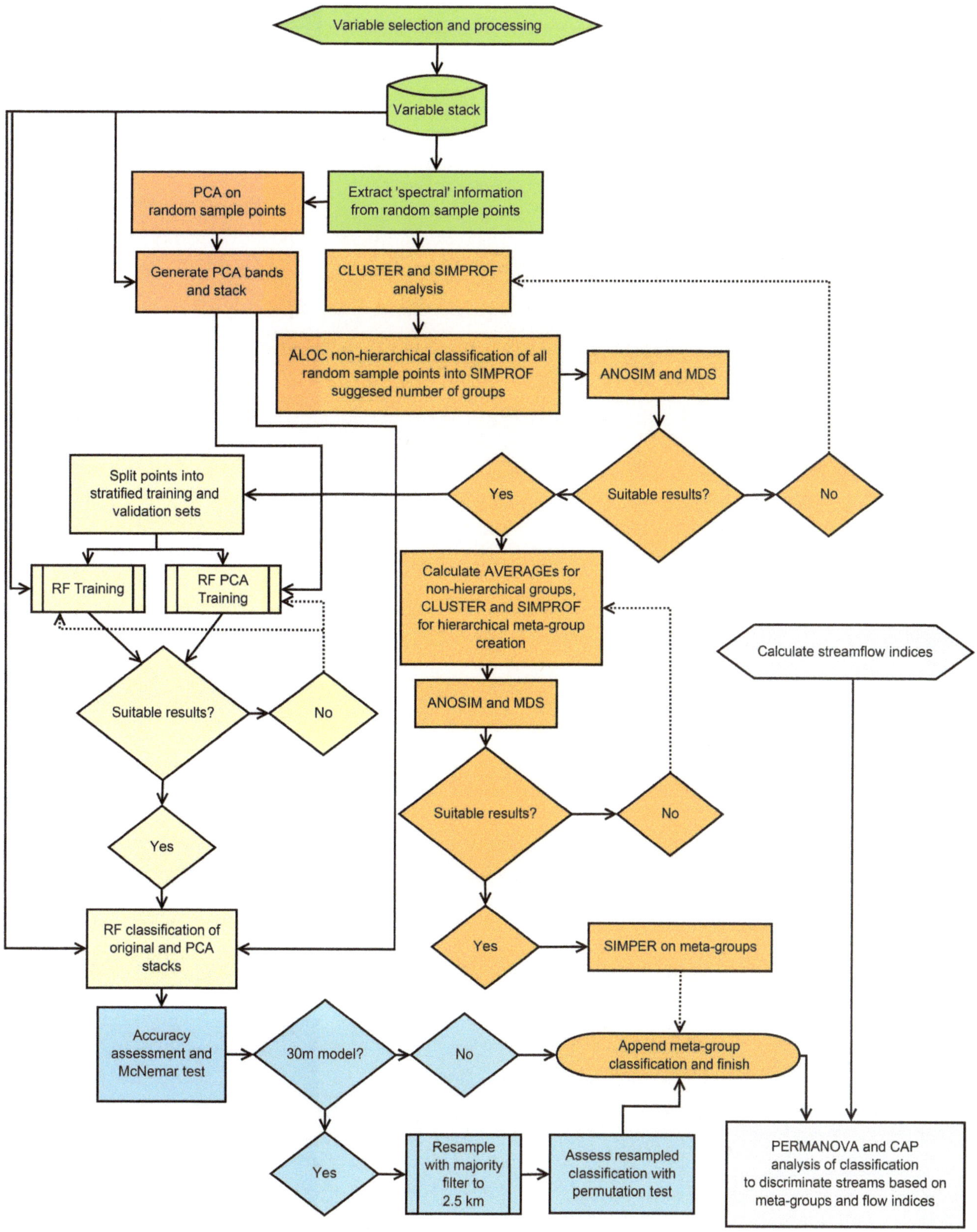

Figure 1. Graphical overview of the methods employed in the creation of the hydrological regionalisation; the training and validation of the RF models used to extend the statistical clustering to the state of Victoria; and of the validation of the classification with hydrological data. Each colour coded section of the figure corresponds to a section in the methods. Green: Variable Selection and Processing; Orange: Development of Classification Groups; Beige: Hybrid Classification with Random Forests; Light Blue: Accuracy Assessment; Grey: Relationship between the Regionalisation and Hydrologic Indices. The process for the ALOC 20 100% models did not involve splitting the ALOC classified random sample points into training and validation subsets and model accuracy was only assessed with OOB accuracy from EnMap-Box.

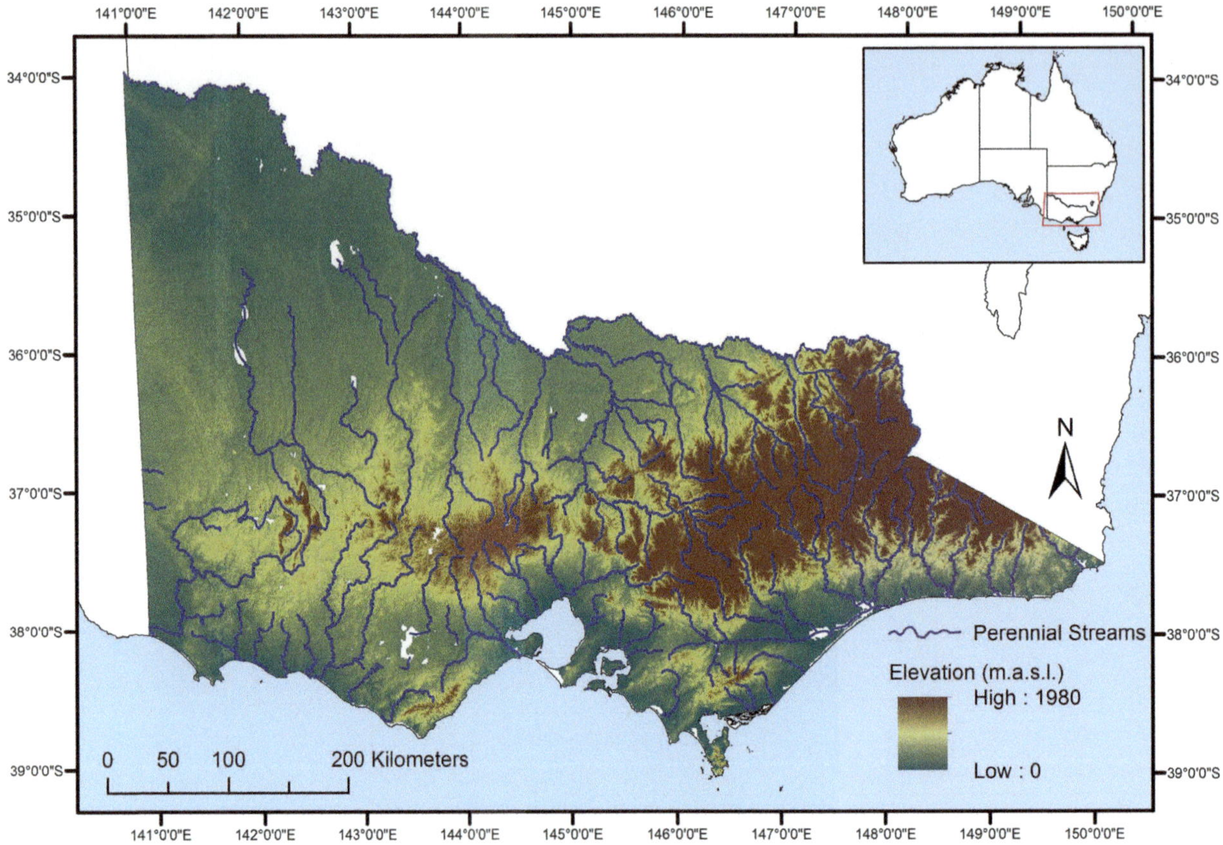

Figure 2. Location of the study area in south-eastern Australia. Dark blue lines represent perennial rivers, while the colour gradient represents elevations, with darker browns indicating higher elevations.

substantial amount of detail and information and can affect the ability of supervised classification methods to successfully classify pixels (See Figure 3 for a comparison between the 30-m and 10-km Landscape Development Index (LDI)). Therefore, in this study, two approaches were used to standardise the scale of our raster data. The first approach was to resample all datasets to the finest resolution (30 m); and the second involved re-sampling all the raster datasets to the coarsest resolution found in our datasets (10 km). All rasters were continuous in their spatial coverage with the exception of the soil hydrological properties (KSAT, PAWC and soil horizon thickness) which had significant gaps where large lakes and wetlands were found. There was also a significant gap in coverage on the eastern headland of Port Phillip Bay. To ensure that all datasets aligned correctly and had the same degree of spatial continuity, the digital elevation model (DEM) was used as a snap raster for the resampling. Once the resampling had been completed using a nearest neighbour algorithm, the now 30-m soil properties were used as a mask to extract all other raster values. The result of this was that all of the datasets used in the analysis had a 30-m spatial resolution and all had corresponding areas of missing data that would be excluded from any analysis.

For the second approach, all of the original datasets were resampled to 10 km using a nearest neighbour algorithm and the mean annual evapotranspiration raster as the snap raster. Two different datasets were used as snap rasters so that the pixels of the resampled rasters (at either 30 m or 10 km) would be aligned correctly at the respective resolution. If this study had been conducted at a continental scale, then a coarser resolution would be more suitable, however as it was conducted on a relatively small scale, we considered that resampling to a finer resolution was both suitable and justifiable. Furthermore, limiting our analysis to a coarser resolution (on the basis of a single coarse-scaled dataset; mean annual evapotranspiration), would have significantly affected the applicability and usefulness of the method presented here, specifically, the ability of the RF model to accurately recover and reproduce allocated class information. Previous research has shown that the accuracy of supervised classifications, at both an overall and per class level, can be affected by the spatial resolution of the input images [39,40]. As such, we assessed this issue through the accuracy of the RF model and the ability of the hybrid classifications to accurately recover the class information using the coarser dataset (see section on accuracy assessment). A layer stack of both sets of variables (30-m and 10-km resolution) was produced in ENVI 4.8 [41] for later use with the RF model.

Two random distributions of sampling points ($n = 10{,}000$ and $n = 410$; with minimum distances between points of 30 m and 10 km respectively) were then created. Ripley's K function [42], which determines whether features are significantly clustered or dispersed over a range of distances, was then used to assess whether both sets of points were distributed across our sampling area. Using the random-sampling points, raster values were extracted from each of the raster layers for later use in clustering, and then the training and validation of the RF model.

Development of classification groups

The selected variables were then statistically analysed to develop the groups (known as 'regions') to be used as the basis for

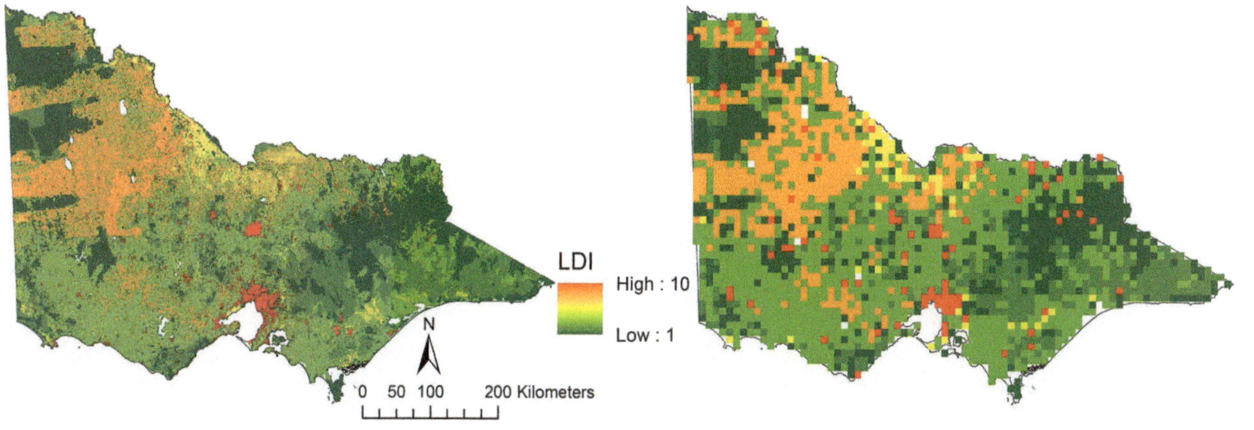

Figure 3. An example of the differences in resolution identified when working in relatively small study areas. The left hand image is the Landscape Development Intensity index (LDI) at a 30-m resolution while the image on the right is the LDI at a 10-km resolution. The accuracy of supervised classifications can be affected by the spatial resolution of the input images and as such we developed models at both resolutions.

classification. The steps involved are outlined in orange in Figure 1. All sampling points that were found to contain missing data were removed prior to analysis, resulting in $n = 9{,}958$ (30-m data) and $n = 406$ (10-km data) for the two sets of data points. For the 30-m data, three 1000-point subsamples were taken for initial statistical analysis, while all data points were used for the 10-km dataset.

In order to develop a bottom-up approach to classification, where groups are an emergent factor of the data rather than defined *a priori*, we cluster-analysed each of the subsamples individually. Due to the large ranges and different scales used across our variable set, Gower similarity matrices were constructed for each of the initial analysis datasets using PRIMER 6 [43]. The CLUSTER [44] function was then used with a SIMPROF [45] test, to identify the number of statistically-significant ($\alpha = 0.05$) groups within the datasets. Essentially, SIMPROF determines the number of significant groups with the assumption of no *a priori* groups by calculating similarities between every pair of samples using the chosen resemblance matrix and a hierarchical cluster dendrogram. Beginning at the top of an already-defined hierarchy (i.e. by the CLUSTER function), progression down the divisions or branches of the dendrogram is only permitted if the current set of samples is deemed to still have statistically significant dissimilarity. Upon encountering a non-significant result (i.e. the samples are similar), no further tests are performed down that branch of the dendrogram and all samples below are considered part of the same group [45]. A limitation of the SIMPROF test is that groups identified by the test may be at too fine a level of detail for practical purposes. However, if the resulting clusters are super-sets of the SIMPROF-defined groups, it is appropriate to define coarser groupings based on an arbitrary slice at some chosen level of similarity [45]. As the 30-m SIMPROF tests were conducted on three 1000-point subsamples, the subsample that produced the largest number of groups was used to determine the number of groups for the 30-m data. Even though SIMPROF uses a hierarchical relationship between sampling points to determine the number of clusters present in the data, we believe this approach is suitable for estimating an appropriate number of non-hierarchical groups as opposed to choosing an arbitrary k number of groups.

We then classified the full datasets ($n = 9{,}958$ and $n = 406$) into the number of classes suggested by the SIMPROF tests using the non-hierarchical clustering algorithm ALOC [26] and the Gower metric in PATN v3.1.2 [46]. Group allocations were then exported from PATN and joined to the original datasets as factors for further analysis. Once group membership information was in PRIMER, the ANalysis Of SIMilarities (ANOSIM; [44]) routine was used to test for statistically-significant differences among sample groups. Based on the R statistic, which is scaled to be between -1 and $+1$, global R values >0 indicate greater dissimilarity between groups than within groups [23]. Group averages were calculated using the AVERAGE [43] tool in PRIMER to visually analyse group separation using MDS. MDS is useful in providing a visual representation of the pattern of similarities between objects or groups while reducing the multidimensional space to be more readily interpretable (i.e. reducing data to two or three dimensions). The ability of MDS to reduce the degree of multidimensional space is measured with a stress value. Essentially, stress is the mismatch between distances between all samples in the plot in multidimensional space and the calculated estimate of their respective locations in two or three dimensions, with lower values indicating better representation [23]. The CLUSTER and SIMPROF routines were then used to hierarchically cluster the ALOC generated group averages into 'meta-groups'. By definition, non-hierarchical groups are not linked based on their hierarchical multivariate relationship to each other, but rather are defined by their multivariate dissimilarity. As such, group x may not be closely related to group y but could be more closely related to group z. By hierarchically clustering our ALOC generated groups, were we able to determine which ALOC groups were more closely related to each other based on their multivariate means. A standardised Euclidean distance similarity matrix was then created and the SIMilarity PERcentage (SIMPER; [44]) routine was then used to analyse variable contribution to each of the meta-groups and to examine between meta-group similarity, while the Kruskal-Wallis [47] statistic was used to assess the ability of each of the variables to differentiate between clusters.

Previous studies (e.g. Wolock et al. [8]) have employed PCA [44] to reduce dimensionality and reduce multi-collinearity among variables. Our method relied firstly on using 'raw' data (i.e. the data was not transformed in any way) to extract spectral information for the statistical clustering and then classification. The results of this method were then compared against a classification based on PCA-transformed data. PCA is a procedure where possibly correlated variables are orthogonally transformed into a new set of linear, uncorrelated variables known as principal components [23]. The transformation results in the first component explaining most of the data variance (i.e. the first component explains as much of the multivariate data as possible) while each additional component in turn is then created to explain the remaining variance. The number of components is less than or equal to the number of original variables and each additional component is created under the condition that it is uncorrelated with all of the preceding components [23]. Here, a PCA was conducted in PRIMER on the same standardised Euclidean distance matrix used for the SIMPER analysis. The eigenvectors, for the first five principal components (eigenvalues ≥ 1) were used in ArcMap to generate PCA bands and new PCA raster stacks were created in ENVI.

Hybrid classification with Random Forests

The regions that were developed in the previous step were then used to classify the raster stack of the variables for the entire study area. Steps in this section are outlined in beige in Figure 1. RF is an ensemble machine-learning method used in classification and regression [27]. The RF method is relatively unknown in land remote sensing and has not been thoroughly evaluated by the remote sensing community, although it has been shown to be more accurate than single decision-tree classifiers [48]. RF requires two parameters for generating a predictive model: the number of trees (k) and the number of variables used for growing the trees (m). Therefore, a dataset can be classified by defining a constant number of m variables, while each of the training samples is classified by k trees. Classification is determined by using the mode of the classes output by individual trees for each training site (x), $\hat{C}_{rf}^{B} = majority\ vote \left\{ \hat{C}_b(x) \right\}_1^B$, where $\hat{C}_b(x)$ is the class prediction of the bth RF tree from a possible B classes [48]. RF increases the diversity of the constituent trees by making them grow from different training data through bootstrap aggregation which involves random re-sampling (without deletion) of the original training dataset [27]. Therefore some data may be used more than once in the training of the model, while some may not be used at all [48]. Being an ensemble method, multiple models (trees) are used allowing the algorithm to obtain better predictive performance than that which would be obtained by using any of the constituent models individually. RF is becoming increasingly popular in data mining, remote sensing and landscape ecology as it is non-parametric, can generate internal, unbiased error estimates and variable importance, is robust to training data reduction and noise, and is highly accurate [27,48].

In order to develop the RF models, the sampling points were randomly split into independent training (80%) and validation (20%) datasets and then stratified using the cluster-membership allocations from PATN. Due to the small sample size of the 10-km dataset, an additional RF model was created using 100% of the sample for RF training. Using ENVI, layer stacks of the raster data (30 m, 10 km, and PCA at both resolutions) were constructed and masked. Layer stacks, masks and training regions of interest were imported in to EN-Map Box 1.4 [28] to permit the building of RF models and the classification of the image stacks. EN-Map Box was set to use 200 trees (k) per sample, and the square root of the number of input variables on the non-PCA transformed data ($m = \sqrt{25} = 5$), or m-1 variables ($m = 4$) for the PCA models. The Gini coefficient [49] was used to calculate impurity, which is one method used to evaluate the best split decision for each tree.

Accuracy assessment

The ability of the classification to accurately represent the information across a range of resolutions was then tested, with the relevant steps outlined in light blue in Figure 1. The accuracy of the RF models in recovering and classifying the image stacks into the ALOC classes was assessed with out-of-bag (OOB) error rates generated in EN-MAP box, and using the independent 20% validation dataset to calculate percent agreement between classified and validation data, user and producer accuracies, and Kappa (κ) coefficients [50] in ENVI and R 3.0.1 [47] with the *psych* [51] and *irr* packages [52]. User's accuracy refers to the probability that a pixel classified into a certain class really belongs to that class, while producer's accuracy refers to the probability that a certain class is classified correctly. The locations of the validation pixels were used to extract class information from the original and PCA classifications and percent agreement and κ coefficients between model types were examined. High levels of agreement between the original and PCA classifications would indicate an insignificant amount of multivariate information loss by PCA and further support the use and application of methods that reduce data dimensionality and multi-collinearity between variables in regionalisation studies. The 10-km sample that used 100% of the data for RF training could not be assessed for accuracy independently and was therefore only assessed with OOB error and class distributions.

We also decided that, due to the resampling of the original data (to 30-m from a range of resolutions), it was worth investigating the effect of resampling the 30-m classifications to a coarser resolution (using the majority filter) to help remove some of the finer-scale variability in the data. A resolution of 2.5 km was chosen as a suitable pixel resolution, to ensure that our resampled assessment points were further apart than the mean distance of the original samples (see below), this meant our resampled classifications were equal to or larger than the resolution of the majority of the datasets while still being finer than the soil erosivity index. To assess agreement between the 30-m and 2.5-km resampled classifications (i.e. original and PCA both at 30-m and resampled 2.5-km), 300 random validation points were selected from the 20% validation datasets, with minimum distances between points of 2.5 km, and used to extract class information from the 30-m and 2.5-km resampled classifications. Agreement between the validation data and the 30-m and 2.5-km resampled classifications was assessed with percent agreement and κ coefficients between model resolutions. Visual inspection of the class distributions was also conducted; for this the 300 random validation points were used for the 30-m and 2.5-km classifications, while all 81 validation points for the 10-km classifications were used. McNemar's chi-square test [53,54] was used to formally test for statistically-significant differences between model types (e.g. 30-m original and 30-m PCA) and resolutions (e.g. 30-m original and resampled 2.5-km original) and the validation dataset using the same 300 random validation points; however we recognised *a priori* that the relatively large number of random validation points would likely make any difference statistically significant and that the resolution tests would not be independent of one another. As such, a permutation- based method ($n = 9999$) was also used to assess for statistically-significant differences in the κ values of the classifications [55,56]. The algorithm worked as follows:

1. If V = validation data; A = classification 1; B = classification 2
2. Let test(x, y), be a function that calculates the test statistic (κ) for the classifications,
3. H_0 = if A and B are approximately equal in their classification accuracy, observations in A and B can be exchanged without affecting κ, given by test(V, A) and test(V, B),

4. Randomly exchange data between A and B, n times, and observe how these changes affect the κ of A and B, i.e. test(V, A) – test(V, B),
5. n permutations would result in N data points. Rank the N data points and observe where the κ from the original test (i.e. not the permutated data) is located among the κ values from the permutated data points. If the original κ is outside the 0.975 percentile (or below the 0.025 percentile) then we can claim that the two classifications are different at $\alpha = 0.05$.

If a high level of agreement was found, as indicated by high percent agreement and high κ, then resampling to 2.5-km *a posteriori* could be considered a suitable and justifiable method for smoothing the finer-scale variability.

Relationships between the regionalisation and hydrologic indices

Finally, we tested the results of our classification against a traditional classification based on hydrologic indices. Steps in this section are outlined in grey in Figure 1. As a preliminary assessment of the ability of the regionalisation to differentiate among streams with differing hydrology, we calculated a range of streamflow indices based on the recommendations of Olden and Poff [57] and then explored the relationships between the regions and streams with a permutation-based ANOVA (PERMANOVA) [58] and a constrained discriminant ordination (Canonical Analysis of Principal Coordinates (CAP); [59,60]).

Streamflow gauge locations were downloaded from the Water Measurement Information System (data.water.vic.gov.au/monitoring.htm). These locations were then overlayed on the regionalisation (30-m ALOC 23 meta-group classification, see results) and had group (region) information appended to them. In regions where there were more than 50 stream-gauges present, 50 were chosen at random to be included in the analysis. Daily streamflow data between 1980 and 2010 was then downloaded for 564 gauges throughout Victoria. A minimum record length of 15 years within the 30-year temporal window was required for a gauge to be included in the analysis [61]. Stream gauges that were potentially subject to modification by weirs, dams or water extractions were not specifically excluded from the analysis. Where there were missing periods of flow information (to a maximum of 20 days in any single event) the record was in-filled using linear interpolation [62] with the Time Series Manager module of the River Analysis Package (RAP) [63]. Gauges that had a single period of missing data greater than 20 days were excluded from the analysis. Thirty-two indices characterising different aspects of the flow regime (Table III, All Streams; [57]) for each stream were calculated using the Time Series Analysis module of RAP. Indices that were related to discharge (i.e. those divided by catchment area) were not included in the analysis.

To test for differences among our groups, the PERMANOVA+ add-on [64] for PRIMER was used. One-way PERMANOVAs, (999 permutations) using Group as a fixed factor, were run for the dataset of flow indices based on the original gauges ($n = 201$) and for an additional dataset consisting of a bootstrapped sample of those flow indices ($n = 383$), based on normalised Euclidean distance matrices. Analyses tested both for main effects and pairwise differences among groups. Traditional Analysis of Variance (ANOVA) is powerful for univariate data however the traditional multivariate analogues (e.g. MANOVA), are too stringent in their assumptions (in particular, that of multivariate normality which is frequently untrue in ecological data [25,58]), for use in ecology. As such, permutation-based non-parametric methods are preferred [58]. PERMANOVA uses permutation methods to test the simultaneous response of one or more variables to one or more factors in an analysis of variance. The use of permutations in PERMANOVA removes the assumption of normal distributions which are required for traditional ANOVA/MANOVA testing and, as such, the only assumption of the test other than independence is that the observations can be exchanged under a true null hypothesis [58]. Another benefit of using a permutation approach is that the permutated P-values provide an exact test of each *individual* null hypothesis of interest, and as such ad-hoc pairwise corrections (e.g. Bonferroni) are not strictly necessary [64].

When data are classified into *a priori* groups, unconstrained ordinations (PCA, MDS) are extremely useful for visualising patterns from a multivariate space. However, the overall dispersion of points (when reduced to two or three dimensions) can often hide the true multivariate differences among those groups and it may be very possible to discriminate among groups through another direction or dimension of the multivariate data [64]. Unlike unconstrained ordinations, constrained ordinations have an *a priori* hypothesis which controls the way the multivariate data can be interpreted in an attempt to relate predictor variables (streamflows indices) to response variables (groups) [60]. In a discriminant analysis, the ordination axes are interpreted in such a way as to maximise the differences between *a priori* groups, while in a canonical correlation, the axes are interpreted to maximise correlations among variables. CAP first calculates the principal coordinate axes (PCO) among N samples, and then chooses an appropriate number of PCO axes (m) for interpretation based on a number of criteria (see Anderson et al. [64] for details), including a leave-one-out cross validation procedure which attempts to maximise classification success. The benefits of CAP over other constrained ordination methods are its ability to use any distance or dissimilarity measure, conduct permutation tests for significance of relationships among variables and predict group membership of new samples [60]. To assess the ability of our regionalisation to discriminate groups based on streamflow indices, three CAPs were conducted using the PERMANOVA+ add-on for PRIMER. Two of the analyses were performed against the group information extracted for each stream gauge using the same normalised Euclidean distance matrices that were used in the PERMANOVA tests (i.e. the original dataset and the bootstrapped dataset of hydrologic indices). The number of axes (m) was not specified and a permutation test ($n = 999$) was conducted to test the strength of the relationship. In addition, a third CAP was conducted using the bootstrap dataset where a stratified random sampling approach was used to remove 20% of the group information as a validation sample. The CAP was conducted as before, with the exception being that the validation cases were allocated groups based on the results of the CAP. Allocation accuracy by CAP was assessed by calculating percent agreement between the CAP allocated group and the original group information, and κ coefficients with the *irr* and *psych* packages in R. Pearson's r [47] was calculated between the between the number of gauges in each group and the number of gauges classified correctly from each group to assess for thresholds at which a specified level of accuracy could be achieved.

Results

Spatial distribution of sample points

Analysis of spatial distribution of the sampling points concluded that the mean, minimum and maximum distances to the nearest neighbour among the 30-m points were 2.4 km, 202 m and 8.8 km, respectively. The corresponding values for the 10-km points were 15 km, 10 km, and 47.4 km, respectively. Ripley's K, based on 999 permutations, indicated significant over-dispersion of the 30-m sample points to a distance of ~250 m and a significant clustering at distances greater than ~650 m, while the 10-km sample points were significantly over-dispersed at distances less than ~12 km, but were significantly clustered at distances greater than ~18 km. Based on these values, the spatial distribution of our sampling points was considered suitable for the analyses as no significant clustering was displayed by the sample points at the resolution of the datasets they would be sampling (e.g. there was no significant over-dispersion or clustering at the mean distance of 15 km for the 10-km points).

Clustering and ordination

Twenty-three groups were identified within the 30-m data points and 20 groups for the 10-km data. As a result, the data were allocated into 23 (ALOC 23: 30-m data) and 20 (ALOC 20: 10-km data) non-hierarchical groups. There were wide variations in the ability of the model variables to differentiate among clusters (Table S2) and the group populations had differing multivariate distributions. The allocated groups were well separated (Figure S1) with global-R values of 0.852 ($p = 0.001$) for the ALOC 23 clustering and 0.908 ($p = 0.001$) for the ALOC 20 clustering, indicating that cluster membership was highly unlikely to be a result of chance alone. This was supported by the fact that neither ANOMSIM resulted in any permutations that had R-statistics greater than or equal to the global-R value.

The allocated groups were then further clustered into hierarchical meta-groups. Using the group averages for ALOC 23 and ALOC 20, 11 and ten meta-groups were generated (Figure S1 and Figure S2). ANOSIM analysis again indicated that cluster membership was highly unlikely to be a result of chance alone and suggested that the meta-groups for both the ALOC 23 and ALOC 20 models were well separated with global-R values of 0.668 and 0.762 ($p = 0.001$). Again no permutations had R-statistics that were greater than or equal to the global-R value. Some variables exhibited no relationship between the observed values and the meta-groups, while other variables show very clear relationships to the meta-groups (Figures S3–S7). For example, the values for the aridity index (low aridity index values represent drier regions) and rainfall decreased from groups A to K. The opposite relationship was observed for maximum and minimum temperatures, again suggesting as we move through regions A to K, the environment becomes drier and warmer. The BioClim variables 8, 9, 15 and 16 also supported this relationship, with increases in BIO08 and BIO09 (mean temperature of the wettest and driest quarter, respectively), and decreases in BIO16 and BIO17 (precipitation of the wettest and driest quarters, respectively). Variable percentage contributions to the meta-groups for both the ALOC 23 and ALOC 20 classifications differed markedly (Table S2, Figure S8) with, for example, saturated hydraulic conductivity of the B soil horizon (B_KSAT) contributing 0% to a number of ALOC 23 meta-groups, while contributing 46% to ALOC 23 meta-group F. The within-group variation was highly variable among groups, with average squared distances of groups ranging from 4.18 to 11.77 for the ALOC 23 meta-groups, and 4.44 to 13.45 for the ALOC 20 meta-groups (Table S2).

Ordination by PCA resulted in the creation of five principal components (PC) for both ALOC 23 and ALOC 20, using a minimum eigenvalue of one. The first PC for each of the ALOC 23 and ALOC 20 clustering explained 45% and 53% of the data variance, with eigenvalues of 11.2 and 13.2, respectively (Table 1). The five PCs combined explained 78% and 87% of the variance, with the final PC having eigenvalues of 1.39 and 1.25, respectively. PC 1 can be interpreted to represent those areas that are wet, cool, heavily vegetated, high elevation environments with steep slopes and low erodibility soils (Table 1, $0.2 < r < -0.2$). With the exception of PC 1, none of the PCs for either of the datasets can be interpreted to represent similar environments across the two classifications. For example, PC 2 of the ALOC 23 dataset can be interpreted to represent areas with thick, weathered, A-horizon soils with high levels of plant available water, with relatively small variations in temperature seasonality and poorly developed B-horizon soils. For the ALOC 20 dataset, on the other hand, PC 2 suggests areas of poorly developed A-horizon soils, but with thick B horizon soils, and cooler, wetter summers.

Random Forests

Training and validation. Two hundred trees were sufficient for the RF models to achieve acceptable accuracies. Exploratory analysis, using models with 5000 trees (not presented) showed little improvement in OOB error (<1%). OOB error rates were low for the ALOC 23 and ALOC 23 PCA RF models, with estimated maximum accuracies of 95% and 92%, respectively. The OOB error rate of the ALOC 20, ALOC 20 PCA and the 100% ALOC 20 and 100% ALOC 20 PCA models (the latter two were created due to the small sample size for the 10-km dataset as described above) was significantly worse, with estimated accuracies of 59%, 56%, 56% and 54%, respectively (Figure 4).

Classification accuracy was 95% ($\kappa = 0.94$) for the ALOC 23 classification and 92% ($\kappa = 0.92$) for the ALOC 23 PCA classification (Table S3) when tested against the validation dataset. The accuracy of the ALOC 20 and ALOC 20 PCA classifications decreased relative to those estimated by the RF OOB error, with accuracies of 46% ($\kappa = 0.42$) and 47% ($\kappa = 0.44$) (Table S3). The producer accuracies differed significantly for each of the classifications (Table S3), with observed minimum producer classification accuracies of 81% for the ALOC 23 classification and 59% for the ALOC 23 PCA classification. Likewise, the ALOC 20 and ALOC 20 PCA classification also exhibited low producer accuracies with minima of 0% observed for a number of classes in each classification. Visual inspection of the resulting classifications showed few obvious differences among the various ALOC 23 classifications (Figure 5), but more differences were apparent among the ALOC 20 classifications (Figure 6).

Comparisons between original, PCA and resampled classifications. The class distributions of the RF and resampled classifications illustrate that the RF classifications at 30-m and resampled to 2.5-km (bottom row, Figure 5) were quite successful in maintaining the distribution of the validation dataset (Figure 7). All 10-km models were missing classes 7, 8, 12, 18, 19 and 20 from the validation dataset (Figure 7), meaning that accuracy assessment of these classes was not possible, although classes 7, 8, 12, and 18 were present in the final classification (Figure 6). Class 1 tended to be over-classified by the ALOC 20 RF models, as evidenced by the large number of validation points classified as such (Figure 7). The ALOC 20 100% classification was also the only classification to have classes 6 and 15 represented at the validation locations.

Agreement between the ALOC 23 and ALOC 23 PCA classifications was high at 93% ($\kappa = 0.93$). There was no statistically-significant difference between the two classifications ($\chi^2 = 2.72$, $p = 0.1$). Agreements for the ALOC 20 and ALOC 20 PCA classifications was lower than that observed for the ALOC 23 classifications, at 72% ($\kappa = 0.67$), which was also non-significant ($\chi^2 = 0.24$, $p = 0.63$). This indicates that the classification based on the five principal components captured the vast majority of the

Table 1. Results of Principal Components Analysis.

ALOC 23						ALOC 20				
	PC1	PC2	PC3	PC4	PC5	PC1	PC2	PC3	PC4	PC5
Eigenvalue	**11.2 (45)**	**3.26 (58)**	**2.07 (66.3)**	**1.58 (72.6)**	**1.39 (78.2)**	**13.2 (52.7)**	**4.04 (68.9)**	**1.68 (75.6)**	**1.58 (81.9)**	**1.25 (86.9)**
Variable										
A_KSAT	0.14	0.14	**−0.42**	**−0.3**	0.18	0.13	**−0.22**	0.15	**0.53**	0.03
A_PAWC	0.01	**0.48**	−0.18	−0.02	−0.12	0.05	**−0.45**	0.12	0.1	−0.07
A_THICK	0.1	**0.45**	**−0.21**	−0.12	−0.02	−0.04	**−0.45**	0.1	−0.04	−0.12
ARIDITY_INDEX	**0.29**	0.01	0.04	0.02	0.03	**0.26**	−0.09	0	−0.03	−0.02
B_KSAT	0.15	0.05	**−0.4**	**−0.31**	0.2	0.16	**−0.22**	**0.26**	**0.29**	−0.08
B_PAWC	0.12	**−0.34**	−0.09	**−0.41**	**0.26**	0.17	0.18	0.01	**0.47**	**0.29**
B_THICK	0.06	**−0.38**	0.14	**−0.38**	0.16	0.1	0.3	−0.1	**0.35**	**0.42**
BIO04	−0.08	**−0.34**	**−0.29**	0.08	**−0.39**	**−0.21**	−0.01	**−0.26**	**0.26**	**−0.36**
BIO08	**−0.22**	0.05	−0.1	0.2	**0.44**	**−0.25**	−0.01	0.19	0.09	0.07
BIO09	**−0.24**	−0.1	−0.17	−0.04	−0.19	0.02	**−0.26**	**0.35**	**−0.26**	**0.52**
BIO15	0.08	0.14	0.16	**−0.43**	**−0.53**	**0.26**	−0.09	0	−0.11	0.1
BIO16	**0.28**	0.03	0.07	−0.05	−0.05	**0.26**	0.01	−0.19	0.03	−0.13
BIO17	**0.27**	−0.02	0.03	0.18	**0.23**	−0.04	**0.38**	**0.38**	−0.08	−0.01
ELEVATION	**0.25**	−0.14	−0.08	0.04	−0.16	**0.24**	0.18	0.11	−0.11	−0.01
ET_ANNUAL	**0.24**	0.14	**0.28**	0.06	0.1	**0.21**	−0.1	**−0.41**	−0.09	0.1
GW_SWL	0.17	−0.08	**−0.29**	0.09	−0.14	**0.23**	0.11	0.2	0.05	**−0.24**
GW_TDS	−0.17	−0.1	**−0.32**	0.08	0.02	−0.16	0.11	**0.41**	0.13	**−0.25**
LDI	−0.11	−0.11	0.16	−0.07	0.05	−0.15	−0.08	**−0.26**	0.18	0.08
MAX_TEMP	**−0.27**	−0.05	−0.18	0	−0.07	**0.26**	−0.09	0.01	−0.03	0
MIN_TEMP	**−0.25**	0.08	−0.05	0	0.15	**0.25**	0.02	0.07	0.08	**−0.21**
RAIN_ANNUAL	**0.29**	0	0.04	0.03	0.02	**0.27**	−0.02	0.04	0.01	−0.07
SLOPE_RAD	**0.23**	−0.1	−0.2	**0.22**	−0.03	**−0.26**	0.09	0.05	0.1	−0.04
SOIL_EROS	**0.26**	−0.07	−0.05	0.09	0.03	**−0.24**	−0.16	−0.07	0.13	0.01
TWI	−0.18	−0.01	0.16	**−0.27**	0.03	**−0.23**	0	0.03	−0.06	0.1
WEATH_IND	−0.16	**0.23**	0.1	**−0.26**	0.14	−0.19	−0.17	−0.03	0.11	**0.27**

Eigenvalues are presented in the top row while the numbers in brackets represents the cumulative % variance explained by the PCA. Bold numbers indicate variables with a correlation ≥0.2 or ≤−0.2 which was used for PC interpretation.

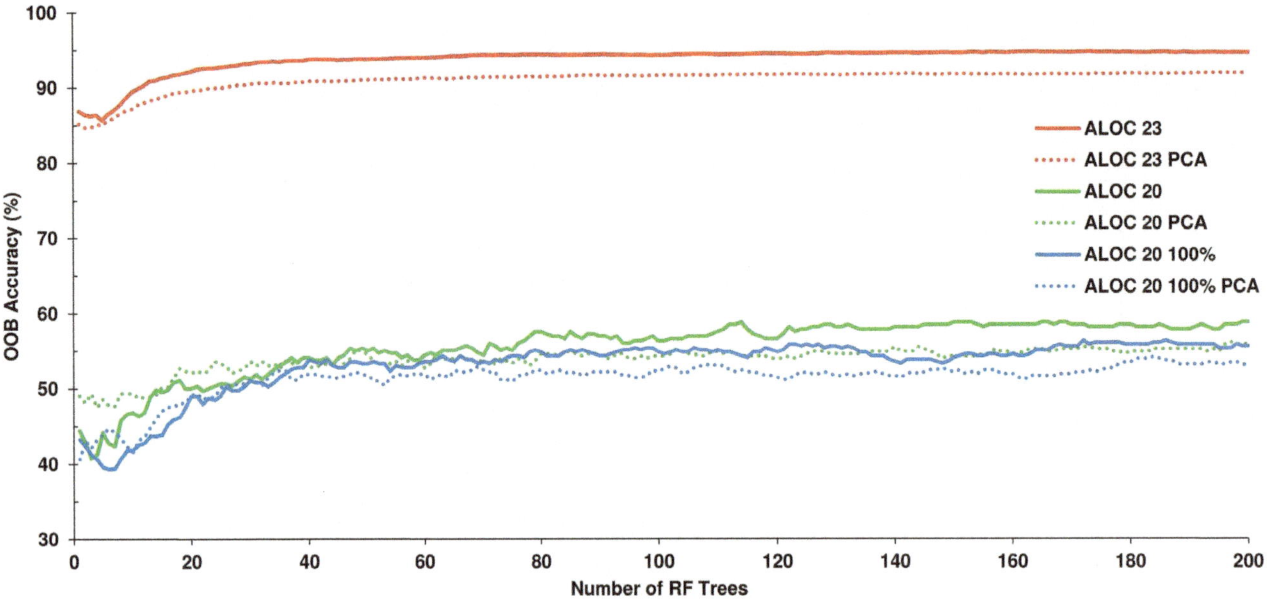

Figure 4. Out-Of-Bag (OOB) percent accuracies for the ALOC clusterings as classified by Random Forests. The 30-m ALOC 23 and ALOC 23 PCA models were significantly more accurate than the 10-km ALOC 20 classifications.

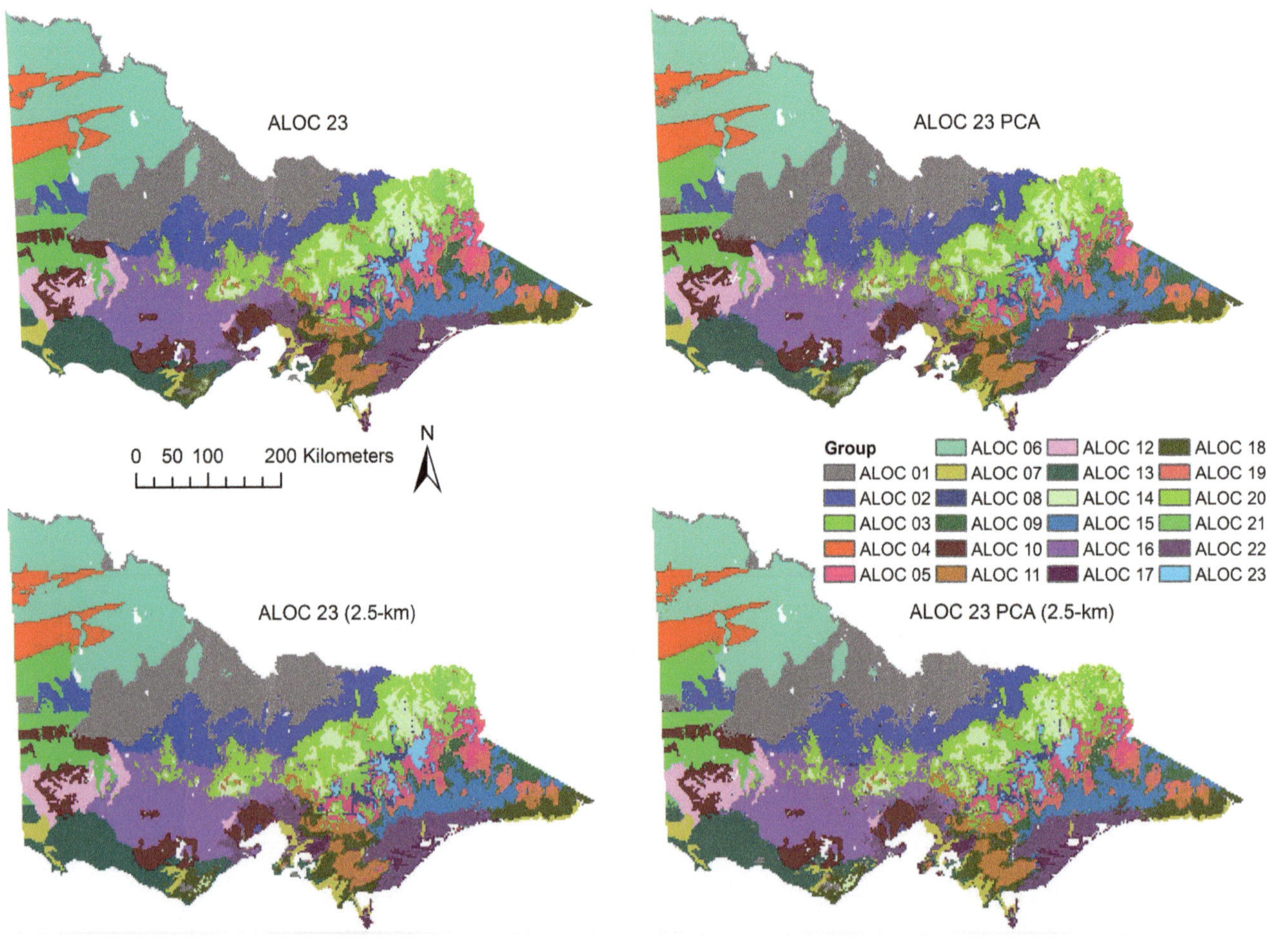

Figure 5. Results of the ALOC 23 (30 m) classifications. Top row - ALOC 23 and ALOC 23 PCA; Bottom row - ALOC 23 and ALOC 23 PCA resampled to 2.5 km. Colours represent each of the ALOC non-hierarchical clusters. Similar colours and cluster numbers do not necessarily represent related groups.

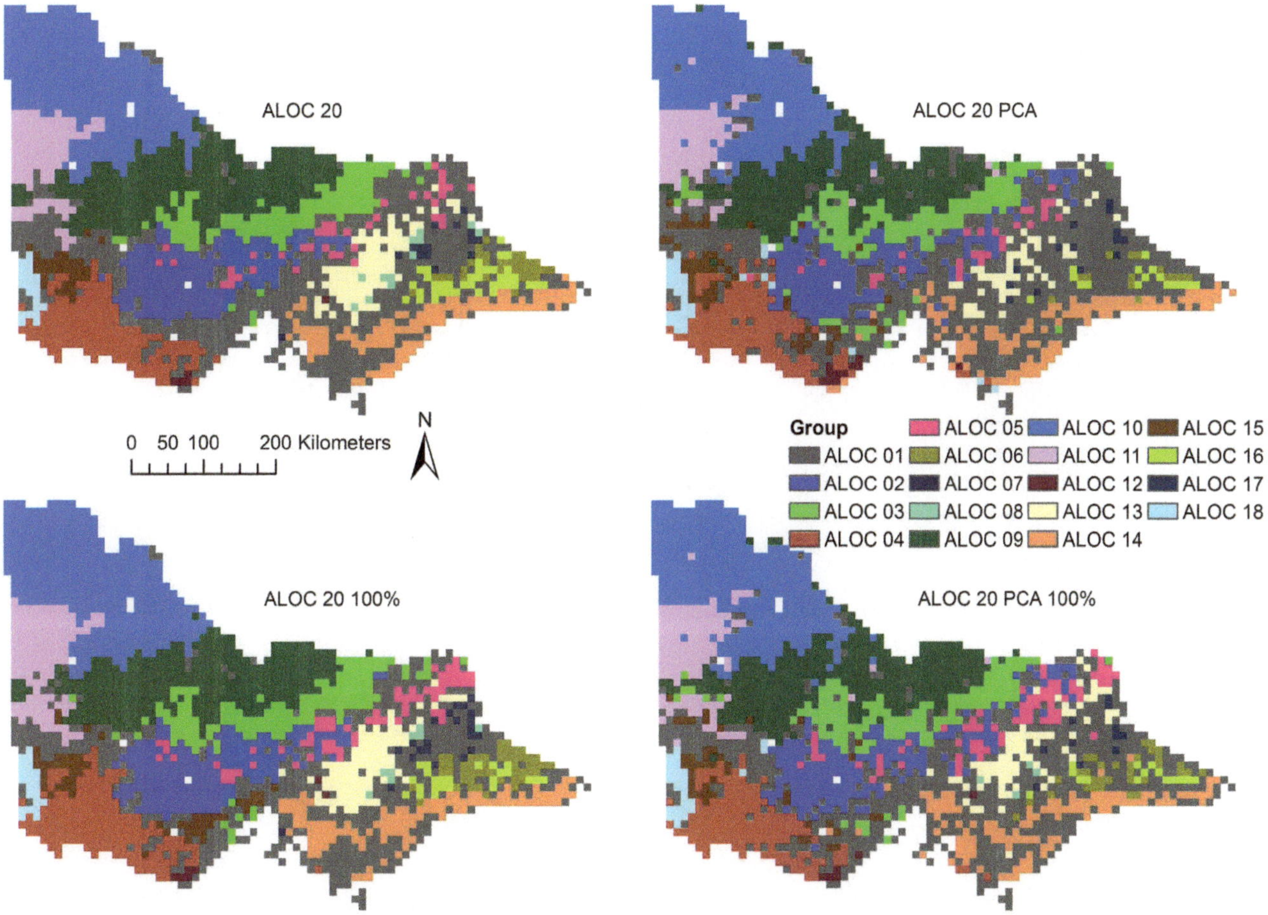

Figure 6. Results of the ALOC 20 (10 km) classifications. Top row - ALOC 20 and ALOC 20 PCA; Bottom row - ALOC 20 (100%) and ALOC 20 PCA (100%). Note that not all ALOC clusters are present in the final classifications. Colours represent each of the ALOC non-hierarchical clusters. Similar colours and cluster numbers do not necessarily represent related groups.

variability among points at the 30-m scale, and most of the variability at the 10-km scale.

Visual inspection of the resampled classifications (bottom row, Figure 5) showed very similar results to the original 30-m classifications. Agreement between the resampled ALOC 23 classification and the original 30-m classification was high (92%, $\kappa = 0.92$) and an accuracy of 91% ($\kappa = 0.9$) was observed with the 300 randomly chosen points from the validation dataset. This was a 4% difference in accuracy to the original 30-m ALOC 23 classification. However, McNemar's test suggested the difference in classification accuracy was statistically significant ($\chi^2 = 12.56$, $p<0.001$). As the two classifications are not independent, the permutation test was also used to compare the two. This supported the results of the McNemar's test and a permutated κ difference of 0.062 was deemed significant ($p<0.001$). The ALOC 23 PCA resampling results were similar, with a relatively high agreement with the original 30-m PCA classification (88%, $\kappa = 0.87$), and 87% ($\kappa = 0.86$) agreement with the validation dataset, a 5% difference compared to the original ALOC 23 PCA classification. These differences were also statistically significant (McNemar's test, $\chi^2 = 11.76$, $p<0.001$; permutation test, κ difference $= 0.073$, $p<0.001$). Agreement was quite high between the 2.5-km resampled ALOC 23 and 2.5-km resampled ALOC 23 PCA classifications (93%, $\kappa = 0.92$), however unlike the 30-m classification results, when comparing the agreement between the two resampled classifications McNemar's test suggested the results were significantly different ($\chi^2 = 5.88$, $p = 0.015$). As these two classifications were independent of one another the permutation test was not required.

Once the classification assessment was finalised, the meta-group allocations were appended to the classifications. The ALOC 23 meta-group allocations were examined visually (Figure 8) and were deemed suitable given that regions having similar groups (e.g. A and B, represented by distinct colours in Figure 8) exist in similar areas among the two classifications and show quite obvious spatial relationships. While the performance of the ALOC 20 models was weaker, the meta-group assignment results were similar to those observed for the ALOC 23 classification in that similar meta-groups existed closer spatially (not presented).

Spatial variability in western Victoria

The variability of model results was not uniform across the entire Glenelg-Hopkins case-study region (Figure 9). While the eastern and central parts of the region appeared to be relatively spatially uniform (i.e. they do not show a significant amount of variation in ALOC groupings); the western and north-western parts of the region are comprised of a number of ALOC classes. This suggests that the hydrological system varies spatially across

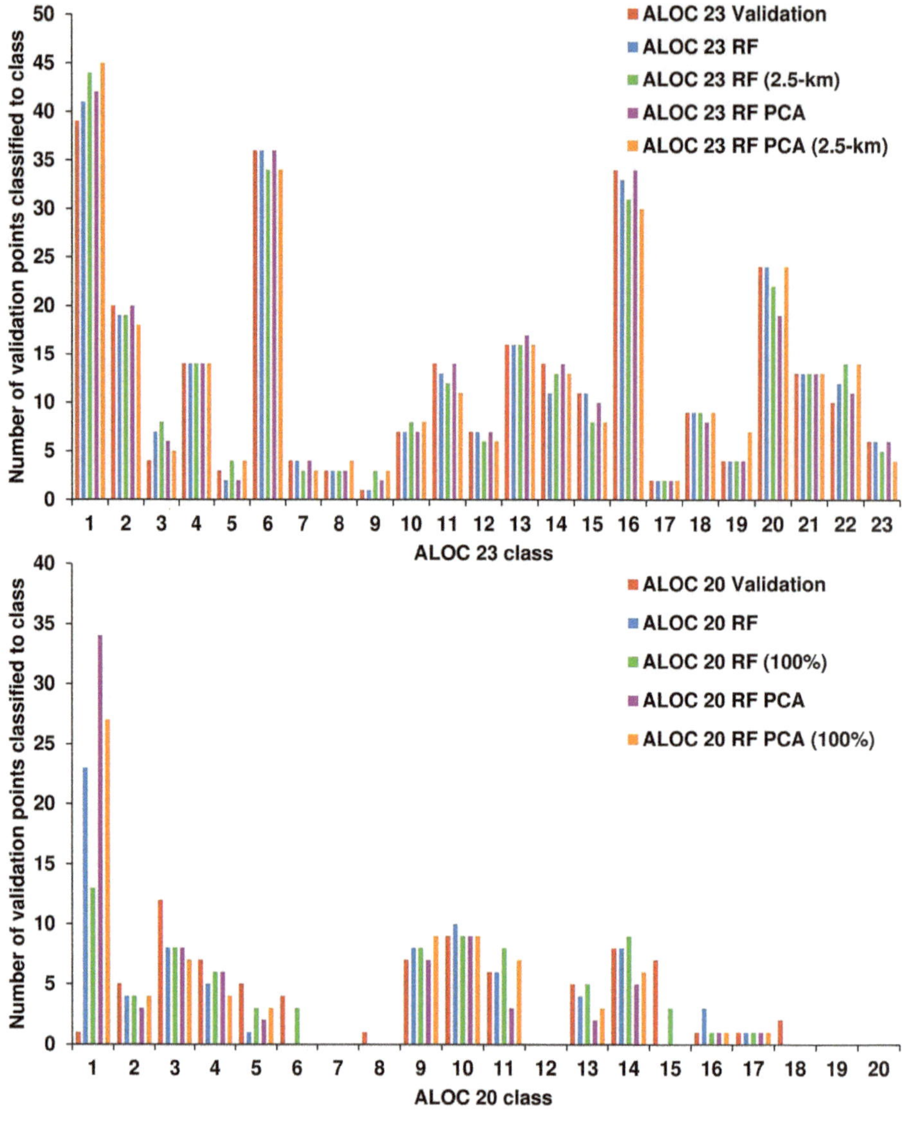

Figure 7. Class distributions from the final classified images. The 20% validation points were used to extract class information from the classified images. The resulting figure shows that the ALOC 23 classifications were much better at maintaining the class distribution of the validation dataset (shown above in red) than the ALOC 20 classifications.

the region, with the most variability likely occurring in the western half of the catchment.

The meta-group classifications support the idea that there was a difference in the hydrological systems of the three major catchments of the Glenelg Hopkins region (Figure 10). Interestingly, the amount of spatial variation did not decrease when the meta-group assignments of the original ALOC classes were examined, suggesting that hydrologic responses could be very different in areas that are quite close together. In the Glenelg River catchment (Figure 10, shown in red), there is obvious spatial variation in the assigned hydrological classes. Of particular interest are the two catchments in the north-east of the catchment that contain Rocklands Reservoir and the majority of the Grampians ranges, as they each consist of five different hydrological meta-groups (E, F, H, I & J). Even though the meta-groups represented in those particular catchments occupy a small area, they are still all present in the resampled classification (Figure 10, bottom row). While the differences were less pronounced in the Hopkins River catchment (Figure 10, outlined in blue), there was some spatial variability in classes in the north (dominated by meta-group I, with some small patches comprised of meta-group E), while most of the catchment belongs to group I and the two southernmost catchments belong to meta-group G. There was even less variation in the classification of the Portland catchment (Figure 10, outlined in green) with meta-group I dominating that catchment. Nonetheless, there were small areas of groups F and J in the south-west catchments of the catchment. Visual examination of the PCA and the resampled classifications showed little difference to that observed in the original 30-m classifications. The most obvious change was the small area in the southern catchment of the Hopkins River catchment (outlined in blue), and the easternmost parts of the Portland catchment (outlined in green), that was classified as meta-group J in the PCA classifications.

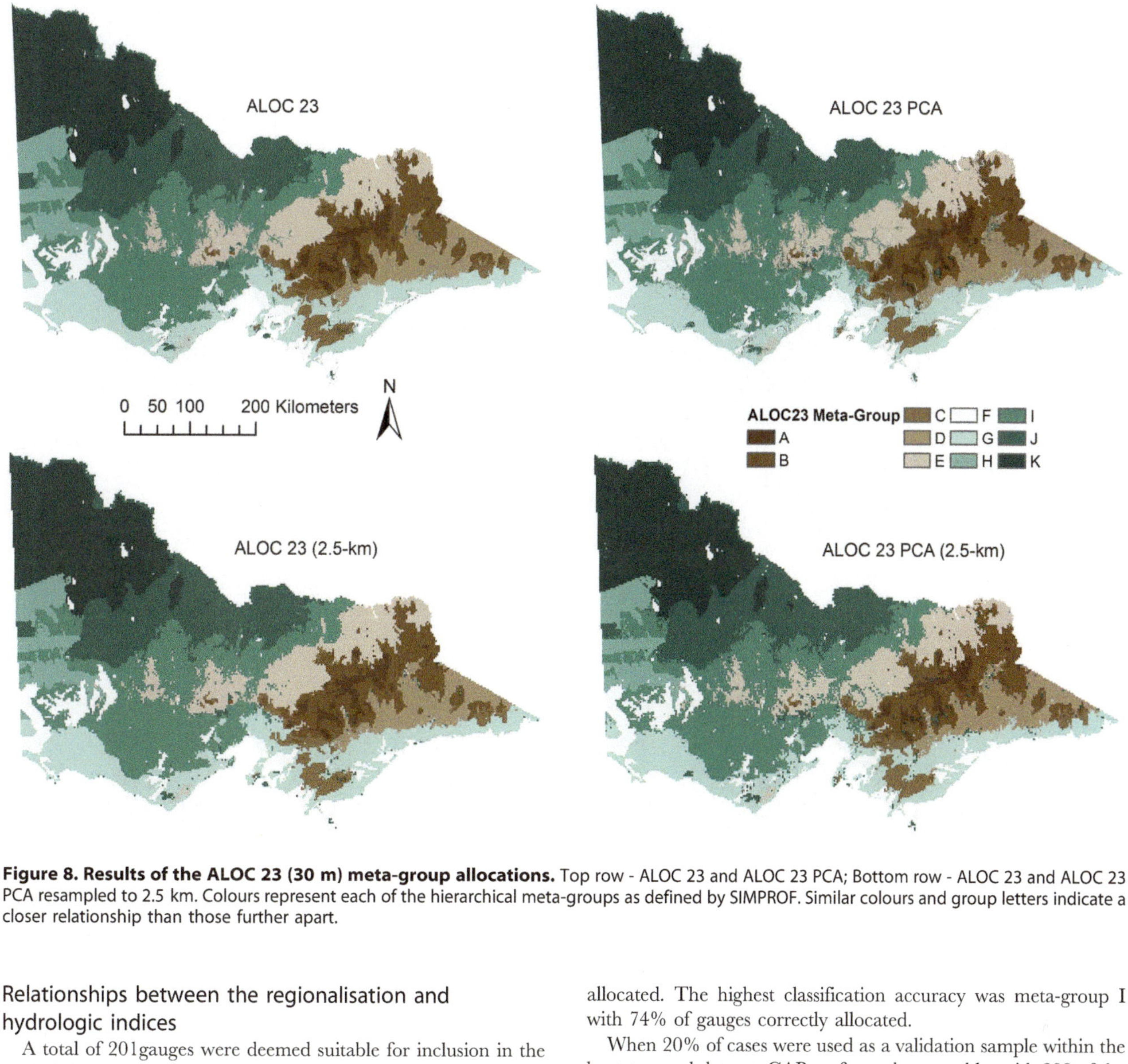

Figure 8. Results of the ALOC 23 (30 m) meta-group allocations. Top row - ALOC 23 and ALOC 23 PCA; Bottom row - ALOC 23 and ALOC 23 PCA resampled to 2.5 km. Colours represent each of the hierarchical meta-groups as defined by SIMPROF. Similar colours and group letters indicate a closer relationship than those further apart.

Relationships between the regionalisation and hydrologic indices

A total of 201gauges were deemed suitable for inclusion in the analysis based on the criteria specified above. With the exception of groups A, K and H, all regions had multiple suitable gauges.

There were significant differences among groups based on their hydrologic indices for both the original (pseudo-$F_{1,7} = 3.655$, $p = 0.001$) and the bootstrapped datasets (pseudo-$F_{1,7} = 9.304$, $p = 0.001$). *A posteriori* pairwise comparisons of groups using the original dataset indicated that the differences were significant ($p \leq 0.05$) between all groups with the exception of pairs B:E, C:D, C:E and F:I. The pairwise comparisons on the bootstrapped dataset were all significant ($p < 0.05$).

The ability of CAP to correctly classify cases within the original dataset based on their hydrologic indices was relatively poor, with only 92 samples correctly classified, but the model was statistically significant (48%, $m = 30$, $p = 0.001$). Stream gauges from meta-group C had the lowest classification accuracy with only 10% of gauges being successfully allocated. The highest classification accuracy was observed for both meta-groups B and G, with 67% of gauges correctly allocated to each. The bootstrap dataset performed better with 253 samples being correctly classified (66%, $m = 28, p = 0.001$). The lowest classification accuracy was observed for meta-group E with only 47% of gauges being correctly allocated. The highest classification accuracy was meta-group I with 74% of gauges correctly allocated.

When 20% of cases were used as a validation sample within the bootstrapped dataset, CAP performed reasonably, with 208 of the gauges correctly classified (67%, $m = 32$, $p = 0.001$). The lowest classification accuracy was observed for meta-group J with 58% of gauges being correctly allocated. The highest classification accuracy was meta-group F with 77% of gauges correctly allocated. Agreement between the samples that were allocated to new groups and their original group was relatively high at 60% ($\kappa = 0.53$). The lowest successful group allocation was observed for meta-group E where only 20% of gauges were correctly classified. The highest allocation accuracy was observed for meta-group D where 100% of gauges were correctly classified.

Pearson's correlation indicated there was a statistically-significant positive relationship between the number of gauges in each group and the number of gauges classified correctly from each group (Figure S9) for the original ($r = 0.91$, $p = 0.002$, $n = 8$), the bootstrapped dataset ($r = 0.93$, $p = 0.001$, $n = 8$), and the bootstrapped dataset with the validation samples ($r = 0.96$, $p < 0.001$, $n = 8$). This indicated that the model was most likely to correctly classify groups that were common in the dataset, with uncommon groups being correctly classified less often. All classifications performed highly favourably compared to a random allocation of

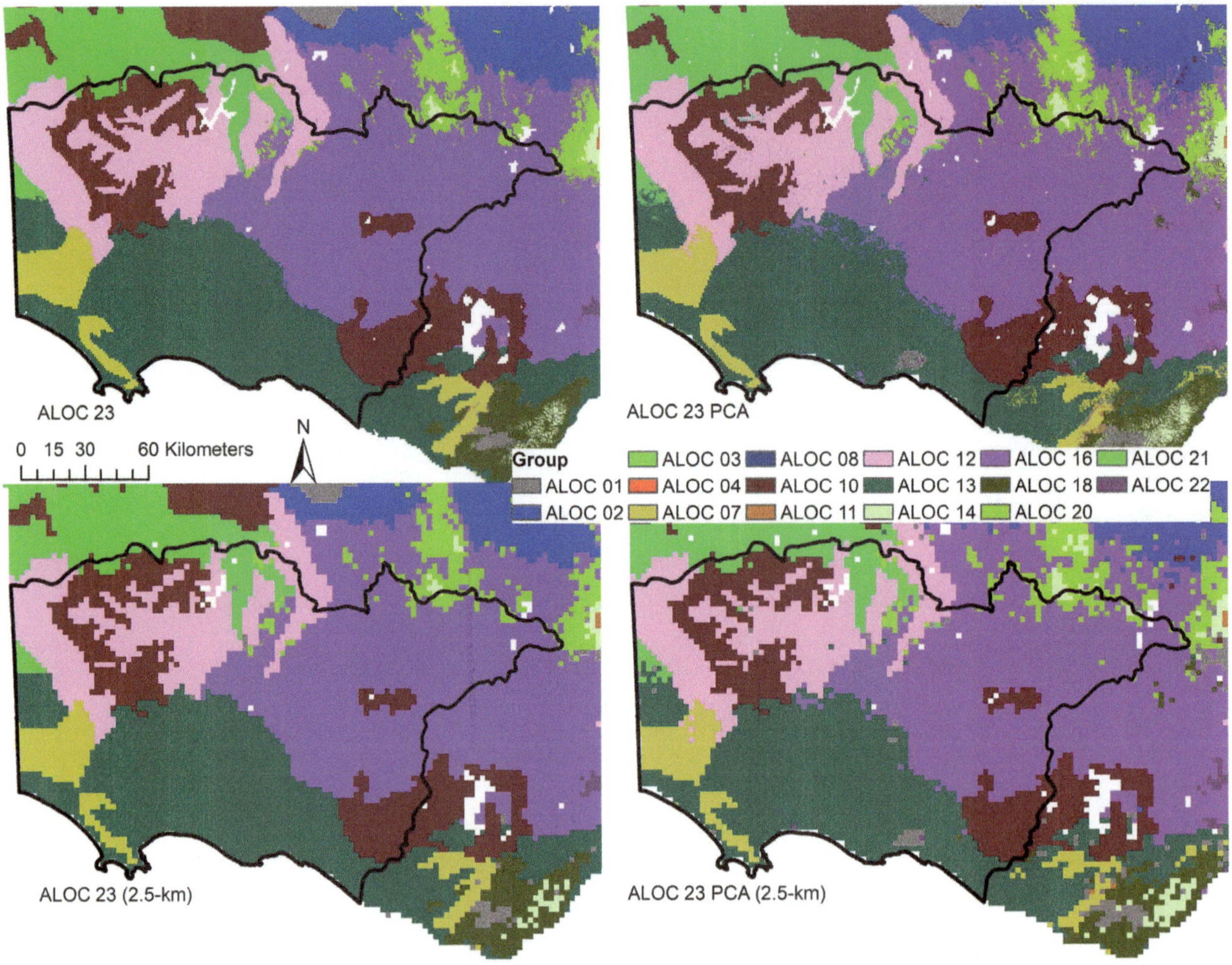

Figure 9. Results of the ALOC 23 (30 m) classifications for the Glenelg Hopkins region. Top row - ALOC 23 and ALOC 23 PCA; Bottom row - ALOC 23 and ALOC 23 PCA resampled to 2.5 km. Colours represent each of the ALOC non-hierarchical clusters. Similar colours and cluster numbers do not necessarily represent related groups. More spatial variability in the ALOC clusters is obvious in the Glenelg catchment (western and NW side), than that in the Hopkins (eastern) and Portland (south central) catchments.

cases among 11 groups which, assuming equal sample sizes, results in 9% of cases being correctly classified.

Discussion

By incorporating a number of environmental variables likely to influence regional hydrology and a range of non-parametric statistical and classification methods, this study aimed to generate a hydrologic landscape classification that did not require the use of an *a priori* selection of a spatial unit such as a catchment. The main objective of the approach was to see whether it was possible to create a classification that could preserve the environmental and hydrological variability that are known to influence streamflows within and between catchments that has typically been lost in previous regionalisation studies. An analysis of the ability of the classification to differentiate between streams from each group based on a number of streamflow indices was also undertaken.

Difference to previous regionalisation studies

Inductive methods of hydrologic regionalisation have been popular in the past (see Table 2 in Olden et al. [9]) and, while there have been a number of studies that have focused on deductive methods (see Table 1 in Olden et al. [9]), the choice of variables, their resolution (temporal and spatial), the classification method, the spatial scale of the classification and the number of groups are all known to influence deductive classifications [9,14]. While the final product of deductive methods is a spatial mosaic of independent hydrologic types, the final classifications do not always only identify hydrologic variation [10,12]. Inter-catchment variability can limit the applicability of hydrologic regionalisations to generalise and predict catchment behaviour as a function of climatic and environmental gradients [14]. As previous studies have relied on catchments, landscape units or stream sections [8,9,14,19], an issue that is more apparent in deductive regionalisation studies is the loss of small-scale spatial hydrologic variability [6,16] as the unit of analysis gets larger. Our method relied on using an accurate supervised image classification method to extend our statistical clustering to an area covering ~228,000 km^2 without the need to rely on catchments, landscape units or stream sections. We opted for this approach as it is well known in the literature that there is significant flow variability within and among catchments and that the variables governing flow

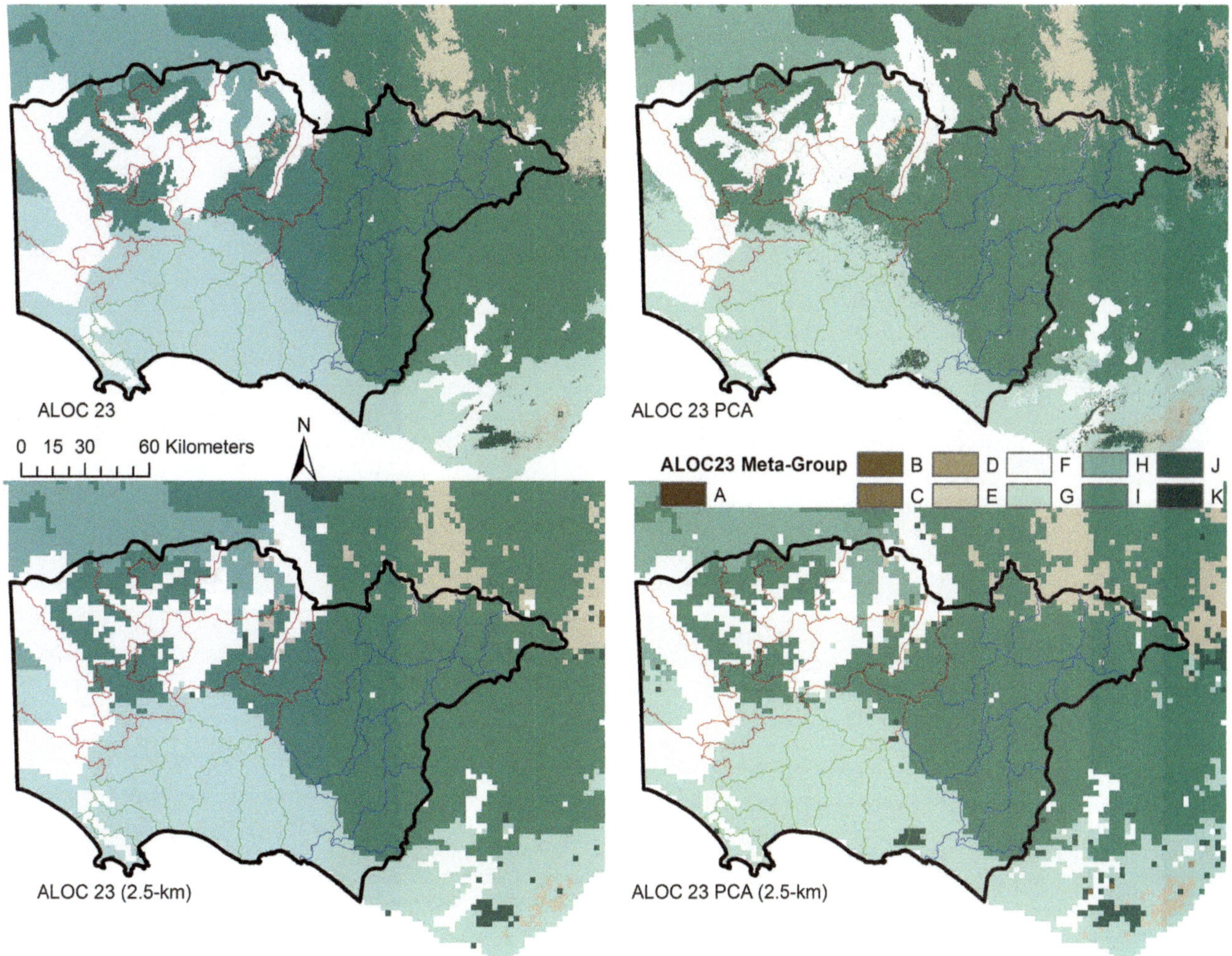

Figure 10. Results of the ALOC 23 (30 m) meta-group allocations for the Glenelg Hopkins region. Top row - ALOC 23 and ALOC 23 PCA; Bottom row - ALOC 23 and ALOC 23 PCA resampled to 2.5 km. Colours represent each of the hierarchical meta-groups as defined by SIMPROF. Similar colours and group letters indicate a closer relationship than those further apart. More spatial variability in the meta-groups is obvious in the Glenelg catchment (shown in red), than that in the Hopkins (blue) and Portland (green) catchments.

variability are dependent on scale [13]. While other regionalisation studies have identified that the primary drivers of catchment function are largely related to climatic gradients [14], our results suggest that a mixture of climatic, geological and environmental functions are driving catchment, and thus hydrologic, variability (Table S2, Figure S8) in our regionalisation. It would be expected, in a traditional approach to hydrologic regionalisations, that some of this variability would be lost – which could explain why other studies have largely identified climatic gradients that vary slowly with space to be the primary drivers of catchment function. Our approach allows for different classes to be represented within a single catchment, thus preserving intra-catchment variability.

Statistical evaluation, clustering, and PCA

Traditional parametric statistical analyses and clustering algorithms such as *k*-means tend to have restrictive assumptions regarding independence of samples, multivariate normality and collinearity. The assumption of samples being distributed normally through multivariate space, for example, is unlikely to be true in most ecological and environmental datasets [25]. The approach that we employed relied on the use of non-parametric and permutation-based statistical methods in conjunction with the RF classification algorithm. This approach had far fewer assumptions relating to data normality and collinearity [27,44,45]. The approach also allowed the decision regarding the number of groups used in the analysis to be statistically-justified, when this decision is typically arbitrary. Our method was supported by the application of both ANOSIM and MDS to assess group separation, with each suggesting that the groups were distinct and that we had chosen an appropriate number of groups for our dataset. While not perfect (SIMPROF, by design, tests for hierarchically-related groups and we were after non-hierarchical groups), we believe this approach to be simpler, more statistically sound and more efficient than methods employed in the past which require large amount of *a posteriori* or *post-hoc* statistical testing [10,19].

The use of PCA-transformed data was shown not to significantly affect the classification accuracy of the model, even though the PCA was only able to explain <80% of the data variability (Table 1, ALOC 23 PCA). This suggests that future classifications could be conducted on PCA-transformed datasets while still producing accurate classification schemes. Our method has essentially shown that it is possible to extract the same number of groups from PCA-transformed data as it was from the non-PCA transformed data. However, using PCA from the beginning has the potential to influence the number of groups identifiable by SIMPROF (as there are fewer data and less variance in the remaining data) and therefore influence the overall classification process. Using PCA could, however, make the process more efficient in that having a reduced number of groups to begin with could remove the need to first use non-hierarchical classification before hierarchically classifying the groups. The major benefit of using PCA-transformed data in this study was that the time to parameterise and classify the raster data with RF was substantially reduced.

Classification by Random Forests

The non-parametric and highly accurate [27,48] RF classifier was very successful in recovering and classifying the ALOC class information of the remaining pixels in the raster datasets. While the ALOC 23 RF models had very high classification accuracies (95%, $\kappa = 0.94$ and 92%, $\kappa = 0.92$), our hypothesis that the classification of the coarser datasets would be inferior was confirmed by the low accuracies of the ALOC 20 classifications (46%, $\kappa = 0.42$ and 47%, $\kappa = 0.44$). As our sampling density was severely limited by pixel size in the 10-km models, this further supports previous research showing that the overall and per-class accuracy of supervised classifications can be limited by pixel size [39,40]. While the RF classifier has been shown to be robust against statistical noise and training data reduction [48], it is possible that, in this case, there were simply not enough training sites to allow for the creation of an appropriate model at the 10-km scale. This analysis used only 325 training sites (80% of $n = 406$ 10-km sample points) to try to produce a classifier for 20 classes and another 33% of the training site data was excluded for OOB accuracy assessment [48]. This left the RF algorithm with only 215 points to generate the required classification trees. The two ALOC 20 100% models used 268 points (66% of $n = 406$ 10-km sample points) to create an RF model but still performed poorly with OOB error rates limiting accuracies to 56% and 54%. As all classes were included in training data for the ALOC 20 100% models, their omission from the final classification (Figure 6 & 7) suggests that they, by chance, happened to be excluded from the training data selected by the bootstrapping step used to calculate OOB error. This could partly explain the high OOB error rates observed for those models. The missing classes from the ALOC 20 classifications (Figure 7, Table S3) could be similarly explained, although it is also possible that they were excluded randomly from the 80% training data at the previous step. As expected, the exclusion of data, either manually for validation purposes or automatically by RF to enable an OOB estimate, appeared to severely limit the classification accuracy when using small amounts of training data [48]. This further supports our hypothesis that limiting the study to a 10-km resolution based on a single, coarse dataset would influence the results presented here, particularly as this study was conducted over a relatively small area. If the study had been conducted at a continental scale, for example, it would be possible to generate more than 15,500 random points at a minimum distance of 10 km and, therefore, resampling the same datasets to a finer resolution (to avoid sub-sampling of pixels) would not be necessary.

The resampling of the ALOC 23 models from a 30-m to 2.5-km resolution appeared to remove some of the finer-scale spatial variability in the classifications (Figures 5 & 8). While not appearing to constitute a significant change between the original and resampled classifications (percent agreements between the resampled and original classifications were 92% ($\kappa = 0.92$) for the ALOC 23 model and 88% ($\kappa = 0.87$) for the ALOC 23 PCA model), the results of both McNemar's test and the permutation test indicated that the resampled classifications were significantly different from their equivalent 30-m classifications. This may seem like a serious drawback to the method, however, when compared to the validation dataset the accuracy of the 2.5-km ALOC 23 classification was only 4% less than that observed for the 30-m model, while the accuracy of the 2.5-km ALOC 23 PCA classification was only 5% less than that of the 30-m PCA model. While statistically the difference may be deemed significant, we contend that in reality a difference of ≤5% would likely not be ecologically or environmentally important and thus, would not affect the ability of the method to create a hydrological landscape classification that could be used to explain spatial differences in streamflow metrics. Additionally, the resampling step was performed *a posteriori* and therefore may not actually be necessary in all cases.

Case study on spatial variability in western Victoria

The spatial variability in the classification of the Glenelg Hopkins region was most evident in the Glenelg catchment, and less so in the Hopkins and Portland catchments (Figure 9 & 10). Spatial hydrologic variability has been observed in a number of studies in the past and the strongest and most variable relationships between environmental factors and water quality and quantity have consistently been found in the Glenelg catchment. Brown et al. [17] found that the Glenelg catchment exhibited the most variability in the relationships between climate, land use and wetland extent, which may help to explain some of the variability observed here. Relationships explored in the past relating the proportion of native vegetation to in-stream salinity [35] showed strong relationships in the Glenelg and Portland catchments, but less evidence for the same relationships in the Hopkins catchment and it was suggested that this may have been due to the degraded nature of the catchment. The degraded nature of the Hopkins catchment could also explain the lack of variability observed in that catchment in this study. A spatially-varying relationship between nutrient exports and land use has also been observed in the Glenelg Hopkins region [34] although, due to the lack of suitable stream-gauge data, it is not clear whether this relationship holds for the streams and rivers of the region. However, the relationships in the region that have been described in the past [17,34,35] are likely to be complicated due to variations in geomorphology, groundwater levels and salt concentrations [36]; conditions that we have attempted to account for in this study. Water resource managers in the region need to take into account possible differences in intra- and inter-catchment hydrology that could drastically affect river management and restoration plans and regionalisation studies such as the one presented here could assist in identifying that variability.

Relationships between the regionalisation and hydrologic indices

Understanding, and being able to accurately predict, streamflow characteristics in ungauged locations is crucial for ecohydrological and other studies [2,6]. Our method set out to test a new approach to hydrological regionalisation that removed the need for catchments as a spatial unit of analysis for our statistical clustering [6,8,14]. However, the ability to link the results of the regionalisation to streamflow indices could have presented an issue given that we did not rely on catchments as a spatial unit.

The results supported our hypothesis that our method would be able to identify and preserve inter- and intra-catchment variability. Pairwise comparisons suggested that, even in the original dataset ($n = 201$), there was enough variability in our 32 streamflow indices to separate all but four pairs (of a total of 28 pairs). A simple bootstrap (with replacement) to $n = 383$ gauges was sufficient for all groups to be easily identified as distinct suggesting there is a minimum number of gauges for the approach to be reliable (likely to relate to the number of each individual class in the dataset). In addition, CAP demonstrated that the regionalisation was able to discriminate among streams from different groups. While the classification was not perfect, the analysis conducted on the bootstrap datasets indicated that the stream gauges could be classified correctly significantly better than chance alone, and that gauges from one class could be correctly classified 100 percent of the time. In general, common classes were correctly classified more frequently than uncommon classes. While the correlation between the number of gauges in each class and the number of correctly-allocated samples was significant, there was no clear threshold (i.e. the relationship was linear), so it is difficult to identify a single minimum number of gauges that could be implemented to ensure that results met pre-defined criteria for reliability. Therefore, future users should interpret results for rare groups with caution. However, we believe that, based on this preliminary assessment of our method, the results illustrate that there is promise in the method for categorising regions, particularly in the absence of comprehensive streamflow data as is the case in many regions in Australia and elsewhere. Further investigation of the validation (i.e. linking the regionalisation to streamflow indices) using more in-depth data mining approaches (e.g. decision trees [49]) is likely to produce even greater classification success.

Conclusions

Hydrologic classifications are increasingly being employed in the management and research of aquatic resources. Our approach differed from inductive hydrological regionalisation where membership is defined quantitatively based on metrics of stream flow, and traditional deductive regionalisation which require the use of catchments or other appropriate spatial units. Instead, membership of pixels was defined qualitatively with the random forest classifier based on a statistical classification of a number of environmental variables that we believe could have a direct influence on the hydrologic cycle. In essence, we present a method that allowed the creation of spatially-independent hydrological regions; these regions represent a series of fundamental hydrologic landscape units that exist in multiple locations depending on environmental similarity rather than a combination of environment and streamflow-metric response similarity. To our knowledge, the application of deductive reasoning and hybrid classification is a novel approach in hydrological regionalisation. This method has removed the need to rely on a spatial unit specified *a priori*, such as a catchment or ecoregion, and has allowed the preservation of intra-catchment variability. Thus, it should be useful in the spatial explanation and prediction of streamflow responses that are known, or suspected, to vary within catchments. The ability of our regionalisation to discriminate among streams from different groups based on their range of flow indices highlights the value of this approach, particularly in regions where streamflow data are lacking.

Supporting Information

Figure S1 MDS Analysis plots for the ALOC 23 and ALOC 20 models. Top row shows the ALOC 23 and ALOC 20 groups, while the bottom row shows the ALOC 23 and ALOC 20 meta-group plots. Legends in the top row represent the meta-groups.

Figure S2 ALOC 23 and ALOC 20 dendrogram demonstrating the hierarchical relationships between the non-hierarchical groups as defined from the ALOC group averages using SIMPROF. Letters in green represent the meta-groups each combination of non-hierarchical groups belongs to.

Figure S3 BioClim variable distributions across each of the ALOC 23 meta-groups.

Figure S4 Groundwater variable distributions across each of the ALOC 23 meta-groups. Note that observations >>30,000 are missing from GW_TDS.

Figure S5 Landscape variable distributions across each of the ALOC 23 meta-groups.

Figure S6 Soil variable distributions across each of the ALOC 23 meta-groups.

Figure S7 Climate variable distributions across each of the ALOC 23 meta-groups.

Figure S8 Variable contribution to each of the hierarchical meta-groups calculated using SIMPER on a standardised Euclidean distance matrix. Any variables contributing <5% to the variance were pooled together and are represented by ALL_OTHER_VARS. Missing groups contained only one ALOC cluster and therefore % variable contribution could not be calculated with SIMPER.

Figure S9 The correlation between the number of samples from each class and the number of samples correctly classified by CAP was highly significant, with the linear relationship between the variables for each sample illustrated in the figure. However, as the relationship was linear there was no clear threshold suggesting a minimum number of gauges needed to guarantee an acceptable level of accuracy. The CAP on the original dataset ($n = 201$) was quite poor (classification accuracy $= 48\%$, $m = 30$, $p = 0.001$), while the bootstrapped dataset ($n = 383$) was a significant improvement (classification accuracy $= 66\%$, $m = 28$, $p = 0.001$). CAP on the bootstrap dataset with a 20% validation sample also performed reasonably (classification accuracy $= 67\%$, $m = 32$, $p = 0.001$).

Table S1 The variables used in the creation of the hydrological regionalisation. A number of variables describing the storage, transport and release of surface water, groundwater and atmospheric water were included in the analysis. * The DTM data was resampled to 30 m to enable geo-TIFF compatibility with ENVI 4.8.

Table S2 SIMPER results from PRIMER 6. Numbers represent % contribution of each of the variables to the ALOC 23 and ALOC 20 meta-groups on a standardised Euclidean distance matrix. Blank columns are meta-groups that only contained 1 ALOC cluster and therefore % contribution could not be calculated using SIMPER. KW = Kruskal-Wallis statistic, with higher values indicating a better ability of that variable to discriminate between clusters. All KW values were significant at $p<<0.001$. See Figure S9 for a graphical representation of this table.

Table S3 Total accuracies and kappa statistics for the 4 RF classified models, and producer and user accuracies for each of the classes defined by the ALOC algorithm as classified by RF. N/A indicates groups that were missing from the classified dataset as a result of exclusion from the samples used to train the RF model. In some cases, groups were absent from the 80% training data, while others were excluded by the bootstrap aggregation step used to train the RF models.

Code S1 S_perm_test.R: R code for running the permutation test for testing for differences between the resampled and original classifications.

Data S1 S_Kappa_ALOC23.csv: CSV file with 4 columns. The first column is the point ID from the random subsample used for comparisons between the original and resampled classifications, the remaining columns are the "true" class from the ALOC clustering, the class extracted from the location of the 30-m RF classification, and the class extracted from the 2.5-km resampled RF classification.

Author Contributions

Conceived and designed the experiments: SCB REL JF. Performed the experiments: SCB. Analyzed the data: SCB REL JF. Wrote the paper: SCB REL VLV JF LL.

References

1. Bunn SE, Arthington AH (2002) Basic principles and ecological consequences of altered flow regimes for aquatic biodiversity. Environmental Management 30: 492–507.
2. Poff NL, Richter BD, Arthington AH, Bunn SE, Naiman RJ, et al. (2010) The ecological limits of hydrologic alteration (ELOHA): A new framework for developing regional environmental flow standards. Freshwater Biology 55: 147–170.
3. Poff NL, Allan JD, Bain MB, Karr JR, Prestegaard KL, et al. (1997) The natural flow regime: A paradigm for river conservation and restoration. BioScience 47: 769–784.
4. Naiman RJ, Latterell JJ, Pettit NE, Olden JD (2008) Flow variability and the biophysical vitality of river systems. Comptes Rendus - Geoscience 340: 629–643.
5. LeRoy Poff N, Allan JD (1995) Functional organization of stream fish assemblages in relation to hydrological variability. Ecology 76: 606–627.
6. Kennard MJ, Pusey BJ, Olden JD, MacKay SJ, Stein JL, et al. (2010) Classification of natural flow regimes in Australia to support environmental flow management. Freshwater Biology 55: 171–193.
7. Winter TC (2001) The concept of hydrologic landscapes. Journal of the American Water Resources Association 37: 335–349.
8. Wolock DM, Winter TC, McMahon G (2004) Delineation and evaluation of hydrologic-landscape regions in the United States using geographic information system tools and multivariate statistical analyses. Environmental Management 34: S71–88.
9. Olden JD, Kennard MJ, Pusey BJ (2012) A framework for hydrologic classification with a review of methodologies and applications in ecohydrology. Ecohydrology 5: 503–518.
10. Snelder TH, Biggs BJF, Woods RA (2005) Improved eco-hydrological classification of rivers. River Research and Applications 21: 609–628.
11. Santhi C, Allen PM, Muttiah RS, Arnold JG, Tuppad P (2008) Regional estimation of base flow for the conterminous United States by hydrologic landscape regions. Journal of Hydrology 351: 139–153.
12. Carlisle DM, Falcone J, Wolock DM, Meador MR, Norris RH (2010) Predicting the natural flow regime: Models for assessing hydrological alteration in streams. River Research and Applications 26: 118–136.
13. McManamay RA, Orth DJ, Dolloff CA, Frimpong EA (2012) Regional Frameworks Applied To Hydrology: Can Landscape-Based Frameworks Capture The Hydrologic Variability? River Research and Applications 28: 1325–1339.
14. Sawicz K, Wagener T, Sivapalan M, Troch PA, Carrillo G (2011) Catchment classification: Empirical analysis of hydrologic similarity based on catchment function in the eastern USA. Hydrology and Earth System Sciences 15: 2895–2911.
15. Poff NL (1996) A hydrogeography of unregulated streams in the United States and an examination of scale-dependence in some hydrological descriptors. Freshwater Biology 36: 71–91.
16. Poff NL, Olden JD, Pepin DM, Bledsoe BP (2006) Placing global stream flow variability in geographic and geomorphic contexts. River Research and Applications 22: 149–166.
17. Brown S, Versace V, Laurenson L, Ierodiaconou D, Fawcett J, et al. (2012) Assessment of Spatiotemporal Varying Relationships Between Rainfall, Land Cover and Surface Water Area Using Geographically Weighted Regression. Environmental Modeling and Assessment 17: 241–254.
18. Mackey BG, Berry SL, Brown T (2008) Reconciling approaches to biogeographical regionalization: A systematic and generic framework examined with a case study of the Australian continent. Journal of Biogeography 35: 213–229.
19. Stein JL, Hutchinson M, Pusey BJ, Kennard MJ (2009) Ecohydrological classification based on landscape and climate data. Appendix 8. In: Pusey BJ, Kennard MJ, Stein JL, Olden JD, editors. Ecohydrological regionalisation of Australia: A Tool for Management and Science. Canberra: Land and Water Australia.
20. Clifford DHT, Stephenson W (1975) An introduction to numerical classification. New York: Academic Press.
21. Cormack RM (1971) A Review of Classification. Journal of the Royal Statistical Society Series A (General) 134: 321–367.
22. Everitt B (1980) Cluster analysis. 2nd Edition. London: Heinemann.
23. Clarke KR, Warwick RM (2001) Change in Marine Communities: An Approach to Statistical Anlysis and Interpretation, 2nd Edition. Plymouth: PRIMER-E.
24. Legendre P, Legendre L (1998) Numerical Ecology, 2nd Edition. Amsterdam: Elsevier Science.
25. Belbin L, McDonald C (1993) Comparing three classification strategies for use in ecology. Journal of Vegetation Science 4: 341–348.
26. Belbin L (1987) The use of non-hierarchical allocation methods for clustering large sets of data. Australian Computer Journal 19: 32–41.
27. Breiman L (2001) Random forests. Machine Learning 45: 5–32.
28. Held M, Jakimow B, Rabe A, van der Linden S, Wirth F, et al. (2012) EnMAP-Box Manual, Version 1.4, Humboldt Universität zu Berlin, Germany. Available: www.enmap.org. Accessed 2013 Aug.
29. Lillesand TM, Kiefer RW, Chipman JW (2008) Remote Sensing and Image Interpretation: 6th edition. New Jersey: Wiley.
30. Department of Environment and Primary Industries (2013) Victorian Resources Online - Temperature, Available: http://vro.dpi.vic.gov.au/dpi/vro/vrosite.nsf/pages/climate-temperature. Accessed 2013 Aug.
31. Department of Environment and Primary Industries (2013) Victorian Resources Online - Rainfall, Available: http://vro.dpi.vic.gov.au/dpi/vro/vrosite.nsf/pages/rainfall. Accessed 2013 Aug.
32. Corrick A (1992) An assessment of Victoria's wetlands. Melbourne: Department of Conservation and Environment.
33. Versace VL, Ierodiaconou D, Stagnitti F, Hamilton AJ (2008) Appraisal of random and systematic land cover transitions for regional water balance and revegetation strategies. Agriculture, Ecosystems & Environment 123: 328–336.
34. Ierodiaconou D, Laurenson L, Leblanc M, Stagnitti F, Duff G, et al. (2005) The consequences of land use change on nutrient exports: a regional scale assessment in south-west Victoria, Australia. Journal of Environmental Management 74: 305–316.
35. Versace VL, Ierodiaconou D, Stagnitti F, Hamilton AJ, Walter MT, et al. (2008) Regional-scale models for relating land cover to basin surface-water quality using remotely sensed data in a GIS. Environmental Monitoring and Assessment 142: 171–184.
36. Yihdego Y, Webb JA (2011) Modeling of bore hydrographs to determine the impact of climate and land-use change in a temperate subhumid region of southeastern Australia. Hydrogeology Journal 19: 877–887.

37. Yihdego Y, Webb J (2013) An Empirical Water Budget Model As a Tool to Identify the Impact of Land-use Change in Stream Flow in Southeastern Australia. Water Resources Management 27: 4941–4958.
38. ESRI (2012) ArcGIS Desktop 10.1, Environmental Systems Research Institute, Redlands. Available: http://www.esri.com
39. Sprintsin M, Karnieli A, Berliner P, Rotenberg E, Yakir D, et al. (2007) The effect of spatial resolution on the accuracy of leaf area index estimation for a forest planted in the desert transition zone. Remote Sensing of Environment 109: 416–428.
40. Mumby PJ, Edwards AJ (2002) Mapping marine environments with IKONOS imagery: Enhanced spatial resolution can deliver greater thematic accuracy. Remote Sensing of Environment 82: 248–257.
41. Exelis Visual Information Solutions (2010) ENVI version 4.8, Exelis Visual Information Solutions, Boulder. Available: http://www.exelisvis.com/ProductsServices/ENVI/ENVI.aspx
42. Dixon PM (2006) Ripley's K function. In: El-Shaarawi A, Piegorsch WW, editors. Encyclopedia of Environmetrics. New York: Wiley.
43. Clarke KR, Gorley RN (2006) PRIMER v6: User Manual/Tutorial. Plymouth: PRIMER-E.
44. Clarke KR (1993) Non-parametric multivariate analyses of changes in community structure. Australian Journal of Ecology 18: 117–143.
45. Clarke KR, Somerfield PJ, Gorley RN (2008) Testing of null hypotheses in exploratory community analyses: similarity profiles and biota-environment linkage. Journal of Experimental Marine Biology and Ecology 366: 56–69.
46. PATN Website. Blatant Fabrications Pty. Ltd. (2009) PATN Version 3.1.2, Available: http://www.patn.com.au/default.htm. Accessed 2012 Apr.
47. R Core Team (2013) R: A language and environment for statistical computing, R Foundation for Statistical Computing, Vienna. Available: http://www.R-project.org/
48. Rodriguez-Galiano VF, Ghimire B, Rogan J, Chica-Olmo M, Rigol-Sanchez JP (2012) An assessment of the effectiveness of a random forest classifier for land-cover classification. ISPRS Journal of Photogrammetry and Remote Sensing 67: 93–104.
49. Breiman L (1993) Classification and regression trees: Chapman and Hall/CRC press.
50. Cohen J (1960) A coefficient of agreement for nominal scales. Educational and Psychological Measurement 20: 37–46.
51. Revelle W (2013) psych: Procedures for Psychological, Psychometric, and Personality Research, Available: http://cran.r-project.org/web/packages/psych/index.html. Accessed 2013 Aug.
52. Gamer M (2012) irr: Various Coefficients of Interrater Reliability and Agreement, Available: http://cran.r-project.org/web/packages/irr/index.html. Accessed 2013 Aug.
53. Fay MP (2010) Two-sided Exact Tests and Matching Confidence Intervals for Discrete Data. R journal 2: 53–58.
54. McNemar Q (1947) Note on the sampling error of the difference between correlated proportions or percentages. Psychometrika 12: 153–157.
55. McKenzie DP, Mackinnon AJ, Péladeau N, Onghena P, Bruce PC, et al. (1996) Comparing correlated kappas by resampling: Is one level of agreement significantly different from another? Journal of Psychiatric Research 30: 483–492.
56. Foody GM (2004) Thematic Map Comparison: Evaluating the Statistical Significane of Differences in Classification Accuracy. Photogrammetric Engineering & Remote Sensing 70: 627–633.
57. Olden JD, Poff NL (2003) Redundancy and the choice of hydrologic indices for characterizing streamflow regimes. River Research and Applications 19: 101–121.
58. Anderson MJ (2001) A new method for non-parametric multivariate analysis of variance. Austral Ecology 26: 32–46.
59. Anderson MJ, Robinson J (2003) Generalized discriminant analysis based on distances. Australian and New Zealand Journal of Statistics 45: 301–318.
60. Anderson MJ, Willis TJ (2003) Canonical analysis of principal coordinates: A useful method of constrained ordination for ecology. Ecology 84: 511–525.
61. Kennard MJ, Mackay SJ, Pusey BJ, Olden JD, Marsh N (2010) Quantifying uncertainty in estimation of hydrologic metrics for ecohydrological studies. River Research and Applications 26: 137–156.
62. Kennard MJ, Pusey BJ, Olden JD, Mackay S, Stein JL, et al. (2009) Ecohydrological classification of Australia's flow regimes. Appendix 5. In: Pusey BJ, Kennard MJ, Stein JL, Olden JD, editors. Ecohydrological regionalisation of Australia: A Tool for Management and Science. Canberra: Land and Water Australia.
63. Marsh NA, Stewardson MJ, Kennard MJ (2003) River Analysis Package. Monash University Melbourne: Cooperative Research Centre for Catchment Hydrology.
64. Anderson MJ, Gorley RN, Clarke KR (2008) PERMANOVA+ for PRIMER: Guide to Software and Statistical Methods. Plymouth, UK: PRIMER-E.

10

A Biodiversity Indicators Dashboard: Addressing Challenges to Monitoring Progress towards the Aichi Biodiversity Targets Using Disaggregated Global Data

Xuemei Han[1,2]*, Regan L. Smyth[1], Bruce E. Young[1], Thomas M. Brooks[1,3,4,5], Alexandra Sánchez de Lozada[1], Philip Bubb[6], Stuart H. M. Butchart[7], Frank W. Larsen[8,9], Healy Hamilton[1], Matthew C. Hansen[10], Will R. Turner[9]

1 NatureServe, Arlington, Virginia, United States of America, **2** Department of Environmental Science and Policy, George Mason University, Fairfax, Virginia, United States of America, **3** International Union for Conservation of Nature, Gland, Switzerland, **4** World Agroforestry Center, International Center for Research in Agroforestry, University of Philippines, Los Baños, Laguna, Philippines, **5** School of Geography and Environmental Studies, University of Tasmania, Hobart, Australia, **6** United Nations Environment Programme World Conservation Monitoring Centre, Cambridge, United Kingdom, **7** BirdLife International, Cambridge, United Kingdom, **8** European Environment Agency, Copenhagen, Denmark, **9** Conservation International, Arlington, Virginia, United States of America, **10** Department of Geographical Sciences, University of Maryland, College Park, Maryland, United States of America

Abstract

Recognizing the imperiled status of biodiversity and its benefit to human well-being, the world's governments committed in 2010 to take effective and urgent action to halt biodiversity loss through the Convention on Biological Diversity's "Aichi Targets". These targets, and many conservation programs, require monitoring to assess progress toward specific goals. However, comprehensive and easily understood information on biodiversity trends at appropriate spatial scales is often not available to the policy makers, managers, and scientists who require it. We surveyed conservation stakeholders in three geographically diverse regions of critical biodiversity concern (the Tropical Andes, the African Great Lakes, and the Greater Mekong) and found high demand for biodiversity indicator information but uneven availability. To begin to address this need, we present a biodiversity "dashboard" – a visualization of biodiversity indicators designed to enable tracking of biodiversity and conservation performance data in a clear, user-friendly format. This builds on previous, more conceptual, indicator work to create an operationalized online interface communicating multiple indicators at multiple spatial scales. We structured this dashboard around the Pressure-State-Response-Benefit framework, selecting four indicators to measure pressure on biodiversity (deforestation rate), state of species (Red List Index), conservation response (protection of key biodiversity areas), and benefits to human populations (freshwater provision). Disaggregating global data, we present dashboard maps and graphics for the three regions surveyed and their component countries. These visualizations provide charts showing regional and national trends and lay the foundation for a web-enabled, interactive biodiversity indicators dashboard. This new tool can help track progress toward the Aichi Targets, support national monitoring and reporting, and inform outcome-based policy-making for the protection of natural resources.

Editor: Julia A. Jones, Oregon State University, United States of America

Funding: This work was supported by awards from the John D. and Catherine T. MacArthur Foundation (URL: http://www.macfound.org/) to NatureServe addressing "Dashboard assessments: proof-of-concept and baselines" (Grant No. 11-98252-000-INP), "Moving graphical presentation of biodiversity conservation monitoring indicators from a static proof-of-concept to a dynamic prototype" (Grant No. 12-100096-000-INP), and the bridge grant between the two (Grant No. 12-102962-000-INP). The funders supported the concept development in study design phase, but had no role in data collection and analysis, decision to publish, or preparation of the manuscript.

Competing Interests: The authors have declared that no competing interests exist.

* Email: xuemei_han@natureserve.org

Introduction

Resource monitoring has long been recognized as a cornerstone of biodiversity and conservation science [1,2,3]. In 2010, at the 10th Conference of the Parties of the Convention on Biological Diversity (CBD), 193 nations agreed to twenty "Aichi Biodiversity Targets", and in doing so committed to updating their National Biodiversity Strategies and Action Plans and developing monitoring programs to assess progress [4]. The Aichi Targets rely upon indicators to report progress towards reducing pressure on biodiversity, maintaining and improving the state of biodiversity, implementing conservation actions to ameliorate biodiversity loss, and providing benefits to human well-being [4]. Many other initiatives and multilateral agreements call for similar indicator-based biodiversity monitoring, including (a) the United Nations Millennium Development Goal #7 [5] and the draft new Sustainable Development Goals [6]; (b) intergovernmental treaties that provide mechanisms for national action and international cooperation, such as the Ramsar Convention on Wetlands [7] and the Convention on Migratory Species [8], (c) science-policy interfaces such as the Intergovernmental Platform on Biodiversity and Ecosystem Services [9]; and (d) partnerships or networks in support of the above mentioned bodies, such as the Biodiversity Indicators Partnership [10,11,12], and the Group on Earth Observations Biodiversity Observations Network Working Group #9 [13].

Monitoring called for by these programs is essential both to document biodiversity change over time [14], to shed light onto key ecological processes [15], and to measure the success or failure of conservation interventions through counterfactual analysis [16,17,18,19]. However, most existing monitoring programs have been designed primarily at localized scales, and often produce information that is disaggregated, heterogeneous, and non-standardized when considered at national or regional scales [20]. Monitoring requirements for measuring conservation performance, of the kind necessary to track the Aichi Targets, require data that transcend the fine temporal, spatial, and organizational scales commonly addressed in current literature [15].

Documentation of conservation impacts and biodiversity response must be accomplished in ways that are scientifically defensible, at appropriate temporal and spatial scales, and simple enough to inform decision-making by the diverse group of individuals and organizations working at the intersection of science and policy. Mounting global evidence shows that biodiversity loss is continuing at alarming rates [21,22], yet currently, two thirds of national reports submitted to the CBD lack evidence-based measures to illustrate changes in the status of biodiversity [23]. National capacity is often insufficient to measure many indicators of interest using on-the-ground methods, particularly in developing countries [24]. Even when national data are available, a lack of standardization across countries can make regional assessment difficult or impossible [20].

To better understand the challenges to effective biodiversity monitoring at national and regional scales, and how finer-scale (e.g. national) data might be integrated into a framework for global monitoring of biodiversity status and trends, we surveyed local conservation experts working in areas of high conservation value on monitoring and capacity needs. Building from the needs identified in those workshops, we then developed the concept for a biodiversity indicators dashboard using indicators derived from global data sets and constructed a dashboard prototype. This is the first operationalized dashboard to date that communicates multiple biodiversity indicators at multiple scales, and directly serves the global need to monitor progress towards Aichi Targets. Full development of the biodiversity indicators dashboard will encompass: (1) identification of appropriate indicators, (2) proof of concept using global data, (3) building the technological infrastructure necessary to host the dashboard, (4) designing the visual interface for multiple platforms (i.e. web and mobile users), and (5) creating systems to support the integration of finer-scale (regional and national) data. Here, we address in detail steps 1 and 2 of the dashboard design, laying the foundation for a web-based tool freely available to all with an interest in biodiversity conservation. A prototype of the tool is now available to the international conservation community at http://dashboarddev.natureserve.org, with steps 3–5 being implemented in an on-going iterative process.

Methods and Results

1. Study Area

We considered three geographically diverse areas with exceptional biodiversity value, that confront a high degree threat and that receive significant investment by international conservation agencies (Figure 1) [25,26]. The Tropical Andes region encompasses the eastern slope of the Andes, containing eight watersheds of headwater rivers (Japura, Putumayo, Rio Maranon, Ucayali, Guapore, Madre de Dios/Beni, Amazon, Magdalena) across Venezuela, Colombia, Ecuador, Peru, and Bolivia. The Great Lakes region of Africa includes five major watersheds (Lake Victoria, Upper Nile, Lake Tanganyika, Lake Malawi/Nyasa, Turkana/Omo) across Ethiopia, South Sudan, Kenya, Uganda, Democratic Republic of Congo, Rwanda, Burundi, Tanzania, Zambia, Malawi, and Mozambique. The Greater Mekong region encompasses the entire Mekong River Basin, spanning China, Myanmar, Vietnam, Lao P.D.R., Cambodia, and Thailand [25].

We delineated regional boundaries for the Tropical Andes, African Great Lakes and Greater Mekong regions using hydrological basins derived from HydroSHEDS and compiled by the UN-FAO, [27,28,29]. We performed analyses at both this regional scale, and at the national scale for the 22 countries that these three regions overlap (including areas outside the focal watershed boundaries).

2. Challenges to Biodiversity Monitoring and Capacity Needs at Regional and National Scales

We conducted seven consultation workshops in the three study regions between September 2011 and August 2012 to (1) better understand the challenges to effective biodiversity monitoring at national and regional scales, (2) identify gaps in current monitoring capacity and potential mechanisms for filling those gaps, and (3) begin to explore mechanisms for integrating local and national monitoring data into future regional and national biodiversity indicators. In total, 260 individuals from 20 countries attended at least one of the workshops, with broad representation from the public, civil-society, and academic sectors. Invitees included those with professional responsibility for National Biodiversity Strategies and Action Plans for monitoring progress towards Aichi Targets, and managers and technical experts responsible for designing and conducting biodiversity monitoring programs at multiple scales.

At each workshop, we solicited multiple-choice feedback on two issues: 1) the spatial scales of monitoring that participants required to guide their work (regional, national, sub-national, watershed, and/or site scales); and 2) the status of monitoring of selected biodiversity indicators for pressure, state, response, and benefits at the national scale, with answer options of "Monitored", "Limited Monitoring" (monitoring that has been conducted in some areas but not systematically done across the country), "Not Monitored", or "Unknown". Of the 260 workshop participants, 132 (51%) submitted answers to these written questionnaires, of which 39% came from the public sector, 45% from civil-society, and 16% from the academic sector. We also recorded and categorized responses to open-ended questions addressing (1) the utility of tracking biodiversity indicators derived from existing global data with a dashboard approach and (2) national challenges in developing sustainable biodiversity monitoring.

To identify the preferred scales of monitoring, we tabulated the frequency of the scales that participants indicated were important. To quantify the existing capacity for monitoring in each of the targeted countries, we calculated a score based on the perceived monitoring status for each biodiversity indicator. The score is scaled 0 (not monitored) to 1 (monitored), and equals $P_1 + 0.5P_2$,

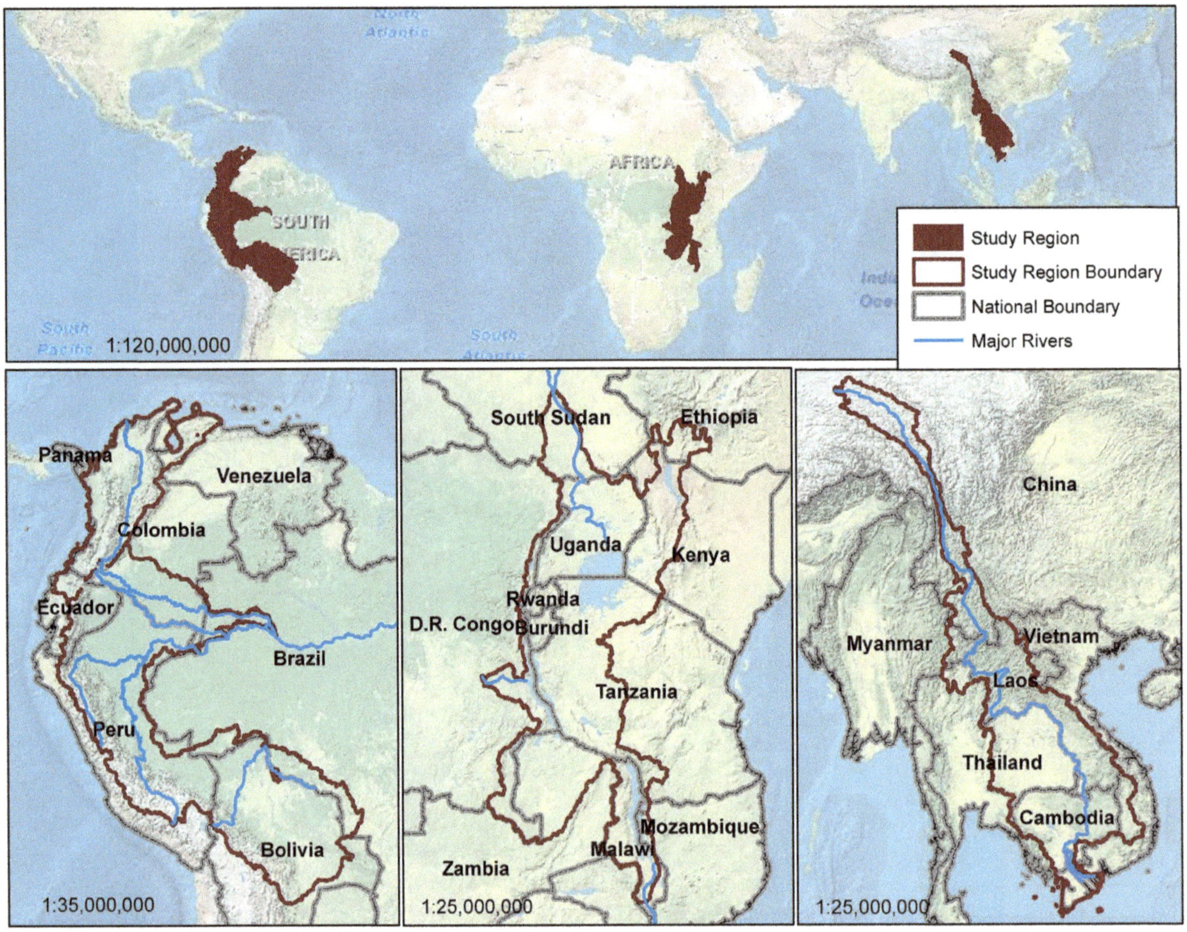

Figure 1. Study area regions. From left to right: the Tropical Andes, the African Great Lakes, and the Greater Mekong.

with P_1 the percent of respondents who answered "monitored" and P_2 the percent of respondents who answered "limited monitoring". We used ANOVA to explore differences of monitoring status between regions, and a repeated-measures ANOVA to examine differences in monitoring status among indicators.

Responses to the questionnaire indicate a strong demand for reliable information on the state of, and pressures facing, biodiversity. Regarding scales of monitoring, participants were most interested in analyses carried out at the site (82%) and national levels (76%), followed by watershed (71%), sub-national (68%) and regional levels (65%).

Our questionnaires revealed significant differences in the degree to which indicators are currently monitored ($p<0.001$), with hydrologic measures (average score $=0.40$) and species extinction risk (0.57) less frequently monitored than deforestation (0.72) and protected area coverage (0.79) (Figure 2). While there were no differences in the average score across regions ($p=0.88$), the status of monitoring differed widely among nations. Of the 22 countries, those with the highest overall scores for existing monitoring were Colombia (0.875), Malawi (0.875), and Thailand (0.75). Countries with very limited monitoring include South Sudan and the D. R. Congo (both ≤ 0.25).

Among the open-ended questions, a third of survey respondents from all sectors expressed high interest in using the dashboard approach, and employing appropriate subsets of global scale data, as a means to gather and share information to assess biodiversity status and threats, assess and improve conservation impacts, and inform policy, planning, and decision-making. Supporting capacity building, promoting stakeholder participation and dissemination of information were also frequently cited by survey respondents as potential benefits of this effort (Figure 3).

Across regions, the challenges to effective monitoring (Figure 4) include the lack of personnel, technology, and financial support for data collection and management (45%), and limited information accessibility and interoperability (40%). Emphasis varies among regions, with African respondents stressing the need for support in data management (24%), and Andean respondents more concerned about scientific standards and methods (25%) and conservation expertise and analysis (39%).

3. Creation of a Biodiversity Indicators Dashboard to Support Monitoring Needs

3.1 The Dashboard Concept. To address the challenges to biodiversity monitoring at regional and national scales identified by the survey, we envision the creation of a biodiversity "dashboard" – a visualization of biodiversity indicators designed

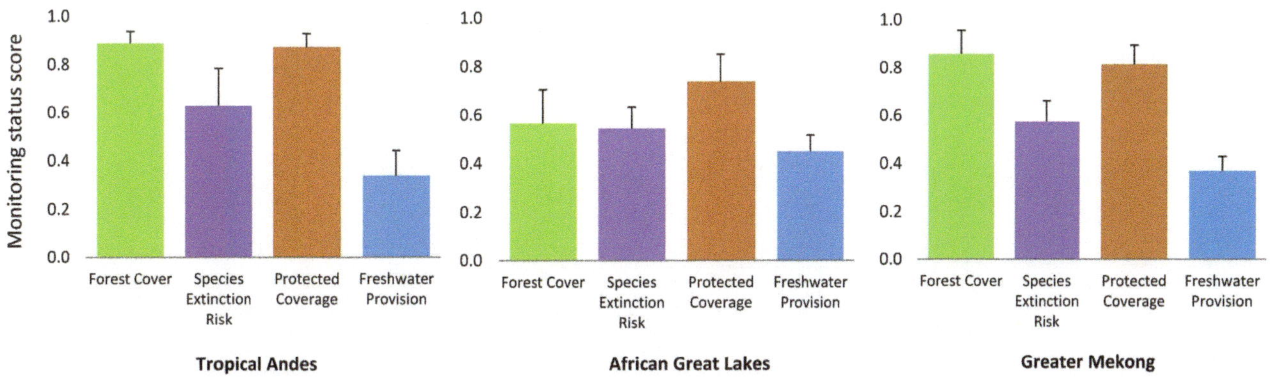

Figure 2. Monitoring status of the indicators, as reported by national experts via questionnaire responses. The mean score and its standard error for each indicator are shown by region. Number of respondent is 36 for Tropical Andes, 46 for African Great Lakes, and 50 for Greater Mekong.

to enable ongoing tracking of biodiversity status and trends and present biodiversity monitoring and conservation performance information in a clear, user-friendly, and unified format that facilitates iterative adaptive management. Using biodiversity indicators of the type developed by Butchart et al. (2010) [22] as the foundation, the dashboard provides a means to disseminate information, promote stakeholder participation, support capacity building, and allow users to better understand the relationship between conservations actions and impact. The utility of the biodiversity indicators dashboard in meeting these needs was confirmed by responses to our survey (Figure 3).

Originating as a business performance monitoring tool, dashboard visualizations are an information management and reporting instrument that has seen increasing use in a variety of contexts to communicate complicated information on current status and historical trends to broad audiences [30,31]. From the World Bank Atlas of Global Development [32] to commercial products used to track stock performance and guide financial investment (e.g., [33]), dashboards distill complicated data by tracking key indicators, usually via a combination of charts and maps. This information is typically served on websites or mobile applications and updated regularly (e.g., annually for World Bank indicators, minute-by-minute for financial markets).

Dashboards have been proposed and employed in various biological and resource management contexts. For example, the CITES Trade Data Dashboard [34] allows users to explore patterns in species exploitation across space, time, and taxonomic affiliation through a dynamic interface [35]. Dashboards also support fisheries management by providing a framework to better visualize relationships among fish populations, socio-economics, and exploitation [36,37].

If a dashboard is to be useful for decision makers, the indicators chosen must present information critical to influencing the decisions to be made. We used the Pressure-State-Response-Benefit (PSRB) framework to guide selection of indicators, following Sparks et al. 2011 [14]. This is derived from the causal-chain Pressure-State-Response and Driver-Pressure-State-Impact-Response frameworks, widely used for reporting on the state of the environment [22,23,38,39,40,41,42,43], and one that has been used by the CBD Ad-Hoc Technical Expert Group of the CBD to guide indicator development for the CBD [44] and recommended for communicating biodiversity indicators [42]. The core elements in PSRB as applied in the dashboard assessments are pressure on biodiversity, its drivers, (e.g., habitat destruction, climate change, invasive species), the state of species and ecosystems (e.g., species extinction risk, animal and plant populations, ecosystem integrity), conservation action or policy responses (e.g., protected areas establishment and management, investment in biodiversity conservation) and benefit to human well-being from the social, economic and cultural impacts of conservation (e.g., maintenance of hydrological functions, climate change mitigation, maintenance of indigenous cultures). By

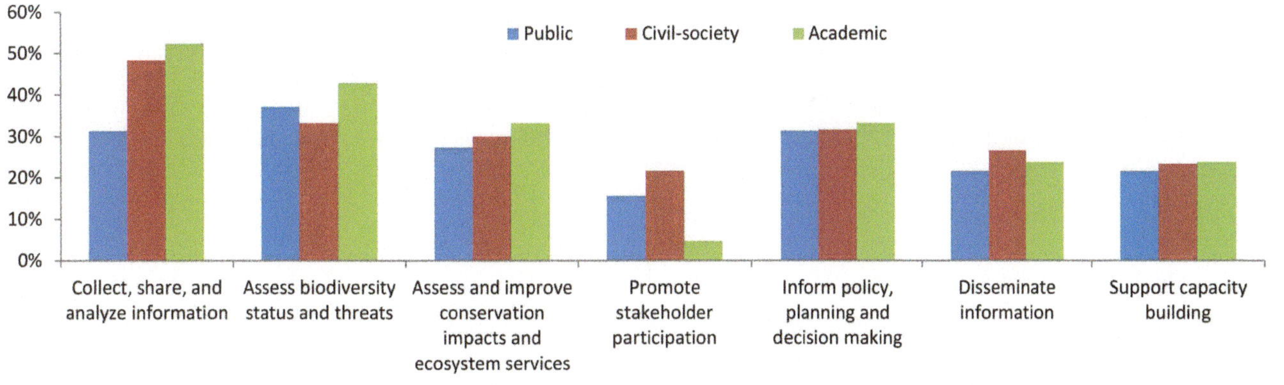

Figure 3. Perceived benefits of using global data within a dashboard approach, by sector. Number of respondent is 51 for public sector, 60 for civil-society, and 21 for academic sector.

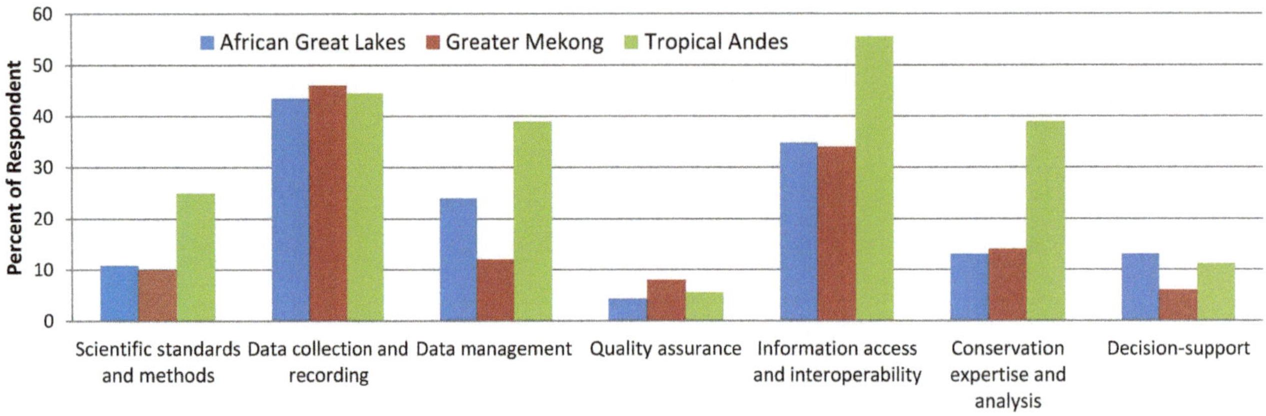

Figure 4. Perceived challenges to biodiversity monitoring by region. Number of respondent is 36 for Tropical Andes, 46 for African Great Lakes, and 50 for Greater Mekong.

viewing dashboard indicators together within the PSRB framework, users can begin to understand interactions between indicators [14].

3.2 Methods for Constructing a Biodiversity Indicators Dashboard using Global Data. We selected one indicator from each PSRB component, with consideration for the availability of global datasets, the degree to which the indicator contributes to evaluating progress made towards the Aichi Targets, the feasibility of trend estimates, and the likely availability of analogous data generated locally for future integration into the dashboard. The four selected indicators are examples of the types of data the biodiversity indicators dashboard can be used to track. At this stage in development of the dashboard, the chosen indicators are not intended to address causal relationships; however, as additional indicators are added, the PSRB approach will facilitate exploration of causal links between indicators. We selected forest cover loss as the pressure indicator, species extinction risk as the state indicator, protected area coverage of key biodiversity areas (KBAs) as the response indicator, and freshwater provisioning to downstream human populations as the benefit indicator. We used data from the first decade of the 21st century to represent current status and provide an initial baseline, or reference point, against which future trends can be assessed. For all but the benefit indicator (freshwater provision), existing data from either previous time steps (i.e. species extinction risk and percent protection of KBAs) or later time steps (i.e. forest cover loss) supporting the tracking of trends.

Global data were disaggregated to provide regional and national indicator values (Table 1). For each indicator, we mapped current condition, charted and mapped trends over time, and generated tabular summaries. All spatial analyses were performed using ArcGIS 10.1 [45] and all statistical calculations were performed in R [46].

a. Pressure Indicator: Forest Loss

The forest loss indicator is derived from the Global Forest Monitoring Project [47,48], which estimated forest cover in 2000 and forest cover loss between 2000 and 2005 using MODIS data [49] calibrated with Landsat [50] imagery, at 18.5-km resolution. Values represent the percent forest cover within each pixel, with forest cover defined as areas with at least 25% cover of trees at least 5 meter in height. The deforestation measure, Gross Forest Cover Loss (GFCL), represents a unidirectional change in forest cover, calculated from the percent forest loss between 2000 and 2005. For each analysis unit (e.g., region, nation) we calculated a mean value for forest cover in 2000 and mean GFCL between 2000 and 2005. We then derived the average annual rate of GFCL for 2000–2005 for each analysis unit, presented as the annual percent forest loss from the 2000 baseline.

While, to our knowledge, the GFCL data provide the best globally consistent spatial representation of deforestation to date, the data are limited in that they do not incorporate information on forest gain from restoration, natural regrowth, and plantation. They also do not address finer resolution forest degradation, as some other regional mapping products do [51,52,53].

b. State Indicator: Species Extinction Risk

The Red List Index (RLI) is a measure of trends in survival probability (the inverse of extinction risk) for sets of species. It is based on the numbers of species within each IUCN Red List category (i.e., Extinct (EX), Extinct in the Wild (EW), Critically Endangered (CR), Endangered (EN), Vulnerable (VU), Near Threatened (NT), Least Concern (LC)) and the changes in these numbers over time resulting from genuine improvement or deterioration in status between assessments [22,54,55,56,57,58].

For this indicator, we used the first and last comprehensive Red List assessment (when all species of a taxonomic group were assessed) for each of three vertebrate groups (1988 and 2008 for birds, 1996 and 2008 for mammals, and 1980 and 2004 for amphibians, noting that the 1980 assessment for amphibians was based on a retrospective assessment), following Butchart et al. (2004, 2005, 2007 and 2010) [22,54,55,56]. We identified all species falling partially or completely within each region and each country using 2010 spatial distribution data for each species [58] (Table 2). For each region and country, we calculated the RLI for each taxonomic group individually and for all taxonomic groups together. This standardized RLI varies between 1 (all species LC) and 0 (all species EX or EW). Following Butchart et al. 2004, 2005 [55,56], species undergoing genuine Red List category changes between assessments contributed to RLI trends only if the driving process of the change (i.e. threat or conservation action) operated within the relevant country or region. For each vertebrate group, we calculated the annual change in aggregate extinction risk by dividing the difference in RLI between the last and first assessment by the number of intervening years. Data Deficient and Not Evaluated species were excluded from this calculation and the annual change value across all taxonomic groups is computed using the mean time difference between assessments for the three groups.

Table 1. Biodiversity indicators summary and data sources.

Framework Component	Pressure/Driver	State	Response	Benefit (Impact)
Indicator	Forest coverage and rate of gross forest cover loss	Red List Index	Protected area coverage in key biodiversity areas	Quality-weighted freshwater provision from natural ecosystems to downstream human population
Aichi Target	Target 5: Loss of habitats is at least halved by 2020	Target 12: Extinctions of known threatened species has been prevented by 2020	Target 11: At least 17% of terrestrial...areas, especially important areas for biodiversity and ecosystem services are conserved by 2020	Target 14: Ecosystems that provide essential services are restored and safeguarded by 2020
What does it show	Spatial data represents the percent of forest cover for each 18.5 km pixel in 2000 and percent gross forest cover loss (i.e., deforestation) from 2000 to2005. Tabular FAO data summarize forest land use coverage and "net forest cover change" by country in 2005 and 2010.	An index of aggregate survival probability of species that occur in the given spatial unit. Values range from 1 (all species Least Concern) to 0 (all Extinct).	Mean percent area of key biodiversity areas covered by protected areas	Quality-weighted delivery of clean freshwater from natural habitats to downstream human populations per unit area
Data Source	Forest cover for year 2000 and gross forest cover loss 2000–2005 through Global Forest Monitoring Project	IUCN Red List assessment for:	- World Database on Protected Areas (UNEP-WCMC) (2010)	- World WaterGAP 2 model runoff map
				- Hydrological drainage direction
		- Amphibians (1980, 2004)	- Global KBAs as represented by Important Bird and Biodiversity Areas (IBAs) and Alliance for Zero Extinction (AZE) sites	
				- Landscan Global Population Database
		- Birds (1988, 2008)		
				- GlobCover land cover
		- Mammals (1996, 2008)		
Time frame	2000–2005	1980–2008	1950–2010	2010
Limitation and caveat	- Resolution is too coarse (18.5 km) to detect deforestation in small areas.	- Differing assessment dates requires interpolation and extrapolation to estimate aggregate trends	- The WDPA omit recently decreed protected areas	- Values are relative, not absolute
	- The gross forest cover loss data shows deforestation only, not taking account afforestation	- Because of the heterogeneous distribution of species, regional extinction risk can skew national indicator values	- The WDPA does not currently document management effectiveness	- Only baseline (2010) data currently available; not able to estimate trend
	- Forest degradation was not quantified	- The proportion of a species' range within a given analysis unit is not considered	- Key biodiversity areas for taxa other than birds that are not endemic to single sites have only been identified in some countries	- Spatial resolution is too coarse (2,5921km^2 pixels) to estimate freshwater provision in small areas
		- Red List categories are necessarily broad classes of extinction risk, so the RLI is moderately sensitive		

c. Response Indicator: Protected Area Coverage of Key Biodiversity Areas

For the Response indicator, we calculated the mean percentage area of key biodiversity areas (sites contributing significantly to the global persistence of biodiversity [59]) falling within protected areas for each analysis unit [60]. We used the World Database on Protected Areas for 2010 [61] to delimit protected areas. Within our study area, key biodiversity areas include 757 Important Bird & Biodiversity Areas (IBAs; the subset of key biodiversity areas important for birds[62]), and 139 Alliance for Zero Extinction sites (AZE; the subset of key biodiversity areas holding effectively the entire populations of highly threatened species, i.e., CR and EN species>95% restricted to single sites [63,64]). We calculated the percentage of each KBA that overlaps protected area boundaries

Table 2. Number of species recorded and analyzed to derive Red List Index (only extant species that are not Data Deficient were included).

	Number of all assessed species				Number of species with changed Red List category			
	Overall	Mammals	Birds	Amphibians	Overall	Mammals	Birds	Amphibians
Assessment year	**2008**	**2008**	**2008**	**2004**	**1980–2008**	**1996–2008**	**1988–2008**	**1980–2004**
Tropical Andes	5357	880	2978	1499	179	13	29	137
African Great Lakes	2483	613	1534	336	30	10	17	3
Greater Mekong	2499	642	1479	378	142	41	71	30

and subsequently generated a national mean. We used the year of establishment to generate time series graphs for each year from 1950 to 2010, assigning an establishment date to those protected areas lacking establishment date by randomly sampling from known dates of designation of protected areas in the same country, and then bootstrapping following the methods of Butchart et al. [60]. We plotted the mean with 95% confidence intervals based on uncertainty arising from the missing data [60]. We also calculated the annual rate of change in protection of key biodiversity areas between 1980 and 2010. The annual rate of change is thus from a time period comparable to that calculated for the State indicator (the Red List Index).

d. Benefit Indicator: Freshwater Provision

Freshwater provision data were developed by Larsen et al. [65,66] using spatially explicit maps of runoff from the global hydrological water model WaterGAP [67], hydrological drainage directions [27,68], downstream human population density [69], and global land cover data (used to weight flow estimates by a quality coefficient, based on information from previous studies [70,71,72,73]). Estimated quality-weighted freshwater provision, reported as a freshwater flow index, was calculated for 2,592 km^2 hexagonal grid cells [66]. Using this grid, we calculated a mean value for each analysis unit. Because the freshwater provision data is currently only available for a single time step (2010), we cannot yet calculate trends.

3.3 Results of Biodiversity Indicators Dashboard with Global Data. The indicators are presented as a series of maps and charts visualized to support a web-enabled biodiversity indicators dashboard (Table 3, Figures 5–7). A web prototype of the dashboard, displaying the results discussed here, is available at http://dashboarddev.natureserve.org.

Together, the dashboard graphics present a picture of the current status (Figure 5) and trends (Figure 6) in biodiversity. Baseline data show large geographical variation in the status of forest cover. National forest coverage is the lowest in Kenya (7.47%), and the highest in D. R. Congo (71.18%) and Lao P.D.R. (69.06%) (Table 3 and Figure 5A). The baseline Red List Index reveals high species extinction risk in Tropical Andes for all taxa (0.89), with variations among countries and among taxonomic groups (Table 3). The conservation response, measured as the average percentage of key biodiversity areas under protection, varies somewhat among regions (44% in the Tropical Andes, 63% in the African Great Lakes region, and 49% in the Greater Mekong as of 2010) but the dashboard shows larger differences at national levels, with lows in Mozambique (20%), Ethiopia (25%), Peru (25%), Vietnam (34%) and Myanmar (35%), and highs in Burundi (100%), Malawi (88%), Venezuela (79%) and Thailand (73%) (Table 3 and Figure 5C). Similarly, baseline data for freshwater provision show large national differences (Figure 5D), with Burundi, Rwanda, Vietnam, and China standing out as areas of high importance.

The trend of forest loss is documented as ongoing in all regions evaluated, with national rates of loss lowest in Peru (0.08%/yr) and D. R. Congo (0.12%/yr), and highest in Kenya (1.2%/yr) (Table 3 and Figure 6A). The Red List Index indicates the worsening status of species, with a decline in Red List Index observed for all nations between 1980 and 2008 (Table 2). Rates of decline were highest in the Tropical Andes, largely driven by amphibians (1.36×10^{-3}/yr), and in the Mekong, due to both mammals (1.29×10^{-3}/yr) and amphibians (0.98×10^{-3}/yr). Protection of key biodiversity areas increased in all regions since 1980, with some regional variation in the rate of increase (0.20% in the African Great Lakes, 0.86% in the Tropical Andes, and 1.43% in the Greater Mekong). Nationally, rates of increase in the

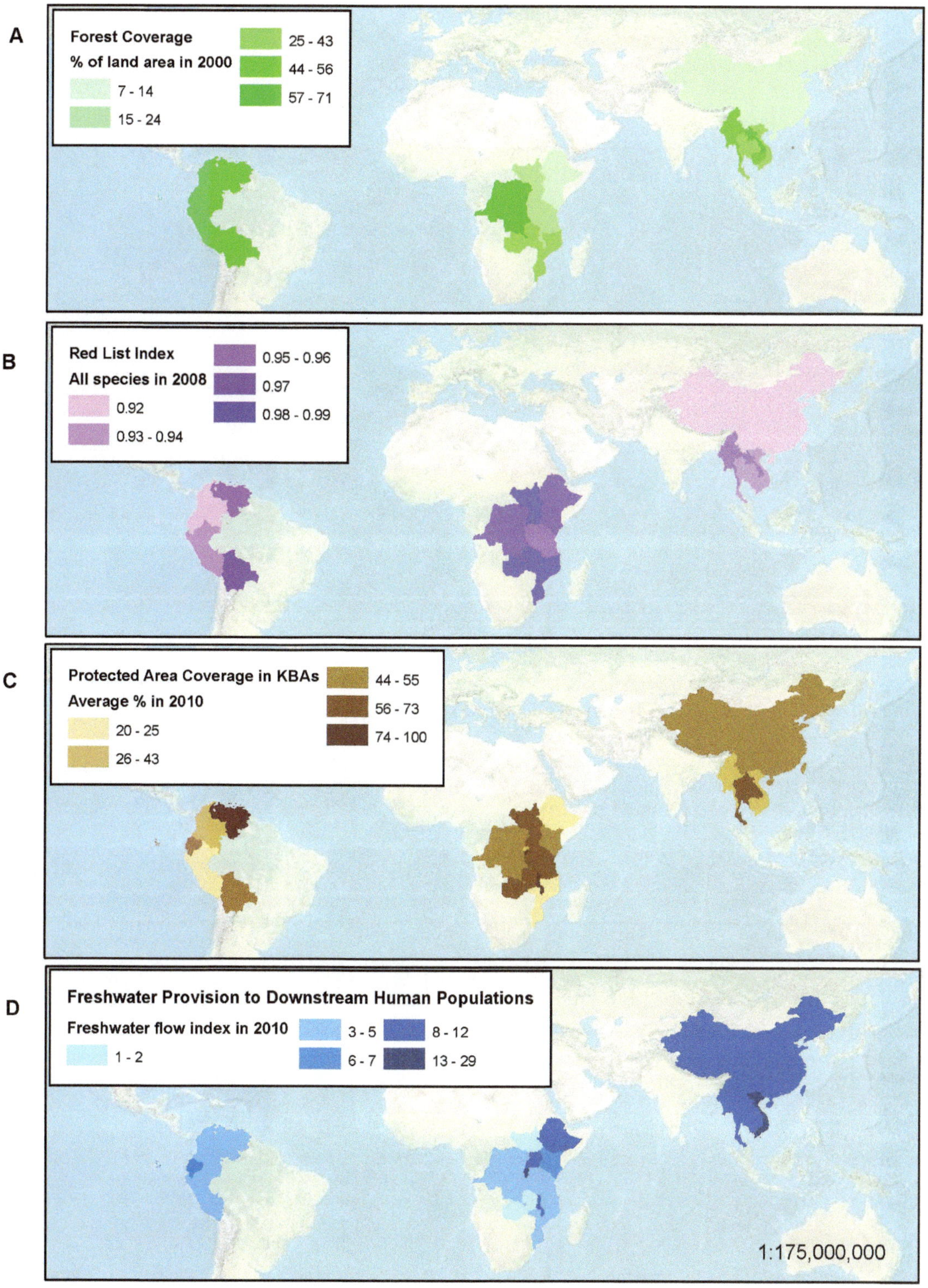

Figure 5. Dashboard indicator baseline results. Results for (A) Forest Cover (2000); (B) Red List Index a measure of change in extinction risk (2008); (C) Protected Area Coverage of Key Biodiversity Areas (2010); and (D) Freshwater Provision (2010).

protection of key biodiversity areas ranged from lows of zero in Burundi, Malawi, Mozambique and Rwanda, and highs of 2.09%, 1.54%, and 1.42% in Lao P.D.R., China, and Cambodia (Table 3 and Figure 6C).

Discussion

1. National and Regional Monitoring Challenges

As approaches to biodiversity management shift towards data-intensive and science-driven methods [74,75], addressing gaps in

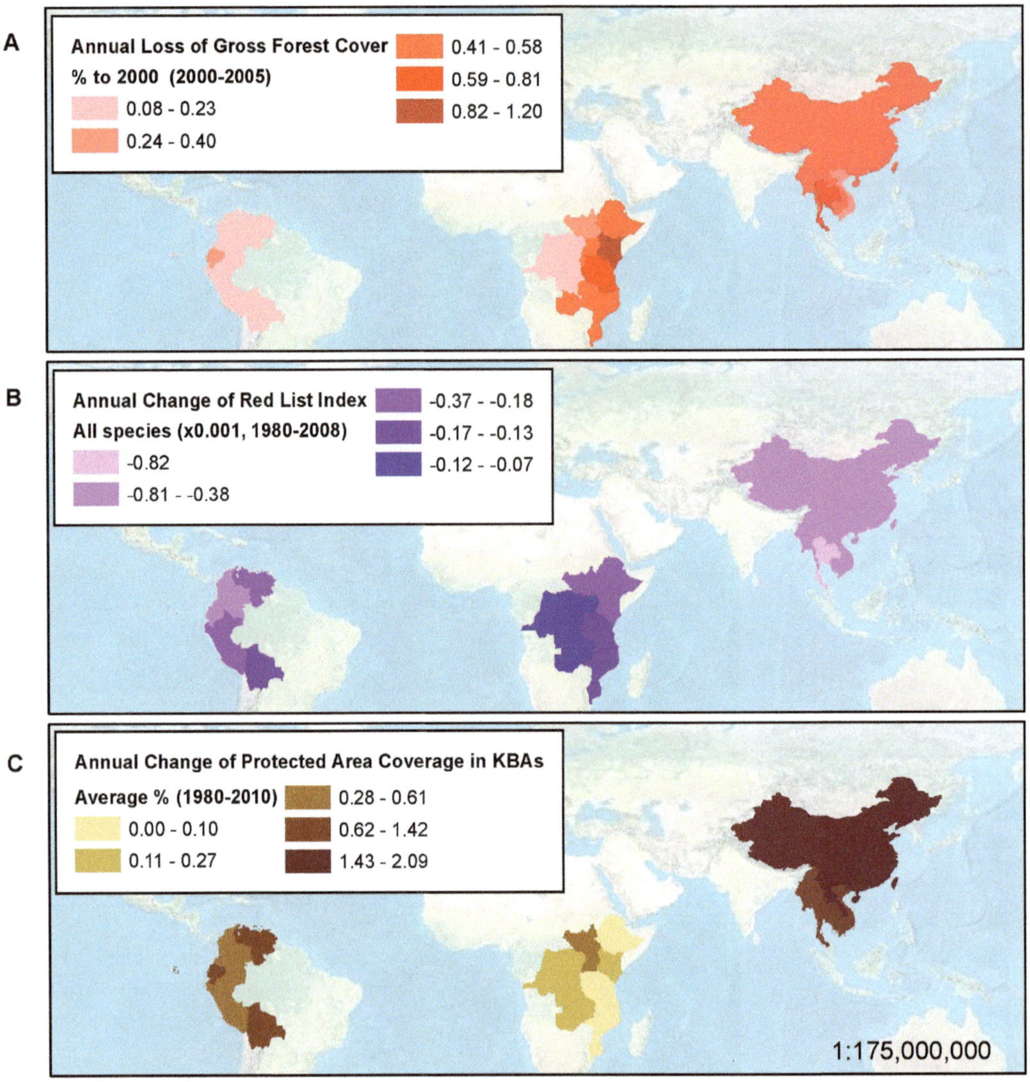

Figure 6. Dashboard indicator trend results. Annual rate of (A) Gross Forest Cover Loss (2000–2005); (B) Change in Red List Index as a measure of extinction risk (change for all species of mammals, birds, and amphibians; 1980–2008); and (C) Change of Protected Area Coverage of Key Biodiversity Areas (1980–2010).

capacity for information generation and dissemination has become increasingly important [76,77,78,79,80]. The prevailing and widely recognized challenges to addressing these gaps include sustaining financial and human resources for on-the-ground monitoring [76,81], overcoming cultural and technical barriers associated with data generation [82,83,84] and information sharing [74,85,86], adopting scientific standards for monitoring and data analysis [79,82,84,87], and developing indicator sets that can effectively inform policy and decision-making processes [10,40,88,89]. Our survey of regional biodiversity experts reaffirms these challenges in generating, managing, and sharing biodiversity information (Figure 4), demonstrates the strong demand for access to biodiversity status and trend data at multiple spatial scales, and indicates that our proposed biodiversity indicators dashboard could be an effective tool to address varied conservation needs (Figure 3).

Recognizing that national and local indicator data are often limited or non-existent, the survey respondents affirmed the value of deriving indicators from global datasets as an intermediate measure necessary to meet current demand. Moving forward, the respondents noted that it will be necessary to augment and validate globally-derived measures with national and local monitoring results, and doing so will require both cost-effective participatory monitoring protocols that ensure sustainable data collection and well-designed standards that ensure data interoperability (Figure 4). A lack of baseline data was the most frequently mentioned monitoring challenge in our survey. The few studies that have systematically evaluated the availability of indicators for monitoring biodiversity targets [90,91] support our findings that biodiversity indicators, particularly indicators of state and benefit, are deficient (Figure 2). Unstable political situations, lack of financial support, and the low priority of biodiversity monitoring culturally or in national development strategies all can impair continuous and systematic data collection. At the same time, the barriers to information access and interoperability, prevent the information that does exist from fully informing conservation efforts. Biodiversity data are generated and kept by different agencies in a fragmented manner. Within our area of study, civil society and

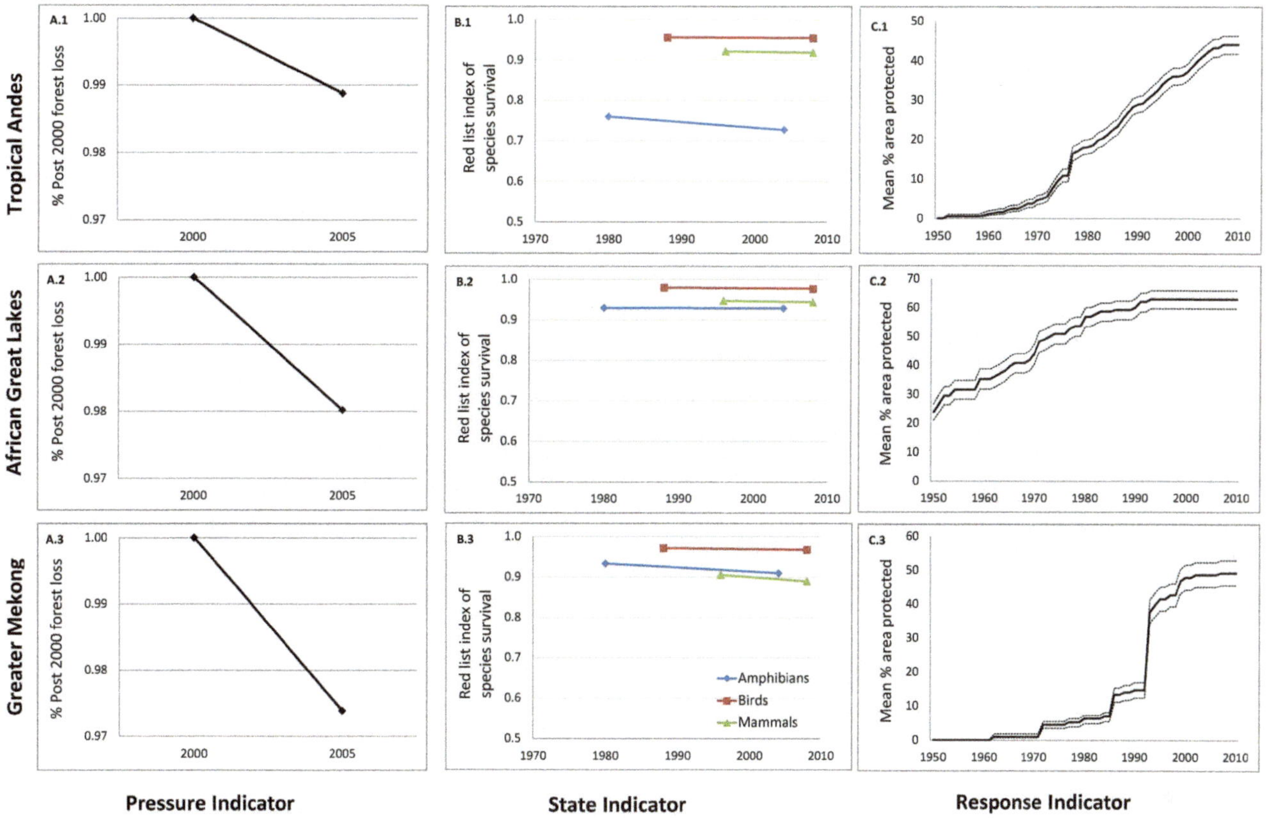

Figure 7. Dashboard indicator trend graphs by region. A.1 – A.3 chart gross forest loss as a percent of forest cover in 2000; B.1-B.3 chart change in Red List Index for mammals (green), birds (red), and amphibians (blue); and C.1-C3 chart change in protected area coverage of key biodiversity areas (1950–2010) with solid lines indicating the mean percent protected across all sites, and dashed line indicating the 95% confidence intervals [60].

academics (such as the Wildlife Conservation Society in Cambodia, NatureKenya in Kenya, NatureUganda in Uganda) accumulate a wide range of site-level monitoring data maintained as project-based resources, while national level monitoring data are typically held by specific government divisions who may or may not share that information with other government entities, much less outside organizations. These challenges are well-known and efforts to address them are being made at global (e.g., Biodiversity Indicators Partnership [12]), regional (e.g., ASEAN Centre for Biodiversity [92], Red Amazónica de Información Socioambiental Georreferenciada [93], Streamlining European Biodiversity Indicators [94], Conservation of Arctic Flora and Fauna [95]), and national levels (e.g., National Biodiversity Data bank in Uganda [96], National Biodiversity Database System in Vietnam). However, those efforts do not cover all countries and regions, and ready access to reliable and geographically consistent biodiversity indicator information at multiple scales remains a confirmed need.

2. Towards a Dashboard

The biodiversity indicators dashboard, as explained here, is designed to address the unmet needs expressed in our survey of biodiversity experts by laying the foundation for better accessibility and interpretability of existing biodiversity trend data within a framework that enhances monitoring capacity and promotes data interoperability and sharing. Many of these indicators are widely used at global scales, but until now have rarely been reported at national scales. While previous studies have demonstrated the utility of biodiversity indicators in monitoring conservation status and trends [14,22], our biodiversity indicators dashboard is the first operationalized online interface that communicates multi-dimensional indicators with spatial representation. By providing easy access to indicator information at national and local scales, it complements global efforts such as the Biodiversity Indicators Partnership [12] and facilitates reporting for Aichi National Biodiversity Strategies and Action Plans (NBSAPs). The intuitive graphics and ease of data access that flow from the dashboard are intended to engage and enhance partnerships at all levels – international, regional, national, and local.

As a data visualization tool, the biodiversity indicators dashboard is designed so that a quick examination communicates the overall status of biodiversity conservation, important trends and patterns, and previously hidden challenges. For example, the graphics and values for the state indicator, such as those presented in Table 3 and Figure 7, communicate a high extinction risk in the Tropical Andes, driven largely by the extinction risk of amphibian species. This finding is consistent with other recent studies [97,98] and highlights the importance of addressing threats to amphibian species if biodiversity is to be maintained.

By using data and methods that are globally consistent, the dashboard facilitates direct comparison of baseline and trends across regions and nations in the three continents targeted for this first stage of dashboard developments. Regional patterns are readily evident, such as the higher rate of decline of the Red List Index in the tropical Andes or the enormous importance of fresh water provisioning in the Greater Mekong countries (Figure 5).

Table 3. Baseline and trend results for all indicators by country and region.

Country/Region	Indicators – Baseline							Indicators – Annual Rate of Change					
	Forest Coverage % (2000)	Red List Index of species survival				Protection Coverage of KBAs % (2010)	Freshwater Flow Index	GFCL % (2000–2005)	Red List Index of species survival (x 0.001)				Protection Coverage of KBAs % (1980–2010)
		3 taxa (2008)	Mammal (2008)	Bird (2008)	Amphibian (2004)				3 taxa (1980–2008)	Mammal (1996–2008)	Bird (1988–2008)	Amphibian (1980–2004)	
Bolivia	56.26	0.97	0.96	0.98	0.90	51.51	1.13	0.21	−0.16	−0.19	−0.06	−0.62	1.03
Colombia	56.06	0.92	0.93	0.97	0.77	42.15	4.82	0.22	−0.38	−0.35	−0.04	−1.09	0.59
Ecuador	50.1	0.92	0.93	0.97	0.73	51.77	6.56	0.29	−0.44	−0.39	−0.03	−1.57	1.24
Peru	52.23	0.94	0.93	0.97	0.82	25.07	2.89	0.08	−0.23	−0.21	−0.05	−0.91	0.61
Venezuela	53.95	0.95	0.95	0.98	0.80	79.34	3.3	0.23	−0.24	−0.32	−0.02	−1.15	1.05
Tropical Andes	***64.04***	***0.89***	***0.92***	***0.95***	***0.73***	***44.10***	***2.06***	***0.23***	***−0.51***	***−0.24***	***−0.07***	***−1.36***	***0.86***
Burundi	19.67	0.96	0.95	0.97	0.88	100	29.13	0.51	−0.07	0.00	−0.10	0.00	0
D.R. Congo	71.18	0.97	0.95	0.98	0.96	55	3.25	0.12	−0.09	−0.27	−0.07	0.00	0.17
Ethiopia	14.27	0.96	0.92	0.98	0.91	24.55	6.76	0.58	−0.20	−0.21	−0.22	0.00	0.1
Kenya	7.47	0.97	0.95	0.98	0.95	54.60	5.71	1.2	−0.18	−0.23	−0.18	−0.10	0.27
Malawi	19.79	0.98	0.98	0.98	0.96	88.42	9.8	0.52	−0.13	−0.20	−0.14	0.00	0
Mozambique	43	0.98	0.97	0.98	0.98	19.70	3.71	0.57	−0.13	−0.17	−0.15	0.00	0
Rwanda	18.92	0.95	0.93	0.97	0.86	38.57	27.48	0.54	−0.10	−0.18	−0.09	0.00	0
South Sudan	22.18	0.99	0.97	0.99	1.00	64	1.39	0.39	−0.18	−0.32	−0.17	0.00	0.45
Tanzania	24.3	0.95	0.93	0.97	0.84	60.97	4.66	0.66	−0.14	−0.17	−0.11	−0.23	0.07
Uganda	21.45	0.98	0.96	0.99	0.95	61.61	8.24	0.45	−0.11	−0.18	−0.11	0.00	0.61
Zambia	38.52	0.98	0.97	0.99	1.00	59.66	2.09	0.47	−0.09	−0.16	−0.09	0.00	0.19
Africa Great Lakes	***25.07***	***0.96***	***0.94***	***0.98***	***0.93***	***62.89***	***7.15***	***0.40***	***−0.14***	***−0.27***	***−0.14***	***−0.05***	***0.20***
Cambodia	49.81	0.94	0.88	0.96	0.96	42.53	8.91	0.57	−0.57	−1.79	−0.37	−0.19	1.42
China	13.17	0.92	0.91	0.96	0.79	53.00	11.36	0.45	−0.42	−0.88	−0.14	−1.03	1.54
Lao P.D.R.	69.06	0.95	0.88	0.98	0.95	62.80	8.51	0.54	−0.40	−1.42	−0.21	−0.33	2.09
Myanmar	50.22	0.95	0.90	0.97	0.97	35.38	9.52	0.55	−0.54	−1.20	−0.47	−0.23	0.92
Thailand	31.77	0.95	0.89	0.96	0.96	73.41	9.37	0.81	−0.82	−1.57	−0.72	−0.56	1.19
Vietnam	38.27	0.94	0.87	0.96	0.91	33.68	19.44	0.4	−0.42	−1.45	−0.16	−0.77	0.99
Greater Mekong	***44.03***	***0.94***	***0.89***	***0.97***	***0.91***	***49.21***	***6.34***	***0.53***	***−0.46***	***−1.29***	***−0.19***	***−0.98***	***1.43***

Stark differences in the efforts of neighboring countries at safeguarding key biodiversity areas, such as for Mozambique and Tanzania, are also easily discernible by a non-scientific audience (Figure 5).

To develop the dashboard concept, we used indicators derived from global data to bypass the many obstacles in obtaining consistent and comprehensive regional and national data. While the reporting of these global indicators at the national level provides new and valuable information, there are limitations in using global data to represent national indicators. While advances in remote sensing technology provide an unprecedented opportunity to gain temporally repetitive information [99], currently available estimates of land cover change can differ substantively among sources (i.e., FAO forest resource assessment [100], Global Forest Cover Loss mapping [48], ESA Global Land Cover and GlobCover products [101,102]) and are often coarse in resolution (342 km^2 and 2592 km^2 respectively for the deforestation and freshwater provision datasets used here). With regard to protected areas, global datasets tend to omit recently decreed areas and often fail to capture important differences in on-the-ground management. Similarly, the use of global distributions and Red List categories may not adequately reflect the conservation status of a given species in a particular region, and because global distribution data is only available for terrestrial vertebrates, the results do not reflect the status of many other components of biodiversity.

The shortcomings of indicators derived from global data can be addressed by integrating nationally and locally derived data into the final dashboard design. Despite the numerous obstacles, data are being generated at a variety of scales deemed useful by survey respondents, both through governmental efforts [92,96,103,104], regional consortiums [92,93,94,105,106,107], and site-specific projects [108,109]. The next stage of the biodiversity indicators dashboard development focuses on building an effective data sharing mechanism the promotes shared identifiers to link data from these different sources [83], digital architecture to coordinate data flow and ensure data ownership, and promoting consent and trust among data contributors. With continued development, we envision the biodiversity indicators dashboard as an interactive, web-accessible platform that can facilitate national reporting towards biodiversity targets while allowing for the integration of localized data to support the type of site-scale monitoring deemed important to survey participants. The dashboard framework has also been designed so that over time, the indicators discussed here can be supplemented with other metrics capturing complementary aspects of each of state, pressures, responses and benefits (e.g. population trends, agricultural intensity, environmental legislation and additional ecosystem service measures).

By serving these data as a web-accessible dashboard, we can put information on status and trends in biodiversity within easy access of users and organizations from all sectors and backgrounds, and facilitate more informed decision-making, enable exploration of patterns among variables, and support the tracking of progress towards conservation goals. The maps presented here and information contained in Table 3 can be depicted in a dashboard format via various means, including as mapped values (Figures 5 and 6), mouse-over boxes displaying the numerical values associated with those maps, tabular data summaries by nation or indicator accessible via interactive menus, and charts of trends that users could generate either by region (Figure 7) or nation.

In agreeing to the Aichi Targets, the nations of the world implicitly committed to developing the data necessary to effectively monitor progress towards meeting biodiversity goals. The challenges in reporting towards those goals are many, but we believe the dashboard approach, as outlined here, provides a valuable framework that can facilitate and advance the type of reporting required by the conservation community. Starting with global indicators and expanding to incorporate additional national and site-scale data identified as important by conservation practitioners on the ground, the biodiversity indicators dashboard can serve as a tool to track progress towards Aichi Targets, support national monitoring and reporting, and inform outcome-based policy-making in the realm of conservation.

Acknowledgments

We thank A Chenery, D. Stanwell-Smith, J Tordoff, A. Ajagbe, C. Josse, the Biodiversity Indicator Partnership, and ASEAN Biodiversity Center for assistance on consulatation workshops; A. S. L. Rodrigues, T. Evans, R. Hoft, and C. Josse for comments on the indicator data and manuscript preparation; the survey respondents; and the many thousands of individuals and organisations who contribute to the compliation of IUCN Red List assessments, the World Database on Protected Areas, and the indentification and documentation of Important Bird and Biodiverseity Areas and Alliance for Zero Extinction Sites.

Author Contributions

Conceived and designed the experiments: TMB XH BEY ASL. Performed the experiments: XH ASL TMB BEY. Analyzed the data: XH ASL SHMB FWL. Contributed reagents/materials/analysis tools: PB SHMB FWL MCH WRT. Wrote the paper: XH RLS BEY TMB ASL PB SHMB FWL HH MCH WRT.

References

1. Holland GJ, Alexander JSA, Johnson P, Arnold AH, Halley M, et al. (2012) Conservation cornerstones: Capitalising on the endeavours of long-term monitoring projects. Biological Conservation 145: 95–101.
2. Lindenmayer DB, Likens GE (2010) The science and application of ecological monitoring. Biological Conservation 143: 1317–1328.
3. Lovett GM, Burns DA, Driscoll CT, Jenkins JC, Mitchell MJ, et al. (2007) Who needs environmental monitoring? Frontiers in Ecology and the Environment 5: 253–260.
4. Diversity CoB (2010) Conference of Parties Decision X/2.Strategic Plan for Biodiversity 2011–2020.
5. United Nations Millennium Development Goals. Available: http://www.un.org/millenniumgoals/. Accessed 2013 January 22.
6. Rio +20 United Nations Conference on Sustainable Development. Available: http://www.uncsd2012.org/. Accessed 2013 January 22.
7. The Ramsar Convention on Wetlands. Available: http://www.ramsar.org/. Accessed 2013 January 22.
8. Convention on Migratory Species. Available: http://www.cms.int/. Accessed 2013 Janurary 22.
9. Intergovernmental Platform on Biodiversity and Ecosystem Services. Available: www.ipbes.net. Accessed 2013 Janurary 22.
10. Walpole M, Almond REA, Besancon C, Butchart SHM, Campbell-Lendrum D, et al. (2009) Tracking Progress Toward the 2010 Biodiversity Target and Beyond. Science 325: 1503–1504.
11. Walpole M, Herkenrath P (2009) Measuring progress towards the 2010 target and beyond: an international expert workshop on biodiversity indicators. Oryx 43: 461.
12. Biodiversity Indicator Partnership (BIP). (2008) Available: http://www.bipindicators.net/.
13. Group on Earth Observations Biodiversity Observations Network (GEO BON). (2008) Available: www.earthobservations.org. Accessed 2013 Janurary 22.
14. Sparks TH, Butchart SHM, Balmford A, Bennun L, Stanwell-Smith D, et al. (2011) Linked indicator sets for addressing biodiversity loss. Oryx 45: 411–419.
15. Pimm SL (1991) The balance of nature?: ecological issues in the conservation of species and communities. Chicago: University of Chicago Press. xiii, 434 p. p.
16. Ferraro PJ, Pattanayak SK (2006) Money for nothing? A call for empirical evaluation of biodiversity conservation investments. Plos Biology 4: 482–488.
17. Pattanayak SK, Wunder S, Ferraro PJ (2010) Show Me the Money: Do Payments Supply Environmental Services in Developing Countries? Review of Environmental Economics and Policy 4: 254–274.

18. Sutherland WJ, Pullin AS, Dolman PM, Knight TM (2004) The need for evidence-based conservation. Trends in Ecology & Evolution 19: 305–308.
19. Ferraro PJ (2009) Counterfactual thinking and impact evaluation in environmental policy. In: Birnbaum M, Mickwitz P, editors. Environmental program and policy evaluation. pp. 75–84.
20. Roberts D, Moritz T (2011) A framework for publishing primary biodiversity data INTRODUCTION. Bmc Bioinformatics 12.
21. Butchart SHM, Baillie JEM, Chenery AM, Collen B, Gregory RD, et al. (2010) National Indicators Show Biodiversity Progress Response. Science 329: 900–901.
22. Butchart SHM, Walpole M, Collen B, van Strien A, Scharlemann JPW, et al. (2010) Global Biodiversity: Indicators of Recent Declines. Science 328: 1164–1168.
23. Bubb P, Chenery A, Herkenrath P, Kapos V, Mapendembe A, et al. (2011) National indicators, monitoring and reporting for the strategic plan for biodiversity 2011–2020. In: UNEP-WCMC, editor. A Report by UNEP-WCMC with IUCN and ECNC for the UK Department for Environment Food and Rural Affairs (Defra) ed. Cambridge.
24. Chandra A, Idrisova A (2011) Convention on Biological Diversity: a review of national challenges and opportunities for implementation. Biodiversity and Conservation 20: 3295–3316.
25. MacArthur Foundation (2011) Conservation & Sustainable Development: international programs strategic framework 2011–2020. Chicago, IL: MacArthur Foundation.
26. Myers N, Mittermeier RA, Mittermeier CG, da Fonseca GAB, Kent J (2000) Biodiversity hotspots for conservation priorities. Nature 403: 853–858.
27. Lehner B, Verdin K, Jarvis A (2008) New Global Hydrography Derived From Spaceborne Elevation Data. Eos, Transactions American Geophysical Union 89: 93–94.
28. HydroSHEDS WWF, USGS (2011) http://www.worldwildlife.org/hydrosheds for general information; http://hydrosheds.cr.usgs.gov for data download and techinical information. March 20, 2008 ed.
29. FAO GeoNetwork (2009) Hydrological Basins derived from Hydrosheds.
30. Eckerson WW, ebrary Inc. (2011) Performance dashboards measuring, monitoring, and managing your business. 2nd ed. Hoboken, N.J.: Wiley.
31. Few S, Safari Tech Books Online. (2006) Information dashboard design. Sebastopol, Calif.: O'Reilly.
32. The World Bank. Available: World Bank Open Data. Accessed 16 February 2013.
33. CNN Money. Available: Accessed 26 February 2014.
34. Convention on International Trade in Endangered Species of Wild Fauna and Flora. Available: CITES trade data dashboards. Accessed 16 February 2014.
35. Chalakkal K (2010) 3. Trade in Endangered Species. Yearbook of International Environmental Law 21: 301–310.
36. Tessier E, Pothin K, Chabanet P, Fleury P-g, Bissery C, et al. (2011) Définition d'Indicateurs de performance et d'un Tableau de bord pour la Réserve Naturelle Marine de La Réunion (RNMR) (Definition of performance indicators and a dashboard to the Natural Marine Reserve of La Reunion (RNMR)). Ifremer, Plouzane, France.
37. Clua E, Beliaeff B, Chauvet C, David G, Ferraris J, et al. (2005) Towards multidisciplinary indicator dashboards for coral reef fisheries management. Aquatic living resources 18: 199–213.
38. Carr ER, Wingard PM, Yorty SC, Thompson MC, Jensen NK, et al. (2007) Applying DPSIR to sustainable development. International Journal of Sustainable Development and World Ecology 14: 543–555.
39. European Environmental Agency. (2007) Available: The DPSIR framework used by the EEA. Accessed Jan 2 2013.
40. Mace GM, Baillie JEM (2007) The 2010 biodiversity indicators: Challenges for science and policy. Conservation Biology 21: 1406–1413.
41. Maxim L, Spangenberg JH, O'Connor M (2009) An analysis of risks for biodiversity under the DPSIR framework. Ecological Economics 69: 12–23.
42. Tscherning K, Helming K, Krippner B, Sieber S, Paloma SGY (2012) Does research applying the DPSIR framework support decision making? Land Use Policy 29: 102–110.
43. UNEP-WCMC (2009) International Expert Workshop on the 2010 Biodiversity Indicators and Post-2010 Indicator Development. Cambridge, UK: UNEP-WCMC.
44. Convention on Biological Diversity (2011) Report of the Ad Hoc Technical Expert Group on Indicators for the Strategic Plan for Biodiversity 2011–2020. High Wycombe, United Kingdom.
45. ESRI (2011) ArcGIS Desktop: Release 10. Redlands, CA: Environmental Systems Research Institute.
46. R Core Team (2013) R: A language and environment for statistical computing.. Vienna, Austria: R Foundation for Statistical Computing.
47. Hansen MC, Stehman SV, Potapov P (2010) Global Forest Monitoring Project. In: Geographic Information Science Center of Excellence tSDSU, editor.
48. Hansen MC, Stehman SV, Potapov PV (2010) Quantification of global gross forest cover loss. Proceedings of the National Academy of Sciences of the United States of America 107: 8650–8655.
49. NASA's Earth Observing System. Available: MODIS: Moderate-resolution Imaging Spectroradiometer. Accessed 2013 June 17.
50. NASA, USGS. Available: LANDSAT Missions. Accessed 2013 June 17.
51. Asner GP, Clark JK, Mascaro J, Galindo Garcia GA, Chadwick KD, et al. (2012) High-resolution mapping of forest carbon stocks in the Colombian Amazon. Biogeosciences 9: 2683–2696.
52. Asner GP, Knapp DE, Balaji A, Paez-Acosta G (2009) Automated mapping of tropical deforestation and forest degradation: CLASlite. Journal of Applied Remote Sensing 3.
53. Asner GP, Powell GVN, Mascaro J, Knapp DE, Clark JK, et al. (2010) High-resolution forest carbon stocks and emissions in the Amazon. Proceedings of the National Academy of Sciences of the United States of America 107: 16738–16742.
54. Butchart SHM, Akcakaya HR, Chanson J, Baillie JEM, Collen B, et al. (2007) Improvements to the Red List Index. Plos One 2.
55. Butchart SHM, Stattersfield AJ, Baillie J, Bennun LA, Stuart SN, et al. (2005) Using Red List Indices to measure progress towards the 2010 target and beyond. Philosophical Transactions of the Royal Society B-Biological Sciences 360: 255–268.
56. Butchart SHM, Stattersfield AJ, Bennun LA, Shutes SM, Akcakaya HR, et al. (2004) Measuring global trends in the status of biodiversity: Red list indices for birds. Plos Biology 2: 2294–2304.
57. Hoffmann M, Hilton-Taylor C, Angulo A, Bohm M, Brooks TM, et al. (2010) The Impact of Conservation on the Status of the World's Vertebrates. Science 330: 1503–1509.
58. IUCN (2010) IUCN Red List of Threatened Species.
59. Eken G, Bennun L, Brooks TM, Darwall W, Fishpool LDC, et al. (2004) Key biodiversity areas as site conservation targets. Bioscience 54: 1110–1118.
60. Butchart SHM, Scharlemann JPW, Evans MI, Quader S, Arico S, et al. (2012) Protecting Important Sites for Biodiversity Contributes to Meeting Global Conservation Targets. Plos One 7.
61. IUCN, UNEP (2010) The World Database on Protected Areas (WDPA). In: UNEP-WCMC, editor. Cambridge, UK.
62. BirdLife Important Bird Area (IBA) Programme (2011) Sites- Important Bird Areas (IBAs). BirdLife International.
63. Extinction AfZ (2010) 2010 AZE Update.
64. Ricketts TH, Dinerstein E, Boucher T, Brooks TM, Butchart SHM, et al. (2005) Pinpointing and preventing imminent extinctions. Proceedings of the National Academy of Sciences of the United States of America 102: 18497–18501.
65. Larsen FW, Londono-Murcia MC, Turner WR (2011) Global priorities for conservation of threatened species, carbon storage, and freshwater services: scope for synergy? Conservation Letters 4: 355–363.
66. Larsen FW, Turner WR, Brooks TM (2012) Conserving Critical Sites for Biodiversity Provides Disproportionate Benefits to People. Plos One 7.
67. Alcamo J, Doll P, Henrichs T, Kaspar F, Lehner B, et al. (2003) Development and testing of the WaterGAP 2 global model of water use and availability. Hydrological Sciences Journal-Journal Des Sciences Hydrologiques 48: 317–337.
68. U.S.Geological Survey (2000) HYDRO1k Elevation Derivative Database In: Sci. CfEROa, editor. Sioux Falls, SD, USA.
69. Oak Ridge National Laboratory (2006) Landscan Global Population Database Oak Ridge, TN.
70. Balmford A, Rodrigues ASL, Walpole M, ten Brink P, Kettunen M, et al. (2008) Review on the economics of biodiversity loss: scoping the science. European Commission.
71. Brauman KA, Daily GC, Duarte TK, Mooney HA (2007) The nature and value of ecosystem services: An overview highlighting hydrologic services. Annual Review of Environment and Resources. pp. 67–98.
72. Bruijnzeel LA (2004) Hydrological functions of tropical forests: not seeing the soil for the trees? Agriculture Ecosystems & Environment 104: 185–228.
73. Dudley N, Stolton S (2003) Running Pure: The importance of forest protected areas to drinking water. World Bank/WWF Alliance for Forest Conservation and Sustainable Use.
74. Hampton SE, Strasser CA, Tewksbury JJ, Gram WK, Budden AE, et al. (2013) Big data and the future of ecology. Frontiers in Ecology and the Environment 11: 156–162.
75. Kelling S, Hochachka WM, Fink D, Riedewald M, Caruana R, et al. (2009) Data-intensive Science: A New Paradigm for Biodiversity Studies. Bioscience 59: 613–620.
76. Danielsen F, Mendoza MM, Alviola P, Balete DS, Enghoff M, et al. (2003) Biodiversity monitoring in developing countries: what are we trying to achieve? Oryx 37: 407–409.
77. Danielsen F, Mendoza MM, Tagtag A, Alviola P, Balete DS, et al. (2003) On participatory biodiversity monitoring and its applicability - a reply to Yoccoz, et al. and Rodriguez. Oryx 37: 412–412.
78. Rodriguez JP (2003) Challenges and opportunities for surveying and monitoring tropical biodiversity - a response to Danielsen, et al. Oryx 37: 411–411.
79. Yoccoz NG, Nichols JD, Boulinier T (2001) Monitoring of biological diversity in space and time. Trends in Ecology & Evolution 16: 446–453.
80. Yoccoz NG, Nichols JD, Boulinier T (2003) Monitoring of biological diversity - a response to Danielsen, et al. Oryx 37: 410—410.
81. Jones JPG, Collen B, Atkinson G, Baxter PWJ, Bubb P, et al. (2011) The Why, What, and How of Global Biodiversity Indicators Beyond the 2010 Target. Conservation Biology 25: 450–457.

82. Bertzky M, Stoll-Kleemann S (2009) Multi-level discrepancies with sharing data on protected areas: What we have and what we need for the global village. Journal of Environmental Management 90: 8–24.
83. Page RDM (2008) Biodiversity informatics: the challenge of linking data and the role of shared identifiers. Briefings in Bioinformatics 9: 345–354.
84. Scholes RJ, Mace GM, Turner W, Geller GN, Jurgens N, et al. (2008) Ecology - Toward a global biodiversity observing system. Science 321: 1044–1045.
85. Chavan VS, Ingwersen P (2009) Towards a data publishing framework for primary biodiversity data: challenges and potentials for the biodiversity informatics community. Bmc Bioinformatics 10.
86. Editorial (2005) Let data speak to data. Nature 438: 531–531.
87. Magurran AE, Baillie SR, Buckland ST, Dick JM, Elston DA, et al. (2010) Long-term datasets in biodiversity research and monitoring: assessing change in ecological communities through time. Trends in Ecology & Evolution 25: 574–582.
88. Rands MRW, Adams WM, Bennun L, Butchart SHM, Clements A, et al. (2010) Biodiversity Conservation: Challenges Beyond 2010. Science 329: 1298–1303.
89. Rudd MA, Beazley KF, Cooke SJ, Fleishman E, Lane DE, et al. (2011) Generation of priority research questions to inform conservation policy and management at a national level. Conservation Biology 25: 476–484.
90. Biala K, Conde S, Delbaera B, Jones-Walters L, Torre-Marin A (2012) Streamlining European biodiversity indicators 2020: building a future on lessons learnt from the SEBI 2010 process. Copenhagen, Danmark: European Environment Agency.
91. Chenery A, Plumpton H, Brown C, Walpole M (2013) Aichi targets passport. Cambridge, UK: Biodiversity Indicators Partnership.
92. ASEAN Center for Biodiversity (ACB). (2005) Available: http://www.aseanbiodiversity.org/. Accessed 2013 Janurary 22.
93. Red Amazónica de Información Socioambiental Geoffeferenciada (RAISG). (1996) Available: Amazonian Network of Georeferenced Socio-Environmental Information, http://raisg.socioambiental.org/. Accessed 2013 Janurary 22.
94. Streamlining European Biodiversity Indicators (SEBI). Available: http://biodiversity.europa.eu/topics/sebi-indicators. Accessed 2013 Janurary 22.
95. Conservation of Arctic Flora and Fauna (CAFF). Available: Accessed 2013 Janurary 22.
96. Uganda National Biodiversity Data Bank. Available: Accessed 2013 Janurary 22..
97. Stuart SN, Chanson JS, Cox NA, Young BE, Rodrigues ASL, et al. (2004) Status and trends of amphibian declines and extinctions worldwide. Science 306: 1783–1786.
98. Young BE, Lips KR, Reaser JK, Ibanez R, Salas AW, et al. (2001) Population declines and priorities for amphibian conservation in Latin America. Conservation Biology 15: 1213–1223.
99. Turner W, Spector S, Gardiner N, Fladeland M, Sterling E, et al. (2003) Remote sensing for biodiversity science and conservation. Trends in Ecology & Evolution 18: 306–314.
100. UN-FAO (2010) Global Forest Resources Assessment. FOOD AND AGRICULTURE ORGANIZATION OF THE UNITED NATIONS,.
101. European Space Agency (ESA) (2010) GlobCover 2009 (Global Land Cover Map). 21st December 2010 ed.
102. European Space Agency (ESA) (2008) Global Land Cover Product (2005–06).
103. National Ecological Observatory Network. (2010)Available: NEON, www.neoninc.org. Accessed 2013 Janurary 22.
104. Hopkin M (2006) Ecology - Spying on nature. Nature 444: 420–421.
105. The Circumpolar Biodiversity Monitoring Programme (CBMP). Available: http://caff.is/monitoring. Accessed 2013 Janurary 22.
106. Natural Heritage Program. (1974) Available: http://www.natureserve.org/visitLocal/. Accessed 2013 Janurary 22.
107. Stein BA, Kutner LS, Adams JS (2000) Precious heritage: the status of biodiversity in the United States. New York: Oxford University Press. xxv, 399 p. p.
108. TEAM Network. Available: Tropical Ecology Assessment & Monitoring Network: Early Warning System for Nature. http://www.teamnetwork.org/. Accessed 2013 Janurary 22.
109. WCS Conservation Support. Available: Status and Impact Monitoring, http://wcslivinglandscapes.com/WhatWeDo/StatusImpactMonitoring.aspx. Accessed 2013 Janurary 22.

11

A Developmental Model for Branching Morphogenesis of Lake Cress Compound Leaf

Akiko Nakamasu[1,2], Hokuto Nakayama[1], Naomi Nakayama[3], Nobuhiko J. Suematsu[2,4]*, Seisuke Kimura[1]*

1 Department of Bioresource and Environmental Sciences Faculty of Life Sciences, Kyoto Sangyo University, Kyoto, Japan, **2** Meiji Institute for Advanced Study of Mathematical Sciences, Meiji University, Tokyo, Japan, **3** Institute of Molecular Plant Sciences, University of Edinburgh, Edinburgh, Lothian, United Kingdom, **4** Graduate School of Advanced Mathematical Sciences, Meiji University, Tokyo, Japan

Abstract

Lake cress, *Rorippa aquatica* (Brassicaceae), is a semi-aquatic plant that exhibits a variety of leaf shapes, from simple leaves to highly branched compound leaves, depending on the environment. Leaf shape can vary within a single plant, suggesting that the variation can be explained by a simple model. In order to simulate the branched structure in the compound leaves of *R. aquatica,* we implemented reaction-diffusion (RD) patterning onto a theoretical framework that had been developed for serration distribution in the leaves of *Arabidopsis thaliana,* with the modification of the one-dimensional reaction-diffusion domain being deformed with the spatial periodicity of the RD pattern while expanding. This simple method using an iterative pattern could create regular and nested branching patterns. Subsequently, we verified the plausibility of our theoretical model by comparing it with the experimentally observed branching patterns. The results suggested that our model successfully predicted both the qualitative and quantitative aspects of the timing and positioning of branching in growing *R. aquatica* leaves.

Editor: Hirokazu Tsukaya, The University of Tokyo, Japan

Funding: This study was supported by Grant-in-Aid for JSPS Fellows (http://www.jsps.go.jp/english/e-grants/index.html) to A.N. and H.N. and by Meiji University Global COE Program "Formation and Development of Mathematical Sciences Based on Modeling and Analysis" (http://gcoe.mims.meiji.ac.jp/index-e.html) to A.N. and N.J.S. In addition, it was partially supported by JSPS KAKENHI Grant Numbers 22870031 and 24770047 and the Science Research Promotion Fund from the Promotion and Mutual Aid Corporation for Private Schools of Japan (http://www.shigaku.go.jp/g_about_pmac.htm) to S.K. The funders had no role in study design, data collection and analysis, decision to publish, or preparation of the manuscript.

Competing Interests: The authors have declared that no competing interests exist.

* Email: suematsu@meiji.ac.jp (NJS); seisuke@cc.kyoto-su.ac.jp (SK)

Introduction

Morphogenesis of multi-cellular organisms requires coordination of growth and developmental pattern formation (i.e., temporal and spatial specification of morphogenic fates). Morphogenesis accompanied by continuous growth requires stable positional information in order to make well-proportioned shapes. From the structure of lungs to the branching of trees, nearly evenly spaced branches represent a major type of morphogenic process and various mathematical models have been applied to explain each phenomenon [1,2].

Branching morphogenesis in plants is often accompanied by growth. One example is compound leaf development. Although plant leaves exhibit striking diversity in their shapes, they can be roughly classified into two classes: simple and compound. Simple leaves have a single blade, whereas compound leaves have multiple blade units. In both types of leaves, a leaf begins as a small bulge, called a leaf primordium, at the tip of the shoot apex. The leaf primordium grows rapidly by cell division and cell expansion, eventually forming the mature leaf shape. In the case of compound leaves, new leaflet primordia emerge iteratively on a growing leaf primordium [3,4]. Recently, a laser ablation experiment on *Eschscholzia californica* leaf primordia revealed that a constant-spacing regulatory mechanism governs leaflet initiation sites [5].

Some geometric features of leaf shapes have been theoretically investigated [6]. The theoretical understanding of leaf shape determination has recently dramatically improved [7,8,9]. For example, Bilsbrough et al. (2011) explained the mechanism underlying the formation of a simple leaf with serrations on its margin by using a mathematical model that utilized a repeated iterative pattern. It has been suggested that the framework of this serration model could also explain the morphogenesis of more complex leaf shapes [8]. The spatially periodic pattern was simulated using the polar auxin transport (PAT) model, which has been widely used to explain various aspects of plant morphogenesis, including phyllotaxis, leaf vascular patterning, and root development [10,11,12,13]. In the leaf morphogenesis, PAT was caused by interactions of the growth-promoting phytohormone auxin, PIN-FORMED1 (PIN1: auxin efflux carrier protein) [14], and CUP-SHAPED COTYLEDON2 (CUC2: transcription factor that negatively regulates local growth rate) [15].

We studied the North American lake cress (*Rorippa aquatica*), which is a semi-aquatic member of the family Brassicaceae, as a model to explain leaf diversity. *R. aquatica* changes the shape of its leaves depending on the growth conditions (e.g., underwater submergence and temperature) [16]. Such morphological plasticity is called heterophylly. The leaf shape can range from simple elliptical to a complex, highly branched compound morphology (Fig. 1A, B). Intermediate forms have also been observed. This plant is a theoretically good model to study leaf morphogenesis since such variation in the leaf shape is observed even within a single plant. There may be a common morphological mechanism that produces the various leaf-shapes in heterophylly. If so, it can be explained by a simple model.

Here, we built the model to explain the morphogenesis of the highly branched compound leaves in *R. aquatica*. Instead of PAT, we utilized RD based patterning [17,18,19] for our model because it is often employed to explain spatial pattern formation in biological systems, such as the pigmenting stripe pattern of marine angelfish [20]. The pattern behavior on a growing domain is well studied and known to be easily controlled [21]. Diffusion-driven instability, which was suggested by Alan Turing in 1952, can create stationary periodic patterns called Turing patterns. They are known to cause the insertion or splitting of reactant peaks to maintain equal intervals when placed on an expanding domain [21,22].

Our model was successful in replicating the growth and branching events in the development of *R. aquatica* compound leaves. It was able to predict the qualitative aspects, such as the positioning, direction, and order of branching, as well as the quantitative aspects, including the increase in the number of branches.

Results and Discussion

Morphological observation

R. aquatica develops smooth-margin simple leaves when grown at 30°C, and compound leaves of a multi-order branched structure when grown at 20°C (Fig. 1A, B). The simple leaf has iterative hydathodes on its margin, which are generally seen at the tip of serrations in *A. thaliana*. The leaves at the earliest stages of development at the shoot apex were dissected and used to examine the branching pattern formation (Fig. 1C). A shoot apex contains leaf primordia at different stages of morphogenesis (Fig. 1D–F). By the observation of dissected shoot apex and from time-lapse movies [Nakayama et al. in preparation] the branched structure of the compound leaf formed through repetitive additions of protrusions that are future leaflets (i.e., leaflet primordia). The number of leaflets increased gradually (Fig. 1D–F). First, primary leaflets appeared on both sides of a leaf primordium (Fig. 1D), and secondary leaflets emerged on the primary leaflets after some elongation (Fig. 1E, F). The increase in the total number of leaflets of all categories (i.e., terminal, primary, secondary, tertiary) was plotted as a progression of the primary leaflet formation (Fig. 1G). The dynamics of repetitive leaflet initiations prompted us to consider that a spatially periodic pattern occurs during compound leaf development in *R. aquatica*.

The total number of leaflets reached a plateau when the leaf became approximately 4 mm in height. At this point, the number of primary leaflets also reached a plateau. Because the number of leaflets on the primary leaflet of mature leaves is almost the same

Figure 1. Morphogenesis of *Rorippa aquatica* leaves. A, B: Mature leaf morphology of the simple leaf that was developed at 30°C (A) and the highly branched compound leaf that was developed at 20°C (B). Scale bar: 1 cm. C: Dissected shoot apex of a plant grown at 20°C, showing the nested group of leaf primordia with indented blade. D–F: Dissected primordial of a plant grown at 20°C for about 2 months. Each primodium has the 32th (D), 35th (E), and 39th (F) leaf primordium from the oldest (i.e. outermost) leaf of a plant. The larger leaf position numbers indicate younger leaves. Scale bar: 1 mm (C) and 200 μm (D–F). G: Comparison of the total number of leaflet primordial between experimentally observed and the theoretically estimated value.

as the number of leaflet primordia on primary leaflets of immature leaves as described later, it is suggested that the branched structures of both mature and immature leaves are geometrically similar. This means that branching is fixed after that specific stage, and the duration of leaflet formation seems constant and determined. From time-lapse recordings of developing leaves, it was observed that new protrusions emerge in the recently elongated regions of a leaf, which exist on the both sides of the tip and the base of the first primary leaflets [Nakayama et al. in preparation].

Modeling for branched structure of compound leaves

In order to explain the branching dynamics in *R. aquatica* leaves, we utilized the framework of a previously developed mathematical model [9] (Fig. 2A). The leaf margin was regarded as a one-dimensional reaction-diffusion domain, which was composed of segmented lines simulating cells. The leaf margin was deformed based on the spatially periodic pattern using the boundary propagation method (BPM), which abstractly achieves this geometric deformation [23]. Briefly, the growth direction of the connection point was vector sum of the normal unit vector of adjacent cells and the growth rate was proportional to the average concentration of reactant in adjacent cells. Segments over a certain length were divided into two daughter cells, which had the reactants state of mother cell.

The growth profile of the BPM ring and the spatiotemporal pattern of the reactant peaks therein are shown in Figure 2. To support the experimental observations, in which new branches were inserted into newly generated gaps between two leaflets, we calculated the RD pattern formation using the parameters that showed insertions of reactant peaks. Expansion at the position of reactant peak makes the peak unstable then it generally induces the peak splitting. However, in appropriate parameter region, regular peak doubling by insertion can be generated (Fig. 2B, Movie S1). For easy selection of pattern behavior on growing domain, the linear activator-inhibitor model was used.

As an initial condition, six cells that had random values of reactants u and v were connected as a ring. The ring grew in one direction based on the location of the first-emerged reactant peak (represented blue in; Fig. 2C–F, Movie S1) and subsequently, the first insertion was generated (Fig. 2C, Movie S1). The second peak emerged on the opposite side of the first peak. When we regarded either of them as a terminal leaflet, the ring was stretched toward each peak to make the main axis (i.e., the proximodistal axis) of leaf growth. Next, tertiary peaks were formed in the center of intervals, which become wider between the first and second peaks, resulting in two secondary axes that corresponded to the primary leaflets (Fig. 2D, Movie S1). As the reactant peaks were inserted in order, new secondary axes joined four by four on the recently extended region of the main axis (Fig. 2E, Movie S1). In addition, ternary axes, corresponding to the secondary leaflets, were generated on the extended region of the secondary axes (Fig. 2F, Movie S1).

The simulated branching dynamics recapitulate the qualitative aspects of the morphogenesis of compound leaves in *R. aquatica*. Hence, the developmental model for indented simple leaves [9] is also applicable to the morphogenesis of branched compound

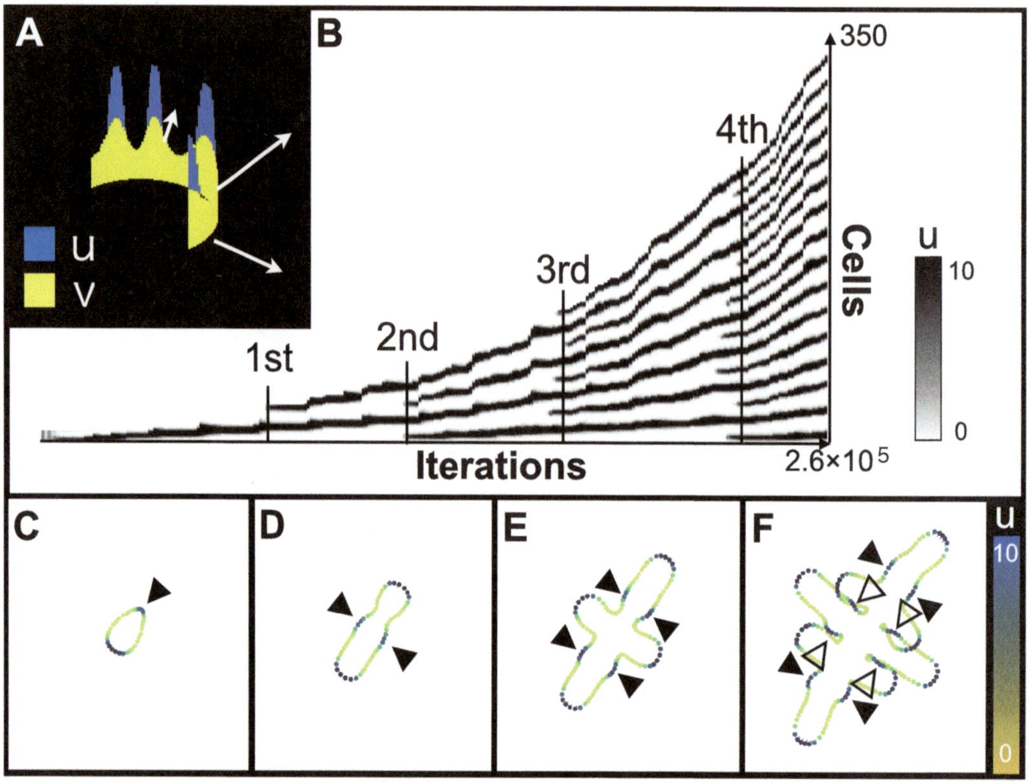

Figure 2. Spatiotemporal plot and growth profiles for the BPM rings by pattern dependent expansion. A: Schematic of the modeling. B: Spatiotemporal plot for peak doubling by insertion. The value of reactant u is represented by the gray scale. Each panel shows the first (C), second (D), third (E), and fourth (F) insertion. Each point indicate the middle point of segmented cell, then the color of points indicate the value of reactant u. Solid arrowheads indicate the points of peak insertion, and empty arrowheads are points of side branch generation.

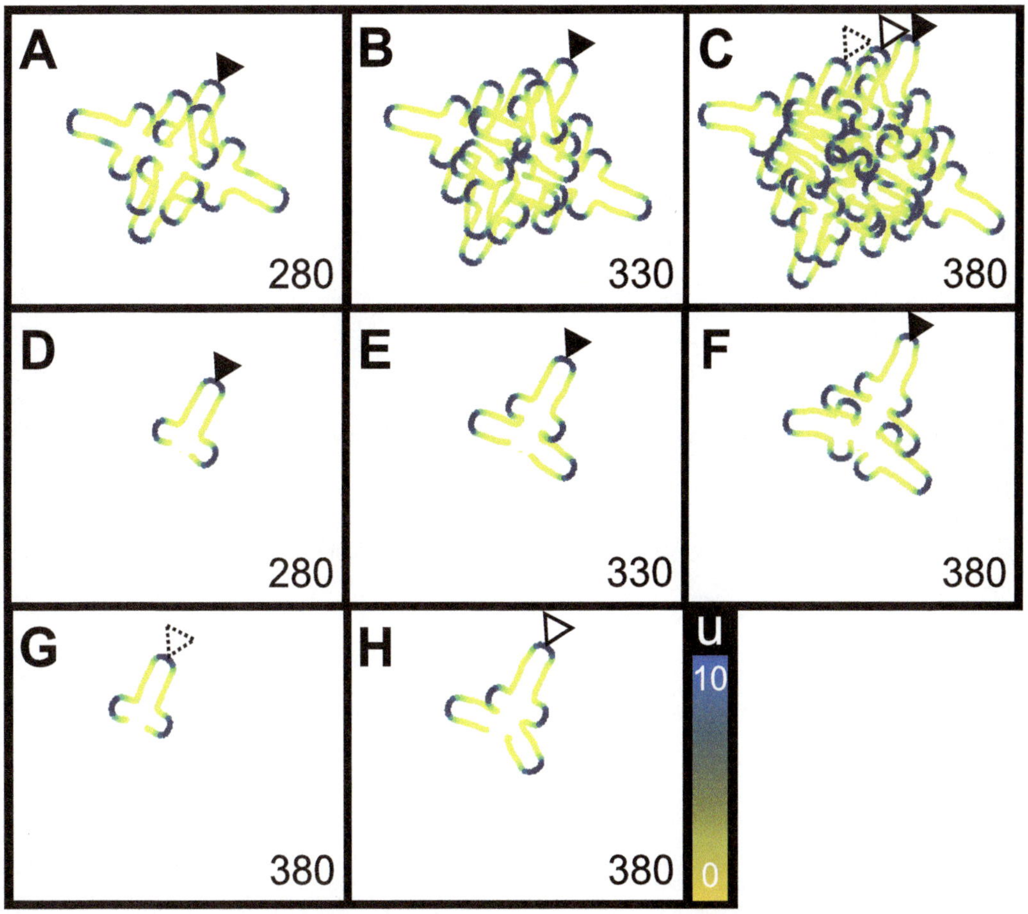

Figure 3. Simulations of leaf primordia and branches. Each panel shows the simulated whole leaves (A–C) and primary leaflets (D–H). The simulated branches were crossover. Each branch was independently formed nested regular branches. The inserted number shows the time of iterative calculations ($\times 10^4$), and the arrowheads indicate each leaflet; filled, flamed, and dotted arrow heads represent the first, second, and third primary leaflet respectively.

leaves. The behaviors of spatially periodic patterns on uniform growing domain have been well studied and analyzed. On the other hand, the study about the behaviors of the patterns on reactant-dependent growing domain is still limited especially in the case of peak insertion. Here, we show that a linear RD pattern can be reduplicated by regular insertion using appropriate parameters. Whatever the mechanism governing the periodic pattern, this nested branched pattern was generated by the simple rule of a continuous growth dependent on spatially periodic pattern.

Comparison between the simulation and actual-leaf branching dynamics

Even with the distinctive character, the snapshot of shapes of organisms can be simulated by various models. Therefore, it is important to compare the spatial-temporal profile of developmental events to reduce the degree of freedom. To test the plausibility of our model, we compared the actual morphogenesis of *R. aquatica* compound leaves to the branched pattern generated by our simulation. Although the simulated branches overlapped and had tangled crossovers, regular and spatiotemporally nested branches were generated as the iterative calculation progressed (Fig. 3, Movie S2). Each branch was formed by a closed ring, and branches were thus independent of each other (Fig. 3F–H and Movie S2). It should be noted that since the actual structure of the leaf primordium varies in three-dimensions, the crossovers of each branch are irrelevant.

When we expressed the number of terminal and primary leaflets where n times insertion had occurred as B_n, we observed an increase in the number of the primary leaflets over time. At first, only the terminal leaflets were produced by first insertion, and thus the number of total leaflets was given by $B_1 = 1$. For every $n \geq 2$, the addition of primary leaflets led to the formula $B_n = 4(n-1) - 1$, thereby ensuring that the additions were proportional to the times of the events. Therefore, we expressed the morphogenetic stage as the number of primary leaflets and placed it on the horizontal axis (Fig. 1G).

When we considered the nature of the peak doubling, the estimation of the total number of leaflets A_n was given by the simple formula $A_n = 2^n - 1$, which was obtained from the subtraction of the one peak corresponding to the base of leaf primordium from 2^n, the total number of the peaks. The theoretical estimation was shown as a solid line marked by squares, and the total number of leaflets from the experimentally observed half primordium was shown as a bar with standard error (Fig. 1G). Despite slight deviations, similar increments were observed in both data sets, suggesting the plausibility of our model.

Next, we compared the generation of secondary leaflets on each primary leaflet. When we put the estimation of the number of secondary leaflets on the ith newest primary leaflet, C_i, and gave

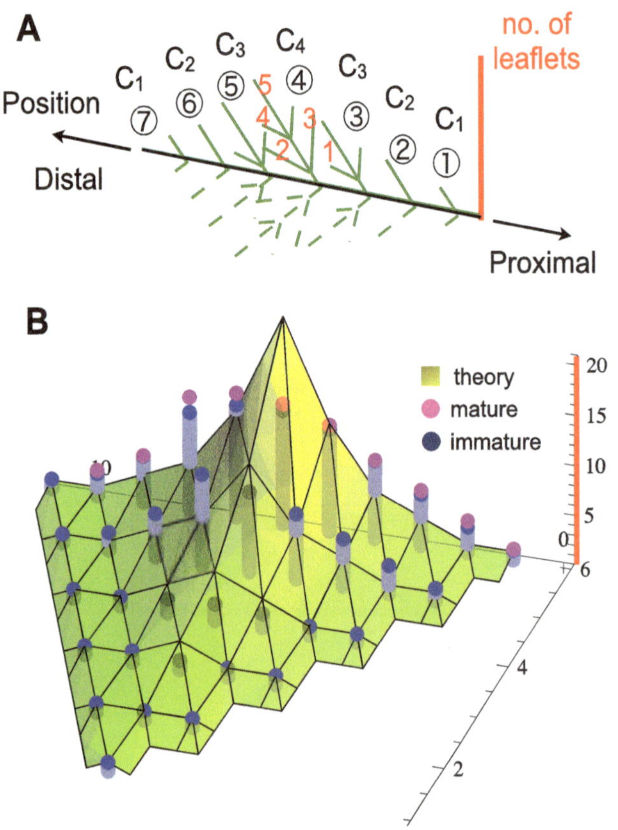

Figure 4. The numbers of leaflet primordia of each primary leaflet. A: Schematic of the branched structure of one half of a *R. aquatica* compound leaf. Circled numbers indicate the positions of primary leaflet (the horizontal axis in B), and theoretically derived recurrence formulas of each primary leaflet are shown by C_i. The red numbers represent the numbers of leaflets formed on the 4th primary leaflet (the vertical axis in B). B: A comparison between the experimentally observed data in actual plants and theoretically estimated numbers derived from mathematical formulae of leaflet on each primary leaflet. The magenta dots show the data from mature leaves. The number of leaflets at each stage was plotted as aligned at the center. The theoretical estimations are represented on a yellow planar graph, and the actual data in developing leaves as blue dots with columns.

$C_1 = 1$ could be represented as $C_i = 2^{i-1} - C_{i-1} (i \geq 2)$. This simple recurrence formula stems from the characteristics of branches being nested and having a regular branching rule that shown by this model (Fig. 3, Movie S2).

The total number of leaflets mentioned above could be represented by the summation of individual leaflets. Therefore, the total numbers of leaflets on n times inserted leaf primordium A_ns which contain a terminal leaflet and $n-19^{th}$ newest two first primary leaflets and following four primary leaflets were given using C_is as

$$A_n = 1 + 2C_{n-1} + 4C_{n-2} \ldots 4C_1 = 1 + 4\sum_{1}^{n-1} C_i - 2C_{n-1} \qquad (C_1 = 1)$$

and thus $A_n = 2^n - 1$.

Since the leaf could be regarded symmetrical along the proximodistal axis, we considered a half of the branched leaf for comparison. As in Figure 4A, we placed the positions of primary leaflets on the half leaf and the numbers of leaflets on each primary leaflet on the horizontal and vertical axis, respectively (Fig. 4B). The spatial branching structure of mature leaves (expanded leaves larger than 5 cm in length at leaf number 10–15) were indicated as magenta dots in Fig. 4B. For immature leaves, we grouped the graphs of blue column corresponding to each primordium at different morphogenic stages. Furthermore, values of C_i against the positions of the primary leaflets were plotted on the planar graph (Fig. 4B).

The analysis that focused on the branching of each primary leaflet showed consistency between the theoretical estimation from the mathematical formula and empirical data in actual plants, although some discrepancies were observed. In the simulation, the number of total leaflets increased from the centrally located primary leaflet, since it was generated first (earliest). On the other hand, the actual location of primary leaflets with the highest number of branches tended to be toward the apical side, rather than the center. Furthermore, the natural leaflet distributions varied mildly compared to the peaked theoretical values.

Branch formation by expansion inhibition

Recently, Vlad *et al.* reported that cell proliferation in the sinus region was inhibited in a gene expression pattern dependent manner at the formation of a compound leaf in *Cardamine hirsuta* [24]. Therefore we attempted a method that was opposite case to that described in the previous sections. The leaf margin was expanded constantly then the expansion was inhibited based on the spatially periodic pattern. As a result, fundamentally equal branched pattern and branching process were observed (Fig. 5B–E, Movie S3), even in the case of expansion inhibition. Because the region without the reactant peaks bulged outward, the branching was generated by repetitive splittings of the pattern (Fig. 5A, Movie S3). This is simply the reverse case of the insertion of Turing-pattern behavior on growing domain [21]. Insertion of peak emerges at the region where the activator is deficient. Thus, it is important to keep the distance between two successive peaks of activators relatively large while domain growth to make insertion happens. To make splitting happen, opposite situation is needed.

Our simulation created symmetrical and regular branched structures, and the timing of peak-doubling was synchronous; however, the actual *R. aquatica* leaf undergoes asynchronous branching. The asynchronous branching can be caused, even in the equal interval pattern, by the introduction of anisotropic growth. In such cases, it is difficult to accurately estimate the increase in the number of leaflets by the methods described in this work. In addition, we assumed interactions among branches in this model; without the interactions, the same results would be calculated using certain timings of branching. Therefore, in order to confirm the biological significance of this model, the existence of interactions among branches should be tested. Recently, the presence of such interactions between leaf tip and leaflet of the compound leaves was shown by a laser ablation experiment in *E. californica* [5].

The model reported here may require further verification; however, it has the potential to simulate the full spectrum of the morphogenic gradient of leaf shapes in *R. aquatica*. It has already been led to a specific new hypothesis to be tested experimentally that will likely reveal new insight into the mechanism behind the diversity of leaf shapes. The molecular mechanisms of compound leaf development have been studied in model plants, such as tomatoes (*Solanum lycoperisicum*) and *Cardamine hirsuta* [3,25], and the differences in the timing and positioning of maturation within the primordia have been related to the species-to-species differences in the direction and position of leaflet initiation [4,26]. The molecular basis of the reactants used in this RD based model has not yet been determined. We would like to determine the molecular aspect of this mechanism in the future.

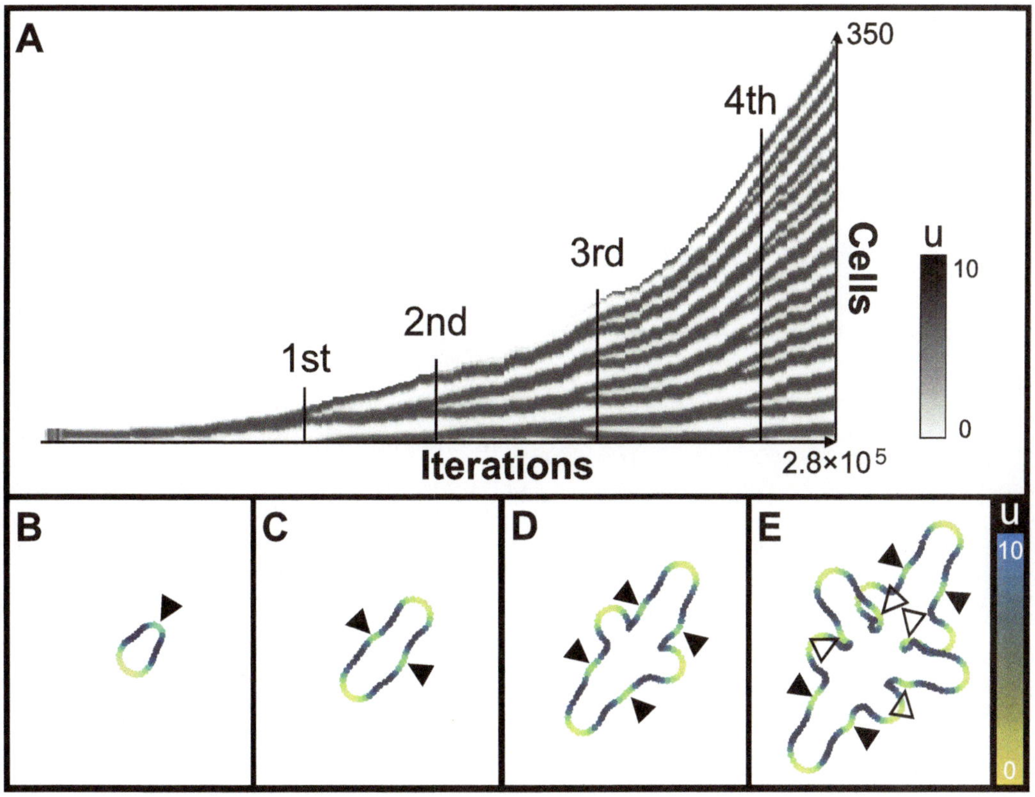

Figure 5. Spatiotemporal plot and growth profiles for the BPM rings by Expansion inhibition. A: Spatiotemporal plot for peak doubling by splitting. The value of reactant *u* is represented by the gray scale. Each panel shows the first (B), second (C), third (D), and fourth (E) splitting. Each point indicate the middle point of segmented cell, then the color of points indicate the value of reactant *u*. Solid arrowheads indicate the points of peak splitting, and empty arrowheads are points of side branch generation.

Materials and Methods

Plant materials and growth condition

Plants were grown in soil for 1.5–2 months in a growth chamber under a continuous light condition with the temperature set at 20°C at a photo flux density of 70 μmol m^{-2} s^{-1}.

Dissection and image acquisition

To investigate the spacing of the branch formation, 12 shoot apices were mounted on a glass slide and dissected under a stereoscopic microscope (LEICA 8AP0) with a LAS EZ digital camera using forceps. The shoot apices were fixed in FAA (4% formaldehyde, 5% acetic acid, and 50% ethanol) and were washed using phosphate buffered saline (PBS) before dissection. Images were captured by a digital camera (Nikon DS-Ri1) mounted on an up-light light microscope (Nikon ECLIPSE 80i).

Simulation

The simulation was performed according to the method described by Bilsbrough et al. (2011). Leaf margin, regarded as a one-dimensional reaction-diffusion domain, was expanded based on the spatially periodic pattern which was generated by the RD model as described in [20]. The differential equations used in the simulation were as follows:

$$\partial u/\partial t = D_u \partial^2 u/\partial x^2 + H(u,v) - eu$$

$$\partial v/\partial t = D_v \partial^2 v/\partial x^2 + I(u) - gv$$

$$H(u,v) = au - bv + c_u \quad I(u) = fu - c_v$$

$$(0 \leq H \leq H_{max}, \ 0 \leq I \leq I_{max})$$

The linear activator-inhibitor model using two variables, an activator "*u*" and an inhibitor "*v*" were chosen for their flexibility of pattern-behavior selection. The parameters used in the simulation were: $D_u = 0.02$, $D_v = 0.8$, $a = 0.08$, $b = 0.08$, $e = 0.03$, $f = 0.1$, $g = 0.06$, $c_u = 0.05$, $c_v = 0.1$, $H_{max} = 0.3$, $I_{max} = 0.5$ in case of pattern dependent expansion then $D_u = 0.04$, $D_v = 0.6$, $a = 0.11$, $b = 0.08$, $e = 0.03$, $f = 0.1$, $g = 0.06$, $_{mi}$, $c_v = 0.2$, $H_{max} = 0.2$, $I_{max} = 0.5$ in the latter case that is expansion inhibition. They were selected to lead stable spatially periodic insertions or splitting of reactant peaks on the expanding reaction-diffusion domain. The model parameters were searched based on empirical knowledge for 2D simulation of Turing pattern by linear model. The parameters making spot and net pattern in 2D tend to form insertions and splittings on growing domain, respectively. Leaf margin was simulated using the boundary propagation method [23], i.e., the propagation of the leaf margin in space over time. This propagation was performed iteratively by updating connection points of arbitrary segments regarded as a cell. Connection points (x_{ij}, y_{ij}) of adjacent cells i and j were displaced at velocities $\overrightarrow{v_{ij}}=(v_{ij}^x,v_{ij}^y)$, which were activated by reactant u as $\overrightarrow{v_{ij}}=k(u_i+u_j)\overrightarrow{N_{ij}}$ or inhibited by u as $\overrightarrow{v_{ij}}=k\overrightarrow{N_{ij}}\Big/\left(1+u_i+u_j\right)$. $\overrightarrow{N_{ij}}$ were the propagation directions, and the summation of normal vectors of the margin of cells i, j pointed outward. u_i, u_j indicate the amount of reactant u of cell i, j respectively. The $k=0.0003dt$ in the case of pattern dependent expansion, then the $k=0.008dt$ in the case of expansion inhibition. A cell divides into two daughter cells when its length exceeds a threshold by updating positions (x_{ij}, y_{ij}). The two daughter cells inherited the reactant state of their mother cell.

Supporting Information

Movie S1 Branch formation by pattern dependent expansion. Pattern dependent expansion of the ring was simulated. Left panel shows the growth profile of the ring. First, second, and third insertions are marked by arrows. Right panel shows the spatiotemporal plot of growing reaction-diffusion domain. Pattern behavior at these parameters shows the insertions of reactant peaks.

Movie S2 Branching profile of first primary leaflet of RD BPM model. Ring was grown by pattern dependently in this simulation. One of the first primary leaflets was picked up for the observation of branch formation. Generated branches crossed but were independent to each other.

Movie S3 Branch formation by expansion inhibition. Constant expansion of the ring was inhibited in a pattern dependent manner in this simulation. Left panel shows the growth profile of the ring. First, second, and third splittings are marked by arrows. Right panel shows the spatiotemporal plot of the growing reaction-diffusion domain. Pattern behavior at these parameters shows the splittings of reactant peaks.

Acknowledgments

We thank Okamoto Y. Kaoru (Kyoto Sangyo University) for helpful discussions.

Author Contributions

Conceived and designed the experiments: AN NJS SK. Performed the experiments: AN HN NN. Analyzed the data: AN. Contributed to the writing of the manuscript: AN HN NN NJS SK.

References

1. Miura T (2008) Modeling lung branching morphogenesis. Curr. Top Dev Biol 81: 291–310.
2. Fisher JB, Honda H (1997) Computer simulation of branching pattern and geometry in Terminalia (Combretaceae), a tropical tree. Botanical Gazette 138(4): 377–384.
3. Kimura S, Koenig D, Kang J, Yoong FY, Sinha N (2008) Natural variation in leaf morphology results from mutation of a novel *KNOX* gene. Curr Biol 18: 672–677.
4. Ikeuchi M, Tatematsu K, Yamaguchi T, Okuda K, Tsukaya H (2013) Precocious progression of tissue maturation instructs basipetal initiation of leaflets in *Chelidonium majus* subsp. *asiaticum* (Papaveraceae). Am J Bot 100(6): 1116–1126.
5. Ikeuchi M, Igarashi H, Okada K, Tsukaya H (2014) Acropetal leaflet initiation of *Eschscholzia californica* is achieved by constant spacing of leaflets and differential growth of leaf. Planta 10.1007/s00425-014-2071-9.
6. Prsinkiewicz P, Lindenmayer A (1990) The algorithmic beauty of plants. New York: Springer.
7. Prusinkiewicz P (2012) Computational models of plants development and form. New Phytol 193(3): 549–569.
8. Prusinkiweicz P, Lane B (2013) Modeling morphogenesis in multicellular structures with cell complexes and L-system. Pattern Formation in Morphogenesis, 137–151.
9. Bilsbrough GD, Runions A, Barkoulas M, Jenkins HW, Hasson A, et al. (2011) Model for the regulation of *Arabidopsis thaliana* leaf margin development. Proc Natl Acad Sci USA 108: 3424–3429.
10. Van Berkel K, De Boer J R, Scheres B, Ten Tusscher K (2013) Polar auxin transport: models and mechanisms. Development 140: 2253–2268.
11. Reinhardt D, Pesce ER, Stieger P, Mandel T, Baltensperger K, et al. (2003) Regulation of phyllotaxis by polar auxin transport. Nature 426: 255–260.
12. Scarpella E, Marcos D, Friml J, Berleth T (2006) Control of leaf vascular patterning by polar auxin transport. Genes Dev 20: 1015–1027.
13. Sabatini S, Beis D, Wolkenfelt H, Murfett J, Guilfoyle T, et al. (1999) An auxin-dependent distal organizer of pattern and polarity in the *Arabidopsis* root. Cell 99: 463–472.
14. Hay A, Barkoulas M, Tsiantis M (2006) *ASYMMETRIC LEAVES1* and auxin activities converge to repress *BREVIPEDICELLUS* expression and promote leaf development in *Arabidopsis*. Development 133: 3955–3961.
15. Nikovics K, Blein T, Peaucelle A, Ishida T, Morin H, et al. (2006) The balance between the *MIR164A* and *CUC2* genes controls leaf margin serration in *Arabidopsis*. Plant Cell 18: 2929–2945.
16. Nakayama H, Nakayama N, Nakamasu A, Sinha N, Kimura S (2012) Toward elucidating the mechanisms that regulate heterophylly. Plant Morphol 24: 57–63.
17. Turing AM (1952) Chemical basis of morphogenesis. Phil Trans R Soc London Ser B 237: 37–72.
18. Meinhardt H (1982) Models of biological pattern formation. London: Academic.
19. Kondo S, Miura T (2010) Reaction-diffusion model as a framework for understanding biological pattern formation. Science 329: 1616–1620.
20. Kondo S and Asai R (1995) A reaction-diffusion wave on the skin of the marine angelfish *Pomacanthus*. Nature 376: 765–768.
21. Meinhardt H (1995) The algolismic beauty of sea shells. Berlin: Springer. p.31.
22. Crampin EJ, Hackborn WW, Maini PK (2002) Pattern formation in reaction-diffusion models with nonuniform domain growth. Bull Math Biol 64: 747–769.
23. Sethian J (1999) Level Set Methods and Fast Marching Methods: Evolving Interfaces in Computational Geometry, Fluid Mechanics, Computer Vision, and Materials Science. Cambridge, UK: Cambridge Univ Press.
24. Vlad D, Kierzkowski D, Rast M I, Vuolo F, Dello Ioio R, et al. (2014) Leaf Shape Evolution Through Duplication, Regulatory Diversification, and Loss of a Homeobox Gene. Science 343: 780–783.
25. Barkoulas M, Hay A, Kougioumoutiz E, Tsiantis M (2008) A developmental framework for dissected leaf formation the *Arabidopsis* relative. *Cardamine hirsute*. Nat Genet 40: 1136–1141.
26. Gleissberg S (2004) Comparative analysis of leaf shape development in *Eschscholzia californica* and other Papaveraceae-Eschscholzioideae. Am J of Bot 91: 306–312.

12

Linking Cascading Effects of Fish Predation and Zooplankton Grazing to Reduced Cyanobacterial Biomass and Toxin Levels Following Biomanipulation

Mattias K. Ekvall[1]*, Pablo Urrutia-Cordero[1,2], Lars-Anders Hansson[1]

1 Lund University, Department of Biology, Ecology Building, SE-223 62, Lund, Sweden, **2** Lund University, Center for Environmental and Climate Research, Ecology Building, SE-223 62, Lund, Sweden

Abstract

Eutrophication has been one of the largest environmental problems in aquatic ecosystems during the past decades, leading to dense, and often toxic, cyanobacterial blooms. In a way to counteract these problems many lakes have been subject to restoration through biomanipulation. Here we combine 13 years of monitoring data with experimental assessment of grazing efficiency of a naturally occurring zooplankton community and a, from a human perspective, desired community of large *Daphnia* to assess the effects of an altered trophic cascade associated with biomanipulation. Lake monitoring data show that the relative proportion of *Daphnia* spp. grazers in June has increased following years of biomanipulation and that this increase coincides with a drop in cyanobacterial biomass and lowered microcystin concentrations compared to before the biomanipulation. In June, the proportion of *Daphnia* spp. (on a biomass basis) went from around 3% in 2005 (the first year of biomanipulation) up to around 58% in 2012. During months when the proportion of *Daphnia* spp. remained unchanged (July and August) no effect on lower trophic levels was observed. Our field grazing experiment revealed that *Daphnia* were more efficient in controlling the standing biomass of cyanobacteria, as grazing by the natural zooplankton community never even compensated for the algal growth during the experiment and sometimes even promoted cyanobacterial growth. Furthermore, although the total cyanobacterial toxin levels remained unaffected by both grazer communities in the experimental study, the *Daphnia* dominated community promoted the transfer of toxins to the extracellular, dissolved phase, likely through feeding on cyanobacteria. Our results show that biomanipulation by fish removal is a useful tool for lake management, leading to a top-down mediated trophic cascade, through alterations in the grazer community, to reduced cyanobacterial biomass and lowered cyanobacterial toxin levels. This improved water quality enhances both the ecological and societal value of lakes as units for ecosystem services.

Editor: Hans G. Dam, University of Connecticut, United States of America

Funding: This work was funded by the European Union Interreg IV A project "Algae be gone!" and by the ERA-Net BiodivERsA, with the national funders FORMAS, part of the 2012 BiodivERsA call for research proposals (LIMNOTIP). The funders had no role in study design, data collection and analysis, decision to publish, or preparation of the manuscript.

Competing Interests: The authors have declared that no competing interests exist.

* Email: mattias.ekvall@biol.lu.se

Introduction

Human impact on aquatic ecosystems, in particular in the form of eutrophication, has been one of the largest environmental problems for aquatic ecosystems during the past decades [1,2]. The subsidy of nutrients, and specifically phosphorus [3,4], to inland waters from the surrounding landscapes lead to increased primary production and facilitates the formation of cyanobacterial blooms during warm summer months [5]. These cyanobacterial blooms, often dominated by different cyanobacterial genera such as *Microcystis spp. and Anabaena spp.*, often have the potential to produce a vast array of toxins, such as microcystins and anatoxin-a [6–9], posing a serious threat to both humans [10,11] and aquatic organisms [12]. Furthermore, cyanobacterial blooms can have considerable impact on both biodiversity and ecosystem functioning, as well as ecosystem services, such as recreation and drinking water supply, making the water resource less desirable [13].

As a counter-action to those alterations, researchers and stakeholders have invested considerable efforts in finding ways to restore reservoirs and lakes to their former glory. Several restoration measures, both in the catchments and in the lakes themselves, have been proposed. Measures in the catchment mainly focus on reducing the external nutrient loading to the lakes, for example by constructing buffer zones around streams in agricultural areas, reducing point sources, and constructing wetlands to catch the nutrients before reaching the lake [14]. In-lake measures include for example chemical treatment of the sediment to reduce internal nutrient loading [15] and different kinds of biomanipulations [16]. Biomanipulations, a term first coined by Shapiro et al. [17], can be of many different types including cyprinid fish removal, as well as zooplankton and piscivorous fish stocking or a combination of both [18–22]. Regardless of biomanipulation type, they all have the common goal of altering the food web and ultimately increase the grazing pressure on phytoplankton, thereby reducing the occurrence of algal blooms through top-down effects [21,23]. Due to high fish predation, the zooplankton communities in eutrophic lakes are generally dominated by small sized species which are less effective in feeding on large phytoplankton compared to large cladoceran zooplankton species, such as *Daphnia magna* [24]. This means that although there might be relatively high numbers of zooplankton, the existing species composition is rarely favoring grazing on the targeted large cyanobacteria, thereby facilitating bloom formation.

In this study we address the question whether alterations in the food chain composition, such as a biomanipulation, can improve the water quality with respect to cyanobacterial biomass and toxin production. To address this we used 13 years of monitoring data from the eutrophic Lake Ringsjön, southern Sweden, of which the last eight years have been subject to cyprinid fish removal (biomanipulation) to track the trophic cascade through the food web. In addition, we performed a field grazing experiment to simulate how an increased abundance of large cladoceran grazers, *Daphnia magna*, would affect grazing efficiency on the phytoplankton community compared to the naturally occurring zooplankton community dominated by small taxa. We hypothesized that an increased amount of large cladoceran grazers would increase the grazing pressure on cyanobacteria relative to the existing zooplankton community and thereby alter the amount of cyanobacterial toxins in the lake.

Materials and Methods

Site description

Lake Ringsjön is a eutrophic lake situated in the southern part of Sweden. The lake consists of three connected basins with a total area of 40 km^2 and has been subject to biomanipulation by cyprinid fish removal, mainly roach (*Rutilus rutilus*) and bream (*Abramis brama*) [25]. The first biomanipulation attempt in Lake Ringsjön was between 1989–1992 [16] and the effects of this biomanipulation lasted until the mid 1990s, when the lake showed signs to return to a turbid state. As a response to the increased turbidity, a new attempt of biomanipulation was initiated in 2005 and was still ongoing at the time of the field study in 2012. Here we focus on this latest effort of biomanipulation in the Western basin of Lake Ringsjön (area: 14.8 km^2, max depth 5.4 m) using available monitoring data from 2000–2004 as a reference for evaluating the changes in the lake during the eight years with biomanipulation (2005–2012).

Lake monitoring data

Lake Ringsjön has been subject to monthly monitoring of water chemistry (total phosphorous), total chlorophyll *a* concentration and phytoplankton- and zooplankton biomass for more than 10 years. Moreover, since 2004 samples for cyanobacterial toxins, specifically microcystins, have been taken on a monthly to bi-weekly basis. The microcystin samples were taken in surface water above the deep-hole of the western basin and were immediately frozen and later analyzed using enzyme-linked immunoassay (Microcystins-DM ELISA Microtiter Plate, Abraxis LLC, Warminster, PA, USA) according to Hansson et al. (2007) [12]. In addition, as a proxy for cyprinid fish abundance, the landings of cyprinid fish from the trawling during the biomanipulation were used to assess the trends in the targeted fish stock. The same type and number of trawlers were used during all years of biomanipulation. As the number of days trawled differed slightly between years due to e.g. weather conditions, catches were normalized to trawling effort and, besides total catch, catch efficiency (i.e. catch per trawling day) were estimated for each year. Predatory fish (mainly pike (*Esox lucius*), pikeperch (*Sander lucioperca*) and large perch (*Perca fluviatilis*)) caught in the trawl were immediately sorted out on the trawling boat and returned to the lake and is hence not included in the landings.

Field grazing-experiment

A field experiment, based on the method described by Lehman and Sandgren (1985) [26], consisting of 12 plastic cubitainers (volume: 10 L) was run once a month between June and August 2012. Cubitainers were divided into two gradients (n = 6) to investigate two different outcomes of lake biomanipulations. The first gradient consisted of a biomass gradient of the natural zooplankton community that exists in the lake today, while the second gradient constituted a gradient representing a scenario where a large cladoceran grazer, here represented by *Daphnia magna*, becomes the dominant herbivore in the lake. Hence, this gradient was designed to represent a future scenario where the fish predation had been reduced to very low levels. The *Daphnia* used were taken from a lab-reared population. These *Daphnia* had been fed with a mixture of phytoplankton during several generations consisting mainly of green algae but also containing low levels of cryptophytes and cyanobacteria.

Each cubitainer received 9 L of 150-μm filtered lake water i.e. water containing the natural phytoplankton community without grazers larger than 150 μm. The zooplankton retrieved on the 150-μm filter were pooled and subsequently added to the cubitainers constituting the natural-grazer gradient in 0.25, 0.5, 1, 2, 4 and 6 times the biomass of zooplankton in the lake at the time of sampling (for absolute biomasses see result section). This grazer community was dominated by copepods and small cladocerans (Chydorids, *Bosmina* spp. and *Ceriodaphnia* spp.), the latter defined as smaller than 400 μm in length. The second gradient received a gradient of cultured *Daphnia magna* (mean ±SD size: 1750.9±376.8 μm) following the same gradient steps as for the natural grazers with 8±4 *Daphnia magna* per liter (mean±SD during the different months) in the "ambient step" (1) depending on the size of the individuals in the culture, thereby successfully generating two different grazer community compositions (Fig. 1). Although these gradients differed in their absolute biomass, with higher total biomasses in the *Daphnia*-dominated gradients due to the large size of *Daphnia magna* and the small size of the individuals in the natural community, the natural grazers always outnumbered the *Daphnia*-gradient. Once filled, 0-samples were taken for phytoplankton enumeration, total microcystin concentrations and extracellular microcystin concentrations. Thereafter the cubitainers were hooked onto a rope and incubated in the lake for 72 hours in the surface water. The cubitainers were then taken out of the lake and samples were taken for phytoplankton and microcystin concentrations. Also, the entire volume in the cubitainer was filtered through a 150-μm mesh and the zooplankton retrieved on the filter were kept as a measure for grazer community composition and biomass determination. Samples for extracellular microcystins were collected by, prior to freezing, removing all cyanobacterial cells from the water by a very low-pressure filtration using GF/C-filters (Whatman). All microcystin samples were stored at −20°C until analyzed in the same way as the lake monitoring samples with the exception that extracellular samples, which contained no cells, were not sonicated. All plankton samples were fixed with Lugols solution and stored in a cooling room at 4°C for later enumeration and biomass determination. Zooplankton samples were counted and measured using a stereoscopic microscope (Olympus SZ40) at 20x magnification and biomasses were estimated using length-weight regressions according to Bottrell et al. [27] and Dumont et al. [28]. Cyanobacteria were counted using an inverted microscope (Olympus CK40) and biomasses were estimated according to the methods described by Ekvall et al. (2013) [6], whereafter net cyanobacterial growth rates (r) were calculated using the equation: $r = \ln(N_t/N_0)/\Delta t$, where N_t is the final cyanobacterial biomass, N_0 the initial cyanobacterial biomass and Δt the running time of the experiment. Calculated growth rates were then correlated to grazer abundance and grazing rates were estimated based on Lehman and Sandgren (1985) [26]. No specific permits were needed to conduct this experiment.

Data analysis and statistics

All lake data were split into either before biomanipulation (2000–2004) or during biomanipulation (2006–2012) and were analyzed on a monthly basis using linear regressions with year as independent variable and the respective response variable as dependent variable. The first year of biomanipulation (2005) were considered a transition year and were not included in any of the two groups. As only one data point on microcystins exists prior to the biomanipulation (2004) changes in microcystin concentration were only analyzed for years with biomanipulation (i.e. changes over time with years of ongoing biomanipulation). The results of the grazer-gradient experiment were also analyzed via linear regressions using net growth rate (r) as dependent- and zooplankton biomass as independent variable. The same approach was used for both the total microcystin concentration, as well as the extracellular concentration although the r in this case represented changes in toxin concentration rather than growth rate. Statistical analyses were made in SPSS 21 for Macintosh (grazing experiment) and Prism 6 for Macintosh (lake monitoring data).

Results

Lake monitoring

In 2005, the first year of biomanipulation, the total catch of cyprinid fish reached around 105 tons (Fig. 2). This was also the year with the highest catch efficiency. Although the total catch has varied among years, the catch efficiency steadily decreased during the period from 2005 to 2008, but thereafter it stabilized at about 800 kg day^{-1} (Fig. 2). This large drop in catchable cyprinid fish suggests that the fish stock, and with that also the predation pressure on zooplankton, decreased with time of biomanipulation. Comparing the effects of the biomanipulation on a monthly basis revealed no significant changes in total zooplankton biomass during any month (Table 1, Fig. 3: a–c). There were no significant trends in the proportion of *Daphnia* spp. before the start of the biomanipulation (Table 1, Fig. 3: d–f). However, in spring (June), the relative contribution of large cladoceran grazers, *Daphnia* spp., increased with time following the biomanipulation ($r^2 = 0.58$; $P = 0.048$; Fig. 3: d), while it remained unaffected in July and August (Table 1, Fig. 3: e–f).

Total cyanobacterial biomass showed a significant decrease in spring (June) with time of biomanipulation (Table 1, Fig. 3: g). Due to this, the cyanobacterial biomass now fluctuates on a considerably lower level in June compared to the period prior to the biomanipulation. These effects were not seen in any of the other summer months investigated (Table 1, Fig. 3: h–i). Although not significant, the cyanobacterial toxins (microcystins) showed similar patterns as the cyanobacterial biomass with much lower concentrations during the biomanipulation compared to before the biomanipulation (Table 1, Fig. 3: j). In July and August, the microcystin concentration still fluctuated around the same level as measured prior to the biomanipulation (2004), while it dropped considerably during June compared to the measured levels in 2004

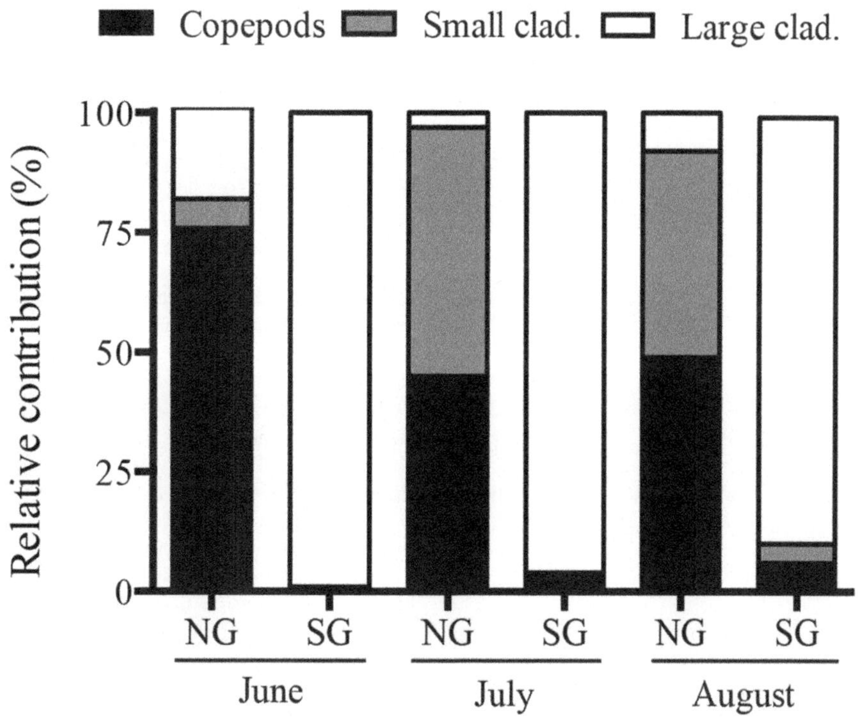

Figure 1. Relative community composition (%), on a biomass basis, of copepods, small cladocerans and large cladocerans in the different grazer communities in June, July and August, respectively. NG = naturally occurring grazer community, SG = standardized (*Daphnia* dominated) grazer community.

Table 1. Results from the linear regressions on total zooplankton biomass, % *Daphnia* spp. (biomass), cyanobacterial biomass, microcystin concentration and total phosphorous concentration in the western basin of Lake Ringsjön before (2000–2004) and during biomanipulation (2006–2012).

	June						July						August					
	Before			During			Before			During			Before			During		
Total phosphorous (unfilt.)	$F_{(1,3)}$	R^2	P	$F_{(1,5)}$	R^2	P	$F_{(1,3)}$	R^2	P	$F_{(1,5)}$	R^2	P	$F_{(1,3)}$	R^2	P	$F_{(1,5)}$	R^2	P
	3.122	0.510	0.175	0.995	0.166	0.364	1.357	0.311	0.328	0.990	0.165	0.365	0.215	0.067	0.675	0.056	0.011	0.822
Σ Zooplankton biomass	$F_{(1,2)}$	R^2	P	$F_{(1,5)}$	R^2	P	$F_{(1,2)}$	R^2	P	$F_{(1,5)}$	R^2	P	$F_{(1,1)}$	R^2	P	$F_{(1,5)}$	R^2	P
	0.006	0.003	0.945	0.029	0.006	0.871	0.017	0.008	0.909	1.158	0.188	0.331	0.153	0.133	0.763	0.124	0.024	0.739
% *Daphnia* spp.	$F_{(1,2)}$	R^2	P	$F_{(1,5)}$	R^2	P	$F_{(1,2)}$	R^2	P	$F_{(1,5)}$	R^2	P	$F_{(1,1)}$	R^2	P	$F_{(1,5)}$	R^2	P
	2.943	0.595	0.228	6.798	0.576	**0.048**	0.080	0.038	0.804	2.072	0.293	0.210	0.015	0.015	0.922	0.281	0.053	0.619
Σ Cyanobacterial biomass	$F_{(1,3)}$	R^2	P	$F_{(1,5)}$	R^2	P	$F_{(1,3)}$	R^2	P	$F_{(1,5)}$	R^2	P	$F_{(1,3)}$	R^2	P	$F_{(1,5)}$	R^2	P
	0.156	0.049	0.719	13.94	0.736	**0.014**	0.506	0.144	0.528	0.461	0.084	0.527	0.104	0.033	0.768	1.235	0.198	0.317
Microcystin concentration	$F_{(d.f.)}$	R^2	P	$F_{(1,5)}$	R^2	P	$F_{(d.f.)}$	R^2	P	$F_{(1,4)}$	R^2	P	$F_{(d.f.)}$	R^2	P	$F_{(1,5)}$	R^2	P
	n/a	n/a	n/a	3.262	0.395	0.131	n/a	n/a	n/a	0.183	0.044	0.691	n/a	n/a	n/a	0.268	0.051	0.627

As only one year of microcystin samples exist prior to the biomanipulation (2004) no analysis could be made on trends in toxins before the biomanipulation (n/a). Significant results (P<0.05) are displayed in bold.

(Table 1, Fig. 3: j–l). Total phosphorous concentrations in the lake showed no trends, neither before nor during the biomanipulation, with similar concentrations between years and months (Table 1, Fig. 3: m–o).

Field grazing experiment

Both natural and standardized (large *Daphnia*) zooplankton were, to some extent, able to graze on the cyanobacterial community. The community dominated by *Daphnia magna* was able to strongly affect the cyanobacterial community in both June and August, while the natural community was only able to exert significant grazing in August (Table 2, Fig. 4). However, in the natural community, not even the highest zooplankton density managed to induce a relative algal growth rate (r) below zero, i.e. their grazing rate was lower than the algal growth rate (Fig. 4). Furthermore, in June, the natural grazer community even had a positive effect on cyanobacterial growth, leading to a net increase in cyanobacterial biomass (Fig. 4). None of the two communities had any significant effect on the cyanobacterial community in July (Table 2).

Total microcystin concentrations (including both intra- and extracellular microcystins, i.e. inside cells and dissolved in the water) were not affected by zooplankton biomass at any of the study occasions (Table 2). However, during June and August, when the *Daphnia* dominated community was able to exert significant grazing on the cyanobacterial community, there was a significant increase in the extracellular microcystin concentration (Table 2, Fig. 5), indicating that the toxins were transferred from the cyanobacteria to the water. No changes in extracellular concentration were seen in any other case (Table 2, Fig. 5).

Discussion

Here we show that cascading effects of a whole lake biomanipulation can be tracked from fish removal all the way down to reduced cyanobacterial toxin levels during years of biomanipulation compared to the time before the biomanipulation. However, the effects of biomanipulation in Lake Ringsjön were only visible in spring (June), and although no differences in total zooplankton biomass were found, the change in cyanobacterial biomass and the trends in microcystin concentration coincided with an upswing in the proportion of large cladoceran grazers, *Daphnia* spp.; a group very sensitive to predation by fish [29]. Furthermore, results from the lake monitoring data are strengthened by the field grazing experiment, showing that *Daphnia* are more efficient in feeding on the cyanobacteria than the zooplankton community now present in the lake. Moreover, despite there were no effects on the total microcystin level, the *Daphnia* dominated gradient made a larger pool of the toxins dissolved in the water and thereby available for bacterial degradation [30]. This may be one of the reasons for the observed decreasing trend in lake toxin levels following an upswing in the proportion of *Daphnia* spp. in June. Total phosphorous concentrations did not change significantly with time and can hence not explain the observed gradual decrease in cyanobacterial biomass. However, the levels of total phosphorous were on a lower and more constant level during the years of biomanipulation. One factor that potentially could explain this pattern could be reduced re-suspension of sediment by bentivorous fish (such as bream) [25].

The effects on lower trophic levels following cyprinid fish removal, but the lack of top-down effects on overall zooplankton biomass, are consistent with the findings from previous biomanipulation attempts in Lake Ringsjön [31]. However, here we argue that the observed effects on lower trophic levels in June are likely a

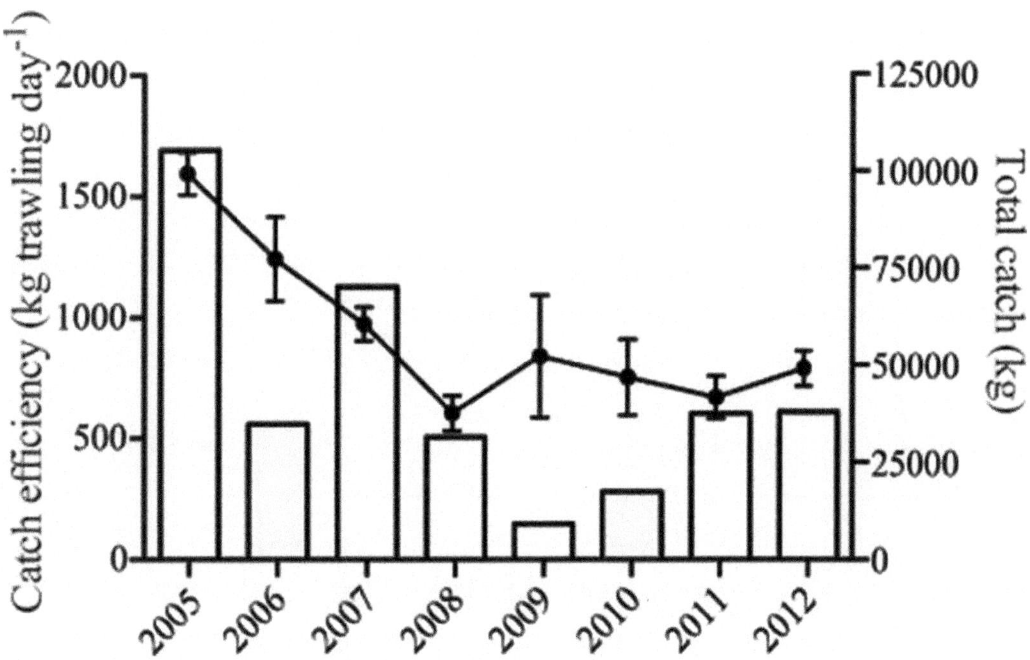

Figure 2. Total reported landings from the trawl per year (bars) and average catch efficiency (kg trawling day^{-1}) ± SE (line).

result of top-down control through a change in the relative zooplankton composition, with a higher proportion of large cladoceran grazers. That none of the other months investigated showed any trends in *Daphnia* spp. proportion, neither before nor during the biomanipulation, could be due to the increased predation pressure in the lake at that time of the year resulting from the hatching and rapid growth of 0+ fish [32]. Recruitment of 0+ fish, and thereby increased predation pressure on zooplankton, is generally found to increase even further following biomanipulation [33], something that was also seen in Lake Ringsjön during the first biomanipulation attempt during the 1990s [34]. Another factor contributing to the collapse of the *Daphnia* spp. in July could be changes in phytoplankton species composition due to natural succession in lakes [35]. Succession towards larger sized cyanobacterial species as the growth season progresses could potentially lead to clogging of the feeding apparatus of the daphnids [24,36]. The combination of these factors would render large zooplankton sandwiched between increased predation pressure and poor food quality, possibly contributing to the low biomass of large cladoceran grazers observed in July and August and, with that, the lack of a trophic cascade following biomanipulation during these months. In addition, although *Daphnia* previously adapted to toxic cyanobacteria have been shown to be able to suppress phytoplankton biomass at high microcystin concentration [37], the higher microcystin levels in July and August, compared to in June, may also have contributed to the reduced performance of the herbivores leading to their population decline [38,39].

The field grazing experiment clearly showed that large *Daphnia* were more efficient in grazing on cyanobacteria than the natural grazer community. While the *Daphnia* strongly affected the cyanobacterial biomass negatively both in June and August, the natural community was only able to affect them negatively in August. In fact, *Daphnia* stocking has previously successfully been used as a restoration tool in combination with piscivorous fish stocking in biomanipulation attempts to control algal blooms [18]. Furthermore, although the natural community was able to exert significant grazing on the cyanobacterial community in August, they were not even at the highest abundances able to reduce the growth rate of the cyanobacteria below zero, i.e. they were not even able to compensate for the cyanobacterial growth during the 72 hour experiment. The fact the natural community were not able to exert significant grazing in June, when it contained a larger proportion of large cladoceran grazers, can be explained by that the biomass of zooplankton found in the lake was considerably lower in June than in August, thereby generating a lower grazing pressure on the cyanobacteria. *Daphnia*, however, generally managed to push the net growth rate of the cyanobacteria below zero i.e. their grazing did not only compensate for cyanobacterial growth during the experiment, but also affected the standing biomass. In June, an increase in biomass of the natural zooplankton community boosted a significant growth of cyanobacteria. A likely explanation is that the natural community then mainly consisted of copepods and small cladocerans, such as *Bosmina* and *Chydorus*, and that these small cladoceran species were not able to graze on the relatively large cyanobacteria. This is in line with the findings by Dawidowicz (1990) [24] who showed that the existing lake community in the moderately eutrophic Lake Ros (Northern Poland) was only able to graze on the smaller sized phytoplankton species. In July, none of the grazer gradients showed significant grazing on the cyanobacterial community. At this point the cyanobacterial community was dominated by large filaments of *Anabaena crassa* and although *Daphnia magna* has been shown to be able to feed on large sized phytoplankton [40], the filamentous character of the *Anabaena crassa* might have led to clogging of the feeding apparatus [41–43].

The cyanotoxin (microcystin) levels remained unaffected by grazers throughout the grazing experiment. However, the *Daphnia* dominated gradient induced an increased concentration of microcystins dissolved in the water, i.e. the extracellular fractions of microcystins. This effect was never observed for the natural grazer community, not even when their grazing rate

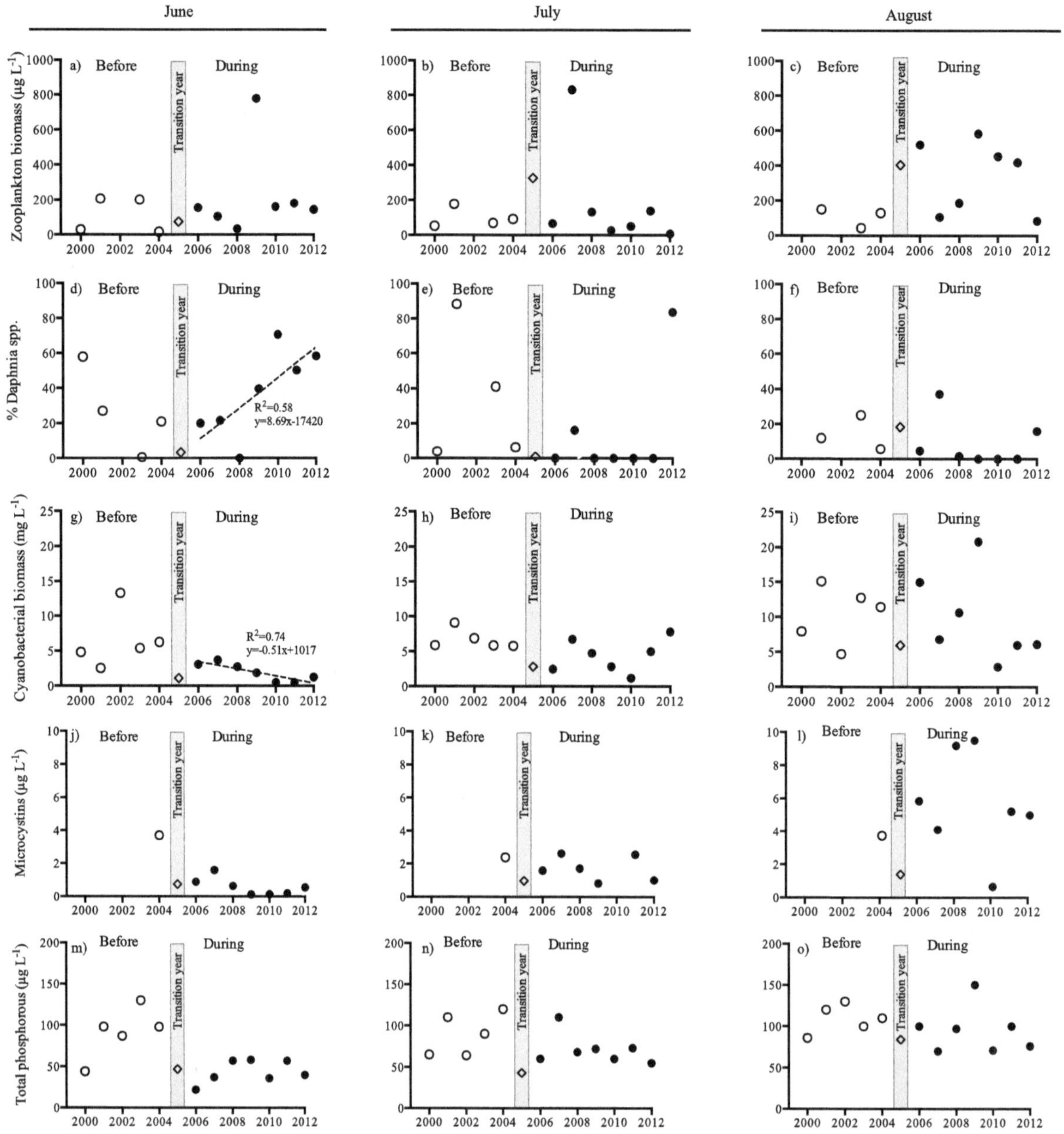

Figure 3. Lake data from Lake Ringsjön (mean + SD) showing total zooplankton biomass (a–c), the proportion of *Daphnia* spp. (%) of total zooplankton biomass (d–f), cyanobacterial biomass (g–i), microcystin concentration (j–l) and total phosphorous concentration (m–o) during June–August before (2000–2004) and during biomanipulation (2006–20012). The first year of biomanipulation (2005) were considered a transition year and were not included in the analysis (gray bar). Regression lines mark significant trends ($P<0.05$).

significantly reduced the cyanobacterial growth in August. *Daphnia*, being a less selective grazer [44], likely ingest more of the toxic cyanobacteria than the natural community, which might feed more selectively, or be too small to feed on large cyanobacteria. If so, the excretion of damaged cells by *Daphnia* is the likely reason for the higher levels of extracellular microcystins in the water. Furthermore, Jang et al. (2007) [45] have suggested that cyanobacteria actively release microcystins into the water as a response to zooplankton. Irrespective of the mechanism behind the increased extracellular toxin level, the toxins are made available for bacterial- and photochemical degradation by large *Daphnia* grazers, but not by the smaller herbivores [30,46]. Although higher dissolved concentrations of cyanotoxins will have an immediate negative effect for ecosystem services, such as drinking water and recreation, it will expose the toxins to degradation and thereby lead to improved water quality in the long run. Furthermore, given that microcystins are not broken down internally in the cyanobacterial cells [47], the breakdown processes of extracellular toxins would likely be the major route for reducing the concentration of microcystins in the water. Hence, a higher portion of large herbivores may not only reduce the abundance of nuisance cyanobacteria, but also reduce the toxin levels, thereby fulfilling one of the main goals with biomanipulation, a notion strengthened by our lake monitoring data. Although our dataset on microcystins only includes one year prior to the biomanipulation it is clear that the levels of

Table 2. Results from regressions on cyanobacterial growth rate (r) and relative change in microcystins concentration (r) in relation to zooplankton biomass for both the a) natural zooplankton community and b) the *Daphnia* dominated community.

a)	**Natural community**								
	June			**July**			**August**		
	R^2	$t_{(d.f.=4)}$	p	R^2	$t_{(d.f.=4)}$	p	R^2	$t_{(d.f.=4)}$	p
Cyanobacteria	0.881	5.453	**0.005**	0.092	0.638	0.558	0.959	−6.734	**0.003**
Total microcystins	0.002	0.094	0.930	0.145	−0.824	0.465	0.563	2.270	0.086
Extracellular microcystins	0.125	0.756	0.492	0.189	0.965	0.389	0.001	0.064	0.952
b)	***Daphnia*-dominated community**								
	June			**July**			**August**		
	R^2	$t_{(d.f.=4)}$	p	R^2	$t_{(d.f.=4)}$	p	R^2	$t_{(d.f.=4)}$	p
Cyanobacteria	0.822	−2.890	**0.045**	0.102	−0.673	0.538	0.835	−4.502	**0.011**
Total microcystins	0.403	−1.644	0.176	0.426	−1.722	0.160	0.239	1.122	0.325
Extracellular microcystins	0.812	4.157	**0.014**	0.442	1.779	0.150	0.797	3.963	**0.017**

Significant results ($P<0.05$) are displayed in bold.

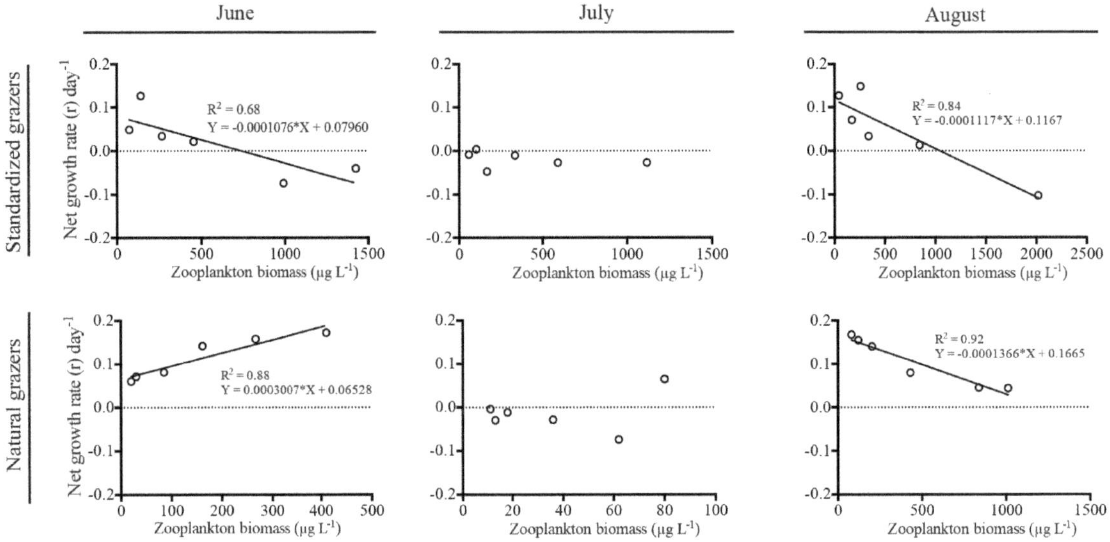

Figure 4. Net growth rates (r) of the cyanobacterial community during June, July and August in relation to grazer abundance. Top panel = *Daphnia* dominated community (standardized grazers) and Lower panel = natural zooplankton community. Regression lines show significant ($p<0.05$) relations.

microcystins in June during that year were much higher than during the years with biomanipulation - when we see a tendency of reduced toxin concentration with time of biomanipulation. During both July and August the levels of microcystins during years of biomanipulation is still at the same level as they were before the biomanipulation, following the patterns in cyanobacterial biomass. This strongly suggests that the biomanipulation, through top down control, has led to a reduction in cyanobacterial toxin levels during early summer. As shown in our field grazing experiment, *Daphnia* grazing did not lead to a drop in total toxin concentration in the water but rather a relative change in the fraction of extracellular vs. intracellular toxins. This could potentially explain the results from the lake as we, as in the field grazing experiment, see a reduction in cyanobacterial biomass but only tendencies to reduced toxin levels as the lake monitoring samples do not discriminate between the different fractions of microcystins.

In conclusion, both monitoring- and experimental data show that biomanipulations, through top-down mediated trophic cascades, can lead to improved water quality expressed both as reduced cyanobacterial biomasses and lowered toxin levels in spring. Following cyprinid fish removal and reduced predation pressure on zooplankton, *Daphnia* spp. became more dominant leading to more efficient top down control on phytoplankton. This led to a reduction in the pool of toxin producing cyanobacteria

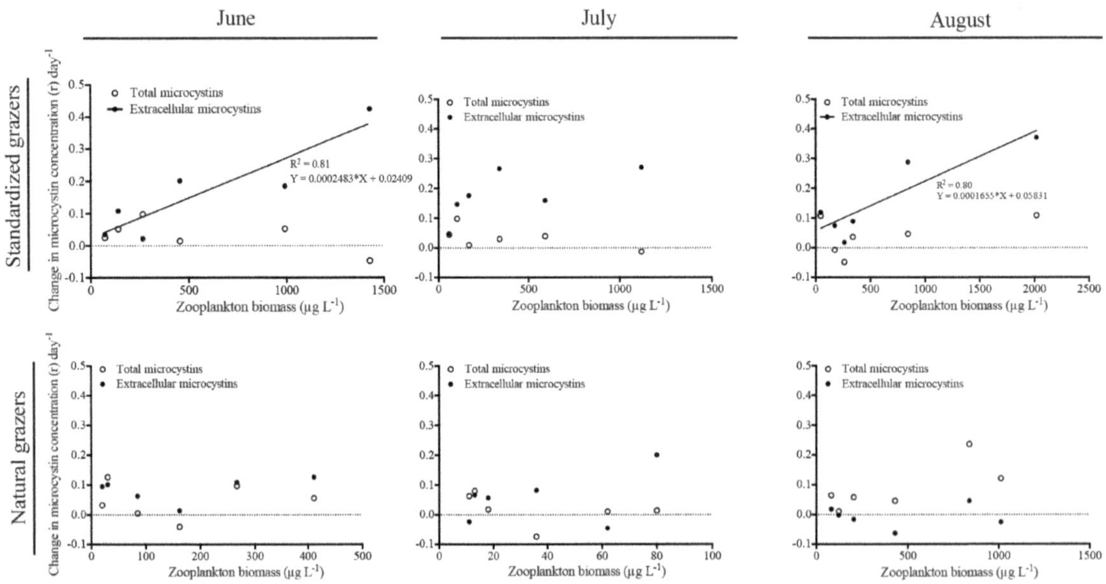

Figure 5. Net change in total microcystin (open circles) and extracellular microcystin concentration (closed circles) in June, July and August in relation to zooplankton biomass. Top panel = *Daphnia* dominated community (standardized grazers) and Lower panel = natural zooplankton community. Regression lines show significant ($p<0.05$) relations.

and microcystins found in June. Furthermore, the grazing experiment revealed that an increased abundance of *Daphnia* leads to higher levels of extracellular toxins making them more susceptible to degradation that would further improve water quality. None of these cascading effects, which ultimately lead to better water quality, were seen when the zooplankton community was dominated by smaller species, neither in the field nor in the grazing experiment. Hence, we here show the importance of fish removal as a tool for top-down mediated cascading effects leading all the way to reduced toxin levels. If the spring clear water phase can be prolonged, as our data suggest, with increased abundance of large and efficient grazers further into the summer season, this would reduce the amount of phytoplankton through bottom-up mediated effects. This would ultimately reduce the formation of nuisance and toxic blooms even further into the season which would improve not only the ecological status of the lake, but also make it more valuable as a resource for ecosystem services, such as recreation and drinking water supply.

Acknowledgments

We would like to thank Marie Svensson for numerous hours spent by the microscope counting zooplankton from Lake Ringsjön. Furthermore, we would like to thank the municipality of Höör for providing monitoring- and fish landing data. We would also like to thank Monika Winder for valuable comments on an earlier version of this manuscript.

Author Contributions

Conceived and designed the experiments: ME LAH. Performed the experiments: ME PUC LAH. Analyzed the data: ME PUC. Contributed reagents/materials/analysis tools: LAH ME PUC. Contributed to the writing of the manuscript: ME LAH PUC.

References

1. Smith VH (2003) Eutrophication of freshwater and coastal marine ecosystems a global problem. Environmental Science and Pollution Research 10: 126–139.
2. Smith VH, Joye SB, Howarth RW (2006) Eutrophication of freshwater and marine ecosystems. Limnology and Oceanography 51: 351–355.
3. Schindler DW (1977) Evolution of Phosphorus Limitation in Lakes. Science 195: 260–262.
4. Schindler DW (1974) Eutrophication and Recovery in Experimental Lakes - Implications for Lake Management. Science 184: 897–899.
5. Paerl HW, Huisman J (2008) Climate - Blooms like it hot. Science 320: 57–58.
6. Ekvall MK, de la Calle Martin J, Faassen EJ, Gustafsson S, Lürling M, et al. (2013) Synergistic and species-specific effects of climate change and water colour on cyanobacterial toxicity and bloom formation. Freshwater Biology 58: 2414–2422.
7. Faassen EJ, Harkema L, Begeman L, Lürling M (2012) First report of (homo)anatoxin-a and dog neurotoxicosis after ingestion of benthic cyanobacteria in The Netherlands. Toxicon 60: 378–384.
8. Codd GA (1995) Cyanobacterial toxins: Occurrence, properties and biological significance. Water Science and Technology 32: 149–156.
9. Codd GA, Bell SG, Brooks WP (1989) Cyanobacterial Toxins in Water. Water Science and Technology 21: 1–13.
10. Falconer IR (2005) Is there a human health hazard from microcystins in the drinking water supply? Acta Hydrochimica Et Hydrobiologica 33: 64–71.
11. Carmichael WW (2001) Health effects of toxin-producing cyanobacteria: "The CyanoHABs". Human and Ecological Risk Assessment 7: 1393–1407.
12. Hansson LA, Gustafsson S, Rengefors K, Bomark L (2007) Cyanobacterial chemical warfare affects zooplankton community composition. Freshwater Biology 52: 1290–1301.
13. Brookes JD, Carey CC (2011) Resilience to Blooms. Science 333: 46–47.
14. Uusi-Kämppä J, Braskerud B, Jansson H, Syversen N, Uusitalo R (2000) Buffer zones and constructed wetlands as filters for agricultural phosphorus. Journal of Environmental Quality 29: 151–158.
15. Reitzel K, Lotter S, Dubke M, Egemose S, Jensen HS, et al. (2013) Effects of Phoslock treatment and chironomids on the exchange of nutrients between sediment and water. Hydrobiologia 703: 189–202.
16. Hansson LA, Annadotter H, Bergman E, Hamrin SF, Jeppesen E, et al. (1998) Biomanipulation as an application of food-chain theory: Constraints, synthesis, and recommendations for temperate lakes. Ecosystems 1: 558–574.
17. Shapiro J, Lamarra V, Lynch M (1975) Biomanipulation: an ecosystem approach to lake restoration. In: Brezonik PL, Fox JL, editors. Proceedings of a symposium on water quality management through biological control, January 23–30, 1975: Gainesville: University of Florida. pp. 85–96.
18. Ha J-Y, Saneyoshi M, Park H-D, Toda H, Kitano S, et al. (2013) Lake restoration by biomanipulation using piscivore and *Daphnia* stocking; results of the biomanipulation in Japan. Limnology 14: 19–30.
19. Theiss J, Zielinski K, Lang H (1990) Biomanipulation by Introduction of Herbivorous Zooplankton - a Helpful Shock for Eutrophic Lakes. Hydrobiologia 200: 59–68.
20. Søndergaard M, Jeppesen E, Mortensen E, Dall E, Kristensen P, et al. (1990) Phytoplankton Biomass Reduction after Planktivorous Fish Reduction in a Shallow, Eutrophic Lake - a Combined Effect of Reduced Internal P-Loading and Increased Zooplankton Grazing. Hydrobiologia 200: 229–240.
21. Shapiro J, Wright DI (1984) Lake Restoration by Biomanipulation - Round Lake, Minnesota, the 1st 2 Years. Freshwater Biology 14: 371–383.
22. Benndorf J, Kneschke H, Kossatz K, Penz E (1984) Manipulation of the Pelagic Food Web by Stocking with Predacious Fishes. Internationale Revue Der Gesamten Hydrobiologie 69: 407–428.
23. Carpenter SR, Kitchell JF, Hodgson JR (1985) Cascading Trophic Interactions and Lake Productivity. Bioscience 35: 634–639.
24. Dawidowicz P (1990) Effectiveness of Phytoplankton Control by Large-Bodied and Small-Bodied Zooplankton. Hydrobiologia 200: 43–47.
25. Bergman E, Hansson LA, Persson A, Strand J, Romare P, et al. (1999) Synthesis of theoretical and empirical experiences from nutrient and cyprinid reductions in Lake Ringsjön. Hydrobiologia 404: 145–156.
26. Lehman JT, Sandgren CD (1985) Species-Specific Rates of Growth and Grazing Loss among Fresh-Water Algae. Limnology and Oceanography 30: 34–46.
27. Bottrell HH, Duncan A, Gliwicz ZM, Grygierek E, Herzig A, et al. (1976) Review of Some Problems in Zooplankton Production Studies. Norwegian Journal of Zoology 24: 419–456.
28. Dumont HJ, Vandevelde I, Dumont S (1975) Dry Weight Estimate of Biomass in a Selection of Cladocera, Copepoda and Rotifera from Plankton, Periphyton and Benthos of Continental Waters. Oecologia 19: 75–97.
29. Brooks JL, Dodson SI (1965) Predation Body Size and Composition of Plankton. Science 150: 28–35.
30. Christoffersen K, Lyck S, Winding A (2002) Microbial activity and bacterial community structure during degradation of microcystins. Aquatic Microbial Ecology 27: 125–136.
31. Bergman E, Bergstrand E (1999) Lack of a top-down effect on the zooplankton community after a cyprinid reduction. Hydrobiologia 404: 77–87.
32. Hansson LA, Nicolle A, Brodersen J, Romare P, Nilsson PA, et al. (2007) Consequences of fish predation, migration, and juvenile ontogeny on zooplankton spring dynamics. Limnology and Oceanography 52: 696–706.
33. Meijer ML, Jeppesen E, Vandonk E, Moss B, Scheffer M, et al. (1994) Long-Term Responses to Fish-Stock Reduction in Small Shallow Lakes - Interpretation of 5-Year Results of 4 Biomanipulation Cases in the Netherlands and Denmark. Hydrobiologia 275: 457–466.
34. Romare P, Bergman E (1999) Juvenile fish expansion following biomanipulation and its effect on zooplankton. Hydrobiologia 404: 89–97.
35. Sommer U, Gliwicz ZM, Lampert W, Duncan A (1986) The Peg-Model of Seasonal Succession of Planktonic Events in Fresh Waters. Archiv Fur Hydrobiologie 106: 433–471.
36. de Bernardi R, Giussani G (1990) Are Blue-Green-Algae a Suitable Food for Zooplankton - an Overview. Hydrobiologia 200: 29–41.
37. Chislock MF, Sarnelle O, Jernigan LM, Wilson AE (2013) Do high concentrations of microcystin prevent Daphnia control of phytoplankton? Water Research 47: 1961–1970.
38. Sarnelle O, Gustafsson S, Hansson LA (2010) Effects of cyanobacteria on fitness components of the herbivore *Daphnia*. Journal of Plankton Research 32: 471–477.
39. Wilson AE, Hay ME (2007) A direct test of cyanobacterial chemical defense: Variable effects of microcystin-treated food on two *Daphnia pulicaria* clones. Limnology and Oceanography 52: 1467–1479.
40. Lebret K, Fernandez MF, Hagman CHC, Rengefors K, Hansson LA (2012) Grazing resistance allows bloom formation and may explain invasion success of *Gonyostomum semen*. Limnology and Oceanography 57: 727–734.
41. Gliwicz ZM (1990) Why Do Cladocerans Fail to Control Algal Blooms. Hydrobiologia 200: 83–97.
42. DeMott WR, Gulati RD, Van Donk E (2001) *Daphnia* food limitation in three hypereutrophic Dutch lakes: Evidence for exclusion of large-bodied species by

interfering filaments of cyanobacteria. Limnology and Oceanography 46: 2054–2060.
43. Webster KE, Peters RH (1978) Some Size-Dependent Inhibitions of Larger Cladoceran Filterers in Filamentous Suspensions. Limnology and Oceanography 23: 1238–1245.
44. Demott WR (1986) The Role of Taste in Food Selection by Fresh-Water Zooplankton. Oecologia 69: 334–340.
45. Jang MH, Jung JM, Takamura N (2007) Changes in microcystin production in cyanobacteria exposed to zooplankton at different population densities and infochemical concentrations. Limnology and Oceanography 52: 1454–1466.
46. Gagala I, Mankiewicz-Boczek J (2012) The Natural Degradation of Microcystins (Cyanobacterial Hepatotoxins) in Fresh Water - the Future of Modern Treatment Systems and Water Quality Improvement. Polish Journal of Environmental Studies 21: 1125–1139.
47. Rohrlack T, Hyenstrand P (2007) Fate of intracellular microcystins in the cyanobacterium *Microcystis aeruginosa* (Chroococcales, Cyanophyceae). Phycologia 46: 277–283.

13

Adaptive Radiation along a Thermal Gradient: Preliminary Results of Habitat Use and Respiration Rate Divergence among Whitefish Morphs

Kimmo Kalevi Kahilainen[1,2]*, William Paul Patterson[3], Eloni Sonninen[4], Chris Harrod[5,6], Mikko Kiljunen[7]

1 Kilpisjärvi Biological Station, University of Helsinki, Kilpisjärvi, Finland, **2** Department of Environmental Sciences, University of Helsinki, Helsinki, Finland, **3** Department of Geological Sciences, Saskatchewan, Isotope Laboratory, University of Saskatchewan, Saskatoon, Saskatchewan, Canada, **4** Finnish Museum of Natural History, University of Helsinki, Helsinki, Finland, **5** School of Biological and Chemical Sciences, Queen Mary University of London, London, United Kingdom, **6** Instituto de Ciencias Naturales Alexander Von Humboldt, Universidad de Antofagasta, Antofagasta, Chile, **7** Department of Biological and Environmental Sciences, University of Jyväskylä, Jyväskylä, Finland

Abstract

Adaptive radiation is considered an important mechanism for the development of new species, but very little is known about the role of thermal adaptation during this process. Such adaptation should be especially important in poikilothermic animals that are often subjected to pronounced seasonal temperature variation that directly affects metabolic function. We conducted a preliminary study of individual lifetime thermal habitat use and respiration rates of four whitefish (*Coregonus lavaretus* (L.)) morphs (two pelagic, one littoral and one profundal) using stable carbon and oxygen isotope values of otolith carbonate. These morphs, two of which utilized pelagic habitats, one littoral and one profundal recently diverged via adaptive radiation to exploit different major niches in a deep and thermally stratified subarctic lake. We found evidence that the morphs used different thermal niches. The profundal morph had the most distinct thermal niche and consistently occupied the coldest thermal habitat of the lake, whereas differences were less pronounced among the shallow water pelagic and littoral morphs. Our results indicated ontogenetic shifts in thermal niches: juveniles of all whitefish morphs inhabited warmer ambient temperatures than adults. According to sampling of the otolith nucleus, hatching temperatures were higher for benthic compared to pelagic morphs. Estimated respiration rate was the lowest for benthivorous profundal morph, contrasting with the higher values estimated for the other morphs that inhabited shallower and warmer water. These preliminary results suggest that physiological adaptation to different thermal habitats shown by the sympatric morphs may play a significant role in maintaining or strengthening niche segregation and divergence in life-history traits, potentially contributing to reproductive isolation and incipient speciation.

Editor: Syuhei Ban, University of Shiga Prefecture, Japan

Funding: Funding of this study was obtained from Ministry of Agriculture and Forestry, Academy of Finland (projects #140903, #1268566, and #134139) and Emil Aaltonen Foundation. The funders had no role in study design, data collection and analysis, decision to publish, or preparation of the manuscript.

Competing Interests: The authors have declared that no competing interests exist.

* Email: kimmo.kahilainen@helsinki.fi

Introduction

A major task in evolutionary biology is to understand the formation of new species. Adaptive radiation is recognised as a key mechanism in speciation [1]. A central tenet of adaptive radiation is phenotypic adaptation to a specific environment [1]. Here, disruptive selection linked to adaptation to different environments is considered an important mechanism, both with regard to the initial divergence, as well as the maintenance of differences [1]. The final outcome of disruptive selection may lead to the formation of new species via ecological speciation [2]. Speciation research conducted over recent decades has demonstrated that many avian, terrestrial and aquatic species are now considered to have arisen via adaptive radiation and ecological speciation [1–3].

Most of this speciation research has focused on ecomorphological divergence, whereas the contribution of thermal adaptations in ecological speciation is generally poorly understood [4]. Many species distributions of closely related species exhibit distinct correlations with ambient temperature rather than ecological niche segregation [5]. Many famous adaptive radiations including Darwin's finches and *Anolis* lizards occur on volcanic island environments displaying pronounced altitudinal and vegetation gradients, resulting in different temperature and precipitation conditions [6,7]. However, even in these model systems most of the research has focused on ecomorphological divergence, largely neglecting the potential selective importance of temperature and related thermal adaptations. Such adaptations are likely most important for poikilotherms, such as amphibians, reptiles and fish.

Polymorphic fishes inhabiting post-glacial lakes have become important models for understanding how disruptive selection acts on phenotypic divergence and how it is related to accumulation of reproductive isolation between diverging species [8–10]. Across different fish lineages, a common pattern displayed is morphological divergence correlated to specific resource utilization in pelagic-littoral resource axis in lakes [e.g. 11]. In more complex cases, morphological divergence can be extended to a third principal lake habitat, the profundal [12–14]. In contrast to morphological divergence, which is well studied, far less is known about physiological adaptations to specific niches. Because freshwater fishes can be divided into cold, cool and warm water adapted species according to species-specific thermal optima, tolerance ranges and habitat selection [15], it is also likely that physiological adaptions should arise early in the divergence process. One of the best experimental studies comes from the salmonid fishes, where the closely related and morphologically similar cisco (*Coregonus fontanea*) and vendace (*Coregonus albula*) from Lake Stechlin in Germany differ in their optimum temperature preferences and standard metabolic adaptations [16]. Beyond this, little is generally known about temperature preferences and metabolic adaptions of polymorphic fish.

Adaptive radiation and ecological speciation is common throughout the natural range of European whitefish (*Coregonus lavaretus* (L.)) [17–19]. Whitefish are known for their ability to diverge, allowing the utilisation of pelagic and benthic habitats in lakes, with up to five sympatric morphs in the same lake [13,17,20]. Ecomorphological divergence of whitefish into morphs using the three major habitats (i.e. pelagic, littoral and profundal zones) in large, deep lakes is particularly pronounced in northern Fennoscandian lakes [12,13,21]. Apart from variation in food resources [22] and basal sources of energy [21], these habitats also differ with regard to their thermal conditions. Thermal stratification of the water column in these lakes means that littoral and pelagic habitats undergo large seasonal changes in water temperature between warm summer months (water temperature range 10–20°C) and cold winter months (0–2°C), when lakes are ice-covered [22]. Profundal fish occupy a cool/cold habitat (2–8°C) throughout the year. Reproductive isolation among whitefish morphs is generally based on spatial and temporal differences in spawning place and time [14,17,19].

Whitefish morphs dwelling in these different habitats differ in the number of gill rakers, which is commonly correlated to trophic ecology and habitat use in each of sympatric whitefish morphs and is also a heritable trait [9,17,23,24]. Gill raker number is closely correlated with overall head and body morphology: pelagic morphs are typically fusiform with a higher number of gill rakers, whereas benthic morphs have more robust body shape and a lower number of gill rakers [21,23]. Much of our understanding of polymorphic whitefish ecology relates to studies conducted during summer [21,23,25], and equivalent studies of habitat selection during winter months (i.e. when lakes in these regions are ice covered) are lacking. Furthermore, data regarding lifetime patterns of thermal niche selection from the point of hatching to maturation are not currently available from polymorphic fish.

Such studies of long-term thermal habitat selection could include the use of electronic data storage tags. However, these approaches are not suitable for delicate and small bodied fishes such as pelagic and profundal whitefish morphs. Here, it is more suitable to use natural anatomical recorders from individual fish such as balancing bones (e.g. otoliths), that document information such as secular variation in seasonal metabolism [26], thermal preferences [27], and growth of fish [28]. Furthermore, otolith microchemistry provides a chemical record of the annual changes in temperature related to metabolism that can be used to quantify variation in the life-time thermal niche via stable isotope analysis [29]. Oxygen stable isotope values ($\delta^{18}O$) have also been used to calculate temperatures encountered during the lifetime of fishes living in the distant past [e.g. 30]. Furthermore, the inorganic carbon isotope value ($\delta^{13}C$) of otoliths is correlated with the specific respiration rate of individual fish [e.g. 26,31,32], and analysed in parallel with $\delta^{18}O$ values. The use of micromilling techniques facilitates sampling of otoliths at a subannual scale throughout the lifetime of the fish by discretely isolating aliquots of carbonate that may represent days to weeks, depending on the growth rate and size of the otolith [29,33]. While otolith $\delta^{18}O$ and $\delta^{13}C$ values have been successfully used to track the lifetime thermal niche and respiration rates for many marine and freshwater species [32,34–36], we are unaware of any comparative studies of polymorphic species.

In this preliminary study, we were interested in putative differences in lifetime thermal niche selection among individuals of different sympatric whitefish morphs inhabiting a subarctic lake and whether such whitefish morphs displayed ontogenetic thermal shifts. If differences in thermal niche selection are apparent, we were interested in determining any correlative differences in respiration rates among whitefish morphs. Finally, we evaluated the likelihood of different hatching times among whitefish morphs using micromilling of otolith nuclei. According to previous data collected during summer sampling campaigns relating to temperature and habitat selection, the profundal whitefish morph was hypothesized to dwell in the coldest and the most annually thermally stable environment relative to the littoral and pelagic morphs [21,25]. If the profundal morph is adapted to a cold and thermally stable environment, we predicted that its respiration rates should be both lower and less variable than morphs that inhabit shallower, warmer habitats. Following the results of a previous otolith isotope study conducted on an alpine monomorphic whitefish population in Lake Annecy, France [32], we hypothesized that all whitefish morphs likely undertake an ontogenetic niche shift from a warmer juvenile thermal niche to a cooler adult thermal niche and that in all morphs respiration rates should decrease with age. Because all sympatric whitefish morphs investigated in this study lake display genetic differences and partial reproductive isolation according to neutral nuclear markers [14,37], we expected to find measurable differences in estimated hatching temperatures among the morphs, reflecting differences in spawning time and inter-morph reproductive isolation.

Materials and Methods

Study area and fish fauna

Samples were collected from the subarctic oligotrophic (total phosphorus 6 µg/L and total nitrogen 160 µg/L) Lake Paadar (68°52′N, 26°35′E), located in the Paatsjoki watercourse in Finnish Lapland (Fig. 1). The drainage area comprises a mixture of peatland, pine (*Pinus silvestris*) and birch (*Betula* spp.) forest. The surface area of the lake is circa 21 km^2, the maximum depth is 56 m, with a mean depth of 11.7 m [40]. Proportions of principal habitat types in this lake are littoral 38% and pelagic/profundal 62% [21]. The fish fauna consists of nine species, whitefish, brown trout (*Salmo trutta* L.), grayling (*Thymallus thymallus* (L.)), pike (*Esox lucius* L.), burbot (*Lota lota* (L.)), perch (*Perca fluviatilis* L.), three-spined stickleback (*Gasterosteus aculeatus* L.), nine-spined stickleback (*Pungitius pungitius* (L.)) and minnow (*Phoxinus phoxinus* (L.)) [40]. Whitefish dominate the fish community, contributing ~90% of the total catch by number [23]. Lake Paadar whitefish stock has evolved into two pelagic and two benthic morphs, each are genetically differentiated from each other [37], according to neutral nuclear DNA (range of F_{st} = 0.02–0.08).

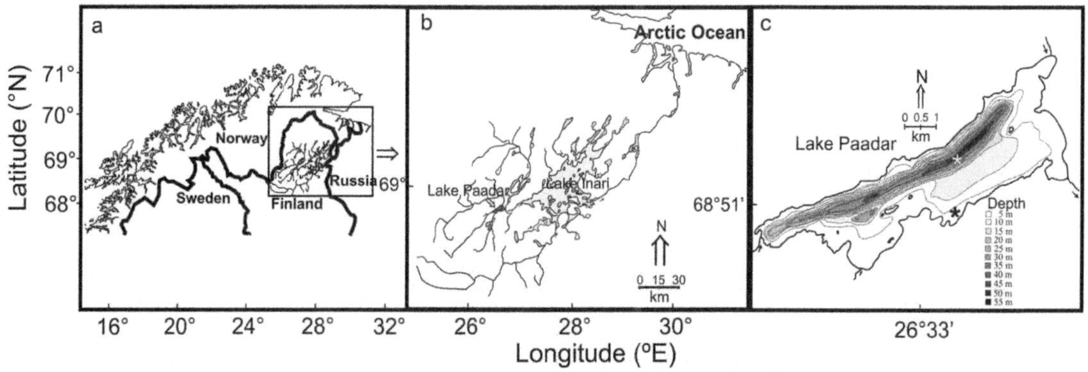

Figure 1. Study area in northern Fennoscandia (a), highlighting the Paatsjoki watercourse (b), and study lake (c). The asterisk in Fig. b indicates the position of water column temperature measurements in 1999–2004 Lake Inari (see Figure 2) and in Fig. c, the position of water column isotope sampling in the pelagic (white asterisk) and littoral (black asterisk) habitats in Lake Paadar (data presented in Figure 3).

Sampling

Whitefish were sampled for research purposes only in September 2004 using a gill net series combined of eight gill nets, each 30 m long and 1.8 m high, with different mesh sizes (12, 15, 20, 25, 30, 35, 45 and 60 mm). Gill net series were set in littoral, pelagic and profundal habitats in the evening and collected the following morning. Fish were immediately removed from gill nets, euthanized by cerebral concussion according to the Finnish Animal Conservation Law (32§ 9.8.2013/584) conducted by lead author (KKK) and subsequently placed on ice for measurements in the laboratory. Fishing rights in Finland belong to the land owner according to the Finnish Fishing Law (5§ 27.5.2011/600) and lead author (KKK) get a permit for research fishing from the land owner (Forest and Park Service, permit number 14.5.2004, 1386/5713-2004). No ethical permission is required for described scientific sampling with gill nets according to the Finnish Animal Conservation Law (7§ 28.6.2013/498).

Fish were divided by species (whitefish by morphs) and total length (± 1 mm) and wet mass (± 0.1 g) were measured from every captured fish. Whitefish were classified into morphs in the field reflecting differences in gill raker, body and head shape [12]. The profundal small sparsely rakered (SSR) benthivorous whitefish morph are small bodied with the lowest relative number (circa 15–20) of short, bent and very widely spaced gill rakers. They also have a characteristic sub-terminal mouth and large eyes positioned high on the head, large fins and a robust body shape [12,21]. The littoral benthivorous morph, that we refer to as the large sparsely rakered (LSR) morph has an intermediate number (c. 20–30) of short and widely space gill rakers, characteristic whitefish coloration and a terminal mouth. A similar body and head shape is typical for the large densely rakered (LDR) morph, but compared to the LSR morph, gill rakers are more numerous (c. 30–36), longer and more densely spaced. The fourth morph is the densely rakered (DR) whitefish, with a slender fusiform body, superior mouth, and numerous (c. 30–40), fine, long and very densely spaced gill rakers [12,21]. A size-based subsample (c. 40–50 individuals) reflecting the length distribution of the catch was frozen (-20°C) for more detailed analysis. The main prey items of the different whitefish morphs were collected for carbon and nitrogen stable isotope analyses [21]. Pelagic zooplankton was collected with vertical hauls using 25 cm diameter (50 μm mesh size) plankton net. Littoral (sampling depths 1–3 m) and profundal (10–40 m) invertebrate samples were collected using Ekman grab. Water samples for oxygen and inorganic carbon stable isotopes were collected at the end of July 2012 from pelagic and littoral habitats (Fig. 1). The oxygen isotope values of water in the pelagic zone is unlikely to change considerably despite the time delay between otolith (2004) and water sampling (2012) due to the long residence time of the large study lake and the sampling time in the middle of summer [38,39]. For inorganic carbon isotopes, there are obviously some seasonal differences between thermally stratified and non-stratified seasons [32] and thus our results from summer thermal stratification should be considered as a snap shot. Temperature and oxygen profiles were recorded at the same time.

Laboratory work

Whitefish samples were thawed in the laboratory and positioned for photographs taken of the left flank of each fish [21]. The first left gill arch was subsequently dissected and gill rakers were counted under a microscope. The middle gill raker length was also measured for comparison of pelagic and benthic morphs [12]. Scales and sagittal otoliths were removed for the determination of age and growth. A section of dorsal muscle was dissected for carbon (δ^{13}C) and nitrogen (δ^{15}N) stable isotope analyses. Muscle and invertebrate samples were dried (60°C) for 48 hours and later ground into a fine powder. A small amount (0.5–0.6 mg) of powder was encapsulated in tin cups. Water samples were collected from littoral (depths 0–3 m) and pelagic/profundal (0–33 m) habitats (Fig. 1) and directly stored in pre-combusted scintillation vials. Water samples for dissolved inorganic carbon were collected and directly injected into pre-phosphoric acid loaded and helium flushed Exetainer vials. All tissue and water samples were analyzed using an isotope mass spectrometer.

Whitefish were aged using both scales and otoliths [41]. According to these estimates of age, we selected a single individual of similar age (5+) from each morph to assess the lifetime thermal habitat use and respiration rates through otolith isotope analysis. Initially, we aimed to run several individuals per morph, but due to time and financial constraints this was not possible. However, of the majority of similar studies typically rely on small sample sizes and an individual-based approach without population-level comparisons [27,32,34]. Due to the lack of this type of data from other polymorphic poikilothermic populations, we hope that our preliminary results would promote more research on thermal adaptation.

One otolith of each morph was glued (Crystalbond©) to a microscope slide, sanded and polished flat on one side until the otolith nucleus was reached. Otoliths were photographed under a stereomicroscope, and the location and spacing of summer and winter circuli determined in order to characterize subsequent sampling paths via computer-controlled micromilling. This was conducted using a custom three-dimensional micromilling system in the Saskatchewan Isotope Laboratory facilities at the University of Saskatchewan [33]. This system permitted the sampling of 33–40 paths concordant with the growth banding in each otolith. The required sample size for isotope analysis (c. 7–25 μg) was obtained from almost all sampling paths (total 140 paths from 4 otoliths), with only four samples yielding insufficient material for analysis. Precision of carbon and oxygen analyses was 0.04 and 0.07‰, respectively.

Temperature Calculations

Isotopic values for fish, otolith, invertebrate and water samples were calculated using:

$$\delta X = \frac{(R_{sample} - R_{standard})}{R_{standard}} \times 1000$$

where δX is the stable isotope ratio of dorsal muscle or otolith carbonate and R is the isotope ratio of carbon, nitrogen or oxygen in the *sample* and the *standard*. As muscle isotope values show variations in C:N ratios, suggesting differences in lipid concentrations, we arithmetically corrected the values [42].

We used the average habitat-specific water and micro-milled otolith $\delta^{18}O$ values to calculate whitefish lifetime thermal niche. Species-specific models are not available for whitefish: hence we estimated values using a general model [29] and models developed for closely related salmonids, brook charr [*Salvelinus fontinalis*; 43] and Arctic charr [*Salvelinus alpinus*; 44]. Here, the latter non-linear model by Godiksen et al. [44] for Arctic charr provided the most similar values compared to annual water column temperature datasets during the lifetime of sample fish from the same watercourse (Fig. 2). The other models gave on the average 0.5–5.0°C higher temperature values (Table S1) than the non-linear model for Arctic charr [44]. Otolith $\delta^{18}O$ values reported relative to Vienna PeeDee Beleminite (VPDB) were transformed to the Vienna standard mean ocean water (VSMOW) scale:

$$\delta^{18}O_{otolith(VSMOW)} = 1.03092 * \delta^{18}O_{otolith(VPDB)} + 30.92$$

We used $\delta^{18}O_{water}$ values from the water samples collected from each habitat type and whitefish otoliths to calculate the fractionation factors (α) for each sample:

$$\alpha = \frac{\delta^{18}O_{otolith(VSMOW)} + 1000}{\delta^{18}O_{water(VSMOW)} + 1000}$$

where $\delta^{18}O_{otolith(VSMOW)}$ is the isotope value of otolith samples and $\delta^{18}O_{water(VSMOW)}$ is the isotope value of water samples (both presented relative to the VSMOW scale). These values were then utilized in the polynomial model [44]:

$$1000\ln\alpha = a + \frac{b}{T} + \frac{c}{T^2}$$

where T is temperature (10^3/K), a, b and c are empirical constants. The empirical constants a, b and c were 1089.24, 89.9 and −617.19 [44].

Respiration calculations

We used photographs to measure and calculate caudal aspect ratio, a morphometric index of fish swimming activity [45]:

$$K_{caud} = \frac{h^2}{s}$$

where h is the height of the opened caudal fin and s is caudal fin surface area. Here, higher values indicate streamlined and active fish such as pelagic cruisers, and low values represent less active, benthivores [45]. In general, K_{caud} can be used successfully to

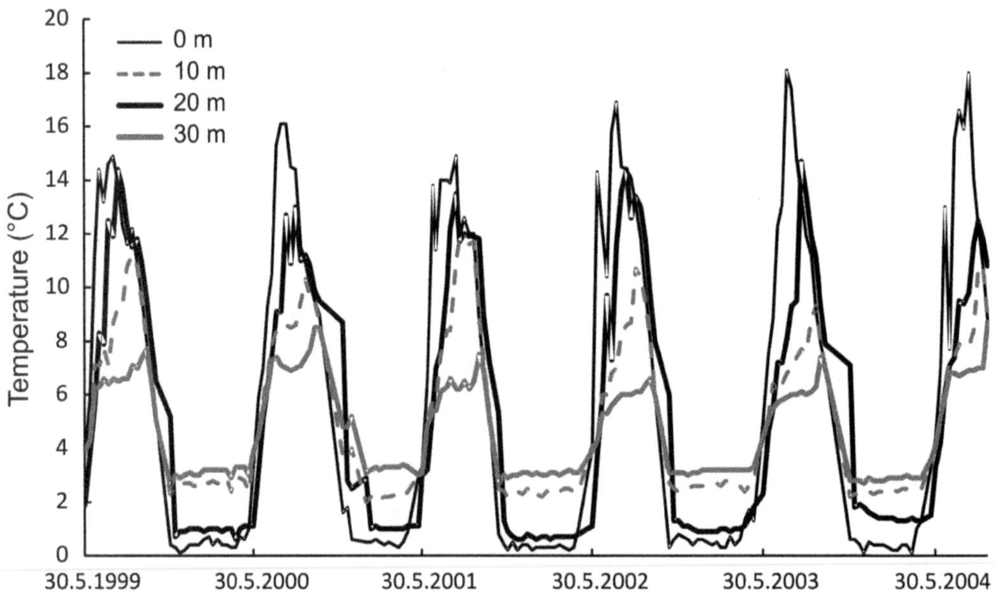

Figure 2. Annual temperature profiles (0–30 m) from Lake Inari, used as a proxy for Lake Paadar annual temperatures, during the life-time (1999–2004) of sampled whitefish morphs.

predict otolith $\delta^{13}C$ values [46]. Solomon *et al.* [31] further developed this relationship to directly estimate proportion of otolith carbon that is metabolically derived (*M*):

$$M = 0.025 + 0.066K_{caud}$$

In addition, estimated of lifetime metabolism of the different whitefish morphs were calculated using the Wisconsin bioenergetics model with parameters for the generalized coregonid model [47]. To model parameters of consumption and to empirically estimate respiration, the model requires data that reflect lifetime temperature, growth in terms of wet mass, proportion of prey items consumed, as well as the energy density of whitefish morphs and invertebrate prey. For the input data, we used temperature estimates from the $\delta^{18}O_{otolith}$ values, i.e. starting the modelling at the estimated time of hatching (1th of June 1999, first modelling day = 150) and continued until the time of death (15th of September 2004 = modelling day 2085). The age and length data from the total sample of each whitefish morph was used to calculate von Bertalanffy non-linear growth curves [48]:

$$L_t = L_\infty\left[1 - e^{-K(t-t_0)}\right]$$

where L_t is the expected total length at time t, L_∞ is the asymptotic total length (i.e. theoretical total length to which a fish would grow if time permitted), K is the rate at which the growth curve approaches the asymptote and t_0 is the hypothetical time at which L_t was zero (i.e. hatching time).

The estimated lengths were transformed into wet mass with corresponding whitefish morph specific mass-length relationships [49]:

$$W = aL^b$$

where W = wet mass (g), L = total length (cm), a = a constant, b = an exponent. Due to a lack of seasonal diet data, we assumed that all whitefish morphs consumed zooplankton prey during the first two months of life and thereafter ate prey associated with their respective habitat, i.e. DR whitefish consumed pelagic zooplankton, SSR whitefish fed on profundal benthic macroinvertebrates and LSR whitefish ate littoral benthic macroinvertebrates [21,25]. As LDR whitefish diet consists of both pelagic zooplankton and insect pupae or adults, we used 50% pelagic and 50% littoral diet proportions. We calculated the energy density (Y) of whitefish morphs and invertebrates using the proportion of elemental carbon (%) derived from the isotope analysis of dorsal muscle and invertebrates [50]:

$$Y = -8.415 + 0.62X$$

where X is the proportion of carbon (%). Energy densities of whitefish morphs and their invertebrate prey were kept as constant. The calculated energy contents for the SSR, LSR, LDR and DR whitefish were 5227, 5134, 5088 and 5150 J/g, respectively. Energy contents for littoral, profundal and pelagic invertebrates were 4694, 4341 and 2747 J/g.

The average gill raker count, gill raker length and isotope values of dorsal muscle ($\delta^{13}C$, $\delta^{15}N$) were statistically compared using analysis of variance and pairwise comparisons with Tukey's test.

Results

The total survey catch was 939 whitefish, of which 179 (40–50 individuals per morph) were subjected to more detailed analysis (Table 1). Whitefish morphs showed differences in gill raker number (ANOVA, $F_{3,175} = 850$, $p<0.001$) and the length of the longest raker (ANOVA, $F_{3,175} = 208$, $p<0.001$): the morphs differed from each other with regard to both measures (Tukey's pairwise comparisons $p<0.05$), except SSR and LSR whitefish which overlapped in gill raker length. There were also significant differences in muscle tissue $\delta^{13}C$ values (ANOVA, $F_{3,175} = 167$, $p<0.001$) and $\delta^{15}N$ values (ANOVA, $F_{3,175} = 200$, $p<0.001$). Here, all morphs differed (Tukey's pairwise comparisons $p<0.05$), except SSR and DR whitefish in $\delta^{13}C$ values. The ecomorphological traits of the individuals selected for the otolith micromilling from each morph were in the range of values for the sample of the larger population (Table 1).

There was a clear variation in $\delta^{18}O_{H2O}$ and $\delta^{13}C_{DIC}$ values in the water column (Fig. 3). As expected, the highest $\delta^{13}C_{DIC}$ values were observed near the surface (−6.4‰ $_{VPDB}$) and the lowest at the deepest sampling depth of 33 m (−12.4‰$_{VPDB}$). $\delta^{18}O_{H2O}$ values were less variable, ranging between −13.7 and −12.6‰$_{VSMOW}$ (Fig. 3). Mean $\delta^{18}O_{H2O}$ values were calculated for each whitefish morph habitat i.e. littoral (mean for depths 0, 1, 2, 3 m, −13.1‰$_{VSMOW}$), pelagic (mean for depths 0, 1, 3, 6, 9 m, −13.3‰ $_{VSMOW}$) and profundal (mean for depths 21, 24, 27, 30, 33 m, −13.6‰ $_{VSMOW}$). At the time of sampling, there was evidence for water column thermal stratification, with water temperatures ranging from 15°C at the surface to 8°C at 30 m (Fig. 3). Oxygen concentrations at this depth interval were *ca.* 9 mg l^{-1}.

$\delta^{18}O_{otolith}$ and $\delta^{13}C_{DIC}$ values of otoliths indicated differences among the whitefish morphs (Fig. 4). The large bodied littoral LSR whitefish had consistently higher otolith $\delta^{13}C$ values than the other morphs and also displayed the most pronounced seasonal variations. The individual of the pelagic LDR whitefish morph also exhibited strong seasonal variation in otolith $\delta^{13}C$ values, as well as some indication of higher otolith $\delta^{13}C$ values with age. The DR whitefish showed seasonal variation in otolith $\delta^{13}C$ and $\delta^{18}O$ values within the two first years of its life, but not in subsequent years. In contrast to the other morphs, the profundal SSR whitefish had the most stable otolith $\delta^{13}C$ and $\delta^{18}O$ values, exhibiting variation in only the first year of life (Fig. 4).

$\delta^{18}O_{otolith}$ values indicated seasonal water temperature variability for the 5+ individuals examined of each of the four of the whitefish morphs, which taken together, represents a time period that extended across six summer and five winter seasons (Fig. 5). Differences between summer and winter habitat temperatures were the most pronounced for the littoral LSR and the pelagic LDR whitefish, suggesting that both inhabit near-surface habitats. The pelagic DR whitefish display pronounced seasonal differences in the two first years of life, with apparently reduced seasonality in subsequent years. The profundal SSR whitefish individual displayed the most distinct thermal history, living at temperatures <11°C throughout their entire lifetime and with a pattern of decreasing temperatures after the first two years of life (Fig. 2 and 5). Although ontogenetic temperature niche shifts were most pronounced for the profundal SSR whitefish and the pelagic DR whitefish, there was a sign of decreased temperature variation for both the LSR and LDR whitefish too. $\delta^{18}O$ values of the kernel (nucleus of otoliths), indicated that the two benthic morphs (profundal SSR whitefish and littoral LSR whitefish) hatched at higher temperatures (11–12°C) than the pelagic (DR and LDR whitefish) morphs (8–9°C).

Table 1. Basic data from whitefish morphs (SSR = small sparsely rakered, LSR = large sparsely rakered, LDR = large densely rakered, DR = densely rakered whitefish).

Variable	SSR whitefish	LSR whitefish	LDR whitefish	DR whitefish
N	40	48	41	50
Total length (cm)	17.5±3.6 11.1–27.4 (16.7)	18.5±4.1 12.2–27.0 (25.9)	29.3±1.9 21.6–32.1 (25.0)	14.0±1.5 11.3–17.8 (15.4)
Total mass (g)	48.1±39.6 10.8–218 (33.6)	52.9±36.4 13.0–160.0 (141.1)	202.9±32.8 85.3–252.6 (124.2)	20.2±6.2 10–37.7 (25.0)
Gill raker number	18.3±1.3 16–21 (19)	24.4±2.2 20–31 (26)	33.7±1.9 30–38 (33)	36.3±2.0 32–40 (38)
Gill raker length (mm)	2.4±0.5 1.5–4.0 (2.6)	2.7±0.6 1.6–4.3 (3.4)	6.3±0.8 4.3–7.6 (5.3)	4.1±0.6 2.8–5.4 (4.6)
$\delta^{13}C$ (‰$_{VPDB}$) dorsal muscle	−28.4±1.1–31.9–26.8 (−28.0)	−24.5±1.5–27.5–21.9, (−22.6)	−26.8±0.7–28.1–25.5 (−26.5)	−28.4±0.2–28.7–27.5 (−28.6)
$\delta^{15}N$ (‰$_{AIR}$) dorsal muscle	9.9±0.4 9.0–10.8 (10.2)	7.1±0.6 6.0–8.1 (6.6)	7.6±0.8 5.5–9.8 (7.4)	8.6±0.5 7.8–9.9 (8.7)
Mean $\delta^{18}O$ (‰$_{VPDB}$) otolith	(−12.4)	(−12.6)	(−12.8)	(−12.8)
Mean $\delta^{13}C_{DIC}$ (‰$_{VPDB}$) otolith	(−15.0)	(−11.6)	(−15.1)	(−16.1)
Invertebrate prey $\delta^{13}C$ (‰$_{VPDB}$)	−29.4±1.1	−23.4±1.9	−30.2±0.6, −23.4±1.9	−30.2±0.6
Invertebrate prey $\delta^{15}N$ (‰$_{AIR}$)	6.6±0.5	2.8±1.2	5.3±0.5, 2.8±1.2	5.3±0.5
M, estimated	0.260	0.286	0.322	0.349

The average (±SD) and range (min–max) for each morph are indicated with the individual used for the otolith micromilling in parenthesis. $\delta^{18}O$ and $\delta^{13}C_{DIC}$ values indicate the average value from all drilled transects. Isotopic values of invertebrate prey of LDR whitefish indicate both pelagic zooplankton and littoral macroinvertebrate values. *M* is the proportion of metabolically derived otolith carbon estimated from the caudal aspect ratio (K_{caud}) [31].

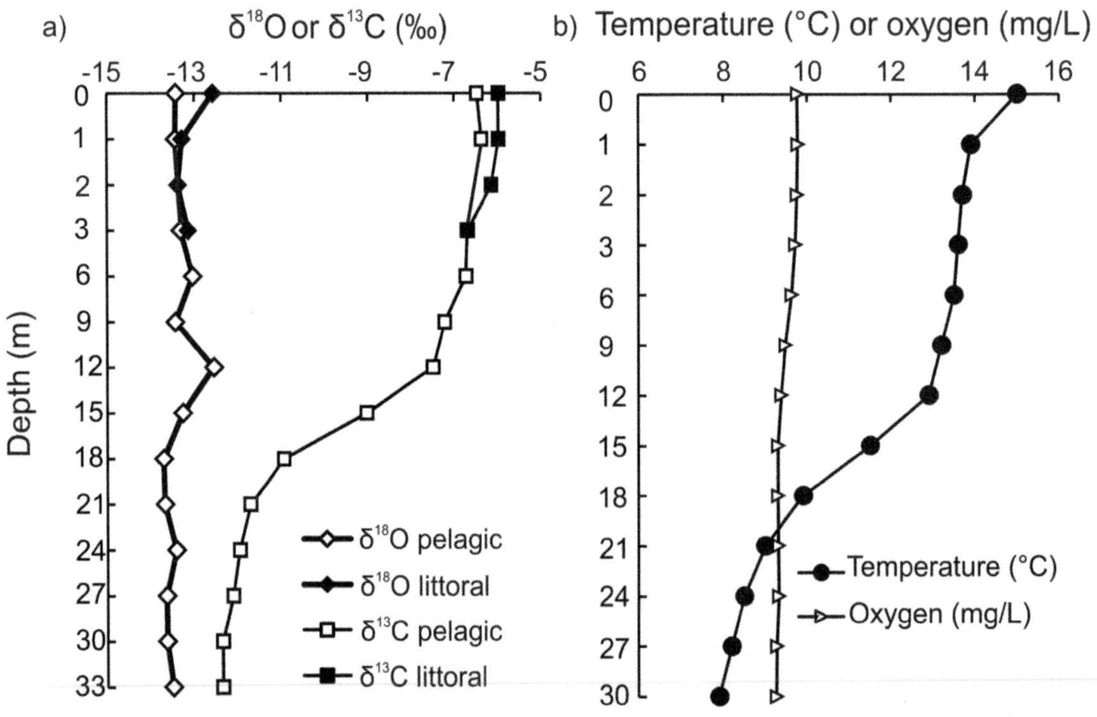

Figure 3. Oxygen ($\delta^{18}O$) and dissolved inorganic carbon ($\delta^{13}C_{DIC}$) stable isotope values and profiles of temperature and oxygen (b) measured from the littoral (0–3 m) and pelagic (0–33 m) water column of Lake Paadar.

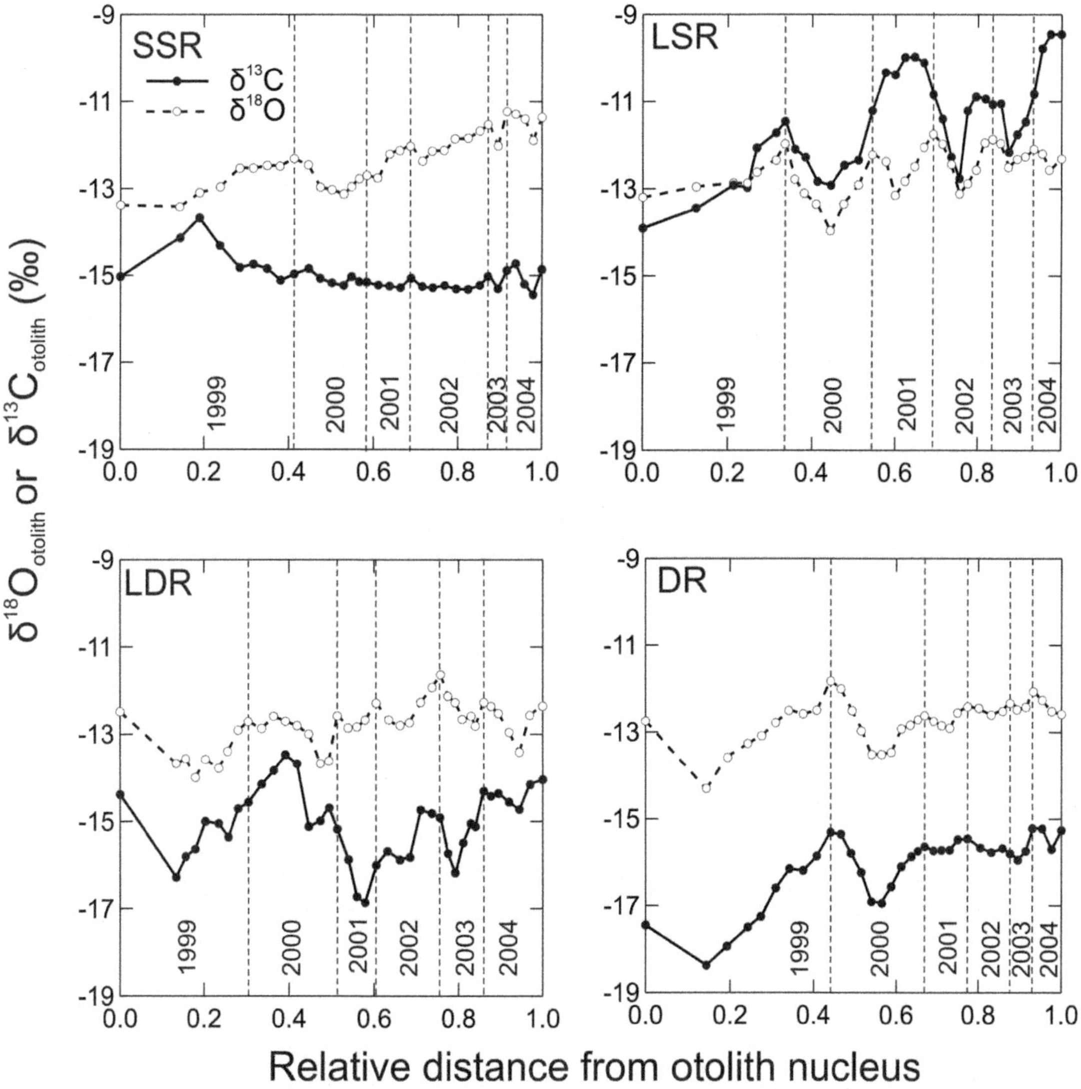

Figure 4. $\delta^{18}O$ and $\delta^{13}C$ values estimated from individual whitefish morph otoliths (SSR = small sparsely rakered, LSR = large sparsely rakered, LDR = large densely rakered and DR = densely rakered whitefish). The first x-axis value indicates the nucleus.

The estimated proportion of metabolically derived otolith carbon (M) calculated from caudal fin morphology was the lowest for profundal SSR whitefish and the second lowest for littoral LSR whitefish (Table 1). The pelagic morphs had markedly higher M, with peak values observed for DR whitefish. Bioenergetic modelling revealed clear growth differences among the morphs. Somatic growth (assessed with wet mass in the model) was the fastest for the large bodied morphs (pelagic LDR whitefish and littoral LSR whitefish), both of which continue growing after six growing seasons (Fig. 6). Somatic growth was much lower for SSR whitefish, which displayed steadily increasing weight. The lowest growth was observed in the pelagic DR whitefish, which attained their estimated maximum size after three years of life. The corresponding estimates of specific respiration rates indicated that seasonal fluctuations were the highest for littoral LSR whitefish and the lowest for profundal SSR whitefish (Fig. 6). The specific respiration rates of pelagic LDR whitefish and DR whitefish were similar, and seasonal fluctuations were generally lower than those estimated from the littoral LSR whitefish.

Discussion

We found support for our predictions that the different whitefish morphs occupied distinct thermal niches associated with differences in estimated respiration rates. The profundal SSR whitefish occupied the coldest, most stable thermal niche and had the lowest respiration estimates, whereas annual fluctuations in the thermal niche were the most pronounced in the littoral LSR whitefish. The thermal niche of the pelagic LDR and DR whitefish also showed clear seasonal fluctuations: respiration rates in individuals from these morphs were generally higher than the SSR whitefish and lower than the LSR whitefish. All morphs used a warmer thermal niche earlier in their ontogeny relative to later life, suggesting ontogenetic movement into colder water. Finally, we demonstrated that pelagic morphs hatched at lower temperatures than benthic morphs, according to the first otolith measurement in the nucleus, which likely relate to temporal and spatial differences in spawning.

In general, our evidence for thermal niche segregation among whitefish morphs provides a further dimension to previously

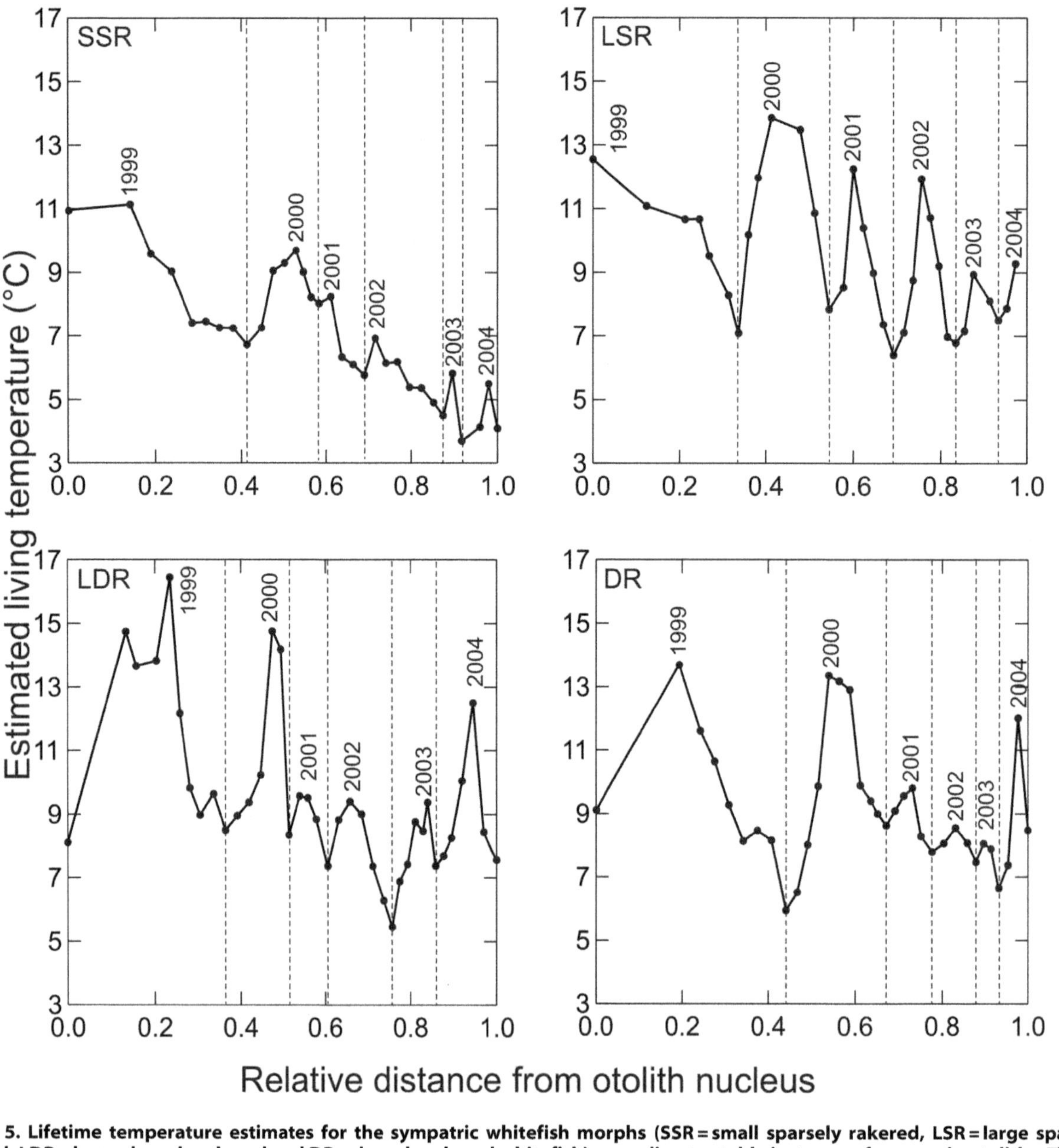

Figure 5. Lifetime temperature estimates for the sympatric whitefish morphs (SSR = small sparsely rakered, LSR = large sparsely rakered, LDR = large densely rakered and DR = densely rakered whitefish) according to stable isotopes of oxygen in otoliths. The first x-axis value indicates the nucleus i.e. estimated temperature at hatching.

documented ecomorphological and life-history divergence of these whitefish morphs [12,21,25]. Sympatric whitefish morphs have diverged and now exploit the three principal habitat types found in lakes (littoral, pelagic and profundal). LSR whitefish are ubiquitous to lakes in the study area, and when found in sympatry with other morphs is a littoral benthivore [21,41]. Otolith $\delta^{18}O$ values indicated that LSR whitefish tolerated pronounced yearly temperature variability, typical for shallow water habitats, suggesting a life-time occupation of littoral habitats. The results from the other benthivorous morph, the SSR whitefish [41] indicate consistent occupation of cold-water profundal niches. Both pelagic morphs displayed evidence for annual variability in their thermal niche, typical for thermally labile pelagic habitats. Results from the LDR whitefish showed more pronounced fluctuations than those shown by the DR whitefish, most likely related to near surface niche occupation in the LDR rather than the DR whitefish [25,51]. In case of whitefish morphs, thermal adaptations are correlated to other ecomorphological divergence, which may be the case with other taxa as well.

There is some recent evidence that adaptive benefit of larger bill size of birds in warm environments is a metabolic benefit via heat loss [52]. Traditionally, bill size is very well known to correlate with foraging, song and mate selection for example in Darwin's finches [6,53]. Different *Anolis* lizard species in the Caribbean Islands utilize distinct parts of trees, reflecting different microclimates [7]. Thermal preferences may have at least partly originated via use of these microclimates and furthermore, may influence relative ecomorphological performance in their respective niche. It is apparent that ecomorphological and thermal adaptations can

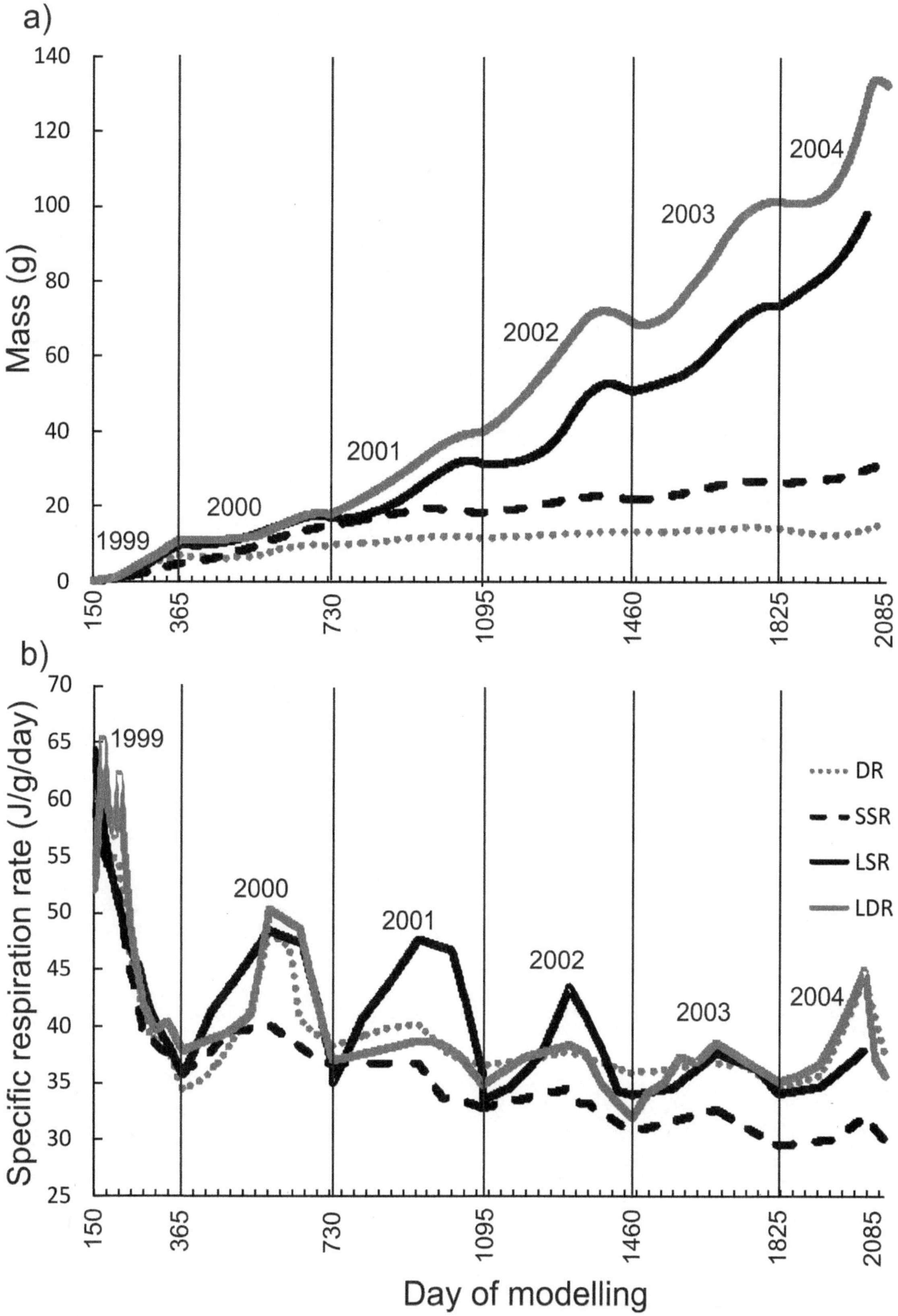

Figure 6. Bioenergetic modelling results of the weight gain (a) and the specific respiration rate (b) of whitefish morphs (SSR = small sparsely rakered, LSR = large sparsely rakered, LDR = large densely rakered and DR = densely rakered whitefish) from hatching (1.6.1999 i.e. day 150) to death (15.9.2004 i.e. day 2085). Years are separated with vertical lines.

both work in the same direction and it is therefore difficult to judge which is more important with regard to the origin of divergence.

Resource competition is considered as an obvious driver of initial divergence of morphs in general [1,54,55] and could be followed by physiological divergence at a later stage. The different habitat and thermal niche selection may result from resource competition among whitefish morphs, in which pelagic zooplankton prey are more efficiently consumed by those pelagic morphs with more gill rakers compared to the sparsely rakered benthic morphs [23]. Previous evidence from allopatric populations has indicated that in the absence of DR whitefish, the pelagic habitat is frequently occupied by LSR whitefish [21,56]. The strong ecomorphological divergence [21 and this study] among the morphs may also suggest physiological adaptations to different thermal niches. For example, in vendace morphs, the deeper water morph is clearly adapted to colder water, with a lower respiration rate relative to that of the shallow water morph [16]. An alternative scenario, especially in allopatric speciation and in the case of secondary contact [57], suggests that physiologically divergent species/morphs colonize and segregate the niches in the same lake or island according to their thermal preferences.

The SSR whitefish inhabiting cold profundal lake habitats exhibited the lowest estimated respiration rates according to calculations based on both morphology and bioenergetics. This is supported by differences in lifespan, as the profundal morph commonly lives for more than 20 years [58]. More variable metabolic rates were calculated for littoral LSR whitefish, which according to morphology (*M*) had the second lowest respiration rate. Both LSR and SSR whitefish forage on benthic macroinvertebrate prey [41], which requires the detection and capture of prey hiding in sediment. Benthic foraging behaviour is less energetically expensive than pelagic foraging, which relies on constant swimming to find zooplankton [45]. However, bioenergetic modelling indicated that the specific respiration rate of LSR whitefish, living in the habitat with the greatest temperature variability, was the highest. The difference between methods likely reflects the strong influence of living temperature on the results of bioenergetic modelling results. Similar differences between approaches were seen in both pelagic morphs: these had higher respiration rates than benthic morphs according to morphology, but were intermediate according to bioenergetics.

Thus, the most consistent finding is that profundal SSR whitefish have the lowest respiration rate, likely related to cold thermal habitats and the lowest risk of predation in profundal zones [59]. In contrast, pelagic morphs are subject to continuous risk of predation by piscivorous brown trout [40,60]. The risk of predation by brown trout is reduced by diel vertical migration by DR whitefish [51], which may drive the highest estimated respiration rate for any of the four morphs. The pelagic DR morph also has the shortest average lifespan, rarely living more than six years [61]. Higher metabolic rates have been estimated for pelagic than benthic morphs through bioenergetic modelling in the closely related lake whitefish (*Coregonus clupeaformis*) [62]. Also, recent genetic results of lake whitefish indicate that metabolic traits are inherited and the pelagic morph consistently exhibits over-expression of genes related to metabolism and oxygen transport compared to benthic morphs, related to higher swimming activity required in pelagic habitats [63,64]. There is an urgent need to evaluate the metabolic differences in other diverging taxa. Reptile and amphibian diversity is correlated with temperature, but very little is known about actual niche-specific metabolic adaptations [7,65].

Individuals of each of the whitefish morphs examined here used warmer water layers during the summer in the first two years of life. This has been observed for many fish species in the first year of life, in both field and experimental studies [35,66]. Dufour *et al.* [32] studied peri-alpine whitefish otoliths using the same techniques as used here and found that whitefish occupied warmer habitats during the first year of life. Whitefish larvae are known to concentrate in near shore areas after hatching and soon start to feed on small zooplankton [67]. Their diet may subsequently shift towards benthic macroinvertebrates or they may continue to consume zooplankton, depending on the individual lake and whitefish morph [68]. In nearby Lake Muddus, surface trawl catches in early autumn frequently contain 0+ DR whitefish in the pelagic habitat, whereas in sheltered shallow littoral bays 0+ LSR whitefish are caught (K. Kahilainen pers. obs.). To date, no 0+ SSR whitefish have been captured during these surveys, suggesting that a shift to morph-specific niches may occur during the first year of life. After the first two summer seasons, all morphs apparently choose cooler waters, shifting to their adult thermal niche. All the specimens studied here had reached or were very close to reaching sexual maturity, which occurs at different ages and sizes at the population level. Here, DR whitefish reach sexual maturity at the lowest age (2–3 years old), SSR whitefish at the highest age (6–7 years old) and the two other morphs at an intermediate 5–6 years old. Little is known about the exact spawning times and places for the different morphs, but field observations of spawning shoals in shallow water close to the main basin during field sampling in September 2004 indicated that DR whitefish were the first morphs to spawn in Lake Paadar. Observations in maturation and gonad state from the other morphs suggest that LDR and LSR whitefish spawned after the DR whitefish, with the SSR whitefish spawning last (K. Kahilainen pers. obs.). In coregonids, such inter-morph differences in spawning times are important factors driving reproductive isolation [17,19,63].

The general differences in spawning time shown by pelagic and benthic morphs may be indirectly related to the role of temperature governing the availability of invertebrate prey resources: pelagic zooplankton are only available in large amounts for a short period each summer, whereas benthic invertebrates are more consistently available year round [22,41,61]. The late spawning time shown by SSR whitefish may reflect temperatures in profundal habitats, which are the highest during the autumnal turnover of the water column. Physiological adaptation to a particular thermal regime may also be an important component of reproductive isolation in many species, but our current understanding is very limited [4]. Recently, niche segregation along thermal gradients associated with variation in both altitude and latitude has been shown to be efficient reproductive isolation barriers even in face of gene-flow in birds, amphibians and fish [69–71].

A recent modelling study also suggests that such physiological adaptations to different thermal habitat may be important in the formation of reproductive isolation and subsequent speciation [72]. Here, adaption to divergent thermal niches is likely to be of crucial importance to the relative capacity to assimilate the energy required to reach sexual maturity and develop gonads. If fish morphs are physiologically adapted to their individual thermal niche, this may cascade affecting spawning times and furthermore hatching times of eggs. The onset of spawning in whitefish is influenced by both temperature and light regimes [73]. An adaptation to lower temperatures may delay spawning times, as observed here for SSR whitefish, the most morphologically and genetically distinct morph [21,37]. The isotope values from the otolith nucleus suggest potentially different hatching temperatures for benthic and pelagic morphs, perhaps related to either temporal or/and spatial differences in spawning. In experiments on lake whitefish morphs, hatching time was demonstrated to be genetically determined, with very small differences between pelagic and benthic morphs [63]. However, little is known about the early development of polymorphic fish eggs in nature, especially with regard to potential differences of degree-day counts among the morphs in relation to their physiological adaptations. These are potentially important thermal adaptations, as in amphibian thermal adaptations in early larval development that apparently separate different populations [71].

Our study indicates that otolith $\delta^{18}O$ and $\delta^{13}C$ analysis provide an excellent tool for understanding lifetime thermal niche and respiration values of whitefish morphs, when $\delta^{18}O_{H2O}$ and $\delta^{13}C_{DIC}$ values are obtained from lake water. Our preliminary study evaluates patterns based on snap shot sampling of water column during the summer stratification, whereas seasonal sampling should be done to track annual differences in $\delta^{13}C_{DIC}$ values [32]. The results also verified the age determination estimates made previously from the same otoliths. Although, larger sample sizes are needed to reveal within-morph variation in thermal niche and respiration between multiple individuals, we hope that our preliminary results based on individual fish promote more studies on thermal adaptation from other polymorphic populations. Even though fine micromilling techniques were utilized to get 33–40 samples from each otolith, the resolution decreased due to declining fish growth rates after the first two years of life. At the very beginning of an individual fish's life, it is unclear at which point during the egg development and what rate otolith growth starts. Reflecting this, do the first measurements taken from the otolith in nucleus indicate hatching temperatures or combine temperatures from the late egg development and hatching periods? Despite the lack of a clear answer, we have observed differences among the morphs, indicating distinct thermal habitats at the very beginning of their lifecycles.

All whitefish morphs can be considered as slow growing fish when compared to the fast growing marine fishes, such as cod (*Gadus morhua*) and halibut (*Hippoglosus stenolepis*) previously examined in similar studies [34,36]. The estimated winter temperatures from otoliths are clearly too high as water temperatures in winter are usually $<4°C$ even in the deepest parts of subarctic lakes [22]. This error is most likely due to the extremely small winter increments in the otoliths and subsequent integration of autumn and spring otolith growth into the same sample. Alternatively, these fish may have a shutdown temperature $<4°C$ that precludes any otolith accretion during the cooler season. Otolith carbonate also records metabolism with winter rings clearly visible in whitefish, including the individuals analysed here. Based on difficulties in estimating timing and duration of a potential winter break in otolith growth and thus thermal estimation, we decided to present the full data in our results and discuss here the uncertainties involved. In addition, it should be noted that we used a fractionation equation developed for Arctic charr, which is adapted to colder temperatures than whitefish [44,74]. An experimental approach using whitefish in general or even whitefish morphs is urgently needed to build species or whitefish morph specific equations between otolith $\delta^{18}O$ values and temperature.

To conclude, we found a clear segregation of thermal niches between profundal and non-profundal whitefish morphs, but our capacity to differentiate between pelagic and littoral morphs was less successful. However, all individuals of the four whitefish morphs displayed evidence for ontogenetic thermal niche shifts, from warm juvenile summer habitats to cooler temperatures. Interestingly, estimated respiration rates were clearly correlated to foraging ecology and life-history, where the profundal SSR whitefish has lower rates than other morphs. These results also provide a useful starting point to conduct further studies of thermal niche life-histories with other polymorphic populations.

Acknowledgments

Authors would like to thank O. Aikio, D. Besic, K. Johansson, P. Jääskeläinen, J. Marttila, K. Mäenpää, J. Niemistö, M. Salonen, S. Timsic and A. Tuomaala for their help during the field and laboratory work. Muddusjärvi Research Station kindly provided needed facilities during the field work. Funding of this study was obtained from Ministry of Agriculture and Forestry, Academy of Finland (projects #140903, #1268566, #134139) and Emil Aaltonen Foundation.

Author Contributions

Conceived and designed the experiments: KKK WPP MK. Performed the experiments: KKK WPP ES CH MK. Analyzed the data: KKK WPP ES CH MK. Contributed reagents/materials/analysis tools: KKK WPP ES CH MK. Contributed to the writing of the manuscript: KKK WPP ES CH MK.

References

1. Schluter D (2000) The ecology of adaptive radiation. Oxford University Press, Oxford.
2. Nosil P (2012) Ecological speciation. Oxford University Press. Oxford.
3. Price T (2008) Speciation in birds. Roberts and Company Publishers, Colorado.
4. Keller I, Seehausen O (2012) Thermal adaptation and ecological speciation. Mol Ecol 21: 782–799.
5. Svensson EI (2012) Non-ecological speciation, niche conservatism and thermal adaptation: how are they connected? Org Div Evol 12: 229–240.
6. Grant PR, Grant BR (2008) How and why species multiply? The radiation of Darwin's Finches. Princeton University Press, New Jersey.
7. Losos JB (2009) Lizards in an evolutionary tree: ecology and adaptive radiation of anoles. University of California Press, Berkeley and Los Angeles.
8. Østbye K, Amundsen PA, Bernatchez L, Klemetsen A, Knudsen R, et al. (2006) Parallel evolution of ecomorphological traits in the European whitefish *Coregonus lavaretus* (L.) species complex during postglacial times. Mol Ecol 15: 3983–4001.
9. Rogers SM, Bernatchez L (2007) The genetic architecture of ecological speciation and the association with signatures of selection in natural lake whitefish (*Coregonus* sp. Salmonidae) species pairs. Mol Biol Evol 24: 1423–1438.
10. Ravinet M, Prodöhl PA, Harrod C (2013) Parallel and nonparallel ecological, morphological and genetic divergence in lake-stream stickleback from a single catchment. J Evol Biol 26: 186–204.
11. Schluter D, McPhail JD (1993) Character displacement and replicate adaptive radiation. Trends Ecol Evol 8: 197–200.
12. Kahilainen K, Østbye K (2006) Morphological differentiation and resource polymorphism in three sympatric whitefish *Coregonus lavaretus* (L.) forms in a subarctic lake. J Fish Biol 68: 63–79.
13. Siwertsson A, Knudsen R, Kahilainen KK, Præbel K, Primicerio R, et al. (2010) Sympatric diversification influenced by ecological opportunity and historical contingency in a young species lineage of whitefish. Evol Ecol Res 12: 929–947.
14. Præbel K, Knudsen R, Siwertsson A, Kahilainen KK, Østbye K, et al. (2013) Ecological speciation in postglacial European whitefish: rapid adaptive radiations into the littoral, pelagic and profundal lake habitats. Ecol Evol 3: 4970–4986.
15. Magnuson JJ, Crowder LB, Medvick PA (1979) Temperature as an ecological resource. Am Zool 19: 331–343.
16. Ohlberger J, Mehner T, Staaks G, Hölker F (2008) Temperature-related physiological adaptations promote ecological divergence in a sympatric species pair of temperate freshwater fish, *Coregonus* spp. Funct Ecol 22: 501–508.
17. Svärdson G (1979) Speciation of Scandinavian *Coregonus*. Rep Inst Freshw Res Drott 57: 1–95.

18. Østbye K, Bernatchez L, Næsje TF, Himberg KJM, Hindar K (2005) The evolutionary history of European whitefish (*Coregonus lavaretus* L.) as inferred from mtDNA phylogeography and gillraker numbers. Mol Ecol 14: 4371–4387.
19. Hudson AG, Vonlanthen P, Müller R, Seehausen O (2007) Review: The geography of speciation and adaptive radiation of coregonines. Arch Hydrobiol Spec Iss Adv Limnol 60: 111–146.
20. Vonlanthen P, Bittner D, Hudson AG, Young KA, Müller R, et al. (2012) Eutrophication causes speciation reversal in whitefish adaptive radiations. Nature 482: 357–363.
21. Harrod C, Mallela J, Kahilainen KK (2010) Phenotype-environment correlations and character displacement in a putative fish radiation. J Anim Ecol 79: 1057–1068.
22. Hayden B, Harrod C, Kahilainen KK (2013) The effects of winter ice cover on the trophic ecology of whitefish (*Coregonus lavaretus* L.) in subarctic lakes. Ecol Freshw Fish 22: 192–201.
23. Kahilainen KK, Siwertsson A, Gjelland KØ, Knudsen R, Bøhn T, et al. (2011) The role of gill raker number variability in adaptive radiation of coregonid fish. Evolutionary Ecology 25: 573–588.
24. Kahilainen KK, Østbye K, Harrod C, Shikano T, Malinen T, et al. (2011) Species introduction promotes hybridization and introgression in *Coregonus*: is there sign of selection against hybrids? Mol Ecol 20: 3838–3855.
25. Kahilainen K, Malinen T, Tuomaala A, Lehtonen H (2004) Diel and seasonal habitat and food segregation of three sympatric whitefish (*Coregonus lavaretus* (L.)) forms in a subarctic lake. J Fish Biol 64: 418–434.
26. Wurster CM, Patterson WP (2003) Metabolic rate of late Holocene freshwater fish: evidence from $\delta^{13}C$ values from otoliths. Paleobiology 29: 492–505.
27. Wurster CM, Patterson WP, Stewart DJ, Bowlby JN, Stewart TJ (2005) Thermal histories, stress, and metabolic rates of chinook salmon (*Oncorhynchus tshawytscha*) in Lake Ontario: evidence from intra-otolith stable isotope analyses. Can J Fish Aquat Sci 62: 700–713.
28. Campana SE (1999) Chemistry and composition of fish otoliths: pathways, mechanisms and applications. Mar Ecol Progr Ser 188: 263–297.
29. Patterson WP, Smith GR, Lohmann KC (1993) Continental paleothermometry and seasonality using the isotopic composition of aragonitic otoliths of freshwater fishes. In: P. Swart P, Lohmann K, McKenzie J, Savin S, editors. Climate change in Continental Isotopic Records. American Geophysical Union,Washington. 191–202.
30. Patterson WP (1998) North American continental seasonality during the last millennium: high-resolution analysis of sagittal otoliths. Palaeogeogr Palaeoclim Palaeoecol 138: 271–303.
31. Solomon CT, Weber PK, Cech JJ, Ingram BL, Conrad ME, et al. (2006) Experimental determination of the sources of otolith carbon and associated fractionation. Can J Fish Aquat Sci 63: 79–89.
32. Dufour E, Gerdeaux D, Wurster CM (2007) Whitefish (*Coregonus lavaretus*) respiration rate governs intra-otolith variation of $\delta^{13}C_{oto}$ values in Lake Annecy. Can J Fish Aquat Sci 64: 1736–1746.
33. Wurster CM, Patterson WP, Cheatham MM (1999) Advances in micromilling techniques: a new apparatus for acquiring high-resolution oxygen and carbon stable isotope values and major/minor elemental ratios from accretionary carbonate. Comput Geosci 25: 1155–1162.
34. Weidman CR, Millner R (2000) High resolution stable isotope records from North Atlantic cod. Fish Res 46: 327–342.
35. Joukhadar Z, Patterson WP, Todd TN, Smith GR (2002) Temperature history of *Coregonus artedi* in the St. Marys River, Laurentian Great Lakes, inferred from oxygen isotopes of otoliths. Arch Hydrobiol Spec Iss Adv Limnol 57: 453–461.
36. Gao Y, Beamish RJ (2003) Stable isotope variation in otoliths of Pacific halibut (*Hippoglossus stenolepis*) and indications of the possible 1990 regime shift. Fish Res 60: 393–404.
37. Couton M (2012) Adaptive radiation of the European whitefish, *Coregonus lavaretus* (L.), in the Pasvik watercourse: the genetic description of a new morph. MSc thesis, University of Tromsø, Tromsø.
38. Forsström L, Sorvari S, Rautio M, Sonninen E, Korhola A (2007) Changes in physical and chemical limnology and plankton during the spring melt period in a subarctic lake. Internat Rev Hydrobiol 92: 301–325.
39. Jonsson CE, Leng MJ, Rosqvist GC, Seibert J, Arrowsmith C (2009) Stable oxygen and hydrogen isotopes in sub-Arctic lake waters from northern Sweden. J Hydrol 376: 143–151.
40. Jensen H, Kahilainen KK, Amundsen PA, Gjelland KØ, Tuomaala A, et al. (2008) Predation by brown trout (*Salmo trutta*) along a diversifying prey community gradient. Can J Fish Aquat Sci 65: 1831–1841.
41. Kahilainen K, Lehtonen H, Könönen K (2003) Consequence of habitat segregation to growth rate of two sparsely rakered whitefish forms (*Coregonus lavaretus* (L.)) in a subarctic lake. Ecol Freshw Fish 12: 275–285.
42. Kiljunen M, Grey J, Sinisalo T, Harrod C, Immonen H, et al. (2006) A revised model for lipid-normalising $\delta^{13}C$ values from aquatic organisms, with implications for isotope mixing models. J Appl Ecol 43: 1213–1222.
43. Storm-Suke A, Dempson JB, Reist JD, Power M (2007) A field-derived oxygen isotope fractionation equation for *Salvelinus* species. Rapid Comm Mass Spec 21: 4109–4116.
44. Godiksen J, Svenning MA, Dempson BJ, Marttila M, Storm-Suke A, et al. (2010) Development of a species-specific fractionation equation for Arctic charr (*Salvelinus alpinus* (L.)): an experimental approach. Hydrobiologia 650: 67–77.
45. Webb PW (1984) Form and function in fish swimming. Sci Am 251: 58–68.
46. Sherwood GD, Rose GA (2003) Influence of swimming form on otolith $\delta^{13}C$ in marine fish. Mar Ecol Prog Ser 258: 283–289.
47. Hanson PC, Johnson TB, Schindler DE, Kitchell JF (1997) Fish Bioenergetics 3.0. University of Wisconsin, Wisconsin.
48. von Bertalanffy L (1938) A quantitative theory of organic growth. Human Biol 10: 181–243.
49. LeCren ED (1951) The length-weigth relationship and seasonal cycle in gonad weight and condition in the perch (*Perca fluviatilis*). J Anim Ecol 20: 201–219.
50. Salonen K, Sarvala J, Hakala I, Viljanen ML (1976) The relation of energy and organic carbon in aquatic invertebrates. Limnol Oceanog 21: 724–730.
51. Kahilainen KK, Malinen T, Lehtonen H (2009) Polar light regime and piscivory govern diel vertical migrations of planktivorous fish and zooplankton in a subarctic lake. Ecol Freshw Fish 18: 481–490.
52. Greenberg R, Cadena V, Danner RM, Tattersall G (2012) Heat loss may explain bill size differences between birds occypying different habitats. PloS One 7: e40933.
53. Podos J (2001) Correlated evolution of morphology and vocal signal structure in Darwin's finches. Nature 409: 185–187.
54. Amundsen PA, Knudsen R, Klemetsen A, Kristoffersen R (2004) Resource competition and interactive segregation between sympatric whitefish morphs. Ann Zool Fenn 41: 301–307.
55. Bolnick DI, Lau OL (2008) Predictable patterns of disruptive selection in stickleback in postglacial lakes. Am Nat 172: 1–11.
56. Kahilainen KK, Malinen T, Tuomaala A, Alajärvi E, Tolonen A, et al. (2007) Empirical evaluation of phenotype-environment correlation and trait utility with allopatric and sympatric whitefish (*Coregonus lavaretus* (L.)) populations in subarctic lakes. Biol J Linn Soc 92: 561–572.
57. Coyne JA, Orr HA (2004) Speciation. Sinauer Associates, Sunderland, Massachusetts.
58. Kahilainen K, Lehtonen H (2002) Habitat use and growth of three sympatric forms of European whitefish, *Coregonus lavaretus* (L.), in subarctic Lake Muddusjärvi. Arch Hydrobiol Spec Iss Adv Limnol 57: 277–290.
59. Kahilainen K, Lehtonen H (2003) Piscivory and prey selection of four predator species in a whitefish dominated subarctic lake. J Fish Biol 63: 659–672.
60. Kahilainen K, Lehtonen H (2002) Brown trout (*Salmo trutta* L.) and Arctic charr (*Salvelinus alpinus* (L.)) as predators on three sympatric whitefish (*Coregonus lavaretus* (L.)) forms in the subarctic Lake Muddusjärvi. Ecol Freshw Fish 11: 158–167.
61. Kahilainen K, Alajärvi E, Lehtonen H (2005) Planktivory and diet-overlap of densely rakered whitefish (*Coregonus lavaretus* (L.)) in a subarctic lake. Ecol Freshw Fish 14: 50–58.
62. Trudel M, Tremblay A, Schetagne R, Rasmussen JB (2001) Why are dwarf fish so small? An energetic analysis of polymorphism in lake whitefish (*Coregonus clupeaformis*). Can J Fish Aquat Sci 58: 394–405.
63. Bernatchez L, Renaut S, Whiteley AR, Derome N, Jeukens J, et al. (2010) On the origin of species: insights from the ecological genomics of lake whitefish. Phil Trans R Soc B 365: 1783–1800.
64. Evans ML, Præbel K, Peruzzi S, Bernatchez L (2012) Parallelism in the oxygen transport system of the lake whitefish: the role of physiological divergence in ecological speciation. Mol Ecol 21: 4038–4050.
65. Cadena CD, Kozak KH, Gómez JP, Parra JL, McCain CM, et al. (2012) Latitude, elevational climatic zonation and speciation in New World vertebrates. Proc R Soc London B 279: 194–201.
66. Hokanson KEF (1977) Temperature requirements of some percids and adaptations to the seasonal temperature cycle. J Fish Res Bd Can 34: 1524–1550.
67. Næsje TF, Sandlund OT, Jonsson B (1986) Habitat use and growth of age-0 whitefish, *Coregonus lavaretus*, and cisco, *C. albula*. Env Biol Fish 15: 309–314.
68. Lindström T (1988) Diet changes during metamorphosis in whitefish. Finn Fish Res 9: 89–105.
69. Culumber ZW, Shepard DB, Coleman SW, Rosenthal GG, Tobler M (2012) Physiological adaptation along environmental gradients and replicated hybrid zone structure in swordtails (Teleostei: *Xiphophorus*). J Evol Biol 25: 1800–1814.
70. Caro LM, Caycedo-Rosales PC, Bowie RCK, Slabbekoorn H, Cadena CD (2013) Ecological speciation along and elevational gradient in a tropical passerine bird? J Evol Biol 26: 357–374.
71. Muir AP, Biek R, Thomas R Mable BK (2014) Local adaptation with high gene flow: temperature parameters drive adaptation to altitude in the common frog (*Rana temporaria*). Mol Ecol 23: 561–574.
72. Ohlberger J, Brännström Å, Dieckmann U (2013) Adaptive phenotypic diversification along a temperature-depth gradient. Am Nat 182: 359–373.
73. Wanzenböck J, Pamminger-Lahnsteiner B, Winkler K, Weiss SJ (2012) Experimental evaluation of the spawning periods of whitefish (*Coregonus lavaretus* complex) in Lake Mondsee, Austria. Fund Appl Limnol Adv Limnol 63: 89–97.
74. Siikavuopio S, Knudsen R, Amundsen PA (2010) Growth and mortality of Arctic charr and European whitefish reared at low temperatures. Hydrobiologia 650: 255–263.

14

Patterns and Potential Drivers of Dramatic Changes in Tibetan Lakes, 1972–2010

Yingkui Li[1,2]*, Jingjuan Liao[2], Huadong Guo[2], Zewen Liu[1], Guozhuang Shen[2]

1 Department of Geography, University of Tennessee, Knoxville, Tennessee, United States of America, 2 Key Laboratory of Digital Earth Science, Institute of Remote Sensing and Digital Earth, Chinese Academy of Sciences, Beijing, P.R. China

Abstract

Most glaciers in the Himalayas and the Tibetan Plateau are retreating, and glacier melt has been emphasized as the dominant driver for recent lake expansions on the Tibetan Plateau. By investigating detailed changes in lake extents and levels across the Tibetan Plateau from Landsat/ICESat data, we found a pattern of dramatic lake changes from 1970 to 2010 (especially after 2000) with a southwest-northeast transition from shrinking, to stable, to rapidly expanding. This pattern is in distinct contrast to the spatial characteristics of glacier retreat, suggesting limited influence of glacier melt on lake dynamics. The plateau-wide pattern of lake change is related to precipitation variation and consistent with the pattern of permafrost degradation induced by rising temperature. More than 79% of lakes we observed on the central-northern plateau (with continuous permafrost) are rapidly expanding, even without glacial contributions, while lakes fed by retreating glaciers in southern regions (with isolated permafrost) are relatively stable or shrinking. Our study shows the limited role of glacier melt and highlights the potentially important contribution of permafrost degradation in predicting future water availability in this region, where understanding these processes is of critical importance to drinking water, agriculture, and hydropower supply of densely populated areas in South and East Asia.

Editor: Hayley J. Fowler, Newcastle University, United Kingdom

Funding: This work was supported by the National Basic Research Program of China (no. 2009CB723906), the National Natural Science Foundation of China (no. 41328001), and the Open Research Fund of Key Laboratory of Digital Earth, Chinese Academy of Sciences (no. 2011LDE004). The funders had no role in study design, data collection and analysis, decision to publish, or preparation of the manuscript.

Competing Interests: The authors have declared that no competing interests exist.

* Email: yli32@utk.edu

Introduction

Rising temperature as a component of climate change has caused most glaciers worldwide to retreat [1,2]. However, the significance of accelerated glacier melt to the water supply of high-latitude and high-altitude areas is still unclear [3]. As the highest and one of the most extensive plateaus on Earth, the Tibetan Plateau, along with its bordering mountains, has the largest number of glaciers outside of the Polar Regions, and these glaciers provide headwaters for most of Asia's great river systems, including the Indus, Ganges, Brahmaputra, Mekong, Yellow, and Yangtze [4,5]. Accelerated glacial retreat has the potential to affect the ecosystems and the livelihoods of more than 1.5 billion people in South and East Asia [4–7].

The Tibetan Plateau also contains >1,000 lakes, most of which belong to inland drainage systems [8], so their dynamics represent integrated effects of climate and associated cryospheric changes within their drainage basins. Recent lake expansions in many areas of the plateau have destroyed roads and infrastructure, flooded grazing and farm lands, and attracted significant attention from the scientific community [9–42]. Investigations of lake changes on the Tibetan Plateau during the past decades have been mainly based on remote sensing due to the lack of long-term systematic observation in this harsh environment [9–42]. Satellite images, such as Landsat MSS/TM/ETM+ imagery, have been used to investigate changes in lake extent [9–36]. Early studies mainly investigated one or a few lake(s) in the past several decades [9–33]. For example, Wang (2007) [14] investigated surface area variations of Nam Co (Tibet: "Co" = lake) and Selin Co from the central Tibetan Plateau, and several lakes in the source area of the Yellow River (northeastern Tibetan Plateau). They found that most lakes on the central plateau were expanding, whereas lakes in the source area of the Yellow River were shrinking. Wu and Zhu (2008) [19] conducted a similar study for Nam Co and indicated a continuous expansion of this lake in the past 40 years. They attributed the lake expansion to glacier retreat, the increase in temperature and precipitation, and the decrease in evaporation in the lake basin. Liu et al. (2009, 2010) [27,28] detected changing patterns of three inland lakes on the central Tibetan Plateau (Nam Co, Zigetang Co, and Co Na) from 1965 to 2005 and correlated lake changes with changes in precipitation, glacial melting, and permafrost degradation. In recent years, more studies have started to investigate lake extent changes in relatively large areas and over the whole plateau [34–36]. For example, Huang et al. (2011) [34] investigated lake extent changes in the source areas of the Yellow and Yangtze Rivers on the northeastern Tibetan Plateau based on three time slices of Landsat images in the 1970s, early 1990s, and around 2004. They founded a widespread declining trend in lake abundance and area in the whole headwater basin of the Yellow River and the southeastern headwaters of the Yangtze River, but lake expansions occurred in the western and northern headwaters of the Yangtze River. Li et al. (2011) [35] conducted a plateau-wide lake change study using Landsat images of four time slices (the 1970s, around 1990, 2000, and 2009) and found that lake expansions dominated on the northern plateau, but lakes shrank in southern regions. The coarse temporal resolution prevented the identification of rapid/slow phases of lake changes and potential spatio-temporal transitions in lake changes.

Variations in lake levels can be monitored by satellite altimetry data [37–42]. Both satellite radar and ICESat altimetry data have been used to investigate recent lake-level variations for several large lakes on the Tibetan Plateau, such as Qinghai Lake [37] and Nam Co [38]. The results indicated that ICESat altimetry data are more precise than satellite radar altimetry and can be used to validate the inter-annual trends of lake-level change. Zhang et al. (2011) [39] examined lake level variations of >100 lakes across the Tibetan Plateau from 2003 to 2009 using ICESat altimetry data. They concluded that most lakes on the plateau were expanding because of glacier retreat in surrounding mountains, whereas the shrinkage of some lakes is mainly caused by decreasing precipitation. Phan et al. (2011) [40] conducted a similar study using the same dataset to examine more lake-level variations across the plateau. Zhang et al. (2013) [41] used ICESat altimetry data from lakes to assess the contribution of lake changes to the increased mass over the Tibetan Plateau derived by the Gravity Recovery and Climate Experiment (GRACE) data. However, as satellite altimetry data are limited to only within the recent decade (for example, ICESat altimetry data are available just from 2003 to 2009), a long-term plateau-wide assessment of lake changes is still lacking.

Although recent changes in lake extents and levels have been investigated using remote sensing techniques, the driving factors for the lake changes still remain unclear. Many studies have argued for the importance of accelerated glacier melting on recent lake changes (especially lake expansion) [10–12,14–26,29,31,39,43,44]. However, only few studies have estimated the contribution of glacier melting to the lake water balance, whereas the estimates have large uncertainties because of the difficulty in determining the volume change of glacial mass [29,31]. Changes in precipitation and evaporation have also been proposed as major drivers for lake changes on the Tibetan Plateau [19,21,26–28,30,33,41]. Some studies suggest that increasing precipitation leads to increasing cloud cover and humidity, decreasing sunshine duration and evaporation, and all these changes favor lake expansions on the central plateau [19,26–28,30]. In contrast, decreasing precipitation and increasing evaporation may lead to lake shrinkages on the southern plateau [20,21]. A few studies also suggest permafrost degradation as a potential factor for lake changes [27,28,35,36], but no detailed studies have been conducted to investigate the role of permafrost degradation on lake changes.

In the work reported here, we document a detailed record of lake extent and level changes since the 1970s across the Tibetan Plateau using the combination of Landsat imagery and ICESat altimetry data. Our aim is to determine if lakes across the Tibetan Plateau behave in a similar fashion or if distinct spatial and temporal patterns can be identified. Then, we test if the pattern of lake changes could be explained by spatial variations in glacial retreat. Finally, we examine the influences of other climate and cryospheric factors on the pattern of lake changes, including changes in precipitation/evaporation and the pattern of permafrost degradation.

Datasets and Methods

We extracted detailed lake-level variations of 94 inland lakes that contain at least four years of measurements between 2003 and 2009 using ICESat altimetry data [45] (Table S1) across the Tibetan Plateau. To systematically examine lake changes beyond 2003–2009, we selected 25 lakes from five regions (five in each region, Fig. 1, Table S2) across the plateau to determine their detailed changes in extent from 1972 to 2010 using Landsat imagery (Tables S3–S7). Comparing the extents of these lakes over nearly four decades allows for an assessment of spatio-temporal patterns of lake change. We then compared these patterns with observed meteorological data from 1970 to 2010 within and around the plateau (Table S8). We also analyzed precipitation trends derived from the Tropical Rainfall Measurement Mission (TRMM) datasets [46] to supplement data from the relatively few meteorological stations on the northwestern plateau. We excluded lakes on ice surfaces and in front of modern glaciers because most of them drain downstream, and variable outflow from the system changes the relationship between changing lake levels and conditions within the drainage basin.

Lake-level extraction based on ICESat altimetry data

We used ICESat altimetry data to extract lake levels in 2003–2009. ICESat was launched in January 2003 and used GLAS (the Geoscience Laser Altimeter System), a laser-ranging instrument, to measure land surface elevation in the world [45]. During its lifetime from 2003 to 2009, the GLAS instrument collected land surface elevations in designated periods each year (typically one in the spring (March) and one in the fall (September to October)). The available GLAS global land surface elevations have a vertical accuracy at the decimeter- to centimeter-level over flat surfaces and within an approximately 70 m diameter footprint [45,47]. This dataset has the potential to monitor the changes in lake levels on the Tibetan Plateau. Because the distances between different ICESat tracks on the Tibetan Plateau are about 70 km [40], ICESat data are only available for the lakes that overlapped with ICESat tracks.

We developed a streamlined workflow to extract lake-level changes (Fig. S1a). We first imported GLAS data into ArcGIS as point features. Then, we extracted GLAS points within each lake using a lake polygon dataset provided by Chinese Academy of Sciences. Because the lake boundary may change over time, this operation may include some GLAS points outside the lakes, introducing outliers. In addition, instrument uncertainty or the impact of cloud or other factors may also introduce outliers. Based on the assumption that surface elevation is relatively constant for any given period across a lake, we designed a mode-mean method to extract lake elevations. As illustrated in Fig. S1b, the mode-mean method first examines the mode of all GLAS elevations based on a ±30 cm window (considering potential influences of waves on lake surface elevation) to exclude outliers and then determine the mean and standard deviation of the lake elevations. After extracting lake elevations for different periods, trend and regression analyses were conducted to determine the overall trends of the lake-elevation changes (the annual lake-level change rate (m/a)) and to identify detailed temporal patterns (Fig. S1c). All 94 inland lakes with at least 4 years of ICESat data are listed in Table S1, including the annual lake-level change rate (m/a), R^2, and p-value. The uncertainty of the extracted lake surface elevation is < 10 cm.

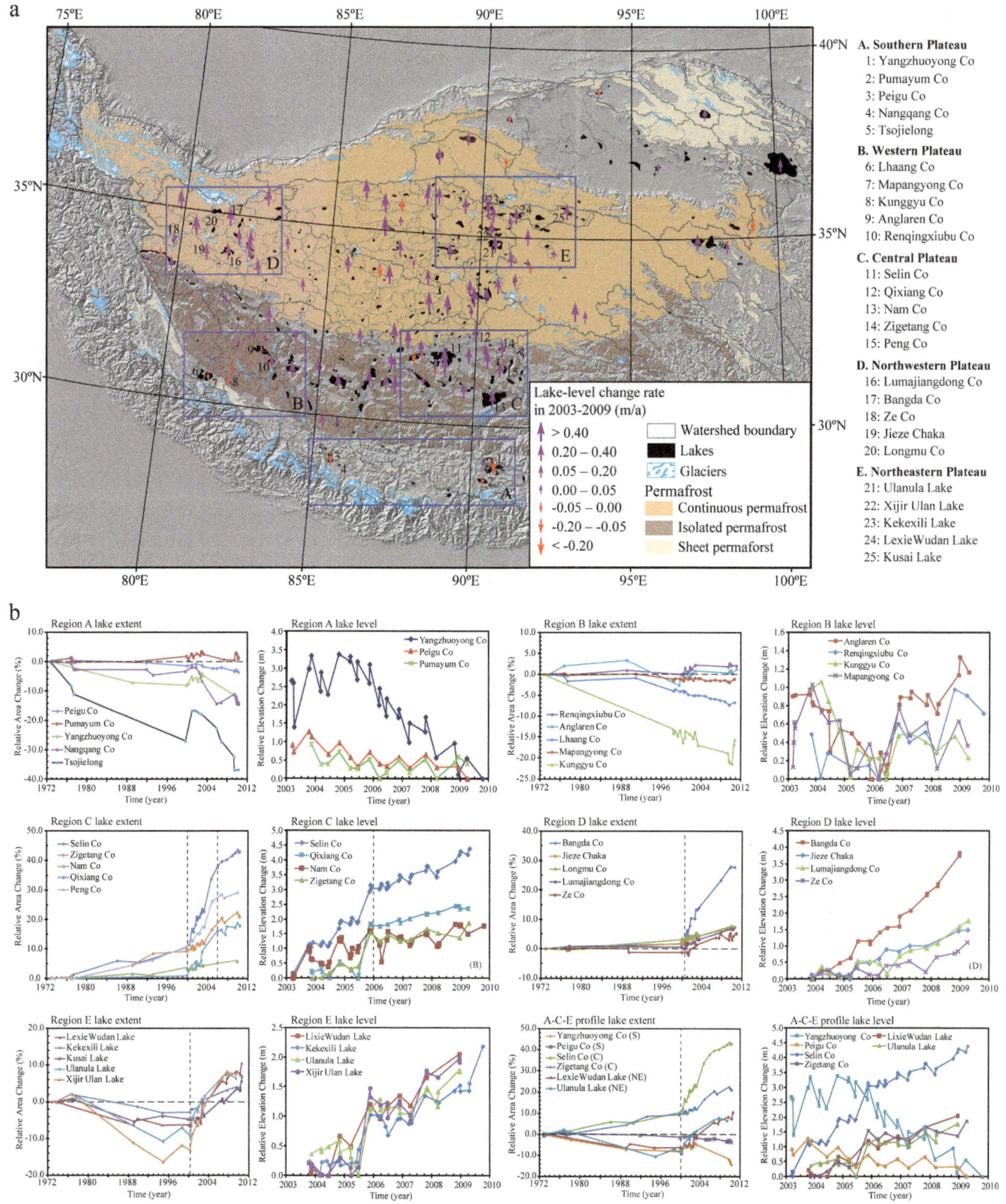

Figure 1. Lake changes across the Tibetan Plateau in recent decades. a, Spatial distribution of lake-level change rates (m/a) of 94 lakes (2003–2009) derived using ICESat altimetry. b, Detailed changes of lake extent since the 1970s (delineated using Landsat imagery) and lake levels in 2003–2009 for 25 lakes. These lakes were selected from five regions marked in panel a (5 lakes in each region: A-southern, B-western, C-central, D-northwestern, E-northeastern Tibetan Plateau). Changes in lake extents and levels along a profile from the southern (A), central (B), to northeastern (E) Tibetan Plateau are also illustrated in panel b.

Lake-extent delineation based on Landsat images

To systematically examine lake changes beyond the period of 2003–2009, we selected 25 lakes from five regions across the Tibetan Plateau to determine detailed changes in lake extent since the 1970s using Landsat MSS and TM/ETM+ images (Table S2). Landsat images were downloaded from the U.S. Geological Survey (USGS)'s Global Visualization Viewer (http://glovis.usgs.gov). To obtain a high temporal resolution of lake-extent changes, we examined and downloaded as many images as possible with a clear view of lake boundaries. In general, about 14–40 images were downloaded and processed for each lake (Tables S3–S7).

Most Landsat images were already geometrically and radiometrically corrected by the USGS EROS Digital Image Processing Center (http://seamless.usgs.gov/faq/nlcd_faq.php). For images that required additional geometric correction, we compared and geo-referenced them with other geometrically corrected images. Landsat MSS imagery has four spectral bands and the last two bands are sensitive to water bodies. TM/ETM+ images have seven and eight bands, respectively, and bands 4 and 5 are sensitive to water bodies [48]. Therefore, lake-boundary extraction was mainly based on these bands. Due to the distinct reflection difference between water and land, we extracted lake boundaries using thresholds automatically defined by the natural break classification in ArcGIS (Fig. S2a). To ensure the quality and accuracy, extracted lake boundaries were visually compared to the composite imagery (bands of 532 or 432 for the TM/ETM+ imagery) and erroneous boundaries were fixed by edit and additional manual digitization. Surface areas of each lake in different time slices were calculated in ArcGIS. Fig. S2b–c illustrates an example of lake boundaries extracted for Selin Co, the second largest lake on the Tibetan Plateau and the composite images that represent detailed lake extent changes in four different years of 1972, 1986, 1999, and 2010.

We assessed the uncertainty of Landsat image-delineated lake extent by comparing the lake boundary delineated using high resolution Google Earth imagery at a similar time slice. Different from the resolution of Landsat imagery (79 m for Landsat MSS and 30 m for TM/ETM+ imagery), the resolution of Google Earth imagery in some regions can reach to 0.6 m [49]. Since Google Earth images are usually a mosaic of multiple high resolution images taken at different times, we selected three relatively small lakes (Laxiong Co, Jieze Chaka, and Daze Co) completely located within the coverage of a single high resolution image to assess the uncertainty. We manually delineated the boundary of each lake from Google Earth (data source of the high resolution imagery: Digital Globe) and then compared it with a Landsat image-delineated lake boundary at a similar time slice (Fig. S3a–c). The comparison indicates that the uncertainty of Landsat image-delineated lake surface area is generally $<1.0\%$ and larger lakes tend to have smaller percentage area differences (Fig. S3d).

Data sources and analysis of climate variables

We collected observed meteorological data (including daily mean temperature, precipitation, solar radiation, wind speed, and vapor pressure) for 1970–2010 from 117 stations within and around the Tibetan Plateau (Table S8) from the China Meteorological Data Sharing Service System (http://cdc.cma.gov.cn/). We calculated annual average temperature and annual precipitation for each station using the observed daily data. Since most stations have no directly observed evaporation data, we estimated the potential evapotranspiration of each station based on the Penman-Monteith model recommended by the Food and Agriculture Organization of the United Nations (FAO) [50]. We then applied the Kriging method to interpret the spatial distribution of annual temperature, precipitation, and potential evapotranspiration in each year across the plateau. A lapse rate of 6°C/1000 m was used to scale temperature to the surface of 4500 m above sea level (most lake levels are around this elevation) to interpret the spatio-temporal distribution of the temperature on this surface (Fig. S4). The spatial distributions of precipitation (Fig. S5) and potential evapotranspiration (Fig. S6) were interpreted without scaling due to the lack of lapse rates for these two variables. Since all climate variables are affected by local topography [51], the interpreted temperature, precipitation, and potential evapotranspiration may not represent the "real" values in each location. Instead, we mainly used the interpretations to examine the temporal trends of these variables across the Tibetan Plateau (Figs. 2, S7, S8).

Given the importance of precipitation in driving lake changes and the fact that few meteorological stations are located on the northwestern Tibetan Plateau, we further analyzed the Tropical Rainfall Measuring Mission (TRMM) Multi-satellite Precipitation Analysis 3B43 (Monthly precipitation data from 1998 to 2012, 0.25×0.25 degree merged TRMM, and other sources estimates) [46] to supplement the assessment of the role of precipitation on lake changes. This dataset combines the estimates generated by the TRMM and other satellites product (3B42) and the global gridded rain gauge data produced by NOAA's Climate Prediction Center and/or the global rain gauge product produced by the Global Precipitation Climatology Center. Recent studies suggest that TRMM data, compared to rain gauge observations, show better performance than other satellite precipitation products with high correlation coefficients and lower root mean square errors [52–54]. We used the precipitation record from 92 stations within the plateau (with more than 10-years of precipitation record since 1998) to assess the quality of TRMM-derived precipitation and to examine the similarities and differences between the trends derived from these two data sources (Fig. S9). We also examined spatial patterns of TRMM-derived precipitation in 1998–2012 across the Tibetan Plateau (Fig. S10).

Data sources and analysis of glacier and permafrost coverage for each lake basin

We delineated the drainage basin for each lake in ArcGIS using the 90-m resolution Shuttle Radar Topography Mission (SRTM) DEMs [55]. Then, we calculated the glacier and permafrost coverage (%) for each lake basin using the GLIMS glacier database [56] downloaded from the National Snow and Ice Data Center (http://nsidc.org/glims/) and the permafrost data [57] of the Tibetan Plateau downloaded from the Environmental and Ecological Science Data Center for West China (http://westdc.westgis.ac.cn/), respectively (Table S1).

Results

Patterns of lake change in 1972–2010

Our results in lake level and areal extent document both seasonal and inter-annual changes of Tibetan lakes (Figs. 1, S11). For example, seasonal changes of Selin Co in surface area can be identified using multiple Landsat TM images from different seasons (Fig. S11). Seasonal changes are also apparent in lake-level variations derived by ICESat altimetry data (Fig. S11). Lake level is usually low during winter or early spring and reaches its maximum in fall (September to October). Seasonal changes can also be observed in other lakes (Fig. 1b).

Beside seasonal changes, our results reveal a distinct southwest-northeast spatio-temporal pattern of lake inter-annual changes

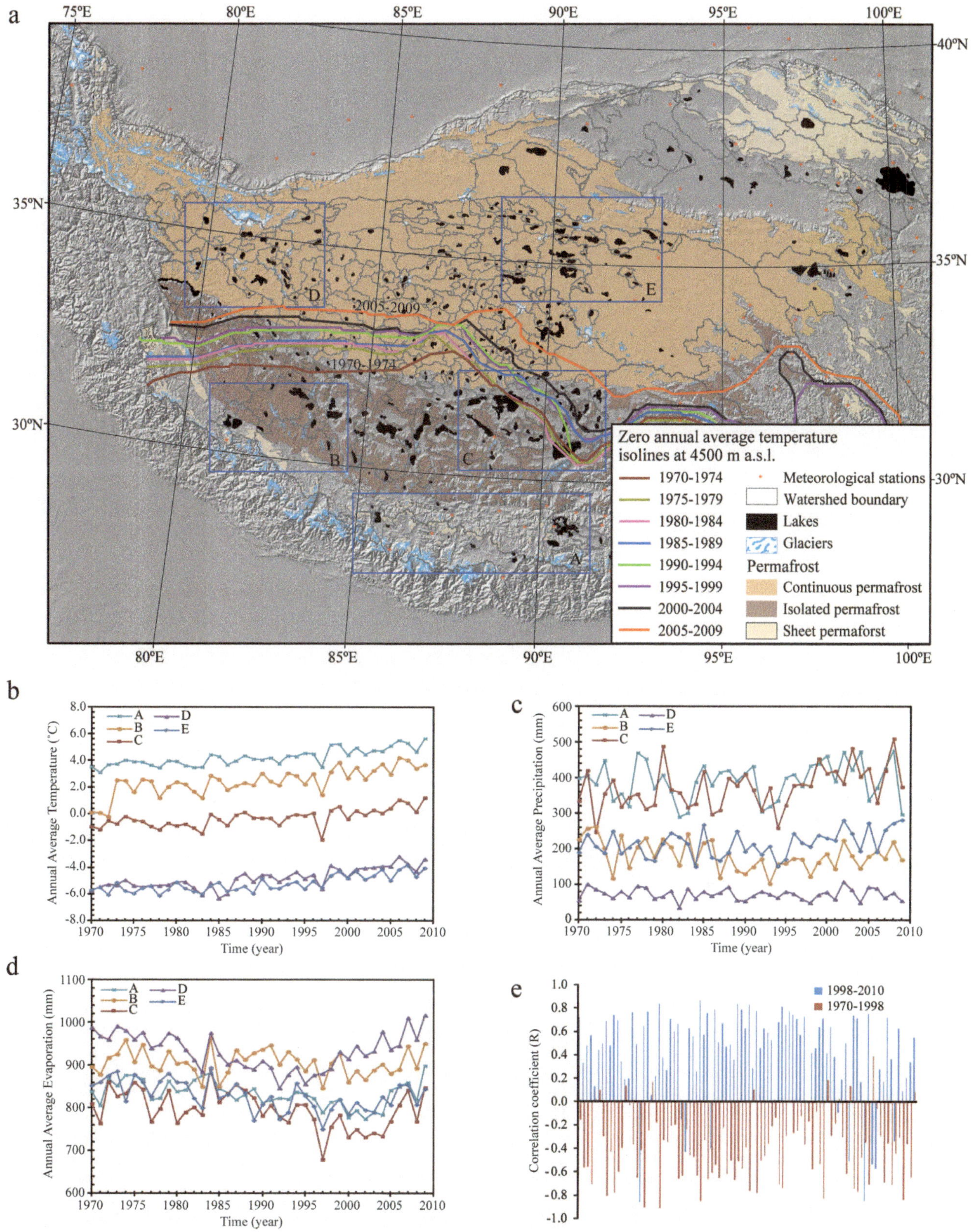

Figure 2. Climate variations on the Tibetan Plateau since the 1970s. a, Map showing the northward migration of 5-year annual average zero-temperature isolines (at 4500 m a.s.l). b, c, d, Variations in annual average temperature (b), precipitation (c), and evaporation (d) in five regions (A–E). e, Potential evapotranspiration trends before and after 1998 derived from meteorological stations within the Tibetan Plateau. The increasing or decreasing trends over time are represented by the correlation coefficient, R (R>0 indicates an increasing trend, and R<0 a decreasing trend).

across the Tibetan Plateau in 1972–2010 from shrinking, to stable, to rapidly expanding (Fig. 1). Lakes on the southern and western plateau (Regions A and B in Fig. 1) have experienced continuous shrinkage or been relatively stable since the 1970s, except for a short period of slight expansion in 2000–2004. The lakes closest to the Himalayas have shrunk, whereas others have been relatively stable. On the central and northern plateau, most lakes expanded rapidly after 2000, but lake changes before 2000 varied with region. Lakes on the central plateau (Region C in Fig. 1) expanded continuously in 1972–2010, with expansion rates relatively limited (<10% in lake-extent change) before 2000, but rapid after 2000 (20–40%). A slowing trend in the expansion rate appeared after 2006. Lakes on the northwestern and northeastern plateau also expanded rapidly after 2000. However, before 2000, lakes on the northwestern plateau (Region D in Fig. 1) were stable or slightly shrinking and lakes on the northeastern plateau (Region E in Fig. 1) experienced various extents of shrinkage. In addition, no slowing trend occurred after 2006 in these two regions, and all lakes continue to expand rapidly. This relatively long-term record of lake changes indicates that lakes were primarily shrinking or relatively stable on the southern and western plateau, but expanding dramatically on the central and northern plateau after 2000, even after different patterns of lake change before 2000.

Temperature variations across the Tibetan Plateau

The annual average temperature on the Tibetan Plateau has continuously increased since the 1970s, with a spatial pattern of much higher than 0°C distributed in the south, close to 0°C in the center, and below 0°C in the north. The temperature zone of close to 0°C (freezing point) has gradually migrated northward (Fig. 2a). The temperature trend, interpreted based on observed meteorological data, is consistent with the pattern of continuous increase in temperature after 2000 revealed using the land surface temperature (LST) product from Moderate Resolution Imaging Spectroradiometer (MODIS) [58].

Precipitation variations across the Tibetan Plateau

Compared to the continuous increase in temperature, patterns of precipitation change are much more complex across the Tibetan Plateau. As illustrated in Figs. S5 and S10, precipitation on the Tibetan Plateau varies significantly among different years, but generally decreases from southeast to northwest. For the precipitation trend in each meteorological station (Fig. S7), during the period of 1970–2010, 75.3% of stations (64 out of 85 stations within the plateau that have >20 years of precipitation records) showed increasing trends, but only 15.3% of the trends (13 stations) were significant at the level of 0.05 ($p<0.05$). The other 21 stations (24.7%) showed decreasing trends, but none of them were significant. Because dramatic lake expansions on the central and northern plateau occurred mainly after 2000, we also analyzed the precipitation trends before and after 1998. For the period of 1970–1998, 58.1% of stations (50 out of 86 stations within the plateau with >15 years of precipitation records) showed increasing trends, but the increase was significant at only one station. The other 36 stations (41.9%) showed decreasing trends, but only one trend was significant. For the period of 1998–2010, 43.5% of stations (40 out of 92 stations within the plateau that have >10 years of precipitation records) showed increasing trends (but only 6 stations reached $p<0.05$). The other 52 stations (56.5%) showed decreasing trends (14 stations reached $p<0.05$). Our results are consistent with conclusions drawn by Yang et al. (2011) [59] that the precipitation trends on the Tibetan Plateau were not significant at $p<0.05$ although slightly increasing trends are observed in the semi-arid and semi-humid zones (central plateau) and decreasing trends in the humid and arid zones (southern and western plateau).

Our results indicate that TRMM-derived and station-observed precipitation are highly correlated ($R^2=0.79$, Fig. S9a). Specifically, 90 out of 92 stations have positive correlations between these two datasets ($p<0.05$ for 76 stations, Fig. S9b). Such statistically significant correlations indicate that TRMM-derived data can be used to evaluate precipitation patterns on the Tibetan Plateau. Comparing precipitation trends derived for each station with its corresponding trend derived from the TRMM dataset also indicates that 72 out of 92 stations (78.3%) within the plateau have similar trends between these two datasets (Fig. S9c). The other 20 stations with opposite trends are located mainly in high-relief mountainous areas, such as several stations close to the Himalayas on the southern plateau and stations close to the eastern marginal mountains of the plateau (Fig. 3a). These results suggest that TRMM-derived data can be used to reflect precipitation trends in most areas of the Tibetan Plateau, especially relatively flat plateau surfaces on the central and northern plateau.

The TRMM-derived precipitation trends from 1998 to 2012 on the Tibetan Plateau and its surrounding regions (Fig. 3a) are similar to the trends derived using station-observed precipitation data. Specifically, precipitation on the southern and southwestern plateau is decreasing, especially close to the Himalayas. Precipitation on the central and northern plateau is slightly increasing in many regions, but decreasing in other regions, such as on the northwestern plateau, where five of our examined lakes are located (Fig. 1a), and in some areas on the central-northern plateau.

Variations in potential evapotranspiration across the Tibetan Plateau

Chen et al. (2006) [60] analyzed the variations in potential evapotranspiration across the Tibetan Plateau using the Penman-Monteith method [50] and concluded that potential evapotranspiration in most areas of the plateau decreased from 1961 to 2000. They argued that the decreasing trend in potential evapotranspiration was mainly caused by the decreasing wind speed on the Tibetan Plateau [60]. We extended this analysis to 2010 using the same method and parameters (Figs. S6, S8a). Our results indicate that 72.6% of stations (61 out of 84 stations within the plateau that have >20 years of record) showed decreasing trends in potential evapotranspiration during the period of 1970–2010 (Fig. S8a) and 34.5% of the trends (29 stations) were significant ($p<0.05$). The other 23 stations (27.4%) showed increasing trends, but only the trends at 8 stations were significant (Fig. S8d). In 1970–1998 (Fig. S8b), 88.2% of stations (75 out of 85 stations within the plateau that have >15 years of record) showed decreasing trends, and the trends at 35 stations (41.2%) were significant. The other 10 stations (11.8%) showed increasing trends, but the increase was significant at only one station. These results are consistent with the conclusions drawn by Chen et al. (2006) [60].

However, in contrast to the decreasing trend before 1998, potential evapotranspiration showed increasing trends in most stations after 1998 (Fig. S8c). Specifically, 87.9% of stations (80 out of 91 stations within the plateau that have >10 years of record) showed increasing trends, with the trends at 42 stations (46.2%) being significant ($p<0.05$) (Fig. S8d). The other 11 stations (12.1%) showed decreasing trends (but only three of those were significant, Fig. S8d).

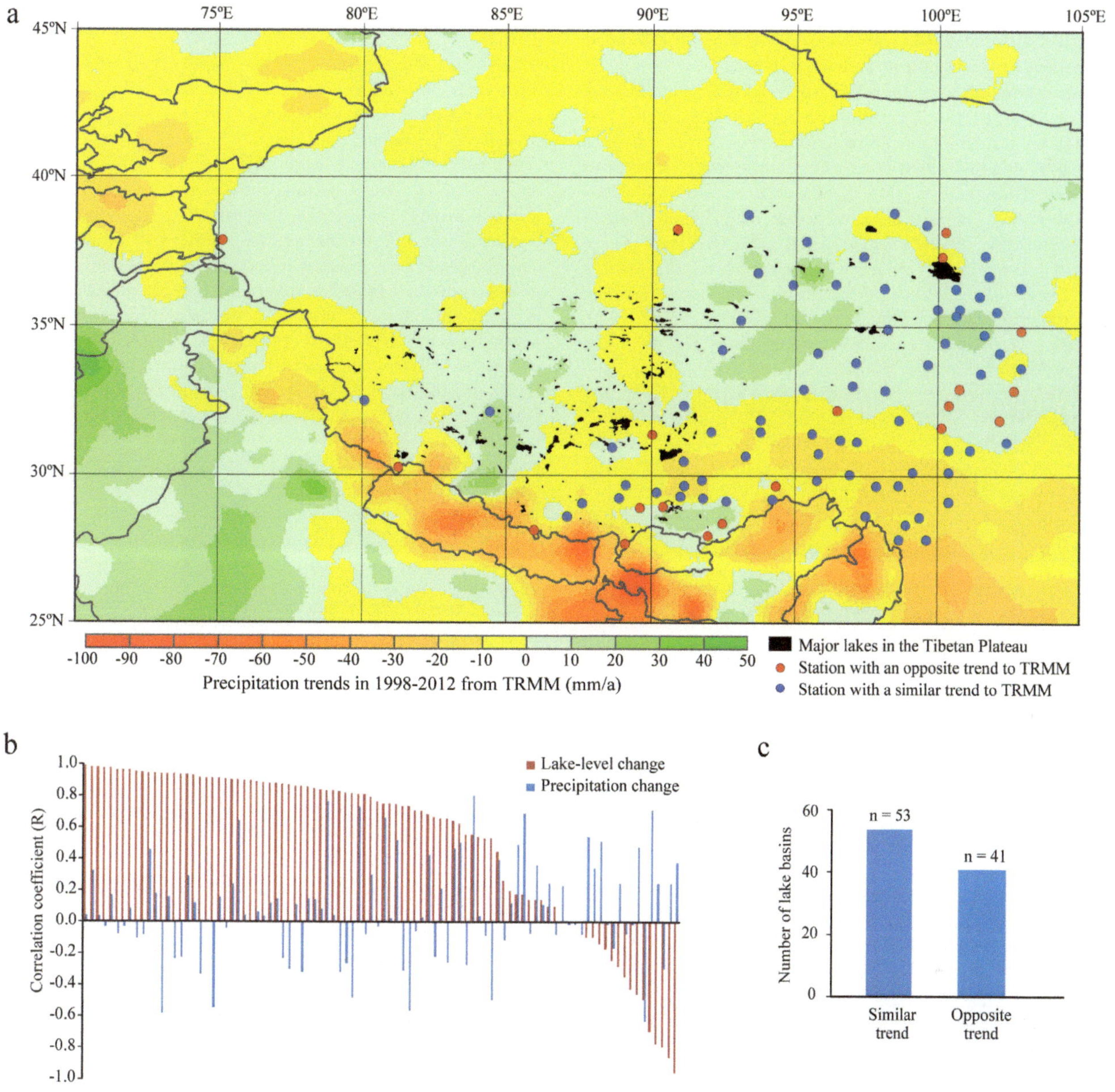

Figure 3. TRMM-derived precipitation trends in 1998–2012 on the Tibetan Plateau and its surrounding regions. a, Spatial distribution of TRMM-derived precipitation trends. b, Comparison between lake-level and precipitation trends for the 94 lake basins listed in Table S1. The increasing or decreasing trends over time are represented by the correlation coefficient, R. c, Numbers of lake basins with similar and opposite trends between lake-level and precipitation changes.

Glacier and permafrost coverage over the Tibetan Plateau

Glaciers distributed on the Tibetan Plateau and surrounding mountains covered about 100,000 km^2 in the 1970s [4]. A study of 7,090 glaciers suggested that the area of these glaciers decreased from ~13,363 km^2 in the 1970s to ~12,130 km^2 in the 2000s, with a total area reduction of 9.2% [4]. Glacial retreat is more apparent in the Himalayas and other marginal mountains, whereas the glaciers on the central plateau are relatively stable [4,6,61]. The glacier coverage of the 94 lake basins we investigated ranges from 0.0 to 10.5% and 29 lakes (mainly from the central and northern plateau) are free of glacier coverage in their drainage basins (Table S1).

The permafrost coverage is about 1.3×10^6 km^2, accounting for ~67% of the total area of the Tibetan Plateau [62,63]. The thickness of the permafrost ranges from 10 m to 312 m [64] and the total volume of ground ice held in the permafrost is estimated at 9,528 km^3, twice the volume of glaciers [65]. The permafrost coverage of the 94 lake drainage basins ranges from 7 to 100%. Fifty-five lake basins with 100% permafrost coverage are located on the central and northern Tibetan Plateau (Table S1).

Discussion

Our derived changes in lake levels and extents are at least one order of magnitude higher than the uncertainties inherited from the Landsat/ICESat data. Therefore, the plateau-wide pattern of lake changes we derived represents the recent changes in water balances of these lakes driven by potential climate and associated cryospheric changes within their drainage basins.

Temperature has continuously risen on the Tibetan Plateau since the 1970s, causing accelerated glacier melt (Fig. 2a–b). The rate of glacial retreat is more apparent in the Himalayas and other marginal mountains, with a decreasing trend from the margin to the center of the plateau [4,6,61]. However, our derived spatio-temporal pattern of lake changes across the Tibetan Plateau (Fig. 1) contrasts distinctly with the spatial characteristics of glacier retreat, suggesting limited influence of glacier melt on lake changes in this region. Specifically, our results indicate no statistically significant correlation between changes in lake levels (2003–2009) and glacier coverage in each lake's drainage basin (Fig. 4a). Most lakes (>79%) in drainage basins containing no glaciers on the central and northern plateau still experienced expansion (Fig. 4b). In contrast, lakes on the southern plateau, fed by rapidly retreating glaciers, were relatively stable or shrinking. Therefore, our results suggest that glacier melt is not the dominant driver for the recent lake expansions across the Tibetan Plateau.

Meteorological records indicate a trend of stable or slightly decreasing precipitation on the southern, western, and northwestern plateau, but a slight increase in precipitation on the central and northeastern plateau (Fig. 2c). TRMM-derived precipitation trends in 1998–2012 show a similar pattern (Fig. 3a), but with more variability within the plateau. The overall pattern of lake changes across the plateau seems to be correlated with changing precipitation patterns, except for the northwestern plateau, where lakes expand rapidly after 2000 even under a slightly decreasing trend in precipitation. In particular, the relatively stable or slightly decreasing precipitation may explain the relative stability or slight shrinkage of lake extents/levels on the southern and western plateau, and the slight increase in precipitation seems to be related to the rapid expansions of most lakes on the central and northern plateau after 2000. Several recent studies [33,41] have suggested the important role of precipitation in recent lake expansions. However, our detailed trend analysis indicates that not all lake-level trends in 2003–2009 of the 94 lakes we analyzed are directly related to the precipitation trends derived from the TRMM dataset. Specifically, only 56.4% of lakes we analyzed (53 out of 94) have similar (increasing or decreasing) trends between lake-level changes and precipitation variations in their corresponding drainage basins after 2000. In contrast, 43.6% of lakes (41 out of 94) have opposite trends (Fig. 3b–c). Although uncertainties exist in the TRMM dataset and TRMM-derived precipitation does not necessarily represent local precipitation observed from meteorological stations for all areas over the plateau, the comparison between these two datasets (Figs. 3a, S9) indicates that their derived precipitation trends are highly correlated (with positive correlations), especially on relatively flat plateau surfaces where our analyzed lakes and their basins are mainly located. Thus, our observation that 43.6% of lakes have opposite trends with precipitation is likely not caused by the TRMM dataset uncertainties. This indicates that the variations in precipitation alone may still not be able to fully explain the overall pattern of recent dramatic lake changes across the whole plateau.

The potential evapotranspiration estimated from meteorological stations decreased before 2000 [60], but showed a slightly increasing trend after 1998 in most stations (Figs. 2d–e, S8). The increasing evapotranspiration after 1998 would reduce surface flow into the lakes and mitigate the impact of potential precipitation increase on lake expansions. Therefore, recent dramatic expansions of most lakes on the central and northern plateau cannot be explained by changes in potential evapotranspiration.

The plateau-wide pattern of lake changes is consistent with the distribution of permafrost on the Tibetan Plateau (Fig. 1a). Due to its high elevation, 67% of the plateau is covered by permafrost that is sensitive to rising temperature [57,62], leading to the increase in active layer thickness, thawing area, and the number of thawing days [66,67]. Thus, permafrost degradation may play an important role in driving lake changes. When ground temperature is much lower than the melting point of the frozen soil, the water contribution from permafrost is limited and lake extents/levels are mainly controlled by other factors, such as the balance between precipitation and evaporation. When temperature increases to the melting point, accelerated permafrost melting will start to contribute to lake changes. However, if temperature continuously increases and remains above the melting point, the water contribution from permafrost will gradually become limited because water held in the frozen soil may already have been released. This supply-limited relationship between the amount of water contributed by permafrost and the ground temperature may explain the spatio-temporal pattern of lake changes across the Tibetan Plateau. The annual average temperature on the southern and western plateau has been much higher than 0°C since the 1970s (Fig. 2a–b); this may indicate that water contribution from permafrost was already limited and lake variations were mainly responding to changes in precipitation and evaporation. On the central plateau, temperature was lower than but close to the melting point before 2000; thus, lakes may have been able to expand slightly due to the contribution from the permafrost that was partially melted. Temperature in most areas of the central plateau might reach the melting point after 2000, increasing the permafrost contribution and causing rapid lake expansions. After 2006, continuously rising temperature may have reduced the permafrost contribution on the central plateau, slowing the trend of lake expansion. On the northern plateau, temperature was much lower than the melting point before 2000. Lake levels during this period remained relatively stable or shrank, depending on the water balance from other factors. Temperature on a significant portion of the northern plateau may have reached the melting point after 2000, causing accelerated permafrost melting and rapid lake expansion. Due to the relatively low temperature on the northern plateau, the permafrost contribution after 2006 may still be increasing, causing continuous lake expansion without a slowing trend.

This explanation is partially supported by the distribution of different permafrost types across the Tibetan Plateau (Fig. 1a). Permafrost in the southern plateau is discontinuous, sporadic and found mainly in higher mountains, whereas permafrost in the central and northern plateau is continuous and covers most plateau surfaces and mountains [57,62]. Long-term permafrost monitoring on the Tibetan Plateau indicated that the permafrost area has been shrinking rapidly in response to rising temperature in recent decades [57,62]. In particular, the lower elevation limit of permafrost moved up by 25 m in the north in the past 30 years and between 50 and 80 m in the south over the last 20 years [57,62]. A previous study indicated that the active layer thickness of the permafrost along the Qinghai-Tibet Highway increased at a rate of 7.5 cm/a from 1995 to 2007 [66]. Furthermore, recent work showed that the active layer thickness increased at a rate of 3.1 cm/a from 1998 to 2010 for the permafrost distributed above 4700 m a.s.l [67]. Based on drilling data, a recent study found that ground ice is widely distributed in the Beiluhe Region, a permafrost monitoring area on the northern plateau [68]. This study also found that ice contents within 1–3 m beneath the

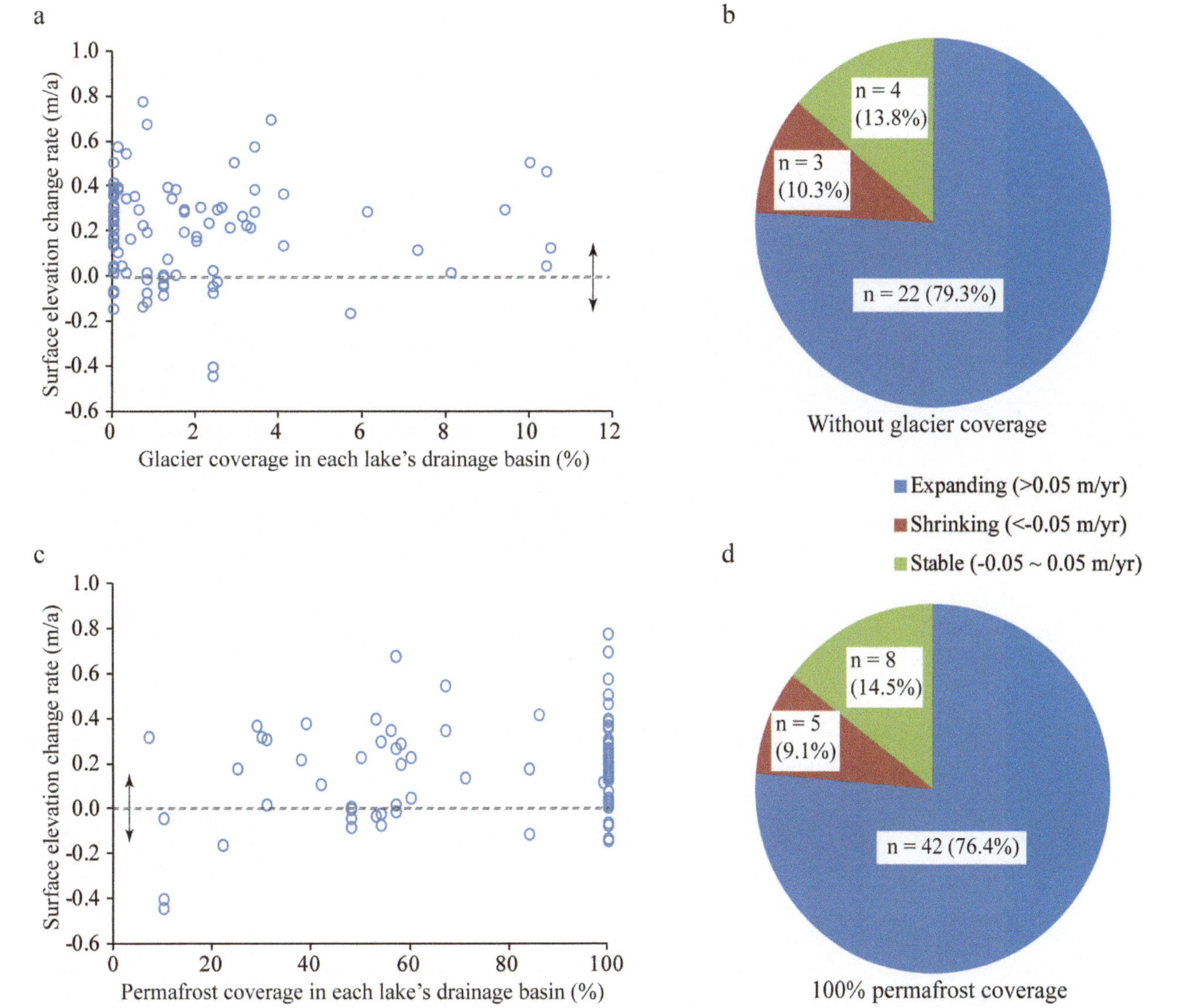

Figure 4. Relations between lake change and glacier/permafrost coverage in each lake's drainage basin. a, Relations between lake-level change rate in 2003–2009 (m/a) and glacier coverage (%). b, Numbers of expanding, stable, and shrinking lakes in drainage basins with 0% glacier coverage. c, Relations between lake-level change rate in 2003–2009 (m/a) and permafrost coverage (%). d, Numbers of expanding, stable, and shrinking lakes in drainage basins with 100% permafrost coverage.

permafrost table, where the increased active layer occurs, are > 50% [68]. With the extensive drainage area feeding each lake (Table S1), permafrost thaw may contribute a significant amount of water to sustain the continuous lake expansion observed in the recent decade. Our results indicate a positive correlation between lake-level changes (2003–2009) and permafrost coverage in each lake's drainage basin (Fig. 4c). In particular, >76% of lakes in drainage basins fully covered by permafrost had a clear trend of increasing lake levels (Fig. 4d). However, we recognized that not all permafrost melt would be converted into runoff flowing into lakes, and detailed studies of permafrost melt are of critical importance to quantify the water contribution from permafrost in the future.

Our observations from the Tibetan Plateau are consistent with the findings from the Arctic that lakes in continuous permafrost regions tend to expand, whereas lakes in isolated permafrost regions are shrinking [69,70]. Our results also suggest that continuously rising temperature may eventually reduce the water contribution from permafrost, causing permafrost-driven lake expansions to cease. For example, the permafrost contribution to lake changes on the southern and western plateau may have already become limited due to high temperatures, and lake variations are mainly responding to the balance between precipitation and evaporation. The space-time transition of lake changes across the Tibetan Plateau suggests that the dramatic lake expansions on the central and northern plateau since 2000 are probably ephemeral and may not be sustained under continuously rising temperature. This finding has important implications for ecosystem, water resource, agriculture, and infrastructure management on the Tibetan Plateau and its surrounding regions.

Supporting Information

Figure S1 Lake-level extraction based on ICESat altimetry data. a, Flowchart of lake surface elevation extraction method. b, The mode-mean method used to extract lake surface elevations. c, Lake surface elevation variations of Selin Co on the Central Tibetan Plateau from 2003 to 2009, representing a continuous increasing trend in surface elevation.

Figure S2 Lake-extent delineation based on Landsat images. a, Flowchart of lake-boundary extraction method. b, Lake-boundary variations of Selin Co representing a continuous lake expansion since the 1970s. c, Landsat images of three marked regions in panel b (A, B, C) around Selin Co, representing lake-boundary changes in four different years (1972, 1986, 1999, 2010).

Figure S3 Accuracy assessment of Landsat image derived lake boundary for three lakes using Google Earth high resolution imagery. a, Laxiong Co. b, Jieze Chaka. c, Daze Co. d, A comparison table representing the percentage difference in lake areas.

Figure S4 Spatial variations of annual average temperature at 4500 m a.s.l. in different years since the 1970s across the Tibetan Plateau. Black dots represent meteorological stations used for temperature interpretation. A lapse rate of 6°C/1000 m was used to scale all observed data to the surface of 4500 m a.s.l. and then the Kriging method was used to interpret the spatial distribution of the temperature in each year at this surface.

Figure S5 Spatial variations of annual precipitation in different years since the 1970s across the Tibetan Plateau. Black dots represent meteorological stations used for precipitation interpretation. The interpretation was simply based on the observed data from each station.

Figure S6 Spatial variations of annual potential evapotranspiration in different years since the 1970s across the Tibetan Plateau. Black dots represent meteorological stations used for interpretation.

Figure S7 Precipitation trends observed in meteorological stations within the Tibetan Plateau. a, Precipitation trends in 1970–2010. b, Precipitation trends in 1970–1998. c, Precipitation trends in 1998–2010. d, Percentage of stations with different precipitation trends and statistical significant levels.

Figure S8 Potential evapotranspiration trends observed in meteorological stations on the Tibetan Plateau. a, Potential evapotranspiration trends in 1970–2010. b, Potential evapotranspiration trends in 1970–1998. c, Potential evapotranspiration trends in 1998–2010. d, Percentage of stations with different potential evapotranspiration trends and statistical significant levels.

Figure S9 Comparison between precipitation data derived from TRMM and observed from meteorological stations within the Tibetan Plateau. a, Regression between these two datasets. b, Number of stations where the correlation between these two datasets with $p<0.05$ vs. $p>0.05$. c, Number of stations where precipitation are with similar vs. opposite trends between these two dataset.

Figure S10 Spatial variations of TRMM-derived annual precipitation in different years from 1998 to 2012 across the Tibetan Plateau. The spatial resolution of the TRMM dataset is 0.25×0.25 degree.

Figure S11 Variations in surface area and elevation of Selin Co from 1999 to 2010 indicating both seasonal and inter-annual changes.

Table S1 Lake-level changes of 94 lakes from 2003 to 2009 derived using ICESat altimetry data and the glacier and permafrost coverage in each lake's drainage basin.

Table S2 Selected lakes to delineate lake-extent changes using Landsat images.

Table S3 Lake-extent changes in the southern plateau (Region A) delineated using Landsat images.

Table S4 Lake-extent changes in the western plateau (Region B) delineated using Landsat images.

Table S5 Lake-extent changes in the central plateau (Region C) delineated using Landsat images.

Table S6 Lake-extent changes in the northwestern plateau (Region D) delineated using Landsat images.

Table S7 Lake-extent changes in the northeastern plateau (Region E) delineated using Landsat images.

Table S8 Meteorological stations (117) within and around the Tibetan Plateau used in this study.

Acknowledgments

We thank the National Climatic Data Center of China Meteorological Administration for providing meteorological data. We appreciate Drs. Carol Harden and Jon Harbor for help in the manuscript preparation, and the Academic Editor, Dr. Hayley J. Fowler, and two anonymous reviewers for their valuable comments and suggestions.

Author Contributions

Conceived and designed the experiments: YL JL HG. Performed the experiments: YL JL ZL GS. Analyzed the data: YL JL ZL GS. Contributed reagents/materials/analysis tools: YL JL ZL GS. Contributed to the writing of the manuscript: YL JL ZL GS.

References

1. Oerlemans J (2005) Extracting a climate signal from 169 glacier records. Science 308: 675–677.
2. Solomon S, Qin D, Manning M, Chen Z, Marquis M, et al. (2007) Climate Change 2007: The Physical Science Basis: Contribution of Working Group I to the Fourth Assessment Report of the Intergovernmental Panel on Climate Change. Cambridge University Press, New York.
3. Parry ML, Canziani OF, Palutikof JP, van der Linden PJ, Hanson CE (2007) Climate change 2007: Impacts, adaptation and vulnerability. Contribution of

Working Group II to the Fourth Assessment Report of the Intergovernmental Panel on Climate Change.
4. Yao T, Thompson L, Yang W, Yu W, Gao Y, et al. (2012) Different glacier status with atmospheric circulations in Tibetan Plateau and surroundings. Nature Clim. Change 2: 663–667.
5. Immerzeel WW, Van Beek LPH, Bierkens MFP (2010) Climate change will affect the Asian water towers. Science 328: 1382–1385.
6. Bolch T, Kulkarni A, Kääb A, Huggel C, Paul F, et al. (2012) The state and fate of Himalayan glaciers. Science 336: 310–314.
7. Scherler D, Bookhagen B, Strecker MR (2011) Spatially variable response of Himalayan glaciers to climate change affected by debris cover. Nature Geosci. 4: 156–159.
8. Wang S, Dou H (1998) Records of China Lakes (in Chinese). Beijing: Science Press.
9. Yang RH, Yu XZ, Li YL (2003) Dynamic analysis of the remote sensing information of the surface enlargement of Serling Lake in Tibet. Remote Sensing for Land & Resources 2: 64–67 (in Chinese).
10. Lu AX, Yao TD, Wang LH (2005) Remote sensing research about typical glacier and lake variation in Qinghai-Tibet Plateau. Journal of Glaciology and Geocryology 27: 783–792 (in Chinese).
11. Lu AX, Wang LH, Yao TD (2006) The study of Yamzho Lake and Chencuo Lake vaiation using remote sensing in Tibet Plateau from 1970 to 2000. Remote Sensing Technology and Application 21: 173–177 (in Chinese).
12. Zhao YY, Zhao XT, Zheng MP (2006) Surface variation of Palgon Lake in Tibet in recent 50 years. Journal of Geographical Sciences 80: 876–884 (in Chinese).
13. Li XY, Xu HY, Sun YL (2007) Lake-level change and water balance analysis at Lake Qinghai, West China during recent decades. Water Resources Management 21: 1505–1516.
14. Wang L (2007) The study of typical glaciers and lakes fluctuations using remote sensing in Qinghai-Tibetan Plateau. IEEE Geoscience and Remote Sensing Symposium: 4526–4529.
15. Wu ZH, Jiang QG, Chen B (2007) Study of dynamic changes of lakes in Qinghai-Tibet Plateau based on remote sensing and GIS. Global Geology 26: 66–70 (in Chinese).
16. Yao T, Pu J, Lu A, Wang Y, Yu W (2007) Recent glacial retreat and its impact on hydrological processes on the Tibetan Plateau, China, and surrounding regions. Arct. Antarct. Alp. Res. 39: 642–650.
17. Ye QH, Zhu LP, Zheng HP, Naruse RJ, Zhang XQ, et al. (2007) Glacier and lake variations in the Yamzhog Yumco basin, southern Tibetan Plateau, from 1980 to 2000 using remote-sensing and GIS technologies. J. Glaciol. 53: 673–676.
18. Niu YF, Li CX, Xi XH, Zong DY (2008) Plateau lake variation monitored by satellite remote sensing and relation to climate change. Arid Land Geography 31: 284–290 (in Chinese).
19. Wu YH, Zhu LP (2008) The response of lake–glacier variations to climate change in Nam Co Catchment, central Tibetan Plateau, during 1970–2000. J. Geogr. Sci. 18: 177–189.
20. Yang L, Chen F, Chun X, Fan Y, Sun Y, et al. (2008) The Jilantai Salt Lake shorelines in northwestern arid China revealed by remote sensing images. Journal of Arid Environments 72: 861–866.
21. Ye QH, Yao TD, Zheng HX, Zhang XQ (2008) Glacier and lake co-variations and their responses to climate change in the Mapang Yumco Basin on Tibet. Geographical Research 27: 1178–1190 (in Chinese).
22. Bian D, Du J, Hu J, Li C, Li L, et al. (2009) Response of lake water level variation to the climate change within Yamdrok Lake Basin in Tibet from 1975 to 2006. Journal of Glaciology and Geocryology 31: 404–409 (in Chinese).
23. Jia SY, Xiao PF (2009) The monitoring of lake change in south Qiangtang Area of Qinghai-Tibet Plateau using multi-temporal Tupu. Remote Sensing for Land & Resources 4: 78–85 (in Chinese).
24. Lei YB, Zhang HC, Wang S, Li SJ, Fan HF, et al. (2009) Change in lake area of Zigetang Co on Central Tibetan Plateau since the 1970s and its mechanisms. Journal of Glaciology and Geocryology 31: 48–54 (in Chinese).
25. Ye QH, Zhong ZW, Kang S, Stein A, Wei Q, et al. (2009) Monitoring glacier and supra-glacier lakes from space in Mt. Qomolangma Region of the Himalayas on the Tibetan Plateau in China. Journal of Mountain Science 6: 211–220.
26. Bian D, Bian BC, La B, Wang CY, Chen T (2010) The response of water level of Selin Co to climate change during 1975–2008. J. Geogr. Sci. 65: 313–319.
27. Liu J, Wang SY, Yu SM, Yang DQ, Zhang L (2009) Climate warming and growth of high-elevation inland lakes on the Tibetan Plateau. Glob. Planet. Change 67: 209–217.
28. Liu J, Kang S, Gong T, Lu A (2010) Growth of a high-elevation large inland lake, associated with climate change and permafrost degradation in Tibet. Hydrol. Earth Syst. Sci. 14: 481–489.
29. Zhu LP, Xie M, Wu YH (2010) Quantitative analysis of lake area variations and the influence factors from 1971 to 2004 in the Nam Co Basin of the Tibetan Plateau. Chin. Sci. Bull. 55: 1294–1303.
30. Zhang B, Wu YH, Zhu LP, Wang JB, Li JS, et al. (2011) Estimation and trend detection of water storage at Nam Co Lake, central Tibetan Plateau. Journal of Hydrology 405: 161–170.
31. Lei Y, Yao TD, Yi CL, Wang WC, Sheng YW, et al. (2012) Glacier mass loss induced the rapid growth of Linggo Co on the central Tibetan Plateau. J. Glaciol. 58: 177–184.
32. Wang WC, Yao TD, Yang XX (2012) Variations of glacial lakes and glaciers in the Boshula mountain range, southeast Tibet, from the 1970s to 2009. Annals of Glaciology 52: 9–17.
33. Lei YB, Yao TD, Bird BW, Yang K, Zhai J, et al. (2013) Coherent lake growth on the central Tibetan Plateau since the 1970s: Characterization and attribution. Journal of Hydrology 483: 61–67.
34. Huang L, Liu JY, Shao QQ, Liu RG (2011) Changing inland lakes responding to climate warming in Northeastern Tibetan Plateau. Clim. Change 109: 479–502.
35. Li J, Sheng Y, Luo J, Shen Z (2011) Remotely sensed mapping of inland lake area changes in the Tibetan Plateau. J. Lake Sci. 23: 311–320 (in Chinese).
36. Liao JJ, Shen GZ, Li YK (2013) Lake variations in response to climate change in the Tibetan Plateau in the past 40 years. International Journal of Digital Earth 6: 534–549.
37. Zhang GQ, Xie HJ, Duan SQ, Tian MZ, Yi DH (2011) Water level variation of Lake Qinghai from satellite and in situ measurements under climate change. Journal of Applied Remote Sensing 5: 053532–1.
38. Kropacek J, Braun A, Kang S, Feng C, Ye QH, et al. (2012) Analysis of lake level changes in Nam Co in central Tibet utilizing synergistic satellite altimetry and optical imagery. International Journal of Applied Earth Observation and Geoinformation 17: 3–11.
39. Zhang G, Xie H, Kang S, Yi D, Ackley S (2011) Monitoring lake level changes on the Tibetan Plateau using ICESat altimetry data (2003–2009). Remote Sens. Environ. 115: 1733–1742.
40. Phan VH, Lindenbergh R, Menenti M (2011) ICESat derived elevation changes of Tibetan lakes between 2003 and 2009. Int. J. Appl. Earth Obs. Geoinf. 17: 12–22.
41. Zhang G, Yao T, Xie H, Kang S, Lei Y (2013) Increased mass over the Tibetan Plateau: from lakes or glaciers? Geophys. Res. Lett. 40: 1–6, doi:10.1002/grl.50462.
42. Zhang G, Xie H, Yao T, Kang S (2013) Water balance estimates of ten greatest lakes in China using ICESat and Landsat data. Chin. Sci. Bull. 58(1): 1–15.
43. Li ZX, He YQ, An WL, Song LL, Zhang W, et al. (2011) Climate and glacier change in southwestern China during the past several decades. Environ. Res. Lett. 6: 045404.
44. Yao T, Thompson LG, Mosbrugger V, Zhang F, Ma Y, et al. (2012) Third Pole Environment (TPE). Environ. Dev. 3: 52–64.
45. Schutz BE, Zwally H, Shuman C, Hancock D, DiMarzio P (2005) Overview of the ICESat Mission. Geophys. Res. Lett. 32: L21S01.
46. Huffman GJ, Adler RF, Bolvin DT, Gu G, Nelkin EJ, et al. (2007) The TRMM Multisatellite Precipitation Analysis (TMPA): Quasi-Global, Multiyear, Combined-Sensor Precipitation Estimates at Fine Scales. J. Hydrometeorol. 8: 38–55.
47. Duong H, Lindenbergh R, Pfeifer N, Vosselman G (2008) Single and two epoch analysis of ICESat full waveform data over forested areas. International Journal of Remote Sensing 29: 1453–1473.
48. Frazier PS, Page KJ (2000) Water body detection and delineation with Landsat TM data. Photogrammetric Engineering & Remote Sensing 66: 1461–1467.
49. Frankl A, Zwertvagher A, Poesen J, Nyssen J (2013) Transferring Google Earth observations to GIS-software: example from gully erosion study. International Journal of Digital Earth 6: 196–201.
50. Walter IA, Allen RG, Elliott R, Mecham B, Jensen ME, et al. (2000) ASCE Standardized Reference Evapotranspiration Equation, p. 209–215. In: Evans RG, Benham BL, Trooien TP (eds.) Proc. National Irrigation Symposium, ASAE, Nov. 14–16, 2000, Phoenix, AZ.
51. Daly C, Conklin DR, Unsworth MH (2010) Local atmospheric decoupling in complex topography alters climate change impacts. Int. J. Climatol. 30: 1857–1864, doi:10.1002/joc.2007.
52. Yin ZY, Zhang X, Liu X, Colella M, Chen X (2008) An Assessment of the Biases of Satellite Rainfall Estimates over the Tibetan Plateau and Correction Methods Based on Topographic Analysis, J. Hydrometeorol. 9: 301–326, doi:10.1175/2007jhm903.1.
53. Gao YC, Liu MF (2013) Evaluation of high-resolution satellite precipitation products using rain gauge observations over the Tibetan Plateau. Hydrol. Earth Syst. Sci. 17: 837–849.
54. Tong K, Su FG, Yang DQ, Zhang LL, Hao ZC (2014) Tibetan Plateau precipitation as depicted by gauge observations, reanalyses and satellite retrievals. Int. J. Climatol., 34: 265–285, doi:10.1002/joc.3682.
55. Jarvis A, Reuter HI, Nelson A, Guevara E (2008) Hole-filled seamless SRTM data V4. International Centre for Tropical Agriculture (CIAT). Available: http://srtm.csi.cgiar.org. Accessed 2012 July 10.
56. GLIMS, National Snow and Ice Data Center. (2005), updated 2012. GLIMS Glacier Database. Boulder, Colorado USA: National Snow and Ice Data Center. Available: http://dx.doi.org/10.7265/N5V98602. Accessed 2012 July 15.
57. Cheng G, Wu T (2007) Responses of permafrost to climate change and their environmental significance, Qinghai-Tibet Plateau. J. Geophys. Res. 112: F02S03.
58. Qin J, Yang K, Liang SL, Guo XF (2009) The altitudinal dependence of recent rapid warming over the Tibetan Plateau. Clim. Change 97: 321–327.
59. Yang K, Ye B, Zhou D, Wu B, Foken T, et al. (2011) Response of hydrological cycle to recent climate changes in the Tibetan Plateau. Clim. Change 109: 517–534.

60. Chen SB, Liu YF, Thomas A (2006) Climatic change on the Tibetan Plateau: Potential evapotranspiration trends from 1961–2000. Clim. Change 76: 291–319.
61. Xu B, Cao J, Hansen J, Yao TD, Joswia DR, et al. (2009) Black soot and the survival of Tibetan glaciers. Proc. Natl. Acad. Sci. U.S.A. 106: 22114–22118.
62. Zhao L, Ping CL, Yang DQ, Cheng GD, Ding YJ, et al. (2004) Changes of climate and seasonally frozen ground over the past 30 years in Qinghai-Xizang (Tibetan) Plateau, China. Glob. Planet. Change 43: 19–31.
63. Nan ZT, Li SX, Cheng GD (2005) Prediction of permafrost distribution on Qinghai–Tibet Plateau in the next 50 and 100 years. Science in China Series D: Earth Sciences 48: 797–804.
64. Wu QB, Zhang TJ, Liu YZ (2010) Permafrost temperatures and thickness on the Qinghai-Tibet Plateau. Global and Planetary Change 72: 32–38.
65. Zhao L, Ding YJ, Liu GY, Wang SL, Jin HJ (2010) Estimates of the reserves of ground ice in permafrost regions on the Tibetan Plateau. Journal of Glaciology and Geocryology 32(1): 1–9 (in Chinese).
66. Wu Q, Zhang T (2010) Changes in active layer thickness over the Tibetan Plateau from 1995 to 2007. J. Geophys. Res. 115: D09107.
67. Li R, Zhao L, Ding YJ, Wu TH, Xiao Y, et al. (2012) Temporal and spatial variations of the active layer along the Qinghai-Tibet Highway in a permafrost region. Chin. Sci. Bull. 57: 4609–4616.
68. Yang Y, Wu Q, Yun H (2013) Stable isotope variations in the ground ice of Beiluhe Basin on the Qinghai-Tibet Plateau. Quaternary International 313/314: 85-91.
69. Yoshikawa K, Hinzman L (2003) Shrinking thermokarst ponds and groundwater dynamics in discontinuous permafrost near Council, Alaska. Permafrost Periglac. 14: 151–160.
70. Smith LC, Sheng Y, MacDonald GM, Hinzman LD (2005) Disappearing Arctic lakes. Science 308: 1429.

15

Further Insights into Metal-DOM Interaction: Consideration of Both Fluorescent and Non-Fluorescent Substances

Huacheng Xu[1,2]*, Jicheng Zhong[2], Guanghui Yu[2,3], Jun Wu[4], Helong Jiang[2], Liuyan Yang[1]*

1 State Key Laboratory of Pollution Control and Resources Reuse, School of the Environment, Xianlin Campus, Nanjing University, Nanjing, China, **2** State Key Laboratory of Lake Science and Environment, Nanjing Institute of Geography and Limnology, Chinese Academy of Sciences, Nanjing, China, **3** College of Resources and Environmental Sciences, Nanjing Agricultural University, Nanjing, China, **4** Key Laboratory of Soil Environment and Pollution Remediation, Institute of Soil Science, Chinese Academy of Sciences, Nanjing, China

Abstract

Information on metal binding with fluorescent substances has been widely studied. By contrast, information on metal binding with non-fluorescent substances remains lacking despite the dominance of these substances in aquatic systems. In this study, the metal binding properties of both fluorescent and non-fluorescent substances were investigated by using metal titration combined with two-dimensional correlation spectroscopy (2D–COS) analysis. The organic matters in the eutrophic algae-rich lake, including natural organic matters (NOM) and algae-induced extracellular polymeric substances (EPS), both contained fluorescent and non-fluorescent substances. The peaks in the one-dimensional spectra strongly overlapped, while 2D–COS can decompose the overlapped peaks and thus enhanced the spectral resolution. Moreover, 2D FTIR COS demonstrated that the binding susceptibility of organic ligands in both NOM and algal EPS matrices followed the order: 3400>1380>1650 cm^{-1}, indicative the significant contribution of non-fluorescent ligands in metal binding. The modified Stern-Volmer equation also revealed a substantial metal binding potential for the non-fluorescent substances ($\log K_M$: 3.57~4.92). As for the effects of organic ligands on metal binding, EPS was characterized with higher binding ability than NOM for both fluorescent and non-fluorescent ligands. Algae-induced EPS and the non-fluorescent substances in eutrophic algae-rich lakes should not be overlooked because of their high metal binding potential.

Editor: Dariush Hinderberger, Martin-Luther-Universität Halle-Wittenberg, Germany

Funding: This work was funded by the National Natural Science Foundation of China (51209192, 51379199 and 41301241), the Natural Science Foundation of Jiangsu province, China (BK2012501), the China Postdoctoral Science Foundation (2013M540438), the Jiangsu Planned Projects for Postdoctoral Research Funds (1301007A), the State Key Laboratory of Pollution Control and Resource Reuse Foundation (PCRRF13011), and the State Key Laboratory of Lake Science and Environment, Chinese Academy of Sciences (2012SKL007). The funders had no role in study design, data collection and analysis, decision to publish, or preparation of the manuscript.

Competing Interests: The authors have declared that no competing interests exist.

* Email: hcxu@niglas.ac.cn (HCX); yangly@nju.edu.cn (LYY)

Introduction

Dissolved organic matter (DOM) in aquatic environments can interact with metals and consequently affect their toxicity, mobility and bioavailability [1,2]. Investigation into the binding properties of DOM is necessary to clearly understand the biogeochemical behavior of metals. DOM in waters is generally characterized with heterogeneous organic compositions including fluorescent (i.e., protein-, humic-, and fulvic-like) and non-fluorescent (i.e., carbohydrates, lipids, and lignins) substances [3–5]. Thus, a comprehensive study on metal-DOM interaction should include related information on both fluorescent and non-fluorescent substances.

Fluorescence quenching titration is a popular method in the study of metal-ligand interaction [6–8]. Saar and Weber [9] compared synchronous fluorescence (SF) quenching titration and ion-selective electrode potentiometry and proposed the former. Ghatak et al. [10] later studied the binding heterogeneity of sediment humic substances and reported more energy binding sites for Fe(III). More recently, through SF quenching titration technique, the heterogeneous metal binding behaviors between leaf litter and soil DOM can be explored [7,8]. Although widely used, SF technique is not applicable for non-florescent substances, even with these making up a significant proportion of aquatic DOMs [11].

By contrast, Fourier transform infrared spectroscopy (FTIR) can be used to explore non-fluorescent substances [12,13]. Binding information on non-fluorescnt substances can be obtained by investigating spectral variations in response to metal addition. Thus, the metal binding properties of both fluorescent and non-fluorescent substances can be obtained through the combination of the SF and FTIR techniques.

However, one-dimensional SF and FTIR spectra usually exhibited many overlapped peaks because of the heterogeneous nature of the investigated DOM [6,10,14]. Hence, further analysis based solely on the traditional visual inspection approach is laborious and difficult. Recent studies showed that two-dimensional correlation spectroscopy (2D–COS) can be used to solve the peak overlapping problems by extending peaks along the second dimension [12,14]. This technique can also be used to explore the sequence of metal-sensitive spectral changes through asynchronous map, which benefits the analysis of binding mechanisms [7,8,15,16]. To date, only one study on the application of 2D–COS for exploring the metal binding properties of soil organic matters has been published [16].

Anthropogenically induced eutrophication in recent years has caused the increasing occurrence of cyanobacterial blooms in lakes, especially in eutrophic shallow lakes [17]. Our previous studies have demonstrated that DOM in such algal-rich lakes should consist of natural organic matter (NOM) and algal extracellular polymeric substances (EPS). EPS can be further fractionated into loosely bound EPS (LB–EPS) and tightly bound EPS (TB–EPS) in terms of its binding strength with algal cells [18,19]. Due to the high reactively, both NOM and algae-induced EPS can bind with metals and consequently play an important role in the toxicity, mobility, and bioavailability of metals.

The present study were therefore aims (1) to investigate the metal-induced spectral variations of both fluorescent and non-fluorescent substances in NOM and algal EPS fractions, and (2) to explore the detailed metal binding behavior and mechanisms by using 2D–COS combined with SF and FTIR techniques. For these purposes, the interactions of a commonly found metal Cu(II) in eutrophic shallow lakes [20] with NOM, LB–EPS and TB–EPS were respectively studied. The results of this study would provide a comprehensive picture of metal-DOM interaction in aquatic environments.

Materials and Methods

Ethics statement

No specific permits were required for the described field studies. The location studied is not privately-owned or protected in any way and our studies did not involve any endangered or protected species.

Sample preparation and EPS extraction

A total of 2.4 L of surface water samples (30 cm from the surface) containing 1.0 g L^{-1} algal biomass were collected from Meiliang Bay, which is located in the northern region of Lake Taihu (Figure S1 in File S1). The samples were collected in the summer-fall period because typical algal blooms occurred during this period [21]. All samples were transported to a laboratory within several hours and firstly filtered through the 0.45 μm polytetrafluoroethylene membranes (Xingya Purification Materials Co., Shanghai, China) for NOM measurement. The residues were carefully scraped and dissolved with a 0.05% NaCl solution to the original volume and centrifuged at 5000 *g* for 15 min, with the liquid collected carefully for measurement of LB–EPS. Afterward, the harvested samples were re-suspended in NaCl solution, heated at 60°C for 30 min and centrifuged at 15,000 *g* for 20 min, with the liquid collected as TB–EPS fraction. More information on sample collection and fractionation procedure can be found elsewhere [18,19].

Cu(II) titration and complexation modeling

Titration experiments were carried out by adding 0.1 mol L^{-1} $Cu(NO_3)_2$ solutions to a series of sealed brown vials containing 50 ml of diluted solution [dissolved organic carbon (DOC): 10 mg L^{-1}] using an automatic syringe [1,2,19,22]. Metal concentrations ranging from 0 to 100 μmol L^{-1} were obtained in the final solutions by adding no more than 50 μl metal titrant. The pH was maintained at 6.0 through adding 0.1 M HCl or NaOH solution under which no precipitate was formed [1]. All solution samples were shaken for 24 h at room temperature to ensure complexation equilibrium [2]. Afterward, the titrated solutions and freeze-dried samples underwent SF and FTIR analyses, respectively. Each titration experiment was performed in duplicate.

The modified Stern-Volmer equation [23] was applied to estimate the conditional stability constants between metals and SF/FTIR-derived peaks:

$$\frac{F_o}{F_o - F} = \frac{1}{f \cdot K_M \cdot C_M} + \frac{1}{f} \quad (1)$$

Here F and F_0 are the measured SF or FTIR intensities at the metal concentration C_M and the beginning of the titration (i.e., no metal addition), respectively. The parameters K_M and f represents the conditional stability constant and the fraction of the initial spectral intensities which correspond to the metal binding.

Two-dimensional correlation spectroscopy

SF was measured using the Hitachi F–7000 fluorescence spectrometer (Hitachi High Technologies, Tokyo, Japan) in synchronous mode at room temperature by ranging the excitation wavelengths from 200 to 450 nm with a constant offset (60 nm) [7,24]. The freeze-dried samples were mixed with KBr (sample to KBr ratio of 1:100, w/w) and then compressed to form a disc. FTIR spectra were obtained by collecting 64 scans at a resolution of 2 cm^{-1} with a Nicolet Nexus 870 FTIR spectrometer [25].

The 2D–COS was produced according to the method of Noda and Ozaki [14]. The metal addition was used as the external perturbation, and thus a set of concentration–dependent SF/FTIR spectra were obtained.

For the perturbation–induced spectral variation y(x, t) as a function of a spectral variable (x) and a perturbation variable (t), the dynamic spectrum $\tilde{y}(x,t)$ is formally defined as follows:

$$\tilde{y}(x,t) = \begin{cases} y(x,t) - \bar{y}(x) \; for \; T_{\min} \le t \le T_{\max} \\ 0 \quad otherwise \end{cases} \quad (2)$$

where $\bar{y}(x)$ denotes the reference spectra, which is typically the average spectrum and is expressed as $\bar{y}(x) = \frac{1}{T_{\max} - T_{\min}} \int_T^{T\max} y(x,t)dt$. The synchronous correlation spectroscopy can be written as follows:

$$\phi(x_1,x_2) = \frac{1}{T_{\max} - T_{\min}} \int_T^{T\max} y(x_1,t) \cdot z(x_2,t)dt \quad (3)$$

The asynchronous correlation spectroscopy can be obtained directly from the dynamic spectrum $\tilde{y}(x_2,t)$and its orthogonal spectrum$\tilde{z}(x_2,t)$.

$$\varphi(x_1,x_2) = \frac{1}{T_{\max} - T_{\min}} \int_T^{T\max} \tilde{y}(x_1,t) \cdot \tilde{z}(x_2,t)dt \quad (4)$$

Prior to 2D–COS analysis, the SF and FTIR spectra were normalized by the summed absorbance from 200 to 450 nm and

400 to 4000 cm^{-1}. The noise components were removed by principal component analysis [16,22]. Then the reconstructed data matrix was progressed using the "2D shige" software (Kwansei Gakuin University, Japan).

Results and Discussion

Determination of the organic matters in the eutrophic algae-rich lake

Figure 1 shows the typical SF and FTIR spectra for the NOM and algal EPS matrices. NOM was characterized with two prominent peaks (230, 280 nm) and one shoulder peak (300~390 nm), whereas the EPS matrix showed only the prominent peaks (Fig. 1a). Peaks located at 230 and 280 nm were ascribed to tyrosine- and tryptophan-like substances, respectively, whereas peaks ranging from 300 to 390 nm were attributed to humic-like substances [7]. The SF results indicated that the algal EPS mainly contained fluorescent protein-like substances, while not only protein-like but also humic-like substances were observed in NOM. In the previous studies aiming at lab-cultured cynobacterium *M. aeruginosa*, Xu et al. [18] detected the protein-like substances in all EPS fractions but the humic-like substances in only the soluble EPS fractions. And Qu et al. [25] demonstrated that the mainly components in bound and dissolved extracellular organic matters were protein-like substances and protein-like + humic-like substances, respectively. These similar results indicated that the distribution patterns of organic compositions were not form-specific.

As for the FTIR spectroscopy, the NOM and algal EPS matrices were both characterized with four obvious absorption peaks at 3400, 1650, 1380, and 1100 cm^{-1} (Fig.1b). We assigned these bands as follows: the band at 1650 cm^{-1} was ascribed to the CO stretching of amide I in protein compounds, the band at 1380 cm^{-1} to the CH deformations in aliphatic groups, and the bands at 3400 and 1100 cm^{-1} to the stretching vibration of OH and CO in carbohydrate-like substances [16,26]. The combined results of SF and FTIR showed that DOM in the eutrophic algae-rich lakes, including NOM and algal EPS, definitely contained both fluorescent (230, 280, and 300~390 nm, and 1650 cm^{-1}) and non-fluorescent (3400, 1380, and 1100 cm^{-1}) substances.

It should be noted that the organic compositions of EPS matrix were highly dependent on the extraction methods. For example, Liu et al. [27] compared five methods for *M. aeruginosa* EPS extraction and found different fluorescent properties among the five methods. A further study also demonstrated that EPS using cation exchange resin extraction contained more aromatic condensed structures with less acidic functional groups when compared with formaldehyde/NaOH extraction [28]. The heterogeneity in organic compositions exhibits different metal binding ability, which will be discussed in the following section.

Metal binding properties of the fluorescent substances

Changes in one dimensional SF spectra with metal addition. The effects of Cu(II) addition on the variations of SF spectra are shown in Fig.2. Fluorescence quenching generally occurred with increasing metal addition. This result revealed strong metal binding and the marked modification of the electronic structures of DOM [7,29]. Further analysis showed that the behavior of spectral changes between NOM and EPS exhibited obvious differences. Monotonic fluorescence quenching was observed for NOM along the entire wavelength region (200~450 nm), whereas both fluorescence quenching (280 nm) and enhancement (230 nm) were identified in the EPS matrix. The complex formation and the fluorophore-quencher collision were responsible for fluorescence quenching [30,31], while the fluorescence enhancement was attributed to the changes in quantum yields of tyrosine-like substances in algal EPS matrix caused by high levels of Cu(II) and/or interactions with inorganic and other organic components [2]. Metal-induced fluorescence enhancement was previously reported in other studies but via fluorescence emission and excitation matrix analysis [1,19,32].

In addition, the quenching rates and degrees were highly related to the organic ligands. Metal addition caused a sharp and gradual reduction at 250~300 nm and 300~370 nm, respectively for NOM. However, the algal EPS matrix exhibited a more pronounced quenching degree than NOM at 250~300 nm. This result indicated that the protein-like substances in NOM might not be as involved in metal binding as those in the algal EPS matrix [7]. The different quenching rates and degrees observed here indicated that the functional sites for metal binding in the NOM and algal EPS matrices were unevenly distributed.

Metal binding sequencing investigated by 2D fluorescence COS. Although SF can capture heterogeneous

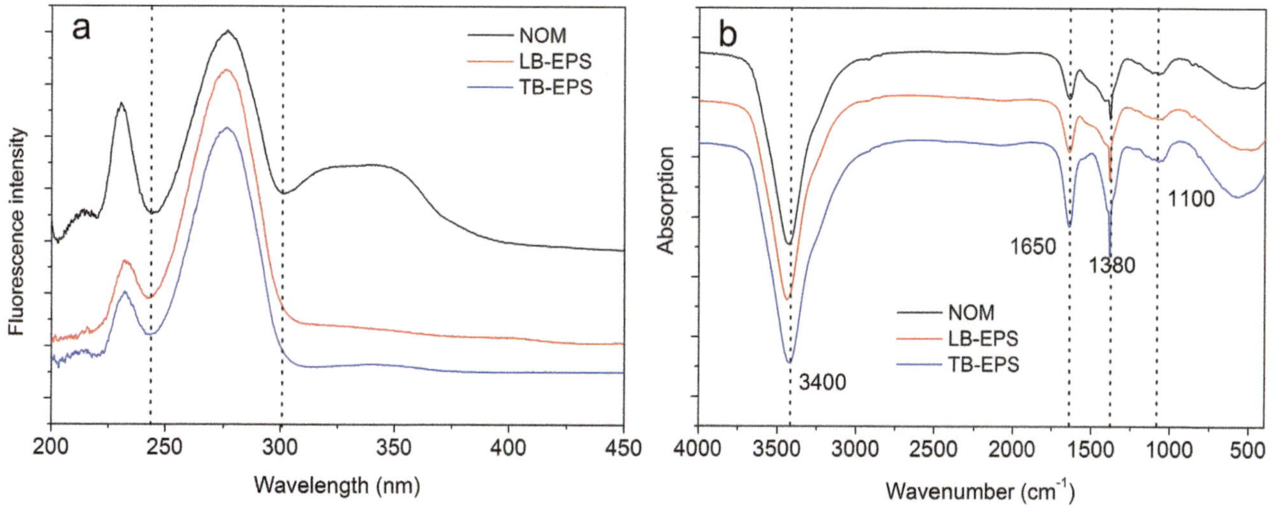

Figure 1. Typical spectral shapes of SF (a) and FTIR (b) for NOM and algal EPS matrix in the eutrophic algae-rich lake.

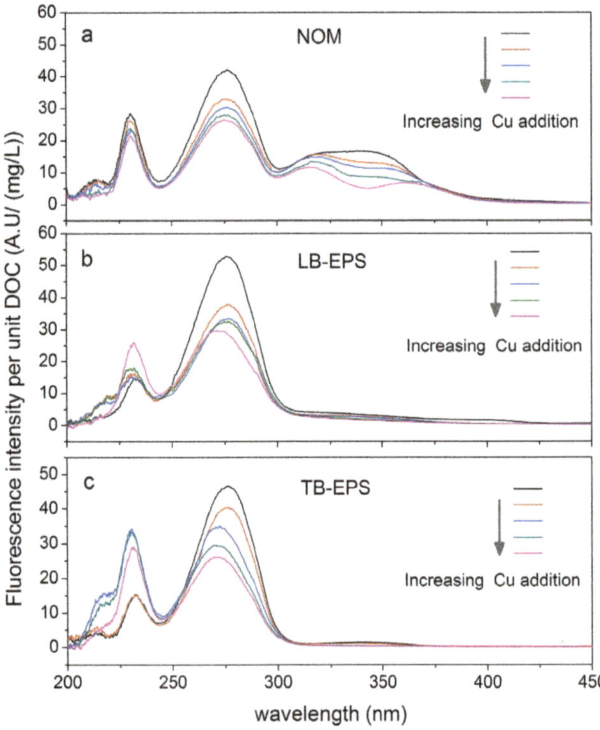

Figure 2. Changes in the one-dimensional SF intensities with Cu(II) addition. The arrows refer to the direction of the increasing metal concentrations.

quenching degrees and rates, extensive information on binding sites and binding sequence cannot be obtained via the one-dimensional spectral technique. 2D–COS was applied in the present study to explore binding behavior in detail (Fig. 3). The synchronous map displayed two autopeaks at 230/230 and 280/280 nm along the diagonal line for both NOM and algal EPS. The synchronous map is a symmetric spectrum with respect to the diagonal line. The correlation peaks include autopeaks and crosspeaks on the diagonal and off-diagonal lines, respectively. An autopeak represents the overall susceptibility of the corresponding spectral region to the change in spectral intensity as an external perturbation is applied to the system [14,16,22]. The intensities of the autopeaks followed the order: 230>280 nm for NOM and 280>230 nm for the EPS matrix. This suggested that tyrosine-like substances in NOM were more susceptible to metal addition compared with the tryptophan-like substances, whereas the tryptophan-like substances in algal EPS matrix exhibited higher metal sensitivity than the tyrosine-like substances [8,14].

A crosspeak in synchronous map represents the simultaneous or coincidental changes of two different spectral variables. NOM and algal EPS were respectively characterized with positive and negative crosspeak at 280/230 nm. This characterization implies that tyrosine- and tryptophan-like substances in NOM covary in their interaction with metal addition, whereas those in algal EPS might vary with opposite directions. Consistent with our previous observations (Fig. 2), NOM demonstrated monotonic quenching effects, whereas the algal EPS matrix demonstrated both fluorescence quenching and enhancement.

An asynchronous map reveals the sequential or successive changes in spectral intensities with increasing metal addition. As shown, one and three negative crosspeaks were observed above the diagonal line for the NOM and algal EPS matrices, respectively. Based on Noda's rule [14], the sequences of binding affinities followed the order: 230>275 nm for NOM, 265>280>225>230 nm for LB–EPS and 270>280>230>232 nm for TB–EPS. The asynchronous maps revealed that metal binding in NOM took place at the shorter wavelengths (~230 nm) prior to that at the longer wavelength range (>265 nm); while for the algal EPS matrix, meals binding first occurred at the longer wavelengths. As indicated by these results, the preferential metal ligands in NOM and algal EPS were tyrosine- and tryptophan-like substances, respectively. In addition, the asynchronous maps also showed that the wavelength 230 nm in the one-dimensional SF spectra overlapped with 230 and 232 nm, and the wavelength 280 nm overlapped with 270, 275 and 280 nm. Unlike the one-dimensional spectral technique, 2D–COS can enhance spectral resolution and yield extensive detailed binding information.

Information on metal binding with both fluorescent and non-fluorescent substances

FTIR can be used to characterize non-fluorescent substances [12,13]. The absorption variations of the normalized one-dimensional FTIR spectra of NOM and algal EPS in response to Cu(II) addition are given in Figure S2 in File S1. Although four absorption peaks were obviously observed at 3400, 1650, 1380 and 1100 cm^{-1}, the one-dimensional FTIR was strongly overlapped and information on the interrelationship among metal-sensitive functional groups cannot be obtained.

Therefore, 2D–COS was applied to explore in detail the binding information on fluorescent and non-fluorescent substances (Fig.4). We mainly focused on the region between 1700~1300 cm^{-1} because this spectral region usually contains the major bands of the amide, carboxylic, ester, and carbohydrate functional groups [12,16]. Both NOM and algal EPS matrices were characterized with two autopeaks located at 1380/1380 and 1650/1650 cm^{-1}, with the former exhibiting higher intensity than the latter. Combined with the synchronous maps of the 4000–400 cm^{-1} region (Figure S3 in File S1), the peak at 3400 cm^{-1} was more susceptible to Cu(II) addition, followed by the peak at 1380 cm^{-1} and 1650 cm^{-1}. This result revealed the substantial contributions of non-fluorescent substances (3400, 1380 cm^{-1}) in metal binding. Zhang et al. [31] also emphasized the metal binding potential of non-fluorescent carboxyl and hydroxyl functional groups in biofilm EPS, although no further evidence was given in their studies.

By contrast, the asynchronous maps revealed that the sequencing of binding affinities of the peaks followed the order: amide I in proteins (1650 cm^{-1})> lignin and aliphatic CH (1380 cm^{-1}) for NOM and LB–EPS, and amide I in proteins (1650 cm^{-1})> lignin and aliphatic CH (1380 cm^{-1})> cellulose CH_2 (1420 cm^{-1}) for TB–EPS [14]. Therefore, metal Cu(II) preferentially bound with the fluorescent protein-like substances, followed by the non-fluorescent liquid- and carbohydrate-like substances, regardless of the difference in DOM fractions (i.e., NOM, LB–EPS and TB–EPS).

Metal binding abilities of both fluorescent and non-fluorescent substances

The conditional stability constants ($\log K_M$) based on the modified Stern-Volmer equation are tabulated in Table 1. The $\log K_M$ values for EPS matrix were calculated within the range from 4.66 to 5.03, which were slightly higher than those reported in Lee et al. [28] but lower than those in McIntyre and Gueguen [33]. The heterogeneous binding capacities were partly attributed to the different organic compositions and functional groups, which

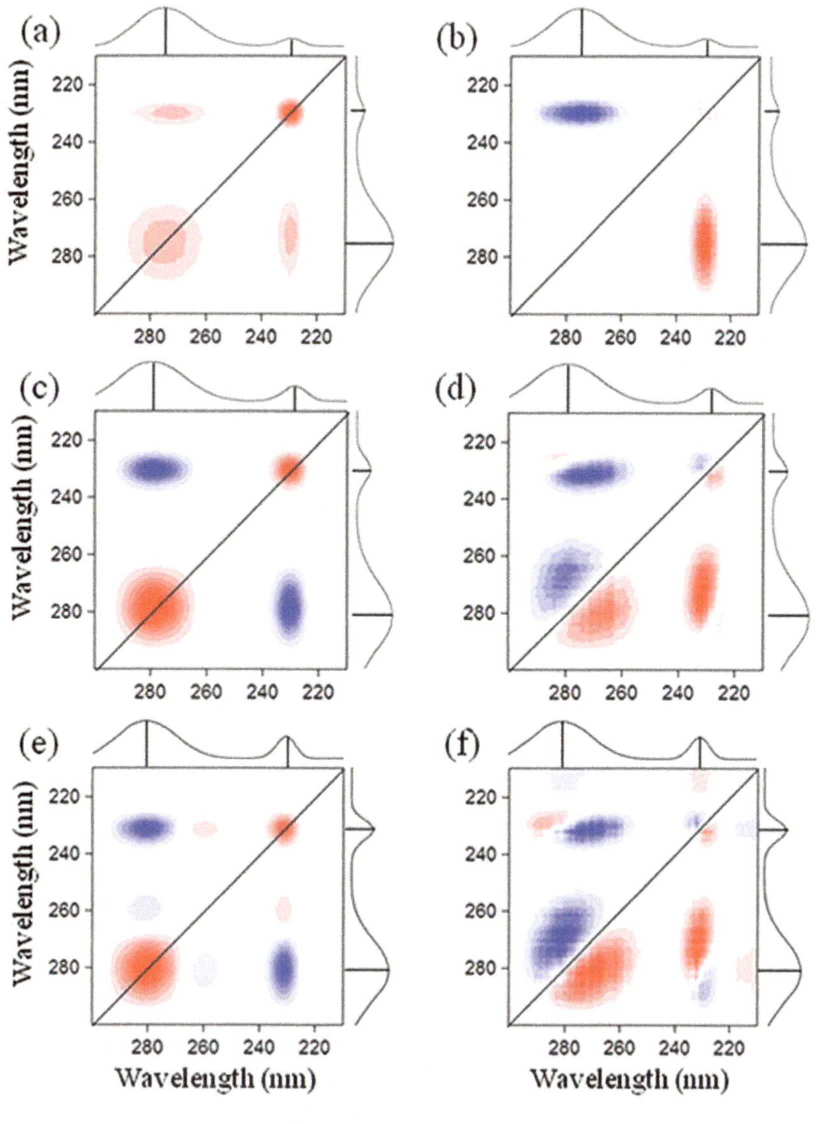

Figure 3. The 2D fluorescence correlation maps generated from 210 to 300 nm region for NOM and EPS matrix with increasing Cu(II) addition. (a) synchronous map for NOM; (b) asynchronous for NOM; (c) synchronous map for LB–EPS; (d) asynchronous for LB–EPS; (e) synchronous map for TB–EPS; (f) asynchronous for TB–EPS. Red represents positive correlations and blue represents negative correlations; higher color intensity indicates a stronger positive or negative correlation.

were highly dependent on the extraction methods [28]. Further analysis showed that the stability constants between NOM and EPS exhibited obvious differences. Specifically, the $\log K_M$ of tryptophan-like substances in NOM (4.51±0.06) was lower than that in the LB–EPS and TB–EPS fractions (>4.66±0.05). The results indicated that the tryptophan-like substances in the algal EPS exhibited higher binding abilities than those in NOM [19].

For the FTIR-derived peaks, the peak at 1650 cm^{-1} showed a higher $\log K_M$ value than the peaks at 1100, 1380 and 1420 cm^{-1} for NOM (4.62±0.08 vs. <4.27±0.05), LB–EPS (4.79±0.04 vs. < 4.59±0.08) and TB–EPS (4.96±0.07 vs. <4.92±0.03). This result implies that the binding ability of fluorescent proteins is stronger than that of the non-fluorescent liquids and carbohydrates. The calculated $\log K_M$ values were consistent with the results of asynchronous correlation spectroscopy, thus providing evidence for the feasibility of 2D–COS in the investigation of metal-DOM interaction. Further analysis showed that the algal EPS matrix exhibited a higher $\log K_M$ value than NOM for both the fluorescence proteins (>4.79±0.04 vs. 4.62±0.08) and the non-fluorescence liquids and carbohydrates (>4.26±0.10 vs. < 3.57~4.27). Therefore, the algal EPS matrix is expected to play a more important role than NOM in the migration and detoxification of heavy metals in eutrophic algae-rich lakes.

Significance of this study

Both fluorescent and non-fluorescent substances were detected in the DOM found in lakes (Fig. 1). Previous studies on metal-DOM interaction mainly focused on fluorescent substances [1,34,35], lacking the related information on non-fluorescent substances. Non-fluorescent substances may sometimes even occupy a significant proportion in aquatic environments [11]. The contribution of non-fluorescent substances in metal binding was investigated for the first time in this study, which is an

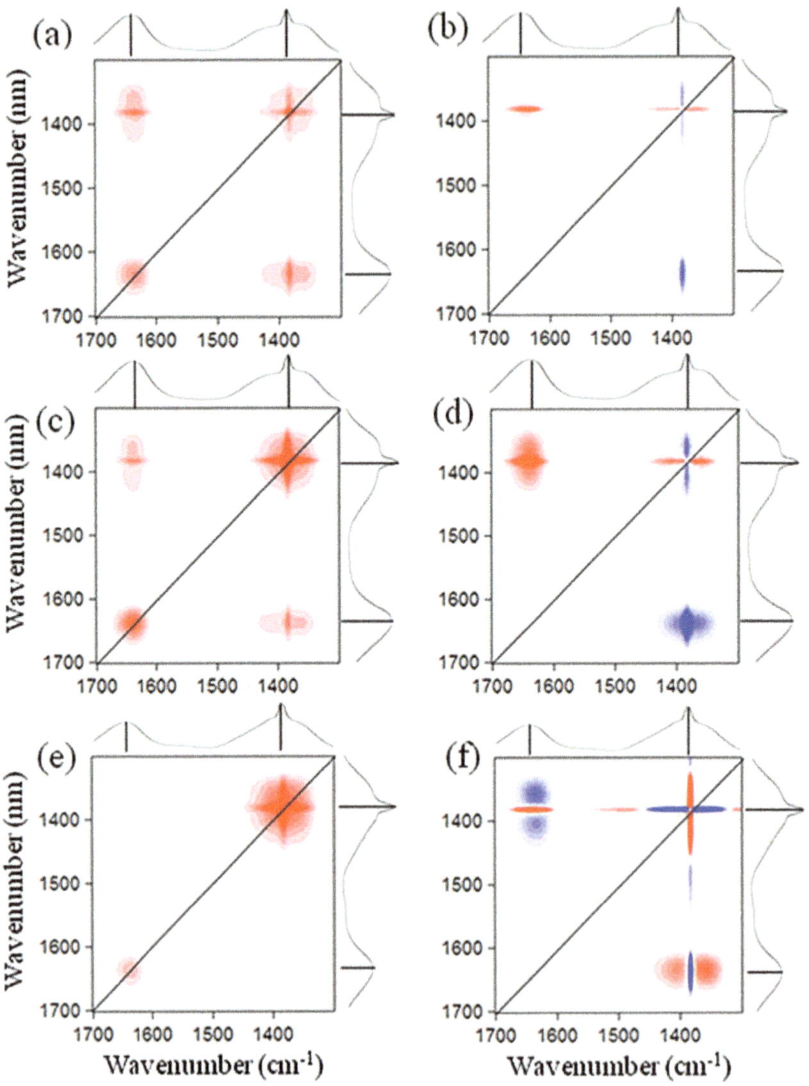

Figure 4. The 2D infrared correlation maps generated from 1300 to 1700 cm^{-1} region for NOM and EPS matrix with increasing Cu(II) addition. (a) synchronous map for NOM; (b) asynchronous for NOM; (c) synchronous map for LB–EPS; (d) asynchronous for LB–EPS; (e) synchronous map for TB–EPS; (f) asynchronous for TB–EPS. Red represents positive correlations and blue represents negative correlations; higher color intensity indicates a stronger positive or negative correlation.

important complement to the research scope of metal-DOM interaction.

Algal blooms have become common in recent years because of climatic changes and nutrient enrichment [17]. This phenomenon has made the algae-induced EPS an important contributor to the DOM pool [33]. Although studies on metal binding with NOM have been carried out [2,8], information on the metal-EPS interactions in eutrophic algae-rich lakes remains insufficient. This information would help provide a comprehensive picture of metal-DOM interactions in aquatic environments. The higher metal-EPS binding capacities observed in this work highlighted the ecological importance of algal EPS in the toxicity, mobility, and bioavailability of metals because the toxicity of metals is related to free iron activity [36].

Although the metal binding degrees and rates can be explored through a one-dimensional spectral analysis, the interrelationships among different organic ligands cannot be obtained. The 2D–COS applied in this study clearly demonstrated the heterogeneous metal binding sites between NOM and algal EPS. More importantly, 2D–COS can capture the binding sequencing of metal-sensitive spectral variations, which could improve our understanding about metal binding mechanisms. This also allows the scope of 2D–COS to be unlimited in the fields of metal binding within lake ecosystems [7]. The interaction of DOM with other objects (i.e., nanoparticles and persistent organic pollutants) in selected aquatic environments (i.e., groundwater and wastewater) can also be explored by 2D–COS in future studies. External perturbation need not be limited to metal concentration but may include other parameters, such as temperature, pressure, and pH.

Conclusions

DOM in eutrophic algae-rich lakes was fractionated into NOM, LB–EPS, and TB–EPS fractions. Both fluorescent and non-fluorescent substances were determined in these fractions. Although fluorescence titration was an effective method for

Table 1. Conditional stability constants ($\log K_M$) for SF/FTIR-derived peaks binding to Cu(II) as calculated using the modified Stern-Volmer equation.

Samples	Wavelength (nm)	$\log K_M$	R^2	Samples	Wavenumber (cm^{-1})	$\log K_M$	R^2
	230	4.69±0.04	>0.87		1100	3.57±0.04	>0.93
NOM	275	4.51±0.06	>0.72	NOM	1380	4.27±0.05	>0.88
					1650	4.62±0.08	>0.71
	230	Not modeled			1100	4.35±0.03	>0.92
LB-EPS	265	4.94±0.07	>0.51	LB-EPS	1380	4.59±0.08	>0.71
	280	4.66±0.05	>0.84		1650	4.79±0.04	>0.86
	230	Not modeled			1100	4.74±0.05	>0.80
	233	Not modeled			1380	4.92±0.03	>0.86
TB-EPS	270	5.03±0.04	>0.88	TB-EPS	1420	4.26±0.10	>0.62
	280	4.74±0.03	>0.95		1650	4.96±0.07	>0.76

investigating metal-ligand interaction, FTIR was able to determine both fluorescent and non-fluorescent substances. 2D–COS can enhance spectral resolution and thus can be used to explore detailed binding properties such as binding sites and binding sequencing. The metal binding properties of non-fluorescent substances were quantified, and the significance of the algal EPS matrix in metal detoxification was emphasized. In sum, 2D–COS combined with SF and FTIR analysis is an effective approach to explore the metal binding properties of both fluorescent and non-fluorescent substances.

Supporting Information

File S1 Supporting figures. Figure S1. The map of Lake Taihu with the location of the sampling station; Figure S2 The variations of the normalized one-dimensional FTIR with metal addition; Figure S3 The 2D synchronous correlation maps generated from 4000–400 cm^{-1} region of the FTIR spectra for NOM and algal EPS.

Author Contributions

Conceived and designed the experiments: HCX LYY. Performed the experiments: JCZ GHY. Analyzed the data: JW. Contributed reagents/materials/analysis tools: HLJ. Wrote the paper: HCX.

References

1. Ohno T, Amirbahman A, Bro R (2008) Parallel factor analysis of excitation–emission matrix fluorescence spectra of water soluble soil organic matter as basis for the determination of conditional metal binding parameters. Environ Sci Technol 42: 186–192.
2. Yamashita Y, Jaffe R (2008) Characterizing the interactions between trace metals and dissolved organic matter using excitation-emission matrix and parallel factor analysis. Environ Sci Technol 42: 7374–7379.
3. Wang Y, Zhang D, Shen ZY, Feng CH, et al. (2013) Revealing sources and distribution changes of dissolved organic matter (DOM) in pore water of sediment from the Yangtze estuary. PLoS ONE 8: e76633.
4. Yao X, Zhang YL, Zhu GW, Qin BQ, Feng LQ, et al. (2011) Resolving the variability of CDOM fluorescence to differentiate the sources and fate of DOM in Lake Taihu and its tributaries. Chemosphere 82: 145–155.
5. Zhang YL, Yin Y, Feng LQ, Zhu GW, Shi ZQ, et al. (2011) Characterizing chromophoric dissolved organic matter in Lake Tianmuhu and its catchment basin using excitation–emission matrix fluorescence and parallel factor analysis. Water Res 45: 5110–5122.
6. Cabaniss SE (1992) Synchronous fluorescence spectra of metalfulvic acid complexes. Environ Sci Technol 26: 1133–1139.
7. Hur J, Lee BM (2011). Characterization of binding site heterogeneity for copper within dissolved organic matter fractions using two-dimensional correlation fluorescence spectroscopy. Chemosphere 83: 1603–1611.
8. Hur J, Lee BM (2011) Comparing the heterogeneity of copper-binding characteristics for two different-sized soil humic acid fractions using fluorescence quenching combined with 2D-COS. Scientific World Journal 11: 1865–1876.
9. Saar RA, Weber JH (1980) Comparison of spectrofluorometry and ion-selective electrode potentiometry for determination of complexes between fulvic acid and heavy-metal ions. Anal Chem 52: 2095–2100.
10. Ghatak H, Mukhopadhyay SK, Jana TK, Sen BK, Sen S (2004) Interactions of Cu(II) and Fe(III) with mangal humic substances studied by synchronous fluorescence spectroscopy and potentiometric titration. Wetlands Ecol Manage 12: 145–155.
11. Leenheer JA, Croue JP (2003) Characterizing aquatic dissolved organic matter. Environ Sci Technol 37: 18–26.
12. Hussain ANA, Elizabeth CM, Patrick GH (2010) Using two-dimensional correlations of ^{13}C NMR and FTIR to investigate changes in the chemical composition of dissolved organic matter along an estuarine transect. Environ Sci Technol 44: 8044–8049.
13. Landry C, Tremblay L (2012) Compositional differences between size classes of dissolved organic matter from freshwater and seawater revealed by an HPLC-FTIR system. Environ Sci Technol 46: 1700–1707.
14. Noda I, Ozaki Y (2004) Two-dimensional correlation spectroscopy: Applications in vibrational and optical spectroscopy. John Wiley and Sons Inc., London.
15. Li XM, Shen QR, Zhang DQ, Mei XL, Ran W, et al. (2013) Functional groups determine biochar properties (pH and EC) as studied by two-dimensional ^{13}C NMR correlation spectroscopy. PLoS ONE 8: e65949.

16. Yu GH, Wu MJ, Wei GR, Luo YH, Ran W, et al. (2012) Binding of organic ligands with Al(III) in dissolved organic matter from soil: Implications for soil organic carbon storage. Environ Sci Technol 46: 6102–6109.
17. Paerl HW, Paul VJ (2011) Climate change: Links to global expansion of harmful cyanobacteria. Water Res 46: 1349–1363.
18. Xu HC, Cai HY, Yu GH, Jiang HL (2013) Insights into extracellular polymeric substances of cyanobacterium *Microcystis aeruginosa* using fractionation procedure and parallel factor analysis. Water Res 47: 2005–2014.
19. Xu HC, Yan ZS, Cai HY, Yu GH, Yang LY, et al. (2013) Heterogeneity in metal binding by individual fluorescent components in a eutrophic algae-rich lake. Ecotox Environ Safe 98: 266–272.
20. Shen J, Liu EF, Zhu YX, Hu SY, Qu WC (2007) Distribution and chemical fractionation of heavy metals in recent sediments from Lake Taihu, China. Hydrobiologia 581: 141–150.
21. Qin BQ, Zhu GW, Gao G, Zhang YL, Li W, et al. (2010) A drinking water crisis in lake Taihu, China: linkage to climatic variability and lake management. Environ Manage 45: 105–112.
22. Xu HC, Yu GH, Yang LY, Jiang HL (2013) Combination of two-dimensional correlation spectroscopy and parallel factor analysis to characterize the binding of heavy metals with DOM in lake sediments. J Hazard Mater 263: 412–421.
23. Hays MD, Ryan DK, Pennell S (2004) A modified multisite Stern-Volmer equation for the determination of conditional stability constants and ligand concentrations of soil fulvic acid with metal ions. Anal Chem 76: 848–854.
24. Xu HC, Jiang HL (2013) UV-induced photochemical heterogeneity of dissolved and attached organic matter associated with cyanobacterial blooms in a eutrophic freshwater lake. Water Res 47: 6506–6515.
25. Qu FS, Liang H, He JG, Ma J, Wang ZZ, et al. (2012) Characterization of dissolved extracellular organic matter (dEOM) and bound extracellular organic matter (bEOM) of *Microcystis aeruginosa* and their impacts on UF membrane fouling. Water Res 46: 2881–2890.
26. Xu HC, Cai HY, Yu GH, Jiang HL (2013) Investigation on extracellular polymeric substances from mucilaginous cyanobacterial blooms in eutrophic freshwater lakes. Chemosphere 93: 75–81.
27. Liu LZ, Qin BQ, Zhang YL, Zhu GW, Gao G (2014) Extraction and characterization of bound extracellular polymeric substances from cultured pure cyanobacterium (*Microcystis wesenbergii*). J Environ Sci 26: 1725–1732.
28. Lee BM, Shin HS, Hur J (2013) Comparison of the characteristics of extracellular polymeric substances for two different extraction methods and sludge formation conditions. Chemosphere 90: 237–244.
29. Plaza C, Brunetti G, Senesi N, Polo A (2006) Molecular and quantitative analysis of metal ion binding to humic acids from sewage sludge and sludge-amended soils by fluorescence spectroscopy. Environ Sci Technol 40: 917–923.
30. Pan XL, Liu J, Zhang DY (2010) Binding of phenanthrene to extracellular polymeric substances (EPS) from aerobic activated sludge: A fluorescence study. Colloids and Surfaces B: Biointerfaces 80: 103–106.
31. Zhang DY, Pang XL, Mostofaa KMG, Chen X, Mu GJ, et al. (2010) Complexation between Hg(II) and biofilm extracellular polymeric substances: An application of fluorescence spectroscopy. J Hazard Mater 175: 359–365.
32. Hernández D, Plaza C, Senesi N, Polo A (2006) Detection of copper(II) and zinc(II) binding to humic acids from pig slurry and amended soils by fluorescence spectroscopy. Environ Pollut 143: 212–220.
33. McIntyre AM, Gueguen C (2013) Binding interactions of algal-derived dissolved organic matter with metal ions. Chemosphere 90: 620–626.
34. Wu J, Zhang H, Shao LM, He PJ (2012) Fluorescent characteristics and metal binding properties of individual molecular weight fractions in municipal solid waste leachate. Environ Pollut 162: 63–71.
35. Chai XL, Liu GX, Zhao X, Hao YX, Zhao YC (2012) Complexion between mercury and humic substances from different landfill stabilization processes and its implication for the environment. J Hazard Mater 209–210: 59–66.
36. Sheng GP, Xu J, Luo HW, Li WW, Li WH, et al. (2013) Thermodynamic analysis on the binding of heavy metals onto extracellular polymeric substances (EPS) of activated sludge. Water Res 47: 607–614.

16

Seasonal Dynamics of Atlantic Herring (*Clupea harengus* L.) Populations Spawning in the Vicinity of Marginal Habitats

Florian Eggers[1,2]*, Aril Slotte[1], Lísa Anne Libungan[3], Arne Johannessen[2], Cecilie Kvamme[1], Even Moland[4], Esben M. Olsen[4,5,6], Richard D. M. Nash[1]

1 Institute of Marine Research, Bergen, Norway, **2** Department of Biology, University of Bergen, Bergen, Norway, **3** Department of Life and Environmental Sciences, University of Iceland, Reykjavík, Iceland, **4** Institute of Marine Research, Flødevigen, Norway, **5** Centre for Ecological and Evolutionary Synthesis (CEES), Department of Biosciences, University of Oslo, Oslo, Norway, **6** Department of Natural Sciences, Faculty of Science and Engineering, University of Agder, Kristiansand, Norway

Abstract

Gillnet sampling and analyses of otolith shape, vertebral count and growth indicated the presence of three putative Atlantic herring (*Clupea harengus* L.) populations mixing together over the spawning season February–June inside and outside an inland brackish water lake (Landvikvannet) in southern Norway. Peak spawning of oceanic Norwegian spring spawners and coastal Skagerrak spring spawners occurred in March–April with small proportions of spawners entering the lake. In comparison, spawning of Landvik herring peaked in May–June with high proportions found inside the lake, which could be explained by local adaptations to the environmental conditions and seasonal changes of this marginal habitat. The 1.85 km^2 lake was characterized by oxygen depletion occurring between 2.5 and 5 m depth between March and June. This was followed by changes in salinity from 1–7‰ in the 0–1 m surface layer to levels of 20–25‰ deeper than 10 m. In comparison, outside the 3 km long narrow channel connecting the lake with the neighboring fjord, no anoxic conditions were found. Here salinity in the surface layer increased over the season from 10 to 25‰, whereas deeper than 5 m it was stable at around 35‰. Temperature at 0–5 m depth increased significantly over the season in both habitats, from 7 to 14°C outside and 5 to 17°C inside the lake. Despite differences in peak spawning and utilization of the lake habitat between the three putative populations, there was an apparent temporal and spatial overlap in spawning stages suggesting potential interbreeding in accordance with the metapopulation concept.

Editor: Brian R. MacKenzie, Technical University of Denmark, Denmark

Funding: These authors have no support or funding to report.

Competing Interests: The authors have declared that no competing interests exist.

* Email: eggersf@t-online.de

Introduction

Typically, fish species may be split into populations based on their degree of reproductive isolation from each other in space and/or time, which could be reflected in genetic or phenotypic differences driven by diverging environmental conditions [1–3]. Under such circumstances exploitation on one population should have little effect on the population dynamics of a neighboring population, and therefore it is also common to assess and manage such populations separately [4,5]. On the other hand, there are also examples where populations are recognized to be separate with diverging spawning season and/or spawning area, but due to mixing in other seasons a separate management of the populations may be difficult [6,7]. The need to identify the different populations, especially where exploitation occurs on mixtures of populations is important for successful management [8,9]. Fisheries biologists therefore often use the term stock instead of population in their fisheries advice; i.e. sometimes a population is harvested and therefore managed as one stock and at other times several separate populations are harvested and managed as one stock. In Begg et al. [10] the concept of a fish stock was simply defined as characteristics of semi-discrete groups of fish with some definable attributes, which are of interest to fishery managers. The definition of ICES [11] for a stock as a part of a fish population usually with a particular migration pattern, specific spawning grounds, and subject to a distinct fishery, will be used hereby. In theory, all individual fish in an area, being part of the same reproductive process, are comprised as a stock. When referring to fisheries management, the term "stock" is used, otherwise the term "population" is preferred.

Atlantic herring (*Clupea harengus* L.) is characterized by highly complex population structure and migration patterns [12]. It is an iteroparous clupeid, becoming sexually mature at two or three years of age, and a total spawner that aggregates at spawning, laying benthic eggs on shells, gravel, coarse sand and small stones at depths down to 250 m [13]. The larvae hatch after 2–4 weeks depending on temperature [14,15]. They drift with currents until metamorphosis [16–18], with vertical migration increasing throughout ontogeny [19,20] and affecting the dispersal trajectories of larvae. The different herring populations are generally classified according to their spawning grounds, which, due to the specific spawning substratum requirements, are fixed geographically and used at a predictable time of the year. Due to physical and geographical barriers, such as prevailing currents and general location of nursery areas, there is often little mixing of larvae, thus tending to isolate the different populations. However, there are occasions where larvae and juveniles may co-occur. Under these circumstances identification of individuals or groups of individuals is undertaken using otolith or meristic characters [1,21–24] as well as genetic markers [25–28]. In the 1950–60s experimental studies [29–31] demonstrated that myotome counts in herring were influenced by both temperature (negatively) and salinity (positively) experienced during the incubation period. The consequence is that mean vertebral count of adult herring is an indicator of spawning ground and spawning times and in some cases also population.

In Norwegian waters some herring populations occupy marginal habitats along the coastline and deep inside fjords, most of which are thought to be stationary with adaptations to local conditions. Hence, they are often phenotypically and, in some occasions, genotypically different from the nearby oceanic population. Examples of such local herring populations are Trondheimsfjord herring [32,33], Borge Poll herring [34], Lusterfjord herring [35], Lindåspollene herring [36], Balsfjord herring [37], Lake Rossfjord herring [38] and the summer/autumn spawners in northern Norway [39]. Despite the discovery of these local populations, the overall research effort targeting marginal areas along the Norwegian coast has been rather low, and it is therefore expected that a number of additional local populations may exist.

Migratory coastal or oceanic populations may occasionally enter the marginal habitats along the Norwegian coast and mix with local herring. This is in accordance with the metapopulation concept, where two or more distinguished subpopulations have variable but moderate interbreeding and significant gene flow [40]. Temporal and spatial overlap during spawning may allow genetic exchange between subpopulations, which is a prerequisite for the existence of metapopulations. An example of such an overlap was demonstrated by Johannessen et al. [41],[42] in the local Lindåspollene herring, where significant changes in life history traits over a 50 year period were linked to genetic exchange with the oceanic population according to the metapopulation concept.

An important mixing area for herring is the northeastern North Sea and Skagerrak, where three different stocks may occur, Norwegian Spring Spawners (NSS), North Sea Autumn Spawners (NSAS) and Western Baltic Spring Spawners (WBSS). Some of these stocks comprise different herring populations, such as coastal Skagerrak spring spawners or more local herring populations, which are not directly subjected to a distinct fishery. The different populations (stocks) can be distinguished by spawning site, spawning season, meristic characters such as the number of vertebrae (VS) and otolith characteristics [23,41].

Of particular interest in the Skagerrak area is a brackish water environment inside Landvikvannet, an inland lake in southern Norway connected to the open sea through an artificial channel. The Institute of Marine Research (IMR) has been sampling herring in Landvikvannet on regular basis since 1984, mainly in May. Data from these investigations demonstrate that herring inside the lake are normally ripe or with running gonads, with a low mean vertebral number (<56.0), slow growth and high fecundity [43,44]. This has led to the hypothesis that the lake is visited on an annual basis by a herring population with specific adaptations to spawning in these brackish water environments. However, in the coastal areas outside the lake, ripe and spawning herring with higher growth and mean vertebral numbers (56.0–57.5) have occurred in samples over the period February–June [43]. This indicates that there may be a mixture of several populations in the area with some temporal and spatial overlap in spawning, which could be linked to spatial seasonal differences in environmental conditions. Such metapopulation dynamics may be revealed by a more detailed seasonal sampling outside the May period normally focused on in IMR's investigations in Landvikvannet. Hence, the principal objective of the present study was to explore the overlap in time, space and maturation stages of phenotypically different herring appearing in Landvikvannet and neighboring fjord areas and their dependence on seasonal changes in environmental conditions.

Material and Methods

Study area

Landvikvannet is a 1.85 km^2 lake located on the Norwegian Skagerrak coast (Figure 1). In 1877 a 3 km long channel (Reddal channel, Figure 1) was constructed, connecting the lake to the open sea. This narrow 1–4 m deep channel transformed Landvikvannet into a brackish system and in addition lowered the water level in the lake by 3 m. At the entrance of the lake there is a small 25 m deep basin. Further into the lake the bottom depth decreases rapidly to 7–10 m. Most of the shoreline is covered by reeds; otherwise the shore is rocky and steep. There is inflow of saltwater over the tidal cycle, whereas freshwater empties into the lake from streams, resulting in a halocline. Oxygen is depleted in the lower layers whereas the surface layer is oxygen rich. In Landvikvannet, herring have been caught by floating gillnets together with trout (*Salmo trutta*) and other freshwater fish since shortly after the channel was opened.

The Reddal channel drains into Strandfjorden (Figure 1), where conditions are estuarine. The outer Strandfjorden is narrow and shallow (1–7 m), whereas the inner part is deeper (10–13 m). Most herring samples were collected in the inner part, close to the mouth of the Reddal channel. The shore is rocky and steep with sparse macroalgae in the upper few meters. At depths >5–6 m the bottom consists of sand and mud. The outermost fjord (Bufjorden, Figure 1) is small with direct connection to Skagerrak. Strandfjorden is connected to the open ocean via Bufjorden (Figure 1). The entrance of Bufjorden is characterized by a 54 m deep basin. The physical environment is similar to Strandfjorden, only less influenced by fresh water runoff. Access to Bufjorden is from the south or east.

Environmental data

To explore whether potential differences in habitat utilization and timing of peak spawning among herring populations were dependent on seasonal changes in environmental conditions, sampling of environmental data was undertaken between March and June 2012 both inside and outside the lake habitat. Note, that no stations could be sampled in February due to ice cover. Water samples were collected at the site where gillnets were moored in the inner part of Strandfjorden and at the entrance of Landvikvannet in the first basin (Figure 1). We measured temperature and salinity at depth with a CTD (STD/CTD – model SD204, SAIV Ltd. Environmental sensors and Systems, Bergen, Norway), while oxygen and hydrogensulfide concentrations were analyzed in the laboratory at the Institute of Marine Research (IMR). In the lake, water samples were collected each 0.5 meter down to the depth of oxygen depletion (hypoxic depth), which was found using the Winkler test [45], thereafter water samples were taken at 5 m

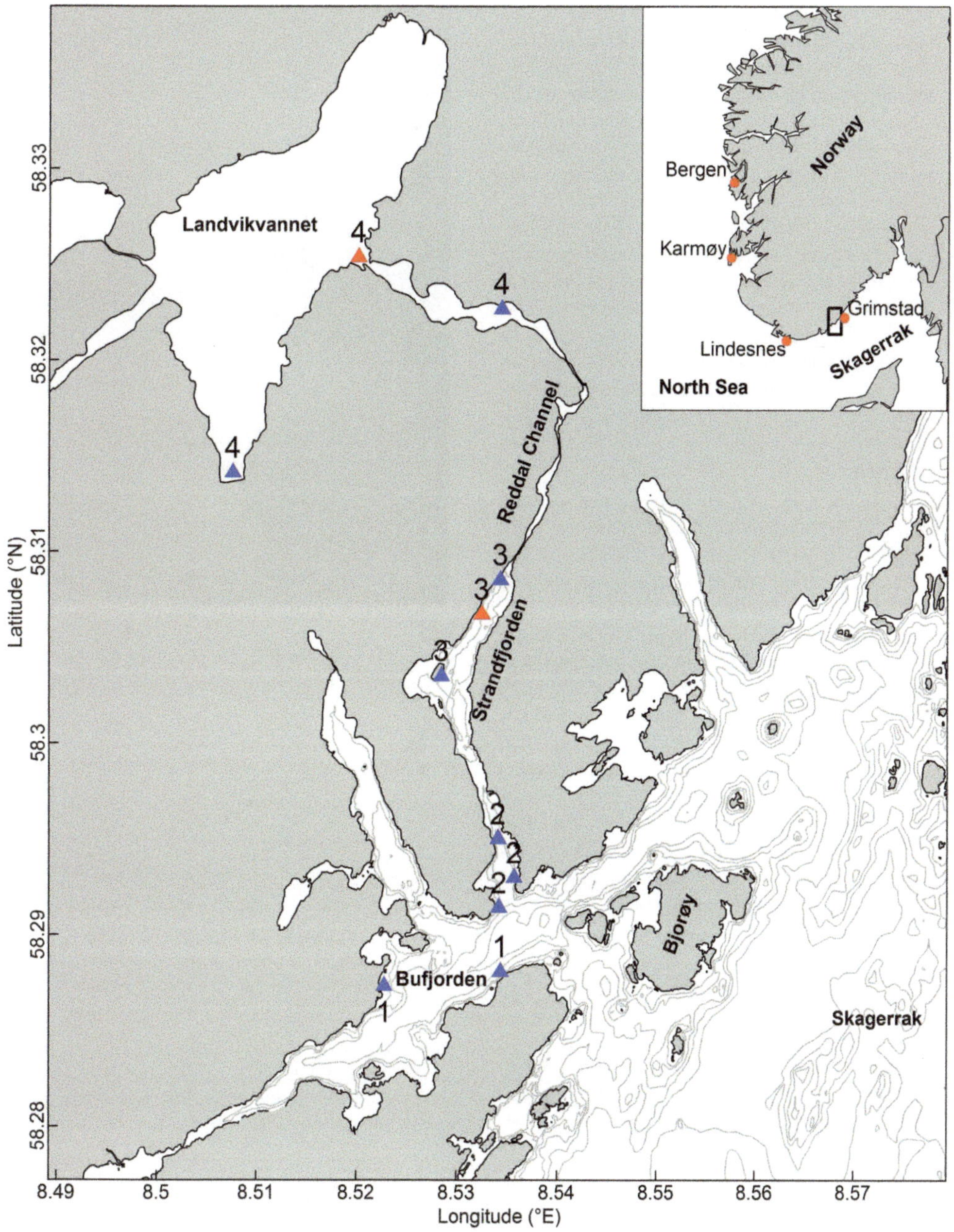

Figure 1. Map of the study area. The map shows CTD-stations (red) and gillnet stations (blue) in 1 = Bufjorden, 2 = Outer part of Strandfjorden, 3 = Inner part of Strandfjorden, 4 = Landvikvannet.

depth intervals. The choice of position for sampling environmental data inside the lake is based on the depth contours of the area. The lake itself is rather shallow, and the bottom depth at most gillnet stations is 2–4 m. However, at the entrance the lake is at its deepest (25 m), which is why this position has been used since investigations started in the area in the 1980s. The environmental conditions at this site between 0 and 10 m have been examined thoroughly over a number of years and are comparable to conditions elsewhere in the lake and as such can be used to characterize the whole lake. These data are therefore representative of all gill net sampling sites.

Biological data

To explore the potential overlap in time, space and maturation stages of phenotypically different herring appearing inside and outside the lake habitat, herring were sampled with gillnet over the full spawning season in 2012 (February–June) concurrently in both habitats (Figure 1, Table 1). In February, due to ice cover both in the lake and inner fjord habitats of Strandfjorden, samples were only taken further out in Bufjorden. The floating gillnets with a mesh size of 26 mm and 29 mm, a depth of 8 m and a length of approximately 10 m were used randomly in all areas. Soak time was 24 hours. This experiment was approved by the Norwegian committee for the use of animals in scientific experiments (FDU). Special permission to fish with floating gillnet inside

Landvikvannet and in the connected fjord system in 2012 was given by the County Governor of Aust-Agder, Department of Climate and Environment, Ragnvald Blakstadsv. 1, Postbox 788 Stoa, 4809 Arendal, Norway. The permission was given to the Institute of Marine Research under the prerequisite that details on the catch were reported when the investigations were finished. The report was delivered to the authorities according to the plan. Our study did not involve endangered or protected species.

Biological samples were analyzed according to IMR standard protocols [46]. The maximum sample size was 100 herring. Biological parameters included in the present study were total length (nearest 0.5 cm below), weight (nearest gram below), sex, stage of maturity, age (otolith readings) and vertebral count (VS). Maturity stages were determined by visual inspection of gonads according to the following scale: immature = 1–2, maturing = 3–4, ripe = 5, spawning/running = 6, spent = 7 and recovering = 8 [46].

Image and shape analyses

Individuals of NSS herring were identified from otoliths, based on a sharper distinction between winter and summer rings compared to local spring spawners (Figure 2). This distinction was also independently tested using image and shape analyses of the otoliths. The rest of the individuals were divided into two populations based on sampling location: local Landvikvannet herring (LV) sampled inside Landvikvannet and coastal Skagerrak spring spawners (CSS) sampled outside Landvikvannet (Table 2). We expected that LV herring would mainly consist of individuals with similar biological characteristics as normally found in May, whereas the CSS herring would mainly consist of spring spawners with characteristics normally found along the Skagerrak coast during February–June. However, some mixture of the two populations would be expected, and this would be evident from results of the biological analyses. To investigate changes in the mixture of NSS, CSS and LV herring in the two habitats, selected biological characters (otolith shape, vertebral count, growth and maturation stage) were analyzed over the full season. The numbers analyzed by month and population are given in Table 2.

Otolith shape was analyzed using the programming language R [47]. Outlines of otoliths were collected from digital images using the package pixmap [48], and applying the conte function [49] to record a matrix of X and Y coordinates (Figure 2a). Mean shape of otoliths differed among the populations, where the modifications in the shape of otoliths mainly were found at the excisura major and antirostrum areas (Figure 2b).

To remove size-induced bias, otolith sizes were standardized to equal area by dividing the coordinates of each otolith with the square root of the otolith area. Equally spaced radiis were drawn from the otolith centroid to the otolith outline, using the regular radius function [49]. Independent Wavelet shape coefficients were obtained by conducting a Discrete Wavelet transform on the

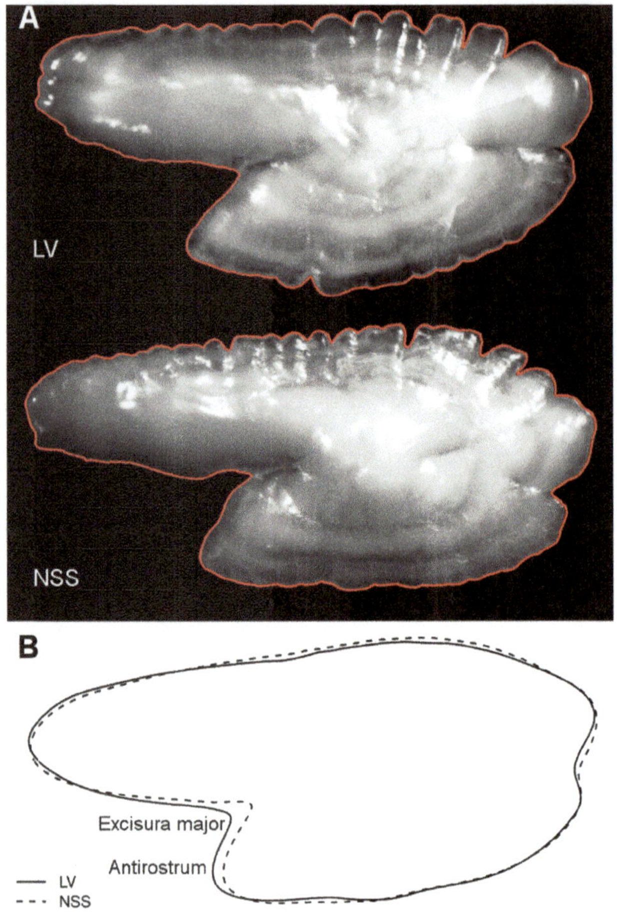

Figure 2. Example of otolith characteristics from two herring populations. A) Example of otoliths used for the shape analysis from Landvikvannet herring (LV) and Norwegian spring-spawning herring (NSS), both at the age of 3 years. Individuals of NSS herring were subjectively identified based on a sharper distinction between winter (dark areas) and summer rings (white areas). Red outline marks the shape of the otolith which was used to compare among populations. B) shows the mean shape of otoliths for the two populations, where the excisura major and antirostrum areas are the most variable areas.

Table 1. Total number of herring caught in the local area for 2012, in brackets number of gillnets; ice = no sampling possible because the area was covered by ice.

Date	Landvikvannet	Inner Strandfjorden	Outer Strandfjorden	Bufjorden
15/2	Ice cover	Ice cover	28 (1)	11 (1)
6/3	4 (3)	129 (1)	119 (1)	
20/3	47 (3)	542 (1)		
26/3	115 (3)	486 (1)		100 (1)
11/4	290 (2)	663 (1)		
14/5	177 (1)	69 (1)		
21/6	82 (1)	66 (1)		
Total	**715**	**1955**	**147**	**111**

Table 2. Total number of herring analyzed in 2012 by month for the three putative herring populations, Norwegian spring spawners (NSS), Coastal Skagerrak spring spawners (CSS) and Landvik herring (LV), in brackets number of NSS inside Landvikvannet.

Month	NSS	CSS	LV
2	7 (0)	32	0
3	108 (38)	440	113
4	32 (14)	68	86
5	8 (5)	61	95
6	0 (0)	66	77
Total	**155 (57)**	**667**	**371**

equally spaced radiuses using the wavethresh package [50]. To determine the number of Wavelet coefficients needed for the analysis, the deviation of the reconstructed Wavelet otolith outline from the original outline was evaluated. To correct for fish length, an ANCOVA was performed on the wavelet coefficients taking fish length as a covariate. Coefficients which could not be adjusted by linear relationships on fish length, due to interaction between the origin and length were excluded from the analysis [51–53]. To adjust the Wavelet coefficients for allometric growth, a normalization technique based on regression was applied to scale the Wavelet coefficients [54].

Data analyses

The number of gillnets varied between Landvikvannet and the neighboring fjord area. Therefore, to estimate the proportions of the LV, CSS and NSS herring, the total catches landed were standardized by catch per unit effort (CPUE), i.e. catch per gillnet.

All statistical analyses were conducted in R (version 3.0.1; [47]). A significance level of $\alpha = 0.05$ was used for all statistical tests. For the plots, mean and standard error (1 SE) are shown. Some samples had very few or no data, and samples with $N<5$ were excluded.

Analysis of Covariance (ANCOVA) was used to test for sex differences in the biological characters (length, age, VS and stage of maturity). Differences in VS among different herring populations were assessed using Analysis of Variance (ANOVA), and a Kruskal-Wallis test for length and age variables as these were not normally distributed. For pairwise comparisons of VS a paired T-test was used, and the Mann-Whitney test for length and age comparisons.

Length-at-age data, used as a proxy for growth of individual herring, were fitted to the von Bertalanffy growth model (VBGM) [55]:

$$L_t = L_\infty(1 - e^{-K(t-t_0)})$$

where L_t is the average length at age t, L_∞ is the asymptotic maximum length, K is the von Bertalanffy growth rate coefficient, i.e. the rate at which length approaches the maximum length asymptote and t_0 is the intercept on the time axis. Growth was compared between the different groups using ANOVA.

Variation in otolith shape, as reflected by the scaled Wavelet coefficients, was analyzed with Canonical Analysis of Principal coordinates (CAP) [56] using the capscale function in the vegan package in R [57]. Using multivariate data to represent otolith shape, an ANOVA like permutation test (vegan package) was used to assess the significance of constraints using 5000 permutations. Variation in otolith shape was analyzed with CAP, while length and VS were compared with ANOVA with respect to herring group: NSS, LV and CSS, the month in which they were caught over the sampling period (Feb–June) and age in years (3–12) using the following models: shape~herring population*month*age, length~herring population*month*age and VS~herring population*month*age. Non-significant interaction terms ($p>0.05$) were excluded from the models. P-values for all posteriori comparisons were corrected with the Bonferroni correction [58]. Possible trends of length and VS within herring populations were tested for significance using linear regression, while the stage of maturity was tested with the Spearman's rank correlation coefficient. For the comparisons of environmental data at time of spawning with the VS of herring, measurements from 3 m were used for Landvikvannet due to the depth of oxygen depletion in combination with previous (2010) acoustic observations of school depth [43]. In Strandfjorden, measurements from 5 m were used, based on acoustic observations of herring school depth during tagging experiments and the gillnet sampling [43].

Results

Environmental conditions

The environmental conditions differed considerably between Landvikvannet and the neighboring fjord, and changed over the spawning season in both locations (Figure 3). Anoxic conditions were found in Landvikvannet at increasing depths from 2.5 m in March to 5 m in June. Salinity ILV at 0–1 m increased over the season from 1‰ in March to 7‰ in June, but was stable around 20–25‰ deeper than 10 m. In comparison, there were no anoxic conditions in Strandfjorden, the salinity at 0–1 m increased from 10‰ in March to 25‰ in June and was stable at 35‰ deeper than 5 m. The temperature at 0–5 m depth increased from March to June from 5 to 17°C in Landvikvannet, and from 7 to 14°C in Strandfjorden.

Population structure

A total of 1260 herring were analyzed during the 2012 spawning season. Total length ranged from 22.0–34.5 cm (mean: 28.3 cm) and age from 2–12 years (mean: 4.2 years). None of the biological characters varied between sexes ($p>0.05$). Hence, all further analyzes were carried out with sexes combined.

Mean length, age and vertebral count (VS) differed significantly among the three herring populations ($p<0.001$, Figure 4). For age and length, pairwise comparisons were also significant ($p<0.001$), with the exception of CSS versus LV for age ($p>0.05$). The vertebral count differed significantly ($p<0.001$) for all pairwise comparisons. The main tendency was a significant increase in

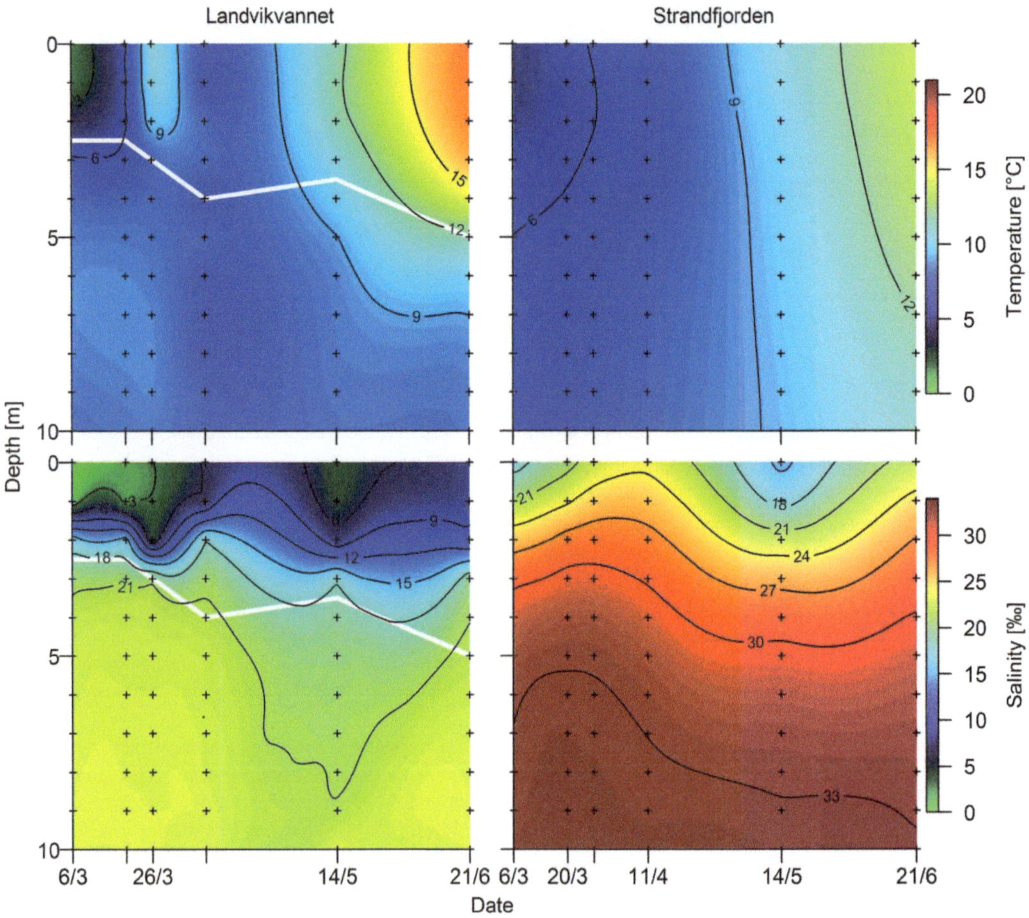

Figure 3. Seasonal change in temperature and salinity by depth. Temperature (upper) and salinity (lower) in Landvikvannet and in Strandfjorden over the study period from March to June. White line indicates the depth of oxygen depletion.

mean body length and VS when moving from LV to CSS to NSS, whereas men age decreased. The most common age was 3 years for NSS, CSS and LV herring. The 4 year olds were also abundant in CSS and LV herring, but hardly present among NSS herring.

Length-at-age data indicated the highest growth for NSS herring, and lowest for LV herring ($p<0.01$) (Figure 5). The von Bertalanffy growth model supported these growth differences (Table 3). Consequently, there were three categories: 'high growth rate' (NSS herring), 'moderate growth rate' (CSS herring) and 'low growth rate' (LV herring).

Between February and June there was a change in the abundance of the different populations (Figure 6). During February–April CPUE was highest for CSS and NSS herring with a low proportion of LV herring (<20%). Also the proportion of NSS herring entering Landvikvannet was insignificant (<10%). The proportion of spawning and spent herring during this period was highest in NSS herring and a little lower for CSS herring, but still indicating peak spawning of two different populations in the fjord habitat during this period. Among the LV herring analyzed in March–April an even lower proportion were in spawning and spent stages than for CSS herring, indicating a later spawning peak for LV herring. This was further demonstrated in the May–June sampling showing a spatial shift in CPUE towards higher abundance of LV than CSS and NSS herring.

Otolith shape differed among the three herring populations ($p<0.001$, Table 4, Figure 7) and also varied though the spawning season ($p<0.001$, Figure 8A). Vertebral count and length differed between the populations ($p<0.001$) and between months ($p<0.001$, Figure 8B, C). Age was a significant factor for all characters ($p<0.001$) and therefore incorporated in the model for all comparisons. Posteriori comparisons showed that LV and CSS differed in otolith shape, VS and length ($p<0.04$, Figure 8, Table 4). NSS and LV ($p<0.001$) as well as NSS and CSS ($p<0.02$) also differed, while no differences were detected for NSS caught inside or outside the lake ($p>0.05$). There was a signifiant ($p<0.001$) negative trend in the mean Canonical scores (CAN1) derrived from the CAP analysis of otolith shape, vertebral count and length for LV and CSS herring at standardized ages over the spawning season, but not for NSS (Figure 8). This indicates that LV herring, characterized by slow growth and low vertebral count, were arriving and mixing with CSS herring.

Maturation and spawning time

Herring in spawning condition were present and overlapped in time for LV, CSS and NSS herring, however, maturation and timing of spawning was delayed in LV compared to NSS and CSS herring (Figure 6). This indicates an adaptation to the environmental conditions and seasonal change in Landvikvannet. Since differences in vertebral count are linked to environmental conditions, the temperature and salinity at depth and time of

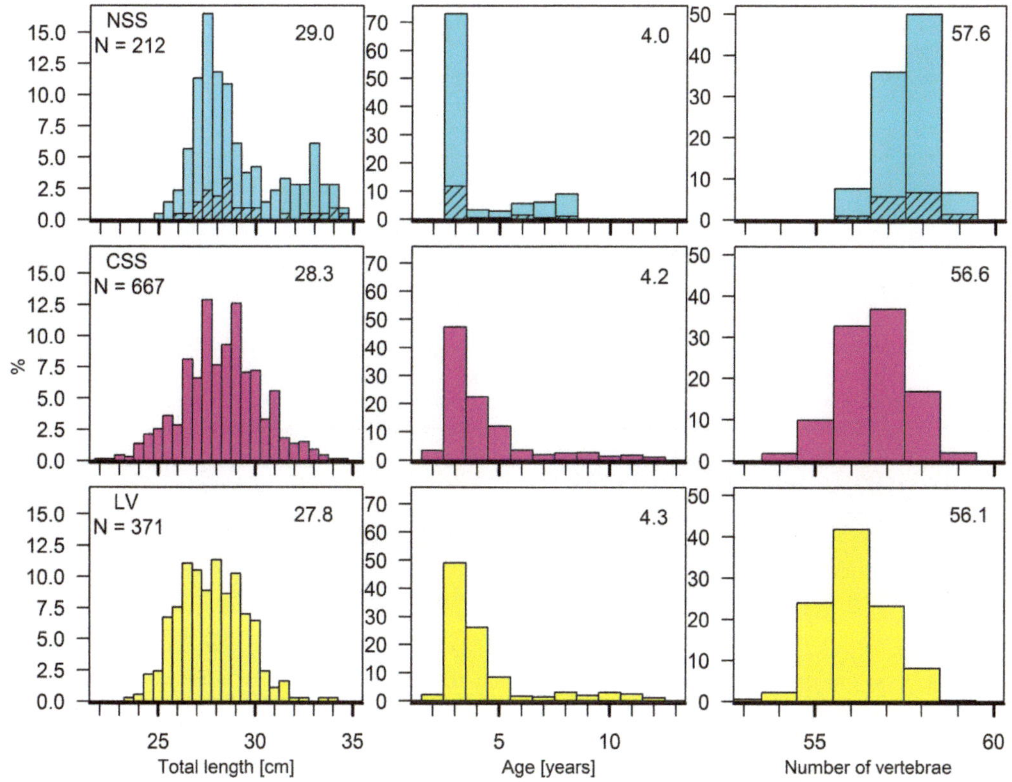

Figure 4. Distribution of length, age and vertebral counts of different herring populations. Comparison between Norwegian spring spawning (NSS), Coastal Skagerrak spring spawning (CSS) and Landvik (LV) herring. Shaded areas are NSS herring inside Landvikvannet. The mean values are included.

spawning affects the vertebral count. The salinity at expected spawning depth in Landvikvannet was distinctly lower (10–15‰) than in the adjacent fjord (>30‰), which could explain the low vertebral count observed in Landvikvannet. The vertebral count was not significantly related to change in salinity over season within habitats; there was negligible change at assumed spawning depth. However, there were significant changes in temperature over season in both habitats, coinciding with a significant decrease in vertebral count at spawning time for both CSS and LV herring ($p<0.05$).

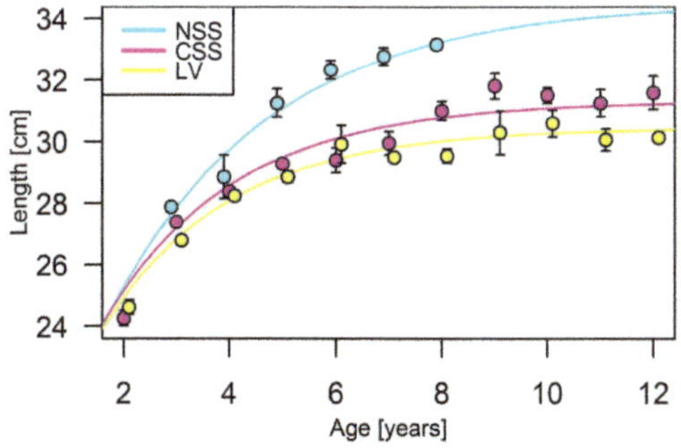

Figure 5. Growth curves of different herring populations. Length-at-age for Norwegian spring spawning (NSS, N = 212), Coastal Skagerrak spring spawning (CSS, N = 667) and Landvik (LV, N = 371) herring in samples pooled over the 2012 spawning season. Means and standard error (1 SE) are given, lines show van Bertalanffy growth models fitted to data.

Discussion

This study reveals strong seasonal dynamics involving three populations of a pelagic migratory fish, the Atlantic herring, in the vicinity of a marginal inland brackish water lake habitat (Landvikvannet) on the Norwegian Skagerrak coast. Gillnet sampling was standardized, implying that the observed differences between herring populations and over season dynamics were not affected by the selectivity normally experienced with gillnet sampling [59]. Three putative herring populations were identified; Norwegian spring spawners (NSS), Landvik herring (LV) and Coastal Skagerrak spring spawners (CSS). Individual NSS herring were identified subjectively based on otolith growth characteristics, and statistically based on otolith shape and mean vertebral count (57.5). NSS herring also had higher growth than the other populations, which is typical for this stock [13,43]. Identification of individual CSS and Landvik herring was not possible. Individuals sampled inside the lake were all classified as LV herring, whereas those sampled outside the channel connecting the lake to the sea were assigned as CSS herring. However, there was a significant decrease in vertebral count over the sampling season in both LV and CSS herring, from levels known as typical for CSS herring (56.5–56.9) in March–April to levels typical for Landvik herring (<56.0) in May–June, again based on historic data [43]. This trend in vertebral count was followed by a decrease in size and change in otolith shape, and a marked change in the relative proportions of the two populations.

The observed seasonal dynamics in biological characters clearly indicate that the assignment of individual fish into CSS and LV herring simply based on sampling location was uncertain, and that

Table 3. Von Bertalanffy growth parameters (L_∞, k, and t_0) of herring populations Norwegian spring spawners (NSS), Coastal Skagerrak spring spawners (CSS) and Landvik herring (LV).

	L_∞	K	t_0
NSS	34.51	0.33	−1.98
CSS	31.31	0.41	−1.98
LV	30.33	0.43	−1.98

the two populations were mixing both inside and outside the lake habitat together with NSS herring showing a different peak occurrence. Early in the season in February–April the biological characteristics indicated that NSS and CSS herring predominated, with only small numbers entering the lake. There was a clear temporal and spatial overlap in spawning individuals from these two populations, although proportions spawning in CSS were comparatively lower than in NSS herring. In May–June there was a significant change with the appearance of a new spawning wave of LV herring, with the highest proportion found inside the lake. Still, the immigration of this population was evident throughout both habitats, where many of the herring found in the fjord would be expected to enter the lake. The data on otolith shape, vertebral count and growth in May tended to differ from the observations in June in both locations, which indicated a spatial and temporal overlap in May between minor proportions of NSS and CSS herring completing their spawning season at the same time as the LV herring was peaking.

All three putative populations were caught at the same location, in the same gillnets, at the same time with running gonads, suggesting that the populations together form a metapopulation [40]. However, there is doubt as to whether interbreeding between distinct populations is occurring despite their proximity in spawning condition. Since breeding was not observed directly,

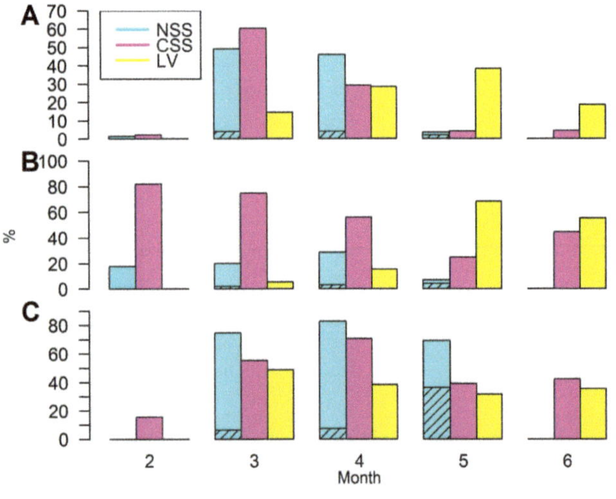

Figure 6. Seasonal change in proportion of different herring populations. Proportion (%), standardized to one gillnet per sample and area, by month of Norwegian spring spawning (NSS), Coastal Skagerrak spring spawning (CSS) and Landvik (LV) herring relative to a) total number analyzed over entire study period (see Table 1 for N), b) total number at month and c) spawning and spent herring (stage of maturity>=6) relative to total number at month (see Table 2 for N). Shaded areas are NSS herring inside Landvikvannet.

one cannot exclude the possibility that the populations separate for spawning events. Such a full separation seems unlikely for NSS and CSS herring because of the high temporal and spatial overlap; whereas it seems more likely for LV herring considering the limited temporal and spatial overlap with the other populations.

The idea that LV herring is reproductively isolated from other populations may be supported by the low vertebral count and concept of natal homing. Differences in vertebral count stem from the incubation phase and thus reflect the origin of the fish at spawning [60]. In general, there is a positive correlation with salinity [31] and negative with temperature [21,29,61] experienced prior to hatching. Hence, the warmer and less saline ambient environment for herring occurring inside Landvikvannet in May–June compared with that experienced by CSS in March–April in the fjord habitat, could result in the observed differences in vertebral count. The low vertebral count of LV herring and the late timing of spawning is an indication of spawning and adaptations to the environmental conditions of the lake habitat. However, this also implies that natal homing [62,63] of Landvik herring occurs on an annual basis. The vertebral number for LV herring in May has been remarkably stable (55.5–55.8) since 1984 [43], supporting natal homing. The principle of natal homing is central to the discrete population concept [12]. Moreover, recent genetic studies support the occurrence of natal homing of herring in the North and Baltic Seas [6,64]. Likewise, Brophy et al. [65] suggested that spawning season and location of Atlantic herring could be predetermined and not learnt from repeated spawning [66]. Support for natal homing and adaptations of Landvik herring to environmental conditions of its marginal habitat also originates from a recent genetic study using 20 microsatellite markers, where Landvikvannet differed from other local herring in Lindåspollene, Lusterfjord and Trondheimsfjord as well as from other herring populations surrounding the Norwegian Sea [67]. Unpublished results on the microsatellite locus Cpa112, which is non-neutral to salinity variability with allele frequencies varying from 45% in the Baltic to 2–4% in the North Sea [27], have shown that Landvik herring is obvious with a frequency of 15% (Carl André, pers. Comm., Department of Biology and Environmental Sciences - Tjärnö, University of Gothenburg, Strömstad, Sweden).

It seems clear from this study that we can refute the hypothesis of a resident local population inside the lake; LV herring definitely migrates into the lake habitat from coastal areas. In this sense the Landvik herring differs from other local herring populations, such as the Trondheimsfjord or Lindås herring, which can be observed throughout the year in their local areas [32,33,36,41]. This may simply be because of the unsuitability of this location as a nursery area for juveniles and feeding grounds for adults. Both CSS and LV herring may still represent more stationary coastal populations not undertaking large scale oceanic migrations. The observed relatively low investment costs in reproduction (low GSI) of NSS compared with that of LV herring supports the assumption that

Table 4. Comparing otolith shape, vertebral count (VS) and length among herring populations Norwegian spring spawners (NSS), Coastal Skagerrak spring spawners (CSS) and Landvik herring (LV).

			Otolith shape				Vertebral count				Fish length			
Comparison		**N**	**df**	**Var**	**F**	**P**	**df**	**Mean Sq**	**F**	**P**	**df**	**Mean Sq**	**F**	**P**
Overall	NSS vs LV vs CSS	897	2	3.28	5.36	<0.001	2	109.95	136.44	<0.001	2	129.80	102.58	<0.001
	Month		1	1.20	3.91	<0.001	1	71.49	88.71	<0.001	1	690.00	545.44	<0.001
	Age		10	4.49	1.47	0.001	10	3.87	4.80	<0.001	10	178.20	140.90	<0.001
	Residuals		883	270.41			867	0.81			867	1.30		
Posteriori	LV vs CSS	745	1	0.69	2.22	0.04	1	32.10	36.69	<0.001	1	13.10	10.08	0.006
	NSS vs LV	500	1	1.45	4.76	<0.001	1	219.80	276.99	<0.001	1	250.45	196.30	<0.001
	NSS vs CSS	549	1	0.84	2.72	0.02	1	115.53	149.39	<0.001	1	178.20	114.88	<0.001
	NSS-ILV vs NSS-OLV	152	1	0.20	0.65	>0.05	1	0.23	0.47	>0.05	1	1.85	1.65	>0.05

NSS herring were also compared between sampling locations, inside (NSS-ILV) and outside (NSS-OLV) Landvikvannet. ANOVA like permutation tests were used to assess the difference in otolith shape and ANOVA for the vertebral count and fish length comparisons. For otolith shape: df: degrees of freedom, Var: Variance among populations, F: pseudo F-value, P: proportion of permutations which gave as large or larger F-value than the observed one. For the vertebral count and fish length: df: degrees of freedom, Mean Sq: Mean Square, F: F-value, P: P-value. P-values for posteriori comparisons have been corrected with a Bonferroni correction. $P<0.05$ indicates a significant effect.

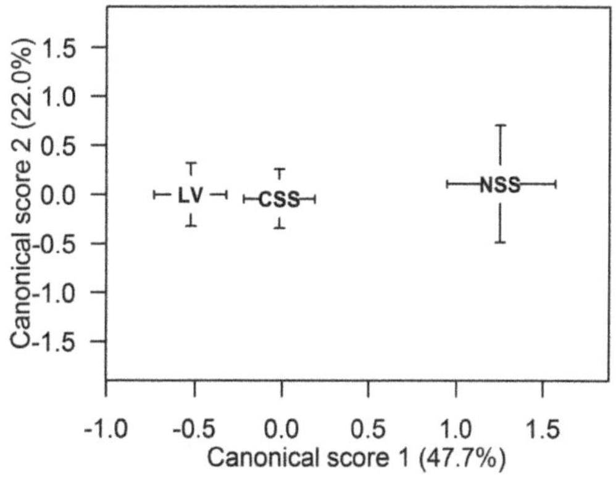

Figure 7. Otolith shape compared for different herring populations. Canonical scores for Norwegian spring spawning (NSS, N = 152), Coastal Skagerrak spring spawning (CSS, N = 397) and Landvik (LV, N = 348) herring are shown on discriminating axes 1 and 2. Black letters represent the mean canonical value for each group with standard error of the mean (1 SE).

NSS is more migratory [44]. The fact that growth of CSS was higher than in LV herring, further suggest that these two populations may not overlap much during the nursery period or at adult feeding grounds. In fact, there is probably little or no spatial overlap for most of the year, with overlap only occurring during the spawning season.

The movements of herring between the fjord and Landvikvannet habitats have also been studied with acoustic telemetry [43,68]. The telemetry study showed that some fish moved in and out of the lake habitat, whereas others stayed inside the lake for more than two weeks. Those fish that arrived and only stayed for a short period of time were interpreted as being NSS or CSS, whereas the ones remaining in the area for extended periods of time were thought to be local LV herring. It is likely that some NSS and CSS herring have short visits to the lake as exploratory migrations searching for good habitats cued by the current from the Reddal channel, but migrate out again to spawn in areas which are more characteristic of their normal spawning habitat. Conversely, fish that stay for two weeks inside the lake before leaving is a reasonably good indication of an established adaptation to the lake and to potential spawning within the lake.

The appearance of NSS herring in the habitats within Landvikvannet and adjacent fjords probably does not represent natal homing. The predominance of 3-year-olds among the NSS stock as well as the high stability of growth and meristic characters over the season, suggest independent selection of spawning grounds, as supported by Slotte and Fiksen [69]. In NSS herring specifically, the use of spawning grounds other than their natal ground is common. NSS herring have a tendency to change their spawning ground as they grow older with larger fish tending to migrate further, in this case southward, and thus potentially increase their life time fitness [69–71]. Such straying from natal spawning grounds results in considerable gene flow [72,73]. The predominance of 3-year-old NSS mixing with CSS and Landvik herring in 2012 may be explained by the relatively unusual spawning migrations of NSS herring in 2009–2010. During these two years a significant proportion of the adult NSS migrated from wintering grounds in the northern Norwegian Sea to areas south of 60°N, resulting in the largest fishery in the fjords (e.g. Boknafjorden) east of the traditional spawning grounds off Karmøy since the 1950s [74]. Based on vertebral count and growth data, it was apparent that the fishery was targeting NSS

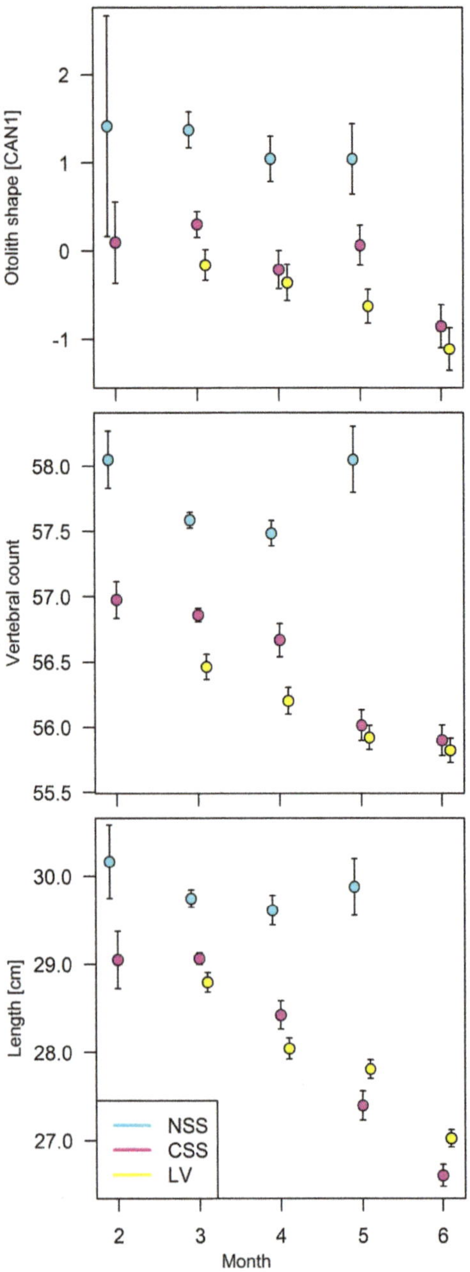

Figure 8. Seasonal changes of otolith shape, vertebral counts and length for different herring populations. For standardized ages. Comparison between Norwegian spring spawning (NSS), Coastal Skagerrak spring spawning (CSS) and Landvik (LV) herring (see Table 2 for N). Values given are means and standard errors (1 SE).

herring [75] and the abundance was high as evaluated by catch levels (Table 5). One hypothesis is that the 3 year old NSS mixing with CSS and Landvik herring in 2012 was a result of this significant spawning at the southern grounds in 2009. Generally, if first time spawners of NSS do not meet older conspecifics and learn to follow their migration towards the spawning grounds then the location of the spawning ground is a chance event [70,71,76,77]. In addition, NSS herring tend to migrate upstream to spawn [69]. Therefore it is not unlikely that NSS from Boknafjorden or further south may have spawned close to their nursery areas or even migrated further south-eastwards against the

Table 5. Commercial catches of herring off Karmøy 2005–2012.

	Year of catch							
Month	**2005**	**2006**	**2007**	**2008**	**2009**	**2010**	**2011**	**2012**
1					0.1			
2	21.2				172.0	3302.9	609.1	897.3
3	24.5	32.6	16.5		19052.0	14877.0	6528.4	6283.2
4	129.2	0.7	1.0	4.8	2301.2	1000.3	52.0	13.4
8	1.0							
9					0.9			
10			0.1					
11		72.8				0.5		
12	0.2							
Total	**176.1**	**106.1**	**17.6**	**4.8**	**21526.2**	**19180.7**	**7189.5**	**7193.9**

Live weight (tons) calculated from landed weight to live weight equivalent for Norwegian spring spawning herring in the Norwegian statistical area 08 (SW coastal Norway) by month and year as registered in the Directorate of Fisheries database.

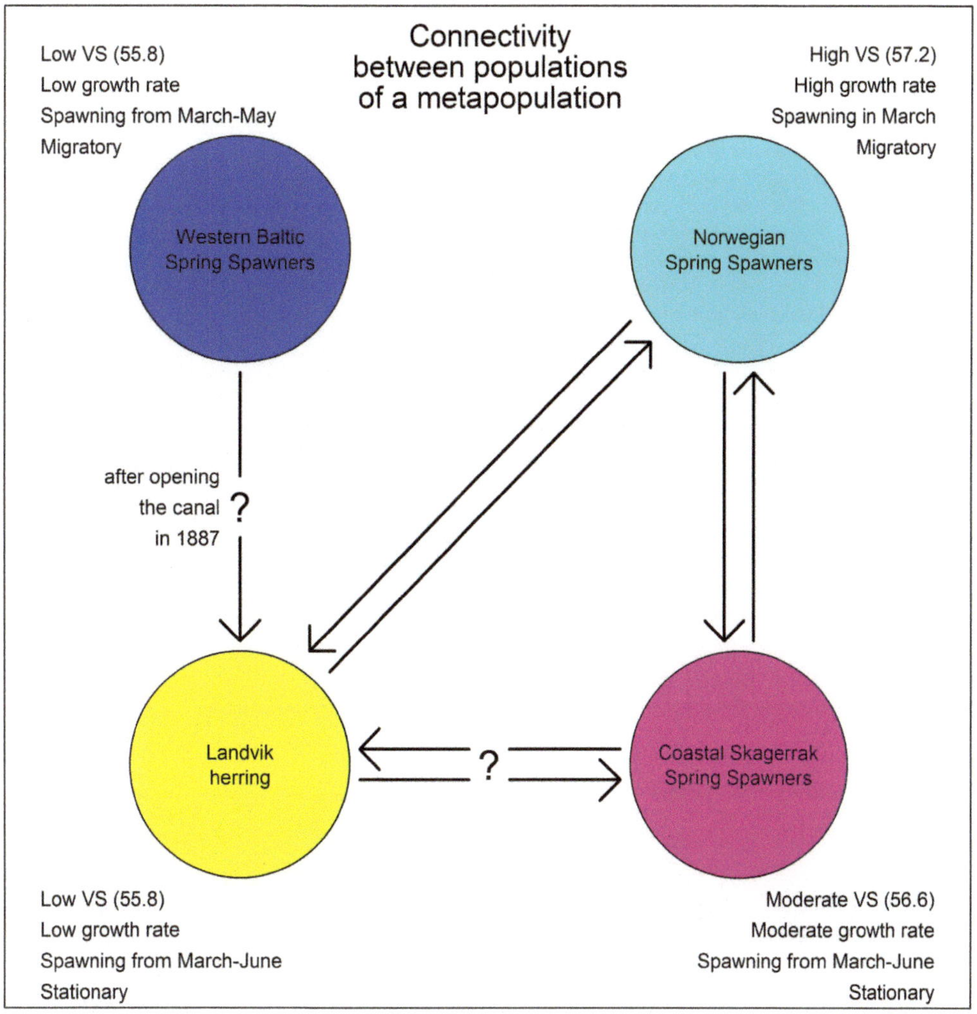

Figure 9. A schematic model of potential metapopulation dynamics in the study area. Potential connectivity between populations of a metapopulation in the study area of Landvikvannet and the connected fjords as hypothesized based on the results of the present study. The biological characteristics (VS = vertebral counts) of the different populations are given.

coastal current to spawn. In addition, school composition tends to involve size-matching among individuals [78], in this case younger, smaller NSS. Three year old NSS (mostly first-time spawners), may have adopted the behavior of the joint local populations with whom they mix during the nursery period as postulated in the adopted-migrant hypothesis [40,79].

From an evolutionary perspective, the Landvikvannet habitat has only been available for marine species for a relatively short period of time. This raises the question of the origin of the herring first colonizing the lake after the opening of the Reddal channel (Figure 9). One possibility is that CSS herring entered the lake sometime after the opening of the channel and successfully spawned there. Due to lower salinity and higher temperature in the lake the offspring developed significantly divergent characters over the years. A strong natal homing effect of herring would lead to the development of a new local population inside Landvikvannet. Hendry and Kinnison [80] concluded that a time span less than 100 years can be sufficient for significant microevolution to develop in response to local agents of selection. Also, Neb [81] demonstrates that such a time interval and differences in salinity are sufficient for herring to diverge in meristic characters. This explanation assumes reproductive isolation during spawning between the original CSS herring and the "new" Landvik herring. A second possibility is that the origin of Landvik herring could be Western Baltic Spring Spawners (WBSS) herring. First time, or even repeated, spawners could have established a new spawning ground in Landvikvannet. The reason for not conducting an annual migration to the original spawning grounds off the island Rügen may be a trade-off between survival of progeny and physiological migration constraints, as shown for NSS by Slotte [70]. WBSS close to their feeding grounds in the Skagerrak could have "discovered" Landvikvannet, cued by similar environmental conditions as those of their original spawning grounds. The continued link to Landvikvannet may have been a result of a fidelity to this site rather than for joining conspecifics in a migration back in to the Baltic region. Huse et al. [76] demonstrate that a high ratio of first-time spawners could lead to the establishment of new wintering grounds. In the case of Landvik herring, it may have led to a new spawning ground.

In conclusion, the present study provides evidence for a distinct small local population of herring associated with Landvikvannet, partly mixing with NSS and CSS herring. This population of LV herring resides, during part of the year in brackish water with many morphometric characteristics indicative of spawning in warm and low salinity environments. Whilst ripe and spent fish have been found in the area, there is no direct evidence of spawning in the lake. If spawning does occur there are no data to indicate likely survival rates or even the residence time of offspring in the lake. There has been one attempt to find eggs with a diver for 1 hour at one of the many bays in the lake, without success. Also, limited plankton net sampling in selected parts of the lake have failed to capture any larvae. The only evidence of potential spawning in the lake, is from two eels with stomachs full of fertilized herring eggs. There is also no clear evidence of the origin of this population, however, they could have arisen from either WBSS or other local CSS. The presence of mixtures of these and other stocks and populations in the Skagerrak area have been shown previously [6,82]. Recent genetic studies using microsatellite DNA [83] have demonstrated differences between Landvik herring and many other stocks, in addition, unpublished results on one microsatellite locus (Carl André, pers. Comm., Department of Biology and Environmental Sciences - Tjärnö, University of Gothenburg, Strömstad, Sweden) suggesting that Landvikvannet herring has not recently immigrated from the Baltic.

The results of the present study may also have some implications for the official ICES stock assessment of herring in the North Sea and Skagerrak area. The present work demonstrates that there can be a fairly complex population structure in the areas with more than one 'stock' which can be mixed. Whilst this may not be a significant problem for the assessment of NSAS or WBSS due to the relatively small abundances of CSS and LV herring, there is a possibility that these smaller populations could be very vulnerable to overfishing [9]. This is probably not unique for coastal areas as there are a number of relatively small populations bordering the North Sea and Skagerrak area [84].

From management point of view, probably the most striking result of the present study is the conclusive evidence of NSS herring as far southeast as in the Skagerrak. This is the first time that individuals from this historically large herring stock have been studied in the Skagerrak area. By definition this stock is not exploited south of 62°N, with exception of the spawning period when they previously have been found as far south as to Lindesnes (Figure 1). This signifies that migration dynamics and population connectivity among herring in the Northeastern Atlantic may be more dynamic than previously assumed, and this must be taken into account in the future development and implementation of new management strategies.

Acknowledgments

Knut Hansen is thanked for his very valuable contribution with running the sampling program throughout the 2012 spawning season and being in charge of biological analyses. The following are also thanked for their contributions to the research: Åse Husebø undertook the initial work on photographing and outlining otoliths for the shape analyses, Jostein Røttingen and Inger Henriksen contributed to sampling and biological analyses, Terje Jåvold analyzed the water samples and Øystein Paulsen undertook photographing and assisted with the sampling. The two anonymous reviewers are thanked for their very valuable suggestions for improvements to the manuscript.

Author Contributions

Conceived and designed the experiments: FE AS LAL. Performed the experiments: FE AS LAL. Analyzed the data: FE AS LAL. Contributed reagents/materials/analysis tools: FE AS LAL AJ EMO EM. Wrote the paper: FE AS LAL RDMN. Reviewing the manuscript: FE AS LAL AJ EMO EM CK RDMN.

References

1. Heincke F (1898) Naturgeschichte des Herings. Berlin: Otto Salle Verlag.
2. Sinclair M, Iles DT (1988) Population richness of marine fish species. Aquat Living Resour 1: 71–83.
3. McPherson AA, Stephenson RL, O'Reilly PT, Jones MW, Taggart CT (2001) Genetic diversity of coastal Northwest Atlantic herring populations: implications for management. J Fish Biol 59: 356–370.
4. Wallace RK, Fletcher KM (1997) Understanding Fisheries Managment: A Manual for understanding the Federal Fisheries Management Process, Including Analysis of the 1996 Sustainable Fisheries Act: Mississippi-Alabama Sea Grant Consortium.
5. Cochrane KL (2002) A fishery manager's guidebook. Management measures and their application. Rome: FAO.
6. Ruzzante DE, Mariani S, Bekkevold D, André C, Mosegaard H, et al. (2006) Biocomplexity in a highly migratory pelagic marine fish, Atlantic herring. Proc R Soc B Biol Sci 273: 1459–1464.
7. Stephenson RL, Melvin GD, Power MJ (2009) Population integrity and connectivity in Northwest Atlantic herring: a review of assumptions and evidence. ICES J Mar Sci 66: 1733–1739.
8. Kell LT, Dickey-Collas M, Hintzen NT, Nash RDM, Pilling GM, et al. (2009) Lumpers or splitters? Evaluating recovery and management plans for metapopulations of herring. ICES J Mar Sci 66: 1776–1783.
9. Hintzen NT, Roel B, Benden D, Clarke M, Egan A, et al. (2014) Managing a complex population structure: exploring the importance of information from fisheries-independent sources. ICES J Mar Sci In press.
10. Begg GA, Friedland KD, Pearce JB (1999) Stock identification and its role in stock assessment and fisheries management: an overview. Fish Res 43: 1–8.
11. ICES (2012) Report of the ICES Advisory Committee 2012.
12. Iles TD, Sinclair M (1982) Atlantic herring: Stock discreteness and abundance. Science 215: 627–633.
13. Runnstrøm S (1941) Quantitative investigations on herring spawning and its yearly fluctuations at the West coast of Norway FiskDir Skr Ser HavUnders 6.
14. Meyer HA (1878) Beobachtungen über das Wachstum des Herings im westlichen Theilen der Ostsee. Wiss Meer 2: 227.
15. Soleim PA (1942) Årsaker til rike og fattige årganger av sild. FiskDir Skr Ser HavUnders 7.
16. Corten A (1986) On the causes of the recruitment failure of herring in the central and northern North Sea in the years 1972–1978. J Cons int Explor Mer 42: 281–294.
17. Dragesund O, Hamre J, Ulltang Ø (1980) Biology and population dynamics of the Norwegian spring- spawning herring. Rapp P-v Réun Cons Int Explor Mer 177: 43–71.
18. Russell FS (1976) The eggs and planktonic stages of British marine fishes. London: Academic Press.
19. Blaxter JHS, Parrish BB (1965) The importance of light in shoaling, avoidance of nets and vertical migration by herring. J Cons int Explor Mer 30: 40–57.
20. Woodhead PMJ, Woodhead AD (1955) Reactions of herring larvae to light: a mechanism of vertical migration. Nature 176: 349–350.
21. Hulme TJ (1995) The use of vertebral counts to discriminate between North Sea herring stocks. ICES J Mar Sci 52: 775–779.
22. Bekkevold D, Clausen LAW, Mariani S, André C, Christensen TB, et al. (2007) Divergent origins of sympatric herring population components determined using genetic mixture analysis. Mar Ecol Prog Ser 337: 187–196.
23. Clausen LAW, Bekkevold D, Hatfield EMC, Mosegaard H (2007) Application and validation of otolith microstructure as a stock identification method in mixed Atlantic herring *(Clupea harengus)* stocks in the North Sea and western Baltic. ICES J Mar Sci 64: 377–385.
24. Geffen AJ, Nash RDM, Dickey-Collas M (2011) Characterisation of herring populations to the west of the British Isles: an investigation of mixing between populations based on otolith microchemistry. ICES J Mar Sci 68: 1447–1458.
25. Bekkevold D, André C, Dahlgren TG, Clausen LAW, Torstensen E, et al. (2005) Environmental correlates of population differentiation in Atlantic herring. Evolution 59: 2656–2668.
26. Jørgensen HBH, Hansen MM, Bekkevold D, Ruzzante DE, Loeschcke V (2005) Marine landscapes and population genetic structure of herring (*Clupea harengus* L.) in the Baltic Sea. Mol Ecol 14: 3219–3234.
27. André C, Larsson LC, Laikre L, Bekkevold D, Brigham J, et al. (2011) Detecting population structure in a high gene-flow species, Atlantic herring (*Clupea harengus*): direct, simultaneous evaluation of neutral vs putatively selected loci. Heredity 106: 270–280.
28. Limborg MT, Helyar SJ, De Bruyn M, Taylor MI, Nielsen EE, et al. (2012) Environmental selection on transcriptome-derived SNPs in a high gene flow marine fish, the Atlantic herring (*Clupea harengus*). Mol Ecol 21: 3686–3703.
29. Hempel G (1953) Die Temperaturabhängigkeit der Myomerenzahl beim Hering (*Clupea harengus* L.). Naturwissenschaften 17: 467–468.
30. Blaxter JHS (1957) Herring rearing III - The effect of temperature and other factors on myotome counts. Mar Res Scot 1: 1–16.
31. Hempel G, Blaxter JHS (1961) The experimental modification of meristic characters in herring (*Clupea harengus* L.). J Cons int Explor Mer 26: 336–346.

32. Broch H (1908) Norwegische Heringsuntersuchungen während der Jahre 1904–1906. Bergen Museums Årbok 1.
33. Runnstrøm S (1941) Racial analysis of the herring in Norwegian waters. FiskDir Skr Ser HavUnders 6.
34. Rasmussen T (1942) The Borge Poll Herring. FiskDir Skr Ser HavUnders 7: 63–73.
35. Aasen O (1952) The Lusterfjord herring and its environment. FiskDir Skr Ser HavUnders 10.
36. Lie U, Dahl O, Østvedt OJ (1978) Aspects of the life history of the local herring stock in Lindåspollene, western Norway. FiskDir Skr Ser HavUnders 16: 369–404.
37. Jørstad KE, Pedersen SA (1986) Discrimnation of herring populations in a northern Norwegian fjord: genetic and biological aspects. ICES CM 1986/H:63.
38. Hognestad PT (1994) The Lake Rossfjord herring (*Clupea harengus* L.) and its environment. ICES J Mar Sci 51: 281–292.
39. Husebø Å, Slotte A, Clausen LAW, Mosegaard H (2005) Mixing of populations or year class twinning in Norwegian spring spawning herring? Mar Freshw Res 56: 763–772.
40. McQuinn I (1997) Metapopulations and the Atlantic herring. Rev Fish Biol Fish 7: 297–329.
41. Johannessen A, Nøttestad L, Fernö A, Langård L, Skaret G (2009) Two components of Northeast Atlantic herring within the same school during spawning: support for the existence of a metapopulation? ICES J Mar Sci 66: 1740–1748.
42. Johannessen A, Skaret G, Langård L, Slotte A, Husebø Å, et al. (2014) The Dynamics of a Metapopulation: Changes in Life-History Traits in Resident Herring that Co-Occur with Oceanic Herring during Spawning. PLoS ONE 9: e102462.
43. Eggers F (2013) Metapopulation dynamics in Atlantic herring (*Clupea harengus* L.) along the coast of southern Norway and in the local area of Landvikvannet. Bergen: University of Bergen. 100 p.
44. Silva FFG, Slotte A, Johannessen A, Kennedy J, Kjesbu OS (2013) Strategies for partition between body growth and reproductive investment in migratory and stationary populations of spring-spawning Atlantic herring (*Clupea harengus* L.). Fish Res 138: 71–79.
45. Winkler LW (1888) Die Bestimmung des im Wasser gelösten Sauerstoffes. Ber Dtsch Chem Ges 21: 2843–2854.
46. Mjanger H, Hestenes K, Svendsen BV, Wenneck TdL (2012) Håndbok for prøvetaking av fisk og krepsdyr. Bergen: Institute of Marine Research.
47. R Core Team (2012) R: A language and environment for statistical computing. R Foundation for Statistical Computing, Vienna, Austria. ISBN 3-900051-07-0, URL http://www.R-project.org.
48. Bivand R, Leisch F, Maechler M (2011) pixmap: Bitmap Images ("Pixel Maps"). R package version 0.4-11. http://CRANR-projectorg/package=pixmap.
49. Claude J (2008) Morphometrics with R. Springer, New York, USA. 316 pp.
50. Nason G (2012) wavethresh: Wavelets statistics and transforms. R package version 4.5. http://CRAN.R-project.org/package=wavethresh.
51. Agüera A, Brophy D (2011) Use of saggital otolith shape analysis to discriminate Northeast Atlantic and Western Mediterranean stocks of Atlantic saury, *Scomberesox saurus saurus* (Walbaum). Fish Res 110: 465–471.
52. Begg GA, Overholtz WJ, Munroe NJ (2001) The use of internal otolith morphometrics for identification of haddock (*Melanogrammus aeglefinus*) stocks on Georges Bank. Fish Bull 99: 1–14.
53. Longmore C, Fogarty K, Neat F, Brophy D, Trueman C, et al. (2010) A comparison of otolith microchemistry and otolith shape analysis for the study of spatial variation in a deep-sea teleost, *Coryphaenoides rupestris*. Environ Biol Fish 89: 591–605.
54. Lleonart J, Salat J, Torres GJ (2000) Removing Allometric Effects of Body Size in Morphological Analysis. J Theor Biol 205: 85–93.
55. Bertalanffy Lv (1934) Untersuchungen über die Gesetzlichkeit des Wachstums. Wilhelm Roux Arch Entwickl Mech Org 131: 613–652.
56. Anderson MJ, Willis TJ (2003) Canonical Analysis of Principal Coordinates: A useful method of constrained ordination for Ecology. Ecology 84: 511–525.
57. Oksanen J, Blanchet FG, Kindt R, Legendre P, Minchin PR, et al. (2013) vegan: Community Ecology Package. R package version 2.0-7. http://CRAN.R-project.org/package=vegan.
58. Sokal RR, Rohlf FJ (1995) Biometry. 3rd edn. New York, NY: W. H. Freeman and Company.
59. Hamley JM (1975) Review of gillnet selectivity. J Fish Res Bd Can 32: 1943–1969.
60. Pavlov DA, Shadrin AM (1998) Development of variation in the number of myomeres and vertebrae in the White Sea herring, *Clupea pallasi marisalbi*, under the influence of temperature. J Ichthyol 38: 251–261.
61. Johnston II, Cole N, Vieira VV, Davidson II (1997) Temperature and developmental plasticity of muscle phenotype in herring larvae. J Exp Biol 200: 849–868.
62. MacLean JA, Evans DO (1981) The stock concept, discreteness of fish stocks, and fisheries management. Can J Fish Aquat Sci 38: 1889–1898.
63. Horrall RM (1981) Behavioral stock-isolating mechanisms in great lakes fishes with special reference to homing and site imprinting. Can J Fish Aquat Sci 38: 1481–1496.
64. Gaggiotti OE, Bekkevold D, Jørgensen HBH, Foll M, Carvalho GR, et al. (2009) Disentangling the effects of evolutionary, demographic, and environmental factors influencing genetic structure of natural populations: Atlantic herring as a case study. Evolution 63: 2939–2951.
65. Brophy D, Danilowicz BS, King PA (2006) Spawning season fidelity in sympatric populations of Atlantic herring (*Clupea harengus*). Can J Fish Aquat Sci 63: 607–616.
66. Haegele CW, Schweigert JF (1985) Distribution and characteristics of herring spawning grounds and description of spawning behavior. Can J Fish Aquat Sci 42: 39–55.
67. Pampoulie C, Slotte A, Óskarsson GJ, Helyar SJ, Jónsson Á, et al. (2014) Stock structure of Atlantic herring (*Clupea harengus* L.) in the Norwegian Sea and adjacent waters: Concordant genetic patterns between neutral and selective microsatellite loci? Submitted to Mar Ecol Prog Ser.
68. Eggers F, Olsen EM, Moland E, Slotte A (2014) Individual habitat transitions of Atlantic herring (*Clupea harengus* L.) in a human-modified coastal system. Submitted to Mar Ecol Prog Ser.
69. Slotte A, Fiksen Ø (2000) State-dependent spawning migration in Norwegian spring-spawning herring. J Fish Biol 56: 138–162.
70. Slotte A (1999) Effects of fish length and condition on spawning migration in Norwegian spring spawning herring (*Clupea harengus* L). Sarsia 84: 111–127.
71. Slotte A (2001) Factors influencing location and time of spawning in Norwegian spring spawning herring: An evaluation of different hypotheses. In: Funk F, Blackburn J, Hay D, Paul AJ, Stephenson R et al., editors. Herring: Expectations for a New Millennium: University of Alaska Sea Grant. pp. 255–278.
72. Hourston AS (1959) The relationship of the juvenile herring stocks in Barkley sound to the major adult herring populations in British Columbia. J Fish Res Bd Can 16: 309–320.
73. Smith PJ, Jamieson A (1986) Stock discreteness in herrings: A conceptual revolution. Fish Res 4: 223–234.
74. Directorate of Fisheries (2013) Landing- and sales documents (Landings- and sluttsedler) from Norwegian vessels landed in Norway and abroad. Statistics Department, Bergen, Norway.
75. Slotte A, Stenevik EK, Kvamme C (2009) A note on NSS herring fishery south of 62°N in 2009. Pelagic Fish Research Group, Institute of Marine Research, Bergen. 3 p.
76. Huse G, Fernö A, Holst JC (2010) Establishment of new wintering areas in herring co-occurs with peaks in the first time/repeat spawner ratio. Mar Ecol Prog Ser 409: 189–198.
77. Petitgas P, Secor DH, McQuinn I, Huse G, Lo N (2010) Stock collapses and their recovery: mechanisms that establish and maintain life-cycle closure in space and time. ICES J Mar Sci 67: 1841–1848.
78. Pitcher TJ, Magurran AE, Edwards JI (1985) Schooling mackerel and herring choose neighbours of similar size. Mar Biol 86: 319–322.
79. Corten A (2002) The role of "conservatism" in herring migrations. Rev Fish Biol Fish 11: 339–361.
80. Hendry AP, Kinnison MT (1999) Perspective: the pace of modern life: measuring rates of contemporary microevolution. Evolution 53: 1637–1653.
81. Neb K-E (1970) Über die Heringe des Wendebyer Noors. Ber Dtsch Wiss Komm Meeresforsch 21: 265–270.
82. Bekkevold D, Clausen LAW, Mariani S, Carl A, Hatfield EMC, et al. (2011) Genetic mixed-stock analysis of Atlantic herring populations in a mixed feeding area. Mar Ecol Prog Ser 442: 187–199.
83. Skírnisdóttir S, Ólafsdóttir G, Helyar S, Pampoulie C, Óskarsson GJ, et al. (2012) A Nordic network for the stock identification and increased value of Northeast Atlantic herring (HerMix). Matís ohf., Reykjavík, Iceland. 50 p.
84. Dickey-Collas M, Nash RDM, Brunel T, van Damme CJG, Marshall CT, et al. (2010) Lessons learned from stock collapse and recovery of North Sea herring: a review. ICES J Mar Sci 67: 1875–1886.

17

Relationships between Bacterial Community Composition, Functional Trait Composition and Functioning Are Context Dependent – but What Is the Context?

Ina Severin*¤a, Eva S. Lindström, Örjan Östman¤b

Department of Ecology and Genetics/Limnology, Uppsala University, Uppsala, Sweden

Abstract

Bacterial communities are immensely diverse and drive many fundamental ecosystem processes. However, the role of bacterial community composition (BCC) for functioning is still unclear. Here we evaluate the relative importance of BCC (from 454-sequencing), functional traits (from Biolog Ecoplates) and environmental conditions for per cell biomass production (BPC; ^{3}H-leucine incorporation) in six data sets of natural freshwater bacterial communities. BCC explained significant variation of BPC in all six data sets and most variation in four. BCC measures based on 16S rRNA (active bacteria) did not consistently explain more variation in BPC than measures based on the 16S rRNA-gene (total community), and adding phylogenetic information did not, in general, increase the explanatory power of BCC. In contrast to our hypothesis, the importance of BCC for BPC was not related to the anticipated dispersal rates in and out of communities. Functional traits, most notably the ability to use cyclic and aromatic compounds, as well as local environmental conditions, i.e. stoichiometric relationships of nutrients, explained some variation in all six data sets. In general there were weak associations between variation in BCC and variation in the functional traits contributing to productivity. This indicates that additional traits may be important for productivity as well. By comparing several data sets obtained in a similar way we conclude that no single measure of BCC was obviously better than another in explaining BPC. We identified some key functional traits for productivity, but although there was a coupling between BCC, functional traits and productivity, the strength of the coupling seems context dependent. However, the exact context is still unresolved.

Editor: Dionysios A. Antonopoulos, Argonne National Laboratory, United States of America

Funding: This study was supported by a grant from the Carl Tryggers Foundation to ESL and ÖÖ, from a grant from the Olsson-Borgh foundation to IS, and by separate grants from the Swedish Research Council to ESL (project number 2009-5172) and ÖÖ (2007-5932). The funders had no role in study design, data collection and analysis, decision to publish, or preparation of the manuscript.

Competing Interests: The authors have declared that no competing interests exist.

* Email: inaseverin@gmx.de

¤a Current address: Marine Biological Section, Department of Biology, University of Copenhagen, Helsingør, Denmark
¤b Current address: Department of Aquatic Resource, Swedish University of Agricultural Science, Uppsala, Sweden

Introduction

Lakes and reservoirs are central compartments in the global carbon cycle [1], through the processing of organic matter by microorganisms, making resources available for higher trophic levels of the grazer chain [2] or contributing to the outgassing of carbon dioxide to the atmosphere [3,4]. It is well known that local environmental conditions such as temperature and availability of nutrients are related to bacterial production [5]; the question is, however, if also bacterial community composition (BCC) plays a role. Results obtained so far are rather inconclusive, both in field [6,7] and experimental studies [8–12].

Disparate results may rise from methodological differences, for instance due to the fact that in natural bacterial communities a substantial proportion of the cells may be inactive or dormant [13], and BCC measures including such cells may obscure BCC-function relationships. Further, results could be affected by the choice of method to classify operational taxonomic units (OTUs), i.e. whether sequence similarities alone or also phylogenetic distances between taxa [14,15] are taken into account. Using a phylogenetic diversity measure may result in a tighter observed coupling between BCC and functioning if closer related taxa are functionally more similar. However, carbon processing traits tend to be dispersed in the 16S rRNA phylogeny [16], i.e. they are shared in phylogenetically shallow clusters all across the bacterial realm. Therefore, any BCC measure may have a limited explanatory power for productivity compared to measurements of key functional traits. For instance, in a previous study [12], we showed that the community productivity of heterotrophic bacteria is dependent on the community's ability to use certain carbon substrates in habitats where these substrates are abundant. Thus, functional trait composition may have a better explanatory power for functioning than BCC.

In addition to the methodological issues, the strength of BCC-functioning relationships can differ due to ecological reasons, depending for instance on the proportion of generalists and specialists in a community, since a great proportion of generalists should result in weaker BCC-functioning relationships [17]. The degree of generalism may in turn depend on community assembly mechanism and environmental heterogeneity. Dispersal may, for example, favor resource generalists [18], and BCC and functioning should therefore be uncoupled if the dispersal rate is high. In contrast, large differences in environmental conditions and low (but not limiting) dispersal among communities could favor resource specialist taxa via species sorting processes and, thus, stronger BCC-functioning relationships are expected [7,19].

For a better understanding of the importance of BCC for community functioning we need a conceptual framework and systematic studies including both methodological awareness and ecological considerations. Here we used six different data sets of aquatic bacteria, from 7–15 communities each, to explore the relative importance of the local environment, carbon processing trait composition and BCC for the productivity of bacterial communities depending on dispersal rates and environmental heterogeneity among communities. Two of the data sets originate from lake sediment communities, two from the epilimnion communities of lakes and two from stream communities. The ecosystem function of interest in this study is bulk bacterial community production (leucine incorporation) and the different functional traits under consideration are the communities' ability to use different carbon substrates (assessed from Biolog-plates). Bacterial taxonomic community composition (BCCt) and phylogenetic community composition (BCCp) were determined by high-throughput DNA sequencing, hypothesizing that BCCp would explain more variation in productivity than BCCt. Further, our BCC measures were based on the 16S rRNA (rBCC) as well as the 16S rRNA gene (dBCC). Assuming that rBCC would better reflect the active part of the community [20] we hypothesize that rBCC will be more closely linked to productivity than measures based on the total community (dBCC). Finally, we hypothesize that BCC (incorporating all aspects of BCC) would explain less variation in productivity with increasing dispersal rates and/or lower environmental heterogeneity among communities in a data set because these conditions are assumed to favor habitat generalism. In contrast, functional trait composition would not be affected by the degree of generalism and should therefore explain relatively more of the variation in productivity with increasing dispersal rates and/or higher environmental heterogeneity.

Material and Methods

All abbreviations are given as an overview in Table S1.

Ethics Statement

No permits were required to sample any of the water bodies in this study. The authors also confirm that the sampling did not affect endangered or protected species.

Sampling

Three freshwater systems in two geographic regions in Sweden were sampled in summer and autumn 2010; pelagic lake water (epilimnion), lake sediments (upper 1 cm) and stream waters. One of the lake systems is situated in the province of Jämtland (approximately 63°N and 13°E) where all the 14 sampled lakes are connected to the river Indalsälven either by an inlet and/or an outlet. Samples were obtained in June from all pelagic lake waters (Jw) and 7 sediments (Js) (due to harsh weather conditions the sediments in the remaining seven lakes could not be sampled). The other lake-system is situated in Uppland (approximately 60°N and 17°E), where water (Uw) and sediments (Us) were sampled in 15 hydrologically unconnected lakes in June. Stream samples were obtained from 15 sites in River Fibyån, Uppland, in July (S I) and September (S II). We assumed that 1) dispersal of bacterial cells among communities was lowest in sediments since it requires both sediment resuspension within lakes and dispersal among lakes; 2) dispersal rates among stream water communities was highest due to shorter water retention time in streams in relation to lakes, but that dispersal was greater in S II compared to S I since water levels indicated a higher water flow in September; 3) among the lakes, dispersal was higher among the Jämtland lakes since they are all part of the same river system while the Uppland lakes were not. Based on these assumptions the data sets were ranked from 1–6 where 6 denotes the highest dispersal rate (S II) and 1 the lowest (Us).

Environmental data

Non-purgeable total organic carbon (hereafter termed TC) in water samples was determined by measuring organic carbon after acidification with HCl (TOC-5000, Shimadzu, Kyoto, Japan). Total nitrogen (TN) in water samples was measured spectrophotometrically (Hitachi U-2000, Hitachi, Ltd., Tokyo, Japan) as nitrate after oxidation at high temperature. Total phosphorus (TP) in water and sediment samples was also measured spectrophotometrically after oxidative hydrolysis of organically bound phosphorus. Total carbon (TC) and total nitrogen (TN) in sediment samples was determined in freeze-dried and ash-free sediments by combustion with oxygen (elemental combustion system, Costech Analytical Technologies, Inc., Valencia, CA, USA).

The Jämtland lakes are generally more oligotrophic with low levels of total organic carbon compared to all Uppland sites (Table 1). However, the environmental variability, measured as coefficient of variation (CV) among sites, was not consistently higher in any data set but differed between environmental variables (Table 1).

Bacterial abundance

Cell abundances were determined flow-cytometrically [21] (CyFlow space, Partec GmbH, Münster, Germany) for water samples and microscopically (Nikon Eclipse E600 fluorescence microscope, Nikon Corporation, Tokyo, Japan) for sediment samples. Following the protocol by [21] water samples were fixed with filtered formaldehyde (3.7% final concentration) and stored at 4°C for a maximum of two days. The cells were stained with SYTO 13 (Invitrogen, Life Technologies Ltd, Paisley, UK). The sediment samples were diluted 10× with filtered lake water (0.2 μm filters, Supor-200 Membrane Disc Filters, 47 mm; Pall Corporation, East Hills, NY, USA) and fixed with filtered formaldehyde (3.7% final concentration). This sediment slurry was then diluted 500× with an 50/50 mix of tap water and deionized water and sonicated at 100 W for 1 min on ice. After settlement of the particles, the cells in the supernatant were stained with DAPI (4′,6-Diamidino-2-Phenylindole, Dihydrochloride, Invitrogen, Life Technologies Ltd, Paisley, UK) for 15 min and filtered onto 0.2 μm black polycarbonate filters (Sorbent AB, Västra Frölunda, Sweden). At least 10 fields with at least 200 cells in total were counted for each filter.

Table 1. Dispersal, environmental conditions and heterogeneity (measured as coefficient of variation, CV) in the data sets.

Data set	Dispersal category	TC	TN	TP	TN/TC	TP/TC	CV TC	CV TN	CV TP
Jw	1	4	0.1	15	0	4.5	43.5	60.5	60.6
Uw	2	21.7	1.1	46.6	0.03	4.46	32.2	29.6	73.3
Js	3	4.3	0.4	0.1	0.05	2.31	54.8	45.3	38.9
Us	4	15	1.1	0.1	0.1	0.02	44.6	39.1	23.8
S I	5	29.4	1.9	93.9	0.08	0.01	44.5	53.2	78.3
S II	6	31.8	2	103.9	0.07	4.08	50.1	33.3	106.2

TC: total organic carbon [mg l^{-1} for water and % dry weight for sediments], TN: total nitrogen [mg l^{-1} for water and % dry weight for sediments], TP: total phosphorus [µg l^{-1} for water and % dry weight for sediments].

Bacterial production

The incorporation of ^{3}H-labelled leucine into bacterial protein in water samples was determined using the modified method from [22]. In short, for each sample two parallels and a blank (immediate addition of a final concentration of 5% TCA) were incubated in a final concentration of 100 nM ^{3}H-leucine for one hour in the dark at close to ambient temperatures (the same temperature for all samples within a data set). The incubation was stopped by adding a final concentration of 5% TCA to the water samples. After washing with 5% TCA and 80% Ethanol, 0.5 ml of the scintillation cocktail (Optiphase Hisafe 2, PerkinElmer, Inc., Waltham, MA, USA) were added and the samples were kept for at least 24 hours before measurement of the incorporated ^{3}H-leucine (Packard Tri-Carb 2100TR Liquid Scintillation Analyzer, GMI, Inc, Ramsey, MN, USA). For sediment samples, homogenized sediment was diluted 1000× with sterile-filtered (0.2 µm-filter) lake water from the same location. This sediment slurry was incubated with a final concentration of 10 µM ^{3}H-leucine for one hour. Samples and blanks were then treated as described for the water samples.

Average per cell productivity (BPC) was inferred as bacterial production (BP)/bacterial abundance (BA). This was done in order to obtain a measure of bulk community productivity independent of abundance.

Functional trait composition

Functional trait composition of communities was assessed as the capacity of the communities to use different carbon sources. 150 µl of either water or a 1000× dilution of the sediment were incubated on Biolog EcoPlates (Biolog, Inc., Hayward, CA, USA). These 96-well plates contain 3 sets of 31 carbon substrates and a water blank. The use of these substrates was followed by absorbance measurement of the colorless tetrazolium dye which is reduced to a violet formazan during oxidation of the substrates by bacterial metabolism. Changes in color development were measured using a microplate reader (TECAN ULTRA 384, Tecan Group Ltd., Männedorf, Switzerland) at 595 nm. Immediately after inoculation, the zero time-point was measured and measurements were repeated daily. The color development was followed until the maximum color development was reached (no further increase in absorbance). The overall color development of each plate was expressed as average well color development (AWCD, [23]) and the absorbance profiles corresponding to the time at which the AWCD was 1.5 AWCD were used. In the analysis substrates were grouped according to their molecular structure into polymers (tween 40 and 80, α-cyclodextrin and glycogen), aromatic compounds (2-hydroxy benzoic acid, 4-hydroxy benzoic acid, L-phenylalanine, phenyletylamine), non-aromatic amino acids (L-arginine, L-asparagine, L-serine, L-threonine, glycyl L-glutamic acid), cyclic compounds other than aromatics (D-cellobiose, α-D-lactose, β-methyl-D-glucoside, D-xylose, N-acetyl-D-glucosamine, glucose-1-phosphate, D-galactonic acid γ-lactone, D-galaturonic acid), and simpler compounds (the rest). The average AWCD-normalized absorbance scores of the substrates in each group were calculated for the PLS analyses (see below).

Nucleic acid extraction

100 ml of the lake and stream water were filtered onto a 0.2 µm filter (Supor-200 Membrane Disc Filters, 47 mm; Pall Corporation, East Hills, NY, USA). The filters were stored in liquid nitrogen in the field and later at −80°C until further processing. Approximately 0.5 ml of the undiluted sediment was frozen directly. Nucleic acids (DNA and RNA) were extracted using the Easy-DNA kit from Invitrogen (Life Technologies Ltd, Paisley,

UK) according to protocol #3. For the extraction of nucleic acids form the filters, glass beads (0.1 mm zirconia/silica, BioSpec Products, Inc., Bartlesville, OK, USA) were added at the beginning of the extraction procedure. The tubes were then shaken in a vortex for 15 min to break filters and cells. No such step was included for the sediment samples. Extracts were quality-checked on a 1% agarose gel. After completion of the protocol, half of the nucleic acid extract was subjected to DNase treatment (DNase I, Invitrogen, Life Technologies Ltd, Paisley, UK) and reverse transcription (RevertAid HMinus First Strand cDNA Synthesis kit, Fermentas Sweden AB, Helsingborg, Sweden) according to the manufacturers' instructions using random hexamer primers. The resulting cDNA as well as the DNA extracts were stored at −80°C until further processing.

PCR amplification and Template Preparation

The bacterial hypervariable regions V3 and V4 of the 16S rRNA (cDNA) as well as its gene (DNA) were PCR amplified using forward primer 341 5′- CCTACGGGNGGCWGCAG-3′ and individually bar-coded reverse primers 805 5′- GACTACHVGGGTATCTAATCC-3′ [24]. Each 20 μL PCR reaction contained 0.4 U Phusion high-fidelity DNA polymerase (Finnzymes, Espoo, Finland), 1× Phusion HF reaction buffer (Finnzymes), 200 μM of each dNTP (Life Technologies Ltd, Paisley, UK), 250 nM of each primer (Eurofins MWG, Ebersberg, Germany), 0.4 mg mL^{-1} BSA (New England Biolabs, Ipswich, UK) and 5–10 ng of extracted nucleic acid. Thermocycling was conducted with an initial denaturation step at 98°C for 30 sec, followed by 25 cycles of denaturation at 98°C for 10 sec, annealing at 50°C for 30 sec and extension at 72°C for 30 sec, and finalised with a 7-min extension step at 72°C. Three to four technical replicates were run per sample, pooled after PCR amplification and quality-checked on a 1% agarose gel. Purification was carried out using the AMPure XP purification kit (Beckman Coulter Inc., Brea, CA, USA). Nucleic acid yields were then checked on a fluorescence microplate reader (Ultra 384; Tecan Group Ltd., Männedorf, Switzerland) applying the Quant-iT PicoGreen dsDNA quantification kit (Invitrogen, Life Technologies Ltd, Paisley, UK). Finally, PCR amplicons were combined equimolarly, i.e. in equal proportions, to obtain a similar number of 454 pyrosequencing reads per sample.

454 Pyrosequencing

The final, pooled amplicon was 454 pyrosequenced with a 454 GS FLX system (454 Life Sciences) at the Norwegian High-Throughput Sequencing Centre, University of Oslo (NSC; Oslo, Norway; http://www.sequencing.uio.no), using Titanium chemistry. Sequences were, prior to analyses, quality-checked and truncated to 400 bases. Each data set was individually processed with AmpliconNoise to reduce the number of PCR and 454 sequencing artifacts and chimeras [25]. 454 pyrosequencing reads have been deposited in the National Center for Biotechnology Information Sequence Read Archive (NCBI-SRA) under accession number SRP016145. For further analysis, singletons as well as sequences belonging to *Archaea* and chloroplasts were removed from the data set. Operational taxonomic units (OTUs) were defined using complete linkage clustering at a level of 99% sequence identity. Computations were performed on resources provided by SNIC through the Uppsala Multidisciplinary Center for Advanced Computational Science (UPPMAX). Taxonomic affiliation (phylum-level) of OTUs was determined by aligning representative sequences to the Greengenes imputed core reference alignment [26] (http://greengenes.lbl.gov) using PyNAST [27] in Qiime [28].

After processing an average of 1450 (median = 774) and 450 (median = 347) reads per sample of the total community (DNA) and active bacterial community (cDNA), respectively, were obtained. Because of the risk of large sampling errors with few numbers of reads per sample we omitted samples with <200 reads (1 sample in S I and 4 samples in S II for the total commuinity (DNA), and 1 sample in Js, 4 samples in Us and 2 each in the stream samples for the active community (cDNA)). The range in number of reads between cDNA samples was 200–2445 and for DNA 200–4719.

Phylogenetic tree

Due to the limited length of the sequenced region (approx. 400 bp) and the large amount of different taxa (over 30 000) we could only construct a robust phylogenetic tree for a subset of the total community. Otherwise the phylogeny would be a random tree without sufficient node support. Therefore we constructed phylogenetic trees for the most abundant and, hence, likely functionally most important taxa. For the 142 taxa of the cDNA data set and the 153 taxa of the DNA data set that had an average relative abundance of 0.1% across all samples or 0.3% in a single data set (18% and 25% of all reads in total and active community, respectively) we constructed 10 000 single phylogenetic trees and the consensus tree from the 16S RNA sequence using a generalized time reversible (GTR) evolutionary model with gamma-distributed rate variation across variable sites in mrBayes 3.2 [29]. The branch length prior was set to a uniform clock. The standard deviations of splits after 10 000 000 generations was < 0.005 indicating most nodes were well supported. A tree was sampled every 1000th time step and the last 5 000 trees from the two runs were saved and used for calculating the consensus trees. From the consensus trees we caluculated phylogentic similarities of communities using the Phylogenetic Community Dissimilarity [30] in the picante-package for R. It calculates the pairwise phylogenetic distance among nonshared taxa between two communities, i.e this measure is based only on the occurrence of different taxa and their phylogenetic distance, not the relative abundance of taxa.

Bacterial community composition

We calculated four different estimates of bacterial community composition for each of the six data sets, i.e. all BCC measures were only calculated for communities within a data set and do not account for between data set compositional differences (which was much larger than within data set differences). For each DNA and cDNA data set we calculated one measure of taxonomic composition without accounting for phylogenetic distance between taxa (dBCCt and rBCCt for total and active community, respectively), and one measure of phylogenetic community composition accounting for phylogenetic distances (dBCCp and rBCCp for total and active community respectively). To extract one measure of rBCCt and dBCCt for each community we used the site scores from the first axis of a Principal Coordinate Analysis (PCoA) of the Morisita-Horn distance matrix (see 'statistical analyses' below) for each data set. In this case the first axis explained 12–42% of the variation in rBCCt and 33–70% in dBCCt, being highest in the Js data set, and lowest in the S II data set (Table S2). Similarly, using the phylogentic distance matrix (see above) sites scores were obtained from the first axis of PCoA both of the active (rBCCp) and total bacterial communities (dBCCp). Communities with similar PCoA site scores have thus closely related non-shared taxa whereas non-shared taxa are more distantly related between communities with different site scores. The first axis explained similar amounts of variation in community

Table 2. Average Bray-Curtis (BC) and Morisita-Horn (MH) dissimilarities.

Data set	BC Biolog	MH rBCCt	MH dBCCt	MH r/dBCCt
Us	0.32 (0.13)	0.82 (0.11)	0.71 (0.06)	0.78 (0.12)
Js	0.22 (0.03)	0.86 (0.17)	0.70 (0.13)	0.84 (0.14)
Uw	0.22 (0.06)	0.71 (0.11)	0.60 (0.12)	0.52 (0.15)
Jw	0.20 (0.08)	0.63 (0.11)	0.45 (0.09)	0.56 (0.07)
S I	0.13 (0.02)	0.71 (0.09)	0.71 (0.09)	0.62 (0.07)
S II	0.13 (0.03)	0.63 (0.07)	0.72 (0.05)	0.70 (0.09)
Average	0.20 (0.06)	0.72 (0.09)	0.66 (0.10)	0.68 (0.12)

Values are calculated between the sampling sites within each data set for carbon substrate use (Biolog, functional trait composition) and BCC, respectively. Standard deviations (SD) are given in parenthesis. MH r/dBCCt is the average of the within sample Morisita-Horn dissimilarity between rBCCt and dBCCt.

composition in all data sets, 19–40% in rBCCp and 22–38% in dBCCp (Table S2), again being highest in the Js data set in both cases.

Statistical analyses

Average beta-diversity of bacterial communities (OTUs of 99% 16S rRNA similarity) within each data set (rBCCt and dBCCt) was calculated as Morisita-Horn dissimilarities. Morisita-Horn dissimilarities close to zero indicate very similar communnities, whereas values close to 1 indicate completely different communities.

Bray-Curtis dissimilarites of carbon substrate use (AWCD scores of all 31 substrates) were obtained as an estimate of between site variation in trait composition. Bray-Curtis dissimilarities close to zero indicate communities that have a very similar relative use of different carbon substrates, whereas values close to 1 indicate a completely different use of carbon substrates. Non-metric multi-dimensional scaling (nMDS) was performed using PAST version 2.17 [31] to show differences in carbon substrate use and bacterial community composition (BCCt) among data sets.

To study associations between community BPC and local environmental factors, functional trait composition of communities and the different aspects of BCC within data sets, we did partial least square regressions (PLS) with SIMCA 12.0 (Simca 12.0.1., Umetrics AB, Umeå, Sweden). BPC was the dependent variable and TC, TP, TN, TN/TC, TP/TC and average use of the different substrate groups and PCoA sites scores of the first axes of BCC (all BCC measures) were used as explanatory variables (Table 2). PLS has the advantages that it can handle co-variation among variables and is not sensitive to the number of explanatory variables relative sample size as explanatory variables are transformed into one or several latent variables that explain the maximum variance of the dependent variable [32]. From the PLS we extracted Variable Importance for the Projection (VIP-scores) that describe the relative importance of a variable for the

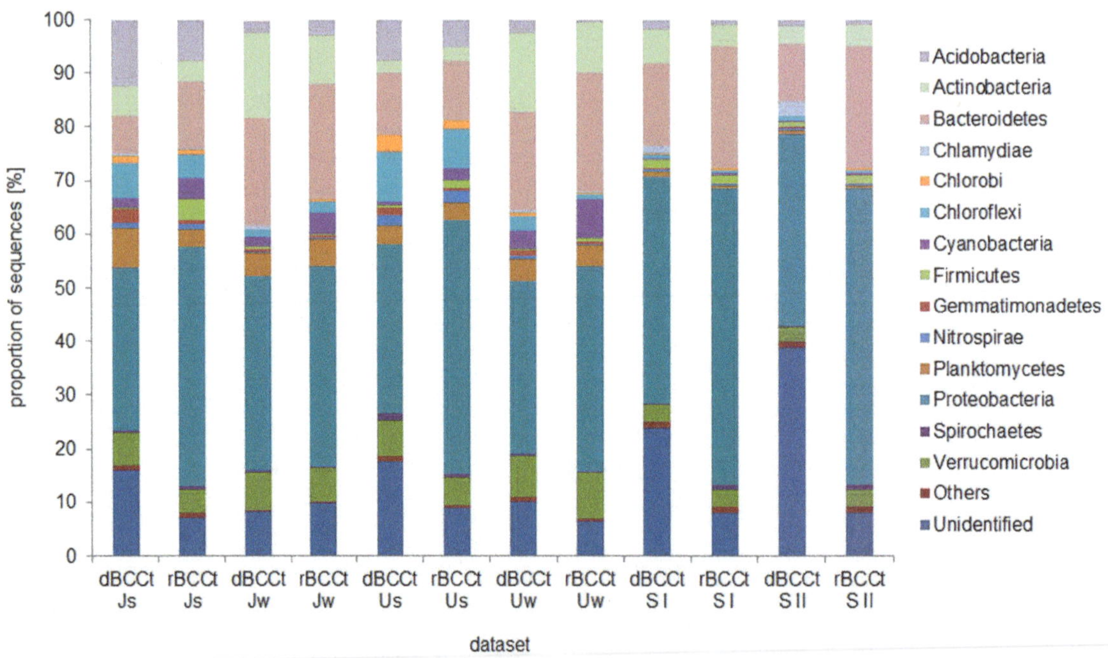

Figure 1. Bar diagrams representing the relative proportion of 16S sequences (rRNA and gene) belonging to the most abundant phyla. 'Others' contain OTUs with a relative abundance below 0.5% in the entire data set. 'Unidentified' denotes OTUs whose taxonomic affiliation is unknown.

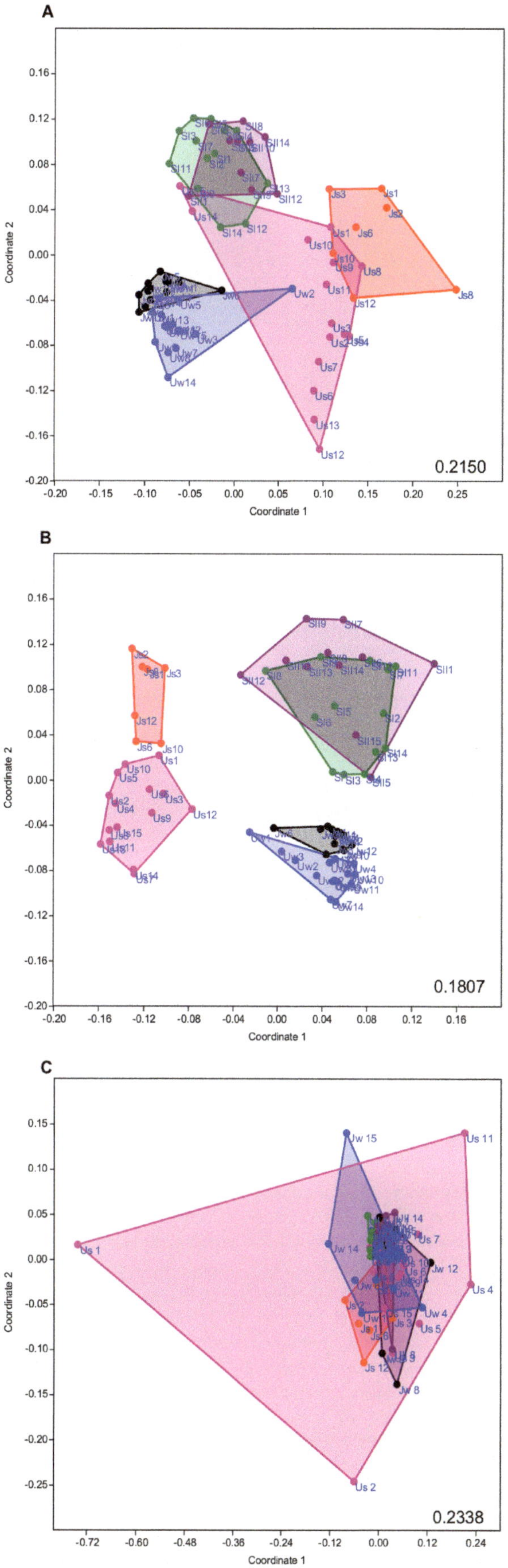

Figure 2. Results from a non-metric multi-dimensional scaling (nMDS) analysis. Depicted are the differences in bacterial community composition between all stations for rBCCt (A) and dBCCt (B), based on Morisita-Horn dissimilarities. The difference in carbon use is based on Bray-Curtis dissimilarities of the Biolog data (C). Stress values are given in the lower right corner.

correlation between the latent explanatory variables and the dependent variable. VIP-scores>1 indicate important variables, and the higher the value the more important is a variable for the correlation between the latent and dependent variable. We used the scaled and centered coefficients between the explanatory variables and the dependent variable to show the direction of the relationship (positive or negative). To infer covariation between different explanatory variables and BPC we plotted the loadings from the PLS. Explanatory variables with high loadings (positively or negatively) explain most of the variation in the latent variables, and explanatory variables close together show high covariation. In the loading plot, explanatory variables close to BPC are positively correlated with BPC whereas explanatory variables distant from BPC are negatively correlated with BPC. However, to get actual values of correlations between explanatory variables we also calculated Pearson correlation coefficients, r. BPC and environmental data was log-transformed prior to the analysis for the data to better fit a normal distribution. We did paired t-tests to compare the explanatory power of active (rBCC) and total (dBCC), respectively phylogenetic based (BCCp) or taxonomic based (BCCt) measures of BCC on BPC.

To study whether dispersal rates and environmental heterogeneity between communities in a data set may contribute to the relative importance of BCC, functional traits and the environment on productivity we did Pearson correlation between the VIP-scores of explanatory variables in a data set and dispersal level or degree of environmental heterogeneity (CV of environmental variables) in each data set. For functional trait composition and environmental variables VIP values could be used directly. For rBCC (active part) and dBCC (total community) we used the highest VIP-value of BCCt and BCCp, respectively.

Results and Discussion

In this field study we investigated the potential importance of bacterial community composition (BCC), functional trait composition (Biolog substrate use), and local environment for functioning (bacterial production per cell, BPC) of freshwater bacterial communities. We hypothesized that these relationships may differ depending on environmental heterogeneity and rates of dispersal among communities. Further, we tested the idea that depending on how BCC was determined, the strength of the relationship to functioning would differ. To enable such an evaluation we obtained six different data sets, each consisting of 7–15 freshwater bacterial communities (water and sediments, lakes and streams), in an identical manner, and statistically evaluated the steering factors for functioning.

The bacterial communities were analyzed by 454 sequencing and their compositions were found to be rather typical for freshwaters [33], e.g., dominated by Proteobacteria and Bacteroidetes (Fig. 1) but there were clear differences in bacterial community composition based both on the 16S rRNA and the 16SrRNA gene (Fig. 2A and B).

The average MH dissimilarity among sites was similar for those two BCC measures, i.e. on average 0.72 among all sites and ranging between 0.63 and 0.86 for rBCCt (Table 2, Fig. 2A) and 0.66 on average and ranging between 0.45 and 0.72 for dBCCt

Table 3. VIP values from PLS analysis between BPC and the explanatory variables for the data sets.

	Data set					
Model fit	**Us**	**Js**	**Uw**	**Jw**	**S I**	**S II**
R2X	23	23	28	56	26	25
R2Y	46	79	54	94	27	51
Explanatory variable						
rBCCt	**1.11 (−)**	0.46 (+)	**1.88 (+)**	**1.76 (−)**	**1.50 (+)**	**1.38 (−)**
rBCCp	**1.02 (−)**	0.36 (+)	0.31 (+)	0.75 (−)	0.11 (−)	0.60 (+)
dBCCt	0.28 (+)	0.56 (+)	0.19 (−)	0.51 (+)	**1.73 (−)**	0.25 (−)
dBCCp	0.95 (−)	**2.25 (−)**	0.45 (−)	0.73 (−)	**2.06 (+)**	1.13 (−)
amino acids	0.47 (+)	**1.28 (+)**	0.56 (−)	0.86 (−)	**1.18 (−)**	0.38 (−)
aromatic	0.43 (−)	**1.21 (−)**	**1.11 (+)**	**1.10 (+)**	**1.24 (+)**	0.94 (−)
simple	**1.13 (−)**	**1.31 (−)**	0.4 (−)	**1.06 (−)**	0.26 (−)	0.28 (+)
polymer	0.33 (+)	0.15 (+)	0.5 (+)	0.63 (−)	0.09 (−)	0.31 (−)
cyclic	**1.92 (+)**	**1.53 (+)**	0.33 (−)	0.62 (+)	0.06 (−)	1.21 (+)
TN	1.10 (−)	0.43 (−)	**1.22 (+)**	0.87 (+)	0.22 (−)	1.51 (−)
TN/TC	0.89 (+)	0.09 (+)	0.67 (+)	**1.25 (+)**	0.80 (+)	0.53 (+)
TC	**1.28 (−)**	0.04 (−)	0.1 (+)	0.64 (−)	0.41 (−)	**1.56 (−)**
TP	0.61 (−)	0.24 (+)	**1.83 (+)**	**1.22 (+)**	0.30 (+)	**1.54 (−)**
TP/TC	**1.13 (+)**	0.92 (+)	**1.65 (+)**	**1.20 (+)**	0.73 (+)	0.58 (−)

VIP>1, identifying variables most relevant for explaining BPC, are shown in bold. The direction of the association between BPC and the explanatory variables is deduced from scaled and centered coefficients (CoeffCS) and given in parenthesis. R2X is the proportion of variation in the explanatory data set explained by the latent factor(s), and R2Y is the proportion of variation in BPC explained by the latent factor(s) from the explanatory data set.

(Table 2, Fig 2B). In contrast, the dissimilarity of functional trait composition (Bray Curtis distance of Biolog substrate use) was much lower, the average being only 0.20 (Table 2). The functional trait composition did also not differ among habitat types in an obvious way (Fig. 2C). This may indicate a redundancy of taxa for processing carbon substrates, i.e. that different taxa can perform similarly on the same carbon substrate.

Assuming that rBCCt should reflect the active community and dBCCt the total community [20] the proportion of active taxa appeared to have been low since the average MH dissimilarity between rBCCt and dBCCt of a given community was 0.68 (Table 2). Especially in sediments, many bacterial cells could have been inactive since MH-distances were on average 0.84 and 0.78 for Jämtland and Uppland respectively, while the lake water communities may have shown a greater proportion of active cells, however still showing MH-distances above 0.5 (Table 2). Therefore it may be expected that rBCCt and dBCCt would be differently related to our functional measure BPC. Our hypothesis was that the active community (rBCC) would be better related to functioning than the total community (dBCC). In our analysis we also included two BCC measures based on phylogenetic distances (rBCCp and dBCCp), hypothesizing that a phylogenetic distance measure would show a stronger coupling between BCC and BCP than the rigid OTU definition used to calculate MH dissimilarities. The potential influential factors on BCP were investigated using PLS which is a statistical method suitable for handling many variables with co-variation among them [32].

The latent variable(s) generated from BCC, carbon substrate use and nutrient levels in the PLS explained most variation in BPC in the Jw data set (94%; R2Y in Table 3) and least variation in S I (27%), and more than 50% in all other data sets. Thus, the latent variables explained a major part of variation in BPC in most data sets. In all data sets the PLS identified BCC to explain an important part of the variation in BPC, since one or several measures of BCC had VIP>1, but other factors contributed as well (Table 3). In four of the data sets (both lake water data sets, Js and S I) a measure of BCC had the highest VIP value, i.e. explained most of the variation in BCP (Table 3, Fig. S1). In the remaining two data sets (Us and S II) functional trait composition of communities and the local environmental conditions, respectively, were better explanatory variables. Since the most important factor thus varied among data sets this result highlights the importance of not relying on a single set of data to draw general conclusions regarding drivers of bacterial functioning in nature. In line with these results are several previous studies showing variable relationships between BCC and functioning [6,7,8,9,34,35].

A methodological consideration arising from these data is that if we had used only one BCC measure instead of four to infer the general importance of BCC for functioning we would have perceived it to be smaller, since "best" BCC measure varied among data sets. However, contrary to our expectations we did not find rBCC (active community) to be more closely coupled to functioning than dBCC (total community) because rBCC did not have higher VIP-values than dBCC (paired t-test: $t_{10} = 0.3$, $p = 0.8$). Further, phylogenetic measures of beta-diversity (BCCp) did not explain more variation than those based on the distance of OTUs (BCCt) (paired t-test: $t_{(10)} = 0.1$, $P = 0.9$). In fact, the BCC measure taking into account the phylogenetic relatedness of the active compartment of the community (rBCCp) performed "worst" with only oneVIP-value>1 (Uppland sediment VIP = 1.02). Methodological reasons for these unexpected results may be that rRNA based methods are poor measures of actual

Table 4. Pearson correlation coefficients between BCC measures (in parenthesis) with VIP-values>1 from the PLS.

Data set	Amino acid	Aromatic	Simple	Cyclic	TC	TN	TP	TN/TC	TP/TC
Us (rBCCt)			−0.31	−0.22	−0.32				
Js (dBCCp)	−0.33	0.46	0.75	−0.69					
Uw (rBCCt)		0.30				0.38	0.63		0.74
Jw (rBCCt)		−0.45	0.41			−0.60		−0.67	−0.62
S I (dBCCp)	−0.16	0.18							
S II (rBCCt)				−0.05	0.20	0.40	0.33		

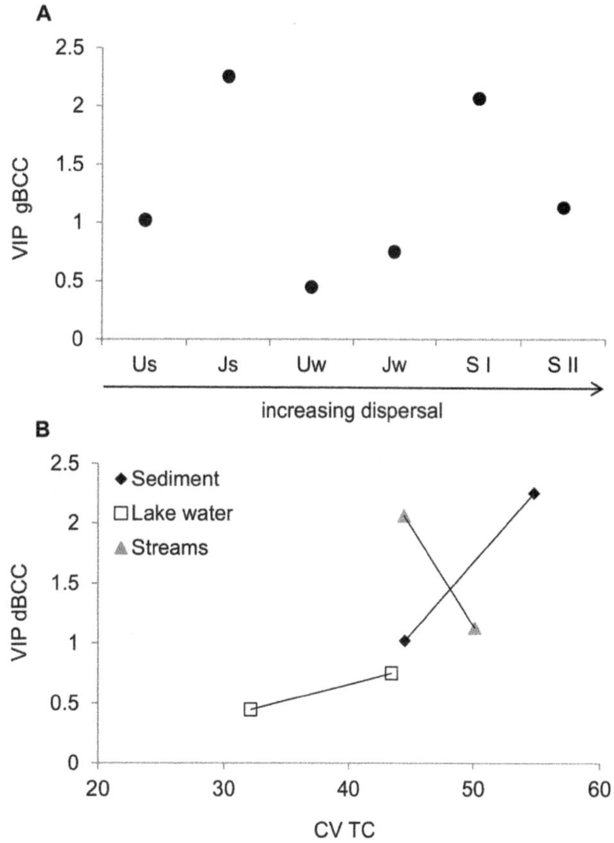

Figure 3. VIP-values of dBCC measures on BPC in relation to dispersal rate (A) and environmental heterogeneity (B). Environmental heterogeneity was measured as coefficient of variation (CV) of total carbon (TC).

activity [36], for instance overestimating the proportion of active cells [37]. Thus, it is unclear what exactly our rBCC measure represents. Moreover, there are limitations in using phylogeny-based estimates of BCC for communities for which the evolutionary history of its constituents is unknown [38]. A problem when assessing phylogenetic diversity of bacterial communities can be that the large number of taxa requires a huge number of variable sites in the gene or genome to get accurate phylogenetic trees to obtain a well-represented estimate of phylogenetic similarity between communities. In our case we could only estimate phylogenetic similarity from less than 25% of total numbers of read (including more taxa would have generated uninformative phylogenetic trees) so our estimates may differ from 'true' phylogenetic similarities between communities. Moreover, the phylogenetic distance estimate we used (PCD) does not account for differences in relative abundance which may be important for community functioning. Thus, we can conclude that methods development is a necessity, both when it comes to accurately define the active proportion of the community, as well as to how phylogenetic diversity should be determined.

At least one of the functional trait groups (i.e. the ability to use certain organic compounds in Biolog plates) showed VIP>1 in each data set (Table 3, Fig. S1). The strongest positive association to any of the functional trait groups was either to aromatic compounds or cyclic compounds (at least three cases with VIP>1 each). This is in line with a previous experimental study where the communities' potential to use aromatic compounds was a key functional trait contributing to variation in bacterial functioning [12]. In contrast, polymers and simple substrates never showed a strong positive association to BPC. Functional traits with high VIP were relatively weakly correlated with BCC measures in both the Uppland and both stream data sets ($r<0.31$), but showed considerably stronger correlations in the two Jämtland data sets ($r>0.45$, Table 4). Weak correlations between functional traits important for BPC (high VIP) and BCC in addition to a large variation in BCC among sites compared to the variation in functional traits strongly imply a functional redundancy of the different bacterial communities. However, in most cases BCC explained more variation in productivity than trait composition. This indicates that there are other taxa specific traits than those measured in Biolog plates that also contribute to productivity, e.g., the use of other resources or a very specific resource. Considering that the aquatic dissolved organic carbon pool is complex and poorly characterized [39] it is not an easy task to pinpoint which organic carbon processing traits should be of interest for a mechanistic understanding of bacterial growth, but should be a subject of future research.

BPC in the second stream sampling was best explained by a negative association with TC, TP, and TN (Table 3, Fig. S1). These changes in the environment were not evidently strongly associated with changes in BCC (Table 4). Also in both the lake water data sets BPC was positively correlated with nutrient levels or ratios between different nutrients and rBCC was well correlated with these environmental changes (Table 4). The influence of local environmental variables on bacterial production was strongly reduced by using BPC, and not gross community production as a response variable in the PLS (results not shown). However, local environmental conditions still showed associations with productivity in all data sets except two (Js and S I). Since a limitation of BPC by C or P alone or the combination of C and P or N is often found in Swedish lakes [40,41], it is not surprising that it was the inorganic nutrient availability and its relation to carbon that explained most variation in productivity. This would indicate that the stoichiometry of nutrients is more important for BPC than nutrient levels per se. In the lake water data sets there was a relatively high covariation between environmental variables and BCC variables contributing to variation in BPC, indicating that the environment partly drives changes in BCC that affect BPC. The negative relation between productivity and nutrient levels in the second stream sampling was surprising and could be a case of negative density dependent interactions, e.g., other nutrients (not considered here) become constraining or there are antagonistic interactions between cells.

Since the strength of the association between BCC and BPC differed between data sets, it seems to be context dependent. A previous modeling study linking diversity and functioning suggested this strength to depend on environmental heterogeneity and species specific traits in relation to the environment [42]. We hypothesized that the dispersal rate among communities would determine the degree of species sorting and in turn the strength of the coupling between BCC and functioning. However, the explanatory power of BCC (rBCC or dBCC) for BPC (VIP-scores) showed no consistent association with the anticipated dispersal rates among the datasets (Fig. 3A). There was a stronger coupling between BCC and BPC in data sets with high environmental heterogeneity in organic carbon but it was not significant ($r = 0.79$; $P = 0.06$; Fig. 3B), maybe partly due to too low sample size. Thus, we could not show a link between different community assembly processes and strength of BCC-functioning relationships. One explanation to this result could be that the range in relative importance of different community assembly processes was not very large in our study. For instance none of the communities were truly isolated and, thus, probably did not experience dispersal limitation. It can also be questioned if dispersal rates were high enough to cause mass effects [10]. Thus, the communities investigated here may all have been formed by local species sorting processes to a similar extent. Future studies aiming to explore the effect of community assembly for the connection between community composition and functioning may, thus, aim for greater gradients in dispersal as well as environmental heterogeneity, also including more disturbed environments than we did, thereby including communities being more likely assembled in different ways. For instance isolated sites such as ground water pockets and environmental gradients of greater range and intensity, such as low pH and high salinity could be included.

In summary, BCC was associated with per cell productivity in all six data sets, but which measure of BCC was best associated differed between data sets. We could not find any consistent difference in BCC-functioning relationship using phylogenetic or non-phylogenetic measures of BCC, nor using BCC from active bacteria or the total bacterial community, highlighting the problem of choosing the best method for BCC measurements in such highly diverse communities. Neither could we find any clear pattern in the relative importance of BCC for functioning between data sets. Thus, the coupling between BCC and functioning in aquatic bacteria seemed context-dependent, but here we could not dissect what the context was. We therefore expect that future research in the area will require methodological considerations on how to measure the active compartment of communities as well as which beta-diversity measures to use. The importance of functional carbon processing traits for productivity was to some extent supported by the compounds tested here, suggesting an importance of the capacity to use especially aromatic and cyclic compounds. However, the results also suggest that important traits still remain to be identified.

Supporting Information

Figure S1 Loading (w*c) bi-plots (of the first and second latent factor) between explanatory variables and per cell bacterial productivity (BPC, filled square) from the PLS for each data set. Explanatory (X) variables are split into functional traits of the community (open circle), environment conditions (open triangle) BCC (cross). Us = Uppland sediment, Js = Jämtland sediment, Uw = Uppland lake waters, Jw = Jämtland lake waters, S I = First stream sampling, and S II = Second stream sampling. For Us, Js and Uw the loadings at the second axis are only shown for illustration purposes as only one latent variable was calculated.

Table S1 Overview of abbreviations used throughout the article.

Table S2 Eigenvalues and variation explained by the first Principal Coordinate Axis (PcoA) of the different BCC for each data set.

Acknowledgments

We thank Jürg Brendan Logue, Joel Segersten and Jan Johansson for help with field sampling and Daniel Lundin for assistance with sequence data processing. Mercè Berga gave valuable comments on the manuscript.

Author Contributions

Conceived and designed the experiments: IS ESL ÖÖ. Performed the experiments: IS ESL ÖÖ. Analyzed the data: IS ESL ÖÖ. Contributed reagents/materials/analysis tools: ESL ÖÖ. Wrote the paper: IS ESL ÖÖ.

References

1. Tranvik LJ, Downing JA, Cotner JB, Loiselle SA, Striegl RG, et al. (2009) Lakes and reservoirs as regulators of carbon cycling and climate, Limnol Oceanogr 54: 2298–2314.
2. Azam F, Fenchel T, Field JG, Gray JS, Meyerreil LA, et al. (1983) The ecological role of water-column microbes in the sea. Mar Ecol Progr Ser 10: 257–263.
3. Jonsson A, Karlsson J, Jansson M (2003) Sources of carbon dioxide supersaturation in Clearwater and humic lakes in northern Sweden. Ecosystems 6: 224–235.
4. Karlsson J, Jansson M, Jonsson A (2007) Respiration of allochthonous organic carbon in unproductive forest lakes determined by the Keeling plot method. Limnol Oceanogr 52: 603–608.
5. Comte J, Fauteux L, del Giorgio PA (2013) Links between metabolic plasticity and functional redundancy in freshwater bacterioplankton communities. Front Microbiol 4: DOI: 10.3389/fmicb.2013.00112
6. Comte J, del Giorgio PA (2009) Links between resources, C metabolism and the major components of bacterioplankton community structure a range of freshwater ecosystems. Environ Micobiol 11: 1704–1716.
7. Lindström ES, Feng XM, Granéli W Kritzberg ES (2010) The interplay between bacterial community composition and the environment determining the function of inland water bacteria. Limnol Oceanogr 55: 2052–2060.
8. Langenheder S, Lindström ES, Tranvik LJ (2005) Weak coupling between community composition and functioning of aquatic bacteria. Limnol Pceanogr 50: 957–967.
9. Leflaive J, Danger M, Lacroix G, Lyautey E, Oumarou C, et al. (2008) Nutrient effects on the genetic and functional diversity of aquatic bacterial communities. FEMS Microbio Ecol 66: 379–390.
10. Lindström ES, Östman Ö (2011) The importance of dispersal for bacterial community composition and functioning. PLoS ONE 10: e25883. 10.1371/journal.pone.0025883
11. Peter H, Ylla I, Gudasz C, Romani AM, Sabater S, et al. (2011) Multi-functionality and diversity in bacterial fiofilms. PLoS ONE 6: e23225. doi: 10.1371/journal.pone.0023225
12. Severin I, Östman Ö, Lindström ES (2013) Variable effects of dispersal on productivity of bacterial communities due to changes in functional trait composition. PLoS ONE 8: e80825 10.1371/journal.pone.0080825
13. del Giorgio PA, Gasol JM (2008) Marine Ecology of the oceans: Physiological structure and single-cell activity in marine bacterioplankton. John Wiley & Sons Inc., Hobo-ken, NJ. p243–298.
14. Srivastava DS, Cadotte MW, MacDonald AM, Marushia RG, Mirotchnick N (2012) Phylogenetic diversity and the functioning of ecosystems. Ecol Lett 15: 637–648.
15. Tan J, Pu Z, Ryberg WA, Jiang L (2012) Species phylogenetic relatedness, priority effects, and ecosystem functioning. Ecology 93: 1164–1172.
16. Martiny AC, Treseder K, Pusch G (2013) Phylogenetic conservatism of functional traits in microorganisms. ISME J 7: 830–838
17. Gravel D, Bell T, Barbera C, Bouvier T, Pommier T, et al. (2011) Experimental niche evolution alters the strength of the diversity-productivity relationship. Nature 469: 89–92.
18. Venail PA, MacLean RC, Bouvier T, Brockhurst MA, Hochberg ME, et al. (2008) Diversity and productivity peak at intermediate dispersal rate in evolving metacommunities. Nature 452: 210–215.
19. Östman Ö, Drakare S, Kritzberg ES, Langenheder S, Logue JB, et al. (2010) Regional invariance among microbial communities. Ecol Lett 13: 118–127.
20. Jones SE, Lennon JT (2010) Dormancy contributes to the maintenance of microbial diversity. Proc Natl Acad Sci USA 107: 5881–5886.
21. del Giorgio P, Bird DF, Prairie YT, Planas D (1996) Flow cytometric determination of bacterial abundance in lake plankton with the green nucleic acid stain SYTO 13. Limnol Oceanogr 41: 783–789.
22. Smith DC, Azam F (1992) A simple, economical method for measuring bacterial protein synthesis in seawater using tritiated leucine. Mar Microb Food Webs 6: 107–114.
23. Garland JL, Mills AL (1991) Classification and characterization of heterotrophic microbial communities on the basis of patterns of community-level sole-carbon-source utilization. Appl Environ Microbiol 57: 2351–2359.
24. Herlemann DPR, Labrenz M, Juergens K, Bertilsson S, Waniek JJ, et al. (2011) Transitions in bacterial communities along the 2000 km salinity gradient of the Baltic Sea. ISME J 10: 1571–1579.
25. Quince C, Lanzen A, Davenport RJ, Turnbaugh PJ (2011) Removing noise from pyrosequenced amplicons. BMC Bioinformatics 12: doi: 10.1186/1471-2105-12-38
26. DeSantis TZ, Hugenholtz P, Larsen N, Rojas M, Keller K, et al. (2006) Greengenes, a chimera-checked 16S rRNA gene database and workbench compatible with ARB. Appl Environ Microbiol 72: 5069–5072.
27. Caporaso JG, Bittinger K, Bushman FD, DeSantis TZ, Andersen GL, et al. (2010) PyNAST: a flexible tool for aligning sequences to a template alignment. Bioinformatics 26: 266–267.
28. Caporaso JG, Kuczynski J, Stombaugh J, Bittinger K, Bushman FD (2010) QIIME allows analysis of high-throughput community sequencing data. Nature Methods doi: 10.1038/nmeth.f.303.
29. Ronquist F, Teslenko M, van der Mark P, Ayres D, Darling A, et al. (2012) MrBayes 3.2: Efficient Bayesian phylogenetic inference and model choice across a large model space. Syst Biol 61: 539–542
30. Ives AR, Helmus MR (2010) Phylogenetic metrics of community similarity. Am Nat 176: 128–142.
31. Hammer O, Harper DAT, Ryan PD (2001) PAST: Paleontological statistics software package for education and data analysis. Palaeontol Electron 4, art. 4: 9pp. (unpaginated)
32. Carrascal LM, Galván I, Gordo O (2009) Partial least squares regression as an alternative to current regression methods used in ecology. Oikos 118: 681–690
33. Newton RJ, Jones SE, Eiler A, McMahon KD, Bertilsson S (2011) A guide to the natural history of freshwater lake bacteria. Microbiol Mol Biol Rev 75: 14–49.
34. Reed HE, Martiny JBH (2013) Microbial composition affects the functioning of estuarine sediments. ISME J 7: 868–879.
35. Sjöstedt J, Pontarp M, Tinta T, Alfredsson H, Turk V, et al. (2013) Reduced diversity and changed bacterioplankton community composition do not affect the utilization of dissolved organic matter in the Adriatic Sea. Aquat Microb Ecol 71: 15–24.
36. Blazewicz SJ, Barnard RL, Daly RA, Firesstone MK (2013) Evaluating rRNA as an indicator of microbial activity in environmental communities: limitations and uses. ISME J 7: 2061–2068.
37. Franklin RB, Luria C, Ozaki LS, Bukaveckas PA (2013) Community composition and activity of estuarine bacterioplankton assessed using differential staining and metagenomic analysis of 16S rDNA and rRNA. Aquat Microb Ecol 69: 247–261.
38. Gravel D, Bell T, Barbera C, Combe M, Pommier T, et al. (2012) Phylogenetic constraints on ecosystem functioning. Nature communications 3: doi: 10.1038/ncomms2123
39. Thurman EM (1985) Developments in Biogeochemistry: Organic geochemistry of natural waters. Martinus Nijhoff/Dr. W. Junk Publishers (Kluwer Academic Publishers Group, Dordrecht). p497
40. Bell RT, Vrede K, Stensdotter-Blomberg U, Blomqvist P (1993) Stimulation of the microbial food web in an oligotrophic, slightly acidified lake. Limnol Oceanogr 38: 1532–1538.
41. Jansson M, Bergström A-K, Lymer D, Vrede K, Karlsson J (2006) Bacterioplankton growth and nutrient use efficiencies under variable organic carbon and inorganic phosphorus ratios. Microb Ecol 52: 358–364.
42. Cardinale BJ, Nelson K, Palmer MA (2000) Linking species diversity to the functioning of ecosystems: in the importance of environmental context. Oikos 91: 175–183.

18

Controlling Cyanobacterial Blooms in Hypertrophic Lake Taihu, China: Will Nitrogen Reductions Cause Replacement of Non-N_2 Fixing by N_2 Fixing Taxa?

Hans W. Paerl[1]*, Hai Xu[2], Nathan S. Hall[1], Guangwei Zhu[2], Boqiang Qin[2], Yali Wu[2], Karen L. Rossignol[1], Linghan Dong[1], Mark J. McCarthy[3], Alan R. Joyner[1]

1 Institute of Marine Sciences, The University of North Carolina at Chapel Hill, Morehead City, North Carolina, United States of America, **2** State Key Laboratory of Lake Science and Environment, Nanjing Institute of Geography & Limnology, Chinese Academy of Sciences, Nanjing, PR China, **3** Marine Science Institute, The University of Texas at Austin, Port Aransas, Texas, United States of America

Abstract

Excessive anthropogenic nitrogen (N) and phosphorus (P) inputs have caused an alarming increase in harmful cyanobacterial blooms, threatening sustainability of lakes and reservoirs worldwide. Hypertrophic Lake Taihu, China's third largest freshwater lake, typifies this predicament, with toxic blooms of the non-N_2 fixing cyanobacteria *Microcystis* spp. dominating from spring through fall. Previous studies indicate N and P reductions are needed to reduce bloom magnitude and duration. However, N reductions may encourage replacement of non-N_2 fixing with N_2 fixing cyanobacteria. This potentially counterproductive scenario was evaluated using replicate, large (1000 L), in-lake mesocosms during summer bloom periods. N+P additions led to maximum phytoplankton production. Phosphorus enrichment, which promoted N limitation, resulted in increases in N_2 fixing taxa (*Anabaena* spp.), but it did not lead to significant replacement of non-N_2 fixing with N_2 fixing cyanobacteria, and N_2 fixation rates remained ecologically insignificant. Furthermore, P enrichment failed to increase phytoplankton production relative to controls, indicating that N was the most limiting nutrient throughout this period. We propose that *Microcystis* spp. and other non-N_2 fixing genera can maintain dominance in this shallow, highly turbid, nutrient-enriched lake by outcompeting N_2 fixing taxa for existing sources of N and P stored and cycled in the lake. To bring Taihu and other hypertrophic systems below the bloom threshold, *both* N and P reductions will be needed until the legacy of high N and P loading and sediment nutrient storage in these systems is depleted. At that point, a more exclusive focus on P reductions may be feasible.

Editor: Brett Neilan, University of New South Wales, Australia

Funding: This study was supported by the US National Science Foundation (0826819, 1230543, and 1240851), the National Science Foundation of China (41003043, 41230744), Natural Science Foundation of Jiangsu Province (BK2012895), and the External Cooperation Program of Chinese Academy of Sciences. The funders had no role in study design, data collection and analysis, decision to publish, or preparation of the manuscript.

Competing Interests: The authors have declared that no competing interests exist.

* Email: hpaerl@email.unc.edu

Introduction

Excessive phosphorus (P) and nitrogen (N) inputs promote hypertrophic conditions in lakes worldwide [1,2], often manifested as harmful (toxic, hypoxia-generating, food-web disrupting) cyanobacterial blooms [3,4]. Some cyanobacteria can convert atmospheric nitrogen (N_2) to biologically-available ammonia via N_2 fixation, providing diazotrophic cyanobacteria with an advantage in waters that are replete in other essential nutrients (e.g., P, Fe, trace metals) but deficient in N [5]. This observation forms a basis for the paradigm that low N:P ratios favor cyanobacterial dominance [6,7]. However, N_2 fixation is energetically unfavorable, and cyanobacteria relying on atmospheric N_2 often do not achieve maximal growth rates [8].

Nutrient loading dynamics have changed, however, since this paradigm was first introduced [6,7]. Agricultural, urban, and industrial expansion in developed and developing regions has dramatically increased reactive N discharges [9], while P loading has stabilized or decreased after recognition that P enrichment influenced eutrophication in freshwater ecosystems [c.f., 10]. Today, unprecedented amounts of biologically-available N are discharged over land, through subsurface aquifers, and via the atmosphere [9]. Freshwater and marine ecosystems are now N:P enriched and more eutrophic [11,12]. Recent studies have shown that these ecosystems can become even more eutrophic when they receive persistent N loading (i.e., they remain sensitive to N inputs) [1,2,13].

The increasing dominance and geographic expansion of non-N_2 fixing cyanobacterial genera, specifically blooms of the widespread, toxin-producing genus *Microcystis*, are a troubling indicator of excessive N loading [14]. This pattern has been observed in large lake systems (e.g., Lake Erie, North American Great Lakes; Lake Kasumigaura, Japan; Lake Okeechobee, Florida; Lake Taihu, China), and estuarine and coastal waters (e.g., Baltic Sea, Europe; Potomac River-Chesapeake Bay and the San Francisco Bay Delta, North America; Peel Harvey and Swan

River Estuaries, Australia) [15]. *Microcystis* spp., and some benthic analogues (e.g., non-N_2 fixing *Lyngbya* and *Oscillatoria*), are indicative of N-over-enriched conditions [16]. The continuing threat from N-driven eutrophication persists because: 1) some N can be 'lost' by denitrification, creating a demand for continued N inputs to sustain eutrophication [17]; 2) historically high P loads accumulate in sediments, providing a reservoir for recycled P to sustain eutrophication; and 3) in nutrient-affected eutrophic lakes, N_2 fixation does not meet ecosystem-scale N requirements [12,18,19]. These studies indicate that eutrophication is accelerated by increasing external N inputs, and they point to the importance of controlling N inputs (in addition to P) for reducing and reversing eutrophication along the freshwater to marine continuum [2,20].

Schindler et al. [21] and others [22,23] suggest that N_2 fixing cyanobacteria will replace non-N_2 fixers once N inputs are reduced, especially if P remains available. This possibility is troubling in lakes and reservoirs used for drinking water, fishing, and recreational purposes, because some N_2 fixers (e.g., *Anabaena* spp., *Aphanizomenon* spp., *Nodularia* spp.) also produce toxins. We evaluated this potential scenario in highly eutrophic (hypertrophic) Lake Taihu, China's third largest freshwater lake, which has experienced massive cyanobacterial blooms dominated by *Microcystis* spp. [24–26]. Previous in situ bioassay work [26,27] in Lake Taihu has shown that P inputs control algal production in the winter-spring, but N-limitation and N and P co-limitation are common during the summer-fall bloom period. Blooms in Lake Taihu remain dominated by non-N_2 fixing *Microcystis*, with potentially N_2 fixing *Anabaena* and *Aphanizomenon* present as sub-dominant genera. The blooms are initiated in late spring (April–May) and persist well into fall (November).

Using in situ 1000 L mesocosms deployed for several weeks during the peak bloom period (June-August), we examined the response of the native phytoplankton community to a range of nutrient enrichments, including one in which only P was administered to enhance N-limitation. The aim of these experiments was to assess the potential for N_2 fixing cyanobacteria to replace non-N_2 fixers under N limited conditions in Lake Taihu.

Methods

Location and field sites

Lake Taihu (Taihu means 'large lake' in Mandarin) is located approximately 150 km west of Shanghai, China (Lake coordinates 31° 10′ N, 120° 9′ E). Taihu is China's third largest freshwater lake at ~2400 km^2 (Fig. 1). It is shallow (mean depth ~2 m), polymictic, and hypertrophic, and it was formed as a former oxbow of the nearby Yangtze River [28]. Taihu's drainage basin is 36,500 km^2 and contains more than 30 input sources, ranging from rivers to small streams and man-made drainage canals. Water exits the southeastern corner of Taihu via the Taipu River, which drains through Shanghai into the East China Sea (Fig. 1). About 40 million people live within Taihu's watershed, which supports approximately 11% (and expanding) of the Chinese economy [28]. The lake is the key drinking water source for ~10 million people, but also serves as a repository for waste from the urban, agricultural, and industrial segments of the local area.

No specific permissions were required for field sampling and studies. These studies did not involve endangered or protected species.

Mesocosm bioassays were located at the interface of one of the northern bays, Meiliang Bay, and the lake proper (Fig. 1). This area was chosen because it is the site of recurring, intensive annual *Microcystis* spp. blooms [24,28], previous nutrient limitation studies [27], and is located near the Nanjing Institute of Geography and Limnology's Taihu Laboratory for Lake Ecosystem Research (TLLER) at Wuxi. Meiliang Bay receives freshwater inputs from the Liangxi and Zhihu Gang rivers, which drain partially-treated wastewater from industrial, residential, and agricultural areas. The named rivers in Fig. 1 contribute more than 85% of the lake's freshwater inflow.

Mesocosms

Twelve cylindrical, 1 m diameter, 1.8 m tall, 1000 L volume, semi-translucent fiberglass mesocosms, which are open at the top and closed at the bottom, were constructed by the shipyard in Wuxi. The mesocosms were installed at the lakeshore near TLLER (Fig. 2), providing ambient natural light and temperature conditions. The mesocosms were outfitted with a flotation flange to ensure that the mesocosms would remain upright and above the lake surface water line. We also monitored daily ambient physical and chemical parameters (temperature, transparency, dissolved oxygen, pH, conductivity), nutrient concentrations (N, P, and carbon species), and phytoplankton community biomass and composition in the mesocosms and nearby lake water. The mesocosm experiment was started on 15 July and ended on 31 July 2013.

We measured phytoplankton community biomass and compositional response to N limitation (i.e., exclusive P enrichment) vs. P limitation (N enrichment), as well as co-limitation, over the two-week period, which allowed the phytoplankton community to adjust to selective nutrient enrichment over numerous generations. The potential for a shift from non-N_2 fixing to actively N_2 fixing cyanobacteria is supported by previous (and ongoing) microscopic observations showing that both groups co-existed during N replete and N-limited periods [24,27].

Prior to deployment, the mesocosms were cleaned and 'conditioned' by filling several times with water from Meiliang Bay (Fig. 1). Once filled for the experiments, the mesocosms were allowed to stabilize for 24 h prior to initiating the following triplicated treatments: 1) Control (no nutrients); 2) nitrogen only (N-only); 3) phosphorus only (P-only); and 4) nitrogen and phosphorus addition (N+P). Nitrogen was added as a solution of 0.3 mg L^{-1} NO_3-N (as KNO_3) and 0.2 mg L^{-1} NH_4-N (as NH_4Cl; final concentration 0.5 mg L^{-1} N). Phosphorus was added as 0.02 mg L^{-1} P (as K_2HPO_4). These nutrient enrichments were based on recent (since 2007) nutrient measurements during spring in Meiliang Bay prior to the bloom initiation [26,27]. Nitrogen and P were supplied every two days throughout the experiment.

While we recognize the importance of sediments as a source or sink for nutrients, we did not add sediments to the mesocosm experiment because: 1) the objective of this study was to evaluate the potential for planktonic N_2 fixers to dominate under well-defined water column N and P regimes; 2) preliminary experiments (in 2012) indicated that the presence or absence of sediments had little effect on phytoplankton community structure and N_2 fixation rates over the course of the mesocosm deployments; and 3) we were concerned that sediments would affect DIN availability via denitrification and excessive microbial (non-phytoplankton) uptake, which could introduce variability, lead to poor replication, and defeat the purpose of the experiment (examining water column utilization of nutrients).

Mesocosms were thoroughly stirred with a plastic paddle following nutrient additions and daily prior to sampling to maintain homogeneous nutrient and seston distributions. Growth of periphyton on the mesocosm walls was controlled by scrubbing the walls of mesocosms every two days with a pre-cleaned brush. This procedure proved effective in minimizing periphyton growth

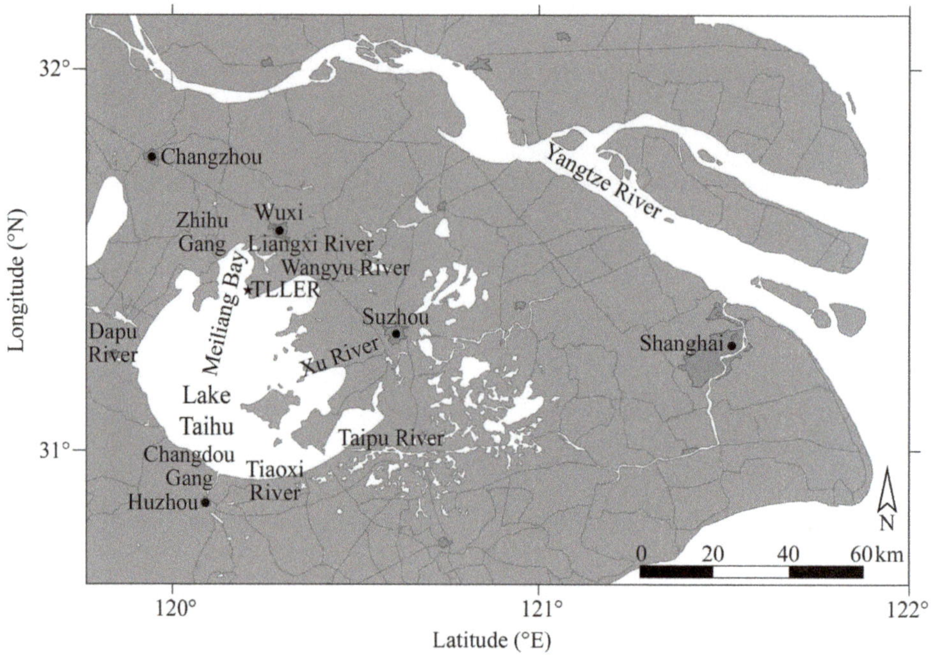

Figure 1. Map of Lake Taihu showing major tributaries and nearby cities. The location of Taihu in China is shown on the inserted map.

during the experimental period. Subsamples were collected every two days for nutrient analyses (dissolved inorganic N and P forms, Total N and P), carbon and N content of the seston and chlorophyll *a* (Chl *a*)). Samples were filtered within 2 h of collection on 25 mm Whatman GF/F filters (pre-combusted at 450°C for elemental C and N analyses). Retained material was used for particulate analyses and filtrate was analyzed for dissolved nutrients. Subsamples for photopigments (fucoxanthin, 9′cis-neoxanthin, violaxanthin, diadinoxanthin, antheraxanthin, myxoxanthophyll, alloxanthin, lutein, zeaxanthin, chlorophyll *b*, *β*-carotene, echinenone), which are diagnostic of major phytoplankton taxonomic groups, N_2 fixation potential, and microscopic identification and enumeration of phytoplankton were collected on GF/F filters every four days. All subsamples were collected prior to nutrient additions. Water temperature, dissolved oxygen, and pH were measured with a YSI Model 6600 multiprobe.

Nutrient and other chemical analyses

Soluble reactive P (SRP) was determined using the molybdenum blue method [29]. NH_4^+-N was measured by the indophenol blue method, and NO_x (NO_3^--N + NO_2^- N) was analyzed with the cadmium reduction method [29]. Total phosphorus (TP), total dissolved phosphorus (TDP), total nitrogen (TN), and total dissolved nitrogen (TDN) were analyzed using a combined persulphate digestion [30], followed by spectrophotometric analysis, as for SRP and NO_3^--N.

Figure 2. Photographs of mesocosm array, located on the shore of Taihu near the Taihu Laboratory for Lake Ecosystem Research (TLLER), at Wuxi, China.

Photopigment determinations

Chlorophyll *a* served as an indicator of total phytoplankton biomass, while diagnostic chlorophyll and carotenoid photopigments were used as indicators of major phytoplankton classes [31]. For Chl *a*, 50–100 mL subsamples were collected from mesocosms and filtered onto 25 mm GF/F filters, which were blotted dry, folded in foil wrappers and frozen until analysis. Chl *a* concentrations were determined spectrophotometrically after extraction in 90% hot ethanol [32].

Diagnostic photopigments were measured using high performance liquid chromatography (HPLC). Approximately 100–150 mL of sample water was vacuum filtered using 25 mm GF/F filters (nominal pore size 0.7 μm). Filters were blotted dry, frozen and transported (frozen) to The University of North Carolina at Chapel Hill, Institute of Marine Sciences for analysis. Filters were extracted in 100% acetone, sonicated, and stored at 20°C for approximately 24 h. Extracts (200 μL) were then injected (using an autosampler) into a Shimadzu Model SIL-20AC HT HPLC equipped with a Model SPD-M10Avp photo diode array detector, following procedures described by Van Heukelem et al. [33] and Pinckney et al. [31]. Pigments were identified according to their absorption spectra, which were determined using commercially obtained pigment standards (DHI-Danish Hydraulic Institute, Denmark).

Contributions of the dominant four algal classes (diatoms, chlorophytes, cryptophytes, and cyanobacteria) to Chl *a* were calculated using Chemtax [34]. The input pigment ratio matrix for the four classes was adapted from Schlüter et al.'s [35] study of lacustrine phytoplankton species. Average ratios for each class were calculated based on species averages within their light availability experiments. With only four dominant classes identified microscopically, the number of pigments used for Chemtax was reduced to twelve. Dominant photopigments, algal classes represented, and input ratios used for the Chemtax analysis are shown in Table 1. Phytoplankton samples were fixed with Lugol's iodine solution (2% final concentration) and sedimented for 48 h. Cell density was measured with a Sedgwick–Rafter counting chamber under microscopic magnification of X200–400. Phytoplankton species were identified according to Hu et al. [36].

Nitrogen fixation

Nitrogen fixation (nitrogenase activity) rates were estimated using the acetylene reduction (AR) assay, as described by Paerl [37], by dispensing 90 mL of sample into 125 mL stoppered serum vials, and adding 7 mL of acetylene generated from calcium carbide (Aldrich). Triplicate light and dark bottles, as well as 0.2 μm filtered (sterile PALL Inc. nitrocellulose filters) lake water blanks were incubated in situ in Taihu for 4 h, after which 4 mL headspace samples were collected in pre-evacuated Vacutainer (Becton Dickinson Inc.) tubes. The tubes were then transported to the laboratory, and 0.2 mL was withdrawn for measurements of ethylene production (from acetylene) by flame ionization gas chromatography using a Shimadzu GC9 gas chromatograph.

Statistical analyses

Phytoplankton responses to the factorial nutrient addition design were assessed using an unsupervised, objective, nutrient-limitation classification model [38]. The model is based on a generalized linear model of treatment contrasts and time effects accounted for by orthogonal polynomials. Aikake information criteria (AIC) was used to determine the model that achieves the greatest degree of parsimony with biomass responses to treatments and time. Separating the time effects from the treatment effects provides a classification of which nutrient, if any, was limiting, or

Table 1. Dominant algal classes and input accessory pigment ratios used for Chemtax determination of class contribution to total chlorophyll *a*.

	Fuco	Neo	Viol	Diad	Anth	Myx	Allo	Lut	Zea	Chl *b*	*β* Car	Ech
Diatom	0.51	0	0	0.074	0	0	0	0	0	0	0.003	0
Cryptophyte	0	0	0	0	0	0	0.37	0	0	0	0.004	0
Chlorophyte	0	0.038	0.026	0	0.016	0	0	0.15	0	0.36	0.003	0
Cyanobacteria	0	0	0	0	0	0.14	0	0	0.28	0	0.097	0.076

Fuco = fucoxanthin; Neo = 9′cis-neoxanthin, Viol = violaxanthin; Diad = diadinoxanthin; Anth = antheraxanthin; Myx = myxoxanthophyll; Allo = alloxanthin; Lut = lutein; Zea = zeaxanthin; Chl *b* = chlorophyll b; *β* Car = *β*-carotene; Ech = echinenone.

whether both nutrients were limiting. Seven classifications of nutrient limitation status are possible: 1-*null*, meaning no nutrient effects; 2 *and* 3- *exclusive N or P limitation*, where only N or P stimulates biomass above the control, and addition of the non-limiting nutrient in the N+P treatment provides no additional stimulation; 4 and 5-*primary N or P limitation*, where addition of one nutrient stimulates growth above the control, but addition of the non-limiting nutrient provides additional stimulation; 6-*combined limitation*, where both N, P, and N+P additions are different from the control; and 7-*exclusive combined limitation*, when neither N nor P additions alone stimulate biomass above the control, but the combined N+P treatment stimulates growth. This classification model was used to analyze responses of total phytoplankton biomass, as Chl *a*, and the biomass responses of specific taxa based on diagnostic photopigments or microscopic enumeration. To determine confidence in each nutrient limitation classification, bootstrapping with replacement was used to produce 1000 resampled data sets from the original triplicated mesocosm treatments. The frequency of bootstrapped nutrient limitation determinations provides a measure of confidence in the determination made from the original experimental dataset [38].

Results

Nutrient addition effects on phytoplankton biomass

For the first two days, N-only additions led to a small but significant stimulation of phytoplankton biomass relative to controls and P additions (Fig. 3). Similarity between N-only and N + P additions showed that the initial community was N limited rather than co-limited. After two days, only the N+P treatment showed continued growth, with Chl *a* concentrations reaching 60 μg L^{-1} or higher, reflecting typical bloom conditions [24–26]. All other treatments showed similar declines in Chl *a* through 27 July 2013.

Overnight on 27 July into early morning 28 July 2013, the TLLER meteorological station recorded 48.6 mm rainfall during a strong thunderstorm. Nutrient concentrations (in mg L^{-1} N or P) in the rainwater were 3.41 TN, 3.39 TDN, 0.46 NH_4^+, 0.69 NO_3^-, 0.032 TP, 0.20 TDP, and 0.025 PO_4^{3-}. If only inorganic nutrients are assumed bioavailable, the N and P added to the 1000 L mesocosms would have increased bioavailable N and P concentrations by ~0.04 and 0.0009 mg L^{-1} (102:1 molar ratio), respectively. If all N and P added from the rain were assumed bioavailable, N and P bioavailability would have increased by 0.12 and 0.0012 mg L^{-1} (221:1 molar ratio), respectively. This N enriched and P poor rain event served as a 'natural experiment' to evaluate the effects of N addition to the experimentally-manipulated mesocosm phytoplankton communities. The atmospheric N addition led to a rapid stimulation of biomass in the P addition treatments and, to a lesser extent, in the controls. N-only and N + P treatments showed no response, indicating that growth in those treatments was not N limited during the time of the storm. Assessments of nutrient limitation status by the unsupervised nutrient limitation classification model were affected by the storm and corroborate the overall N limitation status of the phytoplankton community. For the entire experimental period, the nutrient limitation model best supported by the data was a primary combined N and P limitation (Table 2), but 23% of the resampled datasets supported a primary N limitation determination. When only data prior to the storm were analyzed, confidence in a primary N limitation status for Chl *a* response was high (93% of resampled datasets), with a small portion (6.3%) of primary combined limitation determinations (Table 2).

Phytoplankton community comparisons between mesocosms and ambient lake water

When examined over time for phytoplankton biomass and composition, ambient lake water tended to show elevated levels of total community biomass and major algal groups relative to the control (no nutrient addition) mesocosms (Fig. 4). These differences were observed shortly after the initiation and remained throughout the duration of the experiment (16 d), indicating that nutrient-limited conditions were maintained in the controls throughout the experiment. While total biomass (as Chl *a*) was consistently lower in controls, the phytoplankton composition remained similar between controls and ambient lake water, indicating that potentially confounding issues, such as selective grazing, death, or major shifts in phytoplankton composition due to 'container effects' were minimal. While phytoplankton biomass remained low in control mesocosms, Chl *a* and other diagnostic photopigment concentrations remained detectable throughout the experiment (Fig. 4).

Nutrient addition effects on phytoplankton community composition

Overall, the addition of either N or P alone showed little stimulation of the dominant phytoplankton classes over the control; however, combined N+P additions stimulated all four major classes (Fig. 4). For diatoms, P additions led to small increases in biomass above the control. The nutrient classification status was assessed as primary P limitation, but confidence in this determination was weak with primary combined limitation being nearly as likely (Table 2). Exclusion of the post-storm periods revealed that, prior to this point, the diatom community had likely been co-limited (Table 2). Cryptophytes showed weak stimulation by N-only additions (Fig. 4) and were classified as having a primary N limitation status. However, confidence in this assessment was not strong with primary P limitation, exclusive N + P co-limitation, and primary combined limitation being collectively as likely (Table 2). Chlorophytes were assessed as primary combined limitation, and exclusion of the post-storm period increased confidence in this assessment. Cyanobacteria were assessed as primarily being N limited but, as with the other phytoplankton classes, confidence in this determination was weak. It is instructive that for none of the four phytoplankton classes, in any of the 1000 bootstrapped data sets, was there a single instance where nutrient limitation status was assessed as exclusive N or exclusive P limitation. This underscores the finding that the dominant phytoplankton classes were largely co-limited by N and P during the experiment. N and P availability nearly balanced demand such that, when N or P was added alone, it drove growth limitation to limitation by the other nutrient. In this case, only combined additions of N and P led to continued accumulation of biomass.

The lack of cyanobacterial biomass stimulation by P additions was surprising, since ambient phosphate concentrations were quite low at the start of the experiment (~2–4 μg P-PO_4 L^{-1}). Total dissolved inorganic N (DIN) ($NH_4^+ + NO_x$) concentrations in the P treatments were also quite low (~20–40 μg DIN L^{-1}), providing presumably favorable conditions (DIN:DIP of approximately 10) for selective stimulation of cyanobacterial growth, especially among N_2 fixing genera in response to P additions [7].

N+P additions promoted the growth of all groups relative to controls, with cyanobacteria showing particularly strong stimulation (Fig. 3). This appeared to mirror conditions in ambient lake water, which showed a steady stimulation of cyanobacterial biomass and bloom development in parallel with the experimental

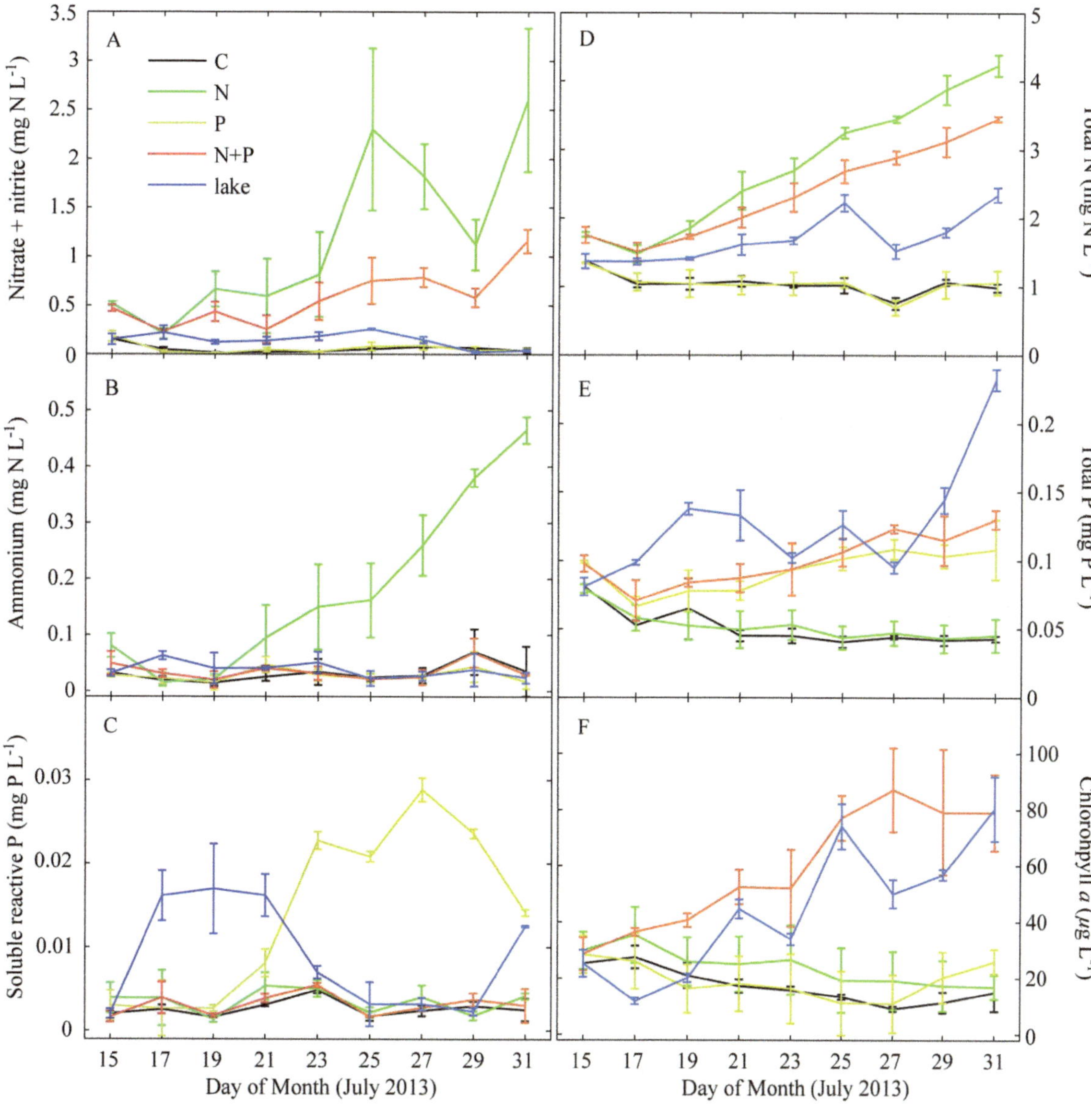

Figure 3. Time series of dissolved and total nitrogen and phosphorous forms and phytoplankton biomass as chlorophyll *a* from the summer 2013 mesocosm experiment. Solid lines connect means of triplicate mesocosms. Error bars are one standard deviation.

period. By the end of the experiment, community composition of the N+P treatments were very similar to ambient lake water (Fig. 5). In comparison, community composition of the controls, P-only, and N-only treatments was similar to the initial community composition (Fig. 5).

Nutrient conditions during the mesocosm experiment

Dissolved inorganic and total N and P concentration showed strong patterns related to nutrient additions and biological utilization (Fig. 3). Control and P only addition mesocosms showed almost no detectable DIN, indicating that N was likely the most limiting nutrient throughout the experimental period. Even with N additions, for the first two and four days, NOx and NH_4^+ concentrations decreased in both the N-only and N+P addition treatments. This indicates that N availability did not meet N demand at the beginning of the experiment and suggests that NH_4^+ was preferentially utilized over NO_x. After this initial period, both NO_x and NH_4^+ increased in the N-only additions. However, the N+P additions treatments showed half the NO_x buildup and no significant NH_4 accumulation, indicating that while N limitation was more evident than P limitation, N+P co-limitation most likely characterized conditions during the experimental period. Total N predictably increased in response to N-only and N+P enrichment, but notably it did not increase in response to P-only enrichment. This served as independent (of acetylene reduction measurements) evidence that N_2 fixation did not respond in a significant manner to the P-only treatments and that microbial production in response to the P-only treatments was likely supported on regenerated sources of N. SRP remained at low levels similar to the control in the N-only and N+P treatments. In the P-only treatment, SRP accumulated after four days and reached a peak on 27 July. After the 27 July storm, SRP dropped by 50%, indicating significant uptake in response to increased N availability from rainwater. Total P increased in response to both

Table 2. Nutrient limitation classifications (N.L.C.) and ambiguity of classification for total phytoplankton biomass and specific algal taxa (Andersen et al. 2005).

Group	N.L.C.	Null	XN	N1	XP	P1	XC	C1
Total	C1	0	0	22.5	0	3.3	1.1	73.1
phytoplankton	(N1)	(0)	(0)	(93.3)	(0)	(0.2)	(0.2)	(6.3)
Diatoms	P1	0	0	1.3	0	51.4	1.6	45.7
	(XC)	(0)	(0)	(20.0)	(0)	(9.9)	(45.5)	(24.6)
Cryptophytes	N1	0	0	54.2	0	18.7	11.5	15.6
	(N1)	(0)	(0)	(64.2)	(0)	(5.2)	(25.3)	(5.3)
Chlorophytes	C1	0	0	31.9	0	0	8.5	59.6
	(N1)	(0)	(0)	(9.7)	(0)	(3.2)	(7.9)	(79.2)
Cyanobacteria	N1	0	0	38.4	0	30.6	12.1	18.9
	(N1)	(0)	(0)	(40.5)	(0)	(29.5)	(9.5)	(20.5)
Microcystis sp.	P1	0	0	1.4	0	87.7	0.3	10.6
	(P1)	(0)	(12.0)	(2.8)	(0)	(69.1)	(0.5)	(15.6)
Anabaena spp.	–	7.8	35.7	18.1	11.5	28.0	1.5	13.7
	(–)	(11.3)	(64.0)	(7.7)	(0)	(5.7)	(9.2)	(2.1)

Total phytoplankton biomass as spectrophotometric Chl *a*. Ambiguity of the classification of the nutrient limitation status is shown by the frequency (%) of different classifications from 1000 bootstrapped data sets. Values in parentheses are results excluding data from the period following the 27 July 2013 rain event. A – indicates that the null model was chosen in greater than 5% of boot-strapped cases.

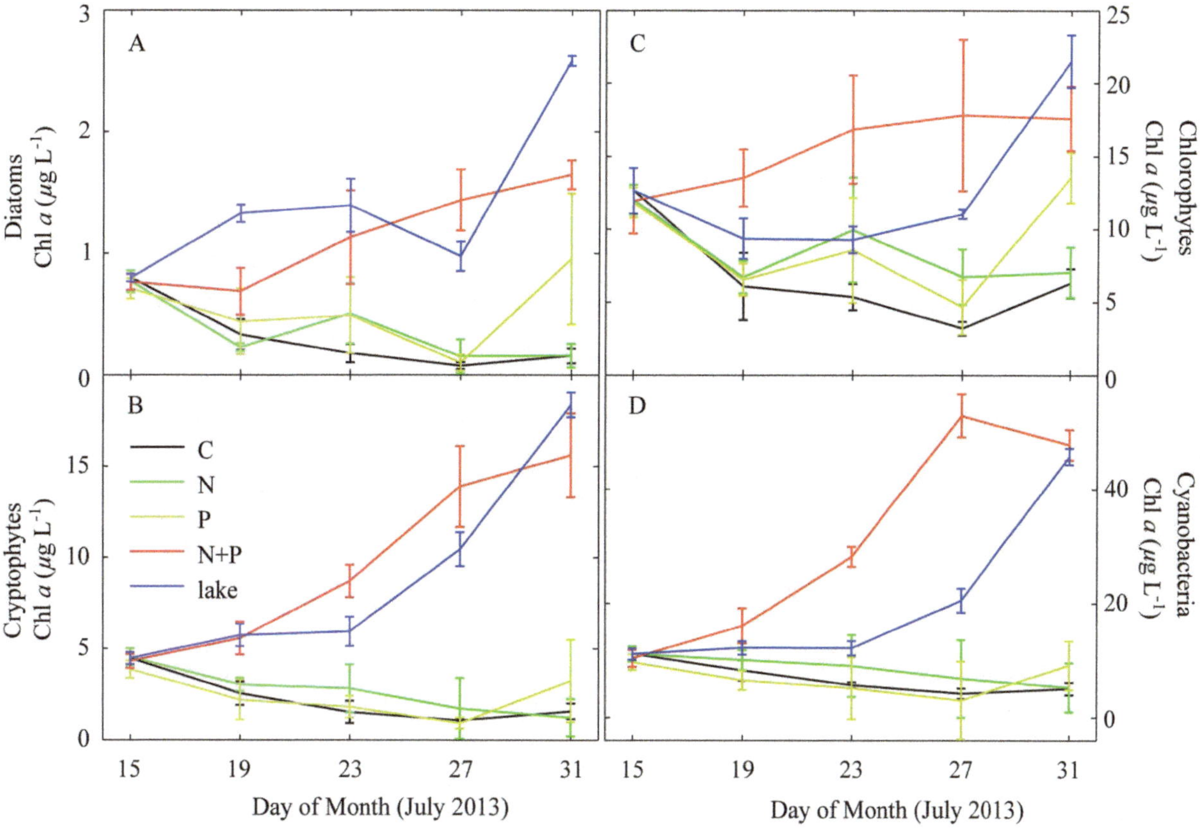

Figure 4. Time series of biomass of the dominant algal classes from the summer 2013 mesocosm experiment. Solid lines connect means of triplicate mesocosm tanks. Error bars are one standard deviation.

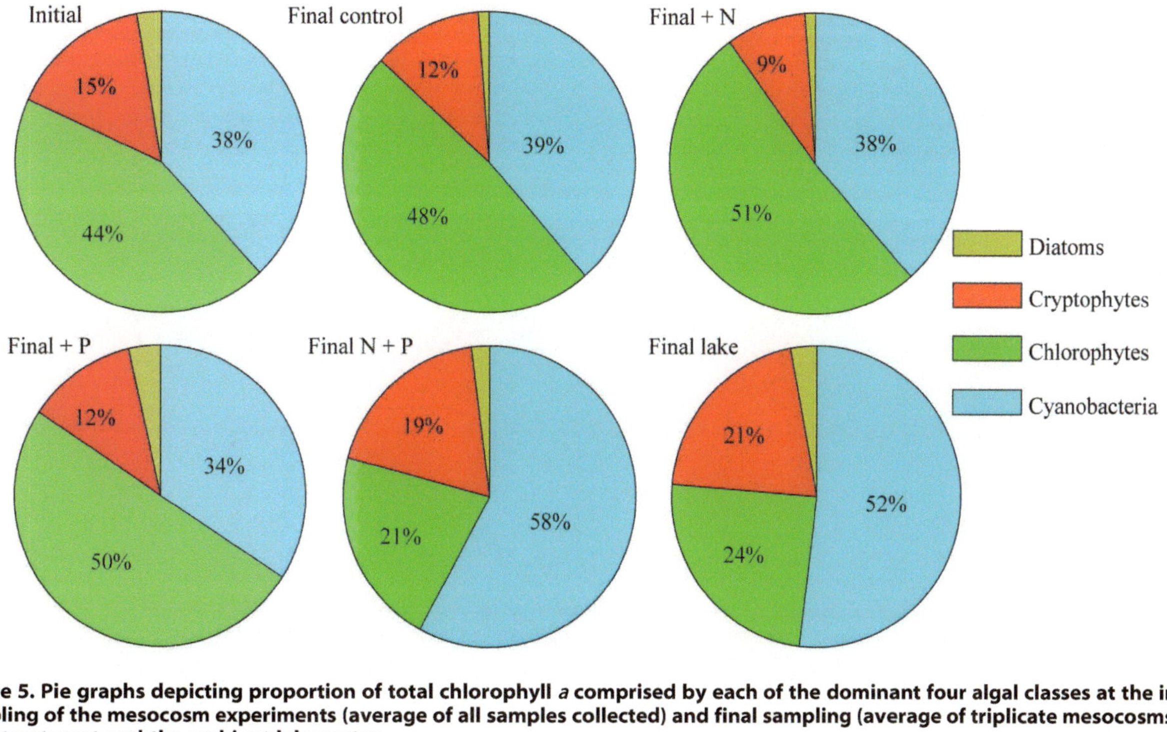

Figure 5. Pie graphs depicting proportion of total chlorophyll *a* comprised by each of the dominant four algal classes at the initial sampling of the mesocosm experiments (average of all samples collected) and final sampling (average of triplicate mesocosms) for each treatment and the ambient lake water.

P-only and N+P additions, but there were no significant differences in response to these treatments. Overall, nutrient concentration results support previous findings that during summer months this region of Taihu is N+P co-limited [26,27]. Exclusive P limitation was not detected throughout the course of these experiments.

Microscopic observations of phytoplankton community dynamics

In general, microscopic observations tended to confirm HPLC diagnostic pigment analyses with respect to the responses of major phytoplankton groups to various nutrient additions. Specific attention was paid to the dominant N_2 fixing (*Anabaena* sp.) vs. non-N_2 fixing (*Microcystis* sp.) cyanobacterial genera responses to various nutrient additions (Figs. 6, 7).

The diatom community was dominated by a mix of solitary centric (e.g., *Cyclotella* sp.), pennate (e.g., *Navicula* sp.), and chain-forming genera (e.g., *Melosira* sp.). Chlorophytes proved to be highly diverse throughout all treatments, but the largest fraction of chlorophyte biomass was comprised of colonial (e.g., *Scenedesmus* spp., *Crucigenia* sp., *Pediastrum* spp., *Eudorina* sp.) and filamentous (*Ulothrix* sp.) forms. Cryptophytes were largely comprised of nanoplanktonic *Cryptomonas* spp. and *Chroomonas* spp. In general, dominant members of these eukaryotic groups were similar to those reported during a long-term monitoring study [24]. Dominant cyanobacterial genera in order of prominence were *Microcystis*, *Oscillatoria*, *Anabaena*, *Aphanizomenon*, and *Merismopedia*.

Among these, *Anabaena* and *Aphanizomenon* are capable of N_2 fixation. *Microcystis* responded in two ways to the nutrient addition treatments: stimulation by N+P additions, and a negative response to P-only additions (Fig. 6). The nutrient limitation classification for *Microcystis* was assessed as primary P limitation because the nutrient classification does not differentiate between positive or negative effects [38]. For more than 8% of the resampled datasets, a null model fit the data equally as parsimoniously as a model that included treatment contrasts for cell abundances of *Anabaena*. This is conceptually equivalent to an insignificant *F*-value in an ANOVA and indicates that high variability within replicates and a high frequency of zero counts precluded an objective assessment of its nutrient limitation status. However, by the end of the experiment, *Anabaena* had highest cell densities in the P-only treatments (Fig. 6). *Aphanizomenon* abundance was also highly variable but showed declines in abundance for all treatments.

Nitrogen fixation responses

Nitrogen fixation (as acetylene reduction) responses were monitored throughout the experimental period. Extremely low (and statistically insignificant) rates of N_2 fixation were observed relative to filtered lake water and reagent blanks in both the lake and among mesocosm treatments (Fig. 8). These results support the above-mentioned microscopic observations indicating that, while filamentous N_2 fixing genera *Anabaena* and *Aphanizomenon* were present, and *Anabaena* increased in the P-only treatment (Fig. 6), heterocyst frequency was very low, and only traces of N_2 fixation were observed at the end of the experiment (Figs. 7, 8). Therefore, while the potential for this process exists, there was no significant activity. These results were supported by direct measurements of total N (TN) in the mesocosms, which showed no evidence that N_2 fixation led to a significant increase in TN in response to P additions. This proved true for all treatments, including the P-only treatment, which was expected to yield the highest potential for N_2 fixation.

We calculated the percentage of *Anabaena's* N requirement (based on growth rates and cell N quotas) that may have been satisfied by N_2 fixation, based on our AR measurements and conversions to N_2 fixed using a 4:1 ethylene production:N_2 fixation ratio. At most, only 20% of *Anabaena's* N needs were met this way. If we extrapolate these results to the entire phytoplankton community, less than 5% of N needs are met by N_2 fixation under P-enriched conditions. There were no significant rates of N_2 fixation in any of the other treatments.

These results indicate that despite the presence of N limited and N+P co-limited conditions, diazotrophic genera must be competing with non-N_2 fixing genera (i.e., *Microcystis* and eukaryotes) for combined N sources to sustain growth. This situation persisted throughout the experimental period.

Interestingly, the P only treatments did not stimulate total cyanobacterial community production. Instead, only diatoms showed selective stimulation by this treatment, and even then, this stimulation only occurred early in the experiment. Overall, enhanced N limitation (through selective P enrichment) did not lead to a significant increase in cyanobacterial biomass or result in

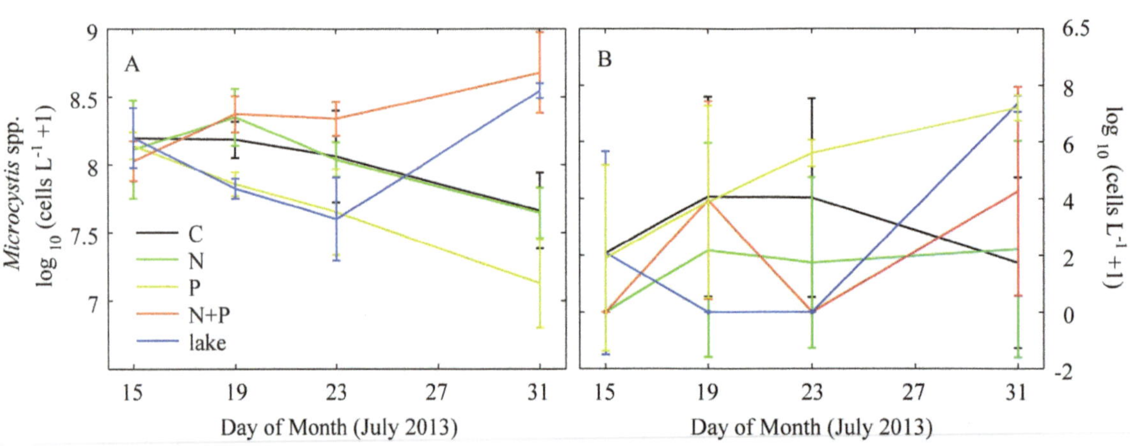

Figure 6. Time series of cell abundances for *Microcystis* spp. and *Anabaena* spp., the dominant non-nitrogen fixing and nitrogen fixing genera during the summer 2013 mesocosm experiment. Solid lines connect means of $\log_{10}$ values of triplicate mesocosm tanks. Error bars are one standard deviation.

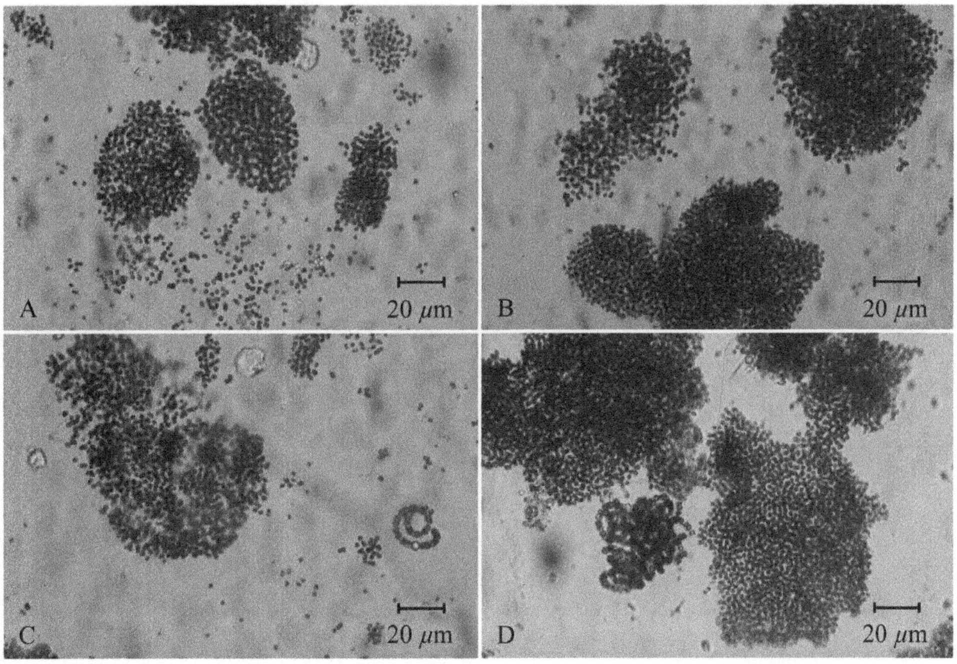

Figure 7. Representative microscopic observations (X200, brightfield) of dominant cyanobacterial genera in the mesocosm experiment conducted during 2013. Samples were collected towards at the end of the experiment, 31 July 2013. (A) Control (no nutrient addition). (B) N-only addition. (C) P-only addition. (D) N+P addition. Note the overall dominance by *Microcystis* spp. colonies. However, an increase in number of filaments of the N_2 fixing genus *Anabaena* was observed in the phosphorus and phosphorus and nitrogen treatments. No significant N_2 fixation was observed in either of these treatments however. Biomass-wise, *Microcystis* spp. clearly dominated the cyanobacterial community in all treatments.

N_2 fixation, but the N+P treatment did enhance cyanobacterial biomass (Fig. 4).

Discussion

Results from the mesocosm nutrient addition experiment confirm prior work based on short-term (<1 week) microcosms, indicating that, during summer bloom periods, growth of Taihu's cyanobacteria-dominated phytoplankton community is primarily controlled by N supply and further stimulated by N and P together [26,27]. Despite summer N limitation, N_2 fixing cyanobacteria did not significantly increase in abundance, heterocyst frequency, or N_2 fixing activity, either in the mesocosm experiments or in ambient lake water. These results do not support Schindler et al.'s [21] conclusion that diazotrophic cyanobacteria should dominate and quantitatively fix N_2 to compensate for N requirements needed to sustain phytoplankton community production rates. The length of mesocosm experiments of up to a month should have provided plenty of time for numerous generations of N_2 fixers to proliferate and for vegetative cells to differentiate into heterocysts [5,8]. Hence, the low N fixation rates were not attributable to short incubation time. The question therefore arises: Why were non-N_2 fixing genera (*Microcystis*) not replaced by active diazotrophs (*Anabaena, Aphanizomenon*) during P enrichment, despite the fact that these genera were observed in the lake throughout the experimental period?

In addressing this question, we note that phytoplankton growth also may be constrained by other factors, including temperature, availability of micronutrients (Fe, trace metals), and adequate light for supporting energy-demanding processes, such as N_2 fixation. Other ecological factors could play additional roles, including the ability to regulate buoyancy (important for accessing light and nutrients), grazing, and beneficial as well as antagonistic associations with other microbes. These factors are known to play a role in the competitive interactions between diazotrophic and non-diazotrophic cyanobacterial species [3].

In shallow, well mixed, highly turbid Taihu, the ability of non-diazotrophs to access and store nutrients, while maintaining high photosynthetic rates and competing for N sources with diazotrophs, is likely important in the initiation and maintenance of blooms. The non-diazotroph *Microcystis* dominated prior to the experiment and remained dominant while N was severely limiting (i.e., P alone treatment). N limitation constrained the ability of a bloom, equal in biomass to the ambient lake, to form. *Microcystis* maintained its dominance over N_2 fixing genera under N limitation and during a decrease in overall biomass. Its ability to effectively compete for dwindling N sources, especially NH_4^+ regenerated from sediments and organic N, provides an advantage [39,40]. *Microcystis* dominance may also be related to: 1) effective cellular N storage; 2) access to sediment and water column regenerated N (as NH_4^+) via buoyancy regulation; 3) close associations with heterotrophic bacteria present around colonies, which enhanced 'phycosphere'-scale nutrient (including N and P) recycling [41]; and 4) more efficient use of light for supporting photosynthetic growth and nutrient sequestration than diazotrophic genera. The latter selective mechanism may be quite important in Taihu's turbid, shallow waters, in which *Microcystis* may be particularly effective in vertically orienting itself [42].

Since N_2 fixation is a very energy-demanding process [5], light availability in highly turbid Taihu may be insufficient for diazotrophs to compete with the effective N sequestration mechanisms that *Microcystis* possesses [39,43]. Secchi depths in Taihu are commonly in the range of 8–85 cm (mean = 38 cm; [44,45], which indicates that a substantial amount of the 2 m deep

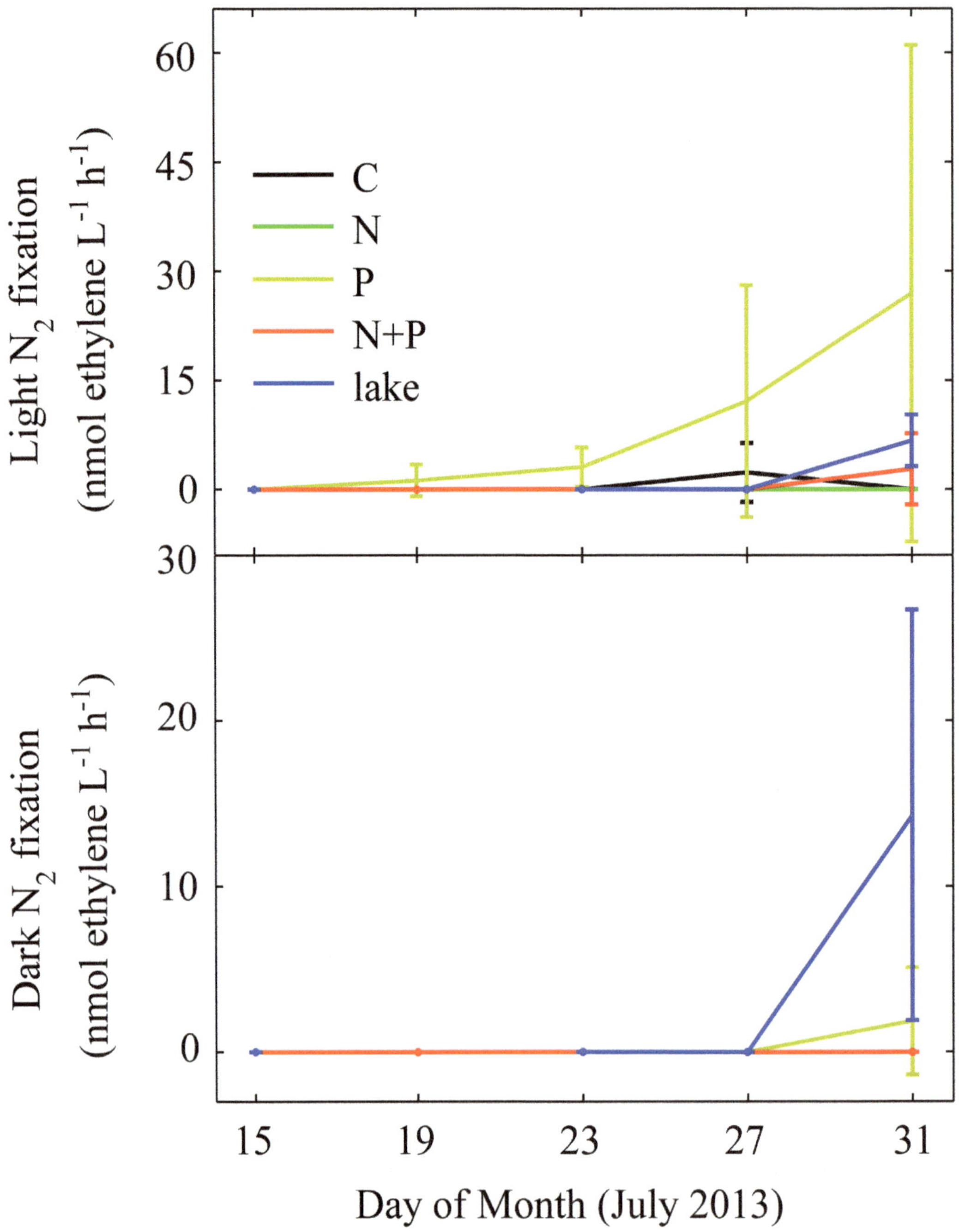

Figure 8. Time series of light and dark acetylene reduction rates from the summer 2013 mesocosm experiment. Solid lines connect means of triplicate mesocosms. Error bars are one standard deviation.

water column experiences very low light levels (<5% of incident light). Suspended solids, and not chlorophyll or chromophoric organic matter, accounts for the vast majority of light attenuation in Taihu [45]. Therefore, we suggest that light limitation may interact with nutrient limitation in this hypertrophic lake to favor non-N_2 fixing cyanobacteria, especially during maximum bloom periods, when surface scums effectively shade the underlying water column. Other shallow, highly turbid lakes (e.g., Lake Okeechobee, FL., Dutch lakes) show similar patterns, with *Microcystis* maintaining dominance over N_2-fixing genera in N and light limited conditions [46,47].

Iron (Fe) limitation could have played a role in the inability of N_2 fixers to assume dominance under the extreme N-limited conditions imposed in these experiments [5]. However, recent examinations indicated Fe-replete conditions during the summer blooms in Taihu [48]. Therefore, this is not a likely explanation for the absence of this process.

From a nutrient management perspective, results from our mesocosm experiments indicate that more N limited conditions induced by selectively enrichment with P did not lead to replacement of non-diazotrophs with diazotrophs in Taihu. This scenario may become more probable as overall nutrient reductions, leaching of stored nutrient supplies in sediments, and increases in transparency take place over time. However, this scenario will likely take several years to decades to emerge in this large, relatively long residence time (~1 year) lake. Therefore, we conclude that N input reductions, in concert with P reduction (i.e., a dual nutrient strategy), is the most effective near-term strategy for reducing phytoplankton biomass, bloom potentials, and improving the overall trophic state of Taihu, as well as downstream N limited Yangtze estuary and South China Sea coastal waters. This approach requires a specific set of nutrient reduction targets, which are being formulated through the combined use of nutrient addition and dilution bioassays [26,27]. At present, it appears that N input reductions would not lead to replacement of *Microcystis* with N_2 fixers, even though this outcome could be considered a positive outcome, given the tendency for lower toxicity, growth rates, and ultimate biomasses of N_2 fixers [8].

The use of microcosm and mesocosm experiments for determining phytoplankton responses to nutrient inputs has been criticized by Carpenter [49] as not reflective of ecosystem-level conditions. These approaches employ small-scale containers over short (days to weeks) incubation times and may not adequately mimic temporal and spatial scales for nutrient cycling and fluxes. However, prior work has shown remarkable agreement among nutrient enrichment and limitation experiments carried out over these time intervals with a wide size range of incubation vessels [50,51]. Furthermore, incubations of at least two weeks should be adequate to elicit taxa-specific (including diazotrophic) responses to nutrient enrichment and other ecophysiological factors (e.g., changes in irradiance, temperature, pO_2) [52].

We previously examined this question in Lake Taihu using 1 L to 4 L polyethylene Cubitainers over a time span of 1 week or less [26,27]. These experiments indicated that N and P co-limitation was prevalent, especially during the critical summer bloom period. Others [51,53] have shown similar results in eutrophic to hypereutrophic lakes. The question persists, however; can these results be extrapolated over longer time scales, and do they reflect natural conditions, such as sediment-water column nutrient exchange and seasonally-variable physical (temperature. light), chemical, and biological (grazing, microbial interactions) conditions? Can species and group succession patterns be replicated over the relatively short time scales that these experiments are conducted? This question is particularly relevant to the issue of replacement of non-N_2 fixing with N_2 fixing cyanobacterial species.

In shallow, hypertrophic Lake Taihu, there was remarkable agreement between ambient lake water and mesocosms deployed in the lake. As N limitation was intensified during the summer period, the non-N_2 fixing cyanobacterial community was not replaced by a N_2 fixing community. Most likely, *Microcystis* remained dominant during P enriched conditions in the lake because it effectively competes for regenerated N (as NH_4^+) from sediments and the water column [27,54], as noted in previous mesocosm experiments in a shallow, eutrophic, Swedish lake [39]. Energetically, *Microcystis* must be able to access regenerated N more efficiently than *Anabaena* and *Aphanizomenon* can fix N_2 to meet their N requirements. The tradeoffs of regenerated N vs. fixed N need further study, especially relative to selective nutrient management strategies aimed at bringing hypertrophic lakes, like Taihu, below bloom thresholds. Our results, combined with the fact that sediment N:P is low in Taihu [55] and most eutrophic lakes, show that *both* N and P reductions are needed in the short-term, in large part due to the legacy of high N and P loading and sediment nutrient storage common in this and other eutrophic lakes [7,47,56]. For this reason, reducing only P loading to Lake Taihu, without parallel N reductions, is unlikely to improve water quality (i.e., reductions in bloom frequency and intensity) on a time scale that will be acceptable to the public and managers.

Recent evidence suggests that N loading may indirectly cause P release from sediments via disruption of the sulfur and Fe cycles [57], providing additional justification for a dual nutrient management approach. Failure to control N loading also has consequences for aquatic systems located downstream [20]. Excessive N loads to coastal and estuarine systems is linked to widespread development of algal blooms, such as red tide, and bottom-water hypoxia (e.g., the northern Gulf of Mexico). A P-only approach to watershed nutrient management for Taihu and other freshwater systems simply shunts N-driven eutrophication issues further downstream to the N sensitive coastal zone [20], an approach that we believe is irresponsible.

Acknowledgments

We appreciate the in depth review of this manuscript by 2 anonymous reviewers. Technical support was provided by Betsy Abare, Randy Sloup, Jianrong Ma, and Jianming Deng. The Taihu Laboratory for Lake Ecosystem Research (TLLER), Chinese Academy of Sciences, provided the environmental monitoring data.

Author Contributions

Conceived and designed the experiments: HWP HX NSH GZ. Performed the experiments: HWP HX NSH GZ LD KLR YW. Analyzed the data: HWP HX NSH ARJ KLR MJM BQ. Contributed reagents/materials/analysis tools: KLR LD ARJ. Contributed to the writing of the manuscript: HWP HX NSH MJM KLR ARJ GZ BQ.

References

1. Elser JJ, Bracken MES, Cleland EE, Gruner DS, Harpole WS, et al. (2007) Global analysis of nitrogen and phosphorus limitation of primary producers in freshwater, marine and terrestrial ecosystems. Ecol. Lett. 10: 1124–1134.
2. Lewis WM Jr, Wurtsbaugh WA, Paerl HW (2011) Rationale for control of anthropogenic nitrogen and phosphorus in inland waters. Environ. Sci. & Technol. 45: 10030–10035.
3. Paerl HW, Fulton RS III, Moisander PH, Dyble J (2001) Harmful freshwater algal blooms, with an emphasis on cyanobacteria. The Scientific World 1: 76–113.
4. Huisman JM, Matthijs HCP, Visser PM (2005) Harmful cyanobacteria. Springer Aquatic Ecology Series 3. Springer.
5. Fogg GE (1974) Nitrogen fixation. In: Stewart WDP, editor. Algal physiology and biochemistry. Blackwell, Oxford. 560–583.
6. Schindler DW (1975) Whole-lake eutrophication experiments with phosphorus, nitrogen and carbon. Verh. Int. Ver. Theor. Angew. Limnol. 19: 3221–3231.
7. Smith VH (1983) Low nitrogen to phosphorus ratios favor dominance by blue-green algae in lake phytoplankton. Science 221: 669–671.
8. Attridge EM, Rowell P (1997) Growth, heterocyst differentiation and nitrogenase activity in the cyanobacteria *Anabaena variabilis* and *Anabaena cylindrica* in response to molybdenum and vanadium. New Phytol. 135: 517–526.
9. Howarth RW, Swaney DP, Billen G, Garnier J, Hong B, et al. (2012) Nitrogen fluxes from large watersheds to coastal ecosystems controlled by net anthropogenic nitrogen inputs and climate. Front. Ecol. & the Environ. 10: 37–43.
10. Likens GE editor (1972) Nutrients and eutrophication, American Society of Limnology Oceanography special symposium 1. Am. Soc. Limnol. Oceanography.
11. Nixon SW (1995) Coastal marine eutrophication: A definition, social causes, and future concerns. Ophelia 41: 199–219.
12. Lewis WM Jr, Wurtsbaugh WA (2008) Control of lacustrine phytoplankton by nutrients: Erosion of the phosphorus paradigm. Inter. Rev. Ges. Hydrobiol. 93: 446–465.
13. Conley DJ, Paerl HW, Howarth RW, Boesch DF, Seitzinger SP, et al. (2009) Controlling eutrophication: Nitrogen and phosphorus. Science 323: 1014–1015.

14. Paerl HW, Otten TG (2013) Harmful cyanobacterial blooms: Causes, consequences and controls. Microb. Ecol. 65: 995–1010.
15. Paerl HW, Hall NS, Calandrino ES (2011a) Controlling harmful cyanobacterial blooms in a world experiencing anthropogenic and climatic-induced change. Sci. Total Environ. 409: 1739–1745.
16. Paerl HW, Fulton RS III (2006) Ecology of harmful cyanobacteria. In: Graneli E, Turner J, editors. Ecology of harmful marine algae. Springer-Verlag. 95–107.
17. Seitzinger SP, Harrison JA, Boehlke JK, Bouwman AF, Lowrance R, et al. (2006) Denitrification across landscapes and waterscapes: A synthesis. Ecol. Appl. 16: 2064–2090.
18. Paerl HW, Scott JT (2010) Throwing fuel on the fire: Synergistic effects of excessive nitrogen inputs and global warming on harmful algal blooms. Environ. Sci. & Technol. 44: 7756–7758.
19. Scott JT, McCarthy MJ (2010) Nitrogen fixation may not balance the nitrogen pool in lakes over timescales relevant to eutrophication management. Limnol. Oceanogr. 55: 1265–70.
20. Paerl HW (2009) Controlling eutrophication along the freshwater–marine continuum: Dual nutrient (N and P) reductions are essential. Estuar. Coasts 32: 593–601.
21. Schindler DW, Hecky RE, Findlay DL, Stainton MP, Parker BR, et al. (2008) Eutrophication of lakes cannot be controlled by reducing nitrogen input: Results of a 37 year whole ecosystem experiment. Proc. Nat. Acad. Sci. USA 105: 11254–58.
22. Bryhn AC, Håkanson L (2009) Coastal eutrophication: Whether N and/or P should be abated depends on the dynamic mass balance. Proc. Natl. Acad. Sci. USA 106: E3, doi:10.1073/pnas.0810905106.
23. Paterson MJ, Schindler DW, Hecky RE, Findlay DL, Rondeau KJ (2011) Comment: Lake 227 shows clearly that controlling inputs of nitrogen will not reduce or prevent eutrophication of lakes. Limnol. Oceanogr. 56: 1545–1547.
24. Chen Y, Qin B, Teubner K, Dokulil MT (2003) Long-term dynamics of phytoplankton assemblages: *Microcystis*-domination in Lake Taihu, a large shallow lake in China. J. Plankt. Res. 25: 445–453.
25. Qin B, Zhu G, Gao G, Zhang Y, Li W, et al. (2010) A drinking water crisis in Lake Taihu, China: Linkage to climatic variability and lake management. Environ. Man. 45: 105–112.
26. Xu H, Paerl HW, Qin B, Zhu G, Gao G (2010) Nitrogen and phosphorus inputs control phytoplankton growth in eutrophic Lake Taihu, China. Limnol. Oceanogr. 55: 420–432.
27. Paerl HW, Xu H, McCarthy MJ, Zhu G, Qin B, et al. (2011b) Controlling harmful cyanobacterial blooms in a hyper-eutrophic lake (Lake Taihu, China): The need for a dual nutrient (N & P) management strategy. Water Res. 45: 1973–1983.
28. Qin B, Liu B, Havens K (2007) Eutrophication of shallow lakes with special reference to Lake Taihu, China. Hydrobiologia 581: 1–327.
29. American Public Health Association [APHA] (1995) Standard methods for the examination of water and wastewater, 19th ed. APHA, American Water Works Association, Water Environment Federation, Washington, DC.
30. Ebina J, Tsutsui T, Shirai T (1983) Simultaneous determination of total nitrogen and total phosphorus in water using peroxodisulfate oxidation. Wat. Res. 17: 1721–1726.
31. Pinckney JL, Richardson TL, Millie DF, Paerl HW (2001) Application of photopigment biomarkers for quantifying microalgal community composition and in situ growth rates. Org. Geochem. 32: 585–595.
32. Papista E, Acs E, Boeddi B (2002) Chlorophyll *a* determination with ethanol–a critical test. Hydrobiologia 485: 191–198.
33. Van Heukelem L, Lewitus A, Kana T, Craft N (1994) Improved separations of phyto-plankton pigments using temperature-controlled high performance liquid chromatography. Mar. Ecol. Progr. Ser. 114: 303–313.
34. Mackey MD, Mackey DJ, Higgins HW, Wright SW (1996) CHEMTAX - a program for estimating class abundances from chemical markers: application to HPLC measurements of phytoplankton. Mar. Ecol. Prog. Ser. 144: 265–283.
35. Schlüter L, Lauridsen TL, Krogh G, Jørgensen T (2006) Identification and quantification of phytoplankton groups in lakes using new pigment ratios – a comparison between pigment analysis by HPLC and microscopy. Freshw. Biol. 51: 1474–1485.
36. Hu H, Li Y, Wei Y, Zhu H, Shi Z (1980) Freshwater Algae in China. Shanghai Science and Technology Press. [in Chinese].
37. Paerl HW (1998) Microbially-mediated nitrogen cycling. In: Burlage R, editor. Techniques in microbial ecology. Oxford University Press. 3–30.
38. Andersen T, Saloranta TM, Tamminen T (2005) A statistical procedure for unsupervised classification of nutrient limitation bioassay experiments with natural phytoplankton communities. Limnol. Oceanogr. : Methods 5: 111–118.
39. Blomqvist P, Pettersson A, Hyenstrand P (1994) Ammonium-nitrogen: A key regulatory factor causing dominance of non-nitrogen-fixing cyanobacteria in aquatic systems. Arch. Hydrobiol. 132: 141–164.
40. Flores E, Herrero A (2005) Nitrogen assimilation and nitrogen control in cyanobacteria. Biochem. Soc. Transact. 33: 164–167.
41. Paerl HW, Millie DF (1996) Physiological ecology of toxic cyanobacteria. Phycologia 35: 160–167.
42. Reynolds CS, Jaworski GMH, Cmiech HA, Leedale GF (1981) On the annual cycle of the blue-green alga *Microcystis aeruginosa* Kütz. emend. Elenkin. Phil. Trans. Roy. Soc. Lond. Ser. B 293: 419–477.
43. Kappers FI (1980) The cyanobacterium *Microcystis aeruginosa* and the nitrogen cycle of the hypertrophic Lake Brielle (The Netherlands), In: Barica J, Mur L, editors. Hypertrophic ecosystems. Dr. W. Junk. 37–43.
44. Zhang Y, Qin B, Zhu G, Gao G, Luo L, et al. (2006) Effect of sediment resuspension on underwater light field in shallow lakes in the middle and lower reaches of the Yangtze River: A case study in Longgan Lake and Taihu Lake. Science in China: Series D Earth Sciences 49: 114–125.
45. Shi K, Zhang Y, Liu X, Wang M, Qin B (2014) Remote sensing of diffuse attenuation coefficient of photosynthetically active radiation in Lake Taihu using MERIS data. Remote Sensing of Environment 140: 365–377.
46. Scheffer M, Rinaldi S, Gragnani A, Mur LR, van Nes EH (1997) On the dominance of filamentous cyanobacteria in shallow, turbid lakes. Ecology 78: 277–282.
47. Havens KE, James RT, East TL, Smith VH (2003) N:P ratios, light limitation, and cyanobacterial dominance in a subtropical lake impacted by non-point source nutrient pollution. Environ. Pollut. 122: 379–390.
48. Xu H, Zhu G, Qin B, Paerl HW (2012) Growth response of the cyanobacterium *Microcystis* spp. to nitrogen, phosphorus and iron enrichment in different regions of Lake Taihu, China. Hydrobiologia 700: 187–202.
49. Carpenter SR (1996) Microcosm experiments have limited relevance for community and ecosystem ecology. Ecology 77: 677–680.
50. Elser JJ, Marzolf E, Goldman CR (1990) The roles of phosphorus and nitrogen in limiting phytoplankton growth in freshwaters: A review of experimental enrichments. Can. J. Fish. Aquat. Sci. 47: 1468–1477.
51. Spivak AC, Vanni MJ, Mette EM (2010) Moving on up: can results from simple aquatic mesocosm experiments be applied across broad spatial scales? Freshw. Biol. 56: 279–291.
52. Paerl HW (1990) Physiological ecology and regulation of N_2 fixation in natural waters. Adv. Microb. Ecol. 11: 305–344.
53. Dodds WK, Johnson KR, Priscu JC (1989) Simultaneous nitrogen and phosphorus deficiency in natural phytoplankton assemblages: Theory, empirical evidence and implications for lake management. Lake and Res. Man. 5: 21–26.
54. McCarthy MJ, Lavrentyev PJ, Yang L, Zhang L, Chen Y, et al. (2007) Nitrogen dynamics and microbial food web structure during a summer cyanobacterial bloom in a subtropical, shallow, well-mixed, eutrophic lake (Lake Taihu, China). Hydrobiologia 581: 195–207.
55. Trolle D, Zhu G, Hamilton D, Luo L, McBride C, et al. (2009) The influence of water quality and sediment geochemistry on the horizontal and vertical distribution of phosphorus and nitrogen in sediments of a large, shallow lake. Hydrobiologia 627: 31–44.
56. Jeppesen E, Søndergaard M, Meerhoff M, Lauridsen TL, Jensen JP (2007) Shallow lake restoration by nutrient loading reduction–some recent findings and challenges ahead. Hydrobiologia 584: 239–252.
57. Smolders AJP, Lucassen ECHET, Bobbink R, Roelofs JGM, Lamers LPM (2010) How nitrate leaching from agricultural lands provokes phosphate eutrophication in groundwater fed wetlands: The sulphur bridge. Biogeochemistry 98: 1–7.

19

Habitat Capacity for Cougar Recolonization in the Upper Great Lakes Region

Shawn T. O'Neil, Kasey C. Rahn¤, Joseph K. Bump*

School of Forest Resources and Environmental Science, Michigan Technological University, Houghton, Michigan, United States of America

Abstract

Background: Recent findings indicate that cougars (*Puma concolor*) are expanding their range into the midwestern United States. Confirmed reports of cougar in Michigan, Minnesota, and Wisconsin have increased dramatically in frequency during the last five years, leading to speculation that cougars may re-establish in the Upper Great Lakes (UGL) region, USA. Recent work showed favorable cougar habitat in northeastern Minnesota, suggesting that the northern forested regions of Michigan and Wisconsin may have similar potential. Recolonization of cougars in the UGL states would have important ecological, social, and political impacts that will require effective management.

Methodology/Principal Findings: Using Geographic Information Systems (GIS), we extended a cougar habitat model to Michigan and Wisconsin and incorporated primary prey densities to estimate the capacity of the region to support cougars. Results suggest that approximately 39% (>58,000 km^2) of the study area could support cougars, and that there is potential for a population of approximately 500 or more animals. An exploratory validation of this habitat model revealed strong association with 58 verified cougar locations occurring in the study area between 2008 and 2013.

Conclusions/Significance: Spatially explicit information derived from this study could potentially lead to estimation of a viable population, delineation of possible cougar-human conflict areas, and the targeting of site locations for current monitoring. Understanding predator-prey interactions, interspecific competition, and human-wildlife relationships is becoming increasingly critical as top carnivores continue to recolonize the UGL region.

Editor: Cédric Sueur, Institut Pluridisciplinaire Hubert Curien, France

Funding: This project was funded and supported by the School of Forest Resources and Environmental Science Ecosystem Science Center at Michigan Technological University. The funders had no role in study design, data collection and analysis, decision to publish, or preparation of the manuscript.

Competing Interests: The authors have declared that no competing interests exist.

* Email: jkbump@mtu.edu

¤ Current address: University of Montana, Missoula, Montana, United States of America

Introduction

Cougars (*Puma concolor*) once spanned North and South America, ranging from south of the boreal forests to Patagonia [1,2]. By the early twentieth century in the United States, human persecution, habitat degradation, and human expansion resulted in the extirpation of cougars from two-thirds of their historic range including eastern and midwestern America [3]. Cougars persisted only in the American west, where populations are increasing for the first time in nearly a century [4]. As a result of this increase, cougars are recolonizing portions of their former range. For example, natural recolonization, aided by changes in cougar protection and prey management, led to the return of viable populations in Wyoming and the Black Hills of South Dakota by 2000 [4]. Recolonization also occurred in the Badlands of North Dakota and in western Nebraska [5,6].

Cougars, particularly young males, will travel hundreds of kilometers in search of new territory [7–9]. In 2011, one individual traveled more than 1,700 km from Minnesota to Connecticut, and may have traveled a straight-line distance of 2,500 km from the Black Hills to the East Coast [10]. In recent years, reports of cougar presence have been more frequent and widespread in the northern Midwest and Great Lakes States. Verified photograph and video evidence from automatic cameras, human-cougar encounters, DNA samples (scat, hair and blood) and track records was recorded each year in Michigan, Wisconsin, and Minnesota between 2008 and 2013. DNA analysis of samples suggested that a minimum of 6 individual cougars visited Wisconsin during this time period [11]. Evidence of cougar presence was also verified in Ontario, Canada [12]. Biologists believe confirmed cougar occurrences are young dispersing males from western populations, as evidenced by the presence of radio collars originating from western-based research and monitoring programs [10, Michigan Department of Natural Resources (DNR) *unpublished data*]. This increase in occurrence is consistent with the expectation that as populations rise in the American west, long-distance dispersals become more frequent, particularly among males [9,13]. Cougars are now exploring much of the Midwest, ostensibly dispersing to expand to new territories, to increase mating opportunities, and to avoid overlap with existing home ranges [9,13,14].

The forests of northern Wisconsin and Michigan appear to exhibit high potential for cougar recolonization, due to favorable configurations of land cover, road density, abundant food resources, and relatively low human population densities [15,16].

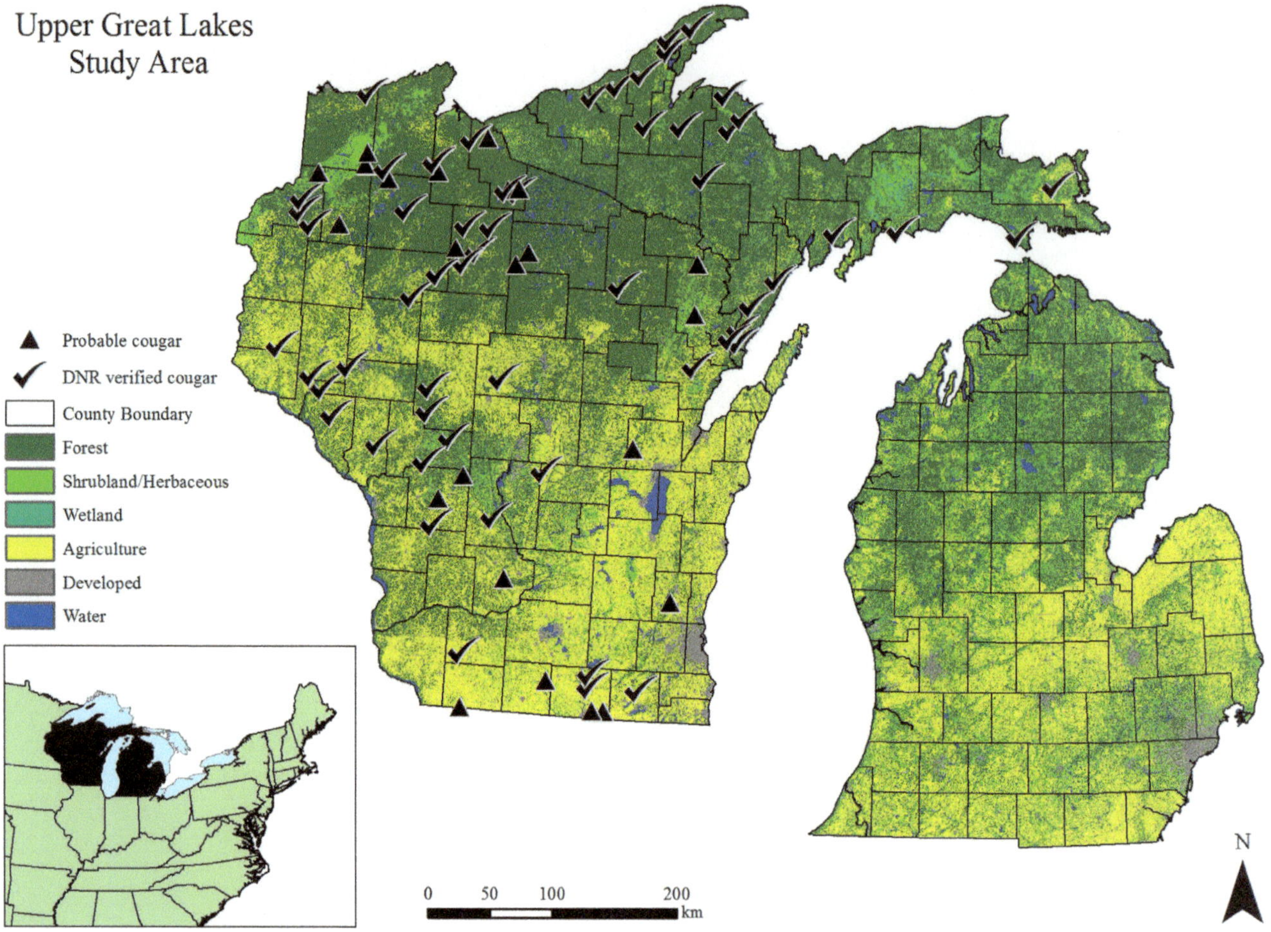

Figure 1. The study area for an analysis of cougar habitat and capacity for the Upper Great Lakes states, USA. Probable and verified cougar locations are represented from 1 January, 2008 to 1 June, 2013.

These same features have also shown to be important habitat features for gray wolves (*Canis lupus)* in this region [17–19]. Evidence of cougar range expansion has future conservation implications for the states of Michigan, Wisconsin, and Minnesota, which likely contain suitable habitat [20]. Despite little evidence of a breeding population [e.g. female cougars, kittens; but see [21,22], these states are preparing for the emergence of a reproducing cougar population at some point in the near future [23]).

Potential cougar range expansion to the Upper Great Lakes (UGL) region raises important ecological questions and may require novel strategies to manage the species. Current research and management needs include mapping the geographic extent of potential cougar habitat in the UGL and estimating its capacity to support cougars. Researchers have modeled potential cougar habitat based on life history requirements to address questions of habitat extent where accurate data on the species' distribution are lacking [16,20]. These models can be combined with prey densities to estimate an area's capacity for the species [24,25]. Here, we address the need for spatially explicit cougar habitat data in Michigan and Wisconsin by extending an expert-assisted GIS spatial model [16,20,26]. We also provide an exploratory validation of the model, and estimate cougar capacity based on prey resources (white-tailed deer [*Odocoileus virginianus*]) and favorable land cover characteristics.

Methods

Study area

Our study area encompassed the states of Michigan and Wisconsin where DNR-confirmed cougar locations were widespread since 2008 (Fig. 1). Land cover types in the northern regions of both states are forest-dominated, with agriculture and human development becoming more predominant in the south. Forested land occurred over approximately 47% of the entire study area. Forests were generally of northern hardwood association, with common types including maple (*Acer* spp.) – birch (*Betula spp.*) – hemlock (*Tsuga canadensis*), aspen (*Populus* spp.) – birch (*Betula* spp.), and oak (*Quercus spp.*). Additional forest types included spruce-fir (*Picea* spp., *Abies balsamea*), and pine (*Pinus banksiana, P. resinosa, P. strobus*). Agricultural land use occurred on approximately 31% of the study area and generally consisted of wheat, corn, and soybean crops. Livestock grazing and dairy farming were common in more southern regions, particularly in Wisconsin. Other land cover types included shrubland/herbaceous, wetlands, and urban/developed (Fig. 1). Elevation ranged from 141 to 602 m and terrain was generally flat to hilly with an overall median slope estimate of 1.07° (Interquartile Range [IQR] = 0.34°–2.73°); however, more rugged terrain was present throughout the region and slopes reached 81.5° in some of these areas. The study area was characterized by

Table 1. Weights for attributes within variables used to model potential habitat suitability for cougars in the Upper Great Lakes states (Adapted from [26]).

Variable	Attribute	Weight (±S.E.)	Percent importance from highest ranking variable
Land cover	Mixed forest	1.92 (0.51)	100
	Deciduous forest	1.61 (0.37)	84
	Evergreen forest	1.59 (0.62)	83
	Shrublands	1.12 (0.85)	58
	Wetlands	0.67 (0.29)	35
	Grasslands	0.61 (0.47)	32
	Agricultural	0.28 (0.17)	15
	Barren/developed/other	0.19 (0.05)	10
Distance to paved roads	Long (>5 km)	1.43 (0.71)	100
	Medium (0.3–5 km)	0.88 (0.34)	62
	Short (<0.3 km)	0.69 (0.73)	48
Distance to water	Short (<1 km)	1.57 (0.41)	100
	Medium (1–5 km)	0.92 (0.27)	59
	Long (>5 km)	0.52 (0.27)	33
Human density	Low (<5 persons/km^2)	2.28 (0.39)	100
	Medium-Low (6–10 persons/km^2)	1.00 (0.18)	44
	Medium-High (11–19 persons/km^2)	0.46 (0.27)	20
	High (>20 persons/km^2)	0.25 (0.07)	11
Slope	Steep (>15°)	1.17 (0.54)	100
	Moderate (5–15°)	1.17 (0.41)	100
	Gentle (<5°)	0.66 (0.53)	56

abundant lakes and streams, with distance to water not exceeding 11 km (Median $=0.35$ km, IQR $=0.15$–0.68 km). Human population density ≤ 5 persons/km^2 characterized much of the northern regions of the study area, as opposed to the southern regions where population centers were more common and population densities commonly exceeded 100 persons/km^2 in these areas.

Overall Modeling Approach

We applied an expert-assisted spatial habitat model for cougars [16,20,26] to our study area. As a means of validation, we summarized predicted habitat around confirmed locations, and investigated the association between model-predicted cougar habitat values and confirmed cougar locations during the study period. We used the results of the habitat model to eliminate areas deemed unsuitable for cougars, incorporated estimates of deer densities and associated deer biomass within the UGL to investigate potential prey biomass, and subsequently combined potential landscape habitat characteristics with prey resources to generate a range of cougar capacity estimates for the UGL region.

Cougar Habitat Model

We used previously established modeling framework and weights [16,20,26] to extend a cougar habitat model to the UGL. This model incorporated five, differently weighted parameters determined to be essential for cougar habitat: land cover, slope, human population density, distance to roads, and distance to water [26] (Table 1). All raster data generated by the habitat analysis were output to 30 m cell size in a GIS and we performed all spatial analysis in ArcGIS 10.1 (Environmental Systems Research Institute, Inc., Redlands, CA).

To represent roads we used Tiger/line shapefiles acquired from the U.S. Census Bureau. We excluded non-paved roads from the analyses; remaining features represented paved roads that presumably impede cougar movement and/or increase risk of mortality [27–29]. We calculated distance to roads and reclassified the distances into three categories for use in the habitat model (Table 1). We calculated human population density (persons/km^2) for each block group using 2010 U.S. Census Bureau data and grouped the results into four density classes (Table 1). We used the 1 arc-second (30 m spatial resolution) digital elevation model (DEM) from the National Elevation Dataset to calculate slopes (°) and reclassified the results into three categories (Table 1). We gathered rivers, streams, and lake features from the Michigan Geographic Data Library and from the ArcGIS Resource Center for Wisconsin. We retained all rivers, streams and water bodies assumed to hold water under normal, non-drought conditions (e.g. all lakes, rivers, and stream features including artificial paths, canals/ditches, intermittent streams, and perennial streams), calculated distance to these features, and reclassified distance to water into three categories (Table 1). We used the IFMAP/GAP Land Cover dataset [30] for Michigan and the National Land Cover Gap Analysis Project for Wisconsin. These datasets were comparable in that they allowed us to reclassify similar cover types into 8 final categories that were then weighted according to the initial habitat model [26] (Table 1).

We weighted each variable in the model according to expert-based variable rankings and weights published previously [16,26] (Table 1). Weights from this model were established via the Analytical Hierarchy Process (AHP) where experts reviewed all

Table 2. Weights for each variable used in the calculation of potential habitat suitability for cougars in the Upper Great Lakes states (Modified from [26]).

Variable	Weights (S.E.)	Percent importance from highest ranking variable (land cover)
Land cover	1.84 (0.59)	100
Human density	1.22 (0.82)	66
Distance to paved roads	0.86 (0.45)	47
Slope	0.61 (0.56)	33
Distance to water	0.47 (0.26)	26

Weights were based on an Analytical Hierarchy Process analysis [31] and represent the relative importance of each variable to cougar habitat as established by a survey of experts [20,26].

possible pairs of environmental attributes and assigned a rating based on each comparison [16,26,31]. Such analyses are useful for making *a priori* predictions in situations where empirical data are lacking, as is the case in our study area where until recently, cougars had not been confirmed present since the early 1900s [11,32]. Next, we completed a weighted summation using the five variables and their respective weights (Table 2). We divided all pixels in the weighted sum raster by the maximum value that a pixel could receive, which resulted in a relative ranking for habitat potential [16]. Resulting pixel values ranged from 0 to 1, with 1 representing greatest habitat potential. To eliminate noise associated with pixel-scale variation, we smoothed the habitat raster using the mean neighborhood statistic within a 13.75 km^2 circular moving window, corresponding to $1/4^{th}$ of the minimum home range size of a female cougar [9]. This size window retains a high level of resolution while also maintaining a realistic spatial scale for cougar perception of habitat, given that scale of selection is complex and likely varies depending on behaviors [33]. For ease of comparison, we scaled final habitat values into percentages and

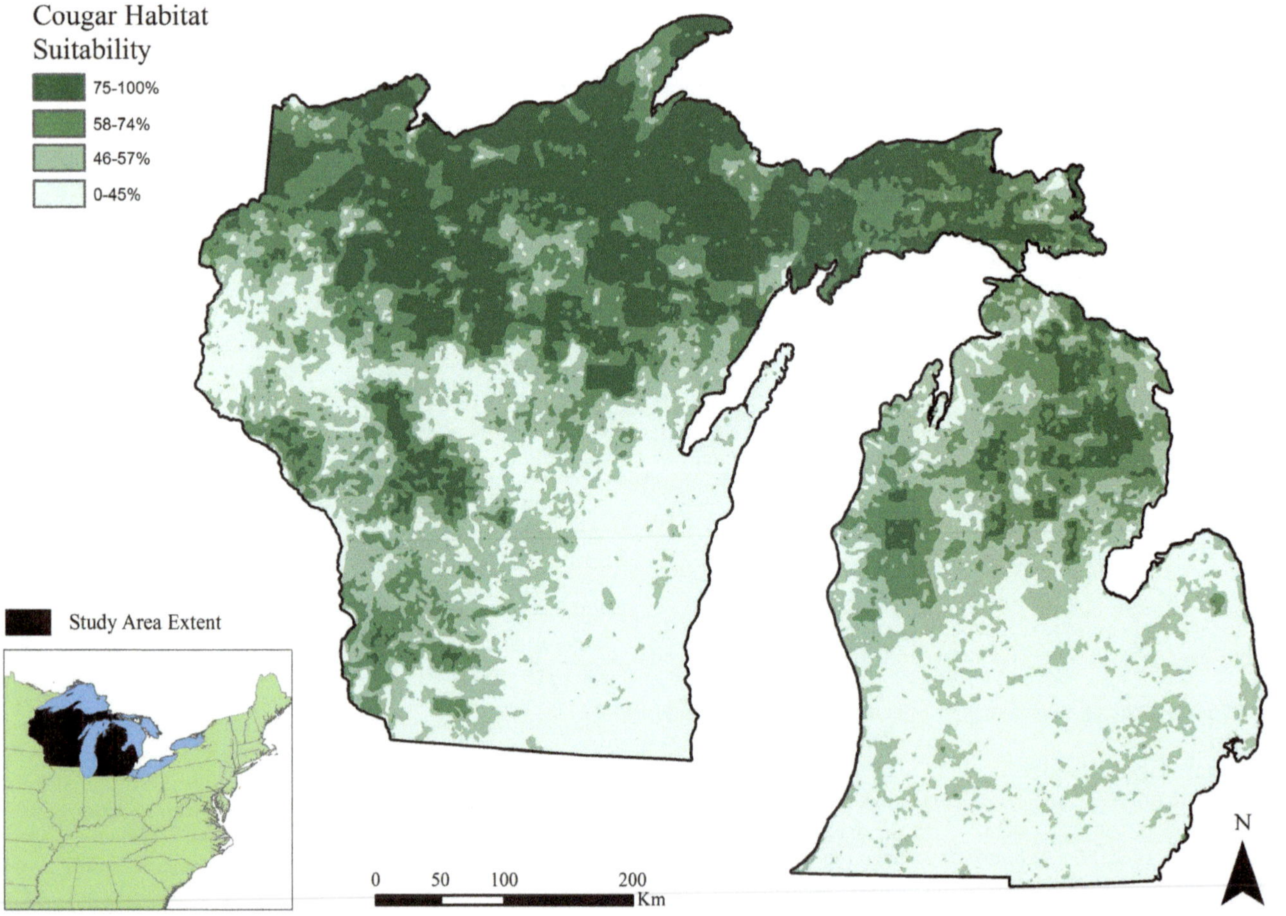

Figure 2. Cougar habitat rankings in the Upper Great Lakes region, USA. The cougar habitat model was generated based on expert-assisted variables and weights from LaRue [20], and LaRue and Nielsen [16,26].

Table 3. Confirmations of cougars by year, 1 January, 2008 to 1 June, 2013, in Michigan and Wisconsin, USA.

	Cougar confirmations			
Year	Michigan	Wisconsin	Probable sightings	Common methods[a]
2008	3	3	6	tracks, photo, DNA
2009	3	8	4	tracks, photo, observed (treed)
2010	1	6	5	photo, tracks, scat
2011	7	7	1	photo, tracks, scat (DNA), video
2012	6	13	9	photo, tracks, video, observed
2013[b]	0	1	NA	observed (treed)
Total	20	38	25	

[a]Ordered from most frequent to least frequent.
[b]Only includes confirmations reported through 1 June, 2013.
Confirmations are those verified by the Department of Natural Resources from either state as cougar (Michigan DNR, *unpublished data*; Wisconsin DNR, *unpublished reports*). Probable sightings were also reported in Wisconsin during the same time period.

reclassified them into the same categories as the original cougar habitat model [16] (Fig. 2).

Cougar Data and Model Validation

We reviewed current literature, archived reports, and press releases from the Michigan and Wisconsin DNRs, to document confirmed (i.e., verified evidence of) cougar occurrences in the study area from 1 January, 2008 to 1 June, 2013. Our review was supplemented by Rare Mammal Observations reported annually by the Wisconsin DNR [10,34–37] and a confirmation database from the Michigan DNR (Michigan DNR, *unpublished data*). These sources of information included all known cougar confirmations documented in the study area since 2008 (Fig. 1). Dates were chosen because the first cougar confirmed in these states since the 1900s occurred in 2008, despite statewide collection of cougar reports from Wisconsin since 1991 [38]. Numbers of verified occurrences increased after 2008 [10,34–37]. Verified evidence included tracks, photographs, scat, video, visual observation by wildlife officials, and DNA gathered from blood, scat, or hair. We recorded the date, township location, and type of evidence (photographs, tracks, scat, encounter, DNA) for each confirmation (Table 3, Fig. 1). Locations classified as "probable" occurrences by biologists based on available evidence [34] were included in Fig. 1 but not used for analysis.

We used the verified location data on the presence of cougars in Wisconsin and Michigan to assess the validity of the habitat model. We first summarized habitat model values around verified cougar locations at four spatial scales (i.e. buffers) to explore modeled habitat potential associated with known cougar space use. Four radial buffers were made around verified locations to represent the area potentially used by the observed animals within the time frame during which they were verified as present. Buffer radius distances were 1 km, 5 km, 10 km, and 25 km, respectively. These distances were chosen to cover a possible range of daily and weekly movement patterns, from smaller and more localized movements typically associated with resident cougars [28,39,40], to larger movements associated with long-distance dispersers [4,8,14]. We summarized mean modeled habitat values within these buffered areas. We then generated a random sample of "pseudo-absence" locations [41] to represent available habitat. We generated 100 random locations for every verified cougar locations over the study area, resulting in 5,800 availability samples [42]. Since no verifications occurred in the Lower Peninsula of Michigan, we limited this sample to Wisconsin and the Upper Peninsula of Michigan. We distributed the locations equally by county to ensure a spatially-balanced sample [42]. Then, for each location, we generated buffers using the same radius distances as the cougar locations and summarized habitat values within each buffer using GIS. To compare habitat values within used buffers to habitat values within available buffers, we assigned a binary indicator variable to each location (1 for used, 0 for available) and fit a resource selection function (RSF; [43,44]) to these data via logistic regression [42]. Accordingly, the RSF assessed the influence that habitat values from our model had on the relative probability of cougar presence, allowing us to explore the validity of our habitat modeling approach. We performed the analysis for each buffer distance and thus fit 4 logistic regression models using 'glm' in R Version 3.0.0 (R Development Core Team, www.r-project.org, accessed 15 June 2013). We scaled the habitat model ranking to a percentage to simply model coefficient interpretation, assessed the resulting fit of models by comparing deviance residuals to the null model, and evaluated the ability of the habitat model to predict cougar occurrences using the Area Under Curve (AUC) statistic [45]. We calculated the AUC and its bootstrapped 95% confidence interval using the 'pROC' package in R [46].

Prey Biomass

Carnivore abundance and density depends on prey availability and biomass as well as available space and favorable land cover. Predicting carnivore density based on estimates of prey biomass has been applied in wolves [17,18], tigers (*Panthera tigris*) [47], and Canada lynx (*Lynx canadensis*) [24]. To estimate potential for cougar prey in the UGL, we used DNR estimates of deer density by deer management unit (DMU) in Michigan and Wisconsin [48,49]. We assumed that white-tailed deer would be the primary prey for cougars in the region [50,51]. To incorporate uncertainty in total deer biomass due to population sex/age structure across the UGL, we used three plausible sex/age structures (buck/doe/fawn) described for the northern Great Lakes and Ontario [52]: 40/30/30 (even/balanced), 25/40/35 (unbalanced, N. Wisconsin), and 15/50/35 (unbalanced, Michigan). We assumed the mean late-autumn deer size within these populations was 100 kg for bucks, 66 kg for does, and 33 kg for fawns [52–54]. Thus, deer biomass estimates (kg) for the three conceivable population structures were approximated within each DMU by the following:

Table 4. Modelled cougar habitat values (0–100%) association with 58 cougar occurrences in Michigan and Wisconsin, USA at 4 spatial scales between 2008 and 2013.

Buffer radius	Range	Mean (± SD)	β Habitat	Odds ratio (95% CI)	AUC (95% CI)
1 km	33–90	64±15	0.033	1.03 (1.02–1.05)	0.66 (0.59–0.72)
5 km	29–92	64±15	0.034	1.03 (1.02–1.05)	0.66 (0.59–0.72)
10 km	29–94	65±15	0.038	1.04 (1.02–1.06)	0.67 (0.60–0.73)
25 km	27–100	64±15	0.042	1.04 (1.02–1.06)	0.68 (0.62–0.74)

Coefficients, odds ratios, and area under curve (AUC) statistics were generated by logistic regression models linking modeled habitat values to verified occurrences.

$$M_{deer} = N_{deer} \times (M_{buck} \times N_{buck} + M_{doe} \times N_{doe} + M_{fawn} \times N_{fawn}) \quad (1)$$

where N_{deer} = estimated deer population/DMU, M_{buck} = presumed average mass of an UGL white-tailed buck, M_{doe} = presumed average mass of an UGL white-tailed doe, M_{fawn} = presumed average mass of an UGL fawn, and N values for buck/doe/fawn were the proportion of deer in the sex/age class given a specified population structure. To estimate capacity for cougars in the study area, we first assumed habitat patches scoring <0.75 in the cougar habitat model would not support a population [16]. We subset the habitat model to only include values ≥0.75, restricted the DMUs to this subset, calculated the area in 100 km^2 units for each DMU, and divided the deer biomass estimates by the area. Carbone and Gittleman [55] found that 10,000 kg/100 km^2 could result in 0.94 cougars/100 km^2; we multiplied deer biomass (10,000 kg/100 km^2) by 0.94 to achieve the potential cougar density for each DMU. Thus, potential cougars/100 km^2 depended on both deer density and available habitat based on favorable landscape characteristics.

Potential cougar density estimates (cougars/100 km^2) corresponding to the three aforementioned deer population structures were summed across the study area. To incorporate geographic variation, we averaged potential cougar density across all available habitats using a 100 km^2 circular assessment window. While densities of up to 13 cougars/100 km^2 have been reported in other locations [56], such densities are uncommon [33,57,58]. Given the potential for cougar-human conflicts leading to higher risk of mortality and competition with wolves and other predators for prey [59], we considered it unlikely for population densities to exceed 3 cougars/100 km^2 in the UGL. Thus, despite high estimates of prey biomass in some areas, we limited the maximum potential cougar density to 3/100 km^2. To incorporate additional uncertainty, we also carried out the analysis using maximum potential cougar densities of 2/100 km^2 and 1/100 km^2.

Results

Cougar Habitat Model

Our model indicated high potential for cougar habitat in Michigan and Wisconsin. Cougar habitat values ranged from 0.27 to 0.92 ($\bar{\chi}$=0.56, SD=0.15). Assuming the value ≥0.75 is a conservative estimate of suitable habitat [16] at least 39% (>58,000 km^2) of our total study area contained suitable cougar habitat. Habitat areas were generally contiguous and concentrated throughout the forested regions of northern Wisconsin, much of the Upper Peninsula of Michigan, and the northern Lower Peninsula of Michigan (Fig. 2). The largest contiguous patch of suitable habitat occurred in northern Wisconsin and the Upper Peninsula. This area contained 85% (49,216 km^2) of all habitat receiving scores of 0.75 or higher. According to our model, patches appeared to decrease in frequency and size and were evidently more fragmented at southern latitudes. Less suitable cougar habitat (suitability values <0.75) often coincided with higher human population densities and agriculture-dominated landscapes; these areas were concentrated throughout the south and south-central regions of the study area (Fig. 2).

Cougar confirmations from the DNR in Michigan and Wisconsin totaled 58 from 1 January, 2008 to 1 June, 2013 (38 and 20 in Wisconsin and Michigan, respectively; Table 3). Model validation results suggested that the habitat model was effective in identifying areas of suitable cougar habitat. Habitat values summarized around cougar locations exhibited a wide range of habitat potentially used by cougars in the UGL, but were generally consistent with the expectation that verified locations would occur within higher-ranked vs. lower-ranked habitat. Observed habitat values within buffered areas ranged from 0.27 to 1.00 depending on buffer size (Table 4) and occurred within higher-ranked habitat than the mean cougar habitat ranking for the study area (0.56). Our cougar habitat model was a significant predictor of the relative probability of cougar occurrence at each buffer distance (Table 4). The RSF deviance residuals at each scale improved upon the null model and odds ratios associated with the habitat predictor were significantly greater than one (Table 4) indicating that the odds of an occurrence increased with increases in habitat ranking. The strength of this effect and the model's predictive performance (AUC) both increased slightly with larger buffer sizes (Table 4). At each scale, the odds of a cougar occurrence increased by 2–6% with each percent increase in cougar habitat ranking (Table 4).

Cougar Capacity

Deer density estimates per DMU ranged from 0.5 to >20 deer/km^2 ($\bar{\chi}$= 11.1, SD = 6.1; Fig. 3a). Using these estimates and three potential age/sex population structures, deer biomass per DMU could conceivably range from approximately 3,000 kg/100 km^2 (unbalanced age structure, lowest deer density) to over 200,000 kg/100 km^2 (balanced age structure, highest deer density). After restricting the DMUs to favorable cougar habitat, the mean deer biomass estimates per DMU within potential cougar habitat were approximately 10,000–165,000 kg/100 km^2 depending on the age/sex structure applied (Table 5). Thus, we estimated that prey biomass within potential cougar habitat was geographically variable and could support up to 15 cougars/100 km^2 (Fig. 3b). However, we also assumed that >3 cougars/100 km^2 anywhere within the study area was unrealistic despite high deer densities in some areas. Using three different maximum viable densities (1, 2, and 3 cougars/100 km^2) and allowing lower estimates to depend on approximations of deer biomass where

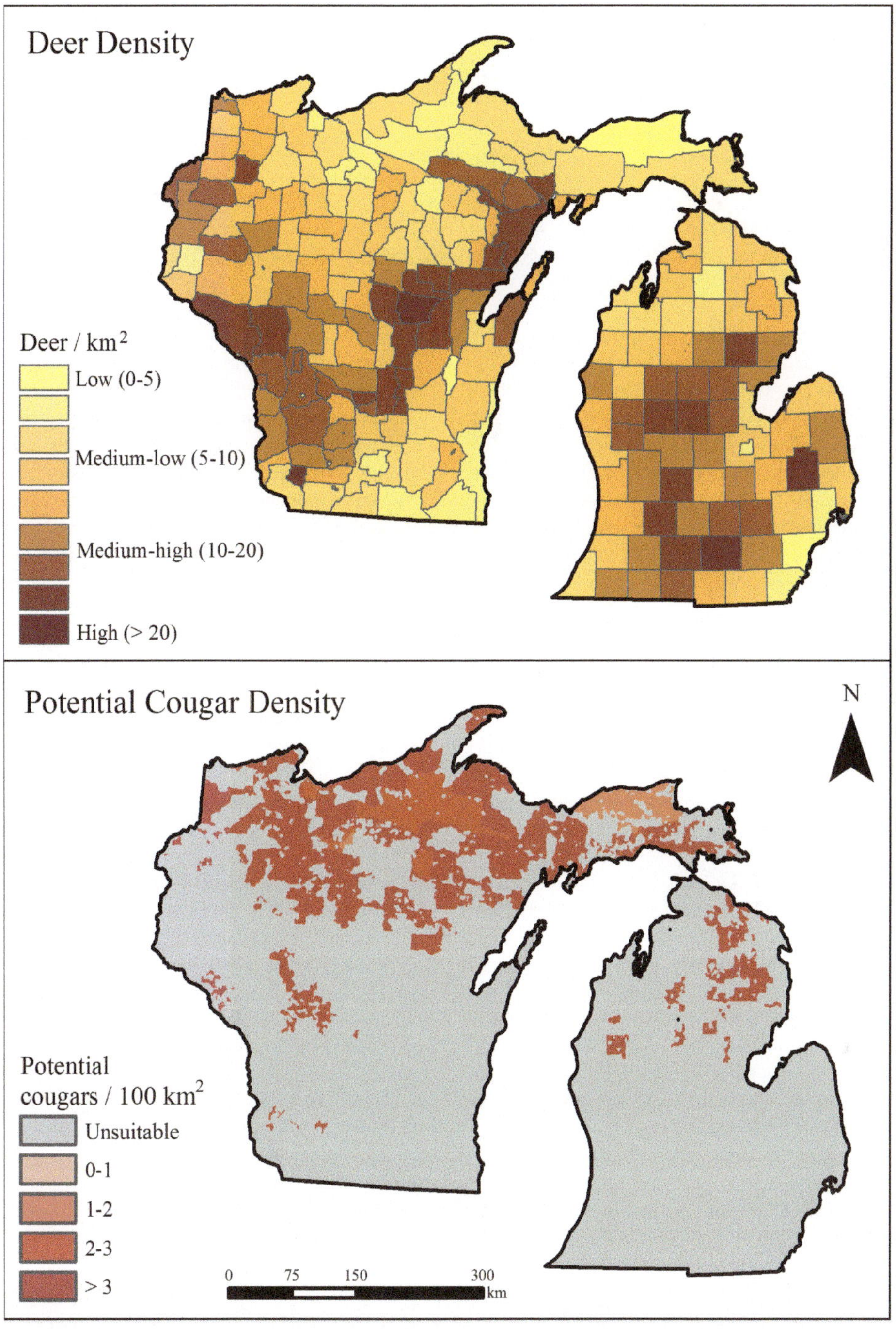

Figure 3. Prey densities and cougar capacity in the Upper Great Lakes region, USA. a) Whitetail deer density estimates were based on best available information from state natural resource management agencies [48,49]. **b**) Geographic variation in cougar capacity based on our most conservative potential deer biomass estimates. Deer biomass estimates were generally high enough to support >3 cougars/100 km^2 across much of the study area.

deer densities were low, we calculated that available resources could sustain 582 to 1,677 cougars within favorable habitat.

Discussion

Our extension of a habitat model for cougars [16,20,26] suggests that Michigan and Wisconsin contain >58,000 km^2 of potential cougar habitat. Based on potential deer biomass estimates, more than 500 cougars could inhabit this overall area. For comparison, cougar range in Washington, USA covers 51% of the state (88,500 km^2) and consistently supports approximately 2,000 cougars [60], and estimates from several models in the northeastern U.S. suggested potential for 322–2,535 cougars depending on the area considered suitable [61]. The northern half of our study area is characterized by dense forest, low human population and road densities, abundant water resources, and

Table 5. Estimates of white-tailed deer biomass within deer management units for Michigan and Wisconsin (the Upper Great Lakes region, or UGL); estimates are based on three conceivable age/sex structures for white-tailed deer and are restricted to favorable habitat as indicated by cougar habitat model values ≥0.75.

Bucks/does/fawns	Range (kg/DMU)	Mean (kg/DMU)	SD (kg/DMU)	Total biomass (kg)
40/30/30	12,494–165,450	76,189	31,812	7.70×10^6
25/40/35	11,284–149,427	68,810	28,731	6.95×10^6
15/50/35	10,674–141,357	65,094	27,179	6.57×10^6

diverse topography. Large expanses of wild land exist that provide contiguous habitat characterized by forested landscapes which support high densities of white-tailed deer. Confirmed location data and associated radio-collar and DNA evidence suggest that multiple cougars occurred in this area between January 2008 and early 2013.

Our investigation of habitat capacity should not be viewed as an attempt to predict the precise distribution and/or exact size of a potential cougar population. Although it would be possible to speculate based on the models we presented (e.g. [18]), the nature of recolonization (i.e. breeding range selection, population size, geographic distribution) will likely depend on additional factors that we could not quantify such as future changes in management, public acceptance [62], the presence and density of competing predators, potential disease components, migratory behavior in deer [63] and other dynamics of prey populations. Both states have experienced similar conservation challenges as wolves have recolonized and recovered in the region [64,65]. Our model validation analysis is exploratory in nature, as the best available information for cougar presence in the UGL is primarily based on incidental, verified observations (Table 3); these location data are assumed to be associated with transient cougars from western populations [13]. Detection and verification probability may vary geographically, and this variation can bias information on cougar presence. Automatic cameras are not evenly spread across the landscape, and are likely underused on large blocks of public lands that contain some of the best cougar habitat in the region. In addition, resource selection by dispersing, transient animals may differ from that of resident individuals [26,66], perhaps with particular regard to human development and activity [15,33]. These reasons likely contributed to our habitat model's relatively low ability to predict cougar occurrences on our study site (model validation AUC scores between 0.59 and 0.74; Table 4). Without knowledge of the specific nature of detection probability across the study site, modeling cougar distribution (e.g. probability and/or likelihood of occurrence) based on presence-only data would require strong assumptions [67,68]. Similarly, it would not be appropriate to base estimates of cougar distribution entirely on locations associated from animals assumed to be non-resident individuals. As such, although our validation used the best available information to provide support for our habitat model, information on cougar habitat use in the UGL remains limited and may not be adequate for making predictions. Given that our objective was to model habitat potential for a future resident population, the expert-assisted model that we implemented is preferable because it only makes assumptions about the most general habitat requirements for cougars. We showed that verified cougar locations in the UGL were consistent with our model, which further suggests that areas of modeled suitable habitat may be able to support a viable cougar population in the future.

Recolonization and recovery of cougars to former ranges such as the UGL will likely require active and adaptive management at both state and federal levels. Natural recolonization is likely to occur eventually [69], providing states with favorable cougar habitat the opportunity to prepare. Public attitudes and their associated influences on public policy are important determinants of large carnivore population viability [62]. Recent evidence suggests that people may be more neutral and are consequently more impressionable in regions where cougars have previously been irrelevant [70], but that humans can be accepting and supportive of cougars, particularly if they have access to good information and are knowledgeable about cougar ecology [62,69,71]. These findings have implications for areas of potential cougar recolonization because there may be a relatively small window of opportunity for outreach and education programs to promote awareness and shape public opinion, as overcoming limited trust once it is instilled is difficult [70]. Large carnivore populations can be polarizing, in part because management tools are often controversial. The appropriateness and ethics of recreational hunting and population control of these animals are hotly debated [72,73] and understanding how public attitudes are influenced by these actions is complicated [74]. Consequently, a natural first step toward preparing for a cougar recolonization would be investigating social acceptance and potential human tolerance of cougars in the UGL states. Public education and outreach could positively shape public opinions and help to avoid agency mistrust [69]. The state of Wisconsin has drafted a management protocol in the event that a breeding population establishes [36]. Additionally, all confirmed cougar observations are publicly available [11]. Anticipating recolonization prior to its occurrence can improve future management and could lay the groundwork for a strong, collaborative cougar conservation program regionally.

In the event that a population becomes established in the UGL, state DNR agencies will need to address concerns of human safety, pet safety, and depredation of livestock [15,75], as well as develop long-term monitoring programs [76]. Strategies necessary to manage a cougar population will likely include plans for education and outreach associated with human safety concerns, potential compensation for livestock losses, mitigation strategies for conflicts with farmers/ranchers, and discussion of possible harvest scenarios [15,75]. Long-term research investigating cougar behavior, habitat use, prey selection, genetic structure of the population, competition with other predators, and impacts on white-tailed deer could eventually be warranted. The addition of another carnivore to the current predator guild in the UGL states could be politically challenging, yet long-term ecological benefits of a viable cougar population [77] could be realized under effective conservation planning scenarios.

Acknowledgments

We thank D. Beyer, Jr., A. Wydeven, B. Roell, J. Wiedenhoeft, S. Roeppke, and M. Hyslop for assistance in data collection. Thanks to the Michigan Department of Natural Resources and Wisconsin Department of Natural Resources for sharing information on cougar verifications. Thanks to D. Beyer, Jr., D. Etter, and A. Wydeven for making improvements to the manuscript.

Author Contributions

Conceived and designed the experiments: STO JKB. Analyzed the data: STO KCR. Wrote the paper: STO KCR JKB.

References

1. Bolgiano C, Roberts J (2005) The eastern cougar: historic accounts, scientific investigations, and new evidence. Mechanicsburg, Pennsylvania, USA: Stackpole Books. 246 p.
2. Laundré JW, Hernández L (2010) What we know about pumas in Latin America. In: Hornocker M, Negri S, editors. Cougar: ecology and conservation. Chicago, Illinois, USA: University of Chicago Press. pp. 76–90.
3. Anderson CR, Lindzey FG, Knopff KH, Jalkotzy MG, Boyce MS (2010) Cougar management in North America. In: Hornocker M, Negri S, editors. Cougar: ecology and conservation. Chicago, Illinois, USA: University of Chicago Press. pp. 41–56.
4. Thompson DJ, Jenks JA (2010) Dispersal movements of subadult cougars from the Black Hills: the notions of range expansion and recolonization. Ecosphere 1: 1–11.
5. Fecske DM, Thompson DJ, Jenks JA, Oehler MW (2008) North Dakota Mountain Lion Status Report. In: Toweill DE, Nadeau S, Smith D, editors. Sun Valley, Idaho, USA. pp. 92–101.
6. Wilson S, Hoffman JD, Genoways HH (2010) Observations of reproduction in mountain lions from Nebraska. Western North American Naturalist 70: 238–240.
7. Sweanor LL, Logan KA, Hornocker MG (2000) Cougar dispersal patterns, metapopulation dynamics and conservation. Conservation Biology 14: 798–808.
8. Thompson DJ, Jenks JA (2005) Long-distance dispersal by a subadult male cougar from the Black Hills, South Dakota. Journal of Wildlife Management 69: 818–820.
9. Logan KA, Sweanor LL (2010) Behavior and social organization of a solitary carnivore. In: Hornocker M, Negri S, editors. Cougar: ecology and conservation. Chicago, Ilinois, USA: University of Chicago Press. pp. 105–118.
10. Wiedenhoeft JE, Wydeven AP, Bruner J (2012) Rare mammal observations 2011. 122–135 p.
11. Wisconsin Department of Natural Resources (2013) Cougars in Wisconsin. Available: http://dnr.wi.gov/topic/wildlifehabitat/cougar.html. Accessed 2013 June 1.
12. Rosatte R (2011) Evidence confirms the presence of cougars (Puma concolor) in Ontario, Canada. Canadian Field-Naturalist 125: 116–125.
13. LaRue MA, Nielsen CK, Dowling M, Miller K, Wilson B, et al. (2012) Cougars are recolonizing the midwest: Analysis of cougar confirmations during 1990–2008. Journal of Wildlife Management 76: 1364–1369.
14. Henaux V, Powell LA, Hobson KA, Nielsen CK, LaRue MA (2011) Tracking large carnivore dispersal using isotopic clues in claws: an application to cougars across the Great Plains. Methods in Ecology and Evolution 2: 489–499.
15. Sweanor LL, Logan KA (2010) Cougar-human interactions. In: Hornocker M, Negri S, editors. Cougar: ecology and conservation. Chicago, Illinois, USA: University of Chicago Press. pp. 190–205.
16. LaRue MA, Nielsen CK (2011) Modelling potential habitat for cougars in midwestern North America. Ecological Modelling 222: 897–900.
17. Mladenoff DJ, Haight RG, Sickley TA, Wydeven AP (1997) Causes and implications of species restoration in altered ecosystems. Bioscience 47: 21–31.
18. Mladenoff DJ, Sickley TA (1998) Assessing potential gray wolf restoration in the northeastern United States: A spatial prediction of favorable habitat and potential population levels. Journal of Wildlife Management 62: 1–10.
19. Potvin MJ, Drummer TD, Vucetich JA, Beyer DE, Peterson RO, et al. (2005) Monitoring and habitat analysis for wolves in upper Michigan. Journal of Wildlife Management 69: 1660–1669.
20. LaRue MA (2007) Predicting potential habitat and dispersal corridors for cougars in midwestern North America: Southern Illinois University.
21. Clark DW, White SC, Bowers AK, Lucio LD, Heidt GA (2002) A survey of recent accounts of the mountain lion (Puma concolor) in Arkansas. Southeastern Naturalist 1: 269–278.
22. Johnson K (2002) The mountain lions of Michigan. Endangered Species UPDATE 19: 27–31.
23. Wisconsin Cougar Working Group (2010) Draft Wisconsin Cougar Response Protocol. Available: http://www.legis.state.wi.us/senate/sen17/news/Press/2010/documents/CougarResponseProtocolforWI2010a.pdf. Accessed 2010 Sept 21.
24. Linden DW, Campa H, Roloff GJ, Beyer DE, Millenbah KF (2011) Modeling habitat potential for Canada lynx in Michigan. Wildlife Society Bulletin 35: 20–26.
25. Laundré JW (2013) The feasibility of the north-eastern USA supporting the return of the cougar Puma concolor. Oryx 47: 96–104.
26. LaRue MA, Nielsen CK (2008) Modelling potential dispersal corridors for cougars in midwestern North America using least-cost path methods. Ecological Modelling 212: 372–381.
27. Dickson BG, Jenness JS, Beier P (2005) Influence of vegetation, topography, and roads on cougar movement in southern California. Journal of Wildlife Management 69: 264–276.
28. Kertson BN, Spencer RD, Marzluff JM, Hepinstall-Cymerman J, Grue CE (2011) Cougar space use and movements in the wildland-urban landscape of western Washington. Ecological Applications 21: 2866–2881.
29. Teichman KJ, Cristescu B, Nielsen SE (2013) Does Sex Matter? Temporal and Spatial Patterns of Cougar-Human Conflict in British Columbia. Plos One 8.
30. Donovan ML, Nesslage GM, Skillen JJ, Maurer BA (2004) The Michigan GAP analysis final report. Lansing, MI: Wildlife Division, Michigan Department of Natural Resources.
31. Saaty TL (1980) The analytic hierarchy process: planning, priority setting, resources allocation. New York, New York, USA: McGraw-Hill International Book Company.
32. Wisconsin Cougar Working Group (2010) Draft Wisconsin Cougar Response Protocol. Available: http://www.legis.state.wi.us/senate/sen17/news/Press/2010/documents/CougarResponseProtocolforWI2010a.pdf. Accessed 2010 Sept 21.
33. Wilmers CC, Wang Y, Nickel B, Houghtaling P, Shakeri Y, et al. (2013) Scale Dependent Behavioral Responses to Human Development by a Large Predator, the Puma. Plos One 8: e60590.
34. Wiedenhoeft JE, Wydeven AP (2009) Rare mammal observations 2008. Wisconsin Wildlife Surveys April 2009, 19: 83–95.
35. Wiedenhoeft JE, Wydeven AP (2010) Rare mammal observations 2009. Wisconsin Wildlife Surveys 20: 103–113.
36. Wiedenhoeft JE, Wydeven AP, Bruner J (2013) Rare carnivore observations 2012. Wisconsin Wildlife Surveys 23: 93–102.
37. Wiedenhoeft JE, Wydeven AP, Bruner J (2011) Rare mammal observations 2010. Wisconsin Wildlife Surveys 21: 99–113.
38. Anderson EM, Wydeven AP, Holsman R (2006) Distribution of cougar observations in Wisconsin 1994–2003. In: McGrinnis HJ, Tischendorf JW, Ropski SJ, editors. Proceedings of the Eastern Cougar Conference, 28-April 2004 -1 May 2004, Morgantown, West Virginia, USA.
39. Beier P (1995) Dispersal of juvenile cougars in fragmented habitat. The Journal of Wildlife Management: 228–237.
40. Beier P, Choate D, Barrett RH (1995) Movement patterns of mountain lions during different behaviors. Journal of Mammalogy: 1056–1070.
41. McDonald L, Manly B, Huettmann F, Thogmartin W (2013) Location-only and use-availability data: analysis methods converge. Journal of Animal Ecology 82: 1120–1124.
42. Northrup JM, Hooten MB, Anderson CR, Wittemyer G (2013) Practical guidance on characterizing availability in resource selection functions under a use-availability design. Ecology 94: 1456–1463.
43. Boyce MS, McDonald LL (1999) Relating populations to habitats using resource selection functions. Trends in Ecology & Evolution 14: 268–272.
44. Johnson CJ, Nielsen SE, Merrill EH, McDonald TL, Boyce MS (2006) Resource selection functions based on use-availability data: Theoretical motivation and evaluation methods. Journal of Wildlife Management 70: 347–357.
45. Conkin JA, Alisauskas RT (2013) Modeling probability of waterfowl encounters from satellite imagery of habitat in the central Canadian arctic. Journal of Wildlife Management 77: 931–946.
46. Robin X, Turck N, Hainard A, Tiberti N, Lisacek F, et al. (2011) pROC: an open-source package for R and S plus to analyze and compare ROC curves. Bmc Bioinformatics 12.
47. Karanth KU, Nichols JD, Kumar NS, Link WA, Hines JE (2004) Tigers and their prey: predicting carnivore densities from prey abundance. Proceedings of the National Academy of Sciences of the United States of America 101: 4854–4858.
48. Wisconsin Department of Natural Resources (2013) Estimated Fall Deer Population, 2012. Madison, Wisconsin, USA: Wisconsin Department of Natural Resources, GIS Services Section. Available: http://dnr.wi.gov/topic/hunt/documents/fallabund.pdf. Accessed 2014 June 1.
49. Michigan Department of Natural Resources (2013) Clickable Map of Proposed 2006–2010 Deer Population Goals. Available: http://www.michigan.gov/dnr/0,1607,7-153-10363_10856_10905-129948-542,00.html. Accessed 2014 June 1.
50. Murphy K, Ruth TK (2010) Diet and prey selection of a perfect predator. In: Hornocker M, Negri S, editors. Cougar: ecology and conservation. Chicago, Illinois, USA: University of Chicago Press. pp. 118–137.
51. DelGiudice GD, McCaffery KR, Beyer Jr DE, Nelson ME (2009) Prey of wolves in the Great Lakes region. In: Wydeven AP, Deelen TR, Heske EJ, editors. Recovery of gray wolves in the Great Lakes Region of the United States: Springer New York. pp. 155–173.

52. Halls LK (1984) White-tailed deer: ecology and management. Harrisburg, Pennsylvania, USA: Stackpole Books.
53. Geist V (1998) Deer of the world: their evolution, behaviour and ecology. Harrisburg, Pennsylvania, USA: Stackpole Books.
54. Verme LJ (1989) Maternal investment in white-tailed deer. Journal of Mammalogy 70: 438–442.
55. Carbone C, Gittleman JL (2002) A common rule for the scaling of carnivore density. Science 295: 2273–2276.
56. Smallwood KS (1997) Interpreting puma (Puma concolor) population estimates for theory and management. Environmental Conservation 24: 283–289.
57. Lambert CMS, Wielgus RB, Robinson HS, Katnik DD, Cruickshank HS, et al. (2006) Cougar population dynamics and viability in the Pacific Northwest. Journal of Wildlife Management 70: 246–254.
58. Quigley H, Hornocker M (2010) Cougar population dynamics. In: Hornocker M, Negri S, editors. Cougar: ecology and conservation. Chicago, Illinois, USA: University of Chicago Press. pp. 59–75.
59. Ruth TK, Murphy K (2010) Competition with other carnivores for prey. In: Hornocker M, Negri S, editors. Cougar: ecology and conservation. Chicago, Illinois, USA: University of Chicago Press. pp. 163–172.
60. Washington Department of Fish and Wildlife (2008) Washington Department of Fish and Wildlife 2009–2015 Game Management Plan. Olympia, Washington, USA: Washington Department of Fish and Wildlife.
61. Glick HB (2014) Modeling cougar habitat in the Northeastern United States. Ecological Modelling 285: 78–89.
62. Smith JB, Nielsen CK, Hellgren EC (2014) Illinois resident attitudes toward recolonizing large carnivores. Journal of Wildlife Management 78: 930–943.
63. Murray BD, Webster CR, Bump JK (2013) Broadening the ecological context of ungulate-ecosystem interactions: the importance of space, seasonality, and nitrogen. Ecology 94: 1317–1326.
64. Beyer DE, Peterson RO, Vucetich JA, Hammill JH (2009) Wolf population changes in Michigan. In: Wydeven AP, Deelen TR, Heske EJ, editors. Recovery of gray wolves in the Great Lakes Region of the United States: an endangered species success story. : Springer New York. pp. 65–85.
65. Wydeven AP, Wiedenhoeft JE, Schultz RN, Thiel RP, Jurewicz RL, et al. (2009) History, population growth, and management of wolves in Wisconsin. In: Wydeven AP, Deelen TR, Heske EJ, editors. Recovery of gray wolves in the Great Lakes Region of the United States: an endangered species success story. : Springer New York. pp. 87–105.
66. Palomares F, Delibes M, Ferreras P, Fedriani JM, Calzada J, et al. (2000) Iberian lynx in a fragmented landscape: predispersal, dispersal, and postdispersal habitats. Conservation Biology 14: 809–818.
67. Pearce JL, Boyce MS (2006) Modelling distribution and abundance with presence-only data. Journal of Applied Ecology 43: 405–412.
68. Hastie T, Fithian W (2013) Inference from presence-only data; the ongoing controversy. Ecography 36: 864–867.
69. Beier P (2010) A focal species for conservation planning. In: Hornocker M, Negri S, editors. Cougar: ecology and conservation. Chicago, Illinois, USA: University of Chicago Press. pp. 177–189.
70. Davenport MA, Nielsen CK, Mangun JC (2010) Attitudes toward mountain lion management in the Midwest: implications for a potentially recolonizing large predator. Human Dimensions of Wildlife 15: 373–388.
71. Corona Research I (2006) Public opinions and perceptions of mountain lion issues. Denver, Colorado, USA: Corona Research, Inc.
72. Treves A (2009) Hunting for large carnivore conservation. Journal of Applied Ecology 46: 1350–1356.
73. Vucetich JA, Nelson MP (2014) Wolf Hunting and the Ethics of Predator Control. In: Kalof L, editor. The Oxford Handbook of Animal Studies. Oxford Handbooks Online: Oxford University Press, Oxford, UK.
74. Treves A, Naughton-Treves L, Shelley V (2013) Longitudinal Analysis of Attitudes Toward Wolves. Conservation Biology 27: 315–323.
75. Cougar Management Guidelines Working Group (2005) Cougar management guidelines. Wild Futures, Bainbridge Island, Washington, USA.
76. U.S. Fish and Wildlife Service (1982) Eastern cougar recovery plan. Atlanta, Georgia, USA: U.S. Fish and Wildlife Service.
77. Ripple WJ, Estes JA, Beschta RL, Wilmers CC, Ritchie EG, et al. (2014) Status and ecological effects of the world's largest carnivores. Science 343: 1241484.

20

Unusual Circulation Patterns of the Rias Baixas Induced by Minho Freshwater Intrusion (NW of the Iberian Peninsula)

Magda Catarina Sousa[1], Renato Mendes[1], Ines Alvarez[1,2], Nuno Vaz[1], Moncho Gomez-Gesteira[2], João Miguel Dias[1]*

1 CESAM, Departamento de Física, Universidade de Aveiro, Campus de Santiago, Aveiro, Portugal, **2** EPhysLab (Environmental Physics Laboratory), Universidade de Vigo, Facultade de Ciencias, Ourense, Spain

Abstract

The Minho River, situated 30 km south of the Rias Baixas, is the most important freshwater source flowing into the Western Galician coast (NW of the Iberian Peninsula). The buoyancy generated by the Minho estuarine plume can reverse the normal circulation pattern inside the Rias Baixas affecting the exchange between the Rias and the ocean, changing the input of nutrients. Nevertheless, this inversion of the circulation patterns is not a well-monitored phenomenon. The only published results based on *in situ* data related to the presence of the Minho River plume inside the Rias de Vigo and Pontevedra correspond to an event measured on spring 1998. In this case unexpectedly higher inflow surface current velocities were found at the Ria de Pontevedra, located further away from Minho River. Thus, the main aim of this study is to research the main factors inducing this unusual pattern on the circulation of the Rias de Vigo and Pontevedra. A numerical model implementation of MOHID previously developed, calibrated, and validated for this coastal area was used. Several scenarios were performed in order to explain the individual effect of the Minho River, rivers discharging into each Rias, and estuarine morphology changes. According to the model results, the Minho River discharge is a key factor in the establishment of the negative circulation, while small rivers inside the Rias slightly attenuate this circulation. The negative circulation was stronger in Ria de Pontevedra independently of the distance of this coastal system from the Minho River mouth, showing that morphologic estuarine features are the main factor justifying the different local circulation patterns.

Editor: Vanesa Magar, Centro de Investigacion Cientifica y Educacion Superior de Ensenada, Mexico

Funding: RM has been supported by the Portuguese Science Foundation (FCT) through a doctoral grant (SFRH/BD/79555/2011). IA and NV are supported by the Ramon y Cajal and Ciência 2008 programs, respectively. This work has been supported by FCT and by the European Union (COMPETE, QREN, FEDER) in the frame of the research project Pest (C/MAR/LA0017/2013) and partially supported by Xunta de Galicia under projects "Programa de Consolidación e Estructuración de Unidades de Investigación Competitivas: Grupos de Referencia Competitiva" (GR2013-001) co-funded by European Regional Development Fund (FEDER), EM2013/003, and "Galician Internal Basin Paleoclimatic Sedimentary Archives: Transport, Provenances, Hydrology and reliability (GALINCLIMARCH - CGL2010-16688/BTE). The funders had no role in study design, data collection and analysis, decision to publish, or preparation of the manuscript.

* Email: joao.dias@ua.pt

Introduction

Major rivers inject freshwater onto the adjacent shelf where mixing of these river plumes takes place, affecting the transport and transformation of dissolved and particulate materials in the coast. The coastal waters freshening due to river runoff is one of the mechanisms that control the density inside the estuaries located up north or south of the river mouth, modifying the circulation and the macronutrients concentrations available for production [1–3]. Numerous studies conducted over diverse coastal systems around the world have revealed a major interest in the ecological consequences of freshwater intrusion on nearby estuaries (e. g. Columbia River [1], Delaware River [3] or Mississippi River [2]). These studies showed that the river plumes from major rivers produce currents in the surface layers of neighbor estuaries, affecting local species and marsh plants.

The NW coast of the Iberian Peninsula is characterized by the presence of several embayments, usually called "Rias". The term "ria" defines a coastal shape correspondent to re-entrance, resultant from the submersion by the sea of the terminal zone of a fluvial network [4]. The Rias Baixas (Figure 1) are located in the northern limit of the Eastern North Atlantic Upwelling System [5]. They are heavy populated ecosystems of enormous potential, both economical (e.g. exploration of marine resources: fisheries, aquaculture and fishing industry) and social (e.g. tourism: beaches and natural beauty). Wind driven coastal upwelling is the main recognized local source of primary production, supporting the high fishery and aquaculture yields [6,7].

Located 30 km southward, the Minho River is the most important river flowing near this coastal region. It has a catchment area of 17080 km^2 and an annual average discharge of 300 m^{3}s^{-1}. The monthly average discharge oscillates between 100 m^{3}s^{-1} in

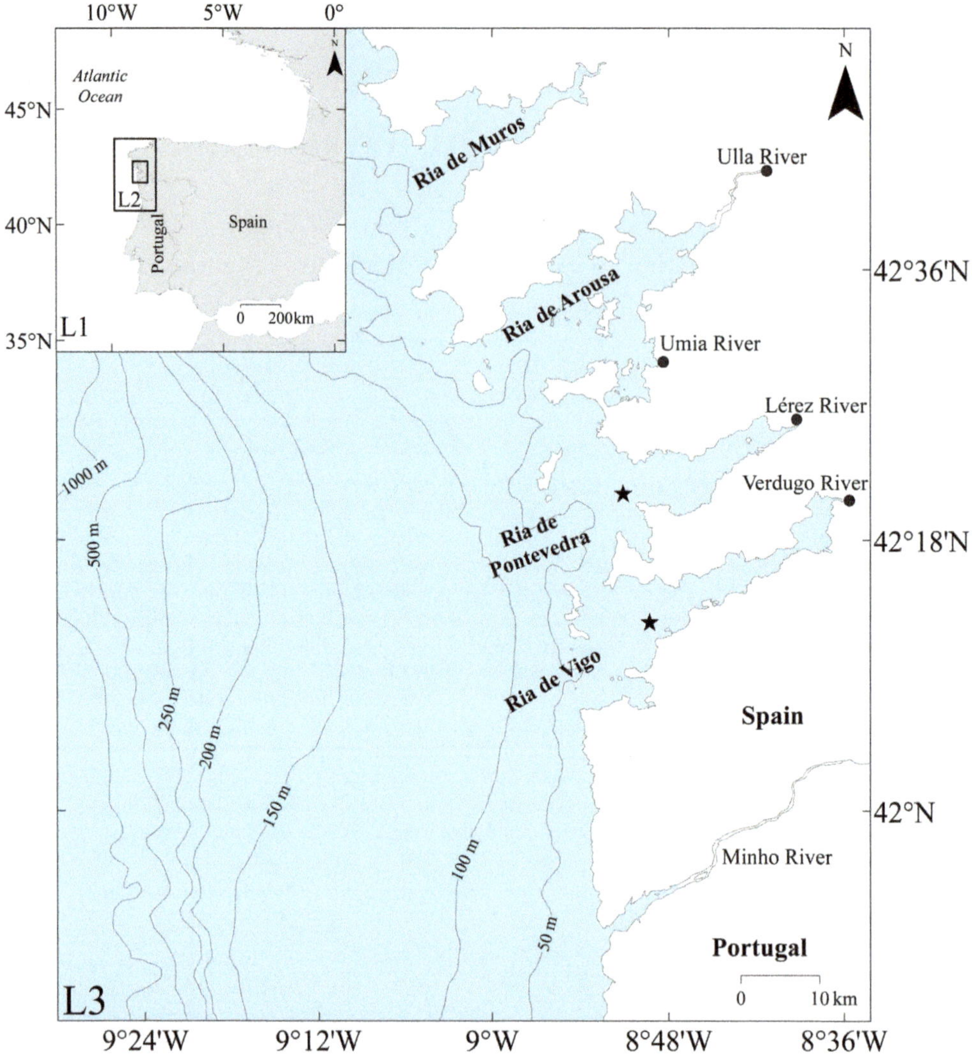

Figure 1. Nested grids and Rias Baixas map.

August and 800 m^3s^{-1} in February [8]. Essentially, during high river discharge and southerly wind conditions, its buoyant plume generates an important salinity decrease at the Rias Baixas mouth, reversing the normal circulation pattern [9–11]. This reverse in the circulation pattern (negative circulation) is characterized by the intrusion of less dense coastal waters at the surface, which accumulate within the Rias and finally flow towards the ocean through the bottom [9,10,12–14]. This inversion of the normal circulation patterns is not a well-documented phenomenon. The infrequent sampling in the inner and outer part of the Rias complicates the characterization of such events through analysis of field measurements. In fact, the only published results related to the presence of the Minho River plume inside the Rias Baixas correspond to an event measured on spring 1998 when a high Minho River discharge was reported as well as southerly wind conditions crucial to spread the river plume towards the Rias Baixas, generating a negative/reverse circulation inside them.

Alternatively, the use of numerical models constitutes a reliable and useful tool to analyze this type of events. The numerical models are used worldwide in order to quantify and understand different processes occurring in coastal waters [15–19]. They allow to investigate hypothetical scenarios (e.g. changes in wind and river discharge), which observations alone cannot provide. In addition, they afford the capability to study processes independently. Using a numerical model, Sousa et al. [14] reproduced the experimental campaign conducted on spring 1998 inside the Rias Baixas. They observed that under northward winds, a continuous moderate Minho River discharge (>700 m^3 s^{-1}) is enough to produce the negative circulation pattern in Rias de Vigo and Pontevedra, reducing the importance of the existence of specific events of high river runoff. However, unexpectedly, higher inflow surface current velocities were found at the Ria located further away from Minho River (at Ria de Pontevedra rather than at Ria de Vigo, which is close to Minho River)). However, the explanation of these velocity differences was not evaluated in this previous study.

Considering these issues the present study aims to research the main reasons that induce this unusual pattern in the circulation in Rias de Vigo and Pontevedra on May 1998 event, through the application of a numerical model using a downscaling approach [11,14]. Several hypotheses were formulated to justify this pattern, namely the impact of different freshwater inflows at the head of each Ria, comparing to the impact of the Minho River and the Rias morphology, that have different width and depth and mouths with different features that may constrain the water exchange with the shelf. To test these hypotheses and explain the different current velocities observed at both Rias were defined several numerical scenarios. In the first, the real conditions corresponding to May 1998 are considered. The second and third scenarios are used in order to evaluate the influence of the Minho River and the rivers flowing into the Rias. The influence of estuarine morphology is evaluated in the last scenario, interchanging the location of the Rias de Vigo and Pontevedra.

The paper is organized as follows. In the second section, the description of study area is presented. The third part presents the general overview of the numerical model as well as the numerical experimental design. Results and discussion can be found in the fourth part. Finally, conclusions are drawn in the fifth part.

Study area

The study region comprises only two Galician Rias: the Rias de Vigo and Pontevedra, which are the closest estuarine regions located up north of the Minho Estuary.

The Ria de Vigo, which is the most meridional of the Rias Baixas is located around 30 km northwards of the Minho River mouth (Figure 1). This Ria is 32.5 km long, presenting an NE-SW direction, 1 km wide in its inner part (NE) and 10 km at the mouth of the Ria (SW). The mean width and depth are 4.8 km and 21 m, respectively. The mouth of this Ria is divided in two entrances defined by a set of islands, which allow the connection to the open sea. The northern entrance is 2.8 km wide and has a maximum depth of 25 m and the southern one is 5.1 km wide and 45 m depth. The Verdugo River is the main tributary of freshwater into Ria de Vigo with a catchment area of 333 km^2. Its annual mean annual discharge is 13 m^3s^{-1} with a greatest seasonal variability (values range from 120 m^3s^{-1} in winter to 1 m^3s^{-1} in summer) [20].

The Ria de Pontevedra is located northward of Ria de Vigo (Figure 1). This Ria is oriented in the NE-SW direction. The mean depth and width are 31 m and 3.8 km, respectively. It is also connected to the ocean by two entrances. The northern one is narrower (3.6 km) and shallower (15 m), while the southern mouth is wider (7.3 km) and deeper (60 m). The Lérez River is the main freshwater source that flows into Ria de Pontevedra. This river has a catchment area of 450 km^2 and an annual mean discharge of 27.5 m^3s^{-1}. The monthly mean discharge oscillates between 2 and 80 m^3s^{-1} and follows a similar pattern to the rainfall, as dams do not control its runoff.

Concerning the tidal forcing, both Rias are estuarine-like systems that have a semi-diurnal tidal regime. The tidal range varies between 2 and 4 m (mesotidal regime) and a significantly low Form Number of 0.25 is found [21]. The Rias behave normally as partially mixed estuaries, with positive residual circulation and two layer circulation pattern, with surface outflow and inflow on the bottom [22–24], although a reverse circulation may be found as previously referred.

Methodology

Numerical model

In this work was used the model MOHID (www.mohid.com), previously implemented and validated for the Rias Baixas by Sousa et al. [14].

The model implementation consists in a set of three-one-way nested models (Figure 1). The first domain is a barotropic tidal driven model using the FES2004 (Finite Element Solution) global solution as forcing [25]. It includes the whole Iberian Peninsula coast (L1, Figure 1), with a horizontal resolution of 0.06°. The ocean boundary conditions were given in cascade starting at the first level. Therefore, the 2D barotropic model is only used to predict the external tidal conditions necessary to feed the L2 (Figure 1) baroclinic model with surface elevation.

The second and third domains are baroclinic models (L2 and L3, Figure 1), being the second domain a coarse representation (horizontal resolution of 0.02°) of the Rias Baixas and adjacent coastal area, while the third includes the same area with a higher resolution (horizontal resolution of 0.005°) and it is directly coupled to L2 at the open boundaries. A *z*-level vertical discretization was adopted, with L2 and L3 having 46 and 42 vertical layers, where the bottom 39 (L2) and 35 (L3) were in Cartesian coordinates and the top 10 m were 7 sigma coordinate layers.

To obtain the initial ocean stratification, L2 and L3 were forced at the open boundaries with monthly mean climatologies profiles from Levitus [26,27].

The surface boundary condition was imposed using hourly high-resolution results from the Weather Research and Forecasting model with a spatial resolution of 4 km. At the surface, heat fluxes were imposed with parameterizations similar to those described by Chapra [28]. The sensible and latent heat fluxes were calculated using the Bowen and Dalton laws, respectively [28]. For the bottom boundary condition, shear friction was imposed, assuming a velocity logarithmic profile.

The freshwater inputs inside Rias Baixas and the Minho estuary outflow were considered as landward boundary conditions (L3). This last was imposed offline in L3 as momentum, water and mass discharges to the coastal model. This outflow was computed with a 0.0065°–0.001° horizontal resolution estuarine model for the inner part of the Minho River [14].

The model was run from November 1997 to May 1998 (the first six-months were considered as spin-up).

A more detailed description of this numerical implementation can be found in Sousa et al. [14].

Model scenarios

Four simulations were carried out in order to evaluate the main factors that induce the unusual pattern on May 1998 (Table 1). In the first scenario, the real conditions corresponding to May 1998 were considered (called reference scenario hereafter). The second simulation was run without Minho River discharge during the whole simulation period (called without Minho River scenarios hereafter), but including the freshwater discharges inside each Ria. In the third scenario, the influence of the rivers flowing into the Rias was analyzed. Thus, no discharges from rivers inside the Rias were considered (called without inside rivers scenario hereafter), but was considered the Minho River discharge. With respect to the influence of estuarine morphology, a new bathymetric grid was built interchanging the position of the entire estuarine areas in study, from the mouth to the easternmost grid point of both Rias (inclusive the islands) (Figure 2). This new grid was considered in the last scenario (called interchanged Rias scenario hereafter), where the Ria de Pontevedra is nearest to Minho River instead of the Ria de Vigo (Figure 2b). Despite this interchanging, each Ria preserves its freshwater input, i.e. Verdugo River discharges to the Ria de Vigo and Lérez River to the Ria de Pontevedra.

Three rivers were imposed as land boundaries for the simulations period: Minho, Verdugo, and Lérez (Figure 1). The Minho River discharge shows an atypical pattern with high values during early May (Figure 3). The maximum value is about 1600 m^3s^{-1}. After this date, a continuous decrease of the river discharge is observed. The Verdugo and Lérez Rivers present a similar pattern with a small runoff decrease in late May, except

Table 1. Model runs overview.

Name	*Minho River discharge*	*Verdugo River discharge*	*Lérez River discharge*
Reference	X	X	X
Without Minho river	-	X	X
Without Inside river	X	-	-
Interchanged Rias	X	X	X

between 29 April and 3 May when the Verdugo River shows high values (250 m^3s^{-1}, Figure 3).

To accomplish the main objective of this work, horizontal maps of the surface density were taken in the instant that Minho River plume reaches the Rias de Vigo and Pontevedra. They were analyzed for the reference and interchanged Rias scenarios. Moreover, the temporal evolution of vertical estuarine circulation at stations located at mouths of Rias de Vigo (~30 m deep) and Pontevedra (~35 m deep) (Figure 1, stars) were calculated for the same scenarios. The residual current was also computed for all scenarios by time averaging the hourly transient values from current velocities at mouth of both Rias (Figure 1, stars) over a period of 13 hours, corresponding to the plume intrusion.

Results and Discussion

In this section, the impact of the Minho River plume and the Rias de Vigo and Ponetevedra main freshwater sources in the establishment of negative circulation in the Rias will be evaluated. In addition, the impact of changes in the position of the Rias (e.g. estuarine morphology *versus* distance to Minho River) on estuarine circulation will be evaluated for the period of May 1998.

Real conditions in May 1998

Surface density maps show a northward spread of the Minho River plume reaching the Ria de Vigo on 11 May (Figure 4a). During the following days, the northward displacement of the plume continued and the plume intrusion is observed in Ria de Pontevedra on 13 May (Figure 4b). This situation generates an unusual surface density pattern influencing the physical properties within the Rias, reversing the normal density pattern characterized by the presence of saltier water at the outer part of the estuary and freshwater at the middle-inner part. These unusual inflows of freshwater are critical from an ecological point of view, affecting the exchange between the Rias and the ocean, and therefore changing the normal input of nutrients [9,10]. Thus, this intrusion into the Rias can impose a control on new production within the estuary environment and changes in the mortality of halophyte organisms inside an estuary [29]. Changes in estuarine density gradients can also affect higher trophic levels, particularly the distribution of fishes in estuaries [2]. These consequences demonstrate the high interest in characterizing these types of events in detail.

In terms of estuarine circulation, the existence of an important external freshwater supply at the mouth of an estuary can generate

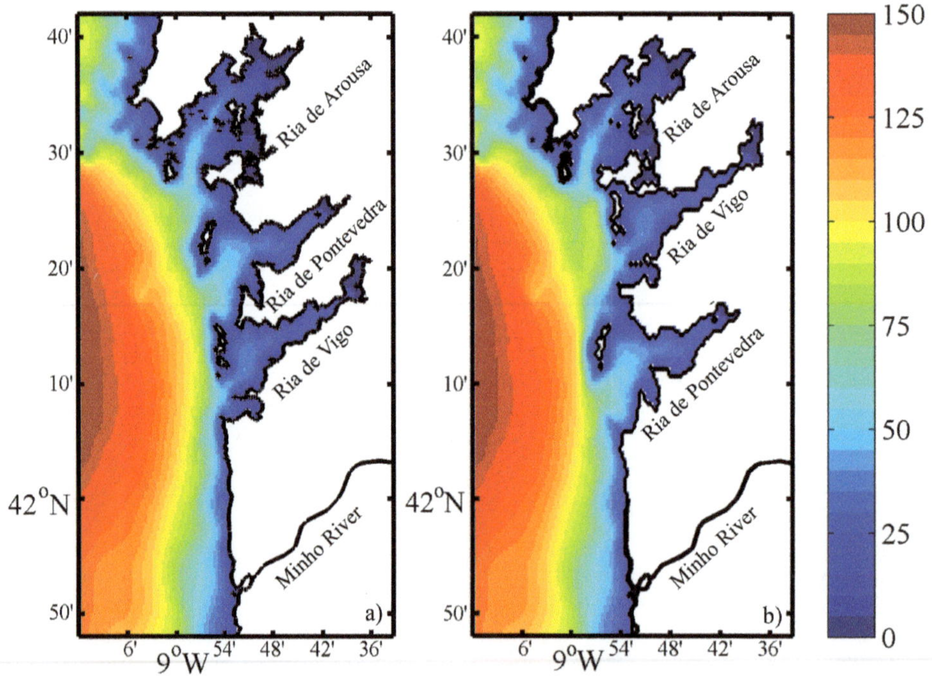

Figure 2. Reference (a) and interchanged (b) bathymetric grids. To simplify visual comparison the Verdugo and Lérez rivers were multiplied by 5.

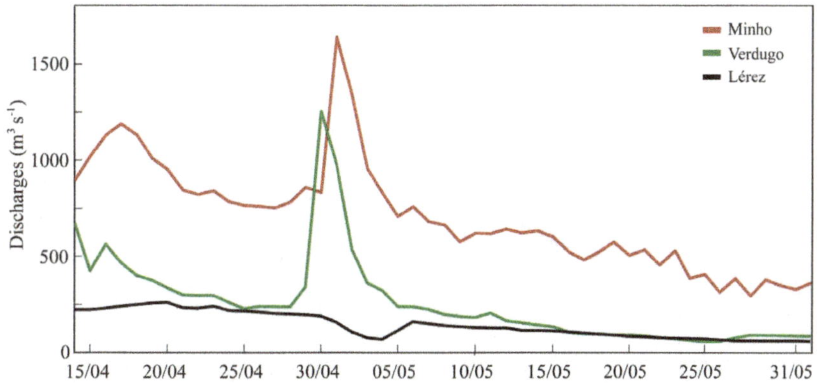

Figure 3. Time series of Minho, Verdugo and Lérez River discharges from 15 April to 31 May 1998.

the reversal of the normal estuarine circulation [14]. Thereby, the temporal evolution of vertical estuarine circulation at stations located at mouths of Rias de Vigo and Pontevedra (Figure 1, stars) between 11 and 16 May 1998 is also studied through the analysis of the along axis velocity component vertical profiles (Figure 5). In Ria de Vigo (Figure 5a), between 11 and 13 May, a negative circulation (upstream circulation) at upper layers (0–8 m) is observed. There is a landward water movement with highest velocities (0.25 m s^{-1}) between 0–2 m depth and a seaward movement at bottom layers. Positive surface velocities are more intense than the negative bottom ones, corresponding to the intrusion of the Minho estuarine outflow into the Ria de Vigo [14].

A similar pattern is recognized in Ria de Pontevedra (Figure 5b), although the surface water moving from ocean to land is observed later (13–15 May). In addition, this unusual circulation is more intense (0.30 m s^{-1}) than in Ria de Vigo, with positive values until 10 m depth. In fact, around 13–14 May surface water moving landward was observed during a tidal cycle. Freshwater supplied by the Minho River generates this unusual circulation, which tends to stop water exchange between the Rias and the shelf, increasing its residence time and consequently changing the water quality [10]. This inversion on estuarine circulation may introduce into the Rias dinoflagellate blooms, which are usually generated in coastal waters, decreasing the abundance of marine species in this area [30].

Due to the ecological importance of estuaries and other coastal ecosystems, external freshwater intrusion inside estuaries has also been studied worldwide, as usually as consequence of remote river plumes spreading. As an example, Monteiro and Largier [31] and Hickey et al. [32] showed that external sources of freshwater can

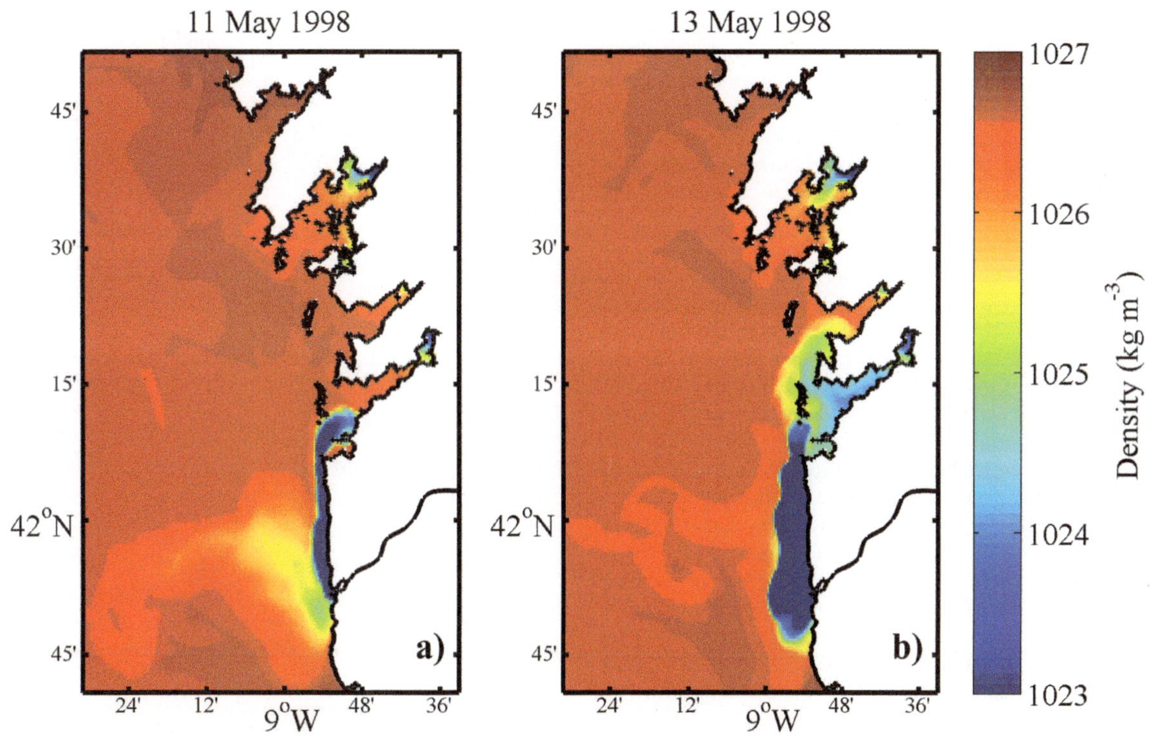

Figure 4. Surface density maps, considering the reference scenario.

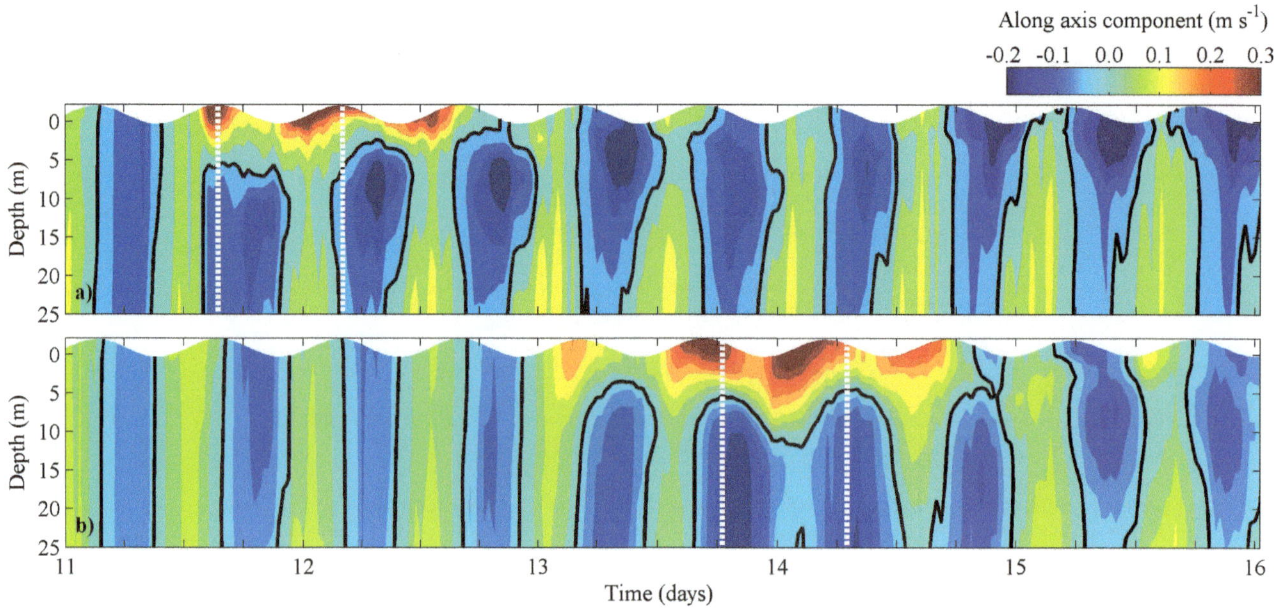

Figure 5. Temporal evolution of vertical profiles of along velocity component at stations located at mouths of Rias de Vigo (b) and Pontevedra (d), considering the reference scenario. Black line corresponds to 0 m s^{-1}.

have an higher impact on the mean baroclinic circulation inside the Saldanha and Willapa bays than usual estuarine processes (ex: changes in river flow or neap-spring variation in mixing). These results are consistent with the results presented here for the Rias de Vigo and Pontevedra.

Effect of Minho River and rivers flowing into the Rias

Residual circulation is crucial in order to determine the long-term transport in estuaries and therefore a key parameter in the dynamical behavior during the intrusion of the Minho estuarine plume on Rias Baixas. Thus, the residual currents induced by the plume intrusion are presented in Figure 6 for the reference, without Minho River, and inside rivers scenarios. The pattern of the vertical structure of velocity is very similar in both Rias concerning the reference scenario (Figure 6, red line). The inversion of circulation occurs at the same depth (~10 m). Both in the surface and in the bottom, the highest velocities are observed at the Ria located further away from Minho River (Ria de Pontevedra). Without Minho River discharge (Figure 6, grey line), lower residual velocities are observed for both Rias, showing that the Minho River discharge is the main responsible for the inversion of the circulation pattern and for the non-tidal estuarine transport of dissolved and particulate materials inside the Rias. For the third scenario (removing the rivers flowing into the Rias, Figure 6, blue line), the residual current profile is similar to the one corresponding to the reference simulation. A greater difference is observed on the surface layer for both Rias. The velocity is higher (about 0.02 m s^{-1}) than for the reference scenario, showing that if rivers flowing into the Rias are neglected, the Minho estuarine plume would affect more significantly the surface circulation of the Rias. These results also demonstrate that the unusual higher surface current velocities found at the Ria de Pontevedra are not induced by the internal freshwater discharges characteristics.

These results show that the Minho River discharge plays an important role in the establishment of the negative circulation inside the Rias. On the contrary, the small rivers that flow through the interior of the Rias do not influence significantly the inversion of the circulation. These findings are in accordance with the results reported for other regions worldwide, as Willapa Bay [1], where under favorable downwelling winds, the plume from Columbia River forces an axial gradient reversal inside the bay.

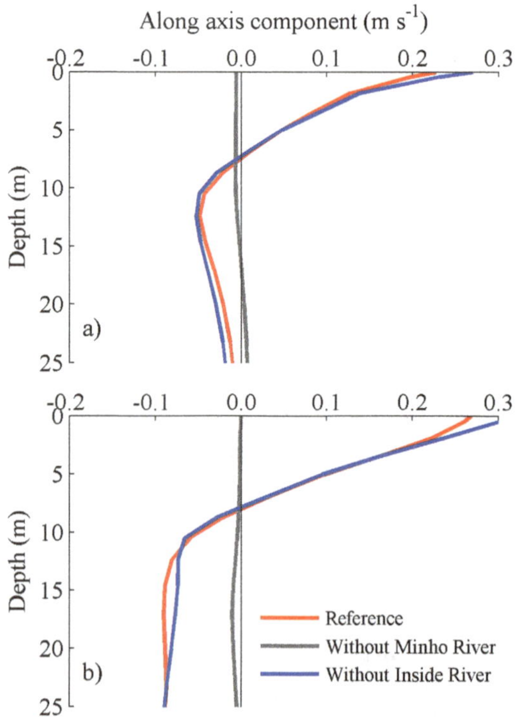

Figure 6. Residual current profiles during the plume intrusion (dashed lines, Figure 4) at stations located at mouths of Rias de Vigo (a) and Pontevedra (b), considering the different scenarios.

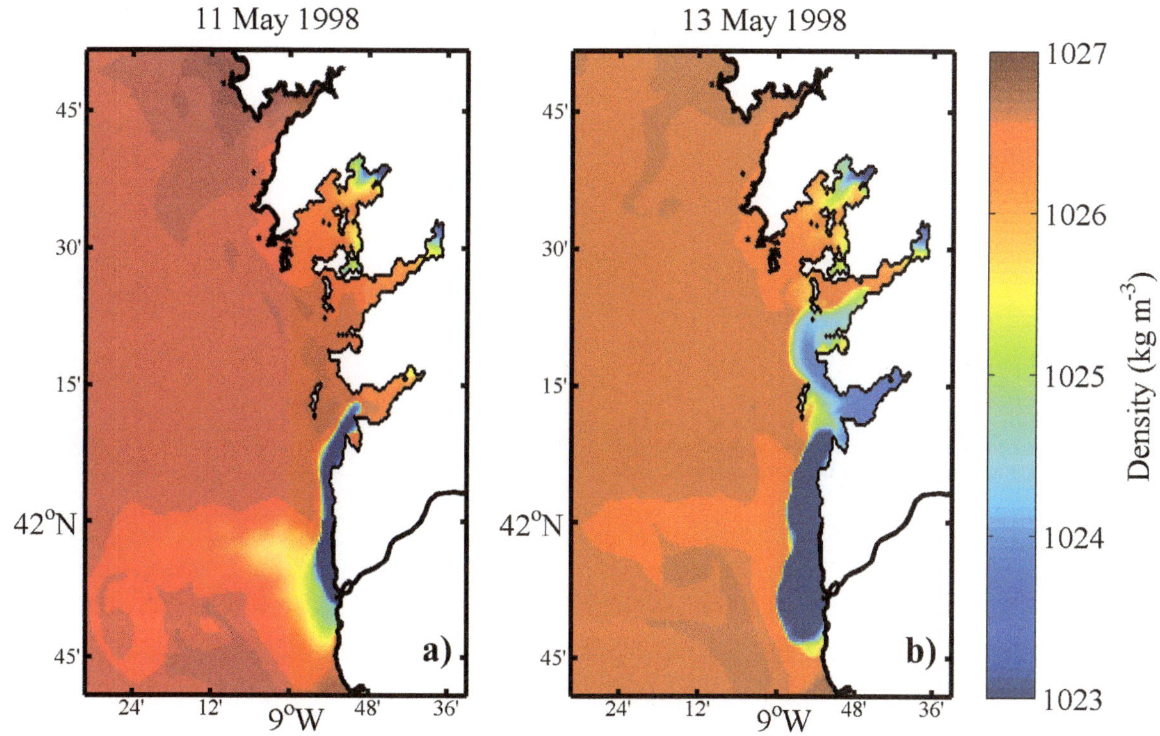

Figure 7. Surface density maps, considering the interchanged Rias scenario.

Effect of estuarine morphology

Comparing the results for each Ria independently, it is observed that the integrated residual currents, both at surface and bottom, are greater in Ria de Pontevedra (that is further way of Minho River) than in Ria de Vigo for all scenarios. For example, in the reference scenario (Figure 6, red line), the surface velocity in Ria de Pontevedra is 0.18 m s^{-1} (Figure 6b) while in Ria de Vigo is 0.12 m s^{-1} (Figure 6a). This could (or not) be related to differences in the morphology of each Ria (width and depth as well as different mouth geometries). In order to better analyze this situation the influence of the estuary morphology was evaluated by interchanging the location of Rias de Vigo and Pontevedra

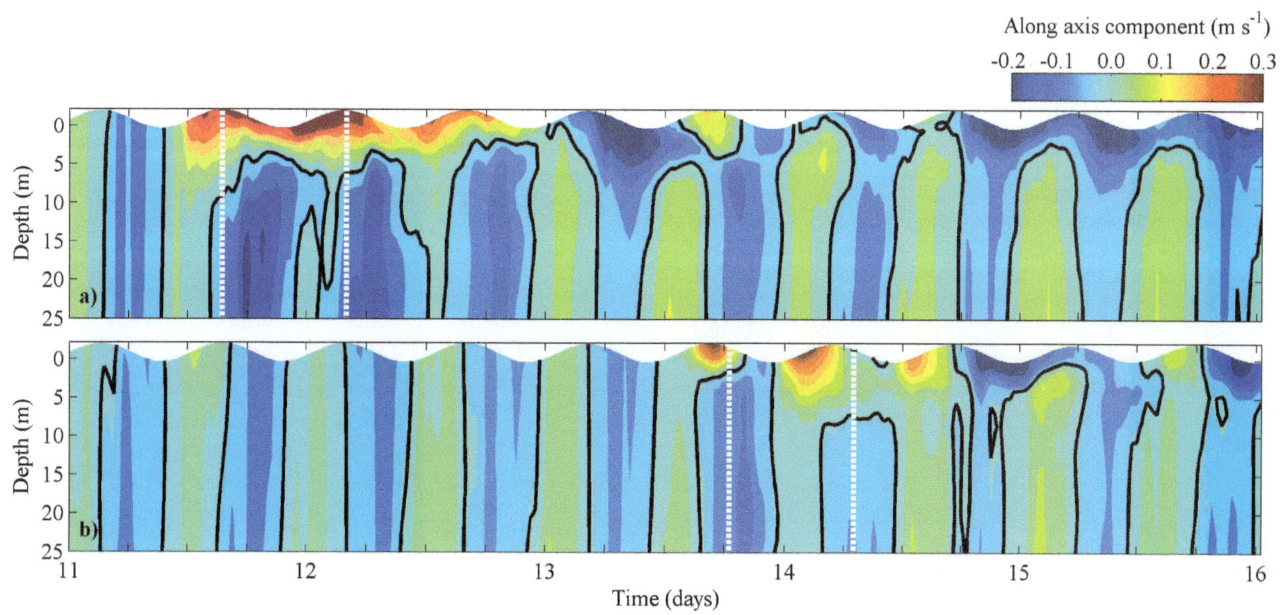

Figure 8. Temporal evolution of vertical profiles of along velocity component at stations located at mouths of Rias de Pontevedra (a) and Vigo (a), considering the interchanged Rias scenario. Black line corresponds to 0 m s^{-1}.

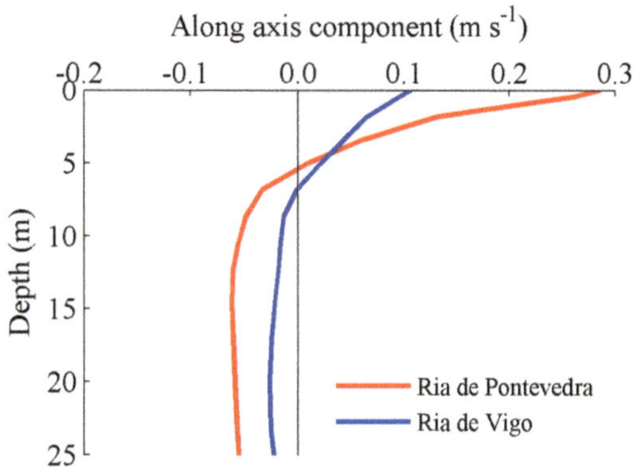

Figure 9. Residual current profiles during the plume intrusion (dashed lines, Figure 7) at stations located at mouths of Rias de Pontevedra and Vigo, considering the interchanged Rias scenario.

(Figure 2b). After that, the methodology adopted for the results presentation in this scenario is analogous to those shown previously.

Generally, the surface density maps are very similar to the reference scenario, with the plume intrusion reaching the Rias at the same period (Figures 7a and 7b).

For this scenario, the temporal evolution of the vertical profiles of along axis velocity component at the mouth of Rias de Pontevedra and Vigo is also calculated (Figures 8a and 8b). The pattern found for Ria de Pontevedra (Figure 8a) is very similar to that found for the real scenario (Figure 5b), with positive velocities at surface layers (0.25 m s^{-1}) and negative values near bottom (–0.10 m s^{-1}), corresponding to the intrusion of the Minho estuarine outflow. In Ria de Vigo (Figure 8d), the surface velocity is weaker (0.15 m s^{-1}) than for the reference scenario (Figure 5a), and in this case the surface water moving landwards only lasts one tidal cycle. In this scenario, the negative circulation is also stronger (in terms of velocity intensity) at Ria de Pontevedra (Figure 8a), indicating that the morphologic estuarine changes are particularly linked to the intensity of the circulation pattern. This is corroborated by the values of the residual currents generated during the plume intrusion in this scenario (Figure 9). Surface velocities are twice higher for Ria de Pontevedra (0.09 m s^{-1}) than for Ria de Vigo (0.04 m s^{-1}), showing that the estuarine morphology influences the intensity of the negative circulation.

As expected, these results reveal that the greater density gradients are visible in the Ria located close to the Minho River mouth for all studied scenarios (including the interchanged one). However, in terms of residual circulation, the morphology of each Ria become a critical factor, since residual flows are highly influenced by depth, width and mouth geometry of an estuary [33,34]. Here, the Ria de Pontevedra always presents higher velocities than the Ria de Vigo, showing the importance of the estuarine morphology. The Ria de Pontevedra is slightly narrower and deeper than Ria de Vigo. As the water content averages for both Rias is similar (3.12 km^3 for Ria de Vigo and 3.45 km^3 for Ria de Pontevedra, Alvarez et al. [35]), the morphology entrances is the crucial factor the plume propagation inside the Rias. Regarding only the Ria de Pontevedra, the residual surface velocity in the interchanged scenario is slightly higher (0.28 m s^{-1}) than the reference one (0.26 m s^{-1}). This difference can be attributed to the higher buoyancy forcing (>density gradients) which squeezes the plume to the surface, increasing the horizontal current velocity towards the interior of the Ria. Warrick and Stevens [36] also showed in their work in Elwha River in the Washington coast that the topographic features influence the buoyant plume behavior, affecting the coastal circulation and sediment transport.

Conclusions

The main aim of this work was to research the main reasons that induce unusual higher inflow surface current velocities at the Ria located further away from Minho River when the normal estuarine circulation is reversed, through a numerical model MOHID. The individual effect of the Minho River discharge, rivers tributaries flowing into the Rias and estuarine morphology in the Rias Baixas circulation was analyzed simulating several scenarios.

The results obtained from this study have shown that the freshwater intrusion of the Minho River into Rias de Vigo and Pontevedra was the main responsible for the negative circulation pattern and unusual density gradient in May 1998.

Without rivers flowing into the Rias, the Minho estuarine plume affects more significantly the Rias surface circulation, showing that small rivers inside Rias slightly attenuate the negative circulation.

Independently of the location of both Rias, the negative circulation was always stronger in Ria de Pontevedra, showing that the proximity with the Minho River is not the main factor justifying this pattern. Instead, the morphologic estuarine characteristics determine the unusual circulation pattern found.

In summary, the model predictions reproduce accurately the circulation pattern on May 1998 event. The obtained model results indicated that the Minho River plays an important role in the density patterns in Rias de Vigo and Pontevedra and that their morphology is the main factor for the establishment of the intensity of the circulation pattern inversion.

Author Contributions

Conceived and designed the experiments: MCS MGG JMD. Performed the experiments: MCS RM. Analyzed the data: MCS RM IA NV MGG JMD. Contributed reagents/materials/analysis tools: MCS RM IA NV. Wrote the paper: MCS RM IA NV MGG JMD.

References

1. Banas NS, Hickey BM, MacCready P, Newton JA (2004) Dynamics of Willapa Bay, Washington: A highly unsteady, partially mixed estuary. Journal of Physical Oceanography 34: 2413–2427.
2. Das A, Justic D, Inoue M, Hoda A, Huang HS, et al. (2012) Impacts of Mississippi River diversions on salinity gradients in a deltaic Louisiana estuary: Ecological and management implications. Estuarine Coastal and Shelf Science 111: 17–26.
3. Wong KC, Lu X (1994) Low-Frequency Variability in Delaware Inland Bays. Journal of Geophysical Research-Oceans 99: 12683–12695.
4. Nonn H (1966) Les régions cotiéres de la Galice (Espagne). Etude géomorphologique [PhD thesis]: Pub. Fac. Lettres. Univ. Strasbourg. 591 p.
5. Wooster WS, Bakun A, Mclain DR (1976) Seasonal Upwelling Cycle Along Eastern Boundary of North-Atlantic. Journal of Marine Research 34: 131–141.
6. Tenore KR, Alonsonoval M, Alvarezossorio M, Atkinson LP, Cabanas JM, et al. (1995) Fisheries and Oceanography Off Galicia, Nw Spain - Mesoscale Spatial and Temporal Changes in Physical Processes and Resultant Patterns of Biological Productivity. Journal of Geophysical Research-Oceans 100: 10943–10966.
7. Alonso-Perez F, Ysebaert T, Castro CG (2010) Effects of suspended mussel culture on benthic-pelagic coupling in a coastal upwelling system (Ria de Vigo, NW Iberian Peninsula). Journal of Experimental Marine Biology and Ecology 382: 96–107.

8. Rio-Barja FX, Rodríguez-Lestegás F (1996) Os ríos, in as augas de Galicia. Consello de Cultura.
9. Alvarez I, deCastro M, Gomez-Gesteira M, Prego R (2006) Hydrographic behavior of the Galician Rias Baixas (NW Spain) under the spring intrusion of the Mino River. Journal of Marine Systems 60: 144–152.
10. deCastro M, Alvarez I, Varela M, Prego R, Gomez-Gesteira M (2006) Mino River dams discharge on neighbor Galician Rias Baixas (NW Iberian Peninsula): Hydrological, chemical and biological changes in water column. Estuarine Coastal and Shelf Science 70: 52–62.
11. Sousa MC, Vaz N, Alvarez I, Dias JM (2013) Effect of Minho estuarine plume on Rias Baixas: numerical modeling approach. Journal of Coastal Research SI65: 2059–2064.
12. deCastro M, Gomez-Gesteira M, Alvarez I, Prego R (2004) Negative estuarine circulation in the Ria of Pontevedra (NW Spain). Estuarine Coastal and Shelf Science 60: 301–312.
13. Mourino C, Fraga F (1982) Hydrography of the Ria of Vigo - 1976-1977 - Abnormal Influence of the Mino River. Investigacion Pesquera 46: 459–468.
14. Sousa MC, Vaz N, Alvarez I, Gomez-Gesteira M, Dias JM (2014) Modeling the Minho River plume intrusion into the Rias Baixas (NW Iberian Peninsula). Continental Shelf Research 85: 30–41.
15. Choi BJ, Wilkin JL (2007) The effect of wind on the dispersal of the Hudson River plume. Journal of Physical Oceanography 37: 1878–1897.
16. García-Berdeal I, Hickey BM, Kawase M (2002) Influence of wind stress and ambient flow on a high discharge river plume. Journal of Geophysical Research-Oceans 107.
17. Jurisa JT, Chant R (2012) The coupled Hudson River estuarine-plume response to variable wind and river forcings. Ocean Dynamics 62: 771–784.
18. Marques WC, Fernandes EH, Monteiro IO, Moller OO (2009) Numerical modeling of the Patos Lagoon coastal plume, Brazil. Continental Shelf Research 29: 556–571.
19. Vaz N, Mateus M, Plecha S, Sousa MC, Leitao PC, et al. (2014) Modeling SST and chlorophyll patterns in a coupled estuary-coastal system of Portugal: the Tagus case study. Journal of Marine Systems. doi: 10.1016/j.jmarsys.2014.1005.1022.
20. Rios AF, Perez FF, Fraga F (1992) Water Masses in the Upper and Middle North-Atlantic Ocean East of the Azores. Deep-Sea Research Part a-Oceanographic Research Papers 39: 645–658.
21. Varela RA, Roson G, Herrera JL, Torres-Lopez S, Fernandez-Romero A (2005) A general view of the hydrographic and dynamical patterns of the Rias Baixas adjacent sea area. Journal of Marine Systems 54: 97–113.
22. Prego R, Dale AW, deCastro M, Gomez-Gesteira M, Taboada JJ, et al. (2001) Hydrography of the Pontevedra Ria: Intra-annual spatial and temporal variability in a Galician coastal system (NW Spain). Journal of Geophysical Research-Oceans 106: 19845–19857.
23. Fraga F, Margalef R (1979) Las rias gallegas. Estudio y Explotacion del mar en Galicia. Santiago de Compostela: 101–121.
24. Prego R, Fraga F (1992) A Simple-Model to Calculate the Residual Flows in a Spanish Ria - Hydrographic Consequences in the Ria-De-Vigo. Estuarine Coastal and Shelf Science 34: 603–615.
25. Lyard F, Lefevre F, Letellier T, Francis O (2006) Modelling the global ocean tides: modern insights from FES2004. Ocean Dynamics 56: 394–415.
26. Locarnini RA, Mishonov AV, Antonov JI, Boyer TP, Garcia HE, et al. (2010) World Ocean Atlas 2009, Volume 1: Temperature. Washington, DC: U.S. Government Printing Office. 184 p.
27. Antonov JI, Seidov D, Boyer TP, Locarnini RA, Mishonov AV, et al. (2010) World Ocean Atlas 2009, Volume 2: Salinity. Washington, DC: U.S. Government Printing Office. 184 p.
28. Chapra SC (1997) Surface Water Quality Modeling. New York: McGraw-Hill.
29. Genio L, Sousa A, Vaz N, Dias JM, Barroso C (2008) Effect of low salinity on the survival of recently hatched veliger of Nassarius reticulatus (L.) in estuarine habitats: A case study of Ria de Aveiro. Journal of Sea Research 59: 133–143.
30. Sordo I, Barton ED, Cotos JM, Pazos Y (2001) An inshore poleward current in the NW of the Iberian peninsula detected from satellite images, and its relation with G-catenatum and D-acuminata blooms in the Galician Rias. Estuarine Coastal and Shelf Science 53: 787–799.
31. Monteiro PMS, Largier JL (1999) Thermal stratification in Saldanha Bay (South Africa) and subtidal, density-driven exchange with the coastal waters of the Benguela upwelling system. Estuarine Coastal and Shelf Science 49: 877–890.
32. Hickey BM, Zhang X, Banas N (2002) Coupling between the California Current System and a coastal plain estuary in low riverflow conditions. Journal of Geophysical Research-Oceans 107.
33. Lee J, Valle-Levinson A (2012) Influence of bathymetry on hydrography and circulation at the region between an estuary mouth and the adjacent continental shelf. Continental Shelf Research 41: 77–91.
34. Valentim JM, Vaz N, Silva H, Duarte B, Cacador I, et al. (2013) Tagus estuary and Ria de Aveiro salt marsh dynamics and the impact of sea level rise. Estuarine Coastal and Shelf Science 130: 138–151.
35. Alvarez I, deCastro M, Gomez-Gesteira M, Prego R (2005) Inter- and intra-annual analysis of the salinity and temperature evolution in the Galician Rias Baixas-ocean boundary (northwest Spain). Journal of Geophysical Research-Oceans 110.
36. Warrick JA, Stevens AW (2011) A buoyant plume adjacent to a headland-Observations of the Elwha River plume. Continental Shelf Research 31: 85–97.

21

High Phylogenetic Diversity of Glycosyl Hydrolase Family 10 and 11 Xylanases in the Sediment of Lake Dabusu in China

Guozeng Wang[1,2]⁹, Xiaoyun Huang[1]⁹, Tzi Bun Ng[3], Juan Lin[1,2]*, Xiu Yun Ye[1,2]*

1 College of Biological Science and Technology, Fuzhou University, Fuzhou 350108, P.R. China, **2** National Engineering Laboratory for High-efficiency Enzyme Expression, Fuzhou 350002, P. R. China, **3** School of Biomedical Sciences, Faculty of Medicine, The Chinese University of Hong Kong, Shatin, New Territories, Hong Kong, China

Abstract

Soda lakes are one of the most stable naturally occurring alkaline and saline environments, which harbor abundant microorganisms with diverse functions. In this study, culture-independent molecular methods were used to explore the genetic diversity of glycoside hydrolase (GH) family 10 and GH11 xylanases in Lake Dabusu, a soda lake with a pH value of 10.2 and salinity of 10.1%. A total of 671 xylanase gene fragments were obtained, representing 78 distinct GH10 and 28 GH11 gene fragments respectively, with most of them having low homology with known sequences. Phylogenetic analysis revealed that the GH10 xylanase sequences mainly belonged to Bacteroidetes, Proteobacteria, Actinobacteria, Firmicutes and Verrucomicrobia, while the GH11 sequences mainly consisted of Actinobacteria, Firmicutes and Fungi. A full-length GH10 xylanase gene (*xynAS10-66*) was directly cloned and expressed in *Escherichia coli*, and the recombinant enzymes showed high activity at alkaline pH. These results suggest that xylanase gene diversity within Lake Dabusu is high and that most of the identified genes might be novel, indicating great potential for applications in industry and agriculture.

Editor: Chunxian Chen, USDA/ARS, United States of America

Funding: This research was supported by the National Natural Science Foundation of China (31301406), the Oceanic Public Welfare Industry Special Research Project of China (201305015) and the Priming Scientific Research Foundation of Fuzhou University (XRC-1326). The funders had no role in study design, data collection and analysis, decision to publish, or preparation of the manuscript.

Competing Interests: The authors have declared that no competing interests exist.

* Email: ljuan@fzu.edu.cn (JL); xiuyunye@fzu.edu.cn (XYY)

⁹ These authors contributed equally to this work.

Introduction

Xylan is the major hemicellulose component of the plant cell wall, which is the second most abundant polysaccharide on earth after cellulose [1]. Xylan is composed of a homopolymeric backbone chain of β-1, 4-linked xylopyranose units with substituted side chains at different positions, and its complete hydrolysis requires a group of enzymes including endo-1,4-β-D-xylanase, β-D-xylosidase, α-D-glucuronidase, α-L-arabinofuranosidase, acetyl xylanesterase, and arylesterase [2,3]. Endo-1,4-β-D-xylanase (EC 3.2.1.8) is a crucial component because it catalyzes the hydrolysis of xylan to short xylooligosaccharides of varying lengths [2,3]. Xylanases have been classified into glycosyl hydrolase (GH) families (http://www.cazy.org/fam/acc_GH.html; [4]) 5, 7, 8, 10, 11 and 43 [3] based on sequence similarities of the catalytic domain. Among these, GH10 and 11 xylanases are the most abundant, which have distinct three-dimensional structures [5], mechanisms of action [6] and substrate specificity to xylan.

Although xylanases are produced by diverse organisms, including bacteria, algae, fungi, protozoans, gastropods and anthropods [3], microbial xylanases are the focus of intense research owing to their significant application in various industrial processes. Specifically, they are used to improve the digestibility of animal feedstock, enhance filterability in brewing, increase dough volume and improve the textural and staling properties of bread, and for fruit juice and wine clarification, the bioconversion of lignocellulosic materials into fermentative products, and facilitation of the release of lignin from the pulp [1,2,7,8]. Many microbial xylanases have been purified and characterized, and the genes encoding xylanases have been cloned and expressed in heterologous systems, and the structures of a number of enzymes have been determined [3,8].

Most microbial xylanases reported to date have acidic or neutral pH optima. Alkaline xylanases are of great interest because xylan is more readily soluble under alkaline pH than neutral pH. Additionally, the application of alkaline xylanases in the paper and pulp industry can reduce the use of chlorine, which is very attractive from an economical and technical point of view [9,10]. Consequently, studies are continually conducted in attempts to identify novel xylanases with potential applications in the pulp and paper industries. Although alkaline xylanases have been reported from microorganisms isolated from nonalkaline environments [11,12], xylanases produced by microorganisms isolated from extreme alkaline environments have attracted increasing attention. These organisms are exposed to hostile environments, resulting in their evolution and accumulation of a variety of adaptive features for activity and stability under these conditions [13–15].

Natural alkaline environments are not common, and soda lakes and deserts are the most stable naturally occurring alkaline ecosystems on earth [16,17]. Soda lakes have high carbonate alkalinity, a pH of 9 to 11, and moderate to extremely high salinity [17]. Despite their extreme environmental conditions, soda lakes harbor extremely productive microbial communities. A number of alkaliphilic microorganisms that play key metabolic roles in soda lakes have previously been studied, and many novel alkaliphilic microorganisms have been isolated from these unique ecosystems [18]. Molecular biological techniques that do not depend on culture have revealed a high diversity and novelty of microbial communities in soda lake environments [19–21].

Although the microbial diversity of soda lakes has been studied extensively, the functional diversity of genes in such systems has not. Lin et al. analyzed the methane monooxygenase genes in Mono Lake and suggested that increased methane oxidation activity was correlated with changes in methanotroph community structure [22]. Kovaleva et al. explored the diversity of RuBisCO and ATP citrate lyase genes in the sediments from six soda lakes of the Kulunda Steppe [23]. However, few studies have focused on the functional gene diversity of plant material dehygrolysis in soda lake environments. We explored the genetic diversity of GH10 xylanase in diverse soil environments using culture-independent molecular methods and found that pH was one of the most important factors influencing the xylan degrading microbial community [24].

In this study, we focused on the genetic diversity of GH10 and GH11 xylanases in the natural Lake Dabusu, an alkaline (pH 10.2) and saline (101 g/liter) lake situated in northeastern China. Partial xylanase genes were amplified directly from the metagenomic DNA of the soda lake sediment. Sequence analysis showed that most xylanase gene fragments had low homologies to known xylanases, suggesting a large number of uncharacterized xylanase genes in the soda lake. Additionally, phylogenetic diversity analysis suggested a surprising diversity of xylanase genes in this harsh environment. A novel GH10 full-length xylanase gene was directly cloned from the metagenomic DNA and expressed in *Escherichia coli*. The recombinant xylanases showed high activity at alkaline pH. Our study provides new insight into the genetic diversity and distribution of microbial xylanases in the soda lake ecosystem, which will help us understand their roles in this microenvironment.

Materials and Methods

Ethics statement

We declare that no living animals were used in this research. No specific permissions were required for the described field studies. The sediment sample was collected from the location that is not privately-owned or protected in any way. The field studies did not involve endangered or protected species because this study only concentrated on the sediment sample.

Sample site

Lake Dabusu is located in the southwestern part of the Qian'an County (Jilin Province, China), in the center of the depressed belt of Songliao Basin. Lake Dabusu is a closed inland alkaline lake located 122 meters above sea level, with an average water depth of about 0.90 meters and an area of approximately 38 km^2 in the rainy season. Because of strong evaporation and a lack of outflow within the closed basin, alkaline materials accumulate in the lake water. Lake Dabusu is a typical soda lake with salinity of 62.34 g L^{-1} to 347.34 g L^{-1} and a pH of 10 to 11 depending on season [25].

Table 1. Physicochemical characters of the alkaline saline lake sediment and xylanase fragment sequences obtained.

Location	T (°C)	pH	Total organic carbon (mg/g)	Total nitrogen (mg/g)	C/N ratio	GH family	Clones sequenced	Sequences recovered	OTUs[a]
44 48 20 N	20	10.2	1.10	0.35	3.14	GH10	550	467	78
123 40 32 E						GH11	250	204	28

[a]5% dissimilarity as cutoff.

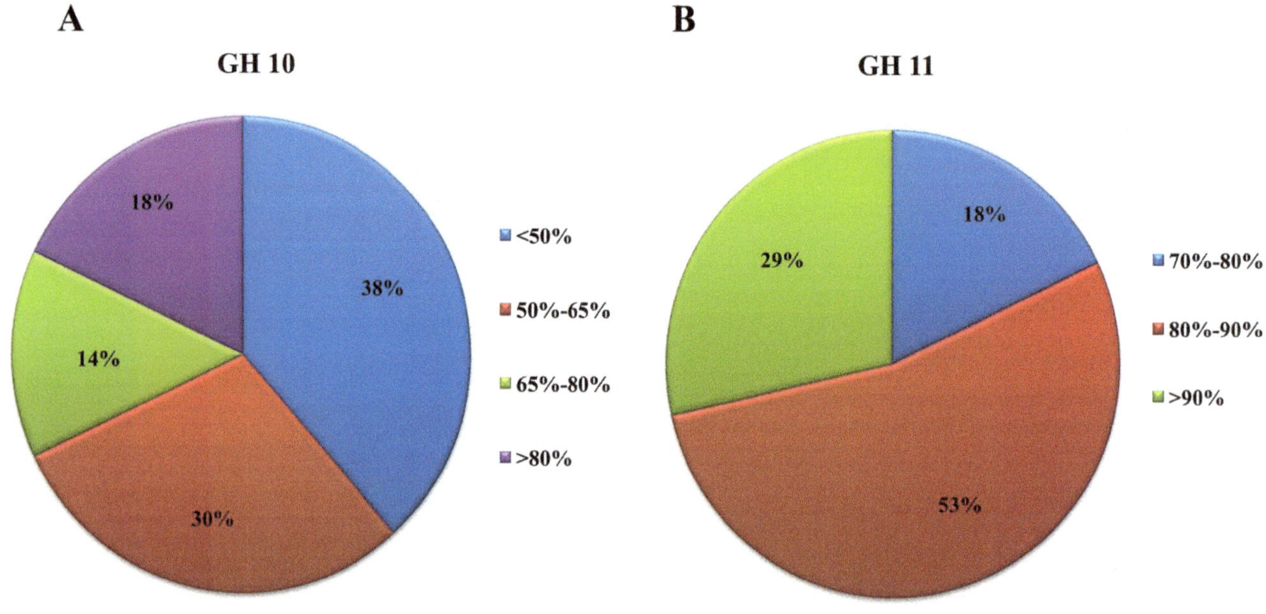

Figure 1. Amino acid sequence homologies of GH10 and GH11 xylanase gene fragments from sediment metagenomic DNA to known xylanases. Each sequence was analyzed by NCBI BLASTp (version 2.2.29) against the GenBank nr database. An E-score (expect value) cutoff of 10^{-10} (default) was applied and the top BLASTp hit to the known xylanases was collected.

Sample collection and DNA extraction

On June 12, 2013, superficial (0 to 10 cm depth) sediment samples were collected and transported into a sterile sampling bag and immediately shipped to the laboratory on ice packs. Upon arrival, portions of the samples were stored at 4°C for physicochemical characterization and −80°C for metagenomic DNA extraction. Details regarding the location and physicochemical properties of the lake sediment are provided in Table 1. By using a modified protocol specific for high molecular weight DNA from environmental samples [26], the metagenomic DNA of the sediment was extracted as described in detail previously [27]. The sediment metagenomic DNA was purified using an Omega Gel Extraction Kit (Norcross, GA) and stored at −20°C until use.

Xylanase gene fragments amplification, library construction and sequencing

GH10 and GH11 xylanase gene fragments were amplified with the purified metagenomic DNA as a template by touchdown PCR using the CODEHOP primers X10-F and X10-R specific for GH10 xylanase and X11-F and X11-R specific for GH11 xylanase [27], respectively. PCR products were then visualized on an agarose gel and purified using a Qiaquick gel extraction kit (Qiagen, Valencia, CA). The purified PCR products were then ligated into the PMD 19-T vector (TaKaRa, Tokyo, Japan) and electroporated into *Escherichia coli* DH5a (TaKaRa, Tokyo, Japan) following the procedure recommended by the manufacturer to construct the clone library for each xylanase family. Positive transformants (white colonies) from each library were randomly picked for further confirmation by PCR with primers M13F (GTAAAACGACGGCCAGT) and M13R (GGATAACAATTTCACACAGGA). They were then sequenced by Life Technologies using the Sanger method with an ABI-3730 automatic sequencer (Life Technologies, Carlsbad, CA).

Phylogenetic analysis

By using the Figaro software [28] (http://sourceforge.net/apps/mediawiki/amos/index.php?title=Figaro), vector sequences introduced by automated Sanger sequencing machines were removed. The sequences were analyzed by NCBI BLASTx (version 2.2.29) searches against the GenBank nr database, with an E-score (expect value) cutoff of 10^{-10}. Nucleotide sequences identified as xylanase gene fragments were translated into amino acids by EMBOSS Transeq (http://www.ebi.ac.uk/Tools/st/emboss_transeq/). Then, they were aligned with known sequences in the GenBank database at the protein level using ClustalW. Redundant amino acid sequences were removed using Cd-hit [29] with a 95% sequence identity cutoff.

The protein sequence similarities were assessed using the BLASTp program (http://www.ncbi.nlm.nih.gov/BLAST/; until January 15, 2014). Phylogenetic trees were constructed with MEGA 4.1 [30] using the neighbor-joining method [31]. Confidence for tree topologies was estimated by bootstrap values based on 1,000 replicates. A total of 39 and 17 representative sequences were selected and used as references for GH10 and GH11 phylogenetic tree constructions, respectively.

Abundance analysis

The abundance of each GH family was estimated using the distance-based operational taxonomic unit and richness determination (DOTUR) software [32]. By using PHYLIP software (http://evolution.genetics.washington.edu/phylip.html), distance matrices of the fragment sequences were calculated at the protein level with the default parameters of protdist. Based on UPGMA (average linkage clustering) implemented in DOTUR with default parameters of precision (0.01) and 1000 bootstrap replicates, sequences were then assigned to OTUs.

Cloning of full-length xylanase gene

Fragment AS10-66, which showed phylogenetic novelty (Figure 1, Table S1) and was most abundant in the GH10 library (see

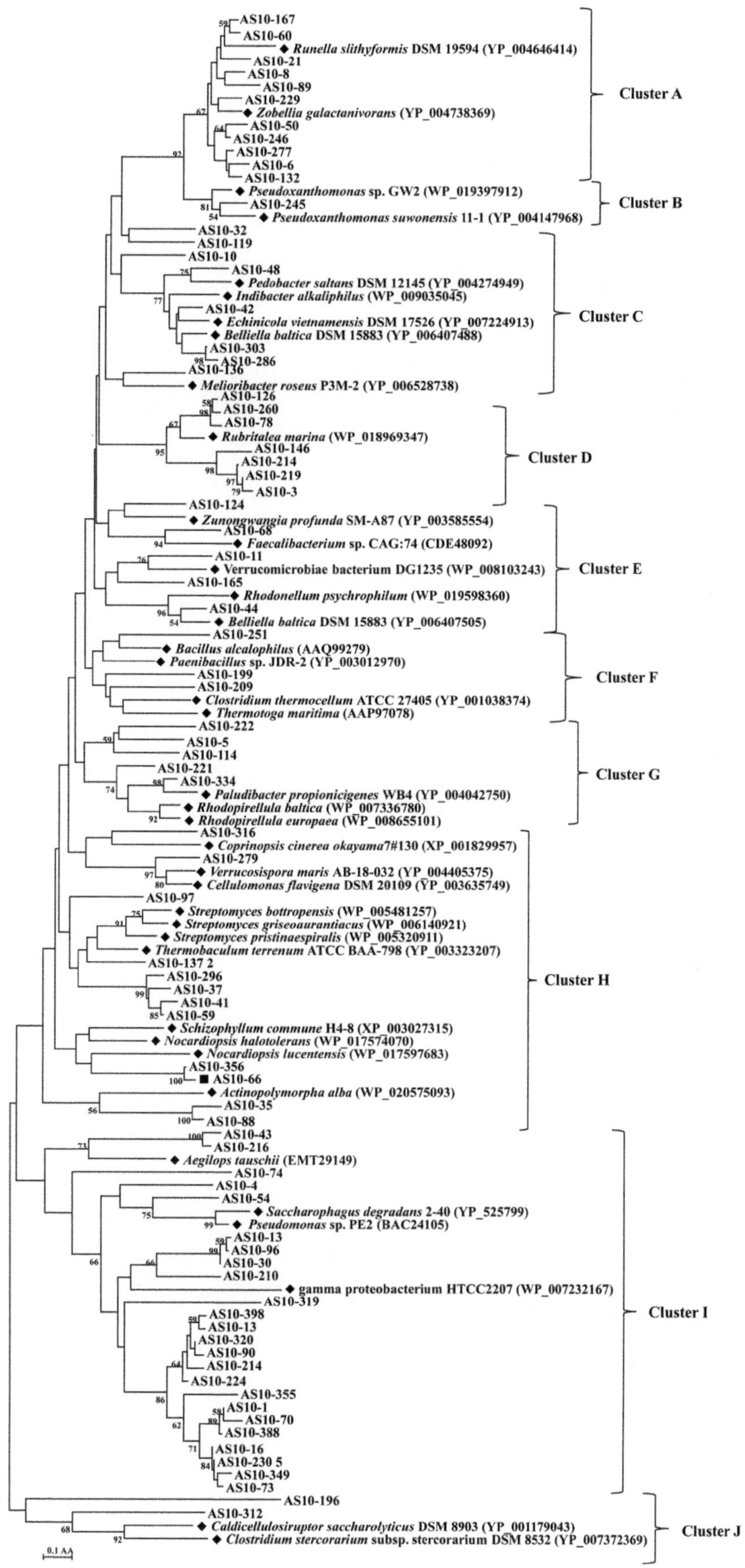
Cluster A
Cluster B
Cluster C
Cluster D
Cluster E
Cluster F
Cluster G
Cluster H
Cluster I
Cluster J
0.1 AA

Figure 2. Phylogenetic analysis based on the partial amino acid sequences of GH10 xylanase genes detected in the Lake Dabusu sediment metagenomic DNA. The tree was constructed using the neighbor joining method (MEGA 4.1). The lengths of the branches indicate the relative divergence among amino acid sequences. The reference sequences are marked with a closed diamond (♦) with source strains and GenBank accession numbers in parentheses. The gene fragments (AS10-66) used for full length cloning were marked with a solid square (■). The numbers at the nodes indicate bootstrap values based on 1,000 bootstrap replications and bootstrap values (>50) are displayed. The scale bar represents 0.1 amino acid substitutions per position.

Table S1), was subjected to full-length gene cloning. The flanking regions of the xylanase gene fragments were cloned by using a modified TAIL-PCR with eight specific primers (Table S3) following the protocol [33]. PCR products of the expected size appeared between the third and fourth rounds of amplification were purified, cloned into the PMD 19-T vectors, sequenced, and then assembled with the known fragment sequence. The full-length xylanase gene was designated as *xynAS10-66*. The signal peptide sequence of XynAS10-66 was predicted with SignalP [34] (http://www.cbs.dtu.dk/services/SignalP/). The DNA and protein sequence identities/similarities were assessed with the BLASTn and BLASTp programs (http://www.ncbi.nlm.nih.gov/BLAST/), respectively.

Xylanase expression and activity assay

Using two primer sets (Table S3), the coding sequence of the xylanase gene *xynAS10-66* without the signal peptide was amplified and cloned into vector pET-22b(+), and then transformed into *E. coli* BL21 (DE3) competent cells for recombinant expression. The positive transformant harboring pET-*xynAS10-66* was grown in LB medium containing 100 μg mL^{-1} ampicillin at 37°C to an A_{600} of 0.6. Protein expression was induced by addition of isopropyl-β-D-1-thiogalactopyranoside (IPTG) at a final concentration of 1 mM, and the culture was incubated at 30°C for additional 12 h.

Xylanase activity was determined by measuring the release of reducing sugar from substrate by the 3, 5-dinitrosalicylic acid method [35]. To accomplish this, reactions containing 0.1 ml of appropriately diluted enzyme and 0.9 ml of 1% (w/v) beechwood xylan as substrate. After incubation at 55°C for 10 min, the reaction was stopped with 1.5 ml DNS reagent and boiled for 5 min, and the absorbance at 540 nm (A_{540}) was measured. Finally, using a standard curve generated with D-xylose, the absorbance was converted into moles of reducing sugars produced. One unit (U) of xylanase activity was defined as the amount of enzyme that released 1 μmol of reducing sugar equivalent to xylose per minute.

Purification and partial characterization of recombinant XynAS10-66

To purify the His-tagged recombinant proteins (rXynAS10-66), culture supernatant was collected after centrifugation (12,000×*g*, 4°C for 15 min). Then the culture supernatant was concentrated with an ultrafiltration membrane (PES5000; Sartorius Stedim Biotech, Germany) and then loaded onto a Ni^{2+}-NTA agarose gel column (Qiagen, Germany) with an imidazole gradient of 20–200 mM in Tris-HCl buffer (20 mM Tris-HCl, 500 mM NaCl, 10% glycerol, pH 7.6). Sodium dodecyl sulfate-polyacrylamide gel electrophoresis (SDS-PAGE) was used to determine the purity and apparent molecular mass of rXynAS10-66.

The optimal pH for xylanase activity of the purified rXynAS10-66 was determined at 37°C with pH ranging from 4.0 to 11.0. The buffers used were McIlvaine buffer (0.2 M Na_2HPO_4/0.1 M citric acid) for pH 4.0–7.0, 0.1 M Tris-HCl for pH 7.0–9.0, and 0.1 M glycine-NaOH for pH 9.0–11.0. The optimal temperature for purified rXynAS10-66 activity was determined over the range of 40–90°C in Tris-HCl buffer (pH 9.0). The K_m and V_{max} values for rXynAS10-66 were determined in Tris-HCl buffer (pH 7.0) containing 1–10 mg mL^{-1} beechwood xylan at 55°C, respectively. K_m and V_{max} were determined from a Lineweaver-Burk plot using the non-linear regression computer program GraFit (Erithacus, Horley, Surrey, UK).

Nucleotide sequence accession numbers

The GH10 and GH11 xylanase gene fragments were deposited into the GenBank database under accession numbers KJ463250–KJ463327 and KJ463328–KJ463355, respectively. Accession number KJ463356 was assigned to the full-length xylanase gene *xynAS10-66*.

Results and Discussion

Soda lakes are one of the most stable naturally occurring alkaline and saline environments that harbor numerous novel microorganisms [16,17,21]. Although the microbial diversity of this unique ecosystem has been thoroughly investigated, only a few studies have considered the functional diversity [22,23]. Functional gene diversity based on the metagenomic sequences can provide insight into functional diversity and metabolic potential at the community level. Xylanases play key roles in the initial steps of plant cell wall breakdown and have great potential for industrial and agricultural applications. Thus xylanase genes were targeted for diversity analysis in this study. Because many microorganisms cannot be cultured in the laboratory owing to the extremely harsh conditions of soda lakes, culture-independent molecular approaches were used to explore the xylanase gene diversity of Lake Dabusu.

Abundance of GH10 xylanase in Lake Dabusu sediment

Using the CODEHOP primers X10-F and X10-R specific for GH10 xylanases [27], PCR product of about 260 bp was amplified from the metagenomic DNA of the sediment. The product was purified and used to construct a clone library. The positive transformants were picked and then confirmed by PCR with primers M13F and M13R. Overall, 550 clones were sequenced, 467 sequences of which were identified as GH10 xylanase gene fragments based on BLASTx analysis and presence of the Asn residue in the protein sequence, which is conserved among GH10 xylanases [36].

After removing redundant sequences using the CD-hit program [29], 78 sequences showed divergence (<95% homology, Table S1). Based on BLASTp analysis, about 68% (53/78) of the sequences had low similarities (<65%) with known xylanases in GenBank (Figure 1), implying that they may be novel xylanases. Abundance analysis using the distance based operational taxonomic unit and richness determination (DOTUR) software [32] showed that AS10-66 was the predominant operational taxonomic unit (OTU), representing 117 sequences. Forty OTUs contained only one sequence (Table S1).

Culture dependent and culture-independent methods revealed the presence of numerous novel microorganisms in soda lake environments [16,21]. In the present study, GH10 xylanase gene

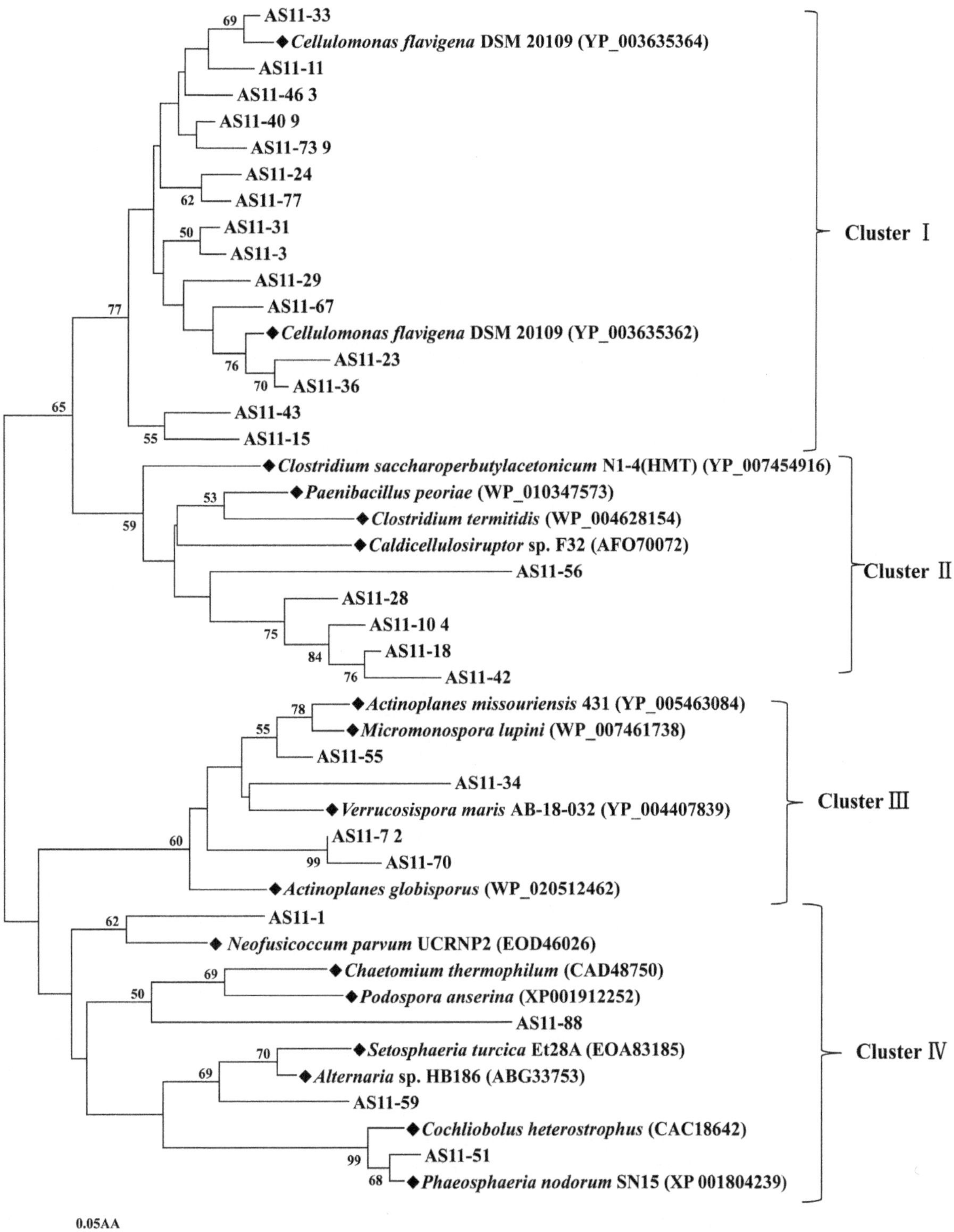

Figure 3. Phylogenetic analysis based on the partial amino acid sequences of GH11 xylanase genes detected in the Lake Dabusu sediment metagenomic DNA. The tree was constructed using the neighbor joining method (MEGA 4.1). The lengths of the branches indicate the relative divergence among amino acid sequences. The reference sequences are marked with a closed diamond (◆) with source strains and GenBank accession numbers in parentheses. The numbers at the nodes indicate bootstrap values based on 1,000 bootstrap replications and bootstrap values (>50) are displayed. The scale bar represents 0.05 amino acid substitutions per position.

fragment sequences obtained directly from the sediment metagenomic DNA had low homology with known xylanases. Specifically, none of the xylanase fragments amplified in this study had greater than 90% homology with known xylanases in the

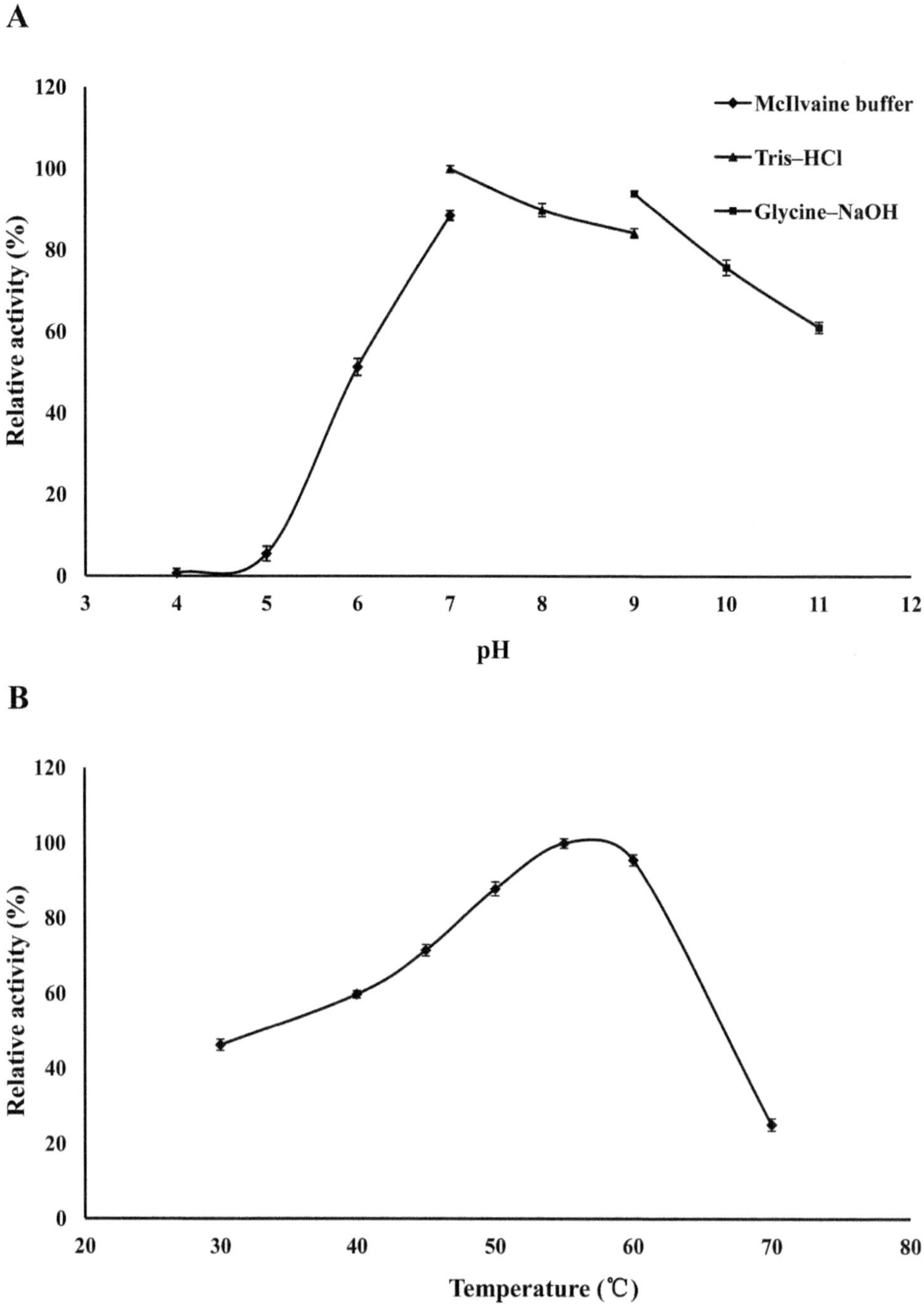

Figure 4. pH and temperature activity profiles of purified recombinant XynAS10-66. A Effect of pH on XynAS10-66 activity. Activities at various pHs were assayed at 30°C in different buffer. **B** Effect of temperature on XynAS10-66 activity in Tris-HCl buffer (pH 9.0).

GenBank database (Table S1). The lowest similarities were observed for AS10-35 and AS10-88, which both had 35% homology with xylanase from *Nocardiopsis* sp. CNS639 (WP_019610727), while AS10-8 showed the greatest homology of 87% with xylanase from *Cellulophaga algicola* DSM 14237 (YP_004163134). Moreover, almost 70% of the obtained sequences had low similarity (<65%) with known xylanases. This is much higher than that from the other soil environments we previously investigated [24]. Overall, these findings suggested that there may be abundant novel xylan-degrading microorganisms in this environment.

High genetic diversity of GH10 xylanase in the alkaline soda lake sediment

Using 78 divergent sequences from the GH10 clone library and 39 reference sequences from the GenBank Database, an unrooted protein-level phylogenetic tree of GH10 xylanases was constructed. All sequences were confined to ten clusters, denoted as Cluster A to Cluster J, indicating substantial diversity among GH10 xylanases in the sediment (Figure 2). The presence of many clades without close relatives suggests their novelty, which might be because of the large portion of unidentified microorganisms in this environment.

Cluster A, C, E and G contained a total of 28 sequences from the soda lake sediment and 14 reference sequences of genera belonging to Bacteroidetes including *Runella slithyformis* DSM 19594, *Pedobacter saltans*, *Indibacter alkaliphilus* and *Rhodopirellula baltica*. Cluster B and cluster I contained the sequences of 23 clones and reference sequences from the phylum Proteobacteria, including gamma proteobacterium HTCC2207, *Pseudomonas* sp. PE2, *Saccharophagus degradans* 2–40 and *Pseudoxanthomonas suwonensis* 11-1. Twelve sequences and 11 reference sequences from different genera in the phylum Actinobacteria, including *Actinopolymorpha*, *Streptomyces* and *Nocardiopsis*, formed cluster H. Three sequences in cluster F and two in cluster J were closely related to xylanase produced by members of the phylum Firmicutes, including *Bacillus alcalophilus*, *Paenibacillus* sp. JDR-2, *Clostridium thermocellum* ATCC 27405, *Caldicellulosiruptor saccharolyticus* DSM 8903 and *Clostridium stercorarium* subsp. stercorarium DSM 8532. Seven sequences in cluster D shared the highest identity with xylanase from *Rubritalea marina*.

As shown in Fig. 2, phylogenetic analysis revealed that Bacteroidetes, Proteobacteria, Actinobacteria, Firmicutes and Verrucomicrobia are the main xylanolytic bacteria to produce xylanases in the sediment of Lake Dabusu. These findings are consistent with the conclusions of previous studies [16,17,20,37]. However, the composition of the xylanolytic community differs from that of the microbial community. Microbial community analysis suggested that Firmicutes are the most abundant microorganisms in soda lake environments, followed by Proteobacteria and Actinobacteria. In the present study, xylanases were found to primarily belong to Bacteroidetes, which harbored more than 35% (28/78) of all GH10 fragment sequences. One reason for this may be that the xylantic microbial community differs from that of the total microbial community. Soda lakes harbor abundant and diverse microorganisms which are critical in decomposition of organic matter and cycling of carbon, nitrogen, phosphorus, and sulphur [38]. Second, the physicochemical properties of Lake Dabusu differ from those of other soda lakes [16], which play important role in the composition of the microbial community [20,39].

Xylanase fragment sequences related to those from Proteobacteria were the second most abundant, accounting for about 30% (23/78) of all GH10 xylanase sequences. All reference sequences of gamma proteobacterium HTCC2207, *Pseudomonas* sp. PE2, *Saccharophagus degradans* 2–40 and *Pseudoxanthomonas suwonensis* 11-1 belonged to the Gammaproteobacteria. This is concurrent with the finding that Gammaproteobacteria is one of the most abundant and diverse groups in soda lakes [37]. Moreover, we found that although these sequences were grouped in a cluster, they were not closely related to each other (Figure 2), suggesting that these sequences might be novel.

Amplification and sequence analysis of GH11 xylanase gene fragments

PCR product with a size of about 210 bp was obtained from the metagenomic DNA of sediment using CODEHOP primers X11-F and X11-R specific for GH11 xylanases [24]. Overall, 250 clones were randomly selected from the clone library constructed using the PCR products and sequenced. As a result, 204 sequences showed 70–98% amino acid identity with known GH11 xylanases (Table S2). Additionally, Glu (catalytic residue), which is highly conserved in GH11 xylanases, was found in all protein sequences (Table S2). Thus, these sequences were considered to be partial GH11 xylanases. After removing the redundant sequences using the CD-hit program, 28 sequences showed divergence (sharing $<$ 95% identity) (Table S2). Abundance analysis using DOTUR software showed that AS11-42 was the predominant OTU, occurring in 39 sequences. Nine OTUs contained only one sequence (Table S2).

Phylogenetic diversity of GH11 xylanase in lake sediment

The 28 distinct partial sequences of GH11 xylanases were used to construct an unrooted phylogenetic tree with 17 reference sequences (Figure 3). Four clusters (I, II, III, and IV) were formed based on high bootstrap values. Some clades formed without closely related references, suggesting that these sequences differed from known xylanases. Nineteen sequences shared the highest identity with xylanases from *Cellulomonas flavigena* DSM 20109, *Actinoplanes missouriensis* 431, *A. globisporus Micromonospora lupine* and *Verrucosispora maris* AB-18-032, and fell into clusters I and III. Five sequences and those of *Clostridium saccharoperbutylacetonicum* N1-4(HMT), *C. Paenibacillus peoriae*, and *Caldicellulosiruptor* sp. F32 were grouped into cluster II. Four sequences closely related to xylanases from *Neofusicoccum parvum* UCRNP2, *Podospora anserine*, *Setosphaeria turcica* Et28A, *Alternaria* sp. HB186 and *Phaeosphaeria nodorum* SN15 fell into cluster IV.

Unlike GH10 family xylanase genes, GH11 xylanase genes from Lake Dabusu were less diverse and showed a different microbial distribution. Sequences of the GH11 xylanase fragments were found to be related to Actinomycetes, Firmicutes and Fungi. Actinomycetes are important degraders of organic matter in most habitats. An alkaline xylanase was purified and characterized from alkaliphilic *Micrococcus* sp. AR-135, which was isolated from an alkaline soda lake in Ethiopia [14]. Upon phylogenetic analysis, more than 67% (19/28) of the sequences were grouped into a cluster containing reference sequences from different genera in Actinobacteria. Along with the 12 sequences of GH10 xylanase, a total 31 fragment sequences were found to be associated with Actinobacteria, indicating the high diversity of xylanolytic actinobacteria in Lake Dabusu.

Using culture and culture-independent methods, the bacterial community in the soda lake was found to be dominated by clones affiliated with Firmicutes. Moreover, several xylanases were characterized from the isolated Firmicutes with high activity at alkaline pH [13,15]. In this study, xylanase fragments related to xylanases of different genera Firmicutes were also found. Specifically, five GH10 xylanase fragment sequences were found to be closely related to *Bacillus alcalophilus*, *Clostridium thermocellum* ATCC 27405 and *Caldicellulosiruptor saccharolyticus* DSM 8903, while five GH11 fragments were closely related to *Caldicellulosiruptor* sp. F32 (AFO70072), *Clostridium termitidis* (WP_004628154) and *Clostridium saccharoperbutylacetonicum* N1-4(HMT) (YP_007454916), suggested the diversity of xylanolytic Firmicutes in this environment.

When compared with Bacteria and Archaea, little is known about the diversity, abundance and activity of micro-eukaryotes in soda lakes. Recently, a high diversity of micro-eukaryotes in soda lakes located in the Ethiopian Rift Valley was revealed by high-throughput sequencing [21]. However, no functional genetic diversity has been reported regarding fungi in soda lake environments. In the present study, four sequences were found to be related to xylanases from Ascomycota, the dominant fungi in soda lakes, suggesting that xylantic fungi are present in these systems.

Direct cloning and novelty of xylanase gene XynAS10-66

Until recently, the majority of known xylanases were obtained from pure cultures. Only a few were cloned using culture-independent approaches, such as construction and screening of metagenomic libraries and PCR-based methods. In our lab, we developed a fast and efficient modified TAIL-PCR method to obtain full-length genes directly from metagenomic DNA based on fragment sequences [33]. In this study, the fragment sequence AS10-66 was selected to obtain the full-length gene because it was the most abundant of all the sequences, representing about one quarter of all sequences (117/467). Moreover, AS10-66 has been reported to have low homology with known xylanases (Table S1).

Based on the fragment sequences of AS10-66 and using modified TAIL-PCR [33], a full-length xylanase gene (*xynAS10-66*) was direct cloned from metagenomic DNA of the soda lake sediment. The complete sequence of *xynAS10-66* contained an open reading frame of 1092 bp encoding a 363-residue polypeptide with a typical signal peptide (residues 1–24). The calculated mass of XynAS10-66 was 38.7 kDa, and the theoretical isoelectric point was 4.74. Sequence similarity searches showed that deduced XynAS10-66 shared the highest homology (44%) with the thermostable GH10 xylanase from *Thermomonospora alba* ULJB1, followed by *Thermobispora bispora* DSM 43833 (44%), an isolate recovered from decaying manure and *Nocardiopsis halotolerans* (43%), an isolate recovered from salt marsh soil in Kuwait [40].

Partial biochemical characterization of purified recombinant XynAS10-66

The gene encoding the mature proteins was expressed in *E. coli* BL21 (*DE3*). Following induction with IPTG at 30°C for 12 h, substantial xylanase activity was detected in the culture supernatant of recombinant cells. Using beechwood xylan as the substrate, rXynAS10-66 showed the highest activity at pH 7.0, while it retained more than 95% activity at pH 9.0, and >60% of the maximum activity at pH 7.0–11.0 (Figure 4A). Using beechwood xylan as the substrate, the K_m and V_{max} values were 1.5±0.04 mg $\cdot$ mL^{-1}, 1102±9.54 μmol $\cdot$ mg^{-1} $\cdot$ min^{-1}, respectively. Although the optimum temperature of the recombinant XynAS10-66 is neutral, it has substantial activity at alkaline pH, with more than 80% and 60% activity being retained at pH 10 and pH 11, respectively. This characterization differs from those of alkaline xylanases obtained from *Bacillus* and *Micrococcus* sp. isolated from other soda lakes, which showed optimum activity at pH 8 to 9 [14,15]. The optimal temperature for the enzyme activity of XynAS10-66 was 55°C (Figure 4B), while more than 60% activity was retained at 45°C to 65°C. The recombinant XynAS10-66 had the same optimum temperature as the alkaline xylanase from alkaliphilic *Micrococcus* sp. AR-135 [14]. However, it had a lower activity than that of xylanases from *Bacillus halodurans* S7 [13] and two thermostable alkaline xylanases from an alkaliphilic *Bacillus* sp. [15]. BLAST analysis revealed that XynAS10-66 shared high similarity with a thermostable endo-beta-1,4-xylanase of *Thermomonospora alba* ULJB1 [41], which showed good activity at up to 95°C. Sequence alignment revealed that XynAS10-66 lacks a carbohydrate binding module (CBM) at the C terminal, which may have resulted in the lower temperature optimum of XynAS10-66. Future studies will be conducted to investigate this phenomenon. Moreover, other fragment sequences will be used to generate more full-length genes directly from the metagenomic DNA, and any genes generated will be characterized in subsequent studies.

In conclusion, culture-independent molecular methods led to recovery of abundant GH10 and GH11 xylanase genes from the sediment of the Lake Dabusu. Similarities among these sequences with known xylanases were low, and they were distantly related based on phylogenetic analysis. These results suggest that xylanase gene diversity within Lake Dabusu is high and most of them might be novel. Our study provides new insight into the genetic diversity and distribution of xylanases in soda lake environments and a rapid culture-independent molecular method for the retrieval of the source of xylanase genes with potential industrial applications.

Supporting Information

Table S1 GH10 xylanase gene fragments detected in the sediment of Lake Dabusu and their closest relative based on amino acid sequence identity and similarity.

Table S2 GH11 xylanase gene fragments detected in the sediment of Lake Dabusu and their closest relative based on amino acid sequence identity and similarity.

Table S3 Primers used for gene cloning and expression.

Author Contributions

Conceived and designed the experiments: JL XY. Performed the experiments: GW XH. Analyzed the data: GW TBN JL XY. Contributed reagents/materials/analysis tools: JL XY. Contributed to the writing of the manuscript: GW XH TBN JL XY.

References

1. Kulkarni N, Shendye A, Rao M (1999) Molecular and biotechnological aspects of xylanases. FEMS Microbiol Rev 23: 411–456.
2. Sunna A, Antranikian G (1997) Xylanolytic enzymes from fungi and bacteria. Crit Rev Biotechnol 17: 39–67.
3. Collins T, Gerday C, Feller G (2005) Xylanases, xylanase families and extremophilic xylanases. FEMS Microbiol Rev 29: 3–23.
4. Cantarel BL, Coutinho PM, Rancurel C, Bernard T, Lombard V, et al. (2009) The Carbohydrate-Active EnZymes database (CAZy): an expert resource for Glycogenomics. Nucl Acids Res 37: D233–238.
5. Biely P, Vrsanska M, Tenkanen M, Kluepfel D (1997) Endo-beta-1,4-xylanase families: differences in catalytic properties. J Biotechnol 57: 151–166.
6. Jeffries TW (1996) Biochemistry and genetics of microbial xylanases. Curr Opin Biotechnol 7: 337–342.
7. Beg QK, Kapoor M, Mahajan L, Hoondal GS (2001) Microbial xylanases and their industrial applications: a review. Appl Microbiol Biotechnol 56: 326–338.
8. Polizeli ML, Rizzatti AC, Monti R, Terenzi HF, Jorge JA, et al. (2005) Xylanases from fungi: properties and industrial applications. Appl Microbiol Biotechnol 67: 577–591.
9. Martin-Sampedro R, Rodriguez A, Ferrer A, Garcia-Fuentevilla LL, Eugenio ME (2012) Biobleaching of pulp from oil palm empty fruit bunches with laccase and xylanase. Bioresour Technol 110: 371–378.
10. Khandeparkar R, Bhosle NB (2007) Application of thermoalkalophilic xylanase from *Arthrobacter* sp. MTCC 5214 in biobleaching of kraft pulp. Bioresour Technol 98: 897–903.
11. Verma D, Kawarabayasi Y, Miyazaki K, Satyanarayana T (2013) Cloning, expression and characteristics of a novel alkalistable and thermostable xylanase encoding gene (Mxyl) retrieved from compost-soil metagenome. PLoS ONE 8: e52459.
12. Simkhada JR, Yoo HY, Choi YH, Kim SW, Yoo JC (2012) An extremely alkaline novel xylanase from a newly isolated Streptomyces strain cultivated in corncob medium. Appl Biochem Biotechnol 168: 2017–2027.
13. Mamo G, Delgado O, Martinez A, Mattiasson B, Hatti-Kaul R (2006) Cloning, sequence analysis, and expression of a gene encoding an endoxylanase from *Bacillus halodurans* S7. Mol Biotechnol 33: 149–159.
14. Gessesse A, Mamo G (1998) Purification and characterization of an alkaline xylanase from alkaliphilic *Micrococcus* sp AR-135. Journal of Industrial Microbiology and Biotechnology 20: 210–214.
15. Gessesse A (1998) Purification and properties of two thermostable alkaline xylanases from an alkaliphilic bacillus sp. Appl Environ Microbiol 64: 3533–3535.
16. Antony CP, Kumaresan D, Hunger S, Drake HL, Murrell JC, et al. (2013) Microbiology of Lonar Lake and other soda lakes. ISME J 7: 468–476.

17. Jones BE, Grant WD, Duckworth AW, Owenson GG (1998) Microbial diversity of soda lakes. Extremophiles 2: 191–200.
18. Zhao B, Chen S (2012) Alkalitalea saponilacus gen. nov., sp. nov., an obligately anaerobic, alkaliphilic, xylanolytic bacterium from a meromictic soda lake. Int J Syst Evol Microbiol 62: 2618–2623.
19. Rees HC, Grant WD, Jones BE, Heaphy S (2004) Diversity of Kenyan soda lake alkaliphiles assessed by molecular methods. Extremophiles 8: 63–71.
20. Foti MJ, Sorokin DY, Zacharova EE, Pimenov NV, Kuenen JG, et al. (2008) Bacterial diversity and activity along a salinity gradient in soda lakes of the Kulunda Steppe (Altai, Russia). Extremophiles 12: 133–145.
21. Lanzen A, Simachew A, Gessesse A, Chmolowska D, Jonassen I, et al. (2013) Surprising prokaryotic and eukaryotic diversity, community structure and biogeography of Ethiopian soda lakes. PLoS ONE 8: e72577.
22. Lin JL, Joye SB, Scholten JC, Schafer H, McDonald IR, et al. (2005) Analysis of methane monooxygenase genes in mono lake suggests that increased methane oxidation activity may correlate with a change in methanotroph community structure. Appl Environ Microbiol 71: 6458–6462.
23. Kovaleva OL, Tourova TP, Muyzer G, Kolganova TV, Sorokin DY (2011) Diversity of RuBisCO and ATP citrate lyase genes in soda lake sediments. FEMS Microbiol Ecol 75: 37–47.
24. Wang G, Meng K, Luo H, Wang Y, Huang H, et al. (2012) Phylogenetic diversity and environment-specific distributions of glycosyl hydrolase family 10 xylanases in geographically distant soils. PLoS ONE 7: e43480.
25. Shen J, Cao JT, Wu YH (2001) Paieoclimatic changes in Dabusu Lake. Chin J Oceanol Limnol 19: 91–96.
26. Brady SF (2007) Construction of soil environmental DNA cosmid libraries and screening for clones that produce biologically active small molecules. Nat Protoc 2: 1297–1305.
27. Wang G, Wang Y, Yang P, Luo H, Huang H, et al. (2010) Molecular detection and diversity of xylanase genes in alpine tundra soil. Appl Microbiol Biotechnol 87: 1383–1393.
28. White JR, Roberts M, Yorke JA, Pop M (2008) Figaro: a novel statistical method for vector sequence removal. Bioinformatics 24: 462–467.
29. Li W, Godzik A (2006) Cd-hit: a fast program for clustering and comparing large sets of protein or nucleotide sequences. Bioinformatics 22: 1658–1659.
30. Tamura K, Dudley J, Nei M, Kumar S (2007) MEGA4: Molecular Evolutionary Genetics Analysis (MEGA) Software Version 4.0. Mol Biol Evol 24: 1596–1599.
31. Saitou N, Nei M (1987) The neighbor-joining method: a new method for reconstructing phylogenetic trees. Mol Biol Evol 4: 406–425.
32. Schloss PD, Handelsman J (2005) Introducing DOTUR, a computer program for defining operational taxonomic units and estimating species richness. Appl Environ Microbiol 71: 1501–1506.
33. Huang H, Wang G, Zhao Y, Shi P, Luo H, et al. Direct and efficient cloning of full-length genes from environmental DNA by RT-qPCR and modified TAIL-PCR. Appl Microbiol Biotechnol 87: 1141–1149.
34. Petersen TN, Brunak S, von Heijne G, Nielsen H (2011) SignalP 4.0: discriminating signal peptides from transmembrane regions. Nat Methods 8: 785–786.
35. Miller GL, Blum R, Glennon WE, Burton AL (1960) Measurement of carboxymethylcellulase activity. Analytical Biochemistry 1: 127–132.
36. Solomon V, Teplitsky A, Shulami S, Zolotnitsky G, Shoham Y, et al. (2007) Structure-specificity relationships of an intracellular xylanase from *Geobacillus stearothermophilus*. Acta Crystallographica Section D 63: 845–859.
37. Ma Y, Zhang W, Xue Y, Zhou P, Ventosa A, et al. (2004) Bacterial diversity of the Inner Mongolian Baer Soda Lake as revealed by 16S rRNA gene sequence analyses. Extremophiles 8: 45–51.
38. Joshi AA, Kanekar PP, Kelkar AS, Shouche YS, Vani AA, et al. (2008) Cultivable bacterial diversity of alkaline Lonar lake, India. Microb Ecol 55: 163–172.
39. Pagaling E, Wang H, Venables M, Wallace A, Grant WD, et al. (2009) Microbial biogeography of six salt lakes in Inner Mongolia, China, and a salt lake in Argentina. Appl Environ Microbiol 75: 5750–5760.
40. Al-Zarban SS, Abbas I, Al-Musallam AA, Steiner U, Stackebrandt E, et al. (2002) *Nocardiopsis halotolerans* sp. nov., isolated from salt marsh soil in Kuwait. Int J Syst Evol Microbiol 52: 525–529.
41. Blanco J, Coque JJ, Velasco J, Martin JF (1997) Cloning, expression in Streptomyces lividans and biochemical characterization of a thermostable endo-beta-1,4-xylanase of *Thermomonospora alba* ULJB1 with cellulose-binding ability. Appl Microbiol Biotechnol 48: 208–217.

22

The Vocal Repertoire of Adult and Neonate Giant Otters (*Pteronura brasiliensis*)

Christina A. S. Mumm[1]*, Mirjam Knörnschild[2,3]

1 Institute of Experimental Ecology, University of Ulm, Ulm, Germany, **2** Institute of Experimental Ecology, Faculty of Natural Sciences, University of Ulm, Ulm, Germany, **3** Smithsonian Tropical Research Institute, Balboa, Panama

Abstract

Animals use vocalizations to exchange information about external events, their own physical or motivational state, or about individuality and social affiliation. Infant babbling can enhance the development of the full adult vocal repertoire by providing ample opportunity for practice. Giant otters are very social and frequently vocalizing animals. They live in highly cohesive groups, generally including a reproductive pair and their offspring born in different years. This basic social structure may vary in the degree of relatedness of the group members. Individuals engage in shared group activities and different social roles and thus, the social organization of giant otters provides a basis for complex and long-term individual relationships. We recorded and analysed the vocalizations of adult and neonate giant otters from wild and captive groups. We classified the adult vocalizations according to their acoustic structure, and described their main behavioural context. Additionally, we present the first description of vocalizations uttered in babbling bouts of new born giant otters. We expected to find 1) a sophisticated vocal repertoire that would reflect the species' complex social organisation, 2) that giant otter vocalizations have a clear relationship between signal structure and function, and 3) that the vocal repertoire of new born giant otters would comprise age-specific vocalizations as well as precursors of the adult repertoire. We found a vocal repertoire with 22 distinct vocalization types produced by adults and 11 vocalization types within the babbling bouts of the neonates. A comparison within the otter subfamily suggests a relation between vocal and social complexity, with the giant otters being the socially and vocally most complex species.

Editor: Dennis M. Higgs, University of Windsor, Canada

Funding: The German Academic Exchange Service (https://www.daad.de/de/index.html) granted a short time stipend for graduate students to CASM, D/10/520, which covered the travelling expenses for the data collection in Peru from September 12th to December 8th in 2011. The Landesgraduiertenförderung Baden-Württemberg (https://www.studieninfo-bw.de/studieren/studienfinanzierung/stipendien/leistungsstipendien/landesgraduiertenfoerderung/) is funding the graduate study of CASM, funding number 1104, 2012-2014. The funders had no role in study design, data collection and analysis, decision to publish, or preparation of the manuscript.

Competing Interests: The authors have declared that no competing interests exist.

* Email: Christina.mumm@uni-ulm.de

Introduction

The complexity of mammalian vocal repertoires is both the sum of several aspects of call structure and function, as well as of evolutionary and selective pressures acting on senders and receivers. Production and perception mechanisms, such as the anatomic structure of the vocal tract or auditory sensitivity, shape and limit the number of distinct vocalizations a species can produce and perceive [1–4]. As described by the source-filter theory [5,6], source (i.e. larynx) induced acoustic parameters include duration, tempo, fundamental frequency and nonlinear phenomena [7,8]. Nonlinear phenomena, being subharmonics, biphonation and deterministic chaos [8], seem to be a rather involuntary by-product of vocal production, nevertheless, increasing the diversity of acoustic signals [9]. Filter (i.e. supralaryngeal vocal tract) induced vocal parameters include formants which, among other cues, indicate the body size of the signaller [10].

Vocalizations can be internally referential, providing information on motivational state, individuality or social origin of the sender, or externally referential, providing information on external events for receivers [1]. According to Morton's motivation-structural rules [11], the physical structure of mammalian and avian vocal signals reflects the motivation and context in which they are produced. Aggressive vocalizations should have lower frequencies and sound harsh, whereas friendly calls should have higher frequencies and sound pure [11]. Bradbury and Vehrencamp [12] refined these motivation-structural rules for the design of mate-attracting, courtship, territorial defence, threat and alarm signals.

The vocalizations within many mammalian repertoires are neither completely discrete, nor completely graded [13,14]. They rather represent a combination of distinct calls, graded signals, variants, and transitions between them [12,15,16]. By varying certain acoustic parameters, animals can use variants of graded vocalizations to signal a more detailed information about their internal state, motivation, or the degree of external danger [12]. Segmental concatenation, being the combination of different vocal cues in one call type [17], as well as a structured combination of vocalizations, resulting in a different or more specific meaning than that of single calls, contribute to the enhancement of a vocal repertoire [18–20].

Cognitively advanced mammals such as primates do not only perceive and recognize distinct meaning from graded call subtypes [18], but may respond differently to apparently similar calls, or, conversely, may be able to derive the same meaning from structurally different vocalizations [21]. This shows that the acoustic structure of vocalizations can, but does not have to be linked with function [21]. Rather, communication is shaped by a combination of the information transmitted, the behavioural context, the relationship of the interacting individuals and, particularly in social groups, by the presence of bystanders [21,22].

The ontogenetic development of a mammalian vocal repertoire undergoes different stages [23]. Some neonate or juvenile calls might disappear during development, whereas the number of vocalizations of the adult repertoire exceeds the number of juvenile calls [16,23]. Structure, usage and the number of vocalizations may change until the full adult and species-specific repertoire is formed [16,23,24]. Typically, the fundamental frequency decreases with age and growth [16,23]. The high frequency and pure tone quality of dependant offspring vocalizations are well suited to elicit parental care behaviour [11]. Babbling behaviour, generally described to facilitate practising the adult vocalizations [25–27], may also direct the attention of caregivers to the vocalizing cub [25,26,28].

Recently, the 'social complexity hypothesis for communication' [29] has been supported by several studies on non-human primates [30,31], birds [32,33], marmots [34] and carnivores [35,36]. This hypothesis states that vocal complexity can be driven by social complexity [29]. Vocal complexity can thereby be classified in terms of vocal repertoire size [31,34,35], information content of the vocalizations or within the vocal system [32], as well as usage diversity of the vocal signals [30,33]. Group size [32,35], or the variety of social roles and relationships [34] can be used to measure social complexity [33]. Effects of habitat, predation pressure or species recognition do not necessarily promote vocal complexity even though they shape the structure of vocalizations [29]. Group size, as a measure for social complexity, is thought to favour the evolution of individual acoustic distinctiveness [37], as well as vocal complexity at the species level [30]. Living in social groups built of closely and frequently interacting individuals with different social roles is demanding and requires a sophisticated communication system (sensu [38,39]). This allows the interacting individuals to deal adequately with challenges originating from their social structure.

With this paper, we want to push forward the formerly poorly noted otters, *Lutrinae*, into the general discussion of acoustic communication and vocal complexity. Generally, otters represent a well-suited taxon for testing the relation of sociality and vocal complexity (as suggested in [40]). They provide a wide variety of social organization from solitary living species to complex social groups [41] and the otter species can be differentiated by their vocalizations [42]. In this paper, we discuss social organization and vocal repertoire size as measures of complexity in otters.

Known for their conspicuous and frequently emitted vocalizations, giant otters (*Pteronura brasiliensis*) are interesting in terms of both their vocal repertoire and their complex social life. This species, endemic to lakes and rivers in Amazonian rainforests and wetlands [41,43], is listed as 'endangered' in the red list of threatened species [44]. Giant otters usually live in family groups with the alpha-couple and their offspring from different years [42–45], resulting in group sizes between three to nine individuals [43,44,46]. A detailed study by Ribas [45] revealed a higher complexity of social organization and variety of relatedness in giant otter groups, differing from the general pattern of a 'parent-brood model' [45] described before. Subadult giant otters may stay as helpers within the family or leave their natal territory when reaching sexual maturity [42,47,48]. The dispersing individuals might stay solitary ('transients') [43,46] or form temporary groups of unrelated members, until they establish their own territory and family group [45]. Group members show a strong association index [46], describing the probability of observing pairs of individuals together (Half-Weight Index, described in [49]). The group cooperates in breeding and territorial defence [43,46,50,51], whereas it is still unclear, whether giant otters truly show cooperative hunting or not [43]. Mostly, group members fish in close proximity. Smaller sized fish are eaten individually, but elder group members might share fish with dependant offspring [41–43]. Giant otters feed together on larger prey like catfish or juvenile caimans ([43], own observation). Giant otter cubs join the fishing bouts of their family at an age of two months. Until they learn to fish on their own, they depend on food sharing from elder group members [43]. Starting with direct provision of small dead fish, elder group members will also provide larger or half-dead fish later on during development of the juveniles, to increase their abilities of capturing and handling prey ([43], own observation). Vocalizations of giant otters have partly been described by previous studies [42,43,52,53]. Leuchtenberger [54] studied the vocal repertoire of wild giant otters in Brazil. She found 15 vocalization categories with several subtypes, resulting in a total of 19 distinct vocalizations. We recently reported vocal individuality in the cohesion calls of giant otters [40] and Leuchtenberger [54] showed sex differences in the giant otter alarm call 'snort'.

To describe the vocal repertoire, we recorded and analysed airborne and underwater calls of wild and captive giant otter groups and provide the behavioural context in which different vocalizations were produced. Furthermore, we described neonate vocalizations and contrast them with the adult vocal repertoire.

The aim of the present study was to give a detailed overview of the giant otters' vocal repertoire and set it into the context of social complexity. Due to the outstandingly social system of giant otters within the *Lutrinae* (compare [41,55–58]), we hypothesized to find a sophisticated vocal system with distinct vocalization types. Moreover, we hypothesized that giant otter vocalizations show a clear relationship between signal structure and signal function according to Morton's motivation-structural rules, and that the neonate vocal repertoire would include precursors of the adult repertoire, as well as age-specific vocalizations.

Materials and Methods

Study sites and study animals

We recorded five wild and three captive giant otter groups. Group size varied from two to fifteen individuals, covering all age classes from new born cubs to adults (giant otter age classes according to [59]). We recorded the vocalizations of wild giant otters at five oxbow lakes, namely Cocha Cashu, Cocha Salvador (September to December 2011), Cocha Tres Chimbadas, Cococha and Cocha Sandoval (April to July 2012) in the Cusco and Madre de Dios Departments, Peru. The lakes Cocha Cashu (N -11°53′3.9984″, E -71°24′28.0008″) and Cocha Salvador (N -11°59′45.9996″, E -71°13′59.0016″) are located within the Manu National Park. Cocha Sandoval (N -12°36′29.4336″, E -69°2′26.9988″) and Cococha (N -12°49′0.624″, E -69°15′36.3456″) are within the Tambopata National Reserve and Cocha Tres Chimbadas (N -12°47′21.7932″, E -69°20′44.0988″) in the reserves buffer zone.

The recordings of the vocalizations of captive giant otters took place in three German zoos, Tierpark Hagenbeck (April 2009, May 2011 and July 2011), Zoo Duisburg (March 2011) and Zoo

Dortmund (March 2009 and April 2011). The enclosures included separable indoor and outdoor areas. The indoor enclosures with artificial light mainly served as retreat areas for the animals and the outdoor enclosures with natural light were always open for visitors. When cubs were born, the zoos inhibited public access to the retreat areas. Giant otters were fed an amount of 2.0 to 6.0 kg of fish (trout, whiting and roach) per day and individual. The animals received supplementary fruit and vegetables as enrichment.

Animal welfare

Since giant otters are listed as endangered by the IUCN red list of threatened species [44], we obtained official research permits (No. 014 S/C- 2011-SERNANP-PNM, 014-2012-SERNANP-JEF, 017-2012-SERNANP-JEF and 0167-2012-DGFFS-DGEFFS) for field work in Peru. The permits provided by SERNANP (Servicio Nacional de Áreas Naturales Protegidas), the Peruvian nature conservation authority and DGFFS (Dirección General Forestal y de Fauna Silvestre), the Peruvian agricultural department, allowed us to record wild giant otter groups. To habituate the giant otters to our presence, we never hid our activities, but kept a minimum observing distance of 10 m–50 m, depending on the giant otters' activities. We increased the distance when new born cubs were present. Wild neonates were recorded coincidentally at one lake, as the group moved the cubs from one den to another.

In Germany, we received research permissions to conduct our study with the captive giant otters from the respective persons in charge (Tierpark Hagenbeck: veterinarian, Zoo Dortmund: zoo director and Zoo Duisburg: curator.). We never separated individual giant otters from the group during recording and abandoned recording when new cubs were born. At Tierpark Hagenbeck, an autonomous recording device was installed at the nesting box inside the indoor enclosure to record the neonates instead.

Recording

We recorded air-borne vocalizations as wave files with a directional microphone (Sennheiser, MKH 416-P48U3) connected to a digital audio recorder (Zoom H2 Handy Recorder; 48 or 96 kHz sampling rate, 24 bit depth resolution). Underwater calls were recorded with a hydrophone (RESON TC4013 Hydrophone, 1 Hz to 170 kHz), amplified by a charge amplifier (#42003, P48), connected to a second audio recorder (Zoom H2 Handy Recorder; 96 kHz sampling rate, 24 bit depth resolution). We documented the behavioural context of air-borne vocalizations by spoken notes and video recordings (Sony, DCR SR-35 camcorder). Underwater behaviour and the behaviour of the captive neonate cubs inside the nesting box could not be recorded. In the wild, we followed the free ranging giant otters with one-person kayaks. The otters' daily activity period from sunrise to sunset (around 5 am to 5 pm), was monitored by two observers with four alternating three hours shifts. In the zoos, we conducted alternating three hours recording sessions, covering the giant otters' activity period in the morning, afternoon or evening.

Classification and analysis of the vocalizations

We classified the adult vocalizations acoustically and visually with respect to the behavioural context. Following Tyack and Miller [60], we named the distinct giant otter calls according to their structure and with reference to the classification by other authors. Subsequently, we assigned them to the behavioural context to avoid a confusion of call structure and function [60]. Only calls having a clear specific function were named accordingly (for instance the begging call and the contact call). Since the giant otter vocal repertoire includes several gradations and variations of the vocalizations [42], we restricted our analysis to vocalizations which could readily be distinguished by structure, context, or a combination of both.

We selected the calls for analysis from the original wave files in Raven Pro (version 1.4, Cornell University, Ithaca, New York). We restricted our selection to vocalizations with a good signal-to-noise ratio, which were not overlapped by other calls and for which we had additional information on the behavioural context for analyses. To reduce an effect of very similar or modulated vocalizations within calling bouts of the same individual [61], we selected the vocalizations from different recording dates whenever possible. Before measuring the acoustic parameters in Raven Pro, we conducted noise reduction (WavePad Sound Editor, version 4.52, NCH Software), erased bird calls or loud background noise (AviSoft SASlab Pro, version 5.1.23, Avisoft Bioacoustics, Berlin, Germany) and normalized the volume of the vocalizations (WavePad Sound Editor). In Raven Pro, we measured frequency variables from a spectrogram and a selection spectrum with high frequency resolution (Fast Fourier transform: 1024-point, window: Hann, overlap: 87.5%, temporal resolution: 2.67 ms, DFT size: 1024 samples, frequency resolution: 46.9 Hz) and time variables from the waveform. In sum, we measured 441 calls, stemming from 8 giant otter groups. Each group consisted of 6–15 individuals (totalling 66 individuals) but many vocalizations could not be attributed to a specific individual. Due to low sample size and low quality of the records, we did not measure the mating call. Neonate calls were not measured due to low recording quality, but are shown in the results to give an overview of the vocal repertoire of giant otters as accurately as possible. The vocalizations and their behavioural context were compared to former findings and descriptions by other authors.

Measurements of acoustic parameters

Aiming for parameters that would best differentiate the vocalizations within the adult vocal repertoire, we analysed the calls both over their entire length and within four subunits of equal duration. Entire vocalizations were measured by selecting a spectrogram section from the highest to the lowest frequency visible (narrower when no harmonics were present and wider when harmonics were present) in Raven Pro. We measured three temporal and three spectral parameters (duration, time to the peak of the dominant frequency, time to peak amplitude, highest and lowest frequency, dominant frequency), and both aggregate and average entropy [62], resulting in 8 parameters for the entire vocalization. The four subunits were measured over the bandwidth of the fundamental frequency, to avoid confounding effects of presence or absence of higher harmonics due to recording conditions. Furthermore, giant otter vocalization types vary strongly in emphasizing different harmonics. We choose to measure the fundamental frequency to gain variables comparable among the vocalization types, as well as among the whole vocal repertoire. In each of these four parts, we measured two temporal parameters (time to the peak of the fundamental frequency, time to peak amplitude), three spectral parameters (lowest, highest and peak frequency of the fundamental frequency), one wave form derived parameter (peak amplitude) and the aggregate and average entropy [62], resulting in 8 parameters for each subunit. We converted the values of peak amplitude within the subunits to percentage in relation to peak amplitude of the entire call.

To describe the frequency contour of the vocalizations, we measured frequency and time at 9 points of equal time distance along the fundamental frequency. The 9 points represented start,

Table 1. Description of the giant otters' vocal repertoire and behavioural context.

Vocalization	Behavioural context	Range	Frequency of occurrence	Age	Wild or captive	Descriptions by other authors
Cohesion: contact and coordination calls						
Bark	Calling or greeting group members, swimming or fishing together, moving together on land, playing	Short – moderate	Occasional	Subadults,Adults	Wild and captive	Coo [54]
Close call	Calling group members, moving together on land, playing	Short – moderate	Occasional	Juveniles, Subadults, Adults	Captive	Coo call [54]
Contact call	Calling group members, swimming or fishing together, moving together on land, playing, begging	Moderate	Very common	Cubs, Juveniles, Subadults, Adults	Wild and captive	Adult call [54], Contact call [40], Grito seco [52], Grito de filhote [52]
Contact call gradation	Calling group members, moving together on land, playing, begging	Moderate	Very common	Cubs, Juveniles, Subadults, Adults	Wild and captive	No description
Hum	Calling group members, swimming or fishing together, moving together on land, change of direction, resting, playing, scent-marking, soothing	Short	Very common	Cubs, Juveniles, Subadults, Adults	Wild and captive	Hum [40,42], Hum and Purr [54], Kontaktsummen [43], Murmúrio [52]
Hum gradation	Calling or greeting group members, swimming together, moving together on land, change of direction or departure, playing	Short – moderate	Very common	Juveniles, Subadults, Adults	Wild and captive	Coo [42,52], Coo-hum [54], Let's go hum [96]
Hum short	Emerging	Short	Occasional	Juveniles, Subadults, Adults	Wild	No description
Isolation call	Calling group members	Large	Occasional	Juveniles, Subadults, Adults	Wild	Cub lost whine [42], Verlorenenruf [43]
Whistle	Calling group members, begging	Short	Occasional	Cubs, Juveniles, Subadults, Adults	Wild and captive	Whistle [42]
Whistle double	Monitoring the surrounding, begging	Moderate	Rare	Cubs, Juveniles, Subadults	Wild and captive	No description
Underwater call	Diving	Unknown	Unknown	Unknown	Wild and captive	Unterwasserlaute [53]
Alarm and threatening vocalizations						
Vocalization	**Behavioural context**	**Range**	**Frequency of occurrence**	**Age**	**Wild or captive**	**Descriptions by other authors**
Growl	Defending a fish, threatening or warning, playing	Moderate	Very common	Cubs, Juveniles, Subadults, Adults	Wild and captive	Adult growl [54], Growl [42], Knurren [43]
Hah!	Warning, submission	Short	Occasional	Cubs, Juveniles, Subadults, Adults	Wild and captive	Hah [42,54]
Snort	Alarm, threatening or warning	Moderate - large	Very common	Cubs, Juveniles, Subadults, Adults	Wild and captive	Snort [42,52,54], Warnschnauben [43]
Wavering scream	Alarm, begging	Large	Occasional	Subadults, Adults	Wild and captive	Adult scream [54], Grito ondulado [52], Schreien [43], Wavering scream [42]

Table 1. Cont.

Cohesion: contact and coordination calls						
Vocalization	**Behavioural context**	**Range**	**Frequency of occurrence**	**Age**	**Wild or captive**	**Descriptions by other authors**
Begging calls						
Vocalization	**Behavioural context**	**Range**	**Frequency of occurrence**	**Age**	**Wild or captive**	**Descriptions by other authors**
Ascending scream	Begging, stealing prey	Moderate - large	Common	Juveniles, Subadults, Adults	Wild and captive	No description
Begging call	Begging	Moderate	Very common	Cubs, Juveniles, Subadults, (Adults)	Wild and captive	Bettelgeschrei [43], Cub call [54], Food begging call [42], Grito de filhote [52]
Begging scream	Begging	Large	Very common	Juveniles, Subadults, (Adults)	Wild and captive	Bettelgeschrei [43], Cub begging scream [54], Food begging call [42], Grito de filhote [52], Squeak [54]
Begging scream gradation	Begging	Large	Common	Cubs, Juveniles, Subadults, (Adults)	Wild and captive	Adult and cub high scream [54], Bettelgeschrei [43], Food begging call [42]
Whine	Begging, Alarm	Moderate - large	Common	Juveniles, Subadults, Adults	Wild and captive	Grito ondulado [52]
Other						
Vocalization	**Behavioural context**	**Range**	**Frequency of occurrence**	**Age**	**Wild or captive**	**Descriptions by other authors**
Mating call	Copulation	Moderate	Rare	Adults	Wild and captive	No description
Suckling call	Nursing	Moderate	Occasional	Cubs, Juveniles, Subadults	Wild and captive	Nursing hum [59], Scream-gurgle [54]

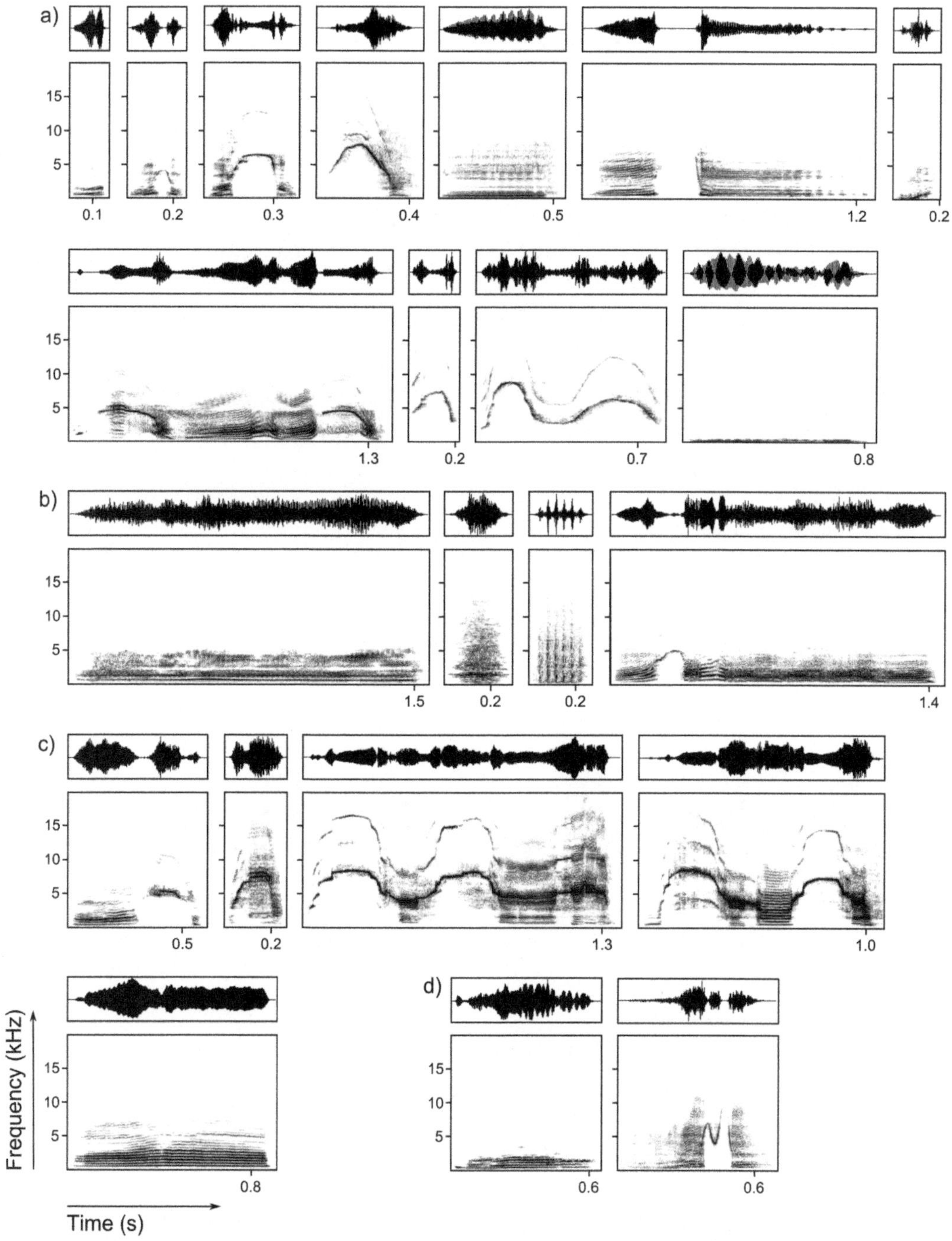

Figure 1. Exemplary calls for the vocal repertoire of giant otters. Calls were obtained from wild and captive individuals. The spectrograms depict frequency over time and were generated using a 1024-point FFT and a Hann window with 75% overlap. The oscillograms show pressure changes over time. a) Cohesion: contact and coordination calls. From left to right: bark, close call, contact call, contact call gradation, hum, hum gradation, hum short, isolation call, whistle, whistle double, and underwater call. b) Alarm calls. From left to right: growl, hah!, snort, and wavering scream. c) Begging calls. From left to right: ascending scream, begging call, begging scream, begging scream gradation, and whine. d) Other calls. From left to right: mating call and suckling call.

mid and end point of each subunit. Since the end points of the subunits one, two and three are also the start points of the subunits two, three and four, those points were only measured once. Time point 9 was redundant with the duration of the entire vocalization and was therefore not included. The onset of each vocalization in time was set to zero seconds and was also not included. Thereby, we measured 9 frequency and 7 time parameters to describe the frequency contour. Additionally, we quantified the occurrence of nonlinear phenomena (NLP). Therefore, we measured the proportion of chaos, subharmonics and biphonation in relation to the duration of the entire call. Overall, we obtained 59 acoustic variables to describe each vocalization.

Table 2. Assessment of model fit of the discriminant function analysis.

Function	Eigenvalue	% of Variance	Test of Function	Wilks' Lambda	Chi-square	df	P
1	4.314	38.2	1 to 9	0.002	2596.028	180	<0.0001
2	2.205	19.5	2 to 9	0.012	1886.126	152	<0.0001
3	1.639	14.5	3 to 9	0.038	1391.124	126	<0.0001
4	1.176	10.4	4 to 9	0.100	978.774	102	<0.0001
5	0.990	8.8	5 to 9	0.218	648.253	80	<0.0001
6	0.447	4.0	6 to 9	0.433	355.809	60	<0.0001
7	0.227	2.0	7 to 9	0.626	198.741	42	<0.0001
8	0.181	1.6	8 to 9	0.769	111.866	26	<0.0001
9	0.102	0.9	9	0.908	41.154	12	<0.0001

Statistics

To reduce the number of measured parameters, we conducted a principal component analyses with varimax-rotation. The principal component analyses fulfilled Kaiser-Meyer-Olkin (KMO) and Bartlett's test criteria and 10 principal components (PCs) with eigenvalues >1 were extracted. We used the 10 PCs to run a stepwise discriminant function analysis (DFA) with leave-one-out cross-validation. Nine of the 10 PCs were included by the DFA. We conducted a one-tailed binomial test to determine whether the DFA classification result differed significantly from a random classification (according to [63]). All statistical tests were performed with SPSS (version 21, SPSS Inc., Chicago, IL, U.S.A.).

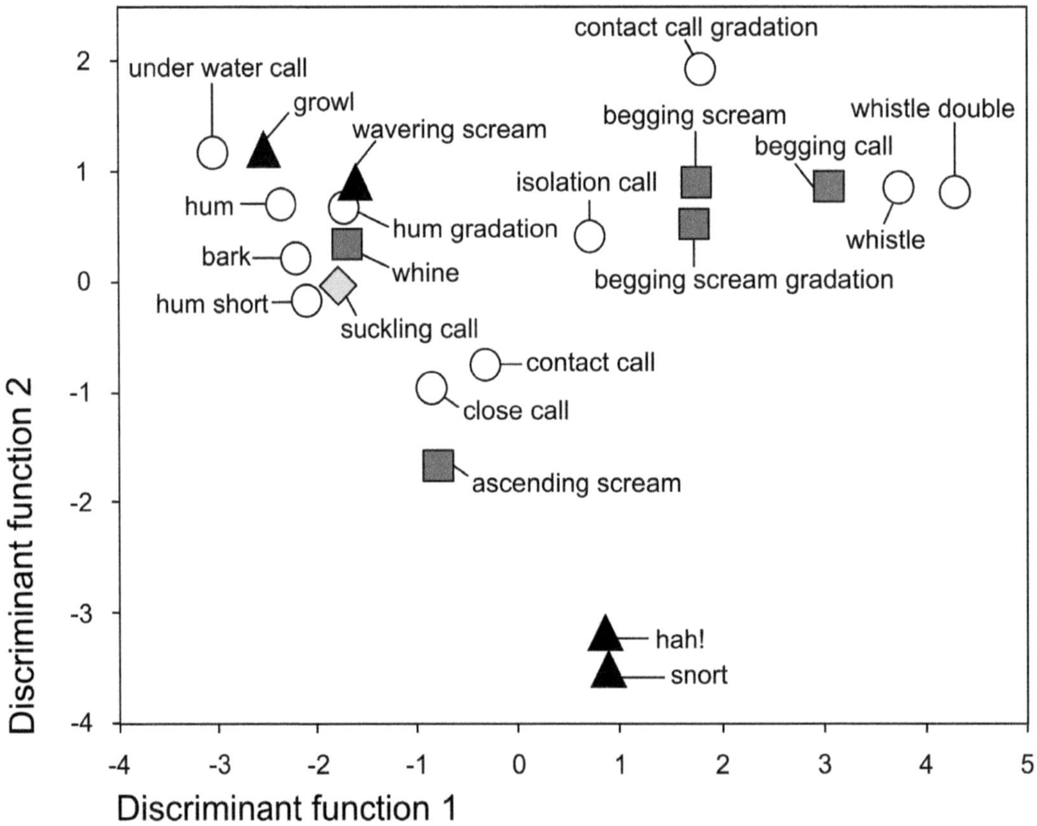

Figure 2. Signal space of the giant otters' adult vocal repertoire defined by the first two discriminant functions. Symbols indicate group centroids. Discriminant function 1 was mainly shaped by the peak frequency within subunit 1 and the fundamental frequency at 1/3 duration of the entire call (combined in one principal component). Discriminant function 2 was mainly shaped by the average entropy in the subunits 2 and 3 (combined in another principal component).

Table 3. Percentage of misclassified cases in the cross-validated DFA, listing the vocalizations with which the misclassified cases have been confused by the DFA*.

Call (n = number of cases)	Cross-validated cases	
	Correctly classified (%)	mainly misclassified to (n = number of misclassified cases)*
Bark (n = 21)	38.1	Hum (n = 8)
Close call (n = 10)	30.1	Ascending scream (n = 2), contact call (n = 2)
Contact call (n = 32)	62.5	**
Contact call gradation (n = 22)	88.0	**
Hum (n = 30)	70.0	**
Hum gradation (n = 29)	65.5	**
Hum short (n = 9)	33.3	Bark (n = 6)
Isolation call (n = 16)	43.8	Begging scream (n = 3), begging scream gradation (n = 3)
Whistle (n = 18)	77.8	**
Whistle double (n = 13)	53.8	**
Underwater call (n = 9)	0.0	Hum (n = 5)
Growl (n = 29)	55.2	**
Hah! (n = 17)	52.9	**
Snort (n = 32)	71.9	**
Wavering scream (n = 11)	18.2	Growl (n = 4)
Begging call (n = 21)	76.2	**
Begging scream (n = 32)	50.0	Begging scream gradation (n = 7)
Begging scream gradation (n = 30)	33.3	Begging call (n = 5)
Ascending scream (n = 25)	60.0	**
Whine (n = 20)	60.0	**
Suckling call (n = 12)	0.0	Growl (n = 4)

* given for calls correctly classified less than 50.0%. ** Original vocalization was correctly classified in >50.0% of the cases.

Results

The vocal repertoire of adult giant otters comprised 22 distinct vocalization types (table 1, figure 1, table S1, table S3, audio S1). Due to its rare occurrence, we did not include the mating call in our statistical analyses. The DFA result supported our classification by assigning the 441 calls to the correct vocalization type in 55.3% of the cross-validated cases (for detailed DFA results, see table 2, figure 2). This classification result differed significantly from a random classification (4.8%; one-tailed binomial test: $p = 0.0001$). The vocalizations were best divided by the peak frequency within subunit 1, the fundamental frequency at 1/3 duration of the entire call (combined in one PC, mainly shaping discriminant function 1), and by the average entropy in the subunits 2 and 3 (combined in another PC, mainly shaping discriminant function 2). Misclassified calls in the cross-validated DFA were mainly misclassified to structurally very similar calls (table 3). For instance, the isolation call was misclassified as belonging to the begging screams in 37.6%. Even though these vocalizations were very similar in structure, only an individual that lost sight of the group emitted isolation calls. Then the group approached the isolated animal, often answering with wavering screams and contact calls. Begging screams were only emitted during intense begging.

We found 11 structurally distinguishable vocalizations within the vocal repertoire of new born cubs (figure 3, table 4, audio S2). The neonate vocalizations included begging call-like and contact call-like vocalizations, distress calls, hum-like vocalizations, whistles, and the suckling call described below. We found no equivalent to the high whistle and the distress calls in the adult repertoire.

Structure and behavioural context of the vocalizations

In table 1, we give a brief overview of the distinct vocalizations within the vocal repertoire. We described the calls found in our study in four major behavioural categories: a) cohesion, including all vocalizations emitted in the context of staying in contact and coordination of the group, b) alarm and threatening vocalizations, referring to warning or threatening behaviour, c) begging vocalizations, and d) other contexts such as mating and nursing. Within these categories, we listed the calls alphabetically. The examination of the behavioural context revealed that giant otters produce structurally similar vocalizations in various situations, as well as using different vocalizations in an apparently similar context.

Cohesion calls

In the context of group cohesion, we found 11 distinct vocalizations, which we defined as bark, close call, contact call, contact call gradation, hum, hum gradation, hum short, isolation call, underwater call, whistle and whistle double. Several of these vocalizations seemed to be structurally derived from the contact call.

Bark. This is a very short, constant frequency vocalization, resembling the constant frequency parts of the contact call. We recorded it in contexts of movement (swimming or fishing together, moving together on land), group cohesion (calling or

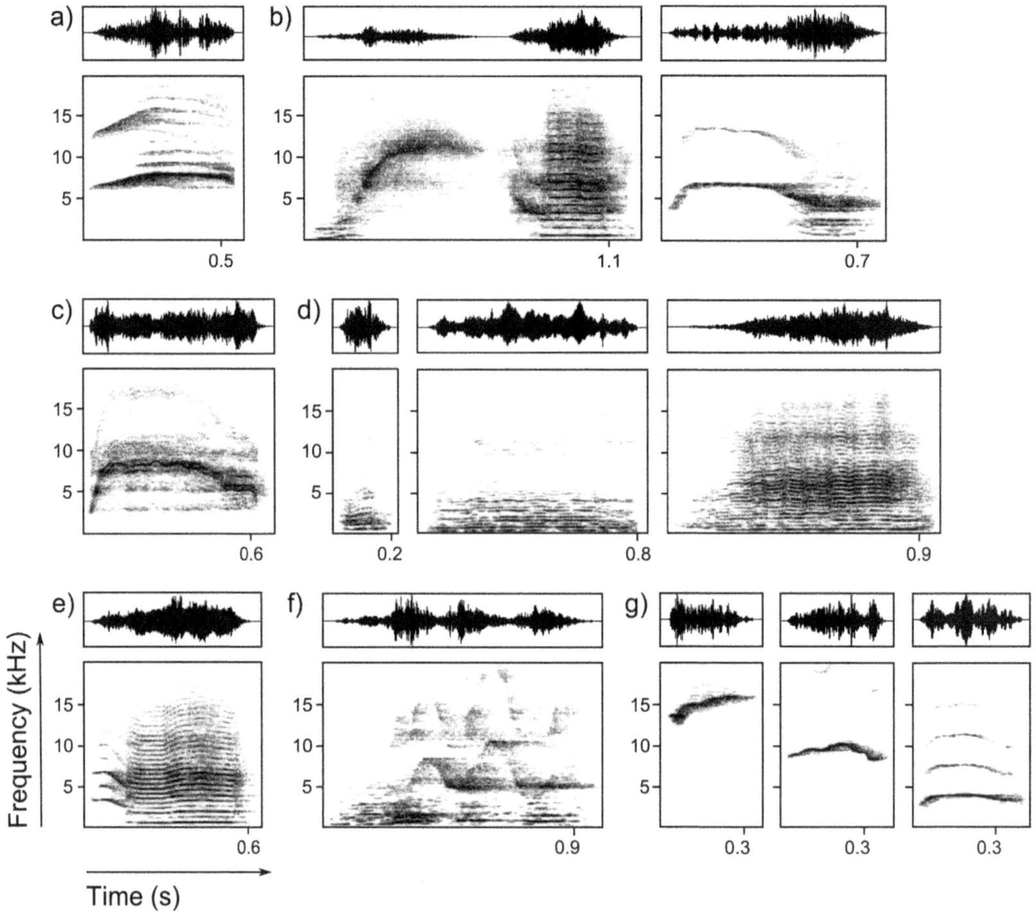

Figure 3. Exemplary calls from neonate giant otters. Vocalizations were recorded from one wild and one captive litter. The spectrograms depict frequency over time and were generated using a 1024-point FFT and a Hann window with 75% overlap. The oscillograms show pressure changes over time. a) Begging call-like vocalization. b) Contact call-like vocalizations. c) Distress call 1. d) Hums. From left to right: bark-like call, hum-like call, and distress call 2. e) Hum gradation-like vocalization. f) Suckling call. g) Whistles. From left to right: high whistle, whistle, and low whistle.

greeting group members) and playing, but could not detect its detailed function. *Close call.* This call is structurally very similar to the contact call, but shorter and the maximum frequency is lower compared to the contact call. It is produced as a cohesion call in close contact (moving together on land, calling group members, playing). *Contact call.* This modulated call starts and ends with a quasi-constant frequency part and has a whistle-like part in the middle. Its main function is to stay in contact with group members and to coordinate group movements. It is produced in situations of visual separation, for instance during fishing by one individual above the surface, while the others are still diving, or when group members are separated over larger spatial distances, as well as in any other context of movement. It may also be included in begging bouts or emitted during play. *Contact call gradation.* This call is structurally very similar to the contact call, but lacks the first quasi-constant frequency part and thus, starts with the whistle-like part. Like the contact call and the close call, it is produced in the context of cohesion (moving together on land, calling group members, playing). It is more often emitted within begging bouts than the contact call. *Hum.* This constant frequency vocalization has a variable duration. It is a short-range contact and cohesion call of low volume, frequently emitted in group activities (swimming or fishing together, moving together on land, resting, playing, scent-marking). It is also used to soothe group members in social interactions or when a group passes the boat of the observer. *Hum gradation.* The structure of the hum gradation reaches from a strongly accentuated hum to a call combination of a hum and a contact call. The contact call part of this vocalization can be included at the beginning, middle or end of the hum. The hum gradation, like hum and contact call alone, was produced in the context of group cohesion. We mainly recorded it when the group changed its direction, or when group members met. The functions of hum (cohesion in close contact and soothing) and contact call (contact, cohesion over larger distances, or reunion after separation) seem thereby combined to the slightly different and more specific meaning of directional change. In terms of group reunion, this call combination can signal individuality and friendly intent by soothing the approaching individuals. *Hum short.* This call is structurally very similar to the bark, but only heard when a giant otter emerges after diving. Other authors have not yet described this call, and we recorded it only from wild giant otters. *Isolation call.* This loud piercing scream is similar in structure to the begging scream gradation and the wavering scream. It is only emitted by individuals who lost sight of the group. The group may answer with wavering screams and contact calls, changing to hums and hum gradations when reaching visual distance. Especially inexperienced juvenile giant otters often get separated from the group during fishing. As long as their fishing and swimming abilities are not fully developed, they tend to eat prey on fallen trees at the lakeshore, whereas the group might move on fishing (own observation). We listed the isolation call in the behavioural context of group coordination and cohesion, since its emission by an isolated individual will result in answering and approaching behaviour of the group. Nevertheless, the call structure applies to distress calls, belonging to assembly alarm vocalizations [12]. The tonal and modulated isolation call is repeated until the group responds and thus, it is designed to attract the attention of group members, likely encoding the location of the lost individual and the urgency of the situation. *Whistle*. This short, modulated and tonal call is less audible than the contact call. Since we recorded it mainly in situations of movement, especially with adults whistling and their cubs following them subsequently, we assume it to be a contact call as well, even though we could not detect its detailed function. *Whistle double*. This rarely emitted vocalization is tonal with two modulations. It was mainly produced while monitoring the surroundings and sometimes emitted during begging. We could not detect its detailed function. *Underwater call*. In the wild, as well as in the zoos, we only recorded one distinct underwater vocalization. This very low frequency call resembles the air-borne hum and we assume it to function as a cohesion call as well, probably to ease coordination of movements during hunting underwater.

Alarm and threatening vocalizations

Alarm and threatening vocalizations were the growl, hah!, snort, and the wavering scream. Especially screams and alarm signals included the 'sound of arousal' [64], i.e. nonlinear phenomena.

Growl. This typical carnivore threat growl has a quasi-constant frequency and is graded according to the intenseness of arousal. Its main function is to defend a fish against begging group members, but it is also emitted in a playing context. In a submissive context, offspring greeting their parents may emit a soft growl after a hah!. *Hah!*. The hah! is a short and sharp exhalation. Giant otters emit the hah! when detecting unusual objects, animals or humans. It might function as a warning of group members against intruders or humans. In a submissive context, offspring greet their parents shaking the head with open mouth and emit the hah!, sometimes followed by a soft growl. *Snort*. The snort is an elongated and accentuated sharp exhalation with a varying number of pulses. It is an alert alarm signal to warn group members against caimans, humans, or any other potential danger, but is also used to threaten conspecifics in conflicts over food or dominance rank. As an alarm call, intermediate between assembly and alert signals, monitoring or periscoping otters emitted these calls to inform group members about potential danger. *Wavering scream*. This call is a loud, sometimes more howl-like, sometimes more piercing modulated scream. It mainly serves as an assembly call, uttered during caiman interactions or acoustic territorial marking and can be heard over a large distance. It is also emitted as an answer to isolation calls. Often group members scream simultaneously with one individual emitting the first scream and the rest of the group joining into a chorus. Single or repeated wavering screams also occur in begging bouts.

Begging vocalizations

The vocalizations emitted in the context of begging behaviour included the ascending scream, the begging call, begging scream and begging scream gradation, and the whine.

According to our observations, begging vocalizations uttered in begging bouts always elicit a response from the individual with a fish. The response might be giving the fish to the begging individual or defending the fish and refusing to provide it. The denial is often accompanied by growls. In case of a denial to share, the begging individual might intensify its begging vocalizations or might even try to steal the fish. All begging vocalizations occurred in the penetrative begging bouts of juvenile giant otters. A bout usually started with single begging calls and was intensified with screams and interspersed with contact call gradations when the cub detected an adult with fish. When receiving or stealing prey from a group member, the calls changed from the begging calls and screams to the ascending scream and the whine, often accompanied by defence displays such as kicking with the hind legs, tail waving and growling at all others in close vicinity. Even adults produced begging vocalizations. In captive groups, all individuals engaged in a piercing begging chorus at feeding time. In the wild group at Cocha Cashu, we recorded begging of the aging mother directed at her youngest son (see video S1), a behaviour previously described by Davenport [65].

Ascending scream. The structure of the ascending scream resembles a contact call with an elongated and ascending first part. This call is emitted within begging bouts or when stealing a fish from a group member, the latter accompanied by defensive body postures like kicking with the hind feet and tail wagging after a successful fish over-take. *Begging call*. This short modulated calls has an elaborate number of subharmonics. Begging bouts often start with one or several begging calls, followed by the other begging vocalizations in alternating or variable order. *Begging scream*. This call is a piercing, modulated scream with subharmonics and often chaotic parts, with a very variable structure. Begging individuals use this scream to intensify their begging behaviour. *Begging scream gradation*. This piercing, modulated screams often has subharmonics or chaotic parts. The structure and frequency modulation is very similar to that of isolation calls. Begging individuals use this scream to intensify their begging behaviour. *Whine*. This howl-like, quasi-constant frequency call has a variable duration and resembles the wavering scream. Like the ascending scream it is produced within begging bouts or when stealing a fish from a group member, the latter accompanied by defensive body postures. In wild groups, the whine was also produced as an alarm call during caiman interactions.

Other contexts

Mating and nursing were the two other behavioural contexts in which we recorded giant otter vocalizations. *Mating call*. This growl-like vocalization has a higher fundamental and more modulated frequency than other growls. We recorded mating calls only from one alpha female, copulating in the water while the male emitted growls. *Suckling call*. This blatant vocalization resembles a hum with additional tonal parts. Generally, only cubs and juveniles emit the suckling call while nursing, but in the Tierpark Hagenbeck group, we recorded suckling calls of subadults nursing after the death of a new born litter.

Discussion

When comparing our findings from Peruvian and captive giant otters to the vocal repertoire of giant otters from Brazil [54], the uniformity of the vocalizations and the related behavioural context is striking. Our classification only slightly differs from Leuchtenberger [54]. Therefore, we think that the vocal repertoire is generally applicable to all local populations. Vocal differences will rather be found within distinct vocalization types among groups, families, sexes or individuals.

Even though we only measured source-induced acoustic parameters, the DFA result supported our classification of the adult vocal repertoire. We interpret the misclassification of

Table 4. Structural description of the neonate giant otter vocal repertoire.

Vocalization	Description	Category	Frequency of occurrence	Wild or captive	Descriptions by other authors
Begging call-like vocalization	Resembles the adult begging call.	'BC-like'	Common	Captive	No description
Contact call-like vocalization	Resembles the adult contact calls.	'CC-like'	Occasional	Wild and captive	No description
Distress call (1)	No equivalent in adult vocalizations. Emitted when cubs are moved from one den to another.	'DC-like'	Rare	Wild and captive	Newborn cub squeaks [42], Jungenfiepsen [43]
Hum gradation-like vocalization	Resembles the adult hum gradation. Combination of a hum and a contact call.	'Hum gradation'	Common	Captive	No description
Bark-like call	Resembles the adult bark vocalization.	'Hums'	Common	Captive	No description
Distress call (2)	No equivalent in adult vocalizations. Hum with accentuation of higher harmonics.	'Hums'	Common	Captive	Cub scream [54]
Hum-like vocalization	Resembles the adult hum.	'Hums'	Very common	Captive	No description
Suckling call	Suckling. Only emitted when nursed.	'Suckling call'	Occasional	Wild and captive	Scream gurgle [54], nursing hum [59]
High whistle	No equivalent in adult vocalizations. Pure tone with F_0 above 10 kHz.	'Whistles'	Common	Captive	No description
Whistle	Resembles the adult whistle. Pure tone with F_0 between 5 kHz and 10 kHz.	'Whistles'	Common	Captive	No description
Low whistle	Resembles the adult whistle. Pure tone with F_0 below 5 kHz.	'Whistles'	Common	Captive	No description

structurally very similar vocalizations being a result of a close structural resemblance, conveying different meaning depending on the situation in which the vocalizations are produced (sensu [21]). We could relate signal structure and function in giant otter vocalizations to the motivational-structural rules described by Morton [11] and Bradbury and Vehrencamp [12]. The vocalizations of giant otters resemble the overall structure of canid [66] and other mammalian [21] vocal repertoires. The giant otter repertoire presents a continuum of distinct and graded vocalizations with variants, encoding information on internal state or external events. While the number of vocalizations a species can produce is limited by anatomical constraints [1,4], vocal complexity can be enhanced by call combinations [67] or segmental concatenation, as has been found in banded mongooses, *Mungos mungo* [17]. Like otters, Herpestids show a wide spectrum of social organization from solitary to highly social group living species [35]. Banded mongooses and meerkats, *Suricatta suricatta*, are obligatory social species, showing a strong group cohesion, cooperative foraging and breeding [68]. Due to the similar structure and function of the giant otters contact call to close calls of banded mongooses [17], we propose that the contact call and structurally similar vocalizations may also include segmental concatenation, which is the combination of different vocal cues in one call type [17]. The individual distinctiveness of giant otters contact calls is mainly encoded in the second constant frequency part and in the frequency contour of the modulated part [40]. In banded mongoose close calls, the discrete individual cues are encoded in the first call segment, whereas the behavioural context is embedded as a graded cue within the tonal part of the call [17]. Since we [40] did not test for other cues than vocal individuality, it may well be that giant otters combine individual signatures and graded behavioural cues in the contact call.

With the results of our underwater recordings, we cannot support the findings of several underwater vocalizations by an earlier study [53]. Air-borne calls can be recorded underwater, since the giant otters' thorax and throat are submerged while swimming. This may lead to misinterpretation of air-borne calls as being produced while diving.

Among snorts, giant otters seem to be able to differentiate the type of signalled danger depending on the situation [43]. The number of pulses and production rate of snorts could refer to the internal state of arousal or the estimation of the external threat perceived by the signaller, as well as signalling the exact type of danger (such as caimans, humans, or non-group members). According to the signal design rules by Bradbury & Vehrencamp [12], the localization of the sender of alert signals can be enhanced by repeated, short and pulsed broadband vocalizations. By varying the repetition rate, senders can inform group members about the urgency of the situation [12]. Reference and emotion can be combined in one call [69]. The content of information and its interpretation depend on the signaller's or the receiver's point of view [69]. By adding unexpected and unpredictable components to alarm calls, nonlinear phenomena could reinforce attention in the receivers, and therefore decrease habituation [64]. The same applies to the wavering scream, but its function needs to be analysed in more detail. It might just signal high arousal, or, as a referential vocalization provide exact information about the external threat perceived. Generally, assembly alarm calls provide information on the spatial location of the group and may encode group specific cues [12]. Several times, we observed giant otter groups screaming in a chorus while patrolling the lakeshore, but we could not detect caimans or any other obvious threat. Therefore, we think that giant otter chorus calling may not always signal caiman presence, but could also represent acoustic territorial signalling. Other group-territorial species like lions or wolves engage in chorus calling, providing information on group size, their presence as territory holders and thereby their resource holding potential to neighbours or potential intruders [12,70–72].

During development, giant otter cubs need to learn how to capture prey on their own. As long as they are depended on food sharing from elder group members, they engage in intense begging for food [43]. A comparable begging system with dependant offspring emitting diverse begging vocalizations can be found in the cooperatively breeding meerkat [73,74]. Meerkat cubs follow their group while foraging and their vocalizations change when observing an adult with food. The intenseness of the vocalizations serves to attract the feeders attention and to 'outcall' siblings [74]. Since this intense calling is energetically costly, the authors hypothesize that the cubs only increase begging when the chance to receive food is optimal [74]. Like meerkat pups, giant otter cubs are mobile and benefit from food provision by adults, from signalling their spatial location and from 'outcalling' their siblings. Furthermore, giant otter cubs may benefit from adding nonlinear phenomena to their begging calls, keeping potential feeders from habituating to the piercing screams [64,75]. Nonlinearities may also provide honest information about the internal state of arousal [64,75], thus denoting condition [76].

Neonate giant otters are vocally active from birth on. Structure and usage of giant otter vocalizations seem to undergo comparable changes as for instance in wolves and wild dogs [16,77]. The distress calls and the high whistle do not persist until adulthood. The other vocalizations we found in new born cubs are likely to be precursors of the distinct adult vocalizations. Giant otter cubs emit the full vocal repertoire not later than at an age of 3 months, whereas the lowering of the fundamental frequency continues until six to 12 months (own observation). The babbling bouts of neonate giant otter cubs very likely fulfil the two main purposes of vocal practise and social care solicitation [25,26,28]. The general structure of the vocalizations in neonate babbling bouts was comparable to adult vocalizations, but showed some subtle differences. For instance, the duration of cub contact call-like vocalizations was extended, and the calls often included gaps. We suggest that the cubs need to practice the shift from the constant frequency part to the modulated frequency and back. Concerning the social organization of giant otters, we expect that babbling also serves to intensify social interactions and caregiving in giant otters. This function of babbling was reported for the social and cooperatively breeding pygmy marmoset, *Cebuella pygmaea* [25,26,28]. Giant otter cubs may switch from babbling to begging vocalizations and isolation calls as care eliciting cues during development.

When comparing social organization among the otter species, we come to the same conclusion as Manser et al. [68] in their comparative study on vocal complexity in Herpestids. Group size alone, as a measure for social complexity, cannot explain the variation of vocal repertoire size in otters. Still, knowledge on the acoustic communication of the 13 otter species is scarce, so that we restrict the comparison to eight out of 13 otter species (table S2, see also [78] on species comparisons). Vocalizations are reported for eight species and for only five of them exist published descriptions of the vocal repertoire [42,55,79–85].

The solitary and solitary foraging Neotropical otter, *Lontra longicaudis* [41,83] has more than four vocalization types [83]. The Eurasian otter, *Lutra lutra*, produces 8 basic vocalization types with some variants [79,80]. Quaglietta et al. [86] recently gave new insight into the social organization of Eurasian otters, revealing a more social lifestyle than previously reported, so that we list it here as semi-solitary. Social interactions among Eurasian otters are enhanced by the overlap of territories, formation of dyads and sharing of resting sites. Quaglietta et al. [86] discuss their findings in relation to environmental factors and otter density. The semi-solitary Cape clawless otter, *Aonyx capensis*, has approximately 13 [55] and the gregarious sea otter, *Enhydra lutris*, 10 vocalization types [81]. North American river otters, *Lontra canadensis*, showing a variable social structure and cooperative foraging [41,87], emit 11 distinct vocalization types (four distinct vocalizations with seven subtypes) [85]. Male groups of North American river otters may include up to 30, and male groups of sea otters may include even up to 2000 animals [41,88,89]. Although the group size of both species exceeds the number of individuals reported for giant otter groups [42], the repertoire of North American river otters and sea otters shows less distinct vocalizations [81,85] than we found in giant otters. Therefore, other factors influencing the social organization of a species need to be taken into account. The large aggregations of male sea otters ('rafts') occur during the nonbreeding season, whereas female groups are much smaller and family units consist of mother and cub. Territorial males may attend mother-cub pairs, but do not help in rearing the offspring [88]. This behavioural pattern is due to the marine habitat of sea otters, as well as due to a conflict between territoriality and gregariousness, resulting in an otter-unusual social organization [41]. In North American river otters, variability of social organization is also based on regional habitat differences [41]. Depending on the region, territorial overlap occurs among females, among males, or among the two sexes. River otters may group together in 'packs' of related or unrelated individuals for several months [41].

The more social species of smooth-coated otter, *Lutrogale perspicillata* [90], has at least four distinct vocalization types, but the vocal repertoire description is only anecdotal [56]. Asian small-clawed otters, *Aonyx cinerea*, showing a social organization very similar to giant otters [41,91–93], were described to have about 12 vocalization types by Timmis [82]. Lemasson et al. [84] classified four main context-dependent vocalization units (7 distinct vocalization types with gradations) for Asian small-clawed otters, resembling the four major behavioural categories we found in giant otters.

Group formation and composition in giant otters are influenced by several factors originating from life history and habitat effects [45,94] and may lead to high variation of relatedness within giant otter groups [45]. Changes in structure of established groups mainly result from the disappearance of one partner of the reproductive pair. Dominant females are thereby replaced by a philopatric female group member, whereas males tend to migrate into another existing group [45,94]. Offspring of the former couple are adopted by the new group member [45,94]. Group size and reproductive success seem to be directly linked to lake size of a territory [94]. If all high quality territories in an area are already inhabited by resident groups, dispersing individuals may form temporary groups in less suitable habitats [45,94].

Besides the plasticity in group composition in giant otters, the basic family group of a breeding pair with offspring of several years likely promotes long-term individual relationships and diversity of social roles in giant and Asian small-clawed otters. These aspects complement to the complexity of a social system [29]. Vocal complexity in giant and Asian small-clawed otters is enhanced by graded vocalizations, and probably by segmental concatenation and call combination. Cohesion calls are individually distinct and both species show vocal individual discrimination [40,84,95], thus, adding an individual scale to their vocal and social complexity.

A more sophisticated vocal repertoire with increasing group size can therefore not fully be supported in otters. On the other hand, the social organization and the strength of social relationships within otter groups may account for their vocal complexity, reflecting an underlying pattern of coevolution of social and vocal complexity in otters, as has been shown for primates [30,31], canids [36,66] and herpestids [35]. Nevertheless, new insight into the social organization and communication of otters may still change our classification (compare [68,78,86]). Consistent with the recent knowledge on otter sociality and acoustic communication, we like to emphasize that giant otters still belong to the socially most complex species and that their social complexity is reflected in their sophisticated vocal system. We hope that this detailed account of the giant otters' vocal repertoire will help to facilitate conservation efforts of this endangered species. By providing the behavioural context and acoustic measures of the vocalizations, the otters' arousal or stress in a given situation may be easier to evaluate.

In conclusion, we found a sophisticated vocal repertoire of adult giant otters, with 22 distinct vocalizations. Thus, the giant otters' social complexity seems to be reflected in their vocal complexity. Neonate vocalizations were also complex and included precursors of the adult repertoire, as well as age-specific calls, supporting our hypotheses on the vocal repertoire of neonate giant otters. Moreover, the relationship of signal structure and function in giant otter vocalizations generally follows the expectations stemming from Morton's motivational-structural rules.

Supporting Information

Table S1 Mean values (±SD) for the most important variables characterizing distinct vocalizations within the giant otters' vocal repertoire.

Table S2 Comparison of social organization and vocal repertoire size in eight of the 13 otter species.

Table S3 List of all 441 adult giant otter vocalizations measured in this study, including origin of the individuals, vocalization type, behavioural context and vocal parameters.

Video S1 Begging, contact, group reunion and contact over larger distance (Lake Cocococha, Peru, 2012). Juveniles are swimming close to the lakeshore and beg, while the family is swimming in the centre of the lake. In sec 5, the family approaches the begging juveniles (sec 38), the adults provide them with fish (sec 44) and emit contact calls (contact calls sec 18.5, 53.5 and 58.0). At min 1 the family is fishing again in the centre of the lake, but still in vocal contact with the juveniles.

Video S2 Group reunion (Lake Cashu, Peru, 2011). The juvenile is swimming alone behind the riparian vegetation, while the family is resting on a trunk (sec 5). At sec 13.5, the family starts to call for the juvenile and approaches him (contact calls sec 13.5, 16.5, 17.5, hum gradation sec 18.5 and 21.5, hums sec 25 and 27).

Video S3 Contact, group cohesion and coordination (Lake Sandoval, Peru, 2012). The giant otter family is swimming together, they emit hums, contact calls and a hum gradation, initializing a change of direction and to go on fishing.

Video S4 Territorial marking (Lake Salvador, Peru, 2011). The alpha couple is emitting hums while marking the entrance of the den by cleaning and smoothing the floor.

Video S5 Soothing hums (Lake Salvador, Peru, 2011). The giant otter family is exploring the hiding place of the observers' boat. The otters are intensely observing and sniffing at the hiding place of our boat and examine the trunk, where we installed a flagging, while emitting hums to soothe each other during this unusual encounter.

Video S6 Wavering screams in an isolation context (Lake Salvador, Peru, 2011). Two juveniles have been separated from the group. One emits a contact call, the other one an isolation call (sec 2), the group answers with wavering screams and approaches the juveniles. When they are closer, they emit contact calls (sec 20).

Video S7 Growling to defend a fish (Lake Sandoval, Peru, 2012). The adult giant otter sitting on the trunk is eating a large fish. He is growling to threaten his group members and to defend his fish against them. At min 1 he emits an accentuated growl against a group member.

Video S8 Minor alarm (Lake Cocococha, Peru, 2012). The alpha male leaves the den and emits hah!s when detecting the human observer (hah! Sec 1.5 and 6).

Video S9 Snorting at each other, hah! in a submissive context (Lake Salvador, Peru, 2011). The eldest daughter of the family is marking branches by scratching the leaves. In second 9, the mother comes and threatens her by body posture, the daughter shows the submissive gesture, exposing her throat and hah!ing with open mouth (sec 10). Then the mother starts to scratch the leaves (sec 14). Sec 33: again, the daughter is marking leaves but she turns away as she sees the mother coming (sec 38). The mother snorts at her (sec 39), the daughter swims away and the mother marks the leaves.

Video S10 Snorting at a caiman (Lake Salvador, Peru, 2011). The group is fighting against juvenile caimans which rest at the lakeshore.

Video S11 Caiman alarm (Lake Cashu, Peru, 2011). One otter detected the very huge caiman, living in this area of the lake, the group members answer his alarming wavering screams with contact calls.

Video S12 Group screaming in alarm (Lake Sandoval, Peru, 2012). The context could not be examined, it could be the close presence of the tourist boat, as well as a caiman detected underwater.

Video S13 Characteristic begging (Lake Tres Chimbadas, Peru, 2012). One giant otter is sitting on a trunk and eating a fish. A juvenile comes close and emits begging calls, interspersed with screams (wavering scream sec 7 and 32, begging scream gradation sec 40 and 42).

Video S14 Begging behaviour of the ageing alpha female directed at her youngest son (Lake Cashu, Peru, 2011). The juvenile is sitting on a trunk and eating a fish. At sec 12, the mother approaches him, climbs the trunk (sec 19) and begs for the fish (begging calls and a begging scream gradation from sec 20). At sec 27 she starts to steal the fish from her offspring. When she emerges with the fish, she emits a growl and an ascending scream (sec 32).

Audio S1 Giant otter adult vocal repertoire, including audio files AS001 to AS441 as listed in table S3.

Audio S2 Neonate giant otter vocal repertoire, including audio files AS442 to AS453, as listed in figure 3 and table S3.

Acknowledgments

Comments from M. Waldmann, N. Duplaix and one anonymous reviewer greatly improved our manuscript. We like to thank E. Kalko (deceased), M. Tschapka, L. Davenport, J. Groenendijk, N. Duplaix, R. Williams, M. C. Urrutia, A. M. L. Chunga Lee, J. M. Camargo Alvarez, P. Toledo Monsonis, Estación Biológica Cocha Cashu, Casa Machiguengka, Posada Amazonas, Explorer's Inn and L. Paypay, Mejía Lodge and family Mejía, SZF Peru and AIDER Peru for their support. We also thank the SERNANP and DGFFS authorities, Zoo Dortmund, Zoo Duisburg and Tierpark Hagenbeck for research permission.

Author Contributions

Analyzed the data: CASM MK. Contributed to the writing of the manuscript: CASM MK. Recordings: CASM.

References

1. Fitch WT, Hauser MD (1998) Unpacking 'honesty': vertebrate vocal production and the evolution of acoustic signals. In: Simmons AM, Fay RR, Popper AN, editors. Acoustic Communication (Springer Handbook of Auditory Research). New York: Springer. 1–44.
2. Seyfarth RM, Cheney DL, Bergman TJ (2005) Primate social cognition and the origins of language. Trends Cogn Sci 9: 264–266. doi:10.1016/j.tics.2005.04.001.
3. Ramsier MA, Cunningham AJ, Finneran JJ, Dominy NJ (2012) Social drive and the evolution of primate hearing. Philos Trans R Soc Lond B Biol Sci 367: 1860–1868. doi:10.1098/rstb.2011.0219.
4. Fitch WT (2000) Skull dimensions in relation to body size in nonhuman mammals: the causal bases for acoustic allometry. Zoology, 103, 40–58.
5. Fant G (1960) Acoustic theory of speech production. The Hague: Mouton de Gruyter.
6. Titze IR (1994) Principles of voice production. Englewood Cliffs N.J.: Prentice Hall.
7. Taylor AM, Reby D (2010) The contribution of source-filter theory to mammal vocal communication research. J Zool 280: 221–236. doi:10.1111/j.1469-7998.2009.00661.x.
8. Wilden I, Herzel H, Peters G, Tembrock G (1998) Subharmonics, biphonation, and deterministic chaos in mammal vocalization. Bioacoustics 9: 171–196.
9. Tokuda I, Riede T, Neubauer J, Owren MJ, Herzel H (2002) Nonlinear analysis of irregular animal vocalizations. J Acoust Soc Am 111: 2908–2919. doi:10.1121/1.1474440.
10. Fitch WT (1997) Vocal tract length and formant frequency dispersion correlate with body size in rhesus macaques. J Acoust Soc Am 102: 1213–1222. doi:10.1121/1.421048.
11. Morton ES (1977) On the occurrence and significance of motivation-structural rules in some bird and mammal sounds. Am Nat 111: 855–869.
12. Bradbury JW, Vehrencamp SL (1998) Signal design rules. In: Bradbury JW, Vehrencamp SL, editors. Principles of animal communication. Sunderland, Massachusetts: Sinauer Associates, Inc. 571–618.
13. Hammerschmidt K, Fischer J (1998) The vocal repertoire of barbary macaques: a quantitative analysis of a graded signal system. Ethology 104: 203–216. doi:10.1111/j.1439-0310.1998.tb00063.x.
14. Green S, Marler P (1975) The analysis of animal communication. In: Marler P, Vandenbergh JG, editors. Handbook of behavioral neurobiology: social behavior and communication. New York: Springer US. 73–147. doi:10.1007/978-1-4615-9116-0.
15. Fischer J, Metz M, Cheney DL, Seyfarth RM (2001) Baboon responses to graded bark variants. Anim Behav 61: 925–931. doi:10.1006/anbe.2000.1687.
16. Schassburger RM (1993) Vocal communication in the timber wolf, *Canis lupus*, Linnaeus. Structure, motivation, and ontogeny Cornell University.
17. Jansen DAWAM, Cant MA, Manser MB (2013) Segmental concatenation of individual signatures and context cues in banded mongoose (*Mungos mungo*) close calls. BMC Biol 10: 97. doi:10.1186/1741-7007-10-97.
18. Cheney DL, Seyfarth RM (2005) Constraints and preadaptations in the earliest stages of language evolution. Linguist Rev 22: 135–159. doi:10.1515/tlir.2005.22.2-4.135.
19. Crockford C, Boesch C (2005) Call combinations in wild chimpanzees. Behaviour 142: 397–421. doi:10.1163/1568539054012047.
20. Zuberbühler K (2002) A syntactic rule in forest monkey communication. Anim Behav 63: 293–299. doi:10.1006/anbe.2001.1914.
21. Seyfarth RM, Cheney DL, Bergman T, Fischer J, Zuberbühler K, et al. (2010) The central importance of information in studies of animal communication. Anim Behav 80: 3–8. doi:10.1016/j.anbehav.2010.04.012.
22. Fichtel C, Manser MB (2010) Vocal communication in social groups. In: Kappeler PM, editor. Animal behaviour: Evolution and mechanisms. Berlin, Heidelberg, Germany: Springer. 29–54.
23. Ehret G (1980) Development of sound communication in mammals. Adv Study Behav 11: 179–225.
24. Seyfarth RM, Cheney DL (1986) Vocal development in vervet monkeys. Anim Behav 34: 1640–1658. doi:10.1016/S0003-3472(86)80252-4.
25. Elowson AM, Snowdon CT, Lazaro-Perea C (1998) "Babbling" and social context in infant monkeys: parallels to human infants. Trends Cogn Sci 2: 31–37. doi:10.1016/S1364-6613(97)01115-7.
26. Elowson AM, Snowdon CT, Lazaro-Perea C (1998) Infant "babbling" in a nonhuman primate: complex vocal sequences with repeated call types. Behaviour 135: 643–664. doi:10.1163/156853998792897905.
27. Knörnschild M, Behr O, von Helversen O (2006) Babbling behavior in the sac-winged bat (*Saccopteryx bilineata*). Naturwissenschaften 93: 451–454. doi:10.1007/s00114-006-0127-9.
28. Snowdon C, Elowson AM (2001) "Babbling" in pygmy marmosets: development after infancy. Behaviour 138: 1235–1248. doi:10.1163/15685390152822193.
29. Freeberg TM, Dunbar RIM, Ord TJ (2012) Social complexity as a proximate and ultimate factor in communicative complexity. Philos Trans R Soc B 367: 1785–1801. doi:10.1098/rstb.2011.0213.
30. Bouchet H, Blois-Heulin C, Lemasson A (2013) Social complexity parallels vocal complexity: a comparison of three non-human primate species. Front Psychol 4: 1–15. doi:10.3389/fpsyg.2013.00390.
31. McComb K, Semple S (2005) Coevolution of vocal communication and sociality in primates. Biol Lett 1: 381–385. doi:10.1098/rsbl.2005.0366.
32. Freeberg TM (2006) Social complexity can drive vocal complexity: group size influences vocal information in Carolina chickadees. Psychol Sci 17: 557–561. doi:10.1111/j.1467-9280.2006.01743.x.
33. Krams I, Krama T, Freeberg TM, Kullberg C, Lucas JR, et al. (2012) Linking social complexity and vocal complexity: a parid perspective. Philos Trans - R Soc B: 1879–1891. doi:10.1098/rstb.2011.0222.
34. Blumstein DT, Armitage KB (1997) Does sociality drive the evolution of communicative complexity? A comparative test with ground-dwelling sciurid alarm calls. Am Nat 150: 179–200. doi:10.1086/286062.
35. Le Roux A, Cherry MI, Manser MB (2009) The vocal repertoire in a solitary foraging carnivore, *Cynictis penicillata*, may reflect facultative sociality. Naturwissenschaften 96: 575–584. doi:10.1007/s00114-008-0506-5.
36. Faragó T, Townsend S, Range F (2014) The information content of wolf (and dog) social communication. In: Witzany G, editor. Biocommunication of Animals. Dordrecht: Springer Netherlands. 41–62. doi:10.1007/978-94-007-7414-8.
37. Pollard KA, Blumstein DT (2011) Social group size predicts the evolution of individuality. Curr Biol 21: 413–417. doi:10.1016/j.cub.2011.01.051.
38. De Waal FBM, Tyack PL (2003) Animal social complexity: intelligence, culture, and individualized societies. Cambridge & London: Harvard University Press.
39. Pollard KA, Blumstein DT (2012) Evolving communicative complexity: insights from rodents and beyond. Philos Trans R Soc Lond B Biol Sci 367: 1869–1878. doi:10.1098/rstb.2011.0221.
40. Mumm CAS, Urrutia MC, Knörnschild M (2014) Vocal individuality in cohesion calls of giant otters, *Pteronura brasiliensis*. Anim Behav 88: 243–252. doi:10.1016/j.anbehav.2013.12.005.
41. Kruuk H (2006) Otters: ecology, behaviour and conservation. Oxford, New York: Oxford University Press.

42. Duplaix N (1980) Observation on the ecology and behavior of the giant river otter *Pteronura brasiliensis* in Suriname. La Terre la Vie - Rev d'écologie 34: 496–620.
43. Staib E (2002) Öko-Ethologie von Riesenottern (*Pteronura brasiliensis*) in Peru Aachen: Shaker Verlag.
44. Duplaix N, Waldemarin HF, Groenedijk J, Evangelista E, Munis M, et al. (2008) *Pteronura brasiliensis*. In: IUCN 2013. IUCN Red List of Threatened Species. Version 2013.2. Available: www.iucnredlist.org. Accessed 10 June 2014.
45. Ribas C (2012) Grau de parentesco e relações sociais em ariranhas (*Pteronura brasiliensis*). Doctor of Philosophy thesis, Instituto Nacional de Pesquisas da Amazônia.
46. Carter SK, Rosas FCW (1997) Biology and conservation of the giant otter *Pteronura brasiliensis*. Mamm Rev 27: 1–26. doi:10.1111/j.1365-2907.1997.tb00370.x.
47. Rosas FCW, Mendes Cabral MM, de Mattos GE, Silva RE (2009) Parental and alloparental care of giant otters (*Pteronura brasiliensis*) (Carnivora, Mustelidae) in Balbina hydroelectric lake, Amazonas, Brazil. Sociobiology 54: 919–924.
48. Evangelista E, Rosas FCW (2011) Breeding behavior of giant otter (*Pteronura brasiliensis*) in the Xixuaú reserve, Roraima, Brazil. IUCN Otter Spec Gr Bull 28: 5–10. Available: http://www.otterspecialistgroup.org/Bulletin/Volume28A/Evangelista_Rosas_2011.html. Accessed 04 September 2014.
49. Cairns SJ, Schwager SJ (1987) A comparison of association indices. Anim Behav 35: 1454–1469. Available: http://linkinghub.elsevier.com/retrieve/pii/S0003347287800180. Accessed 29 August 2014.
50. Leuchtenberger C, Mourão G (2008) Social organization and territoriality of giant otters (Carnivora: Mustelidae) in a seasonally flooded savanna in Brazil. Sociobiology 52: 257–270.
51. Leuchtenberger C, Mourão G (2009) Scent-marking of giant otter in the Southern Pantanal, Brazil. Ethology 115: 210–216. Available: http://doi.wiley.com/10.1111/j.1439-0310.2008.01607.x. Accessed 14 April 2014.
52. Machado MRB (2004) Bioacústica de ariranhas (*Pteronura brasiliensis*) (Carnivora: Mustelidae): discriminação vocal e repertório. Doctor of Philosophy thesis, Universidad Federal do Amazonas - UFAM.
53. Schenck C, Staib E, Yasseri AM (1995) Unterwasserlaute bei Riesenottern (*Pteronura brasiliensis*). Zeitschrift fuer Saeugetierkd 60: 310–313.
54. Leuchtenberger C (2012) Ecologia espacial e comunicação vocal de ariranhas (*Pteronura brasiliensis*) no Pantanal. Doctor of Philosophy thesis, Universidade Federal do Rio de Janeiro. Available: http://busca.ibict.br/SearchBDTD/search.do?command=search&q=+assunto:%22Ariranhas%20Comportamento%22. Accessed 17 June 2014.
55. Larivière S (2001) *Aonyx capensis*. Mamm Species 671: 1–6.
56. Hwang YT, Larivière S (2005) *Lutrogale perspicillata*. Mamm Species 786: 1–4. doi:10.1644/786.1.
57. Ben-David M, Blundell GM, Kern JW, Maier JAK, Brown ED, et al. (2005) Communication in river otters: creation of variable resource sheds for terrestrial communities. Ecology 86: 1331–1345. doi:10.1890/04-0783.
58. Larivière S (2001) *Aonyx congicus*. Mamm Species 650: 1–3.
59. Sykes-Gatz S (2005) International giant otter studbook husbandry and management information and guidelines (2005). Dortmund, Germany: Zoo Dortmund.
60. Tyack PL, Miller EH (2002) Vocal anatomy, acoustic communication and echolocation. In: Hoelzel AR, editor. Marine Mammal Biology: An Evolutionary Approach. Oxford, England: Blackwell Science, Vol. 1, 142–184.
61. Siemers BM, Kerth G (2006) Do echolocation calls of wild colony-living Bechstein's bats (*Myotis bechsteinii*) provide individual-specific signatures? Behav Ecol Sociobiol 59: 443–454. doi:10.1007/s00265-005-0068-x.
62. Charif RA, Waack AM, Strickman LM (2010) Raven Pro 1.4 User's Manual. Ithaca, NY: Cornell Laboratory of Ornithology.
63. Mundry R, Sommer C (2007) Discriminant function analysis with nonindependent data: consequences and an alternative. Anim Behav 74: 965–976. doi:10.1016/j.anbehav.2006.12.028.
64. Blumstein DT, Récapet C (2009) The sound of arousal: the addition of novel non-linearities increases responsiveness in marmot alarm calls. Ethology 115: 1074–1081. doi:10.1111/j.1439-0310.2009.01691.x.
65. Davenport LC (2010) Aid to a declining matriarch in the giant otter (*Pteronura brasiliensis*). PLoS One 5: 1–6. doi:10.1371/journal.pone.0011385.
66. Tembrock G (1976) Canid vocalizations. Behav Processes 1: 57–75.
67. Arnold K, Zuberbühler K (2008) Meaningful call combinations in a non-human primate. Curr Biol 18: 202–203. doi:10.1016/j.cub.2008.01.040.
68. Manser MB, Jansen DAWAM, Furrer RD, Roux A (2014) Vocal complexity in meerkats and other mongoose species. Advances in the Study of Behavior. Elsevier Inc., Vol. 46. 281–310. doi:10.1016/B978-0-12-800286-5.00006-7.
69. Seyfarth RM, Cheney DL (2003) Signalers and receivers in animal communication. Annu Rev Psychol 54: 145–173. doi:10.1146/annurev.psych.54.101601.145121.
70. Mech LD, Boitani L (2003) Wolf social ecology. In: Mech LD, Boitani L, editors. Wolves: behavior, ecology, and conservation. Chicago, London: University of Chicago Press. 1–34.
71. Mech LD, Boitani L (2003) Wolves: behavior, ecology, and conservation. Chicago, London: University of Chicago Press.
72. McComb K, Packer C, Pusey A (1994) Roaring and numerical assessment in contest between groups of female lions, *Panthera leo*. Anim Behav 47: 379–387. doi:10.1006/anbe.1994.1052.
73. Kunc HP, Madden JR, Manser MB (2007) Begging signals in a mobile feeding system: the evolution of different call types. Am Nat 170: 617–624. doi:10.1086/521233.
74. Manser MB, Avey G (2000) The effect of pup vocalisations on food allocation in a cooperative mammal, the meerkat (*Suricata suricatta*). Behav Ecol Sociobiol 48: 429–437. doi:10.1007/s002650000248.
75. Blumstein DT, Richardson DT, Cooley L, Winternitz J, Daniel JC (2008) The structure, meaning and function of yellow-bellied marmot pup screams. Anim Behav 76: 1055–1064. doi:10.1016/j.anbehav.2008.06.002.
76. Riede T, Arcadi AC, Owren MJ (2007) Nonlinear acoustics in the pant hoots of common chimpanzees (*Pan troglodytes*): Vocalizing at the edge. J Acoust Soc Am 121: 1758–1767. doi:10.1121/1.2427115.
77. Robbins RL, McCreery EK (2003) African wild dog pup vocalizations with special reference to Morton's model. Behaviour 140: 333–351. doi:10.1163/156853903321826666.
78. Hauser MD (1996) The evolution of communication. Sebeok TA, editor MIT Press.
79. Gnoli C, Prigioni C (1995) Preliminary study on the acoustic communication of captive otters (*Lutra lutra*). Hystrix 7: 289–296. doi:10.4404/hystrix-7.1-2-4083.
80. Rogoschik B (1989) Vokalisation und Lautentwicklung des europäischen Fischotters. In: Stubbe M, editor. Populationsökologie marderartiger Säugetiere. Halle (Saale): Universität Halle-Wittenberg. 213–221.
81. McShane LJ, Estes JA, Riedman ML, Staedler MM (1995) Repertoire, structure, and individual variation of vocalizations in the sea otter. J Mammal 76: 414–427.
82. Timmis WH (1971) Observations on breeding the Oriental short-clawed otter at Chester Zoo. Int Zoo Yearb 11: 109–111. doi:10.1111/j.1748-1090.1971.tb01868.x.
83. Larivière S (1999) *Lontra longicaudis*. Mamm Species 609: 1–5.
84. Lemasson A, Mikus M-A, Blois-Heulin C, Lodé T (2014) Vocal repertoire, individual acoustic distinctiveness, and social networks in a group of captive Asian small-clawed otters (*Aonyx cinerea*). J Mammal 95: 128–139. doi:10.1644/12-MAMM-A-313.1.
85. Almonte C (2011) The vocal behaviors of North American river otters (*Lontra canadensis*) individual differences and shared repertoires. Doctor of Philosophy thesis, the City University of New York.
86. Quaglietta L, Fonseca VC, Mira A, Boitani L (2014) Sociospatial organization of a solitary carnivore, the Eurasian otter (*Lutra lutra*). J Mammal 95: 140–150. doi:10.1644/13-MAMM-A-073.1.
87. Blundell GM, Ben-David M, Bowyer RT (2002) Sociality in river otters: cooperative foraging or reproductive strategies? Behav Ecol 13: 134–141. doi:10.1093/beheco/13.1.134.
88. Garshelis DL, Johnson AM, Garshelis JA (1984) Social organization of sea otters in Prince William Sound, Alaska. Can J Zool 62(12): 2648–2658. doi:10.1139/z84-385.
89. Estes JA (1980) *Enhydra lutris*. Mamm Species 133: 1–8.
90. Hussain SA (1996) Group size, group structure and breeding in smooth-coated otter *Lutra perspicillata* Geoffroy (Carnivora, Mustelidae) in National Chambal Sanctuary, India. Mammalia 60: 289–297. doi:10.1515/mamm.1996.60.2.289.
91. Sivasothi N, Nor BHM (1994) A review of otters (Carnivora: Mustelidae: Lutrinae) in Malaysia and Singapore. Hydrobiologia 285: 151–170. doi:10.1007/BF00005663.
92. Larivière S (2003) *Amblonyx cinereus*. Mamm Species 720: 1–5. doi:http://dx.doi.org/10.1644/720.
93. Hussain SA, Gupta SK, de Silva PK (2011) Biology and ecology of asian small-clawed otter *Aonyx cinereus* (Illiger, 1815): a review. IUCN Otter Spec Gr Bull 28: 63–75. Available: http://www.otterspecialistgroup.org/Bulletin/Volume28/Hussain_et_al_2011.html. Accessed 04 September 2014.
94. Groenendijk J, Hajek F, Johnson PJ, Macdonald DW, Calvimontes J, Staib E, & Schenck C (2014) Demography of the giant otter (*Pteronura brasiliensis*) in Manu National Park, South-Eastern Peru: implications for conservation. PloS One, 9(8): e106202. doi:10.1371/journal.pone.0106202.
95. Lemasson A, Mikus M-A, Blois-Heulin C, Lodé T (2013) Social partner discrimination based on sounds and scents in Asian small-clawed otters (*Aonyx cinereus*). Naturwissenschaften 100: 275–279. Available: http://www.ncbi.nlm.nih.gov/pubmed/23397263. Accessed 11 March 2013.
96. Hajek F, Groenendijk J (2008) Giants of the Madre de Dios. Lima: Frankfurt Zoological Society.

23

Discerning Fragmentation Dynamics of Tropical Forest and Wetland during Reforestation, Urban Sprawl, and Policy Shifts

Qiong Gao, Mei Yu*

Department of Environmental Sciences and Institute for Tropical Ecosystem Studies, University of Puerto Rico, Rio Piedras, San Juan, PR, United States of America

Abstract

Despite the overall trend of worldwide deforestation over recent decades, reforestation has also been found and is expected in developing countries undergoing fast urbanization and agriculture abandonment. The consequences of reforestation on landscape patterns are seldom addressed in the literature, despite their importance in evaluating biodiversity and ecosystem functions. By analyzing long-term land cover changes in Puerto Rico, a rapidly reforested (6 to 42% during 1940–2000) and urbanized tropical island, we detected significantly different patterns of fragmentation and underlying mechanisms among forests, urban areas, and wetlands. Forest fragmentation is often associated with deforestation. However, we also found significant fragmentation during reforestation. Urban sprawl and suburb development have a dominant impact on forest fragmentation. Reforestation mostly occurs along forest edges, while significant deforestation occurs in forest interiors. The deforestation process has a much stronger impact on forest fragmentation than the reforestation process due to their different spatial configurations. In contrast, despite the strong interference of coastal urbanization, wetland aggregation has occurred due to the effective implementation of laws/regulations for wetland protection. The peak forest fragmentation shifted toward rural areas, indicating progressively more fragmentation in forest interiors. This shift is synchronous with the accelerated urban sprawl as indicated by the accelerated shift of the peak fragmentation index of urban cover toward rural areas, i.e., 1.37% yr^{-1} in 1977–1991 versus 2.17% yr^{-1} in 1991–2000. Based on the expected global urbanization and the regional forest transition from deforested to reforested, the fragmented forests and aggregated wetlands in this study highlight possible forest fragmentation processes during reforestation in an assessment of biodiversity and functions and suggest effective laws/regulations in land planning to reduce future fragmentation.

Editor: Ben Bond-Lamberty, DOE Pacific Northwest National Laboratory, United States of America

Funding: This work was supported by The NASA ROSES Land Cover and Land Use Change Program Grant # NNX12AE98G, and The Puerto Rico NASA IDEAS-ER fund. The funders had no role in study design, data collection and analysis, decision to publish, or preparation of the manuscript.

Competing Interests: The authors have declared that no competing interests exist.

* Email: meiyu@ites.upr.edu

Introduction

Land use and land cover have been changing dramatically across the globe and significantly impact both natural and human systems [1,2,3]. Urbanization, deforestation, and reforestation in abandoned agricultural areas are considered major drivers of the spatial distribution and pattern changes of urban areas, forests, and wetlands in recent decades [2,4,5,6].

Although deforestation as a means to meet the food demands of the growing population has been a major topic [7,8,9,10,11,12], reforestation has been significant in regions with an economic shift from agriculture to industry or to service during economic globalization, which is discussed in the Forest Transition Theory [2,5,13,14]. Reforestation via secondary succession may greatly alter the ecosystem composition, structure, and function [15] and thus add uncertainties to the large-scale carbon budget and water dynamics in the context of global change.

Urbanization is often coupled with the reforestation during the economic shift. People migrated from rural to urban areas after the abandonment of agriculture [16]. The global urban population comprised 10% of the total population in the 1900s and is predicted to comprise 60% in 2030 [17]. The rapid urbanization may accelerate the process of reforestation in rapidly developing countries, such as China. Meanwhile, urban areas expanded at twice the rate of the population increase due to rapid suburban development [3]. While the global urban population is predicted to reach 5 billion in 2030 from 2.86 billion in 2000, the projected urban area in 2030 is nearly triple that in 2000 [3]. Urban expansion drives environmental changes at local, regional, and global scales and impacts biodiversity, biogeochemical cycles, hydrology, climate, and ecosystem services [3,17].

Wetlands are vulnerable to land use/cover changes globally [18]. Because 70% of the global population lives within 80 km of coastlines [19], coastal wetlands are disappearing rapidly due to real estate development and cultivation. Wetlands are further affected by the "coastal squeeze" due to sea level rise [20]. The average annual coastal wetland loss in the US increased from

60,000 acres (24,281 ha) during 1998–2004 to 80,000 acres (32,375 ha) during 2004–2009 [21].

The patterns of land cover changes are often characterized by fragmentation [22,23]. The extension of urban sprawl to distant suburbs and the deforestation processes have been found to cause landscape fragmentation [7,24]. Landscape fragmentation is considered one of the major threats to biodiversity and a major driver of changes in ecosystem structure, function, and disturbance regimes [25,26,27,28,29,30]. Enhanced edge effects due to fragmentation reduce mass, energy, and information flows within a land type, but increase the exchanges among different land types [31,32]. The biophysical environment and socioeconomic factors are often used to explain fragmentation dynamics, such as the analysis of urban sprawl in Maryland, US [24].

Compared with a large volume of literature on deforestation and landscape fragmentation [26,33,34], the role of reforestation in fragmentation has rarely been investigated. Few studies address the composite impacts of urbanization and reforestation on landscape fragmentation dynamics. Although deforestation is still the dominant topic in the studies of land use land cover changes, economic globalization, intensified agriculture, and environmental protection indicate that reforestation, and often associated urbanization, may become widespread in heavily agriculture-dependent regions in the future. The interactions between urbanization and reforestation and their consequences on landscape fragmentation are especially important in tropical regions that contribute significantly to global biodiversity and carbon sequestration [9,35].

Puerto Rico, an island in the Caribbean, is an ideal place to investigate the coupled impacts of urbanization and reforestation on landscape dynamics in tropical regions. Due to the economic shift from agriculture to industry and services, Puerto Rico has been experiencing significant urbanization, reforestation, and urban sprawl since the 1940s [16]. Forest cover increased from 6% of Puerto Rico in the 1940s to 42% in 2000. Urban areas increased from 1.7% in the 1950s to 11% in 1990 and to 14% in 2000, following the global trend of urbanization [17,36]. The human population doubled during this period and reached 3.73 million in 2010 (US Census 2010). The population density of 425 people km^{-2} is greater than that in all 50 US states except for New Jersey. A trend of urban-to-suburban migration (urban sprawl) was revealed [13,37] with the expansion of the low-density residential area from 1991 to 2000, which may fragment natural ecosystems [24]. On the other hand, secondary succession transforms abandoned croplands to pastures, shrublands, and forests [13,38]. Reforestation may favor the aggregation of forests and thus reduce fragmentation. The composite impacts of the two drivers, viz. reforestation and urban sprawl, determine the changes in the patterns of natural systems, e.g., forests and wetlands. However, the quantitative relationship between changes in landscape fragmentation and the composite drivers is still unknown.

Coastal wetlands are especially vulnerable at this heavily populated area. Many wetlands were historically drained for sugar cane production [39] and were recently filled for coastal development [13,40]. The distribution of mangroves declined by 45% during the cultivation era, and coastal urbanization in the 1960s induced a further decrease [41]. The trend reversed in the 1970s due to the policy shift for wetland protection [38]. However, the patterns and underlying mechanisms of the wetland dynamics are still unknown.

Our objectives are to investigate the spatiotemporal patterns of the fragmentation of forests, urban areas, and wetlands in Puerto Rico at local and island-wide scales and to provide insights into the relationship between land cover patterns and the drivers of urban sprawl, reforestation, and policy change. We tested the hypotheses that, despite the overall reforestation, forests were still fragmented by the process of urban sprawl; additionally, the coastal wetlands were aggregated in the study period due to policy changes, e.g., the implementation of The RAMSAR Convention on Wetlands and the US Clean Water Act.

Methods

Study Area

Our study area is the tropical island of Puerto Rico, which is a US territory that lies at the eastern tip of the Greater Antilles (18°15″N and 66°30″W). Puerto Rico has a total area of 9,104 km^2 (Fig. 1). The prevailing trade wind comes from the northeast and traverses the central mountain range that extends east-west. The highest peak is 1,338 m. The northern windward side receives much more rainfall than the southern side, resulting in the high spatially-heterogeneous landscapes within the relatively small area. The rainforest at El Yunque Mountain in the northeast receives an annual rainfall of over 4,000 mm, whereas the dry Guanica forest in the south receives less than 800 mm. Approximately 59% of Puerto Rico is subtropical moist forest. Wet forests and rainforests are located in the mountains, and dry forests are mostly located in the southwest.

Island-wide landscape analysis of urban sprawl, reforestation, and deforestation

To detect the fragmentation dynamics of forests, urban areas, and wetlands, we analyzed the land cover maps for 1977, 1991, and 2000 that were developed by the USDA Forest Service, the Puerto Rico Department of the Natural and Environmental Resources, and the US Geological Survey Caribbean Division [36,38]. We chose these years due to the map availability, the consistency in the map classification and spatial resolution, and the reflection of urban expansion and forest regrowth [38]. All three maps are at the spatial resolution of 30 m. The maps for 1991 and 2000 were derived from the Landsat TM/ETM images using computerized classification, while the map for 1977 was interpreted visually from the aerial photos [38]. To minimize the effects of different data sources and processing methods, the spatial patterns of fragmentation were compared among all the three years; however, the absolute values of the fragmentation indices were only compared for 1991 and 2000. The original categories of the maps were reclassified into urban, forest, wetland (vegetated), and pasture to singularize the interactions between these categories (File S1). Four landscape indices, the mean patch area (AREA_MN), the edge density (ED), the edge-to-area ratio (EDGE_AREA), and the largest patch index (LPI) were calculated using the FRAGSTAT software [42] for forests, urban areas, and wetlands, respectively (Table 1).

Most fragmentation indices contain information of both landscape composition and configuration [23,42]. Generally, ED increases with the number of patches. A larger ED reflects a more fragmented landscape only when the composition of a focal type is fixed. On the other hand, ED becomes a measurement of composition when the patch size and shape are fixed. AREA_MN and EDGE_AREA do not distinguish between a single patch and multiple patches with similar mean sizes and shapes. In fact, AREA_MN only measures the average patch size, and EDGE_AREA measures the average size and shape of the patch(s). These indices mutually compensate each other to measure the fragmentation of landscapes.

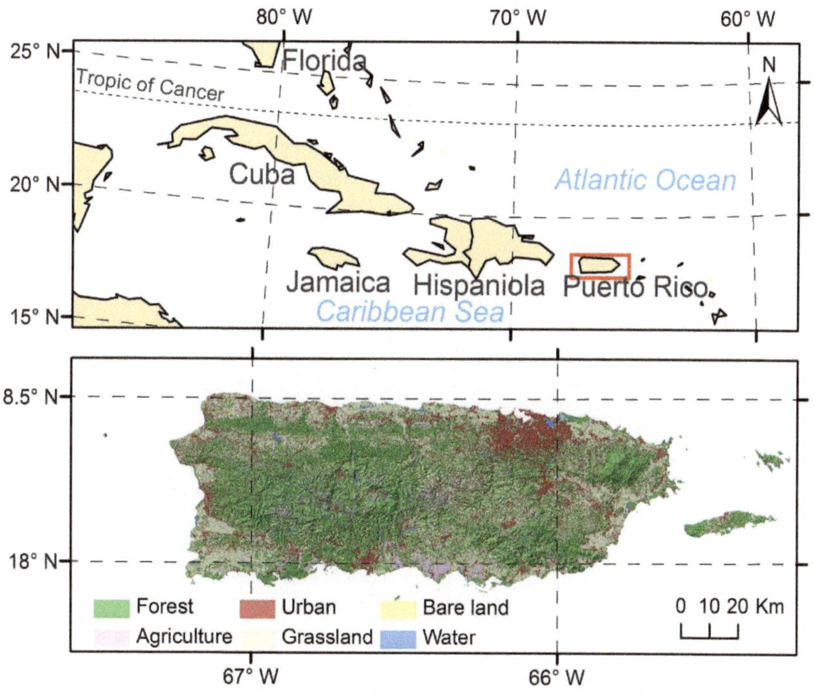

Figure 1. Location, geomorphology, and land cover in 2000 of Puerto Rico (Land cover information from Kennaway and Helmer, 2007).

To decouple the impacts of urban sprawl on forest fragmentation between 1991 and 2000, the union of the urban areas in both years was used as a mask, and the fragmentation indices of forest were recalculated with the mask for both years. This approach allowed us to assess the forest fragmentation when the effect of urban change was removed. The result was compared to the forest fragmentation with urban changes.

To separate the impacts of reforestation from those of deforestation on forest fragmentation, we first compared the forest distribution of 1991 to that of 2000 to detect the reforestation and deforestation sites. A deforestation-only scenario was then created from the forest distribution of 2000 by eliminating the reforested sites. Similarly, a reforestation-only scenario was formed from the union of the forest distributions in both years (therefore, no deforestation occurred in this scenario). The forest fragmentation indices were calculated for the deforestation-only and the reforestation-only scenarios. The results were compared with those derived from the true forest distributions in 1991 and 2000.

Spatiotemporal fragmentation analysis at a 3-km scale using spatial error models

To explore the mechanisms of landscape fragmentation dynamics during urban sprawl and reforestation, we aimed to quantify the relationship between the spatiotemporal pattern of fragmentation and biophysical and socioeconomic factors. The biophysical and socioeconomic variables of elevation, slope, distance to urban centers, and population density were considered as the potential causes of fragmentation dynamics. We used a spatial resolution of 30 m for the elevation (www.edc.usgs.gov), the population density (www.census.gov), the slope, and the distance to urban areas in 1977 (created using the Spatial Analyst extension of ArcGIS software, Esri, Redlands, CA).

Table 1. The fragmentation indices used in the landscape analysis.

Index	Equation	Description
LPI	$LPI = 100\frac{\max(a_i)}{A_T}$ (%)	Largest patch index, percentage of the largest patch of the focal type in the landscape, mainly reflecting configuration. a_i, area of the ith patch (m^2), A_T, total landscape area (m^2).
ED	$ED = 10{,}000\frac{\sum e_i}{A_T}$ ($m\,ha^{-1}$)	Edge density, edge length of the focal type per hectare of landscape area, reflecting both configuration and composition. e_i, edge of the ith patch (m).
AREA_MN	$AREA\_MN = \frac{1}{10{,}000}\frac{\sum a_i}{n}$ (ha)	Mean patch area, mainly reflecting configuration. n, the number of patches of the focal type.
EDGE_AREA	$EDGE\_AREA = \frac{\sum e_i}{A}$ ($m\,m^{-2}$)	Edge to area ratio, edge length per focal type area, mainly reflecting configuration. A, area of the focal type (m^2).

We divided the main island of Puerto Rico into 1,088 non-overlapping large grid cells of 3 km×3 km such that each large grid cell contained 10,000 pixels of 30 m×30 m. Cells with less than 2,500 land pixels (25% of the whole cell) were discarded, resulting in 1,004 cells in the subsequent analyses. For each large cell, we calculated the above fragmentation indices for urban, forest, and wetland classes for 1977, 1991, and 2000. The averages and standard deviations of the distance to urban centers, elevation, geomorphological slopes, and population density, were also calculated for each large cell. Among these variables, the distance to urban centers and population density were considered the composite variables that reflect socioeconomic status. Slopes, including their heterogeneity, have a strong control on urban development and reforestation.

We used a spatial error model to regress the fragmentation indices of the large cells onto the biophysical and socioeconomic variables. The spatial error model represents a linear relationship between the dependent (y) variable and a number of independent variables ($\bar{x}$),

$$y=\bar{x}^{\mathrm{T}}\bar{\beta}+\varepsilon \quad (1)$$

$$\varepsilon=\lambda w\varepsilon+\xi \quad (2)$$

where y refers to any of the fragmentation indices, $\bar{x}^{\mathrm{T}}$ is a transposition of the biophysical and socioeconomic factors, $\bar{\beta}$ is the vector of coefficients, and ε is the error. Unlike an ordinary linear regression model, the spatial error model assumes that the error is spatially correlated, as indicated in Eq. (2), where λ is a scalar of the spatial correlation coefficient, w is a weight matrix that describes the spatial correlation structure, and ξ is the error component without spatial dependence. The w matrix is usually derived from the neighborhood, and the eight nearest neighbors are used for this analysis. The spatial regressions were conducted in the R environment (version 3.0.1, package spdep) to estimate $\bar{\beta}$, λ, and ξ [43].

To assess the average changes in the fragmentation over the island, increments of the fragmentation indices from 1991 to 2000 were computed for each large cell, and the spatial error model was applied to fit an intercept-only model to the increments. The results allow us to determine whether there is a significant change in the fragmentation between 1991 and 2000 at the 3-km scale.

We chose to regress EDGE_AREA and ED for urban and ED and AREA_MN for forest classes on the distance-to-urban (D), the mean and standard deviation of the slopes (s and σ_s), and the mean population density (d_p). These three indices were chosen because scatterplots of their values versus those of the independent variables showed relatively concentrated patterns. The regressions were intended to derive the spatiotemporal relationships between the landscape fragmentation and the major drivers. Retention of the independent variables in the regression was guided by the minimum Akaike Information Criterion (AIC). Due to the strong topographic restriction on the distribution, we did not attempt to apply such regression to wetlands.

To investigate the interactions of urbanization/urban sprawl and reforestation, we regressed the increments of the three fragmentation indices for urban and forest classes and the EDGE_AREA for the wetland class between 1991 and 2000 onto the changes in the total areas of the different land cover types and the biophysical/socioeconomic variables. The analysis allowed us to assess how the changes in the composition of a particular land cover type affect the fragmentation pattern of various land cover types.

Results

Fragmentation analysis at the island-wide scale

Although both forests and wetlands experience total area increase from 1991 to 2000 (both approximately 2%), the island-wide patterns of the two land cover types showed a distinctively different trend: forests are becoming more fragmented, but wetlands are becoming more aggregated (Table 2). The mean forest patch size (AREA_MN) decreased and the forest ED and EDGE_AREA increased, showing increased forest fragmentation. However, the mean wetland patch size more than doubled and the wetland ED and EDGE_AREA decreased, indicating a dramatic reduction in wetland fragmentation.

The number of urban patches increased during 1991–2000 with an 8% increase in total area. Although the largest urban patch increased, the AREA_MN decreased and the EDGE_AREA and ED increased, which indicated slight urban sprawl (Table 2). The land cover transition showed that as to the new urban/suburban areas in 2000, 55% of them were pastures and 31% were forests in 1991.

The impact of urbanization/urban sprawl on forest fragmentation is quantified by comparing the scenarios of 'forests' and 'forests without urban changes' (Table 2). The AREA_MN of forests without urban changes slightly decreased from 7.2 to 7.12 ha between 1991 and 2000, which is much smaller than the 6% decrease observed when urban changes were imposed. Although the urban changes dominated the forest fragmentation, a slight trend of forest fragmentation was detected even without urbanization/urban sprawl.

The analysis of forest transitions revealed that approximately 82% of the forest in 1991 remained in 2000 with the other 18% deforested, mostly to pastures and urban areas. Reforestation also occurred, which largely transformed pastures and low-density urban areas into secondary forests. The combined outcome of deforestation (754 km^2) and reforestation (836 km^2) resulted in a 2% increase in forests during this period.

The spatial configuration of deforested and reforested sites (Fig. 2) indicated that the deforested sites tend to be located in the forest interiors, while the reforested sites tend to be located along the edges of the forests. The average distance of the reforested sites to the unchanged forests (forests in 1991 remained in 2000) is 66.7 m, while the average distance of the deforested sites to the unchanged forests is 55.8 m (16.3% closer to the interior of the forests compared to the reforested sites).

The fragmentation indices of the deforestation-only and reforestation-only scenarios revealed that the AREA_MN for the deforestation-only scenario is 6.07 ha, which is smaller than that of the true forest scenarios in 1991 and 2000 (i.e., 8.03 and 7.56 ha, respectively). However, the AREA_MN is much higher for the reforestation-only scenario (10.73 ha, Table 2). Deforestation decreased the AREA_MN by 24.4%, but reforestation increased it by 33.6%. The coupled deforestation and reforestation processes (the true forest scenario in 2000) resulted in a decrease of 5.9%, indicating a stronger role of deforestation in the forest fragmentation than that of reforestation due to their different spatial configurations. The same pattern occurred for the index of EDGE_AREA (Table 2), which increased by 16.1% due to deforestation, decreased by 19.4% due to reforestation, but still showed a 3% increase when the two processes were coupled.

Table 2. Island-wide fragmentation analyses for urban, forest, and wetland areas in 1991 and 2000.

	AREA_MN	LPI	ED	EDGE_AREA
Urban				
1991	1.184	1.39	25.9	415.6
2000	1.182	1.44	28.1	417.7
Wetland				
1991	3.54	0.08	1.76	246.3
2000	7.63	0.09	1.31	185.5
Forest				
1991	8.03	10.3	37	169.2
2000	7.56	14.1	39	174.8
Forest outside urban areas				
1991	7.20	10.4	29	182.6
2000	7.12	14.2	31	181.3
Forest scenarios				
Deforestation only	6.07	7.8	35	196.4
Reforestation only	10.73	19.2	36	136.3

AREA_MN, mean patch area (ha); LPI, largest patch index (%); ED, edge density (m ha^{-1}); EDGE_AREA, edge to area ratio (m m^{-2}).

Spatiotemporal pattern of landscape fragmentation at the 3-km scale

The fragmentation indices showed different spatial and temporal patterns for urban, forest, and wetland areas (Fig. 3). The fragmentation of urban cover is low within the cities' expansion and tends to be high at the suburban areas, indicating urban sprawl (Fig. 3a). In contrast, forest fragmentation tends to be low in the rural areas, such as the central mountains (Fig. 3b). In general, urban sprawl and forest fragmentation increased from 1991 to 2000, i.e., ED increased and AREA_MN decreased; however, the decreased wetland fragmentation during 1991–2000, especially in the eastern part of the island, indicates wetland aggregation (Fig. 3c) and agrees with the island-wide analysis.

The fitted intercepts of the increments of the fragmentation indices from 1991 to 2000 (Table 3) further confirmed that the enhanced ED (significant for forests and urban areas) and reduced AREA_MN (significant for forests) are strong indicators of increased urban sprawl and forest fragmentation. The results suggest a general trend that a small number of large patches of urban and forest in the large cells may become more aggregated, as reflected in the increased LPI and decreased EDGE_AREA. However, small and moderate patches were more numerous and became more fragmented, causing the reduced AREA_MN and increased ED. The fragmentation of forests is especially prominent, as seen by the 43% decrease in AREA_MN. The significantly reduced ED, increased LPI, and increased AREA_MN (49%) highlight the wetland aggregation.

Relationships between landscape fragmentation and biophysical and socioeconomic drivers

The best-fit spatial error models for the urban EDGE_AREA produced a quadratic function of the distance to urban centers, D; the index increased with the slope, s, and decreased with the heterogeneity of the slope, σ_s, and the population density, d_p (Table 4). The predicted urban EDGE_AREA versus D (Fig. 4a) with other variables fixed at their mean values showed that the

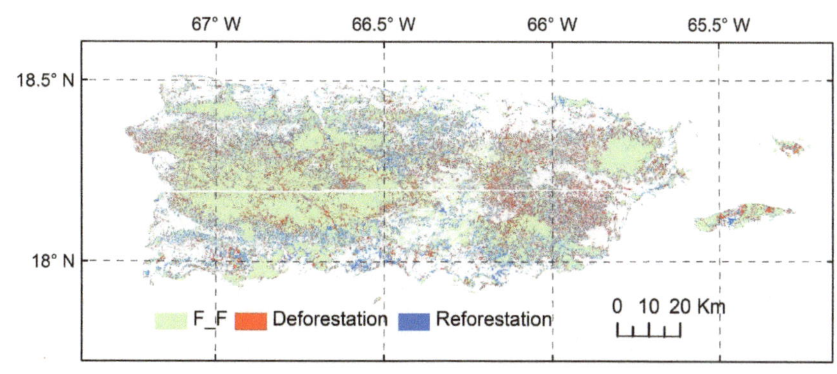

Figure 2. Spatial distributions of reforestation sites, deforestation sites, and unchanged forests (F_F) during the period of 1991–2000. The forest distribution in the deforestation-only scenario includes the F_F only, while that in the reforestation-only scenario includes all the three types. The true distribution in 1991 includes deforestation sites and unchanged forests; the true distribution in 2000 includes unchanged forests and reforestation sites.

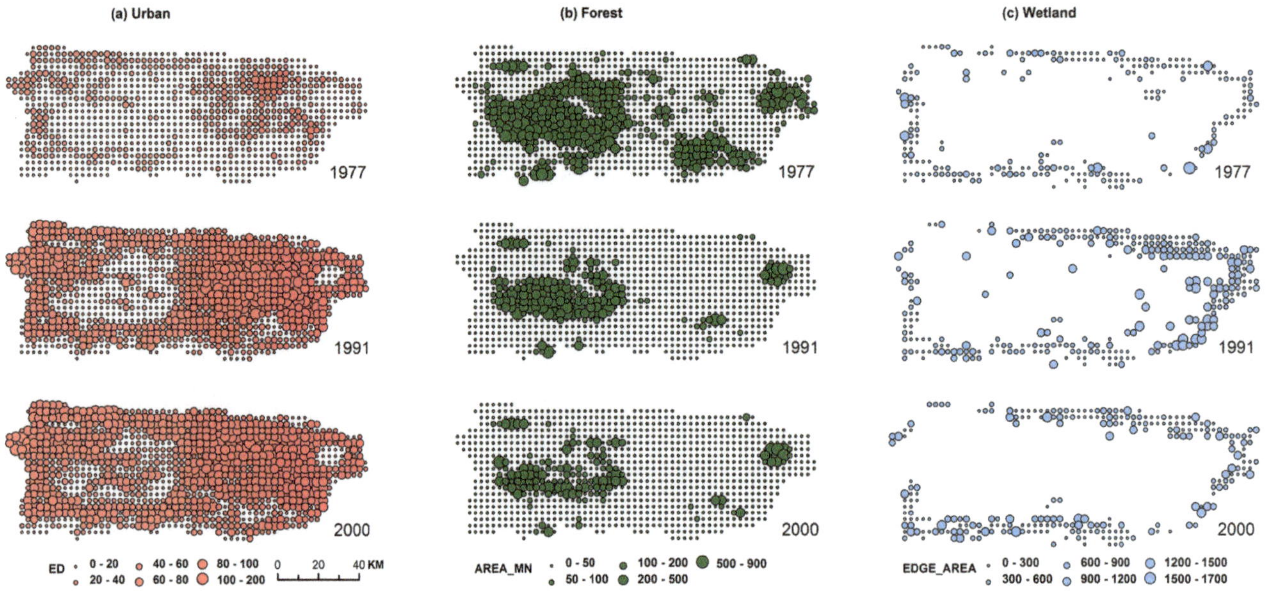

Figure 3. Fragmentation dynamics of (a) urban edge density (ED, m ha^{-1}), (b) forest mean patch area (AREA_MN, ha), and (c) wetland edge-to-area ratio (EDGE_AREA, m m^{-2}) in 1977, 1991, and 2000.

change from 1977 to 1991 is relatively even across the range of D; however, the change from 1991 to 2000 is more complicated. The EDGE_AREA of 2000 is slightly smaller than that of 1991 when D is lower than 7.2 km; but greater when D is larger. The peaks of the curves shifted from 7.3 in 1977 to 8.7 in 1991 and to 10.4 km in 2000. Because EDGE_AREA is a measure of size and shape, this shift indicates small patches developed progressively further away from the urban centers and implies significant urban sprawl toward suburban and rural areas.

The urban edge density decreased with D, but increased with slope, slope standard deviation, and population density when these variables are small (Table 4). However, the second order terms of D, s, and σ_s, tend to reduce the effects of the first order terms. The maximum ED occurred at s of 9.5–10.5° for the three years and at σ_s of 3.9–4.5° in 1991 and 2000. The predicted urban ED with D (Fig. 4b) indicated that the increase in ED was uniform across the range of D from 1977 to 1991. However, ED was more enhanced for smaller D than larger D from 1991 to 2000. Considering the decreased EDGE_AREA over the small distances, the greater increase in ED means increased large patches in these areas.

Similar to the case of EDGE_AREA for urban areas, the ED of forest increases with the first order of D and s, but decreases with σ_s (Table 4). The negative second order term of D indicates the existence of a maximum edge density for both 1991 and 2000 (Fig. 4c). When the distance is short, only a few forest patches exist and the ED is low. When the distance is very large, the forest tends to be less fragmented and therefore has fewer edges. These imply a maximum ED in the moderate distances. Indeed, the peak is found at 6.0 and 7.0 km for 1991 and 2000, respectively, singularizing the shift of the maximum forest fragmentation from suburban toward rural areas. Similarly, the negative second-order term of the slope indicates a peak edge density at approximately 15.6–16.5° for the three years.

The effects of these explanatory variables on the forest AREA_MN are opposite those of the ED, except for the lack of the first order term of D and an additional term of σ_s^2 (Table 4).

Table 3. Assessment of average increments of the fragmentation indices from 1991 to 2000 by fitting the intercept of the spatial error model.

Fragmentation index		Urban	Forest	Wetland
AREA_MN	Intercept	−0.2	−22.2**	3.4**
	Relative change	−0.08	−0.43	0.49
LPI	Intercept	0.73**	0.57*	0.42*
	Relative change	0.10	0.016	0.10
EDGE_AREA	Intercept	−8.5*	−10.3	−17.0
	Relative change	−0.01	−0.04	−0.03
ED	Intercept	3.8*	3.3*	−6.19**
	Relative change	0.07	0.04	−0.18

'**' and '*' indicate the levels of significance with p values less than 0.01 and 0.05, respectively. Relative changes are calculated by dividing the intercept by the mean fragmentation indices of 1991 and 2000.

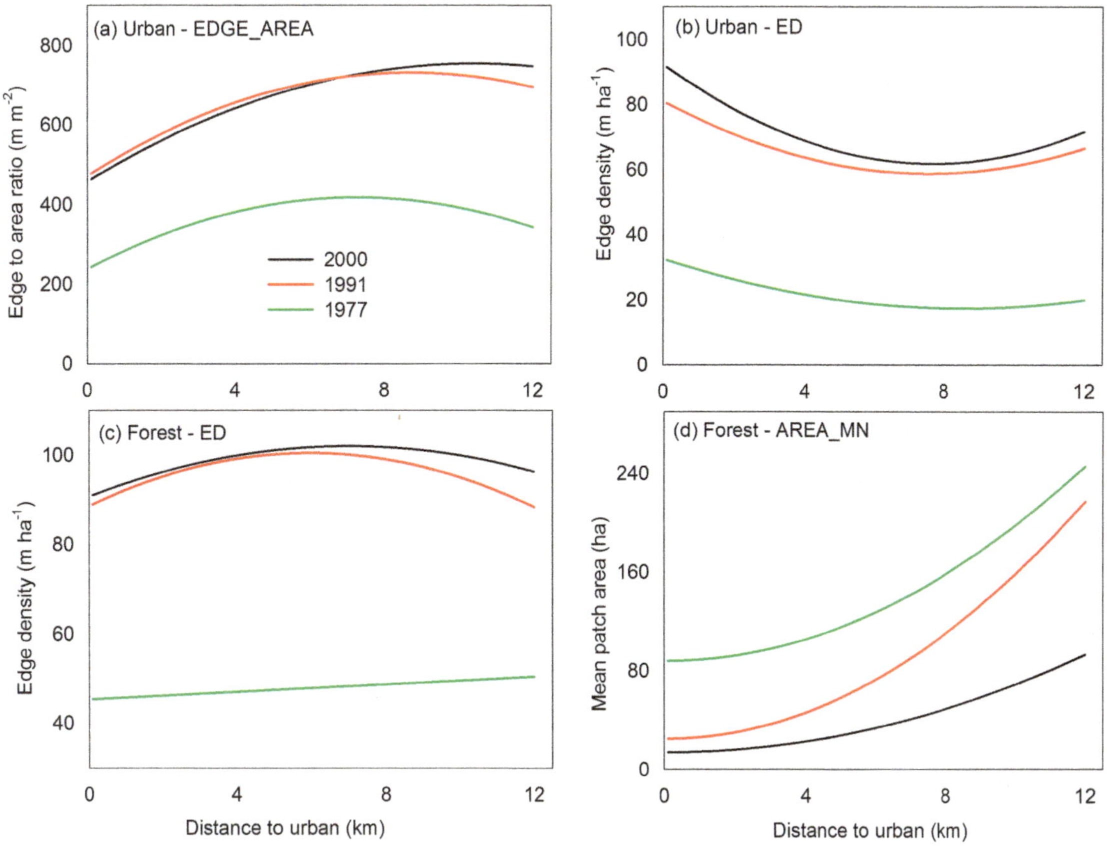

Figure 4. Predicted changes in urban and forest fragmentation index with the distance to urban centers in 1977, 1991, and 2000. (a) the edge-to-area ratio of urban areas; (b) the edge density of urban areas; (c) the edge density of forests; and (d) the mean patch area of forests.

The forest mean patch area increases with D^2, and the curves are all concave upward (Fig. 4d). The AREA_MN consistently decreases from 1977 to 2000 across the range of D. However, two different phases are apparent: fragmentation of the forest mostly occurred close to urban areas from 1977 to 1991, as indicated by the great decrease in the AREA_MN for small D; further fragmentation from 1991 to 2000 mostly occurred far from urban areas, and the reduction in the AREA_MN from 1991 to 2000 increased substantially with D. The AREA_MN also decreases with the first order of s for 1991 and 2000 but increases with the first order of σ_s for the three years. The nonlinear effects of both produce a minimum AREA_MN at the slopes of 5.8 in 1991 and 7.1° in 2000 and a maximum AREA_MN at σ_s of 3.8° in 2000 and 1991 and of 1.2° in 1977.

Interactions of the incremental land cover changes and their impacts on the landscape fragmentation

The regression of the changes in the urban EDGE_AREA and ED from 1991 to 2000 yielded the following equations:

$$\Delta \mathrm{EDGE\_AREA}_U = -0.0093\Delta U_h + 0.010\Delta P - 1.67s + 8.57\sigma_s - 1.074\sigma_s^2 \quad (3)$$

$$\Delta \mathrm{ED}_U = -1.19 + 0.032\Delta U - 0.0061\Delta F - 0.0048\Delta P + 0.518D \quad (4)$$

where $\Delta\mathrm{EDGE\_AREA}_U$ and $\Delta\mathrm{ED}_U$ are the incremental changes of urban EDGE_AREA and ED from 1991 to 2000. ΔU, ΔU_h, ΔF, and ΔP are the corresponding increments of 30 m pixels (proportional to changes in area) for urban, high density urban (class in original land cover maps), forest, and pasture, respectively. Increases in high-density urban and decreases in pasture may both reduce the increment of urban EDGE_AREA. The increase in urban areas may enhance the increment of urban edge density, but increases in forest and pasture tend to reduce the increment of urban edge density.

The increment of the forest edge density (Eq. 5) is enhanced by the increases in urban areas, forests, and pastures, but lowered by the increase in wetlands (ΔW, the increase in wetland pixels from 1991 to 2000). On the other hand, the increase in urban area reduces the increment of the forest mean patch area, and the increase in forest pixels tends to enhance the increment (Eq. 6). The incremental EDGE_AREA for wetlands (Eq. 7, $\Delta\mathrm{EDGE\_AREA}_W$) is reduced by the increase in wetlands and the decrease in agricultural land (ΔA).

$$\Delta ED_F = 6.407 + 0.010\Delta U + 0.0062\Delta F - 0.0049\Delta W + 0.0040\Delta P + 0.076D^2 - 0.923s + 0.031s^2 \quad (5)$$

$$\Delta \mathrm{AREA\_MN}_F = -0.020\Delta U + 0.0257\Delta F - 0.884D^2 - 0.048s^2 \quad (6)$$

Table 4. Spatial error models of land fragmentation indices regressed on biophysical and socioeconomic drivers.

Fragmentation Index	D	D^2	s	s^2	σ_s	σ_s^2	d_p	intercept
$EDGE\_AREA_{U,00}$	56.9	−2.73	19.98		−38.47		−0.119	455.2
$EDGE\_AREA_{U,91}$	59.6	−3.42	20.85		−35.91		−0.121	446.0
$EDGE\_AREA_{U,77}$	49.6	−3.41	6.29		−9.31		−0.063	233.5
$ED_{U,00}$	−7.96	0.519	3.58	−0.188	9.18	−1.02	0.00321	54.9
$ED_{U,91}$	−5.87	0.389	3.54	−0.172	5.30	−0.68	0.00463	51.7
$ED_{U,77}$	−3.56	0.21	1.38	−0.066			0.00631	22.8
$ED_{F,00}$	3.20	−0.23	9.34	−0.285	−4.42			40.96
$ED_{F,91}$	3.97	−0.33	10.20	−0.308	−4.71			33.23
$ED_{F,77}$	0.42		4.25	−0.136	−1.87			22.93
$AREA\_MN_{F,00}$		0.55	−6.68	0.47	14.45	−1.88		
$AREA\_MN_{F,91}$		1.33	−4.41	0.38	12.63	−1.65		
$AREA\_MN_{F,77}$		1.10	2.59	0.62	6.93	−2.93		

Values are significant intercepts or coefficients for the corresponding variables. The first subscript of the fragmentation index indicates urban (U) or forest (F), and the second indicates the year of 2000 (00), 1991 (91), or 1977 (77). D, distance to urban centers (km); s, slope; σ_s, standard deviation of slope; d_p, population density (per km^2).

$$\Delta EDGE\_AREA_W = -0.128\Delta W + 0.037\Delta A \qquad (7)$$

Discussion

The deforestation-to-reforestation process of subtropical Puerto Rico is common in many developing regions [44] and supports the Forest Transition Theory [6,14,45,46] with an economic shift from agriculture to industry and service as the main driver. However, our findings indicate that despite the progressively increased forest cover over the course of the reforestation, the forests are becoming more fragmented, primarily due to two drivers: urban sprawl and deforestation in the forest interiors during and after reforestation.

Accelerated urban sprawl coupled with a forest fragmentation shift to rural areas

The urban sprawl in Puerto Rico is characterized by a shift of the peak edge-to-area ratio from near urban centers to suburban and rural areas and an increased edge density from 1977 to 2000 (Fig. 4, a and b). If we divide the difference between the peaks by the length of the time interval, we find that the average annual peak shift is 0.10 and 0.19 km yr^{-1} for the periods of 1977–1991 and 1991–2000, respectively. The nearly doubled speed of the peak shift in 1991–2000 implies accelerated urban sprawl. It was reported that the peak fragmentation in the state of Maryland, US, shifted from 40 km away from urban centers in 1973 to 55 km away from urban centers in 2000 [24]; the average annual rate was 0.56 km yr^{-1}, which is faster than that in Puerto Rico. Considering the difference in the size of land area between Maryland and Puerto Rico, we divided the absolute annual shift by the peak distance of the previous years to compute the relative peak shift rate of 1.39% yr^{-1} for Maryland and 1.37 and 2.17% during 1977–1991 and 1991–2000, respectively, for Puerto Rico. From 1991 to 2000, the peak of the forest edge density shifted from 5.8 to 7.0 km away from urban centers (Fig. 4c); the annual shift was 0.13 km yr^{-1}. Hence, the accelerated urban sprawl is coupled and approximately synchronized with the increased forest fragmentation toward rural areas, despite the large increase in forest coverage during 1991–2000. The relationship between urban sprawl and forest fragmentation is also demonstrated in Equations 5 and 6, where increased urban area is shown to cause increased forest edge density and decreased forest mean patch area.

Moreover, the accelerated urban sprawl toward farther distances was also accompanied by urbanization near suburbs. The edge-to-area ratio lacks the ability to discern many small patches from a single small patch; however, the edge density compensates for this. The urban edge density in 2000 is greater than that in 1991 across the entire range of distances to urban centers (Fig. 4b). Hence, this finding is an additional indicator of urban sprawl. However, the increase in the urban edge density is greater at shorter distances than at farther distances due to the increased large urban patches near suburbs, which lowers the edge-to-area ratio and indicates urbanization at shorter distances (Fig. 4a). It is interesting to see that when urbanization (ΔU) is considered, the increment of the urban edge density from 1991 to 2000 increased with the distance (Eq. 4), as indicated by the positive coefficient in front of D in the equation (opposite of the trend displayed in Fig. 4b). In other words, when the change of the urban composition is incorporated into the regression, the changes in the urban edge density mainly reflect changes in the landscape configuration, and the increase in the increment of the edge density with D indicates smaller and more irregular urban patches in distant suburbs.

Forest fragmentation is also caused by deforestation in forest interiors during reforestation

The forest fragmentation is mainly caused by urban sprawl, as indicated by the synchronized shifts of peak fragmentation of both forest and urban cover, and the fact that the increase in the urban area (ΔU) significantly reduced the increment of the forest mean patch area and enhanced the increment of the forest edge density (Eq. 5 and 6). The forest fragmentation analysis with and without urban impacts also supports this finding: a 6% decrease in the forest mean patch area was computed under urban change compared with only a 1% decrease without urban influence. On the other hand, the increase in the forest area (reforestation) enhanced the increments of both the mean patch area and edge density of the forest, implying that reforestation mostly formed large patches (Eq. 5 and 6). Among all the causal factors of the forest edge density increment (Eq. 5), urban development is the strongest, as shown by the greatest coefficient of 0.01 in front of ΔU, compared with other land type changes.

In addition to urban sprawl, the greater increase in the forest edge density (Fig. 4c) associated with the greater reduction in the mean patch area at farther distances (Fig. 4d) than those at shorter distances to urban areas from 1991 to 2000 consistently signaled forest fragmentation far from urban areas, i.e., at locations closer to the forest interior. For comparison, previous forest fragmentation mostly occurred near urban areas (1977–1991). The distribution of deforestation sites is nearer to the unchanged forests or forest interiors, compared with that of reforestation sites. The island-wide analysis of the deforestation-only and reforestation-only scenarios demonstrates the stronger role of deforestation, compared with reforestation, on forest fragmentation due to its unique spatial configuration.

The deforestation in the forest interior is mostly associated with the development of pastures from 1991 to 2000. About 75% of deforested areas were converted to pastures and 21% were converted to urban areas in 1991–2000. The increase in the forest edge density is positively associated with the increase in the pasture (Eq. 5). Evidence can be easily found in the western part of the central mountains, where forests were converted to many small patches of pasture for various purposes, such as dairy industry development and land for real estate.

Role of the socioeconomic and biophysical variables in landscape fragmentation

The distance to urban centers, geomorphological slope, and population density played important roles in the urban and forest fragmentation dynamics. The effects of these variables are mostly nonlinear as the second order terms are found in the equations. In a particular year, the fragmentation of forest and urban areas generally increases with the distance to urban centers due to the urban sprawl, reaches a peak, and decreases due to the reduced anthropogenic activities. We found a systematic shift of the peak fragmentation to greater distances, and this trend has been observed in many other places around the world [24].

Slope has been found to be a control variable for both urban and forest areas [24,47]. Our results showed their fragmentation index increased with gentle to moderate slopes. A steep slope cannot be used to build large urban patches, and the scattered construction further fragments both forests and urban cover. When slopes become too steep, urban development is limited by high costs, low commercial value, and risk of landslides, one of the primary natural hazards in Puerto Rico [48]. Therefore, the forests in steep areas are less likely to be disturbed and form large patches with reduced edge density. The slopes with peaks of urban edge density have been found within the range of 9.5–10.5°. The peak edge density of forests has been found at slopes of 15.6–16.5°, which are higher than those for urban edge density, partly because steeper terrain contains more forests and less urban areas. However, the mean patch area of forests has a minimum at 5.8 and 7.1° for 1991 and 2000, respectively, indicating a slight shift of the fragmentation to steeper terrain; the slopes in these ranges are critically vulnerable to further fragmentation.

The impact of the heterogeneity of the slope, as described in the standard deviation, is seldom discussed in the literature. In this study, when the heterogeneity is small or moderate (within 4°), the increase of the slope standard deviation lowered the edge-to-area ratio but increased the edge density of urban areas. A possible consequence of meeting both conditions is increased large patches of urban areas with slope standard deviations in the range of 0–4°. Within the same range, the forest edge density decreased, but the mean patch area increased. Both findings indicate the decreased forest fragmentation with the heterogeneity of the slope. Beyond this range, the high spatial heterogeneity of the terrain limits urban development and fragments the forests. The shift of the peak mean patch area of forests from 1977 (at $\sigma_s = 1.2$) to 1991 and 2000 (at $\sigma_s = 3.8$) indicates the large forest patches appeared in more heterogeneous areas in 1991 and 2000 than in 1977.

Population density has been shown to be a control variable for urban sprawl, but not for forest fragmentation. The urban edge-to-area ratio decreases with the population density because great population densities are generally associated with large urban patches. However, the urban edge density increases with the population density because large urban patches enhance the edge density in terms of composition. A lack of population density in the equations of the forest fragmentation indices (Table 4) suggests that the role of population density is mostly indirect in forest fragmentation.

Significant wetland aggregation despite urban sprawl

Unlike the fragmenting dynamics of forests, wetlands are becoming much more aggregated at both island-wide and local scales. However, the spatial variation of wetland fragmentation is independent of changes in either urban areas or forests (Eq. 7) because wetland dynamics are mainly controlled by the implementation of laws and regulations on wetland protection. The international RAMSAR Convention on Wetland, the Clean Water Act by US Congress, and other relevant state regulations have been enforced since the early 1970s [41]. By comparing the distributions of protected areas [49], approximately 44% of wetlands are protected. Specifically, 56% of forested wetlands (e.g., mangroves and *Pterocarpus* forests) and 29% of herbaceous wetlands are protected. The implementation of wetland regulations directly leads to wetland aggregation, as indicated by the greater mean patch area within the reserves (6.8 ha) compared to that outside the reserves (5.1 ha).

A reduction in agriculture led to decreased wetland fragmentation, as indicated in Eq. 7, because the recovery of wetland is mostly from abandoned agriculture, e.g., sugarcane cultivation and other coastal agriculture. As an example, the Caño Tiburones Nature Reserve, which is the largest herbaceous wetland in Puerto Rico, was formed after the pumps used to drain the land for agriculture broke during Hurricane Georges in late September of 1998. The natural recovery of the wetlands was appreciated and a request to establish a nature reserve was approved by law on Oct.

16, 1998. A similar transformation occurred in the largest *Pterocarpus* forest wetland in Puerto Rico.

Implications of landscape fragmentation during urban sprawl and reforestation

Recent research on global forest cover changes [5] clearly showed that in the period of 2000–2012, deforestation in the forest interiors of Puerto Rico still exists and reforestation continues, especially in the south (map available at http://earthengine partners.appspot.com/science-2013-global-forest). The global urbanization trend with a faster rate of urban expansion than rate of population increase implies that low-density urban or urban sprawl is continuing in many regions, especially in developing countries, such as China and those in South America [3,50]. Globalization, intensified agriculture, and economic shifts also caused forest transitions in many areas of the world (i.e., deforestation to reforestation) [6,51]. Our study indicates that even during the process of reforestation, forests are very likely to become fragmented, primarily due to urban sprawl and deforestation in the forest interiors; the dynamics of forest fragmentation are synchronized with urban sprawl. Because fragmented forests have modified microclimates, disturbance regimes, and biodiversity structures, their functions and services, such as carbon sequestration capabilities, should be evaluated differently from those of continuous forests. Wetlands are becoming aggregated in our study regardless of the strong interference of urbanization along the coast because of the wetland protection regulations. The contrasting trend in fragmentation between forests and wetlands implies that effective regulations and rational land planning shall be implemented for forest protection, especially for the forest interiors.

Acknowledgments

We greatly appreciate the International Institute of Tropical Forestry of USDS Forest Service for sharing the maps of land cover and protected areas in Puerto Rico. We thank Mr. Luis Villanueva for the discussions on wetland issues.

Author Contributions

Conceived and designed the experiments: QG MY. Performed the experiments: QG MY. Analyzed the data: QG MY. Contributed reagents/materials/analysis tools: QG MY. Wrote the paper: QG MY.

References

1. Field CB, Lobell DB, Peters HA, Chiariello NR (2007) Feedbacks of terrestrial ecosystems to climate change. Annual Review of Environment and Resources 32: 1–29.
2. Foley JA, DeFries R, Asner GP, Barford C, Bonan G, et al. (2005) Global consequences of land use. Science 309: 570–574.
3. Seto KC, Gueneralp B, Hutyra LR (2012) Global forecasts of urban expansion to 2030 and direct impacts on biodiversity and carbon pools. Proceedings of the National Academy of Sciences of the United States of America 109: 16083–16088.
4. Gibbs HK, Ruesch AS, Achard F, Clayton MK, Holmgren P, et al. (2010) Tropical forests were the primary sources of new agricultural land in the 1980s and 1990s. Proceedings of the National Academy of Sciences of the United States of America 107: 16732–16737.
5. Hansen MC, Potapov PV, Moore R, Hancher M, Turubanova SA, et al. (2013) High-Resolution Global Maps of 21st-Century Forest Cover Change. Science 342: 850–853.
6. Rudel TK, Coomes OT, Moran E, Achard F, Angelsen A, et al. (2005) Forest transitions: towards a global understanding of land use change. Global Environmental Change 15: 23–31.
7. Chambers JQ, Asner GP, Morton DC, Anderson LO, Saatchi SS, et al. (2007) Regional ecosystem structure and function: ecological insights from remote sensing of tropical forests. Trends in Ecology & Evolution 22: 414–423.
8. Laurence WR, Lovejoy TE, Vasconcelos HL, Bruna EM, Didham RK, et al. (2002) Ecosystem decay of Amazonian forest fragments: A 22-year investigation. Conservation Biology 16: 605–618.
9. Harris NL, Brown S, Hagen SC, Saatchi SS, Petrova S, et al. (2012) Baseline Map of Carbon Emissions from Deforestation in Tropical Regions. Science 336: 1573–1576.
10. Rudel TK (2013) The national determinants of deforestation in sub-Saharan Africa. Philosophical Transactions of the Royal Society B: Biological Sciences 368.
11. DeFries RS, Rudel T, Uriarte M, Hansen M (2010) Deforestation driven by urban population growth and agricultural trade in the twenty-first century. Nature Geosci 3: 178–181.
12. Hansen MC, Stehman SV, Potapov PV, Loveland TR, Townshend JRG, et al. (2008) Humid tropical forest clearing from 2000 to 2005 quantified by using multitemporal and multiresolution remotely sensed data. Proceedings of the National Academy of Sciences.
13. Pares-Ramos IK, Gould WA, Aide TM (2008) Agricultural Abandonment, Suburban Growth, and Forest Expansion in Puerto Rico between 1991 and 2000. Ecology and Society 13.
14. Rudel TK (1998) Is There a Forest Transition? Deforestation, Reforestation, and Development1. Rural Sociology 63: 533–552.
15. Lugo AE, Helmer E (2004) Emerging forests on abandoned land: Puerto Rico's new forests. Forest Ecology and Management 190: 145–161.
16. Grau HR, Aide TM, Zimmerman JK, Thomlinson JR, Helmer E, et al. (2003) The ecological consequences of socioeconomic and land-use changes in postagriculture Puerto Rico. Bioscience 53: 1159–1168.
17. Grimm NB, Faeth SH, Golubiewski NE, Redman CL, Wu JG, et al. (2008) Global change and the ecology of cities. Science 319: 756–760.
18. Fernandez-Prieto D, Finlayson CM (2009) Earth Observation and wetlands. Journal of Environmental Management 90: 2119–2120.
19. Michener WK, Blood ER, Bildstein KL, Brinson MM, Gardner LR (1997) Climate change, hurricanes and tropical storms, and rising sea level in coastal wetlands. Ecological Applications 7: 770–801.
20. Doody JP (2004) 'Coastal squeeze'— an historical perspective. Journal of Coastal Conservation 10: 129–138.
21. Dahl TE, Stedman SM (2013) Status and trends of wetlands in the coastal watersheds of the Conterminous United States 2004 to 2009. U.S. Department of the Interior, Fish and Wildlife Service and National Oceanic and Atmospheric Administration, National Marine Fisheries Service. 46 p. p.
22. Uuemaa E, Mander U, Marja R (2013) Trends in the use of landscape spatial metrics as landscape indicators: A review. Ecological Indicators 28: 100–106.
23. Turner MG (2005) Landscape ecology: What is the state of the science? Annual Review of Ecology Evolution and Systematics 36: 319–344.
24. Irwin EG, Bockstael NE (2007) The evolution of urban sprawl: Evidence of spatial heterogeneity and increasing land fragmentation. Proceedings of the National Academy of Sciences 104: 20672–20677.
25. Saunders DA, Hobbs RJ, Margules CR (1991) Biological consequences of ecosystem fragmentation - a review. Conservation Biology 5: 18–32.
26. Fahrig L (2003) Effects of Habitat Fragmentation on Biodiversity. Annual Review of Ecology, Evolution, and Systematics 34: 487–515.
27. Fischer J, Lindenmayer DB (2007) Landscape modification and habitat fragmentation: a synthesis. Global Ecology and Biogeography 16: 265–280.
28. Laurance WF, Ferreira LV, Merona JMRD, Laurance SG, Hutchings RW, et al. (1998) Effects of Forest Fragmentation on Recruitment Patterns in Amazonian Tree Communities. Conservation Biology 12: 460–464.
29. Laurance WF (2008) Theory meets reality: How habitat fragmentation research has transcended island biogeographic theory. Biological Conservation 141: 1731–1744.
30. Cochrane MA (2003) Fire science for rainforests. Nature 421: 913–919.
31. Gao Q, Yang X (1997) A relationship between spatial processes and a partial patchiness index in a grassland landscape. Landscape Ecology 12: 321–330.
32. Gao Q, Yu M, Wang J, Jia H, Wang K (2004) Relationships between regional primary production and vegetation patterns. Ecological Modelling 172: 1–12.
33. Ewers RM, Didham RK (2006) Confounding factors in the detection of species responses to habitat fragmentation. Biological Reviews 81: 117–142.
34. Swift TL, Hannon SJ (2010) Critical thresholds associated with habitat loss: a review of the concepts, evidence, and applications. Biological Reviews 85: 35–53.
35. DeFries RS, Houghton RA, Hansen MC, Field CB, Skole D, et al. (2002) Carbon emissions from tropical deforestation and regrowth based on satellite observations for the 1980s and 1990s. Proceedings of the National Academy of Sciences of the United States of America 99: 14256–14261.
36. Helmer EH, Ramos O, Lopez TD, Quinones M, Diaz W (2002) Mapping the forest type and land cover of Puerto Rico, a component of the Caribbean biodiversity hotspot. Caribbean Journal of Science 38: 165–183.

37. Martinuzzi S, Gould WA, Gonzalez OMR (2007) Land development, land use, and urban sprawl in Puerto Rico integrating remote sensing and population census data. Landscape and Urban Planning 79: 288–297.
38. Kennaway T, Helmer EH (2007) The forest types and ages cleared for land development in Puerto Rico. Giscience & Remote Sensing 44: 356–382.
39. Lugo AE, Brown S (1988) The Wetlands of the Caribbean Islands. Acta Cientifica 2: 48–61.
40. Brush R (2008) Claiming coastal property in Puerto Rico. Anthropology News. pp. 13–14.
41. Martinuzzi S, Gould WA, Lugo AE, Medina E (2009) Conversion and recovery of Puerto Rican mangroves: 200 years of change. Forest Ecology and Management 257: 75–84.
42. McGarigal K, Cushman SA, Ene E (2012) FRAGSTATS v4: Spatial Pattern Analysis Program for Categorical and Continuous Maps. University of Massachusetts, Amherst.
43. R Core Team (2013) R: A language and environment for statistical computing. R Foundation for Statistical Computing. Vienna, Austria.
44. Rudel TK, Perez-Lugo M, Zichal H (2000) When Fields Revert to Forest: Development and Spontaneous Reforestation in Post-War Puerto Rico. The Professional Geographer 52: 386–397.
45. Turner BL, Lambin EF, Reenberg A (2007) The emergence of land change science for global environmental change and sustainability. Proceedings of the National Academy of Sciences 104: 20666–20671.
46. Mather AS, Needle CL (1998) The forest transition: a theoretical basis. Area 30: 117–124.
47. Butler BJ, Swenson JJ, Alig RJ (2004) Forest fragmentation in the Pacific Northwest: quantification and correlations. Forest Ecology and Management 189: 363–373.
48. Jibson RW (1989) Debris flows in southern Puerto Rico. Geological Society of America Special Papers 236: 29–56.
49. Gould WA, Quinones M, Solorzano M, Alarcon C, Alcobas W (2011) Protected natural areas of Puerto Rico. Scale 1:240,000. IITF-RMAP-02. In: US Department of Agriculture Forest Service IIoTF, editor. San Juan, PR.
50. United Nations Human Settlements Programme (2011) Cities and climate change: policy directions. London; Washington, DC: Earthscan.
51. Redo DJ, Grau HR, Aide TM, Clark ML (2012) Asymmetric forest transition driven by the interaction of socioeconomic development and environmental heterogeneity in Central America. Proceedings of the National Academy of Sciences 109: 8839–8844.

24

The Relationship between the Distribution of Common Carp and Their Environmental DNA in a Small Lake

Jessica J. Eichmiller*, Przemyslaw G. Bajer, Peter W. Sorensen

Department of Fisheries, Wildlife, and Conservation Biology, Minnesota Aquatic Invasive Species Research Center, University of Minnesota, Twin Cities, St. Paul, Minnesota, United States of America

Abstract

Although environmental DNA (eDNA) has been used to infer the presence of rare aquatic species, many facets of this technique remain unresolved. In particular, the relationship between eDNA and fish distribution is not known. We examined the relationship between the distribution of fish and their eDNA (detection rate and concentration) in a lake. A quantitative PCR (qPCR) assay for a region within the cytochrome *b* gene of the common carp (*Cyprinus carpio* or 'carp'), an ubiquitous invasive fish, was developed and used to measure eDNA in Lake Staring (MN, USA), in which both the density of carp and their distribution have been closely monitored for several years. Surface water, sub-surface water, and sediment were sampled from 22 locations in the lake, including areas frequently used by carp. In water, areas of high carp use had a higher rate of detection and concentration of eDNA, but there was no effect of fish use on sediment eDNA. The detection rate and concentration of eDNA in surface and sub-surface water were not significantly different ($p \geq 0.5$), indicating that eDNA did not accumulate in surface water. The detection rate followed the trend: high-use water > low-use water > sediment. The concentration of eDNA in sediment samples that were above the limit of detection were several orders of magnitude greater than water on a per mass basis, but a poor limit of detection led to low detection rates. The patchy distribution of eDNA in the water of our study lake suggests that the mechanisms that remove eDNA from the water column, such as decay and sedimentation, are rapid. Taken together, these results indicate that effective eDNA sampling methods should be informed by fish distribution, as eDNA concentration was shown to vary dramatically between samples taken less than 100 m apart.

Editor: Arga Chandrashekar Anil, CSIR- National institute of oceanography, India

Funding: Funding for this project was provided by the Minnesota Environment and Natural Resources Trust Fund as recommended by the Legislative-Citizen Commission on Minnesota Resources (LCCMR). The funders had no role in study design, data collection and analysis, decision to publish, or preparation of the manuscript.

Competing Interests: The authors have declared that no competing interests exist.

* Email: eich0146@umn.edu

Introduction

Methods to quantify the abundance of fish populations, such as mark-recapture and electrofishing, are costly and time-consuming. In addition, fish are often difficult to capture and detect at low densities, and capture methods themselves can lead to behavioral changes of the target species [1–3]. Molecular methods to detect the DNA released by aquatic organisms into their environment are non-invasive, rapid, and potentially more sensitive than traditional census techniques [4–6]. This environmental DNA (eDNA) is released through processes such as cell sloughage, mucus excretions, and defecation [7]. Notably, eDNA is currently used to monitor the presence of invasive Bigheaded carps (often called 'Asian carps') (*Hypophthalmichthys* spp.) in the Chicago Area Waterway System and the Mississippi River [8]. Although initially developed as a detection tool, molecular techniques that utilize eDNA are evolving to answer more complex questions. For example, several studies have established relationships between eDNA concentration and biomass in aquatic habitats [9–11]. Next-generation sequencing approaches have successfully identified multiple species simultaneously [11,12].

Despite the immense potential for eDNA technology to revolutionize monitoring programs for fish and other aquatic species, little is known about the production, fate, and distribution of eDNA in the natural environment. The distribution of eDNA is of particular importance for development of effective monitoring methods [6]. Surprisingly, Pilliod et al. [9] found that time of day, sampling location, and distance from the target organism (salamanders) had no apparent effect on eDNA concentration in small streams. In contrast, eDNA from snails was more abundant in the middle of a river channel relative to the channel margins [13]. Surface water samples are widely used for eDNA studies [8,9,14]. The rationale for this approach has only been confirmed in one study done in experimental ponds [15]. The possibility that eDNA concentration within a water body may be influenced by fish distribution was initially posed by Takahara et al. [10]. In a lagoon in winter, the concentration of eDNA from common carp (*Cyprinus carpio*, hereafter 'carp') was positively correlated with water temperature and was spatially heterogeneous. The cause of this pattern, and in particular whether it was due to the distribution of carp or higher metabolic activity of fish in warmer waters, was not examined as the distribution of carp was not measured. From the few studies that address the question of eDNA distribution in water, it is clear that it varies among habitat types, and more conclusive explanations of eDNA distribution patterns are needed. Also of interest is the distribution of eDNA in sediments, as sediments likely retain eDNA for long periods of time [16].

To determine whether fish distribution affects eDNA concentration and detection rate in lake water and sediment, we examined the distribution of carp eDNA in a small, shallow lake and compared it to known patterns of carp distribution, which had been monitored for several years. We were interested both in detection rate (percentage of samples in which eDNA levels were present above detection threshold) as well as concentration, because the former is commonly used to assess the likely distribution of invasive Bigheaded carps while the latter measure, if understood, might add more resolution and value to the technique. First, a qPCR assay specific for *C. carpio* eDNA was developed and validated in the lab. Next, since eDNA is often assumed to accumulate in surface water and sediment, surface, sub-surface, and sediment samples were taken throughout the lake. Finally, the concentration and detection rate of eDNA was compared between areas of low- and high-fish use identified from radiotelemetry data. Results of this study provide insights into optimal eDNA sampling methods for small lakes as well as information on how eDNA is distributed in aquatic systems in relation to the distribution of target organisms.

Materials and Methods

Quantitative PCR marker development and validation

Although two *C. carpio* qPCR assays had been developed prior to this study [10,17], a screen against the NCBI database indicated potential non-specific amplification of non-target fish species (Table S1). Therefore, a qPCR assay was developed for the current study. Four genes were considered in the development of a novel qPCR marker specific to the common carp: (1) mitochondrial gene cytochrome *b*, (2) mitochondrial gene cytochrome *c* oxidase subunit 1, (3) mitochondrial gene control (D-loop) region, and (4) the nuclear gene recombination activating gene 1 (RAG1). Candidate primer sets were identified by NCBI Primer-BLAST using sequences under GenBank accession number X61010.1 [18] for mtDNA and EF458304.1 [19] for the RAG1 gene. Specificity was initially screened against the BLASTn database sequences for 15 fish species (Table S1). Minor groove binder (MGB) probes were manually designed using the Primer Probe Test Tool in Primer Express Software v3.0.1 (Life Technologies, Grand Island, NY). Assays with amplification efficiency outside the range of acceptable values of 90–110% or a limit of detection above 300 copies per reaction were not considered. We defined the limit of detection (LOD) as the lowest value at which three replicate reactions would successfully amplify with a quantification cycle (*Cq*) value of less than 40 cycles within the linear range of the standard curve.

Candidate markers were screened for specificity for carp by testing for amplification of 15 ng of fin clip DNA from carp and 34 native and non-native fish species (Table S1). Fin clip samples for genetic marker specificity testing were extracted using the DNeasy Blood and Tissue Kit (Qiagen, Hilden, Germany) and assayed as described below.

Next, we tested markers using aqueous samples. Three 340 L flow-through tanks were set up to confirm the ability of the marker to detect carp eDNA. Prior to this experiment, all tanks were treated with 10% bleach for 30 minutes to remove all traces of DNA. The flow through rate was set at 600 mL/min, and temperature was maintained at 18°C. The first tank was stocked with 10 carp (35 g), and the second tank was stocked with 10 goldfish (*Carassius auratus*) (50 g), while the third tank was stocked with five fish of both species. These stocking levels corresponded to a biomass of 0, 438, and 875 mg/L of carp. Fish were fed once daily *ad libitum* a combination of flake feed (Color Tropical Marine Flake, Pentair Aquatic Eco-systems, Inc., Apopka, FL) and 2.5 mm pellet feed (Oncor Fry, Skretting USA, Tooele, UT) that did not contain target genetic markers. After 6 days, 4 1 L water samples were collected from each tank, immediately stored at 4°C, and filtered within 4 h. Molecular analyses followed protocols described below. This study was carried out in strict accordance with the recommendations in the Guide for the Care and Use of Laboratory Animals of the National Institutes of Health. The protocol for care and holding of laboratory fish was approved by the University of Minnesota's Institutional Animal Care and Use Committee (IACUC) (Protocol: 1407-31659A). No anesthesia or euthanasia was required as part of this study.

Study site

The study site was Lake Staring, a small freshwater lake located in the Upper Mississippi River Basin (44°50'14" N, −93°27'18" W). Lake Staring is a small, shallow lake that experiences frequent mixing due to wind and is typical of high carp density lakes in this region [20]. The surface area of the lake is 65.7 ha, consisting mostly of littoral zone with a depth of less than 2 m. The maximum depth is 4.8 m, and the lake bed is composed of fine sediment. Due to high carp density, the lake lacks aquatic vegetation except for white water lily (*Nymphaea odorata*), which covers less than 10% of the lake area.

Carp population abundance in Lake Staring was estimated in 2011 using a mark-recapture analysis [20]. This analysis showed that the lake was inhabited by approximately 26,000 carp, 95% CI [21,000, 31,000], or approximately 400 carp/ha. The mean body length of carp was 444 mm, indicating that the population was primarily composed of adults [20]. Approximately 14,000 fish were removed from the lake in the winters of 2012 and 2013, and a new population estimate was generated in the fall of 2013 by conducting a mark-recapture analysis. For the mark-recapture analysis, 46 carp were marked with individually numbered tags in November 2013, of which 22 were recaptured during the following four months among 5,457 carp that were captured and examined for marks. Using these data we estimated that Lake Staring was inhabited by 11,153 carp, 95% CI [7,972, 14,334] in the fall of 2013. The biomass decreased only slightly, from 490 kg/ha in 2011 to 397 kg/ha in 2013 because the mean body length of carp increased to 559 mm over this time frame. The biomass at the time of this study was approximately 20 mg/L, assuming an average lake depth of 2 m.

Since 2011, the distribution of carp has been regularly assessed by locating 10–20 carp equipped with internal radio tags (F 1850, Advanced Telemetry Systems, Isanti, MN). Carp location was determined by identification of signal directionality (bearing) with a hand-held antenna and a compass while positioned within 200 m of the radiotagged fish. Two bearings were measured for each fish, each from a different location, and their intersection calculated (LOAS, Ecological Software Solutions, CA) to estimate fish location. Mean measurement error (30 m) was estimated using dummy tags.

Using previously determined carp locations, the pattern of carp habitat usage was examined for the warm season (June–October) when carp maintain stable summertime distributions [21]. A total of 12 radiotelemetry surveys within the 2011–2013 time frame were conducted. Individual locations (N = 135) were pulled across fish and years. Areas of high carp use were estimated by calculating kernel density (search radius = 35 m, approximately one SE of carp location estimate; output cell size = 4.2 m) using spatial analyst in ArcMap (10.0, Esri, Redlands, CA).

Carp showed well-defined areas of habitat usage in Lake Staring (Figure 1A). A density of 800 radiotagged carp/km^2 was used as the cut-off between high- and low-use areas. As such, areas with lower values were considered to be low-use areas, whereas areas with higher densities were classified as high-use areas. The value of 800 radiotagged carp/km^2 corresponded to approximately 1,248 carp/ha. This cut-off value was chosen because lower densities were associated with relatively isolated radiotagged carp observations. The high-use areas were located within or near the patches of lilies, which the carp most likely use for cover as the lake lacks other physical structure. In addition, all high-use areas were within 200 m of the shoreline and less than 2 m deep.

Field sampling

Field sampling took place on 8 October 2013. Average wind speed was 20 km/h from the S [22], while air and water temperatures were 17.8°C and 19.8°C, respectively. Water and sediment samples were collected from 24 locations within the lake. Samples were taken at 18 points at 4 to 5 locations along three N to S transects of the lake and at one location at the E and W ends. Six additional sampling points were added within three patches (two samples per patch) of *N. odorata* where we knew carp were generally found. At each sampling site, surface water, sub-surface (0.5 m depth) water, and sediment were sampled. A surface water sample was also taken in both the inflow and outflow of the lake.

Water samples were collected in 1-L HDPE bottles (Nalgene, Rochester, NY) that had been previously soaked in 10% bleach for at least 30 min to remove all traces of DNA. Bottles were subsequently rinsed with distilled water to remove residual bleach. Surface water samples were taken by partially submerging a

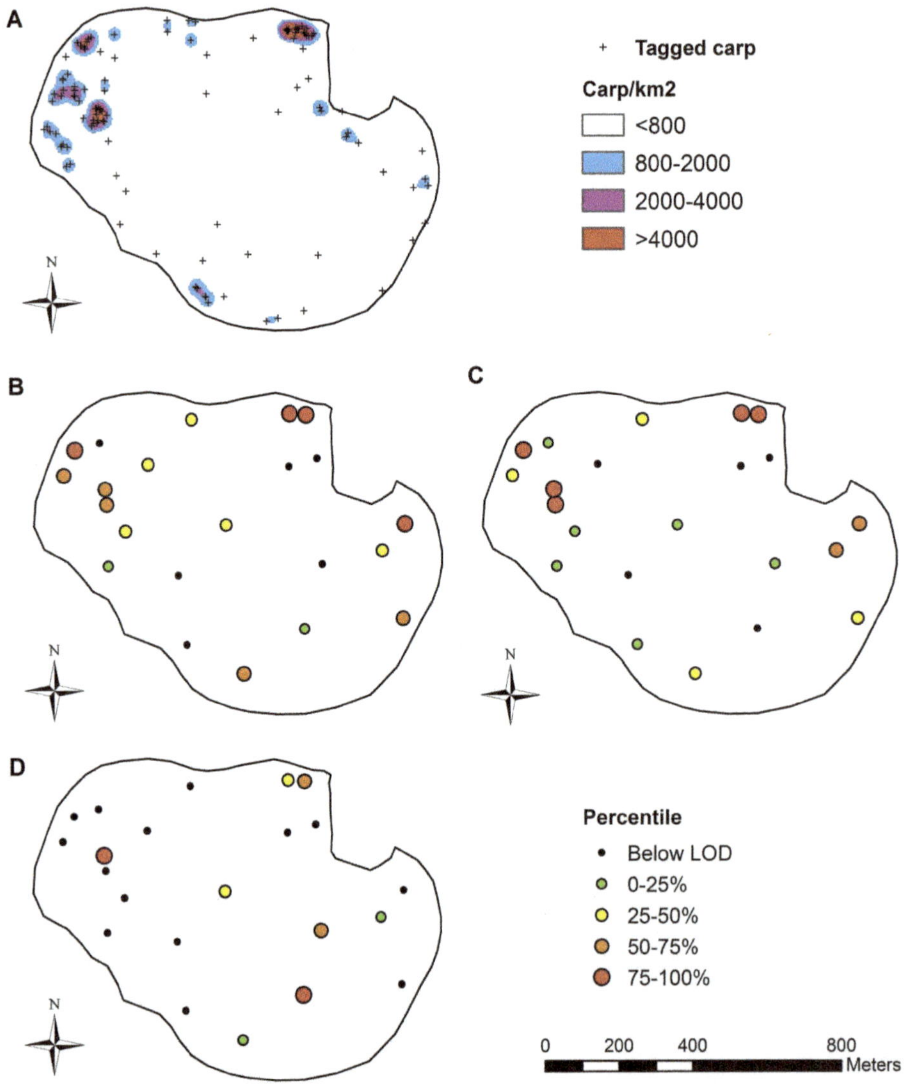

Figure 1. Carp use and distribution of eDNA in Lake Staring. Panel A shows locations of radiotagged carp and high- and low-use areas. Density categories represent the average number of locations of radiotagged carp/km^2. The high- and low-use area cut-off value of 800 radiotagged carp/km^2 corresponded to approximately 1,248 carp/ha. Panels B–D show the pattern of eDNA detection and concentration in surface water (B), sub-surface water (C), and sediment (D). All figures have the same scale. The symbol legend in the upper right refers to panel A, whereas lower right refers to panels B–D.

sample bottle to collect water from the top few cm. Sub-surface samples were taken using a stainless steel Van Dorn sampler (Wildlife Supply Company, Yulee, FL). Sediment samples were collected using a stainless steel Petite Ponar grab sampler (Wildlife Supply Company, Yulee, FL). Sediment was transferred to a sterile Whirl-Pak bag (Nasco, Fort Atkinson, WI) using a sterile polystyrene spatula (Bel-Art, Wayne, NJ). Once collected, samples were immediately placed on ice. Water and sediment samples were stored at 4°C and were filtered within 24 h. No specific permissions were required for access to the study site or collection of samples as part of this study. No animals were collected as part of this study.

Molecular Analyses

Water samples were filtered through Whatman 934-AH 1.5 μm glass microfiber filters (GE Whatman, Fairfield, CT) using a polyphenylsulfone filter funnel (Pall Corporation, Port Washington, NY). Filter funnels and forceps were soaked in 10% bleach and rinsed in distilled water prior to use and between samples. For tank samples, 1 L of water was filtered per sample. For field samples, only 200 mL could be filtered per sample due to clogging from the high amount of suspended solids. Filters were stored at −80°C until DNA extraction.

Sediment samples were homogenized, and a 0.1 g subsample was stored at −80°C for DNA extraction. Preliminary experiments showed that extraction of greater than 0.1 g of sediment lead to reduction in eDNA yield and inhibition of qPCR, regardless of post-extraction inhibitor removal protocols or inclusion of PCR adjuvants. For moisture content analysis, a 10 g subsample of sediment was weighed and then dried at 100°C for 24 h.

DNA was extracted using the QIAamp DNA Stool Mini Kit (Qiagen, Hilden, Germany) using the human DNA analysis protocol. Frozen filters were sliced into 1 mm×5 mm fragments with a sterile razor blade and then transferred to extraction tubes. For sediment samples, extraction buffer was directly added to frozen sediment. Before extraction, 50 ng of UltraPure Salmon Sperm Solution (Life Technologies, Grand Island, NY) was added to adjust for extraction efficiency of DNA as previously described [23]. DNA was eluted in a final volume of 50 μL. To further remove potential inhibitors, all DNA extracts were processed with the Wizard Genomic DNA Purification Kit (Promega, Madison, WI).

A multiplex qPCR assay was designed to amplify both the CarpCyt*b* and the extraction control targets, and the oligonucleotide concentrations were optimized for this study (Table 1). CarpCyt*b* standard was created by cloning PCR product amplified from carp fin clip DNA for the CarpCyt*b* genetic marker (Table 1) using the StrataClone PCR kit (Stratagene, Santa Clara, CA). Purified plasmid DNA was quantified by using a QuantiFluor-ST Fluorometer (Promega, Madison, WI). For the extraction control, standards were created by diluting UltraPure Salmon Sperm Solution (Life Technologies, Grand Island, NY). CarpCyt*b* and extraction control standards were combined prior to preparation of five qPCR standards, ranging from 50 to 300,000 CarpCyt*b* copies and 1.6 to 10,000 pg control DNA per 5 μL.

The assay used iTaq Universal Probes Supermix (Bio-Rad, Hercules, CA). Reactions contained 12.5 μL mastermix, 10 μg bovine serum albumin (New England Biolabs Inc., Ipswich, MA) primers and probe, and water or sediment DNA in a final reaction volume of 25 μL. The volume of water and sediment DNA added to the qPCR reaction was adjusted by testing a dilution series of a subset of 5 samples from water and sediment to confirm that inhibition was not present. For sediment samples, 2.5 μL of DNA extract was added to the reaction, and for water samples 5 μL of DNA extract was added. Reaction conditions consisted of an initial denaturation at 95°C for 3 min, followed by 40 cycles of denaturation at 95°C for 15 s and an annealing and extension step at 60°C for 1 min. Each qPCR run contained triplicate reactions of standards, non-transcript controls, and samples. Amplifications were performed using the StepOnePlus Real-Time PCR System (Life Technologies, Grand Island, NY), and Cq values were automatically determined using the system software. Sample marker concentrations were calculated on a per-run basis. All sediment values are reported per dry g.

Statistical analyses

Detection rate of eDNA was defined as the proportion of samples that were above the qPCR assay LOD. To analyze the effects of water sample depth (surface, sub-surface), carp usage (low-use, high-use), and matrix type (water, sediment) on eDNA detection rate, the number of detections and non-detections were statistically compared using Fisher's exact test. Fisher's exact test was used due to low expected values (<5) in some cells.

Concentration of eDNA in water was analyzed using a three-way Analysis of Variance (ANOVA). Main effects and 2-way interactions of carp usage (low-use, high-use), water sample depth (surface, sub-surface), and lake depth (m) were examined. Lake depth was included to determine the potential for suspended sediment to affect water column eDNA concentration, with shallower depths more likely to be affected by sediment mixing into the water column. The 3-way ANOVA was restricted to sites greater than 0.5 m and less than 2 m depth (i.e. omitting 2 shallower and 7 deeper sampling sites) because of a partial confound (high-use areas were not found at depths greater than 2 m, and no low-use areas were sampled at depths less than 0.5 m). Finally, student's t-test was used to determine whether there was a significant difference between eDNA concentration of sites included in the ANOVA and those excluded for the low-use areas. Since only 2 sites within the high-use area were excluded, no statistical comparison was done.

For all parametric descriptive analyses and statistical tests, eDNA concentrations were $\log_{10}$ transformed to achieve normal data distribution. Values below the LOD were given a value of half the LOD prior to analysis in order to reduce skewing of data. For graphical representation of eDNA concentrations across sampling points, data above the LOD were divided into four equal percentile categories. Percentiles were determined independently for the two sample types: (1) surface and sub-surface water and (2) sediment samples. All statistical tests were conducted in JMP, Version 10 (SAS Institute Inc., Cary, NC).

Results

Carp genetic marker development and laboratory validation

A carp-specific genetic marker (CarpCyt*b*) was developed for a 149 bp region in the cytochrome *b* gene (Table 1). This assay had an R^2 of over 0.99 and an average PCR efficiency of 92%. The lowest copy number that all three replicate reactions reliably and successfully amplified was determined to be the assay LOD. The assay LOD was 50 copies per reaction, which corresponded to 2.0×10^4 copies/L for water samples and 1.0×10^5 copies/g for sediment samples (calculated from the amount of DNA extract analyzed and the volume filtered or weight extracted). The sample LOD varied slightly for individual samples depending on extraction efficiency. Further details regarding qPCR calibration curves can be found in Table S2. The average extraction efficiency

Table 1. Primers and probes used for multiplex quantitative PCR.

Assay	Target	Locus	Primer/Probe	Sequence (5′ to 3′)[a]	Conc. (nM)	Ref.
CarpCytb	*Cyprinus carpio*	Cytochrome *b*	CCcytbF	CTAGCACTATTCTCCCCTAACTTAC	200	This study
			CCcytbR	ACACCTCCGAGTTTGTTTGGA	200	
			CCcytbP	(6FAM) CCCTCTAGTTACACCACC (MGBNFQ)	200	
Extraction control	*Oncorhynchus keta*	ITS[b] region 2	SketaF2	GGTTTCCGCAGCTGGG	200	[36]
			SketaR3	CCGAGCCGTCCTGGTCTA	200	
			SketaP2	(JOE) AGTCGCAGGCGGCCACCGT (BHQ-1)	100	

[a]BHQ1, black hole quencher-1; 6FAM, 6-carboxyfluorescein; JOE, 6-carboxy-4′,5′-dichloro-2′,7′-dimethoxyfluorescein.
[b]ITS, internal transcribed spacer.

was 11% for water samples and 6% for sediment samples. The assay reliably quantified up to 3.0×10^5 copies per reaction, the highest standard tested. No amplification of non-transcript controls was observed. The assay did not amplify DNA from a selection of 34 native and non-native fish species, including Bigheaded carps and other related Cyprinids (Table S1).

In validation tests using laboratory tanks which were stocked with combinations of carp and goldfish, no CarpCyt*b* markers were detected in the tank that contained only goldfish. In the tank with 10 carp, the concentration of markers averaged 1.3×10^7 copies/L 95% CI [1.0×10^7, 1.6×10^7]. In the tank with 5 each of carp and goldfish, the concentration of markers was 4.5×10^6 copies/L, 95% CI [3.1×10^6, 6.7×10^6] which was significantly lower ($p=0.007$, Student's t-test) than that of carp alone, indicating that the presence of more individuals yielded more eDNA. On average, individual carp contributed 3.1×10^{11} copies of CarpCyt*b* in the mixed species tank and 4.4×10^{11} copies per individual in the carp only tank.

Detection rates of eDNA in Lake Staring

The overall detection rate of CarpCyt*b* in water samples was 75% (Table 2). The distribution of eDNA was patchy, with carp eDNA detected tens of meters from sites where eDNA was not detected (Fig. 1B, C). The detection rate was not statistically different between surface and sub-surface samples ($p = 1.00$, Fisher's exact test). However, detection rate was significantly higher in water samples collected in high-use areas ($p=0.009$, Fisher's exact test), a difference of nearly 40% (Table 2).

The overall detection rate of CarpCyt*b* in sediment was only 36% (Table 3). Similar to the water samples, detection pattern was patchy (Fig. 1D). The detection rate of CarpCyt*b* in sediment was slightly higher in high-use areas relative to low use areas, however, the difference between these use areas was not significant ($p = 1.00$, Fisher's exact test). The detection rate of CarpCyt*b* in low-use areas in the sediment was approximately 30% less than for the average value for water samples, however, there was no difference between the detection rate ($p=0.11$, Fisher's exact test). The detection rate of eDNA in sediment within high-use areas was nearly 60% less than water within high-use areas, and the difference was statistically significant ($p=0.005$, Fisher's exact test).

Concentration of eDNA in Lake Staring

CarpCyt*b* concentration in water ranged from below the LOD (2.0×10^4 copies/L) to 1.7×10^6 copies/L (Table S3). The mean CarpCyt*b* concentration across all water samples was 5.7×10^4 copies/L, 95% CI [3.9×10^4, 8.3×10^4] (Table 2). Most samples (84%) had less than 3.0×10^5 copies/L. Only three samples had a marker concentration above 5.0×10^5 copies/L, Surface water samples from the inflow and outflow streams had CarpCyt*b* concentrations of 6.0×10^4 and 3.4×10^4 copies/L, respectively.

Three-way ANOVA showed that carp use pattern had a significant (main) effect on eDNA concentration (Table 4). There was no significant effect of lake depth or sample depth on eDNA, and no 2-way interactions were significant (Table 4). For the subset of sampling locations considered in the ANOVA (0.5 to 2 m lake depth), in low-use areas the average CarpCyt*b* concentration was 3.3×10^4 copies/L, 95% CI [1.8×10^4, 5.9×10^4], and in high use areas CarpCyt*b* concentration averaged 1.6×10^5 copies/L, 95% CI [1.1×10^5, 2.4×10^5]. There was no significant difference in eDNA concentration of the shallow water sites included in the ANOVA and those excluded for the low carp use areas ($p=0.75$).

For sediment samples, the concentration of eDNA ranged from below the LOD (1.0×10^5 copies/g) to 5.4×10^5 copies/g (Table S3). For sediment samples above the LOD, the concentration of CarpCyt*b* was slightly higher in high-use areas (Table 3), but the difference was not significant ($p=0.3$, Student's t-test). On a per mass basis, the lowest measureable sediment concentration of CarpCyt*b* was nearly two orders of magnitude greater than the water sample with the highest concentration of CarpCyt*b*.

Discussion

This study found that both the detection rate and concentration of carp eDNA strongly correlated with the distribution of carp in lake water. In water, the concentration of the carp genetic marker CarpCyt*b* was over 7 times greater in high-use areas as opposed to low-use areas, and detection rate rose from 63% to 100%. The detection rate and concentration of eDNA did not differ between surface and sub-surface water samples. Detection rate was comparably low in sediment, at 36%. The distribution of eDNA is fundamentally important in the design of eDNA sampling schemes and accurate interpretation of eDNA data [6]. Thus, we have shown that the distribution of a target organism must be carefully considered in the design of eDNA sampling schemes and accurate interpretation of eDNA data. Specifically, eDNA is patchily distributed in the environment, and the probability of detecting a target organism may drastically decline tens of meters from areas that are frequently inhabited. As part of this study, a highly-specific qPCR assay for common carp, an invasive and broadly distributed fish, was developed and validated.

The patchiness of eDNA distribution within water and sediment samples taken from Lake Staring was unexpected, given that the lake is small, shallow, and has a high biomass of carp. However,

Table 2. Concentration and detection rates of CarpCyt*b* in water.

Carp usage	Surface			Sub-surface			Total		
	Mean (copies/L) [95% CI]	Detection rate (%)	N	Mean (copies/L) [95% CI]	Detection rate (%)	N	Mean (copies/L) [95% CI]	Detection rate (%)	N
Low-use	3.8×10^4 [2.2×10^4, 6.8×10^4]	60	15	2.7×10^4 [1.8×10^4, 4.2×10^4]	67	15	3.1×10^4 [2.1×10^4, 4.5×10^4]	63	30
High-use	2.1×10^5 [9.8×10^4, 3.7×10^5]	100	7	2.6×10^5 [1.3×10^5, 3.7×10^5]	100	7	2.4×10^5 [1.4×10^5, 3.8×10^5]	100	14
Total	6.6×10^4 [3.7×10^4, 1.2×10^5]	73	22	5.5×10^4 [3.1×10^4, 9.9×10^4]	77	22	5.7×10^4 [3.9×10^4, 8.3×10^4]	75	44

Pilliod et al. [9] also observed high variation in amphibian eDNA concentration of replicate water samples in small freshwater streams. The authors hypothesized that variation was due to downstream pulses of eDNA due to activity of the target organism or variation in the cell type or form (free or cellular) of eDNA. An alternative, and perhaps complimentary explanation, is that mechanisms of eDNA removal from the water column, such as sedimentation and decay, are rapid. Lake Staring is a eutrophic lake, with high productivity and turbid, nutrient rich water. Carp eDNA is primarily contained in the particle size fraction ranging from 1.0–10 μm [24]. Particulate eDNA is continuously settling into lake sediments, but suspended particles are also hot spots of microbial degradation in aquatic systems [25]. Although decay of eDNA was not measured in this study, microcosm studies suggest that eDNA decays rapidly in the environment, and eDNA degrades nearly 90% within several days in water [12,26,27,28]. Taken together, it is likely that rapid removal of eDNA from the water column through processes of decay and sedimentation prevented its accumulation and diffusion from release points in Lake Staring, leading to significantly higher concentrations of eDNA in areas where carp were present.

Regardless of the cause of eDNA's patchy distribution, it has implications for the optimal sampling of eDNA. In low-use areas, the detection rate of eDNA was 40% lower in low-use areas as opposed to high-use areas. Distances between high-use and low-use sampling sites were tens or hundreds of meters apart, and that small distance, in some instances, affected whether eDNA was detected or not. Differences on a fine spatial scale have been observed in streams [9,29], experimental ponds [15], and lakes [10]. Therefore, we conclude that eDNA sampling should be conducted on small spatial scales. Only after extensive testing and consideration should sampling intervals of greater distances be used.

To the authors' knowledge, only one other study has attempted to correlate fish eDNA distribution within a lentic system. A positive relationship between carp eDNA concentration and temperature was noted by Takahara et al. [10] within a Japanese lagoon using a different marker on a slightly coarser scale without explicit information of carp distribution. The authors posited that the distribution corresponded to temperature because carp prefer warmer water; however, the data were not compared to the actual distribution of carp, as in the present study. Therefore other factors, such as higher metabolic activity of fish within warmer waters, could not be ruled out. Nevertheless, there are concordances with patterns observed between these studies. For example, excluding the lagoon channel, areas with higher eDNA appeared to have been located near the shoreline [10]. Similarly, hot spots of eDNA near shore was observed within Lake Staring, due to carp aggregation.

The overall detection rate of carp eDNA in Lake Staring water was 77%, similar to other studies conducted in areas of high fish abundance. For example, the eDNA of Bigheaded carps in the surface waters of a reach of the Mississippi River with a high target population was detected in 64% of samples [30]. Similarly, a 90% of samples were positive for common carp eDNA in a lagoon used for breeding purposes by Takahara et al. [10].

Despite the assumption of eDNA accumulation in surface waters, water depth did not affect concentration or detection rate of carp eDNA in the study lake. Therefore, both surface and sub-surface samples were equally effective for eDNA sampling in the present study. Although most eDNA samples are taken from the water's surface [11,14,26,31], only one study, done in experimental ponds, has confirmed that eDNA is most frequently detected in surface waters [15]. The convention of surface water sampling

Table 3. Concentration and detection rates of CarpCyt*b* in sediment.

Carp usage	Conc. of samples above LOD Mean (copies/g) [95% CI]	N	Detection rate (%)	N
Low-use	1.2×10^5 [7.5×10^4, 1.8×10^5]	5	33	15
High-use	2.3×10^5 [1.1×10^5, 4.8×10^5]	3	43	7
Total	1.5×10^5 [1.1×10^5, 2.1×10^5]	8	36	22

may be a holdover from early methods, wherein eDNA was used to detect floating feces of large marine mammals [32]. Our results indicate that the level of eDNA accumulation in surface water may differ among species, and accumulation likely depends on the relative proportion of eDNA sources, buoyancy of fecal material, and site-specific factors.

Although eDNA is hypothesized to accumulate in sediment, the detection rate was unexpectedly low, at 36%. This is likely due to the high LOD of the CarpCyt*b* marker in sediment. The LOD in sediment was 100,000 copies/g, nearly 4 orders of magnitude higher than for water samples on a per g basis. The high LOD is likely partly due to the limited amount of sediment that could be extracted. DNA extracts prepared with more than 0.1 g of sediment were observed to inhibit the qPCR reaction; therefore, greater amounts of sediment were not capable of being processed. Although the extraction efficiency of sediment was within the range previously observed for commercial DNA extraction kits [33], it was approximately half that of water samples. We do not know the cause of this discrepancy, but we hypothesize that sample chemistry can differentially affect extraction efficiency. Due to the difficulty of DNA recovery from sediment and the potential for qPCR inhibition, water sampling was more efficient for detection of carp in the study lake.

Regardless of the low detection rate of eDNA in sediment, its importance as a reservoir of carp eDNA cannot be disregarded. The concentration of CarpCyt*b* was high in sediment locations where eDNA was detected, but as the majority of the samples (63%) were below the LOD, we were unable to reliably calculate a mean sediment eDNA concentration. The accumulation of eDNA in sediment has been suspected based on the high concentration of microbial DNA in sediment [34,35], but measurements of fish eDNA in sediment have not been previously published. Hot spots of eDNA in sediment did not correlate with carp use. Therefore, there is a need for future studies to measure factors that may control eDNA distribution in sediment, such as deposition, resuspension, and degradation rates.

Table 4. Results of a 3-way ANOVA for CarpCyt*b* marker in water samples.

Effects	df	F ratio	*P* value
Sample depth (surface, sub-surface)	1	0.07	0.79
Lake depth (m)	1	0.08	0.78
Carp use (low-use, high-use)	1	5.77	0.03*
Sample depth×Lake depth	1	0.18	0.67
Lake depth×Carp use	1	0.62	0.44
Sample depth×Carp use	1	0.26	0.62
Error	19		

Conclusions

Sampling design has been previously identified as one of the four critical aspects that must be optimized in a DNA-based monitoring program [6]. The present study showed that common carp distribution led to spatial patterns in both eDNA concentration and detection rate in a small, shallow lake. Our results show that while eDNA is relatively evenly distributed in the water column, eDNA is patchily distributed horizontally. The large variation of eDNA on a small spatial scale, of tens to hundreds of meters, indicates that sampling for aquatic species using eDNA should use a similarly fine scale, at least for initial surveys. The results of this study also indirectly suggest that mechanisms of eDNA removal from the water column are rapid and may partially control eDNA distribution. Although the observations of the current study may not be universally applicable to all species and habitats, our results indicate that eDNA sampling schemes should be critically evaluated for the specific organism and the type of aquatic environment they inhabit. Future research is needed to examine the role of decay, sediment re-suspension, and eDNA release on eDNA distribution in aquatic habitats.

Supporting Information

Table S1 List of fish species tested for marker specificity.

Table S2 Quantitative PCR calibration data.

Table S3 Coordinates, lake depth, fish use, and eDNA concentration at sampling sites.

Acknowledgments

The authors thank Sendréa Best for field and laboratory assistance. Fish telemetry data was collected by Mary Headrick, Joseph Lechelt, Brett Miller, Robert Mollenheuer, and Tracy Szela. Erik Smith and Justine Koch provided assistance with ArcGIS. Mark Hove graciously provided fin clips for use in specificity testing. Commercial fisher Tim Adams and crew also provided valuable assistance obtaining fin clips for specificity testing from the Mississippi River.

Author Contributions

Conceived and designed the experiments: JJE PGB PWS. Performed the experiments: JJE PGB. Analyzed the data: JJE PGB PWS. Contributed reagents/materials/analysis tools: JJE PGB PWS. Wrote the paper: JJE PGB PWS.

References

1. Mesa MG, Schreck CB (1989) Electrofishing mark-recapture and depletion methodologies evoke behavioral and physiological changes in cutthroat trout. Trans Amer Fish Soc 118: 644–658.
2. Cross DG, Stott B (1975) The effect of electric fishing on the subsequent capture of fish. J Fish Biol 7: 349–357.
3. Bayley PB, Austen DJ (2002) Capture efficiency of a boat electrofisher. Trans Amer Fish Soc 131: 435–451.
4. Lodge DM, Turner CR, Jerde CL, Barnes MA, Chadderton L, et al. (2012) Conservation in a cup of water: estimating biodiversity and population abundance from environmental DNA. Mol Ecol 21: 2555–2558.
5. Olson ZH, Briggler JT, Williams RN (2012) An eDNA approach to detect eastern hellbenders (*Cryptobranchus a. alleganiensis*) using samples of water. Wildl Res 39: 629–636.
6. Darling JA, Mahon AR (2011) From molecules to management: adopting DNA-based methods for monitoring biological invasions in aquatic environments. Env Res 111: 978–988.
7. Ficetola GF, Miaud C, Pompanon F, Taberlet P (2008) Species detection using environmental DNA from water samples. Biol Lett 4: 423–425.
8. Jerde CL, Mahon AR, Chadderton WL, Lodge DM (2011) "Sight-unseen" detection of rare aquatic species using environmental DNA. Conserv Lett 4: 150–157.
9. Pilliod DS, Goldberg CS, Arkle RS, Waits LP (2013) Estimating occupancy and abundance of stream amphibians using environmental DNA from filtered water samples. Can J Fish Aquat Sci 70: 1123–1130.
10. Takahara T, Minamoto T, Yamanaka H, Doi H, Kawabata Z (2012) Estimation of fish biomass using environmental DNA. PLoS One 7: e35868.
11. Thomsen PF, Kielgast J, Iversen LL, Wiuf C, Rasmussen M, et al. (2012) Monitoring endangered freshwater biodiversity using environmental DNA. Mol Ecol 21: 2555–2558.
12. Thomsen PF, Kielgast J, Iversen LL, Møller PR, Rasmussen M, et al. (2012) Detection of a diverse marine fish fauna using environmental DNA from seawater samples. PLoS One 7: e41732.
13. Goldberg CS, Sepulveda A, Ray A, Baumgardt J, Waits LP (2013) Environmental DNA as a new method for early detection of New Zealand mudsnails (*Potamopyrgus antipodarum*). Freshw Sci 32: 2555–2558.
14. U.S. Fish and Wildlife Service (2013) Quality assurance project plan (QAPP): eDNA monitoring of bighead and silver carps. 89 pp.
15. Moyer GR, Díaz-Ferguson E, Hill JE, Shea C (2014) Assessing environmental DNA detection in controlled lentic systems. PLoS One 9: e103767.
16. Bohmann K, Evans A, Gilbert TP, Carvalho GR, Creer S, et al. (2014) DNA for wildlife biology and biodiversity monitoring. Trends Ecol Evol 29: 358–357.
17. Mahon AR, Jerde CL, Galaska M, Bergner JL, Chadderton WL, et al. (2013) Validation of eDNA surveillance sensitivity for setection of Asian carps in controlled and field experiments. PLoS One 8: e58316.
18. Chang Y, Huang F, Lo UT (1994) The complete nucleotide sequence and gene organization of carp (*Cyprinus carpio*) mitochondrial genome. J Mol Evol 38: 138–155.
19. Mayden RL, Tang KL, Conway KW, Freyhof J, Sudkamp M, et al. (2007) Phylogenetic relationships of *Danio* within the order Cypriniformes: A framework for comparative and evolutionary studies of a model species. J Exp Zoo B Mol Dev Evol 308: 642–654.
20. Bajer PG, Sorensen PW (2012) Using boat electrofishing to estimate the abundance of invasive common carp in small midwestern lakes. N Amer J Fish Manag 32: 817–822.
21. Bajer PG, Chizinski CJ, Sorensen PW (2011) Using the Judas technique to locate and remove wintertime aggregations of invasive common carp. Fish Manag Ecol 18: 497–505.
22. (NOAA) National Oceanic and Atmospheric Administration Global Historical Climate Network. Available: http://www.ncdc.noaa.gov/cdo-web/datasets#GHCND. Accessed 11 February 2014.
23. Haugland RA, Siefring SC, Wymer LJ, Brenner KP, Dufour AP (2005) Comparison of *Enterococcus* measurements in freshwater at two recreational beaches by quantitative polymerase chain reaction and membrane filter culture analysis. Water Res 39: 559–568.
24. Turner CR, Barnes MA, Xu CCY, Jones SE, Jerde CL, et al. (2014) Particle size distribution and optimal capture of aqueous macrobial Edna. Method Ecol Evol doi: 10.1111/2041-210X.12206.
25. Simon M, Grossart H-P, Schweitzer B, Ploug H (2002) Microbial ecology of organic aggregates in aquatic ecosystems. Aquat Microb Ecol 28: 175–211.
26. Pilliod DS, Goldberg CS, Arkle RS, Waits LP (2014) Factors influencing detection of eDNA from a stream-dwelling amphibian. Mol Ecol Res 14: 109–116.
27. Dejean T, Valentini A, Duparc A, Pellier-Cuit S, Pompanon F, et al. (2011) Persistence of environmental DNA in freshwater ecosystems. PLoS One 6: e23398.
28. Barnes MA, Turner CR, Jerde CL, Renshaw MA, Chadderton WL, et al. (2014) Environmental conditions influence eDNA persistence in aquatic systems. Environ Sci Technol 48: 1819–1827.
29. Jane SF, Wilcox TM, McKelvey KS, Young MK, Schwartz MK, et al. (2014) Distance, flow, and PCR inhibition: eDNA dynamics in two headwater streams. Mol Ecol Res doi: 10.1111/1755-0998.12285.
30. Amberg JJ, McCalla SG, Miller L, Sorensen P, Gaikowski MP (2013) Detection of environmental DNA of Bigheaded carps in samples collected from selected locations in the St. Croix River and in the Mississippi River: U.S. Geological Survey Open-File Report 2013–1080, 44p.
31. Takahara T, Minamoto T, Doi H (2013) Using environmental DNA to estimate the distribution of an invasive fish species in ponds. PLoS One 8: e56584.
32. Tikel D, Blair D, Marsh HD (1996) Marine mammal faeces as a source of DNA. Mol Ecol 5: 456–457.
33. Mumy KL, Findlay RH (2004) Convenient determination of DNA extraction efficiency using an external DNA recovery standard and quantitative-competitive PCR. J Microbiol Meth 57: 259–268.
34. Anno AD, Corinaldesi C (2004) Degradation and turnover of extracellular DNA in marine sediments: ecological and methodological considerations. Appl Environ Microbiol 70: 4384–4386.
35. Pietramellara G, Ascher J, Borgogni F, Ceccherini MT, Guerri G, et al. (2008) Extracellular DNA in soil and sediment: fate and ecological relevance. Biol Fertil Soils 45: 219–235.
36. Domanico MJ, Phillips RB, Oakley TH (1997) Phylogenetic analysis of Pacific salmon (genus *Oncorhynchus*) using nuclear and mitochondrial DNA sequences. Can J Fish Aquat Sci 54: 1865–1872.

Permissions

All chapters in this book were first published in PLOS ONE, by The Public Library of Science; hereby published with permission under the Creative Commons Attribution License or equivalent. Every chapter published in this book has been scrutinized by our experts. Their significance has been extensively debated. The topics covered herein carry significant findings which will fuel the growth of the discipline. They may even be implemented as practical applications or may be referred to as a beginning point for another development.

The contributors of this book come from diverse backgrounds, making this book a truly international effort. This book will bring forth new frontiers with its revolutionizing research information and detailed analysis of the nascent developments around the world.

We would like to thank all the contributing authors for lending their expertise to make the book truly unique. They have played a crucial role in the development of this book. Without their invaluable contributions this book wouldn't have been possible. They have made vital efforts to compile up to date information on the varied aspects of this subject to make this book a valuable addition to the collection of many professionals and students.

This book was conceptualized with the vision of imparting up-to-date information and advanced data in this field. To ensure the same, a matchless editorial board was set up. Every individual on the board went through rigorous rounds of assessment to prove their worth. After which they invested a large part of their time researching and compiling the most relevant data for our readers.

The editorial board has been involved in producing this book since its inception. They have spent rigorous hours researching and exploring the diverse topics which have resulted in the successful publishing of this book. They have passed on their knowledge of decades through this book. To expedite this challenging task, the publisher supported the team at every step. A small team of assistant editors was also appointed to further simplify the editing procedure and attain best results for the readers.

Apart from the editorial board, the designing team has also invested a significant amount of their time in understanding the subject and creating the most relevant covers. They scrutinized every image to scout for the most suitable representation of the subject and create an appropriate cover for the book.

The publishing team has been an ardent support to the editorial, designing and production team. Their endless efforts to recruit the best for this project, has resulted in the accomplishment of this book. They are a veteran in the field of academics and their pool of knowledge is as vast as their experience in printing. Their expertise and guidance has proved useful at every step. Their uncompromising quality standards have made this book an exceptional effort. Their encouragement from time to time has been an inspiration for everyone.

The publisher and the editorial board hope that this book will prove to be a valuable piece of knowledge for researchers, students, practitioners and scholars across the globe.

List of Contributors

John Gichimu Mbaka, Celia Somlai, Denis Köpfer, Andreas Maeck, Andreas Lorke and Ralf B. Schäfer
Institute for Environmental Sciences, University of Koblenz-Landau, Landau, Rhineland-Palatinate State, Germany

Jiali Wang
State Key Laboratory of Simulation and Regulation of Water Cycle in River Basin, China Institute of Water Resources and Hydropower Research, Beijing, China
Institute of Shandong River Wetlands, Laiwu, China

Fang Wang and Hao Wang
State Key Laboratory of Simulation and Regulation of Water Cycle in River Basin, China Institute of Water Resources and Hydropower Research, Beijing, China

Limin Chu
State Key Laboratory of Simulation and Regulation of Water Cycle in River Basin, China Institute of Water Resources and Hydropower Research, Beijing, China
College of Hydrology and Water Resource, Hohai University, Nanjing, China

Zhiping Zhong and Zhipei Liu
Institute of Microbiology, Chinese Academy of Sciences, Beijing, China

Jianyong Gao and Hairong Duan
Genewiz, Inc, Beijing, China

Jason W. Marion, Cheonghoon Lee, Chang Soo Lee and Timothy J. Buckley
Division of Environmental Health Sciences, College of Public Health, The Ohio State University, Columbus, Ohio, United States of America

Qiuhong Wang and Linda J. Saif
Food Animal Health Research Program, Ohio Agricultural Research and Development Center, Department of Veterinary Preventive Medicine, The Ohio State University, Wooster, Ohio, United States of America

Stanley Lemeshow
Division of Biostatistics, College of Public Health, The Ohio State University, Columbus, Ohio, United States of America

Jiyoung Lee
Division of Environmental Health Sciences, College of Public Health, The Ohio State University, Columbus, Ohio, United States of America
Department of Food Science and Technology, The Ohio State University, Columbus, Ohio, United States of America

Tara Wahab, Ingela Hedenström, Sven Löfdahl and Marika Hjertqvist
Public Health Agency of Sweden, Department of Microbiology, Stockholm, Sweden

Dawn N. Birdsell, Cedar L. Mitchell, David M. Wagner and Paul S. Keim
Northern Arizona University, Center for Microbial Genetics and Genomics, Flagstaff, AZ, United States of America

Argaw Ambelu and Seblework Mekonen
Department of Environmental Health Sciences and Technology, Jimma University, P.O. Box 378, Jimma, Ethiopia

Magaly Koch
Center for Remote Sensing, Boston University, 725 Commonwealth Avenue, Boston, Massachusetts, United States of America

Taffere Addis
Ethiopian Institute of Water Resources, Addis Ababa University, Akaki Kifleketema, Addis Ababa, Ethiopia

Pieter Boets, Gert Everaert and Peter Goethals
Laboratory of Environmental Toxicology and Aquatic Ecology, Ghent University, J. Plateaustraat 22, 9000 Ghent, Belgium

Qiuyuan Huang
Department of Geology and Environmental Earth Science, Miami University, Oxford, Ohio, United States of America

Department of Microbiology, University of Georgia, Athens, Georgia, United States of America

Brandon R. Briggs
Department of Geology and Environmental Earth Science, Miami University, Oxford, Ohio, United States of America

Hailiang Dong
Department of Geology and Environmental Earth Science, Miami University, Oxford, Ohio, United States of America
State Key Laboratory of Biogeology and Environmental Geology, China University of Geosciences, Beijing, China
State Key Laboratory of Biogeology and Environmental Geology, China University of Geosciences, Wuhan, China

Hongchen Jiang and Geng Wu
State Key Laboratory of Biogeology and Environmental Geology, China University of Geosciences, Wuhan, China

Christian Edwardson
Department of Microbiology, University of Georgia, Athens, Georgia, United States of America

Iwijn De Vlaminck and Stephen Quake
Departments of Bioengineering and Applied Physics, Stanford University and the Howard Hughes Medical Institute, Stanford, California, United States of America

Sarah J. Ivory
Brown University, Providence, Rhode Island, United States of America

Michael M. McGlue
University of Kentucky, Lexington, Kentucky, United States of America

Geoffrey S. Ellis
U.S. Geological Survey, Denver, Colorado, United States of America

Anne-Marie Lézine
LOCEAN, CNRS, Paris, France

Andrew S. Cohen
University of Arizona, Tucson, Arizona, United States of America

Annie Vincens
CEREGE, CNRS, Aix-en- Provence, France

Muye Gan, Xinyu Zheng and Ke Wang
College of Environmental and Resource Sciences, Zhejiang University, Hangzhou, Zhejiang Province, China

Jinsong Deng
College of Environmental and Resource Sciences, Zhejiang University, Hangzhou, Zhejiang Province, China
School of Civil Engineering and Environmental Sciences and School of Meteorology, University of Oklahoma, Norman, Oklahoma, United States of America

Yang Hong
School of Civil Engineering and Environmental Sciences and School of Meteorology, University of Oklahoma, Norman, Oklahoma, United States of America
State Key Laboratory of Hydroscience and Engineering, Department of Hydraulic Engineering, Tsinghua University, Beijing, China

Stuart C. Brown, Rebecca E. Lester and Laurie Laurenson
School of Life and Environmental Sciences, Deakin University, Warrnambool, Victoria, Australia

Vincent L. Versace
Greater Green Triangle University, Department of Rural Health, Flinders University, Warrnambool, Victoria, Australia
Centre for Environmental Management, Faculty of Science and Technology, Federation University Australia, Ballarat, Victoria, Australia

Jonathon Fawcett
Sinclair-Knight-Merz, Bendigo, Victoria, Australia

Xuemei Han
NatureServe, Arlington, Virginia, United States of America
Department of Environmental Science and Policy, George Mason University, Fairfax, Virginia, United Statesof America

Regan L. Smyth, Bruce E. Young, Healy Hamilton and Alexandra Sánchez de Lozada
NatureServe, Arlington, Virginia, United States of America

Thomas M. Brooks
NatureServe, Arlington, Virginia, United States of America

International Union for Conservation of Nature, Gland, Switzerland
World Agroforestry Center, International Center for Research in Agroforestry, University of Philippines, Los Baños, Laguna, Philippines
School of Geography and Environmental Studies, University of Tasmania, Hobart, Australia

Philip Bubb
United Nations Environment Programme World Conservation Monitoring Centre, Cambridge, United Kingdom

Stuart H. M. Butchart
BirdLife International, Cambridge, United Kingdom

Frank W. Larsen
European Environment Agency, Copenhagen, Denmark
Conservation International, Arlington, Virginia, United States of America

Matthew C. Hansen
Department of Geographical Sciences, University of Maryland, College Park, Maryland, United States of America

Will R. Turner
Conservation International, Arlington, Virginia, United States of America

Akiko Nakamasu
Department of Bioresource and Environmental Sciences Faculty of Life Sciences, Kyoto Sangyo University, Kyoto, Japan
Meiji Institute for Advanced Study of Mathematical Sciences, Meiji University, Tokyo, Japan

Hokuto Nakayama and Seisuke Kimura
Department of Bioresource and Environmental Sciences Faculty of Life Sciences, Kyoto Sangyo University, Kyoto, Japan

Naomi Nakayama
Institute of Molecular Plant Sciences, University of Edinburgh, Edinburgh, Lothian, United Kingdom

Nobuhiko J. Suematsu
Meiji Institute for Advanced Study of Mathematical Sciences, Meiji University, Tokyo, Japan
Graduate School of Advanced Mathematical Sciences, Meiji University, Tokyo, Japan

Mattias K. Ekvall and Lars-Anders Hansson
Lund University, Department of Biology, Ecology Building, SE-223 62, Lund, Sweden

Pablo Urrutia-Cordero
Lund University, Department of Biology, Ecology Building, SE-223 62, Lund, Sweden
Lund University, Center for Environmental and Climate Research, Ecology Building, SE-223 62, Lund, Sweden

Kimmo Kalevi Kahilainen
Kilpisjärvi Biological Station, University of Helsinki, Kilpisjärvi, Finland
Department of Environmental Sciences, University of Helsinki, Helsinki, Finland

William Paul Patterson
Department of Geological Sciences, Saskatchewan, Isotope Laboratory, University of Saskatchewan, Saskatoon, Saskatchewan, Canada

Eloni Sonninen
Finnish Museum of Natural History, University of Helsinki, Helsinki, Finland

Chris Harrod
School of Biological and Chemical Sciences, Queen Mary University of London, London, United Kingdom

Instituto de Ciencias Naturales Alexander Von Humboldt, Universidad de Antofagasta, Antofagasta, Chile

Mikko Kiljunen
Department of Biological and Environmental Sciences, University of Jyväskylä, Jyväskylä, Finland

Yingkui Li
Department of Geography, University of Tennessee, Knoxville, Tennessee, United States of America
Key Laboratory of Digital Earth Science, Institute of Remote Sensing and Digital Earth, Chinese Academy of Sciences, Beijing, P.R. China

Jingjuan Liao, Huadong Guo and Guozhuang Shen
Key Laboratory of Digital Earth Science, Institute of Remote Sensing and Digital Earth, Chinese Academy of Sciences, Beijing, P.R. China

Zewen Liu
Department of Geography, University of Tennessee, Knoxville, Tennessee, United States of America

Key Laboratory of Digital Earth Science, Institute of Remote Sensing and Digital Earth, Chinese Academy of Sciences, Beijing, P.R. China

Huacheng Xu
State Key Laboratory of Pollution Control and Resources Reuse, School of the Environment, Xianlin Campus, Nanjing University, Nanjing, China
State Key Laboratory of Lake Science and Environment, Nanjing Institute of Geography and Limnology, Chinese Academy of Sciences, Nanjing, China

Jicheng Zhong and Helong Jiang
State Key Laboratory of Lake Science and Environment, Nanjing Institute of Geography and Limnology, Chinese Academy of Sciences, Nanjing, China

Guanghui Yu
State Key Laboratory of Lake Science and Environment, Nanjing Institute of Geography and Limnology, Chinese Academy of Sciences, Nanjing, China
College of Resources and Environmental Sciences, Nanjing Agricultural University, Nanjing, China

Jun Wu
Key Laboratory of Soil Environment and Pollution Remediation, Institute of Soil Science, Chinese Academy of Sciences, Nanjing, China

Liuyan Yang
State Key Laboratory of Pollution Control and Resources Reuse, School of the Environment, Xianlin Campus, Nanjing University, Nanjing, China

Florian Eggers
Institute of Marine Research, Bergen, Norway
Department of Biology, University of Bergen, Bergen, Norway

Aril Slotte, Cecilie Kvamme and Richard D. M. Nash
Institute of Marine Research, Bergen, Norway

Lísa Anne Libungan
Department of Life and Environmental Sciences, University of Iceland, Reykjavík, Iceland

Arne Johannessen
Department of Biology, University of Bergen, Bergen, Norway

Even Moland
Institute of Marine Research, Flødevigen, Norway

Esben M. Olsen
Institute of Marine Research, Flødevigen, Norway
Centre for Ecological and Evolutionary Synthesis (CEES), Department of Biosciences, University of Oslo, Oslo, Norway
Department of Natural Sciences, Faculty of Science and Engineering, University of Agder, Kristiansand, Norway

Ina Severin, Eva S. Lindström and Örjan Östman
Department of Ecology and Genetics/Limnology, Uppsala University, Uppsala, Sweden

Hans W. Paerl, Karen L. Rossignol, Linghan Dong, Alan R. Joyner and Nathan S. Hall
Institute of Marine Sciences, The University of North Carolina at Chapel Hill, Morehead City, North Carolina, United States of America

Hai Xu, Guangwei Zhu, Boqiang Qin and Yali Wu
State Key Laboratory of Lake Science and Environment, Nanjing Institute of Geography & Limnology, Chinese Academy of Sciences, Nanjing, PR China

Mark J. McCarthy
Marine Science Institute, The University of Texas at Austin, Port Aransas, Texas, United States of America

Shawn T. O′Neil, Kasey C. Rahn and Joseph K. Bump
School of Forest Resources and Environmental Science, Michigan Technological University, Houghton, Michigan, United States of America

Magda Catarina Sousa, Renato Mendes, João Miguel Dias and Nuno Vaz
CESAM, Departamento de Física, Universidade de Aveiro, Campus de Santiago, Aveiro, Portugal

Ines Alvarez
CESAM, Departamento de Física, Universidade de Aveiro, Campus de Santiago, Aveiro, Portugal
EPhysLab (Environmental Physics Laboratory), Universidade de Vigo, Facultade de Ciencias, Ourense, Spain

Moncho Gomez-Gesteira
EPhysLab (Environmental Physics Laboratory), Universidade de Vigo, Facultade de Ciencias, Ourense, Spain

Guozeng Wang, Juan Lin and Xiu Yun Ye
College of Biological Science and Technology, Fuzhou University, Fuzhou 350108, P.R. China
National Engineering Laboratory for High-efficiency Enzyme Expression, Fuzhou 350002, P. R. China

Xiaoyun Huang
College of Biological Science and Technology, Fuzhou University, Fuzhou 350108, P.R. China

Tzi Bun Ng
School of Biomedical Sciences, Faculty of Medicine, The Chinese University of Hong Kong, Shatin, New Territories, Hong Kong, China

Christina A. S. Mumm
Institute of Experimental Ecology, University of Ulm, Ulm, Germany

Mirjam Knörnschild
Institute of Experimental Ecology, Faculty of Natural Sciences, University of Ulm, Ulm, Germany
Smithsonian Tropical Research Institute, Balboa, Panama

Qiong Gao, Mei Yu
Department of Environmental Sciences and Institute for Tropical Ecosystem Studies, University of Puerto Rico, Rio Piedras, San Juan, PR, United States of America

Jessica J. Eichmiller, Przemyslaw G. Bajer and Peter W. Sorensen
Department of Fisheries, Wildlife, and Conservation Biology, Minnesota Aquatic Invasive Species Research Center, University of Minnesota, Twin Cities, St. Paul, Minnesota, United States of America

Index

N

P

Q

R

S

T

U

W

Z